LOHP
LABOR OCCUPATIONAL HEALTH PROJECT
CENTER FOR LABOR RESEARCH AND EDUCATION
INSTITUTE OF INDUSTRIAL RELATIONS
UNIVERSITY OF CALIFORNIA, BERKELEY

TOXIC
TOXIC
TOXIC
TOXIC
SUBSTANCES CONTROL
SOURCEBOOK

EDITORS

Alexander McRae
The Center for Compliance Information

Leslie Whelchel
The Center for Compliance Information

CONSULTING EDITOR

Howard Rowland
The Rowland Company

THE CENTER FOR
COMPLIANCE INFORMATION

AN ASPEN PUBLICATION

Library of Congress Cataloging in
Publication Information

Aspen Systems Corporation, Center for Compliance
    Information
Toxic substances control sourcebook.

Includes index.
1. Industrial hygiene—United States.
2. Industrial toxicology. I. McRae, Alexander, 1933—
II. Welchel, Leslie. III. Rowland, Howard S. IV. Title

HD7263.A76   1978   362.1   78-2286
                              ISBN 0-89443-033-5

Library of Congress Catalog Card Number: 78-2286
ISBN: 0-89443-033-5

Printed in the United States of America

1   2   3   4   5

CONTENTS

PREFACE   xi

TOXIC CHEMICALS — THE PRESIDENT'S MESSAGE   xiii

CHAPTER ONE — HOW ARE TOXIC SUBSTANCES
REGULATED?   1

   A SUMMARY OF THE TOXIC SUBSTANCES CONTROL ACT   3

      Recognizing Chemical Risks   3

      Scope of the Law   3

      Testing of Chemicals   4

      Premarket Notification   4

      Regulation of Hazardous Chemical Substances and Mixtures   5

      Record-Keeping and Reporting   5

      Relationship to Other Federal Laws   6

      Disclosure of Data   6

      Civil and Criminal Penalties   6

      Effect on State Laws   6

      Actions by Citizens   6

      Exports and Imports   6

      Research, Monitoring and Data Systems   7

      State Programs   7

      Enforcement of the Act   7

      Judicial Review   7

      Employee Protection   7

      Administration of the Act   7

EPA'S IMPLEMENTATION PLAN 9

Overall Program Direction 10

Acquisition of Information and Assessment of Risks to Health and the Environment ... 18

Use of TSCA Regulatory Authorities When Necessary to Control New Chemicals ... 26

Use of TSCA Regulatory Authorities When Necessary to Control Existing Chemicals 29

Dissemination of Information and Assessments to Other Programs and Interested Parties 33

Types of Anticipated Impacts from Implementation Activities 36

REGULATION OF TOXIC AND HAZARDOUS SUBSTANCES UNDER OTHER FEDERAL LAWS 39

Consumer Product Safety Commission 39

Department of Transportation 42

Department of Health, Education, and Welfare 45

Environmental Protection Agency 47

Department of Defense 55

CHAPTER TWO—TOXIC AND HAZARDOUS SUBSTANCES 57

PRELIMINARY LIST OF CHEMICAL SUBSTANCES FOR FURTHER EVALUATION 59

Chemical or Category List 60

Methodology Used to Develop the Preliminary List 76

Types of Comments Sought and Recommended Response Formats 78

FIFTEEN TOXIC CHEMICALS SUBJECT TO IMMINENT REGULATION 81

Arsenic 81

Asbestos 83

Benzene 84

Benzidene 86

Cadmium 87

Ethylene Dibromide 89

Hexachlorobenzene 91

Hydrogen Sulfide 92

Mercury 94

Platinum 95

Polybrominated Biphenyls 96

Polynuclear Aromatic Hydrocarbons 98

Trichloroethylene 99

Tris 101

Vinylidene Chloride 102

HEALTH AND FIRE HAZARDS FROM COMMON INDUSTRIAL CHEMICALS

HEALTH AND FIRE HAZARDS FROM COMMON INDUSTRIAL
CHEMICALS 105

Explanation of Table Headings 105

Table of Chemical Hazards 111

TOXICITY TESTING PROCEDURES

TOXICITY TESTING PROCEDURES 121

Acute Studies 122

Prolonged Studies 126

Chronic Studies 128

Special Tests 129

Research 131

Submission of Test Materials 131

CHAPTER THREE—IMPACT ON INDUSTRY

CHAPTER THREE—IMPACT ON INDUSTRY 133

COST ESTIMATES FOR COMPLIANCE 135

The EPA Report 135

The Dow Study 135

The MCA Study. 137

Component Costs of TSCA 137

Direct Costs of Screening and Testing 137

Administration and Reporting 138

Delays. 138

Bans and Restrictions. 139

Other Costs 139

Conclusions 140

ECONOMIC IMPACT OF TOXIC SUBSTANCES CONTROL
LEGISLATION 141

General Analysis 141

Results by Category 141

Intrepretation 142

Toxicological Testing Costs. 142

Other Related Factors 143

A Forward-Looking Analysis of Production Limitation 143

Chemical Process Industries 143

Macroeconomic Effects. 143

Summary of Economic Impact 144

THE LABORATORY COSTS FOR TOXICITY TESTING.   147

REPORTING REQUIREMENTS UNDER THE TOXIC SUBSTANCES
CONTROL ACT 151

The Chemical Substances Inventory Reporting Regulations 151

Inventory Reporting Forms 165

REREGISTRATION REQUIREMENTS UNDER THE AMENDED
PESTICIDES ACT 179

Instructions for Reregistering Pesticide Products. 179

Reregistration Forms for Pesticide Products 193

CHAPTER FOUR—MANAGEMENT TECHNIQUES FOR TOXIC SUBSTANCES 201

CONTROL PROCEDURES IN TEN MAJOR INDUSTRIES 203

Analyis of Techniques 203

Recommendations. 279

Summary of Use/Exposure Data for Chemicals 295

CHAPTER FIVE—TOXIC SUBSTANCES AND WORKER HEALTH 323

TOXIC SUBSTANCES POSING CANCER RISK TO WORKERS 325

The Regulatory Dilemma 326

Identification, Classification and Regulation of Toxic Materials Posing a Potential
Occupational Carcinogenic Risk to Humans 332

Category I Toxic Materials 333

Category II Toxic Materials. 334

Category III Toxic Materials 335

Model Standards 336

OCCUPATIONS AND DISEASES. 345

A Guide to Toxic Exposures by Occupation 345

Asbestos and Occupational Disease 367

Inorganic Lead and Occupational Disease 371

Case History of Chronic Arsenic Intoxication 377

CHAPTER SIX—SIGNIFICANT CASES INVOLVING TOXIC SUBSTANCES 381

SECRETARY OF LABOR v. THE PROKO COMPANY OF TEXAS, INC.—
Aiborne Asbestos 383

GAF CORPORATION v. OSHRC AND SECRETARY OF LABOR—
Asbestos Fibers.   385

SECRETARY OF LABOR v. GOODYEAR TIRE & RUBBER
COMPANY—
Asbestos Fibers.   391

SECRETARY OF LABOR v. RESEARCH-COTTRELL, INC.—
Asbestos Fibers.   395

SECRETARY OF LABOR v. AMERACE CORPORATION—
Chromic Acid.   399

BULK TERMINALS ET AL. v. THE ENVIRONMENTAL
PROTECTION AGENCY ET AL. —
Hydrochloric Acid and Silicon Dioxide.   403

NATURAL RESOURCES DEFENSE COUNCIL, INC. v. TRAIN—
Lead   407

SECRETARY OF LABOR v. REPUBLIC STEEL CORPORATION—
Sodium Hydroxide.   413

SECRETARY OF LABOR v. SPENCER LEATHERS—
Toxic gases   417

APPENDIX A   — CURRENT FEDERAL RESEARCH ON
TEN MAJOR TOXIC SUBSTANCES.   423

APPENDIX B   — TOXIC SUBSTANCES CONTROL ACT. . .   437

APPENDIX C   — FEDERAL INSECTICIDE, FUNGICIDE
AND RODENTICIDE ACT   487

APPENDIX D   — MARINE PROTECTION, RESEARCH
AND SANCTUARIES ACT   511

APPENDIX E   — SAFE DRINKING WATER ACT.   519

APPENDIX F   — THE FEDERAL WATER POLLUTION
CONTROL ACT.   523

APPENDIX G   —   THE CLEAN AIR ACT   547

APPENDIX H   —   RESOURCE CONSERVATION AND
RECOVERY ACT   553

APPENDIX I   —   REPORTING FOR THE CHEMICAL
SUBSTANCE INVENTORY   561

INDEX   597

PREFACE

Toxic substances control has become a major issue of concern to industry and the public as a whole. There appears to be a rapidly developing perception that many chemical products have harmful effects. Unless this perception can be changed the risks of progress could eventually seem to outweigh the benefits. A climate which is still receptive to innovation and to experimentation might begin to alter.

Evidence of this trend can already be found in the increasing complexity of the federal laws and regulations that are designed to control toxic emissions and the use of toxic chemicals in the manufacturing process. At least a dozen major federal laws, enforced by seven different agencies, address the problem of toxic substances control in various ways.

The culmination of this legislative effort came with the enactment of the Toxic Substances Control Act. Through this statute, American industry is now faced with complex regulations that attempt to cope with the health and environmental effects, safety and toxicity of its products. And the law introduces a new concept—regulation at the very start of research and commercial development.

The Center for Compliance Information was formed to serve the growing need of industry for easily-accessible information on vital regulatory compliance issues. In this volume the Center has compiled information of great value to all those involved with the planning and implementation of toxic substances control and the compliance with relevant regulations. The Toxic Substances Control Sourcebook embodies a wealth of material drawn from a wide spectrum of original sources including government agencies, business associations and private industry.

The volume is divided into logically sequenced, major subject matter sections. The federal laws and regulations governing toxic substances control are presented first with emphasis on the Toxic Substances Control Act of 1976 and its numerous registration, reporting and other requirements. This section is followed by detailed information on the definition, identification, nature and status of toxic and hazardous substances subject to these regulations. Next, detailed aspects of the economic impact on industry of toxic substances control are examined. The fourth section identifies and classifies recommended management techniques for the control of toxic substances posing risk to worker health. The last section contains significant case studies which illustrate actual outcomes in litigation under the toxic substances control laws.

The volume's appendixes contain the full text of the Toxic Substances Control Act of 1976. They also provide the text of other federal laws appearing throughout the book, as well as details on the most current federal research being conducted on toxic substances of vital concern.

In the coming months and years, more and more decision-makers in industry, business, government and other concerned groups will become actively involved with regulations governing toxic substances. The Toxic Substances Control Sourcebook has been designed to guide such decision-makers through the complexities of law, implementation, compliance and effective management.

Alexander McRae   Leslie Whelchel

March 1978

TOXIC CHEMICALS

The presence of toxic chemicals in our environment is one of the grimmest discoveries of the industrial era. Rather than coping with these hazards after they have escaped into our environment, our primary objective must be to prevent them from entering the environment at all.

At least a dozen major federal statutes, implemented by seven different agencies, address this problem in various ways. With the enactment last year of the Toxic Substances Control Act, no further comprehensive federal legislation should be necessary. Now we must inaugurate a coordinated federal effort to exclude these chemicals from our environment.

I am therefore instructing the Council on Environmental Quality to develop an interagency program (1) to eliminate overlaps and fill gaps in the collection of data on toxic chemicals, and (2) to coordinate federal research and regulatory activities affecting them.

The Toxic Substances Control Act enables the federal government, for the first time, to gather the information on chemical substances needed to determine their potential for damaging human health and the environment, and to control them where necessary to protect the public. My FY 1978 budget provides nearly $29 million—a threefold increase over Fiscal 1977—for the Environmental Protection Agency to implement this important Act.

I have instructed the Environmental Protection Agency to give its highest priority to developing 1983-best-available-technology industrial effluent standards which will control toxic pollutants under the Federal Water Pollution Control Act, and to incorporate these standards into discharge permits. My Administration will be seeking amendments to this Act, including revision of Section 307(a), to permit the Environmental Protection Agency to move more decisively against the discharge of chemicals potentially injurious to human health.

Finally, I have instructed the Environmental Protection Agency to set standards under the Safe Drinking Water Act which will limit human exposure to toxic substances in drinking water, beginning with potential carcinogens.

Jimmy Carter
Executive Office Of
The President May, 1977

CHAPTER ONE

HOW ARE TOXIC SUBSTANCES REGULATED?

A SUMMARY OF THE TOXIC SUBSTANCES CONTROL ACT*

Chemicals are all around us—in our air, our water, and our food, and in the things we touch. Many of these chemicals have become essential to our lives, and their production contributes significantly to our national economy. However, for many of these substances, we have little knowledge of the ill-effects they might cause after many years of exposure. The Toxic Substances Control Act, which was signed into law in October 1976, is designed to improve our understanding of the chemicals around us and to provide controls on any chemical which may threaten human health or the environment.

Background: Recognizing Chemical Risks

There has been a dramatic surge in the development of synthetic organic chemicals since World War II. Presently there are an estimated two million recognized chemical compounds. Chemical sales now exceed $100 billion per year, with over 30,000 chemical substances in commerce. To these, a thousand new ones may be introduced each year.

Chemicals play an important role in protecting, prolonging, and enhancing our lives. Synthetic fibers are used to replace human tissue and to create our easy-to-wear wardrobe. Plastics have been molded for use in almost every phase of our activities—in transportation, communication, and industrial and consumer goods. Our leisure time has been enhanced, for example, by durable, low-maintenance pleasure boats and other recreational equipment. Also, the chemical industry makes a significant contribution to the national economy, with sales representing more than six percent of our Gross National Product. Millions of workers are employed by the chemical industry or chemically-dependent industries.

While we have enjoyed the extensive economic and social benefits of chemicals, we have not, however, always realized the risks that may be associated with them.

In the last few years, many chemicals that have been commonly used and widely dispersed have been found to present significant health and environmental dangers. Vinyl chloride, which is commonly used in plastics, has caused the deaths of workers who were exposed to this chemical. Asbestos has long been known to cause cancer when inhaled. Mercury has caused debilitating effects in Japan.

Perhaps the most vivid example of the danger of uncontrolled chemical contaminants is polychlorinated biphenyls, or PCBs. It was not until after tens of millions of pounds of PCBs were produced and released into the environment, however, that scientists realized how toxic and persistent they were. Despite limited restrictions imposed in the early 1970's by industry to reduce the production and to restrict use of PCBs to electrical equipment where escape to the environment would be minimal, high levels of PCBs continue to persist in the Great Lakes and other major waters across the nation. Over the past few years, we have found PCBs in our bodies and even in the milk of our nursing mothers. Recently, a close relative of PCBs, polybrominated biphenyls, or PBBs, has posed a similarly grave threat to human health and the environment. Accidental use of PBBs in animal feed led to the contamination of thousands of Michigan cattle which had to be slaughtered. The health effects of PBBs on the Michigan farming families who were exposed to PBBs and consumed PBB-contaminated products are still uncertain.

In 1971, the President's Council on Environmental Quality developed a legislative proposal for coping with the increasing problems of toxic substances. Over the next five years, Congress held many days of hearings, debated and amended the provisions in committee, failed to resolve the differences between the House and Senate bills in the 92nd and 93rd Congresses, and finally enacted the present legislation in the fall of 1976. The law grants EPA significant new authorities which should greatly improve our ability to anticipate and address chemical risks before it is too late to undo the damage.

The following sections briefly describe the major provisions of the new law. The discussion is intended to familiarize the public with the provisions of the law but not to constitute an authoritative legal statement of the law.

Scope of the Law

The Toxic Substances Control Act authorizes the Environmental Protection Agency to obtain from industry data on the production, use, health effects, and other matters concerning chemical substances and mixtures. If warranted, EPA may regulate the manufacture, processing, distribution in commerce, use, and disposal of a chemical substance or mixture. Pesticides, tobacco, nuclear material, firearms and ammunition, food, food additives,

---

* Office of Toxic Substances, U.S. Environmental Protection Agency, October 1976.

drugs, and cosmetics are exempted from the Act. These products are currently regulated under other laws.

Testing of Chemicals

The Administrator of EPA may require manufacturers or processors of potentially harmful chemicals to conduct tests on the chemicals. Testing may be directed to evaluating the characteristics of a chemical, such as persistence or acute toxicity, or to clarifying its health and environmental effects, including carcinogenic, mutagenic, behavioral, and synergistic effects. Before requiring testing, the Administrator must set forth the need for such testing. Specifically, the Administrator must find that (1) the chemical may present an unreasonable risk to health or the environment, or there may be substantial human or environmental exposure to the chemical; (2) there are insufficient data and experience for determining or predicting the health and environmental effects of the chemical; and (3) testing of the chemical is necessary to develop such data. The manufacturers or processors of a chemical must bear costs of testing that chemical.

Testing requirements under this section must be promulgated by regulation, after opportunity for comments and an oral hearing. A manufacturer or processor of a proposed new chemical may petition the Administrator to develop testing standards for the chemical.

An interagency committee of government experts on chemical substances will advise the Administrator concerning chemicals which should be tested, but his actions are not limited to those recommended by the Committee. The eight committee members will represent the Departments of Labor, Commerce, and Health, Education, and Welfare (including the National Cancer Institute, the National Institute for Occupational Safety and Health, and the National Institute of Environmental Health Sciences); the National Science Foundation; the Council on Environmental Quality; and the Environmental Protection Agency. The committee may designate, at any one time, up to 50 chemicals from its list of recommended substances for testing. Within one year, the Administrator must either initiate testing requirements for those designated chemicals or publish in the Federal Register his reasons for not initiating such requirements.

Premarket Notification

Manufacturers of new chemical substances must give the Administrator 90 days notice before the manufacture of the chemicals. Any chemical which is not listed on an inventory of existing chemicals to be published by the Administrator by November 1977 will be considered "new" for purposes of the premarket notice requirement.

The Administrator may designate a use of an existing chemical as a significant new use, based on consideration of the anticipated extent and type of exposure to human beings or the environment. Any person who intends to manufacture or process a chemical for such a significant new use must also report 90 days before marketing the chemical for that use.

In both of the above cases, the Administrator may extend the 90-day premarket review period for an additional 90 days for good cause.

Notices submitted by the manufacturers for new chemicals or significant new uses of existing chemicals are to include the name of the chemical, its chemical identity and molecular structure, proposed categories of use, an estimate of the amount to be manufactured, the byproducts resulting from the manufacture, processing, and disposal of the chemical, and any test data related to the health and environmental effects which the manufacturer has. In addition, if a rule requiring testing of the chemical or members of its chemical class has been promulgated the manufacturer must submit test data developed from that testing along with the other information.

The Administrator may publish a list of new chemicals or classes of new chemicals which present or may present an unreasonable risk of injury to health or the environment if they are introduced into commerce. When a manufacturer of a new chemical which is on that list submits his premarket notice, he must submit data which he believes show that the chemical will not present such a risk.

The Administrator may determine that there is inadequate information to evaluate the health or environmental effects of a new chemical whether or not it is on the list described above. In that event, he may issue an order 45 days before the expiration of the premarket review period to prohibit or limit the manufacture, processing, distribution in commerce, use, or disposal of a chemical pending acquisition of additional data. If the manufacturer files objections to the order, the Administrator may seek a court injunction to prevent marketing of the chemical until data are developed which indicate that the substance does not present an unreasonable risk.

The Administrator may find that there is a reasonable basis to conclude that a new chemical presents or will present an unreasonable risk of injury to health or the environment. In that event, he may follow similar procedures involving an order, and, if appropriate, court action to prohibit the manufacture of the chemical. If a total ban does not appear necessary, the Administrator may propose a rule which becomes immediately effective limiting the manufacture or use of a proposed chemical or regulating its distribution in commerce, use, or disposal.

For certain new chemicals, the Administrator must publish his reasons in the Federal Register if he does not initiate regulatory action during the premarket review period. These include chemicals subject to testing requirements, those listed by the Administrator as possibly presenting an unreasonable risk, and chemicals whose use was designated as a significant new use.

Exempt from the premarket notification requirement are chemicals: (1) included in categories of chemical substances listed on the inventory of existing chemicals; (2) produced in small quantities solely for experimental or research and development purposes; (3) used for test marketing purposes; and (4) determined by the Administrator not to present an unreasonable risk. A person must apply for an exemption described by (3) and (4) on a chemical-by-chemical basis.

In addition, a person may apply for an exemption from premarket notification requirements if a chemical exists only temporarily and there is or will be no human or environmental exposure. This exemption is directed to chemicals which exist as the result of a chemical reaction in the manufacture or processing of a mixture or another chemical substance.

Regulation of Hazardous Chemical Substances and Mixtures

The Administrator may prohibit or limit the manufacturing, processing, distribution in commerce, use, or disposal of a chemical substance or mixture if he finds that these activities or any combination of them presents or will present an unreasonable risk of injury to health or the environment. Labelling may be required for a chemical or any article containing the chemical. A manufacturer may be required to make and keep records of the processes used in manufacturing a chemical and to conduct tests to assure compliance with any regulatory requirements. Further, the Administrator may require a manufacturer to give notice of any unreasonable risk of injury presented by his chemical to those who purchase or may be exposed to that substance. A manufacturer may also be required to replace or repurchase a substance which presents an unreasonable risk.

In proposing regulatory actions, the Administrator must provide an opportunity for comments by all interested parties, including an oral hearing, and in certain instances, cross-examination. For those unable to afford the costs of participating in rulemaking proceedings, the Administrator may provide compensation.

A rule limiting, but not banning, a chemical may become immediately effective when initially proposed in the Federal Register if the Administrator determines that the chemical is likely to present an unreasonable risk of

serious or widespread injury to health or the environment before normal rulemaking procedures could be completed. However, in the case of a rule prohibiting the manufacture of the chemical, the Administrator must first obtain a court injunction before a rule can be made immediately effective.

If a chemical contains a hazardous contaminant as the result of a certain manufacturing process, the Administrator may order the manufacturer to change his process to avoid such contamination. If a chemical contains a contaminant which may present an unreasonable risk of injury to health or the environment, the manufacturer may be required to give public notice and to repurchase or recall that product.

In addition to these authorities, the law requires that the Administrator take action to regulate polychlorinated biphenyls, by issuing labelling and disposal regulations by July 1977, restricting the chemical's use to closed systems by January 1978, prohibiting all production by January 1979, and distribution in commerce by July 1979.

For those chemicals which present an imminent and unreasonable risk of serious or widespread injury to health or the environment, the Administrator may ask a court to require whatever action may be nesessary to protect against such risk.

Record-Keeping and Reporting

The law authorizes the Administrator to issue rules requiring manufacturers and processors of selected chemicals to report to EPA the name of each chemical, its identity, its proposed uses, estimates of production levels, description of byproducts, adverse health and environmental data, and number of workers exposed to the chemicals. Manufacturers of chemical mixtures and research chemicals are exempt from these requirements unless the Administrator determines such reporting is necessary to enforce the Act. Similarly, in the absence of a determination that reporting is necessary because of an unreasonable risk, small manufacturers are exempt from reporting except for chemicals that are subject to proposed or promulgated testing requirements or limitations under the regulatory provisions of the Act.

The Administrator is required to publish a list of all existing chemicals by November 1977. This list, which will be continually updated, will contain all chemicals manufactured or processed for commercial purposes in the United States or imported into the United States within the last three years.

The law requires any person who manufacturers, processes, or distributes in commerce any chemical substance or mixture to keep records of significant adverse reactions to health or the environment that allegedly were caused by the chemical. Records concerning health effects on employees must be kept for 30 years; other records for five years.

The Administrator may request any health and safety studies on specific chemicals known or available to any person who manufacturers, processes, or distributes such chemicals in commerce. In addition, if such a person has information which indicates that a chemical presents a substantial risk of injury to health or the environment, he must notify the Administrator.

Relationship to Other Federal Laws

The Administrator may determine that an unreasonable risk presented by a chemical may be prevented or sufficiently reduced by action under a Federal law not administered by EPA. If he does, the Administrator will request the agency administering the other law to determine whether the risk exists and if the agency's actions would sufficiently reduce the risk. If the agency finds no risk or takes action directed to the risk, EPA may not take any regulatory action directed to the same risk.

The law directs the Administrator to use other laws administered by EPA to protect against unreasonable risks, such as the Federal Water Pollution Control Act or the Clean Air Act, unless the Administrator determines that it is in the public interest to protect against such risks under the Toxic Substances Control Act.

Disclosure of Data

Confidential data, such as trade secrets and privileged financial data, will be protected from disclosure by the Administrator. All health and safety information on chemicals in commerce submitted under the Act is subject to disclosure. A person submitting other types of data to EPA may designate any part of it as confidential. If the claims of confidentiality are subject to question or if the release of such data is essential for the protection of health or the environment, the Administrator shall notify the person who submitted the data in advance of any contemplated release.

Civil and Criminal Penalties

Any person who fails or refuses to comply with any requirement made under the law may be fined up to $25,000 for each day of violation of the law. Persons who knowingly or willfully violate the law, in addition to any civil penalties, may be fined up to $25,000 for each day of violation, imprisoned up to a year, or both.

Effect On State Laws

With certain exceptions, as set forth below, the Act will not affect the authority of any State or political subdivision to establish regulations concerning chemicals. If EPA issues a testing requirement for a chemical, a State may not establish a similar one. If EPA restricts the manufacture or otherwise regulates a chemical under the Act, a State may only issue requirements which are identical, mandated by other Federal laws, or prohibit the use of the chemical. In response to a request by a State, the Administrator may grant an exemption under certain conditions. Specifically, the Administrator may grant exemptions if the State requirement (1) would not cause a person or activity to be in violation of a requirement under the Act, and (2) would provide a greater degree of protection and not unduly burden interstate commerce.

Actions by Citizens

Any person may bring a civil suit to restrain a violation of the Act by any party or to compel the Administrator to perform any nondiscretionary duty required by the Act. In addition, any person may petition the Administrator to issue, amend, or repeal a rule under the testing, reporting, or restrictions sections of the Act. The Administrator has 90 days to respond to a petition. If he takes no action or denies a petition, the party has the opportunity for judicial review in a U.S. district court. In both civil suits and citizens' petitions, the court may award reasonable legal costs and attorney fees if appropriate.

Exports and Imports

In the case of a chemical produced for export which presents an unreasonable risk to health or the environment of the United States, the Administrator may regulate the chemical. He may also require testing of any exported chemical if such testing is necessary to determine whether there is such a risk to the United States. If a person is exporting, or intends to export, a chemical for which data are required to be submitted under the testing or premarket notification sections, he must notify the Administrator. The Administrator is responsible for notifying the governments of the importing countries of the availability of such data. Similarly, if a person is exporting a chemical subject to a regulatory order or action, he must notify the Administrator who in turn will notify the appropriate governments.

With respect to imports, no chemical substance, mixture, or article containing a chemical substance or mixture will be allowed into the customs territory of the United

States if it fails to comply with any rule or is otherwise in violation of the Act.

Research, Monitoring, and Data Systems

There are several provisions in the Act which call for expanded research and related activities by EPA and other Federal agencies. The Administrator, in consultation with the Secretary of Health, Education, and Welfare (HEW), may enter into contracts and make grants for research, development, and monitoring. In addition to establishing a data system within EPA for data submitted under the Act, the Administrator is responsible for designing and establishing a system for toxicological and other scientific data accessible to all Federal agencies. The Council on Environmental Quality is responsible for studying standard classification and storage systems for chemical substances.

Other research required by the Act to be conducted by the Administrator in consultation with the Secretary of HEW concerns techniques for screening chemical effects and for monitoring chemicals. In addition, the Secretary of HEW has special responsibility related to inexpensive and efficient methods for determining the health and environmental effects of chemicals.

State Programs

The Act authorizes $1.5 million each year for grants to assist States to prevent or eliminate risks associated with toxic substances when the Administrator is unable to take action. The amount of the grant shall be no greater than 75 percent of the costs of the program. In awarding grants, the Administrator shall take into account the seriousness of the health effects which are associated with the chemical substances and the extent of human and environmental exposure to them in the State.

Enforcement of the Act

For purposes of administering the Act, the Administrator or his representative may inspect any establishment in which chemicals are manufactured, processed, stored, or held before or after their distribution in commerce. No inspection shall include financial, sales, pricing, personnel, or research data unless specified in an inspection notice.

The Administrator may subpoena witnesses, documents, and other information as necessary to carry out the Act.

Civil actions concerning violations on lack of compliance with the Act may be brought in a U.S. district court for judicial review. Any chemical substance or mixture which was manufactured, processed, or distributed in commerce in violation of the Act may be subject to seizure.

Judicial Review

Not later than sixty days after a rule is promulgated, any person may file a petition for judicial review of such rule with the U.S. Court of Appeals for the District of Columbia Circuit or with the U.S. Court of Appeals for the circuit of his residence or business.

Employee Protection

If an employee believes that his employer has discriminated against him because of the employee's participation in carrying out the Act, he may file a complaint with the Secretary of Labor. The Secretary shall investigate the alleged discrimination and, if warranted, may order the employer to remedy the effects of any such discriminatory action. Employees and employers may obtain judicial review in the U.S. Courts of Appeals.

The Administrator will evaluate the potential effects on employment of regulatory actions under the Act. In response to a petition by an employee, the Administrator may investigate and hold public hearings concerning job losses or other adverse effects allegedly resulting from a requirement under the Act. The Administrator will make public his findings and recommendations.

Administration of the Act

There are several provisions which deal with general administration of the Act. Among them is a provision authorizing the Administrator to require fees to defray the cost of reviewing testing data and premarket notifications. Such fees shall not exceed $2,500 or, in the case of small business concerns, $100.

The Administrator will establish an office to provide technical and other nonfinancial assistance to manufacturers and processors respecting the requirements of the Act.

Finally, the Administrator will waive compliance with any provision of the Act if requested by the President for national defense purposes.

EPA'S IMPLEMENTATION PLAN*

Summary

The goal of the Toxic Substances Control Act (TSCA) is to protect human health and the environment from unreasonable risks—now and in future generations. To achieve this goal, TSCA implementation activities will emphasize not only control of specific problems under TSCA regulatory provisions, but also use of TSCA authorities to support other Governmental and non-Governmental programs to control toxic substances. These programs include other activities of EPA and other Federal and State agencies, activities of environmental and public interest groups and of professional societies, policies of the financial and investment communities, and efforts of individual companies and trade associations.

During the first several years, EPA will give top priority to the following implementation activities:

Establishment and Implementation of a Pre-market Review System.—The system will emphasize (a) the responsibility of industry to develop adequate data for meaningful chemical assessments, (b) categorization of new chemicals by broad chemical classes and broad uses with particular attention to those categories of greatest environmental concern, and (c) procedures for rapid decisions and adequate documentation of these decisions.

Establishment of Initial Testing Requirements.—Testing will be required in a hierarchical manner on selected categories of both new and existing chemicals. Industry will be required to develop data concerning both toxicity and exposure and to conduct risk assessments of the data. Quality assurance of the data that are developed will be stressed.

Regulatory Actions to Control a Limited Number of Environmental Problems Associated with Existing Chemicals.—In addition to early action on PCB's and selected chlorofluorocarbons, a limited number of serious chemical problems for which adequate data are currently available will be selected for intensive review and for regulatory action as appropriate. Concurrently, a systemmatized approach for identifying, characterizing, and controlling toxic substances under TSCA will be developed and implemented as rapidly as possible.

Assessment and Control of Unanticipated Problems of Urgent Concern.—Unexpected problems will inevitably arise and provisions will be made to respond to such problems without unduly disrupting other priority activities.

In these four priority areas, as well as in other areas, continuing attention will be directed to several overarching concerns. Activities will be oriented to serve the interests of both EPA and other organizations, particularly with regard to data dissemination. Data will be gathered on a highly selective basis to serve specific purposes. Confidentiality aspects will be a major factor influencing data collection, use, and dissemination strategies and activities.

Background

The Problem and TSCA

An estimated 1,000 new chemical substances are introduced into commerce each year in addition to the 30,000 or more which are currently used in many ways. The problems posed by the presence of some of these chemicals in the environment are too well known. Assessing and dealing with chemical problems involves the complex tasks of measuring their presence, estimating their effects, and evaluating the economic and social costs and benefits of their use and control. At the same time, there exists a wide variety of often overlapping and uncoordinated regulatory authorities and support activities directed to toxic substances at the national, State, and local levels. TSCA is designed to help reduce scientific uncertainties concerning toxic substances and to add coherence to the national effort to protect man and the environment from unreasonable risks without unnecessarily blunting a dynamic sector of our economy.

The Congressional Interest

Summary of the Legislative Deliberations.—Early in 1971, the Administration transmitted to Congress a proposal for toxic substances legislation. The report of the Council on Environmental Quality accompanying the proposal, *Toxic Substances,* documented some of the environmental problems which such legislation should address. The two Houses of the 92nd Congress passed different versions of the legislation late in the session. However, there was not time to appoint a conference committee to resolve the differences prior to adjournment.

---

* An Approach to Implementing the Toxic Substances Control Act, U.S. Environmental Protection Agency, February 1977.

Five different bills were introduced during the 93rd Congress, and by July 1973, the two Houses had passed different versions. However, the conference committee could not resolve several of the key differences between the two bills, including provisions concerning premarket screening and the relationship of TSCA to other laws.

In March 1976, in the 94th Congress, the Senate passed a somewhat revised version of the bill it had considered previously. In August 1976, the House passed still a different version. A conference committee met in early September, and the compromise bill that it developed was passed by both the House and the Senate in September 1976, with an effective date of January 1, 1977. The President signed the legislation on October 11, 1976.

During the deliberations of both the House and Senate in the 94th Congress, a number of technical aspects of the legislation were considered in detail. Thus, the committee reports of both houses, together with the report of the conference committee, provide considerable guidance concerning Congressional intent.

Findings.—In proposing toxic substances legislation, the Congress found that human beings and the environment are exposed to many chemicals, some of which may present unreasonable risks due to their manufacture, processing, distribution, use, or disposal and that to address this problem, regulation of both interstate and intrastate commerce is necessary [2(a)].

Policy and Intent.—The Policy of the United States enunciated in the Toxic Substances Control Act is that (a) adequate data on the effects of chemical substances should be developed as the responsibility of those who manufacture and process them; (b) authority should exist to regulate such chemicals which pose unreasonable risks and act on those which are imminent hazards; and (c) exercising this authority should assure that chemical substances will not present unreasonable risks yet not unduly impede technological innovation [2(b)]. The Congress intends that the Act be carried out in a reasonable manner, taking into account the environmental, economic, and social impact of actions taken under the law [2(c)].

Highlights of TSCA

TSCA authorizes EPA to obtain information about existing and new chemicals and take appropriate action against those which may present unreasonable risks. Manufacturers or processors of chemicals may be required to conduct tests and submit to EPA data on the effects and behavior of chemicals. EPA must be notified 90 days in advance of the manufacture of new chemicals and supplied with information necessary to evaluate their effects. When necessary, EPA is authorized to take steps to limit manufacturing, processing, use, or disposal of a chemical substance which may present an unreasonable risk.

TSCA contains several explicit authorities to promote better coordination among Federal agencies in identifying, assessing, and controlling toxic substances. If the Administrator determines that an unreasonable risk may be prevented or sufficiently reduced under a law not administered by EPA, he will request the relevant agency to evaluate the problem and take appropriate action. Similarly, other laws administered by EPA will be used in preference to TSCA when these authorities can adequately address the problems.

Overall Program Direction

General Goal in Implementing TSCA

The general goal in implementing TSCA can be stated as follows:

TO PROTECT HUMAN HEALTH AND THE ENVIRONMENT FROM UNREASONABLE RISKS PRESENTED BY CHEMICAL SUBSTANCES.

The program to be implemented pursuant to TSCA is one of a number of efforts, within and outside Government, directed toward this common goal. These other efforts include not only other programs of EPA and other Federal agencies, but also programs of the States, activities of environmental and public interest groups and of professional societies, policies of the financial and investment communities, and efforts of individual companies and trade associations.

However, the TSCA implementation program is somewhat unique in view of its breadth of coverage and its wide ranging authorities as contrasted to the narrower scope of other programs in the toxic substances area. Thus, there are greater expectations that the program will be able to work toward the common goal on a far broader basis than has been possible in the past.

The number of adverse chemical incidents can and must be reduced. But accidents will continue to occur, and chemicals posing environmental problems will undoubtedly slip through the net of assessment and control. Development of efficient and effective means to minimize the likelihood that such adverse incidents will occur—now and in the future—is the challenge of this legislation.

Purposes of TSCA Implementation Activities

The purposes in implementing the regulatory and non-regulatory provisions of TSCA are two-fold, namely

— TO CONTROL TOXIC SUBSTANCES DIRECTLY, and

— TO SUPPORT OTHER GOVERNMENTAL AND NON-GOVERNMENTAL PROGRAMS TO CONTROL TOXIC SUBSTANCES.

Inherent in the concept of control is the necessity to match the control requirements with the problems. In many cases, chemicals do not pose a threat to health or the environment, and no control is warranted. At the other extreme, there may be a necessity to ban a chemical altogether. In each case, judicious use of control alternatives must undergird successful implementation.

The dual Purposes of direct control through the regulatory authorities of TSCA on the one hand and, on the other hand, indirect control through the use of TSCA authorities to support the efforts of other programs in a position to control toxic chemicals reflect the clear intent of Congress. Implementation of this legislation must rely on achieving a "multiplier effect" through the efforts of many organizations if a significant number of commercial chemicals are to be addressed in the near future. Furthermore, the dual Purposes recognize the many common interests among a large variety of public and private organizations and the necessity to share and conserve resources whenever possible in addressing similar problems. Indeed, more than any other environmental legislation, TSCA should be considered as a mechanism to serve the interests of many organizations.

Several principles will guide implementation activities directed to these dual Purposes.

— The Environmental Protection Agency will assume a position of national leadership in the field of toxic substances control. At the same time, other organizations will be encouraged to utilize the full range of their authorities and capabilities to control toxic substances. In those cases when there are overlapping authorities or capabilities, EPA will encourage the use of the most expeditious and effective approach for addressing urgent environmental problems.

— Information obtained under the law will be made available as promptly and as widely as feasible to enable the expertise of other Federal, State, and local agencies and of the private sector to be utilized as fully as practical in meeting the purposes of the Act. Similarly, relevant information will be actively solicited from all available sources and used appropriately. International exchange of experiences and of data concerning toxic substances assessment and control will be strongly supported.

— All interested parties will have adequate opportunity to participate in development of regulatory requirements. The basis for regulatory decisions will be clearly stated at the time the decisions are made. To the extent possible, such statements will distinguish among scientific facts and uncertainties, scientific judgements, and value judgements.

Operationally, it may be desirable to establish formal agreements with other Federal agencies, and perhaps with other Governments and with international organizations. Also, systemmatic approaches to regularly obtaining up-to-date views on the priorities of other organizations will be developed.

Major Functional Areas

The activities to be conducted under TSCA have been divided into four major functional areas and several supporting areas (see Figure 1), recognizing that all of these areas are interrelated as discussed below. In particular, the acquisition and assessment of information provides the basis for regulatory and related decisions by both EPA and other organizations.

Major Functional Area No. 1: Acquire Information and Assess Risks to Health and the Environment. — TSCA provides several new authorities by which EPA may require industry to develop and provide to EPA information concerning chemical substances. These activities include submittal of information concerning the production, by-products of production, uses, and effects of chemicals. In some cases, such data may be readily available to industry. Sometimes, it will be necessary for industry to develop the data.

EPA will obtain information from other sources as well, including other Governmental organizations and private institutions. Also, EPA will, as necessary, carry out laboratory and field studies to supplement or to confirm data available from external sources. However, the emphasis will be on requiring industry to develop data necessary to assess the environmental acceptability of chemicals since TSCA explicitly places this responsibility squarely on industry [2(b)(1)].

The Agency considers that industry has the responsibility not only of gathering and assembling data but also of assessing the data to determine possible environmental risks. This responsibility is particularly important with regard to test data. Industry will be expected to prepare risk assessments in accordance with guidelines issued by EPA on data submitted pursuant to premarket notification and testing requirements. Obviously, the Agency will review such assessments and also conduct independent assessments when necessary.

In some cases when EPA is using TSCA authorities to acquire for other agencies data which are not of priority interest to EPA, the assessment process will be left to the other agencies. More often, there will probably be a congruence among the interests of EPA and other agencies in

FIGURE 1—GOAL, PURPOSES, ACTIVITY AREAS

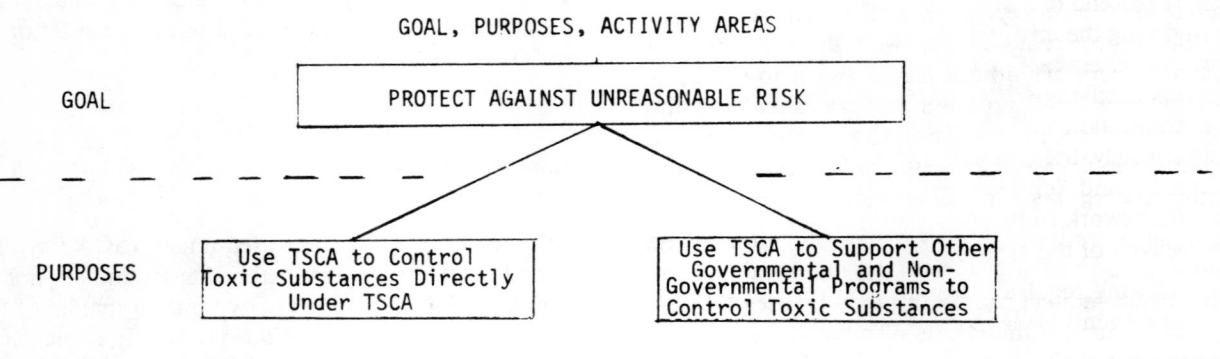

GOAL, PURPOSES, ACTIVITY AREAS

GOAL

PROTECT AGAINST UNREASONABLE RISK

PURPOSES

Use TSCA to Control Toxic Substances Directly Under TSCA

Use TSCA to Support Other Governmental and Non-Governmental Programs to Control Toxic Substances

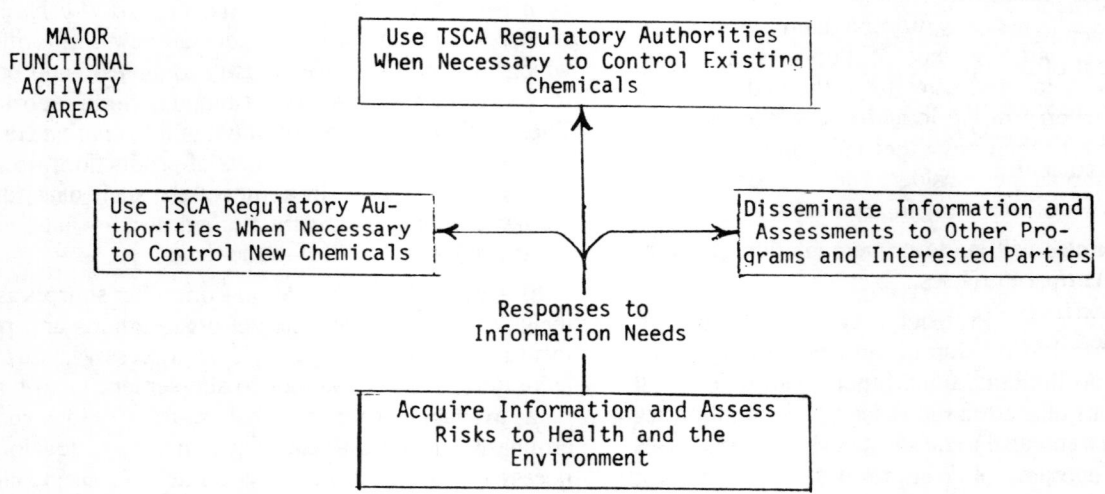

MAJOR FUNCTIONAL ACTIVITY AREAS

Use TSCA Regulatory Authorities When Necessary to Control Existing Chemicals

Use TSCA Regulatory Authorities When Necessary to Control New Chemicals

Disseminate Information and Assessments to Other Programs and Interested Parties

Responses to Information Needs

Acquire Information and Assess Risks to Health and the Environment

SUPPORTING ACTIVITY AREAS

Conduct Research

Assist Interested Parties

Implement TSCA Procedural Aspects

the same environmental problems and joint risk assessments among agencies will be in order.

Major Functional Area No. 2: Use of TSCA Regulatory Authorities when Necessary To Control New Chemicals.—TSCA calls for establishment from scratch by the end of 1977 of a premarket review system for considering the environmental acceptability of all new commercial chemicals. The system includes (a) final regulations clarifying which chemicals are subject to premarket notification and the notification requirements, (b) an appropriately trained staff to receive and assess the notifications, and (c) an internal EPA policy and procedural framework to help insure consistency, efficiency, and objectivity of the reviews.

The following regulatory mechanisms are available to control new chemicals when appropriate:

—Premarket notifications will not be accepted by EPA unless they include all information required by regulations [5(a) and 5(b)].

—If additional information is needed to conduct an adequate assessment, and such information is not available, the manufacture of the chemical will be delayed pending development of the information [5(e)].

—Should a proposed new chemical be determined to present a significant risk, regulatory steps may be taken to control the manufacture, use, or disposal of the chemical [5(f)].

Major Functional Area No. 3: Use of TSCA Regulatory Authorities when Necessary To Control Existing Chemicals.—Among the types of regulatory actions provided for in TSCA are [6(a) and 6(b)]:

—banning or limiting manufacture, processing, distribution, or use of a chemical.
—requiring warning labels.
—requiring specified disposal methods.
—requiring specified quality control measures during the manufacturing process.

There are many other statutes available for controlling toxic chemical problems. The rather comprehensive authority of TSCA is intended to be used when necessary to fill the gaps among these other authorities. Thus, the starting point in determining when and how TSCA regulatory authorities are to be used is an overall assessment of environmental problems associated with a chemical or group of chemicals and a determination as to which regulatory approach will most effectively reduce the problems.

If the most appropriate statute is administered by another Agency, EPA may request the other Agency to take action or explain the basis for non-action within a specified period of time [9(a)]. Close collaboration with the other regulatory agencies beginning with the initial dis-

covery of a chemical problem is essential to insure that this system of formal "referrals" does not unnecessarily disrupt the priority activities of these agencies.

Major Functional Area No. 4: Disseminate Information and Assessments To Other Programs and Interested Parties.—A key to achieving "indirect" control of toxic chemicals through other programs is an effective system to aggressively disseminate to the broadest possible audience information obtained through TSCA. To the extent possible the quality and reliability of the data should be clear, and information should be in a format that will be most meaningful to the users [10(b) and 25(b)].

An effective dissemination program is intimately linked to an earlier determination of the information needs of the users and the timing of these needs so that the appropriate data can be generated in the first place. Thus, the first step is a continuing program to assess user interests and needs and, to the extent possible, to shape the data collection efforts to meet their needs.

Much of the data collected for one user, including EPA, will undoubtedly be of interest to many other users as well. Thus, information will be placed in the public domain as rapidly as possible after receipt by EPA.

Some of the data collected under TSCA possibly will be subject to claims of confidentiality. Rapid and efficient mechanisms for segregating data which are confidential from other data are essential if the user community is to be adequately serviced.

Interrelationships Among the Major Functional Areas.—As shown in Figure 2, the four major functional areas are integral components of the overall system of assessing and controlling toxic substances. Figure 3 shows in more detail crosswalks among specific provisions of TSCA and indicates the reinforcing character of these provisions.

In general, the information acquisition activities (Area No. 1) are determined by the needs for data to (a) provide a basis for decisions concerning the control of new or existing chemicals under TSCA (Areas #2 and #3) and (b) service the interests of others (Area #4). In some cases, the data lead directly to risk assessments within EPA; in other cases, the raw data are forwarded for assessment to other programs.

Initial Priorities

During the initial implementation phase, it will be necessary to establish priorities among the many possible TSCA activities. These priorities should reflect continuing concerns of the Congress, the Agency, and the public that there may be a large number of currently unattended environmental problems which should be addressed very

**FIGURE 2—INFORMATION
FLOW AMONG FUNCTIONAL AREAS**

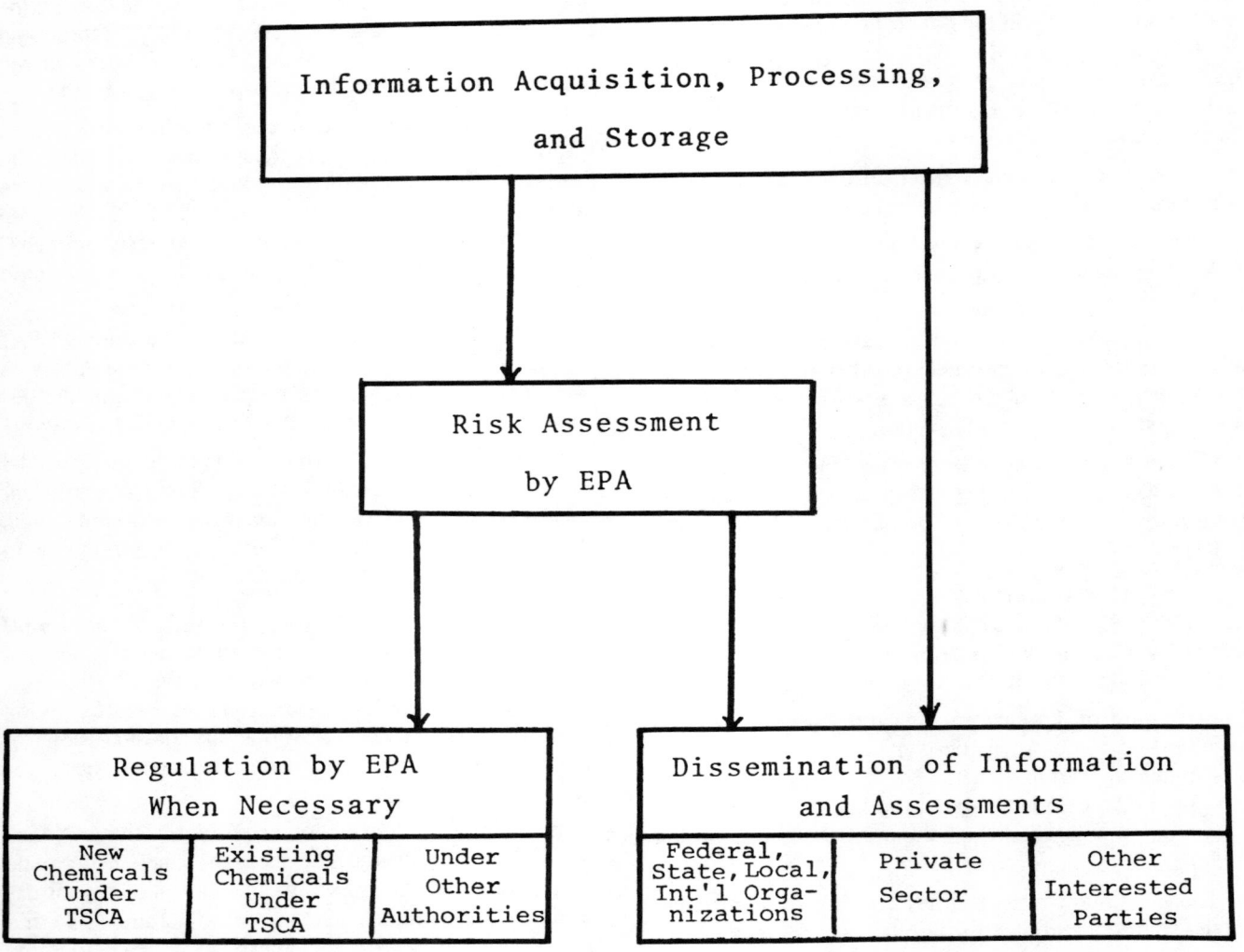

**FIGURE 3—SELECTED CROSSWALKS AMONG
PREMARKET TESTING, REPORTING, AND REGULATORY REQUIREMENTS**

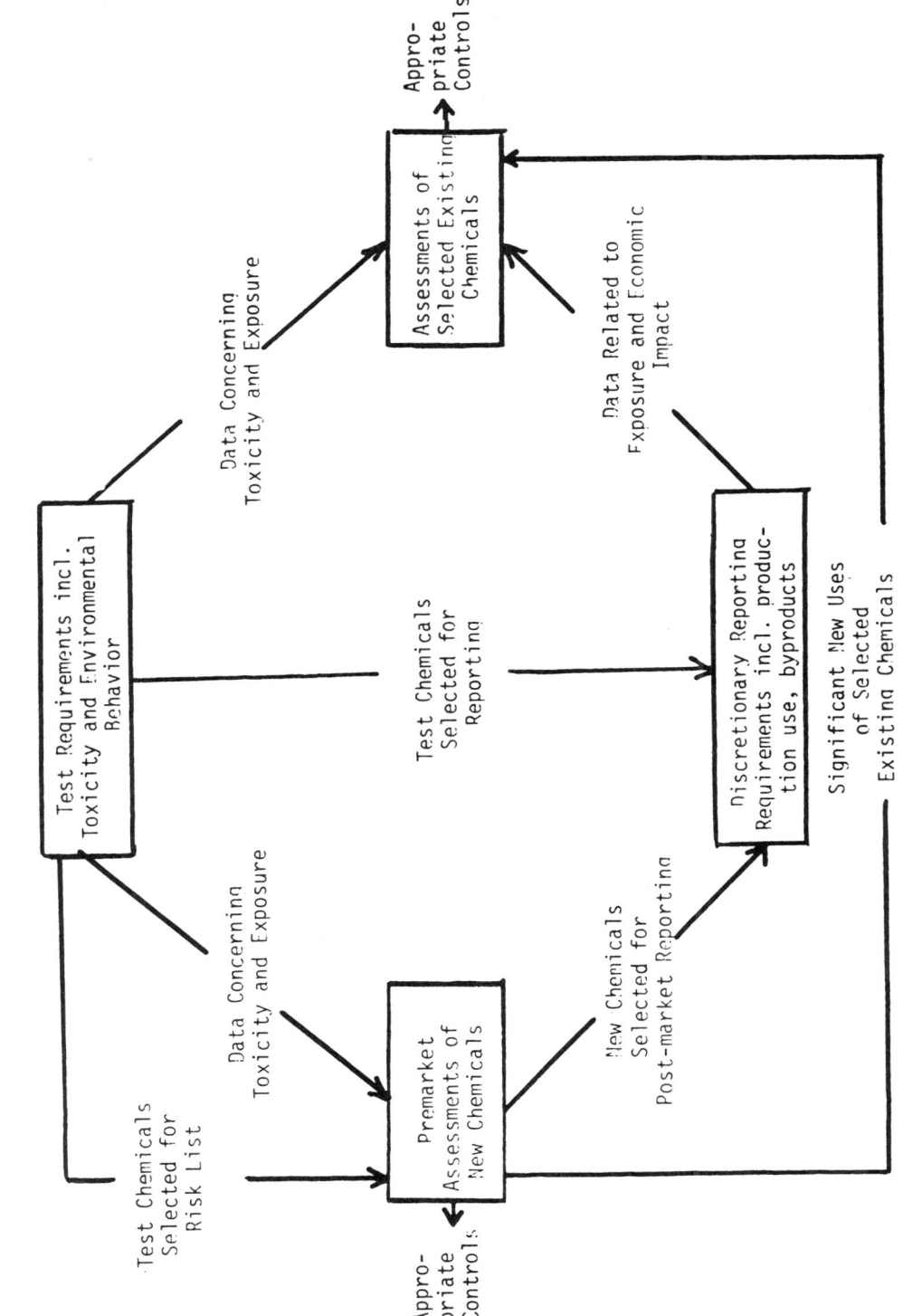

promptly. At the same time, early action must be taken to establish the policy and procedural framework for a long-term program, with the ground rules clearly understood by all interested parties. This is particularly true with regard to the premarket review system which should be environmentally meaningful and administratively efficient from the outset. And, of course, the legislative deadlines should be met.

With regard to regulatory activities, newly identified problems that present a significant risk to health or the environment must be dealt with immediately. Also, the chemical problems cited during the five years of legislative hearings should be promptly reviewed to determine whether regulatory action is needed. Finally, selection of early regulatory activities should include consideration of utilizing a variety of approaches that will not only reduce environmental risks but also clarify the dimensions, the strengths, and the weaknesses of TSCA.

Information is the lifeblood of the overall system. Early development of policies and procedures to establish a broadly based and technically sound data base, drawing on domestic and foreign sources and readily accessible to all interested parties, is essential if future regulatory actions are to be soundly conceived. Given the lead time necessary to develop health and environmental data, steps should be promptly initiated to begin to develop such data on those chemicals which could present serious problems. The information on the health and environmental aspects of commercial chemicals which is currently available only to individual agencies or companies should be promptly made more widely available.

In view of the unusual opportunity provided by TSCA to respond to these general concerns, EPA will give top priority to the following operational activities during the initial three years of TSCA implementation:

Establishment and Implementation of a Pre-market Review System.—The system will emphasize the (a) responsibility of industry to develop adequate data for meaningful chemical assessments, (b) categorization of new chemicals by broad chemical classes and broad uses with particular attention to those categories of greatest environmental concern, and (c) procedures for rapid decisions and adequate documentation of these decisions.

Establishment of Initial Testing Requirements.— Testing will be required in a hierarchical manner on selected categories of both new and existing chemicals. Industry will be required to develop data concerning both toxicity and exposure and to conduct risk assessments of the data. Quality assurance of the data that are developed will be stressed.

Regulatory Actions To Control a Limited Number of Environmental Problems Associated with Exist-

ing Chemicals.—In addition to early action on PCB's and selected chlorofluorocarbons, a limited number of serious chemical problems for which adequate data are currently available will be selected for intensive review and for regulatory action as appropriate. Concurrently, a systemmatized approach for identifying, characterizing, and controlling toxic substances under TSCA will be developed and implemented as rapidly as possible.

Assessment and Control of Unanticipated Problems of Urgent Concern.—Unexpected problems will inevitably arise and provisions will be made to respond to such problems without unduly disrupting other priority programs.

In these four priority activity areas, as well as in other areas, continuing attention will be directed to several overarching concerns. Activities will be oriented to serve the interests of both EPA and other organizations, particularly with regard to data dissemination. Data will be gathered on a highly selective basis to serve specific purposes. Confidentiality aspects will be a major factor influencing data collection, use, and dissemination strategies and activities.

There are many uncertainties as to the number of chemicals subject to premarket notifications and the availability of data concerning these chemicals, the number and extent of unanticipated problems associated with existing chemicals that will emerge, and the difficulties in gaining a scientific consensus concerning the most appropriate approaches to testing. Thus, there is little basis for estimating at this time the relative emphases that should be placed on each of these four types of activities. However, in setting priorities among the many types of potential environmental problems falling within these activities, the following principles will be considered:

—Toxic substance problems of national or global dimensions, and particularly those problems affecting many people or extensive ecological resources, will receive priority over localized problems which affect smaller populations.

—Special attention will be given to the effects of toxic substances on human health, recognizing that ecological impacts can also directly and indirectly affect human health.

—Toxic chemicals which are discharged into the environment in significant quantities and which persist and/or bioaccumulate will be of particular concern.

Protecting against Unreasonable Risk

The concept of unreasonable risk is the central element in relating TSCA implementation activities to the overall goal of TSCA. This concept recognizes that some level of risk must be accepted in our activities involving chemicals. Also, the welfare of future generations as well as our present population must be of concern.

In general, the most severe risks trigger the imminent hazard provision of TSCA [7(a)], and the possibility of less severe risks may trigger testing or other information gathering activities. In some instances the term "unreasonable risk" is used in the Act, in one case "substantial risk" is used [8(e)], and in other cases elaborations of "unreasonable risk" are presented.

The burden of proof for establishing the degree of risk varies in different sections. For example, the proponent of use must establish that selected new chemicals which are on a "risk" list [5(b)(4)(A)] do not present an unreasonable risk. On the other hand, EPA must make explicit findings on risk to take regulatory actions [6(a)]. Meanwhile, there is an overarching requirement on EPA to consider social and economic impacts, as well as environmental impacts, in taking action under any provision of TSCA [2(c)].

The most detailed elaboration in the Act as to the types of considerations involved in assessing risk is set forth in the guidance provided to the Interagency Testing Committee, but this guidance is still very general [4(e)(1)(A)]. Several other provisions of the law, as well as the legislative history, also emphasize the importance of considering carcinogenesis, teratogenesis, and mutagenesis when evaluating risks. While there are no explicit references in the law to acute toxicity concerns, there is no reason for not considering such risks.

Recent court opinions, such as those concerning lead in gasoline and asbestos in drinking water, provide some indication as to the important factors in risk assessment. The 1975 NAS report *Decision Making for Regulating Chemicals in the Environment* also makes some observations concerning those factors that should be considered in assessing risks and in balancing costs, risks, and benefits in decision making.

To foster a degree of consistency in the approach to risk assessments, the following steps will be taken:

—Minimum data requirements for conducting risk assessments will be developed for selected categories of chemicals (e.g., chemical classes, use categories).

—Guidelines for use by industry and the agency concerning risk assessments and the factors involved in cost-risk-benefit decisions will be developed and used.

—Risk assessments of the same chemical or same environmental problems by different programs and different agencies will to the extent feasible be consistent even though the determination as to whether regulatory action is warranted may vary depending on the differing requirements of different statutes.

—Summaries of relevant court opinions related to unreasonable risk will be prepared periodically and made available to risk assessors and decision makers.

—The feasibility of developing criteria for triggering different types of TSCA actions when the chemicals meet the criteria will be explored in detail.

Supporting Functional Activities

Research.—TSCA authorizes a broad range of research activities [10 and 27], and explicitly calls for EPA to direct attention to:

—development of screening techniques to help assess health and ecological effects [10(c)],

—development of monitoring techniques and instruments [10(d)],

—basic research to provide the scientific basis for screening and monitoring developments [10(e)], and

—training of Federal personnel to utilize the new techniques [10(f)].

HEW and other agencies are expected to intensify their research efforts directed to toxic substances as well.

EPA will be developing a more detailed framework for its initial five-year research program during the next few months. EPA intends to give high priority to research both in its own programs and in encouraging a more broadly based and more effectively coordinated national effort involving research by many organizations. Also, the Agency will encourage efforts to expand the pool of technical manpower needed by many organizations to carry out TSCA requirements.

Assistance to Interested Organizations.—Given the limited authorization of $1.5 million for each of the first three years of TSCA for support of a program of state grants (28), the initial grants will be limited to programs in only a few states. The states will be selected on the basis of the likelihood that the proposed programs will significantly upgrade local capabilities to address environmental problems. Should the initial grants prove to be successful in providing an important new dimension to toxic substances control, EPA will consider seeking additional funding in later years and will assess the desirability of incorporating the program into the broader concept of bloc grants.

The Agency will not be in a good position for some time to provide responses to petitions from industry requesting specification of test requirements for individual new chemicals [4(g)]. Within a few years the Agency should have available testing guidelines covering a broad range of chemicals, effects, environmental behavior patterns, and routes of exposure. At that time, authoritative responses to such petitions should be relatively easy to provide. In the interim, heavy reliance will be placed on using, whenever possible, the relevant portions of testing guidelines already prepared by EPA and other agencies for other programs such as those directed to pesticides, food additives, and drugs, taking into account the different types of exposures involved with industrial chemicals.

With regard to EPA assistance in defraying the costs of certain attorneys and witnesses in regulatory proceedings [6(c)(4)(A)], only limited funding will be available in FY 1977, but significant additional funding will be sought in FY 1978. Meanwhile, the criteria for determining eligibility are being developed. Available resources will be distributed on an equitable basis among all bona fide claimants at the end of each fiscal quarter for services rendered during that quarter.

An area of Congressional concern has been the capability of small and medium industrial firms to understand the legal and policy complexities of toxic substances control in relation to specific chemical concerns. Thus, in addition to exempting small business from certain reporting requirements, TSCA calls for an EPA Assistance Office to provide clarification on regulatory and related requirements [26(d)]. The Office has been established. In addition, each of the ten EPA Regional Offices will have an appropriate contact point to help clarify TSCA requirements for the small businessman in particular, and the public in general. The Assistance Office will not be in a position in the near future to provide advice on the environmental acceptability of specific commercial chemicals of interest to individual parties. However, the Office will help identify some of the parameters that should be considered in such assessments and will of course assist in obtaining available information relevant to the particular chemicals of interest.

Procedural Aspects.— A number of TSCA provisions call for development of procedural rules. Also, more explicit guidance will help facilitate implementation of other provisions. Some of EPA's earliest activities will be directed to:

—procedures governing the hearings to be conducted under TSCA and particularly with regard to regulatory actions [6(a)].

—the requirements for selected EPA and HEW employees to disclose their financial interests [26(e)].

—clarification of the requirements concerning claims of confidentiality and requests for information under the Freedom of Information Act [14].

—preemption of State laws [18].

Several TSCA provisions are of special interest to another Federal agency and will receive attention in the near future. For example:

—national defense waivers (Department of Defense) [22].

—employee protection (Department of Labor) [23].

—definition of small manufacturers and processors (Small Business Administration) [8(a)(3)(B)].

—notice to foreign governments of exports (Department of State) [12].

—imports (Department of Treasury) [13].

—reimbursement for use of test data generated by another party (FTC/Department of Justice) [4(b)(3) and 4(c)].

Finally, special studies and reports are called for as follows:

—study of the indemnification aspects of all laws administered by EPA [25(a)].

—annual reports to Congress on TSCA actions [30].

Acquisition of Information and Assessment of Risks To Health and the Environment

General Policy Framework

The gathering, processing, and storing of information will be designed to support specific requirements of EPA and other organizations and will not be considered an end in itself. Data requirements may be very specific or quite broad. For example, information activities may be oriented to:

—Development of specific regulations to control new or existing chemicals under TSCA or other EPA authorities;

—Identification and prioritization of problem chemicals for attention under TSCA or other authorities;

—Supporting specific activities of other agencies and other organizations to assess and control problem chemicals.

—Informing the public of chemical activities and associated problems.

The intended use of the information will determine the type and extent of data that are needed, the timing, the sources, and the most appropriate mechanisms for acquiring the data. When appropriate, data already available in Government files will be used. When industrial data are required, data already available to industry should be used to the extent possible. However, in a number of cases, and particularly with regard to new chemicals, additional data must be developed by industry on a routine basis.

Given the interests of many parties in data that can be developed under TSCA, and the multiplicity of data acquisition efforts already in place, the development of new TSCA data requirements will be coordinated widely within and outside the Government. Such coordination should assist in (a) reducing the likelihood of gaps in information needed for regulatory purposes, (b) conserving resources of EPA and other agencies, and (c) limiting the total reporting burden on industry.

Particular attention will be given to assuring that information that is acquired, and particularly industrial data, is accurate, complete, and current. Among the steps to be considered are:

—Clarity and precision of record keeping and data submission requirements;

—Procedures for assuring appropriate quality control in the acquisition and reporting of information;

—Interim reports of progress in fulfilling long-term data requirements to avoid discovery only at a very late date that unacceptable data have been generated;

—Spot checks and audits of data generation activities conducted in the United States or abroad;

—Development of confirmatory data to check on accuracy and reliability of submissions;

—Prompt and vigorous enforcement against violators of data requirements;

—Encouragement of open communication with industry to discuss implementation problems in satisfying data requirements.

Disclosure of Data

Assertions of trade secrecy and related confidentiality matters could cause many implementation problems and must be addressed promptly. Prior to gathering information which is likely to include confidential data, the need for the information should be very clear. Submitters will be afforded the opportunity to make confidentiality claims at the time of submission. A clear explanation of the consequences of failing to assert confidentiality and the procedure for resolving disputes will be a part of the request for information.

TSCA specifically excludes from claims of confidentiality health and safety studies on chemicals offered for commercial distribution and on chemicals subject to premarket notification and/or testing requirements [14(b)]. Other data which the manufacturer, processor, or distributor consider confidential may be so designated but must be segregated from other non-confidential data. Thus, EPA reporting forms will require confidentiality designations for individual items, and a general confidential stamp for an entire form will not be accepted.

If there is a Freedom of Information request or other action concerning release of data designated as confidential, the originator of the data will be given an opportunity to justify the claim of confidentiality in detail prior to the Administrator's decision on the release of the data. Also, he will have an opportunity to seek judicial relief if there is disagreement with the Administrator's determination. However, if the release of data is necessary to protect health or the environment against an imminent, unreasonable risk, the advance notice of release can be as short as 24 hours [14(c)].

The information system for receiving and storing TSCA data will insure appropriate protection of confidential information.

TSCA Information Gathering Authorities

Figure 4 identifies some of the information that will be received as a result of both non-discretionary and discretionary provisions of TSCA. Non-discretionary provisions include:

—Reporting of chemicals to be included on the initial inventory of existing chemicals [8(a) and 8(b)]
—Premarket notifications [5(a)]
—Industrial reporting of substantial risks associated with commercial chemicals [8(e)]
—Citizen's petitions [21]

Some of the provisions to be implemented at the discretion of EPA are:

—Testing requirements [4(a) and 4(b)]
—Industrial reporting of production and related activities [8(a)]
—Industrial submission of records of adverse reactions to health or the environment [8(c)]
—Industrial submission of health and safety studies [8(d)]
—Inspections and subpoenas [11]
—Acquisition of information from other Agencies [10(b)(2)]

Figure 5 sets forth the implementation timetable for the principal activities related to TSCA testing provisions and the provisions concerning industrial recordkeeping and reporting. These activities are also discussed below.

Risk Assessments

Risk assessments are required under a number of TSCA provisions. Risk assessments are inevitably conducted with less than optimal data—either in terms of quantity or quality. However, a minimum level of data is essential if assessments are to be meaningful, and EPA

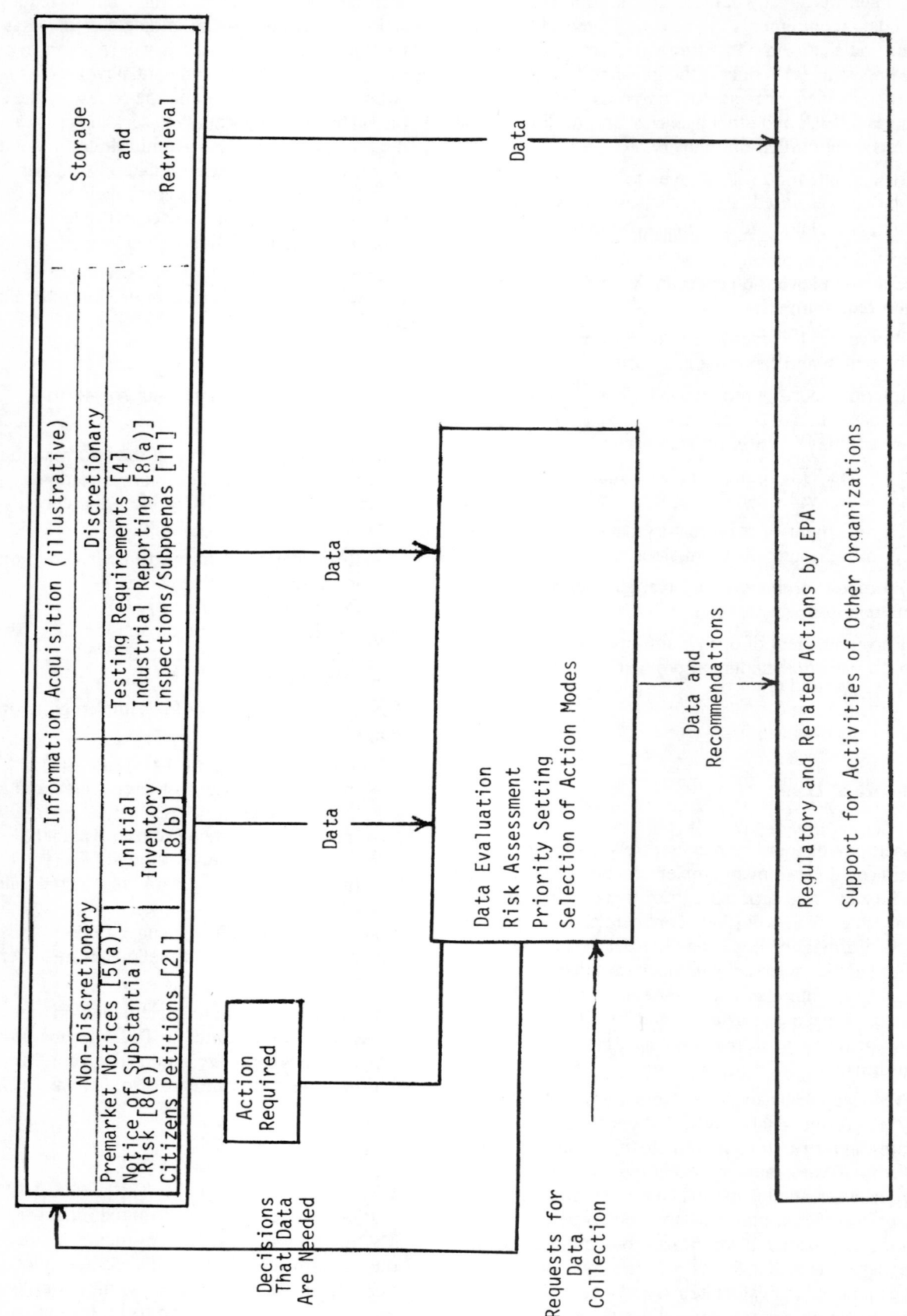

FIGURE 4—INFORMATION ACQUISITION AND ASSESSMENT

will develop guidelines concerning such minimum data requirements.

Manufacturers will be expected to conduct and submit risk assessments, prepared in accordance with EPA guidelines, when developing data under TSCA testing regulations and when submitting premarket notices. Also, risk assessments may be required in connection with information provided under other TSCA provisions. The assessments will utilize not only data developed by the manufacturer but also other available data on the chemicals, and when appropriate, on closely related chemicals. In addition to providing a better basis for EPA evaluation of the likely environmental problems associated with toxic substances, this requirement should stimulate a much broader industrial effort to better understand the environmental acceptability of many chemicals being produced and used.

While much remains to be done to clarify, even in a preliminary way, what is involved in risk assessments, the following types of considerations are important:

—Chemical structure, physical properties, contaminants
—Source assessment and exposure potential: quantities produced, production processes, uses, reactions involved in uses, and types and frequency of environmental discharges
—Environmental behavior and fate, including degradation rates and products, chemical reactions in the environment and inadvertent products, bioaccumulation and biomagnification potential, and possible synergistic effects
—Acute, chronic, and subacute effects on man; absorption, excretion, and metabolism
—Effects on vertebrates, invertebrates, microorganisms, and plants
—Effects on inanimate objects and structures

Testing Requirements

Almost all agencies and organizations in a position to assess and control toxic substances have stressed the need for more timely and more reliable test data on a wide range of chemicals. Test data will be used to support specific regulatory actions and to identify problems that need attention. TSCA test requirements that are developed can be pacesetters for the entire field of environmental assessment techniques. Thus, priority will be given to implementation of the testing provisions of TSCA, recognizing that scientific uncertainties and overlapping organizational interests will complicate rapid progress in this area.

The initial implementation activities related to testing will involve (a) establishing the procedures and format for submitting test data under a variety of TSCA provisions, including premarket notifications which will begin to arrive in December 1977, (b) determining general policies which will provide the framework for TSCA test requirements for determining specific effects of specific chemicals, including procedures for assuring the quality of test data, and (c) developing the initial TSCA test requirements for selected categories of both new and existing chemicals.

In view of the number of chemicals of potential interest, test requirements directed to categories of chemicals is the most appropriate approach. The categories will be based on similarities in chemical structure, chemical use, and/or levels and routes of likely exposure, recognizing that production volume may often be an appropriate surrogate for exposure potential. Not only will such an approach simplify the sorting of priorities, but it will also provide a means for anticipating problems with new chemicals and new uses. In general, EPA will develop general testing requirements for each selected category of chemicals and the manufacturer will in turn propose a detailed protocol for each chemical for EPA approval.

Given the current limitations on the availability of testing facilities and personnel and the scientific uncertainties concerning some types of test methods, the testing requirements during the first several years will be developed on a selective basis directed to some of those chemical effects, types of chemical behavior, and routes of exposure of immediate concern. At the same time, a more comprehensive effort will be undertaken to develop testing approaches on a much wider range of chemicals, effects, behavior, and exposure routes. The recommendations of the Interagency Testing Committee and of other interested parties will be considered within this framework. In short, the categorization scheme can be "tuned", in terms of number and breadth of categories and types of effects and exposure, in accordance with the capability to generate sound data and to use the data effectively.

In general, TSCA testing requirements will:

—require only such data as are necessary to reach regulatory and related decisions.

—provide the flexibility to enable initial state-of-the-art protocols to be effected immediately while permitting modifications and improvements to be made as the science of testing progresses. EPA will update protocols when appropriate without jeopardizing the acceptability of testing already underway in accordance with earlier protocols.

—document the laboratory analytical procedures and the requirements for estimating risks so that industry will clearly know what is required and how the information

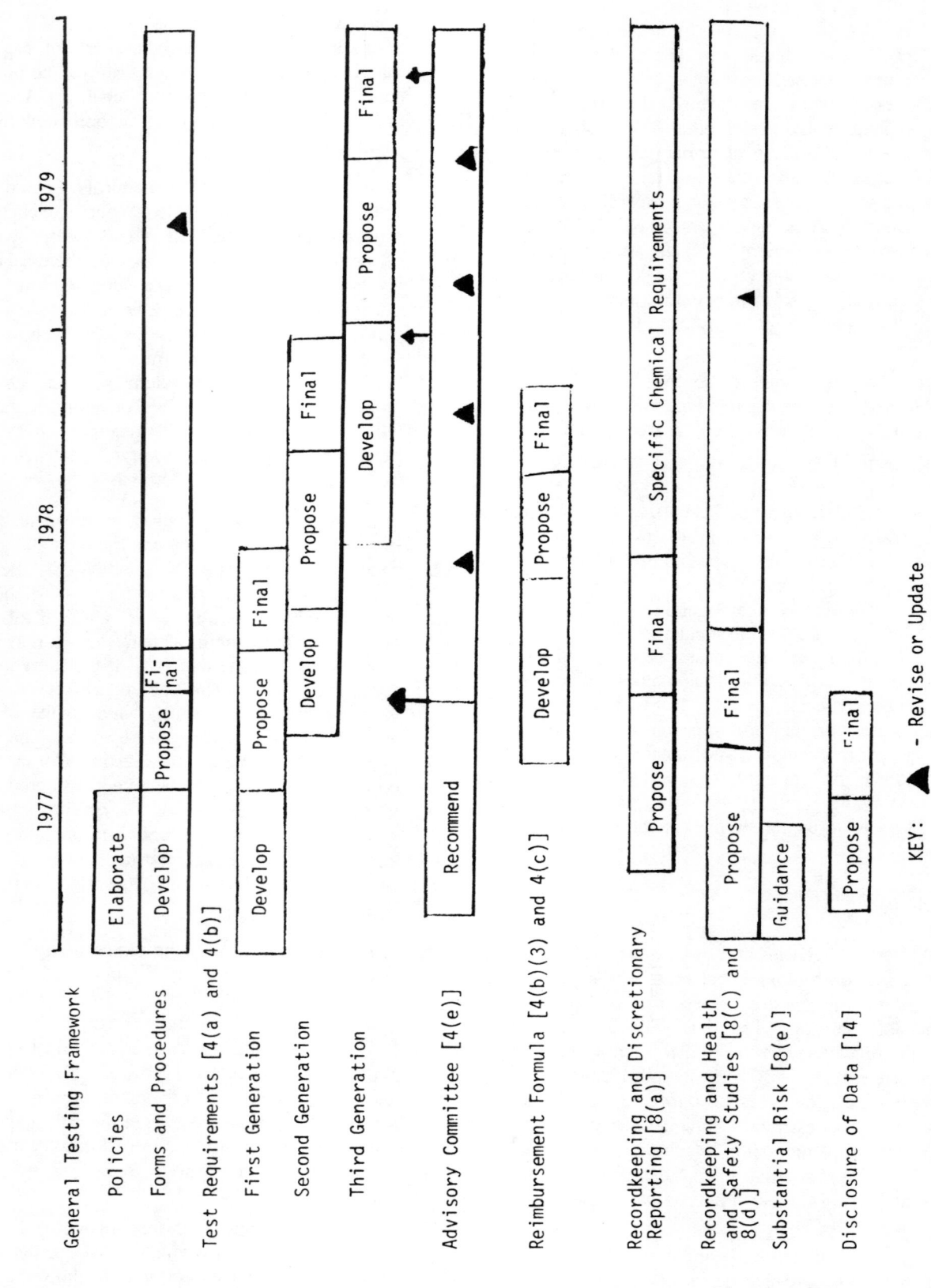

FIGURE 5—IMPLEMENTATION TIMETABLE: TESTING AND REPORTING

FIGURE 6—ORIGIN OF TEST REQUIREMENTS

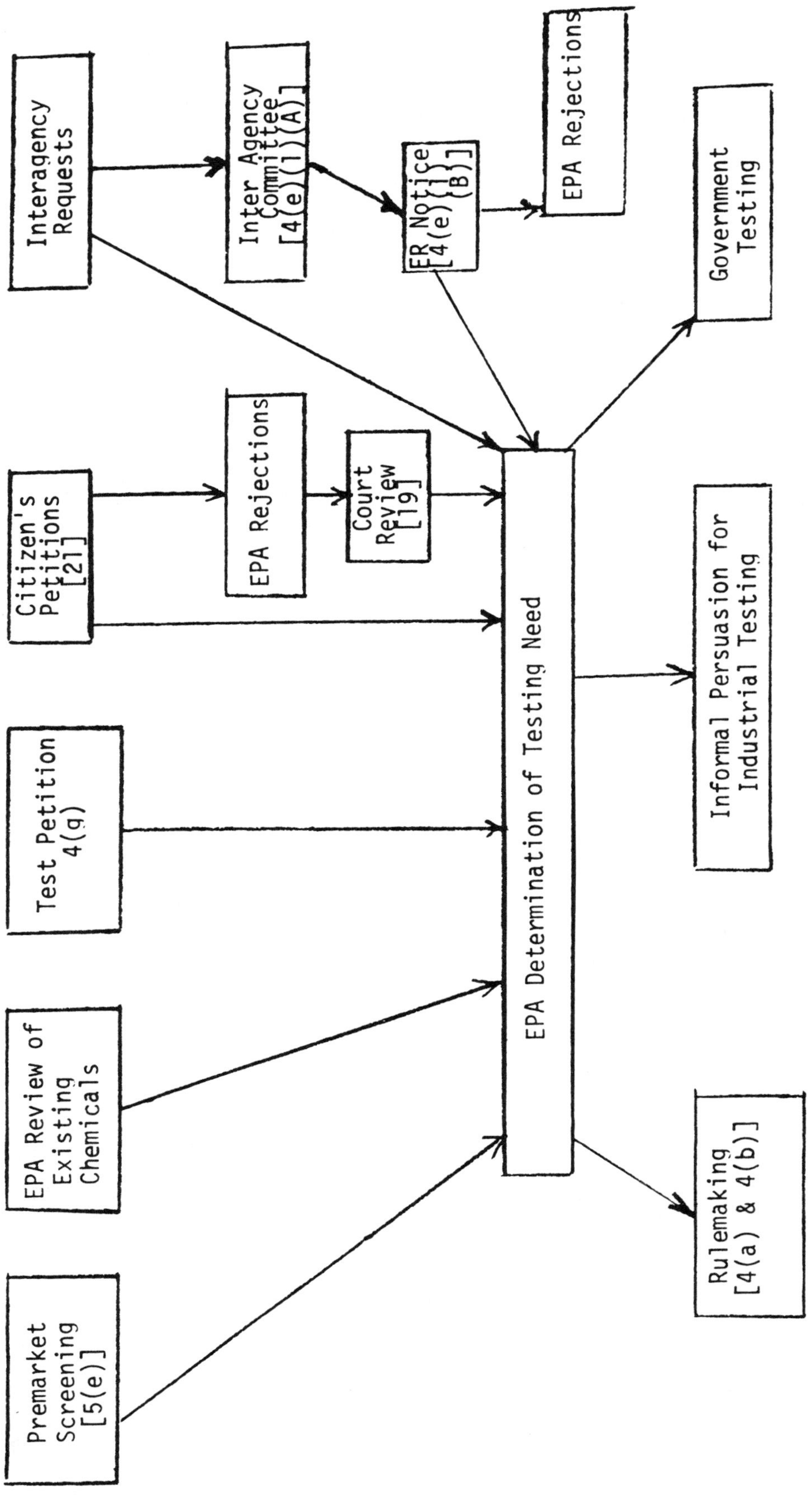

will be assessed. Industry should be able to conduct the same assessments of their data as EPA.

—provide an approach for developing test protocols in a hierarchical manner so that the degree of testing and evaluation is related to the likely exposure and the projected economic and social benefit.

Consistency of test requirements issued by different programs directed to the same classes of chemicals is important. Also, compatability with recommendations of international organizations and with requirements of other Governments is important. However, two pitfalls must be avoided. First, overstandardization of requirements could stifle advances in the state-of-the-art. Secondly, the types and levels of exposures associated with industrial chemicals usually differ significantly with the exposures associated with drugs, food additives, and pesticides. Therefore, the types and extent of testing of industrial chemicals may differ greatly from the more comprehensive approaches used in these other programs.

In developing the near-term requirements, as well as the more comprehensive long term approach, the Congressional concern with the following aspects will be kept in mind:

—carcinogenicity, mutagenicity, teratogenicity
—levels of environmental exposure
—behavioral effects
—synergistic and cumulative effects

With regard to assuring the quality of data that are submitted by industry, EPA will explore the feasibility of expanding the FDA laboratory inspection program and the use of the FDA Good Laboratory Practices manual to encompass TSCA concerns. A counterpart system for environmental testing will be explored within EPA. In the longer term, other approaches may be more appropriate to assure that data are obtained and presented in a technically credible manner [3(12)(B)].

During 1977 the Interagency Advisory Committee will designate up to 50 priority chemicals for testing. The Agency will have up to one year to respond to these recommendations [4(e)(1)]. Meanwhile, as indicated in Figure 6, other testing recommendations will be received from a variety of sources. A sorting of the various recommendations, taking into account the availability of testing facilities, will be a major activity during 1977-78. While such prioritizing is important, promulgation of initial test requirements will not be delayed pending development of an elaborate sorting system which may never be feasible. Early EPA action will demonstrate the Agency's intent to consider all suggestions but then to move in the direction of the priorities as viewed by the Agency.

Policies and procedures will be prepared for enabling the developer of test data to receive fair reimbursement from another party who wishes to use the data [4(b)(3) and 4(c)].

Industrial Reporting and Retention of Information

The recordkeeping and reporting provisions of TSCA are illustrated in Figure 7.

Discretionary acquisition of industrial data concerning production, by-products, and uses [8(a)] will be used for a number of purposes, including:

—to keep track of the commercial development of new chemicals after they have been marketed for the first time. Although the initial production and use patterns might not warrant regulatory intervention in the premarket period, changes in these patterns could become of environmental concern.

—to provide information related to likely types of exposure on existing chemicals which are subject to testing requirements.

—to provide data related to exposure and also related to economic impact of controls on those chemicals which are being considered for regulatory action.

—to provide information on the impact of controls which limit but do not ban chemicals.

—to help pinpoint the types and locations of exposure potential when unexpected chemical problems of urgent concern arise.

—to provide trend data concerning newly emerging chemical technologies which should be of special concern.

The initial emphasis will be on requirements for the establishment within individual industrial firms of complete, up-to-date, and easily accessible records concerning each of the individual chemicals which is used by the firm. Such records will include information concerning the chemical identity, production levels, uses, by-products, health and safety studies, alleged adverse reactions, and other factors of environmental significance [8(a) and 8(c)]. These records will be available for onsite reviews which might trigger requests for the data to be reported to EPA. Also, as individual chemicals are being analyzed in detail by EPA and other agencies, information concerning these chemicals will be requested. In order to know which manufacturers to approach concerning specific chemicals, EPA will require each manufacturer to identify all the chemicals he manufacturers in connection with compiling the initial inventory of existing chemicals [8(b)].

The responsibilities of importers and their suppliers concerning recordkeeping require special attention.

The general procedural requirements for reporting will also be established promptly with the specific chemicals to be subject to reporting identified subsequently [8(a)].

**FIGURE 7—INDUSTRIAL RECORDKEEPING AND
REPORTING FOR EXISTING CHEMICALS**

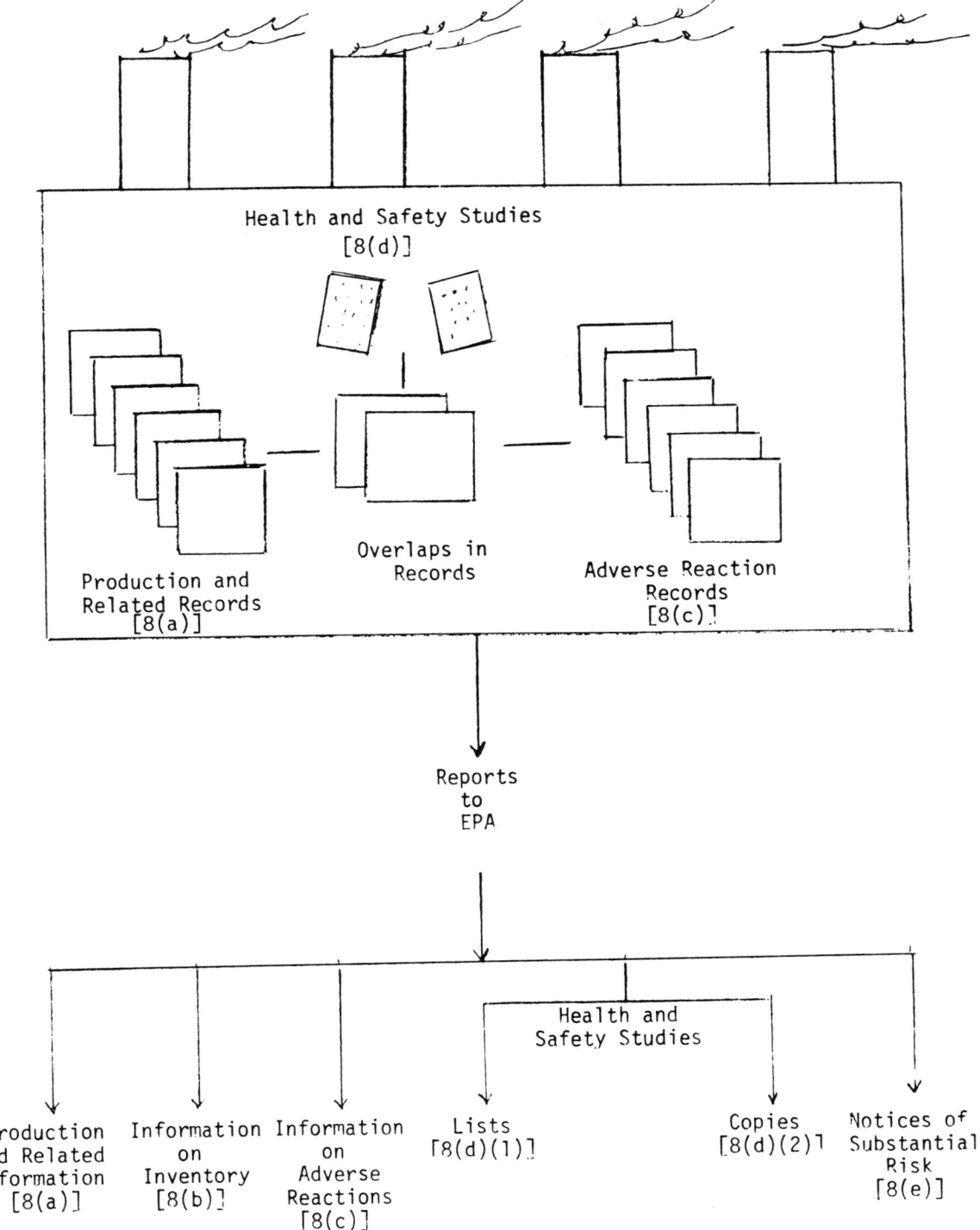

Data on specific chemicals could then be obtained with a minimum of delay. In the early years, this authority will be used selectively to help assess chemical problems under active investigation. At a later date, it may be desirable to build up within EPA a broader base of industrial data. However, before comprehensive reporting requirements are developed, more thorough investigations of the availability of comparable data through other Governmental and non-Governmental mechanisms will be carried out to avoid duplicative reporting.

During the Congressional hearings, concern was repeatedly voiced that considerable amounts of data related to the health and environmental acceptability of commercial chemicals were available in industry files. While the value of such data are unknown, the extent of the backlog of data will be promptly ascertained [8(d)]. Should a manufacturer become aware of data generated by himself or others concerning the likelihood of substantial risks due to chemical exposures [8(e)], notifications to EPA of this data will trigger the activities described under the response section below.

Use of TSCA Regulatory Authorities When Necessary to Control New Chemicals

TSCA Authorities To Control New Chemicals

Priority will be given to establishing the program for premarket screening which must be in place by the end of 1977. TSCA requires notification of all new chemicals at least 90 days prior to their manufacture to provide the Agency an opportunity for determining whether such chemicals pose a risk. If the Agency believes that a new chemical poses an unreasonable risk, of if information is lacking to make an evaluation, the Agency may initiate action leading to the following:

—delay in the manufacture of the new chemical pending development of additional information needed to evaluate the risk [5(e)],

—ban or marketing limitation on the new chemical to protect against an unreasonable risk [5(f)], or

—requirements to track the commercial development of the new chemical after initial marketing through periodic reporting by industry of its production levels, uses, by-products, and related aspects [8(a)].

Also linked to the premarket review are several TSCA authorities which can be used to require certain types of information to be submitted at the time of notification, namely:

—production, use, by-product and related data [5(a)(1)]

—data required by testing regulations [5(b)(1)]

—health and environmental data on chemicals on a "risk" list [5(b)(4)].

If required data are not submitted, the notification will not be accepted.

Preparation of the Inventory of Existing Chemicals

The first order of business to implement premarket screening for new chemicals is the compilation of an inventory of existing chemicals. This inventory will be regularly updated and will serve as the baseline for determining which chemicals to be manufactured after publication of the inventory are "new" and, therefore, subject to premarket notification.

Each manufacturer will be required to report for inclusion in the inventory all chemicals produced after January 1, 1977. Manufacturers may report other chemicals produced between July 1, 1974, and January 1, 1977, if they wish to have these chemicals included; chemicals which are not reported will be subject to premarket notification requirements. To assist in standardizing the nomenclature used in reporting for the inventory, the Agency will publish a candidate list of about 30,000 chemicals, with appropriate names and CAS registry numbers. Also, instructions will be provided for reporting other chemicals not on the candidate list.

The initial inventory will not differentiate among various technical grades of chemicals. The problems related to impurities will be addressed at a later date. The initial inventory will include some categories of chemicals since it is not practical in the short time available, and it may not be desirable, to list every variation of all existing chemicals. Raw agricultural products, for example, will be considered as a single category. Also, minerals will be included on the basis of relatively broad categories.

The exemption from the inventory and from premarket notification of "small" quantities for research purposes will include those amounts no greater than what is reasonably necessary for scientific experimentation, testing, analysis, or research, including such research or analysis necessary for the development of a product [5(h)(3)]. "Test marketing" will be limited to distribution of a chemical to a defined number of potential customers for purposes of evaluating particular uses of that chemical during a predetermined evaluation period [5(h)(1)].

Many intermediate chemicals are often involved in the synthesis of industrial chemicals and are subject to TSCA. Only those intermediate chemicals which cannot be isolated in a practical sense from the immediate vicinity of the reaction process will be exempt from inclusion on the initial inventory and from premarket notification. Thus, for example, those chemicals which normally exist in only a pipeline would be included.

The Agency is considering requiring that all premarket notifications be accompanied by a fee [26(b)]. This requirement would not only help defray costs of administration but would also help insure that Agency efforts are directed to environmental assessments of serious commercial endeavors and are not diverted to address theoretical curiosities submitted by parties who have no intention to manufacture the chemicals commercially. Also, the Agency is considering placing a limit of perhaps one year on the time between premarket notification and initiation of manufacture of the chemical. If the time is exceeded, a second notification would be required.

Data Requirements for the Review of New Chemicals

A meaningful review of new chemicals requires adequate data to review. Thus, EPA reviewers will be provided with internal guidance on minimum data requirements for new chemicals based on chemical categorization considerations. Data requirements will vary among categories. This guidance will be made publicly available.

The guidance for each category will identify the types of data which should normally accompany notification of a new chemical in that category including in some instances data not explicitly required by regulations. A notification submitted without the data will be a candidate for possible action to delay its commercialization. In that event, the Administrator may issue a proposed order to delay manufacture, based on insufficiency of information to "permit a reasoned evaluation of the health and environmental effects" of the new chemical [5(e)]. The guidance will merely serve to alert Agency reviewers to possible problems, and the absence of data will not automatically cause the Agency to issue a proposed order. The Agency is prepared, however, to issue proposed orders when appropriate.

In some cases, it may be appropriate to convert portions of the internal guidance into formal testing requirements. In that event, premarket notifications would not be accepted in the absence of such data. As the guidance is developed, the desirability of such testing requirements will be actively explored.

The Agency does not plan to establish a risk list of new chemicals which pose "an unreasonable risk of injury to health or the environment" at the outset [5(b)(4)]. The internal guidance by chemical categories will help insure a meaningful review of premarket notices, and in a sense perform much of the same function as a risk list. Also, the Agency does not plan to activate reporting of significant new uses of existing chemicals in the short term because of the urgency and complexity of the task of instituting

premarket notification for new chemicals and the difficulty of preparing a meaningful categorization of uses of environmental significance [5(a)(2)].

Limiting Manufacture or Marketing of New Chemicals

The internal guidance on data needs will also assist in determining the key factors underlying decisions as to whether to restrict new chemicals [5(f)]. However, each regulatory decision must be made on a case-by-case basis, and decision-making criteria beyond a check list of the factors to be considered does not appear feasible at this time.

There are often many uncertainties about the commercial viability of new chemicals. In some cases, the likelihood that the potential environmental problems uncovered during premarket review will become actual problems may be questionable. Therefore, alternatives to the resource-intensive process of formal rulemaking and, when necessary, court intervention to limit manufacture or marketing will be considered. For example, simply releasing adverse data to interested parties may in certain instances result in limitations on the type of commercialization which poses environmental problems. However, when necessary, EPA will intervene in a regulatory mode. If worker exposure is the principal concern, referral to OSHA might be the most appropriate course [9(a)].

Review Procedures for New Chemicals

As indicated in Figure 8, the Agency will initially determine if the premarket notification contains the information required by regulations. If such information is missing, the notification will not be accepted. An FR notice indicating receipt of a complete notification will be published within five days [5(d)]. The notification will be accepted even if it is not accompanied by the additional data specified in the internal EPA guidance for the new chemical's category.

The internal guidance will also indicate (a) those categories of chemicals which should in all cases be subjected to in-depth reviews, and (b) those categories which should be screened to determine if in-depth reviews are warranted. In the latter case, technical reviewers will identify the specific chemicals for which no action appears warranted and those that deserve in-depth review, along with all the new chemicals in the first group.

The in-depth review will routinely be completed in 30 days and will include consideration of:

—Physical and chemical properties of the chemical and its byproducts;

—Health and ecological effects;

FIGURE 8—PREMARKET REVIEW SYSTEM

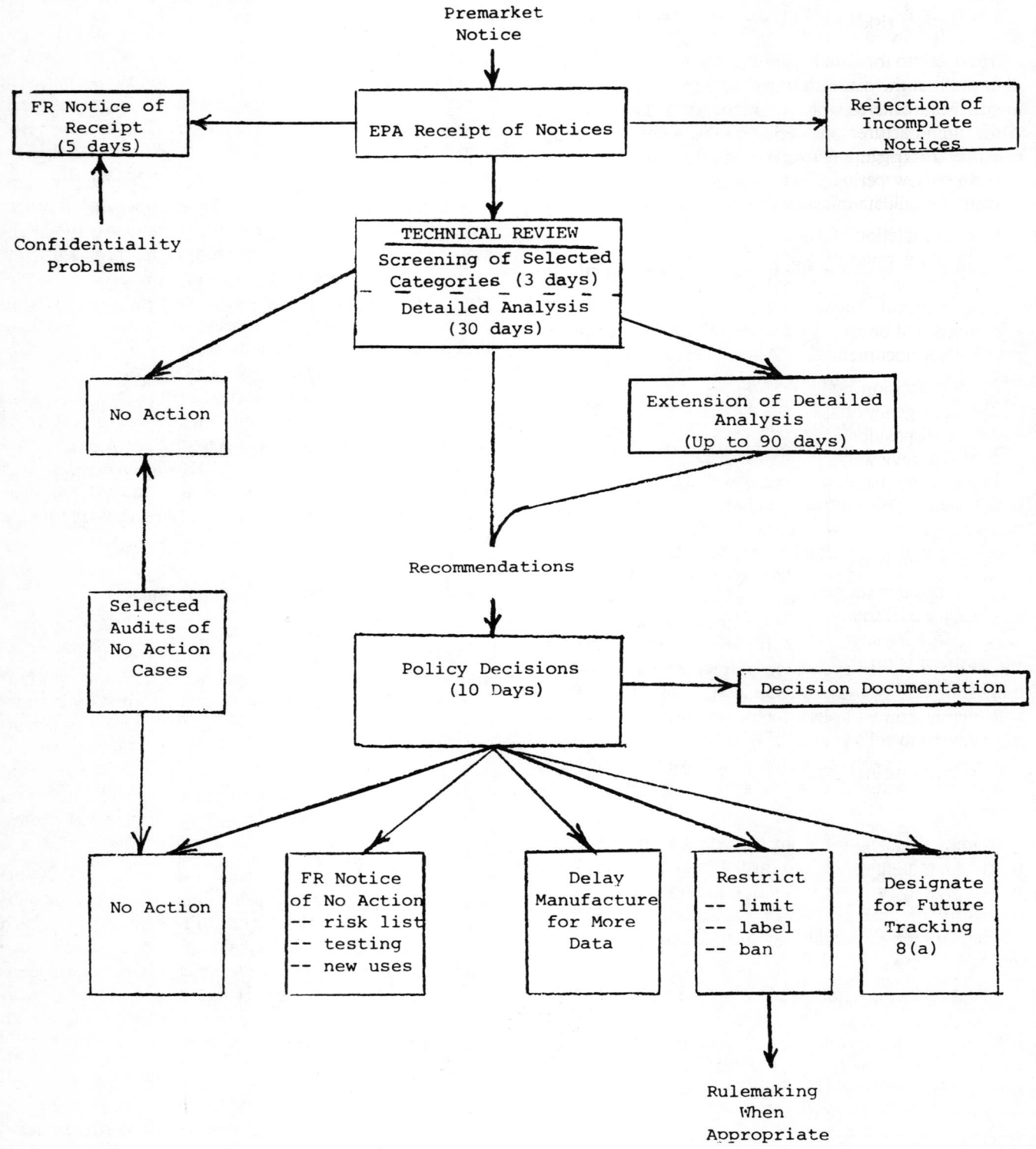

—Environmental behavior and fate including persistence and bioaccumulation;

—Likely sources of environmental discharges and exposed populations;

—Technological and economic factors;

—Industry's risk assessment.

In addition to the data included in the premarket notice, other data on the same chemical and, when appropriate, on chemical relatives which may not have been available to the manufacturer will be reviewed. Criteria will be developed for triggering extensions of up to 90 days of the in-depth review period. These criteria might include the necessity to validate questionable data.

At the completion of the in-depth review, including the supplementary review when appropriate, recommendations will be prepared as to the action that should be taken on each chemical. The recommendations and supporting justification will be concisely summarized in a short decision-making document.

A policy decision body including representatives from appropriate Agency offices will meet on a regular basis and decide the disposition of each of the recommendations of the technical reviewers. The decision will be appropriately documented so that at a later date the basis for the decision is clear. The following types of decisions are envisaged:

1. No action is warranted.

2. No action is warranted but an FR notice of the decision is required if the chemical is subject to testing requirements. Also, at a later date an FR notice will be required if the chemical is on the risk list or is subject to significant new use reporting [5(g)].

3. Manufacture will be delayed pending development of additional data [5(e)].

4. Manufacturing and marketing will be regulated [5(f)].

5. While no action is taken during the premarket period, the chemical will be ear-marked for future reporting after it becomes commercialized [8(a)].

"No action" can be decided by technical screeners on chemicals in selected categories and by the policy body on other chemicals. All "no action" decisions will be subject to a later audit on a selective basis to help insure that the necessity to screen out quickly the less worrisome problems does not result in inappropriate chemicals inadvertently slipping through the review process.

The Agency will give prompt attention to development of a data recovery system for premarket submissions. This will allow use of the Agency reviewers' comments and prior analysis of one new chemical when later reviewing a different, although similar new chemical. The data system

will also permit use of premarket data for other purposes under TSCA.

The premarket screening implementation timetable is set forth in Figure 9.

Use of TSCA Regulatory Authorities When Necessary To Control Existing Chemicals

General Regulatory Approach

TSCA is designed to fill the void in the regulatory span of other authorities. It is intended both to prevent problems and to correct problems.

As indicated in Figure 10, EPA plans to initiate a limited number of regulatory actions directed to existing chemicals in the near future. In addition to addressing important environmental problems, such early actions should stimulate a greater degree of introspection within industry concerning preventive and corrective measures in anticipation of future environmental problems or additional regulatory actions. Also, this early attention to specific problem chemicals should enable EPA to play a lead role in establishing regulatory priorities rather than having external developments become the dominant factor in determining the Agency's priorities. Finally, initial actions directed to a variety of environmental situations should help clarify the scope and limitations of TSCA and its interfaces with other regulatory programs.

The regulatory strategy will emphasize overall assessments of the causes of the environmental problems associated with chemicals and the most effective measures to address these underlying causes. A key factor will be a determination of the most effective regulatory approach using TSCA, another statute, or a combination of statutes. In many cases, authorities other than TSCA will provide the appropriate means for reducing the problems. If the most appropriate statute is administered by another Agency, EPA may request the other Agency to take action or explain non-action within a specified time [9(a)]. However, a collaborative rather than adversarial interagency approach is essential, and close interaction with other agencies beginning with the initial discovery of a chemical problem will help insure that this system of formal referrals does not unnecessarily disrupt the priority activities of these agencies.

When formally referring chemical problems to other agencies for action, EPA will provide all available information concerning potential risk. While EPA may not have in hand complete documentation concerning all aspects of the risk, there should be a reasonably good indication that preventive or corrective action in the near term deserves serious attention. EPA will follow up with

FIGURE 9—PREMARKET SCREENING IMPLEMENTATION TIMETABLE

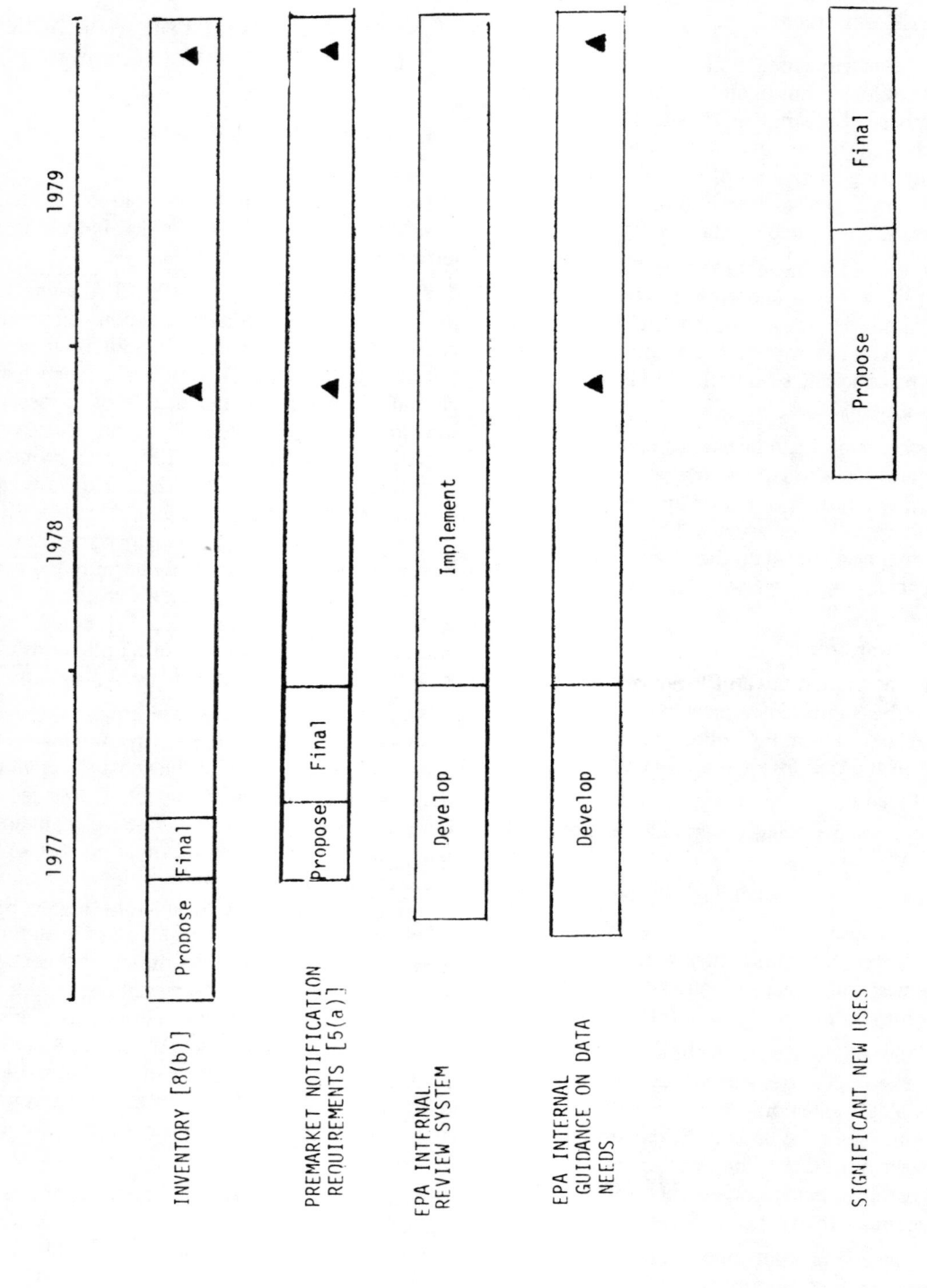

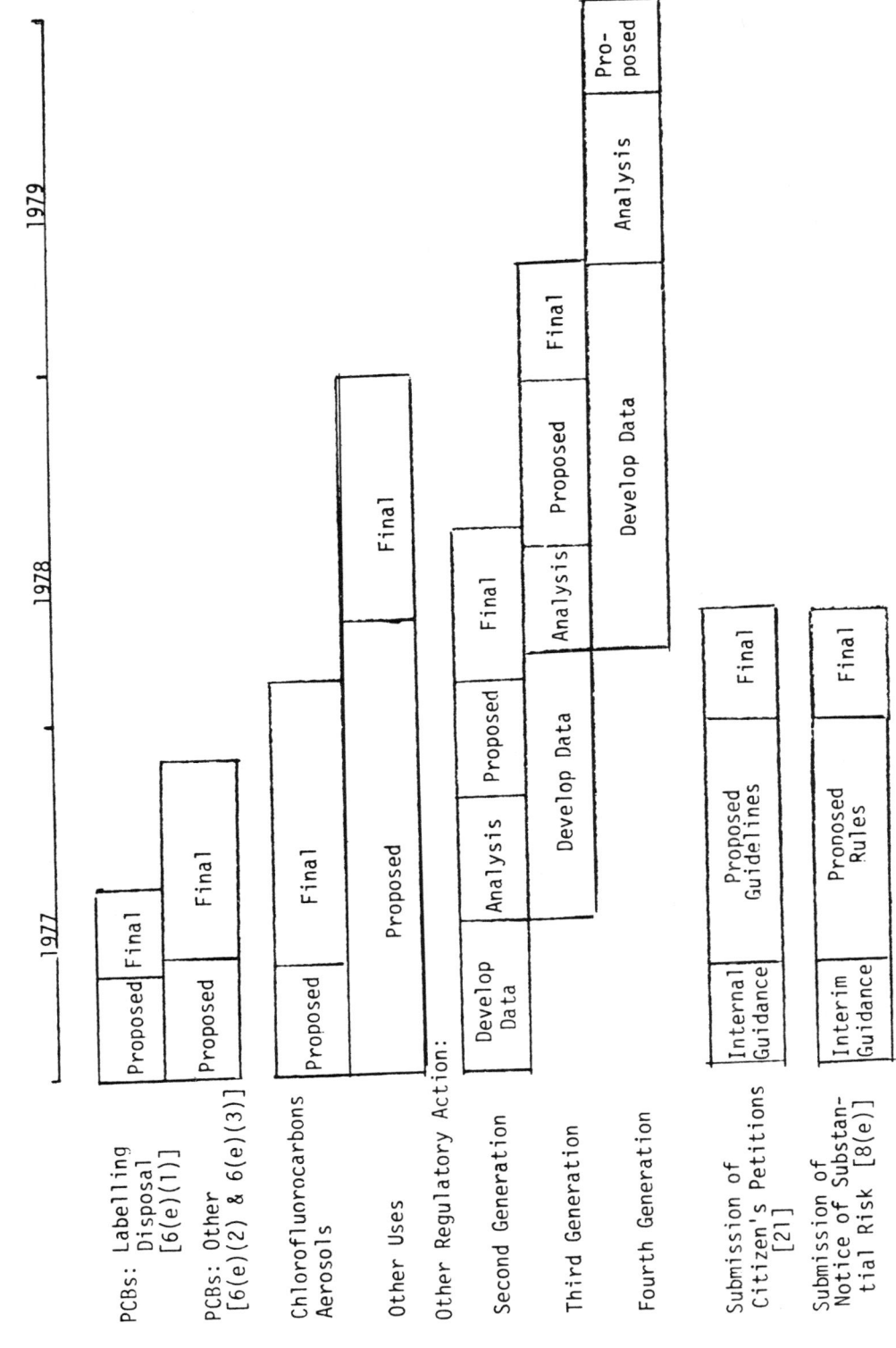

FIGURE 10—USE OF TSCA REGULATORY AUTHORITIES TO CONTROL EXISTING CHEMICALS

the other agencies. If additional data are subsequently developed, and the follow through by the other agency appears inadequate, EPA will not hesitate to reveiw again the situation and to determine appropriate additional actions.

Policy Considerations

The basic consideration for all regulatory decisions is the risk to man or the environment of the problems being addressed. The appropriateness and character of the regulation must then take into account the possibility of actions under other laws [6(c) and 9(a)], the least burdensome approach under TSCA [6(a)], and the overarching requirement to take economics and social factors, as well as environmental concerns, into consideration [2(c)]. The environmental acceptability of the substitute chemicals or alternative technologies that are likely to emerge as the result of the regulatory action will be given special attention. Sometimes, the likely alternatives may be less than optimal from the environmental viewpoint and the incremental environmental gains in adopting the substitutes will be considered.

Often the regulatory action taken under TSCA will be in addition to regulatory actions that have been or could be taken by other programs. Different programs must by law give different weights to the economic and to some of the other factors concerning regulations, and there may be some differences in the orientation of the regulatory decisions made by different programs.

Regulatory actions directed to existing chemicals will not be limited only to correcting problems which are known or believed to be having adverse health and ecological effects but will also be directed to preventing such problems in the future. On occasion, TSCA action may address only a portion of the environmental problem posed by a chemical. Nevertheless, such limited steps can be important, and TSCA action to correct a piece of the problem need not be delayed until the entire problem can be effectively addressed.

Orientation of Initial Activities

Polychlorinated biphenyls and selected chlorofluorocarbons have been of special concern to the Congress for several years, and work is well underway to develop appropriate regulations for limiting their discharge into the environment. The other unattended chemical problems cited during the Congressional hearings are also being examined to determine whether additional regulatory actions are needed immediately.

Many agencies already have lengthy lists of chemicals of particular environmental concern. These lists are being reviewed before reaching a judgement as to whether they can provide useful guidance or whether they should be supplemented with still additional lists. Also, the list to be developed by the Interagency Testing Committee on priorities for testing may be useful in determining regulatory priorities as well.

Expanded efforts will be directed to more systematic procedures for screening and establishing priorities among chemical problems for regulatory attention. While establishing priorities will inevitably involve a number of judgmental decisions, it may be possible to develop improved techniques for assisting in the setting of priorities. Also, the utilization of chemical classes or categories of chemical use might assist in narrowing the vast array of chemicals to a more manageable number.

Underpinning the entire regulatory effort must be a technically sound program of chemical assessment. Chemical assessments have traditionally involved (a) hazard assessment, (b) source assessment, (c) identification of substitutes and alternative technologies, (d) development of control options, and (e) evaluation of the environmental, economic, and related impacts of controls. In some cases, much of the needed data is at hand; more often, supplementary data must be developed. The assessment program will include analyses of individual chemicals, categories of chemicals, chemical technologies, and geographic problems.

Responses to Urgent Problems

Response to the uncovering of chemical problems which might pose urgent risks to health or ecological resources will receive the highest priority, and an on-call response capability will help minimize diversion of resources from other priority activities. In the past such problems have usually come to light as the result of new toxicity test results, new monitoring data, or identification of human or ecological victims of chemical exposure. The discovery of real or alleged urgent problems often results from the conduct of Government programs (e.g., PCB's in the milk of nursing mothers), findings of the scientific community (e.g., nitrosamines in the atmosphere), industrial revelations (e.g., worker deaths from vinyl chloride), and press investigations (e.g., cancer death rates in the Little Elk Valley of Maryland).

TSCA provides additional mechanisms for bringing to light urgent problems. Two of these mechanisms—citizens' petitions [21] and risks uncovered by test data [4(f)]—require a response within a specified time limit. A third principal TSCA mechanism—notification by industry of substantial risk [8(e)]—doesn't have a mandated response time. Regardless of the source of the discovery, the urgency of the problem must drive the response timing.

TSCA also provides several new regulatory mechanisms for limiting chemical exposures quickly if warranted (e.g., imminent hazard [7], regulatory action immediately effective [6(d)(2)], referral to other agency with short deadline for action by the other agency [9(a)]). In the past, local agencies and industry itself have often been willing to take immediate corrective steps in the face of chemical crises. In any event, prompt and effective action to prevent additional damage using TSCA or other programs will be the immediate objective.

The EPA response capability will include available technical specialists to assess the discovery, on-call field and laboratory capability to confirm and supplement the data concerning the discovery, and coordinative and organizational mechanisms—involving a variety of programs within and outside EPA—for implementing prompt and effective corrective actions.

While each new problem will have its own peculiarities, there are usually some common concerns relating to toxicity, exposure, chemical behavior, related commercial activities, and routes of environmental discharges. Generalized checklists of typical concerns are being developed to help insure that technical and policy analyses do not overlook important factors in the face of tight timetables. Also, steps will be taken to insure that interested parties are continuously informed of developments—and particularly acquisition of additional data—in view of the broad political interests in these types of problems.

A very general framework for the response activity is:

(a) Identification of problems associated with chemical activities as the result of

—Systemmatic screening of available information
—Monitoring, toxicological, and epidemiological screening programs
—Ad hoc environmental incidents, research findings, and allegations
—Discoveries submitted under TSCA

(b) Characterization of the problems with particular attention to

—Health and ecological effects and environmental behavior
—Current and projected sources, environmental levels, and exposed population
—Substitutes, control technology, and related cost and economic factors
—Actions to date and actions underway to clarify and control the problems

(c) Development and stimulation of preventive and corrective approaches including consideration of

—Role of relevant authorities of EPA and other agencies

—Alternative approaches to voluntary or regulatory redress
—Environmental and economic impact of approaches
—Implementation of appropriate approach

Dissemination of Information and Assessments to Other Programs and Interested Parties

The Broad Interests in Toxic Substances

Throughout the Congressional consideration of TSCA, there was a recognition of a TSCA role for many of the regulatory and non-regulatory programs of a number of organizations that were in place directed to the assessment and control of toxic substances. Since the enactment of the new law, such programs have already increased in number and in the scope of their interests.

At the Federal level, for example, a large number of chemicals will be explicitly regulated under the Federal Water Pollution Control Act. Many more will be affected by more general standards under that authority and also under the Clean Air Act. The National Academy of Sciences will recommend a large number of chemicals to be considered for possible regulation under the Safe Drinking Water Act. NIOSH has several hundred criteria documents completed or in preparation which will add to the list of chemicals already regulated by OSHA. The Department of Transportation and the Mining Enforcement and Safety Administration similarly have regulations in place or under development affecting many toxic substances. All of these activities must be based on assessments of environmental and related data.

A number of states have taken steps in the toxic substances area. Virginia and Illinois, for example, are particularly interested in data reporting systems. Several states in the Great Lakes area have taken steps concerning PCB's and phosphates. Several states are concerned with chlorofluorocarbons. New Jersey, Texas, and California have very broad concerns over the heavy concentrations of the chemical industry and can be expected to expand efforts in the near future.

Central to the way industry does business are the policies and the attitudes of the financial community concerning investment capital. TSCA, in effect, adds one more risk dimension in the investment world. As this community begins to focus on toxic substances, it needs access to reliable and timely data. The information must be packaged in an understandable and usable form. Meanwhile, the insurance industry is rapidly expanding its interests in toxic substances, particularly with regard to product liability. The issue of substitutes for PCB's also sharpened concern over hazards associated with acceptability of substitutes. Both access to data and early awareness of possible TSCA regulatory actions are important to this side of the commercial community.

Other forces influencing the future directions of the chemical industry are the labor unions, environmental and public interest groups, and the consumer. In all of these cases the specter of possible harmful effects of chemicals can have a direct impact on industrial behavior. TSCA can provide important "early warnings" to these groups who in turn can provide the Government with other early warning signals.

A handful of the larger chemical companies who are responsible for a large proportion of chemical sales have for a number of years conducted sizeable programs to assess the environmental aspects of industrial chemicals. Although there have been many soft spots in these efforts, they nevertheless provide a good foundation for expanded activities. Complimentary efforts have also been supported by industry through a number of trade associations and most recently through the Chemical Industry Institute of Toxicology.

However, relatively few companies have adequate data available to conduct environmental assessments of the chemicals they buy and sell. Despite these shortcomings of the past, the greatest potential impact resulting from TSCA in terms of the number of chemicals that are addressed may lie in the expanded internal assessments and procedures of individual companies, activities that must rest on a solid base of environmental data.

Finally, as the Congress and the courts deepen their involvement in this area, the availability of experts that they can call upon and the credibility of scientific data take on added importance. TSCA will be an important tool for developing the information base which will undergird many major decisions of the future.

Policy Considerations

Several explicit TSCA mechanisms enable interested parties to obtain information. Perhaps the most far-reaching mechanism is the Interagency Testing Committee which provides for seven agencies, in addition to EPA, to set forth the highest priority needs for testing [4(e)]. The provision for citizen's petitions allows any interested party to seek information under the testing and reporting sections of the law [21]. The requirement to place in the Federal Register notices of receipt of premarket data will alert parties concerned with new chemicals reaching the marketplace [5(e)]. The provision concerning disclosure of data clarifies some of the uncertainties concerning the public access to health and safety studies [14(b)]. Of course the overarching requirements of the Freedom of

Information Act are designed to enable all interested parties to obtain information collected under TSCA and other laws. All of these explicit provisions of TSCA underscore the clear intent of the Congress that this legislation service the interests of many organizations in a variety of ways, and particularly with regard to acquisition and dissemination of data.

Therefore, EPA will emphasize coordinated approaches with other Federal agencies to the assessment and control of toxic substances, with particular attention to identifying common information requirements that can be best satisfied through TSCA. The Agency will actively solicit the views of other interested parties as well as to information needs. Information acquired under TSCA will be made available as widely and as promptly as possible. In this regard the interests of the international community are particularly important as we consider the possibility of an international convention to deal on a global basis with the control of toxic substances.

A major effort will be made at the outset to insure that sound procedures are in place which will facilitate a flow of data to all interested parties. Since establishment of these procedural aspects will take priority, it will not be possible to respond to all requests for information to be collected under TSCA until adequate systems are in place to handle the flow. Even in the long run, it will continue to be necessary to prioritize the competing claims from many parties for information services under TSCA.

The Establishment and Operation of Data Systems

System for Receiving and Storing TSCA Data. — Data received pursuant to TSCA will be retained in a discrete data system module with entry into and withdrawal from the system controlled, at least initially, by the Office of Toxic Substances. The TSCA data module obviously will be but one additional component to the much larger Governmental data systems covering toxic substances. To the extent feasible, the TSCA module will be made compatible with and linked to the other existing systems. Figure 11 sets forth the implementantion timetable for data systems.

While the volume of data received under TSCA may be relatively small at the outset, this volume will probably grow rapidly. Therefore, from the outset the system will incorporate automated components as rapidly as possible. The system will be designed and time-phased so that data received in the initial implementation stages are compatible with data received several years into the future, thus avoiding the potential problem of recoding. Development in sub-modules is envisaged, with components (personnel, hardware, and software) added as implementation progresses and data volumes grow. At the outset,

FIGURE 11—DATA SYSTEMS IMPLEMENTATION

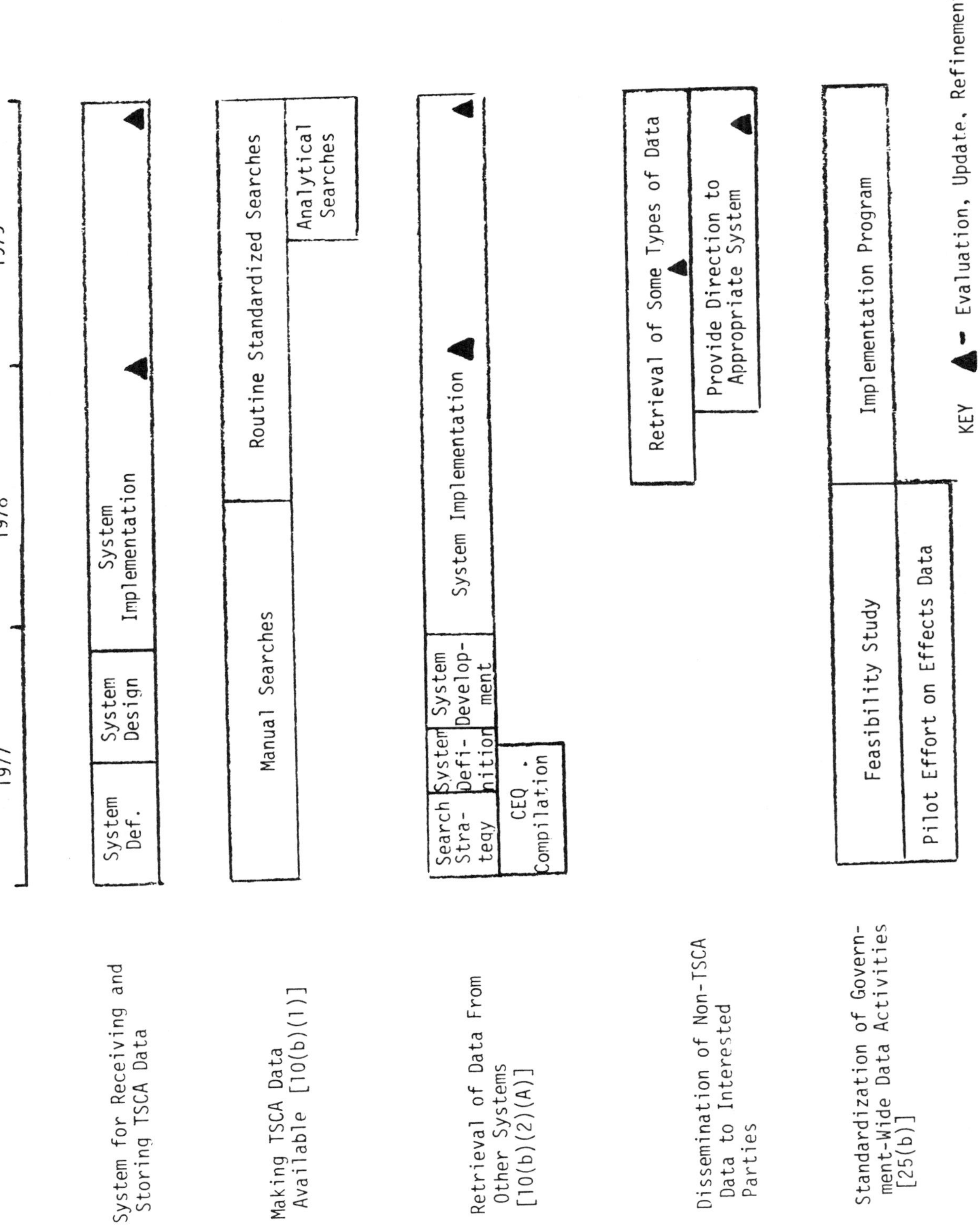

1977	1978	1979

System for Receiving and
Storing TSCA Data

| System Def. | System Design | System Implementation ▲ | ▲ |

Making TSCA Data
Available [10(b)(1)]

| Manual Searches | Routine Standardized Searches |
| | Analytical Searches |

Retrieval of Data From
Other Systems
[10(b)(2)(A)]

| Search Strategy | System Definition | System Development | System Implementation ▲ |
| | CEQ Compilation | | |

Dissemination of Non-TSCA
Data to Interested
Parties

| Retrieval of Some Types of Data |
| Provide Direction to Appropriate System ▲ |

Standardization of Government-Wide Data Activities
[25(b)]

| Feasibility Study | Implementation Program |
| Pilot Effort on Effects Data | |

KEY ▲ - Evaluation, Update, Refinement

however, methods for coding, cataloging, and retrieving data will be established so that consistent ground rules will guide data requirements resulting from rulemaking activities.

The chemicals included in the initial inventory of existing chemicals [8(b)] will provide a core index for the system which will be expanded as the number of chemicals in the system grows. In general, production and use data obtained pursuant to premarket notification [5(a)] and discretionary reporting [8(a)] requirements will be entered from the outset into an automated system. Initially, test data will be retained in hard copy with a locator tag incorporated into the automated system. A test data summary format which could be automated is currently being explored with the National Library of Medicine.

With regard to coding, CAS numbers and manufacturer identification will provide two keys. A chemical use coding system is currently under development to provide another key.

Access to confidential data will be strictly controlled and initially limited to the examination by authorized officials of available information *in situ*.

Making TSCA Data Available.—EPA plans to aggressively develop and carry out procedures for disseminating to the public information obtained under TSCA. For example, test data will be submitted to EPA in sufficient copies so that one copy can be made publicly available without delay through the National Technical Information Service or other appropriate mechanism. Studies of the feasibility of establishing on-line terminal access to non-confidential data and of periodically publishing such data will be soon initiated.

Retrieval of Data from Other Systems.—There are many sources of external technical, scientific, and economic data which should be utilized in implementing TSCA. These sources are located both within and outside the Federal government. CEQ is presently conducting a survey of such data bases. Meanwhile, some states are initiating additional toxic materials data banks. These data activities in particular will be examined in some detail with an eye toward reducing reporting requirements, on the one hand, and avoiding duplicate data searches during examinations of chemical for potential hazards, on the other.

Dissemination of Non-TSCA Data to Interested Parties.—Many parties are interested in improved ways for tapping the multiplicity of Governmental data banks concerning chemicals. As part of the overall data system effort, practical means for facilitating such access will be explored. Initially, an inventory of existing data bases is being conducted. Hopefully, meaningful road maps to Governmental data can be provided to all interested users in the future.

Standardization of Government-Wide Data Activities.—Data standardization is particularly important when considering interlocking data systems. Data required by TSCA rulemaking should be requested in a format which can be used readily by a multiplicity of interested parties. As an initial step towards improved standardization of Government-wide approaches, emphasis will be given to standardizing the formatting and dissemination of newly collected data on health and ecological effects. At a later date, the effort will be expanded to include other types of data, and as time and resources permit, efforts will be directed to achieving compatibility of data already in the files of many agencies with the newly collected and standardized data.

Types of Anticipated Impacts From Implementation Activities

The types and extent of the impacts that will result from implementation activities are speculative at best, given the many uncertainties concerning the effects of chemicals and their behavior in the environment, the large number and variety of chemical products, and the continuing rapid growth of the chemical industry. The five years of Congressional testimony included many general statements concerning the environmental benefits that are likely to ensue. A number of unfortunate chemical incidents were cited as examples of the types of problems that can be avoided. On the economic side, the EPA report *Draft Economic Assessment of the Impact of the Toxic Substances Control Act* of June 1975, as well as economic assessments prepared by industry and by the General Accounting Office, attempted to identify some of the types of direct impact that will undoubtedly occur. However, those discussions were very limited, and little effort was made to address indirect impacts.

The following discussion provides but a very superficial framework for the impact evaluation effort that will accompany implementation activities.

Health and Environment

A primary legislative concern is reducing adverse health effects, and particularly chronic effects. The legislation should offer an opportunity to clarify the health effects of many chemicals and, over time, to reduce the number of deaths and the disease rates attributable to such effects. Also, the possibility of preventing acute effects will be addressed.

Data obtained under TSCA should enhance the capability of OSHA, MESA, and state authorities to reduce the incidence of worker deaths and diseases. Other agencies and organizations which obtain information and support under this legislation should also assist in preventing adverse health impacts at the national and local levels.

The likely impact of TSCA on reducing ecological damage is even more difficult to predict. The early experience in addressing PCB's (i.e., destruction of aquatic resources) and chlorofluorocarbons (i.e., depletion of the ozone layer) clearly demonstrates the importance of the legislation in this regard. A number of toxic chemicals end up in the aquatic environment where ecological damage can be extensive. Prevention of such damage will largely depend on the specific regulatory actions. Ecological test requirements should assist in clarifying the impact of chemicals on the eco-system and in setting quality standards for water and other media under other authorities. This aspect of environmental assessment has been largely neglected by industry in the past.

Commerce and the Economy

Implementation requirements will add a new dimension in financial planning within industry for the development, manufacture, and marketing of chemicals. For example, there will be far greater reluctance to expand commercial investments in chemicals of questionable toxicity, and the search for broader applications of chemicals which are environmentally acceptable will intensify. Some marginal products for which testing is required may give way to substitutes which become commercially competitive. Many firms will be far more cautious in purchasing or selling products of unknown chemical composition. The ripple effect of such adjustments in current marketing practices will impact on a broad range of downstream processors and users.

There will probably be a tendency among some of the larger companies toward greater self reliance on in-house chemical assessments of old and new chemicals and on conducting their own synthesis of small batches of highly reactive chemicals previously purchased from small suppliers. Given the concern over quality control of test data and the shortage of laboratory facilities, in-house toxicological and ecological testing laboratories should become more commonplace. Meanwhile, small firms may tend to move away from product lines that become targets for TSCA attention.

The international development and marketing strategies of multinational firms will also be impacted. Test marketing may be more heavily concentrated in countries where premarket requirements are minimal. Also, there may be a surge of new chemical imports into the United States to establish them as "existing" chemicals and, thus, eliminate the notification period for future imports. In general, more careful planning of international shipment of chemicals will be required.

Industrial Research and Development

Premarket notification requirements and premarket testing requirements will cause some adjustments in the R&D cycle. In a few cases, the testing requirements may in large measure codify existing industrial practices. In most cases, however, the new requirements will alter substantially the time phasing, the types of expertise, and the review processes involved in developing new chemicals. These adjustments will in turn impact on (a) the decisions as to which new chemicals and products should be explored and then developed, (b) the criteria for investing in R&D when there is an increased risk for commercial introductions, and (c) the efforts to "design around" potentially troublesome chemicals from the environmental viewpoint.

The number and quality of environmental assessments conducted by industry should increase markedly. More qualified technical personnel will be attracted to the field, methodological approaches will be significantly upgraded, and the quality assurance procedures will be improved. However, should Governmental requirements "overstandardize" assessment techniques, there is a danger that industrial creativity in improving the state-of-the-art of environmental assessment could be stifled.

In the early years of implementation, the number of new chemicals reaching the marketplace may decline due to uncertainties as to future regulatory requirements, technical and financial difficulties in adjusting to the new procedural requirements, and the increased R&D costs and lead times for some products with limited market potential. However, in the longer run innovation in introducing new products need not be stifled. More intensive investigations of environmentally acceptable chemicals, coupled with incorporation of premarket requirements into a routine R&D cycle, should continue to allow ever expanding benefits to the consumer from the uses of chemicals.

The Scientific Base

The heavy emphasis in the legislation on improving the scientific methodologies undergirding chemical assessments, and the attendant implications for strengthening the technical manpower base, should have a major impact on the chemical and biological sciences. Not only will the legislation give impetus to the advancement of these individual disciplines, but it should stimulate a closer coupling of these disciplines with engineering, economics, and other areas of imporance to toxic substances control. This general impetus to the broad spectrum of sciences may far overshadow the scientific impact of individual regulatory actions.

The inadequacy of science to provide clear answers for regulatory decisions will probably be subject to frequent criticism by many impatient parties. However, carefully documented scientific investigations will be a key to many actions. There is no doubt that the importance of credible technical data, albeit inconclusive, will be widely recognized.

All should benefit from the expanded sharing of scientific data. The importance of common procedures and common formats in carrying out and reporting scientific investigations will take on added importance as the problems involved in exchanging noncompatible data bases become clear.

Social Concerns

Many of the impacts cited above are tied to social concerns, such as employment effects, increased costs of products, and rights of inspection. However, there are even more fundamental social concerns which will be affected by TSCA implementation such as:

—How much effort should be directed to protecting the welfare of future generations?

—To what extent should the public participate in decision-making that has previously been the exclusive domain of private industry?

—How are health concerns to be balanced with economic costs in determining "unreasonable risk"?

There is little experience in measuring the types of social impact that could far outweigh in importance the other more narrow impacts of this legislation. A continuing effort to identify and understand these impacts is the key to determining the value of TSCA as an instrument of public policy.

REGULATION OF TOXIC AND HAZARDOUS SUBSTANCES UNDER OTHER FEDERAL LAWS*

Introduction

This paper describes the criteria used by 23 systems to define a "hazardous substance," primarily for regulatory purposes. The predominance of Federal systems is explained by the number of Federal statutes mandating regulation of hazardous substances in food, in transport, in the work place, and in the air and water environment. In addition, the supply component of the military services uses hazard ratings to help determine precautions needed for storage and shipment of materials. Only the State of California system has been developed for the purpose of comprehensively managing hazardous wastes, including their transport and disposal.

The set of criteria used by each of the hazardous substance classification systems is shown in Table A—Classification Criteria. The only systems not shown in this table are the NIOSH priority list, and the systems of the EPA Sewage Sludge Working Group and the U.S. Coast Guard. The priority list had no criteria that could be categorized, the working group is still drafting guidelines for heavy metal content in sewage sludge, and the Coast Guard regulations are essentially identical to Title 49, Code of Federal Regulations (CFR), Parts 100-199.

Every classification system studied uses toxicological criteria in determining whether a substance should be classified as hazardous. Many systems utilize the toxicology criterion as the primary screen for identifying hazardous substances, although their focus is on acute rather than chronic toxicity data. Acute toxicity data are more extensive than chronic toxicity data, and the results are more clear-cut (i.e., death versus some level of bodily impairment). The acute toxicity tests most frequently used in these systems are the lethal doses of a substance administered orally and dermally, and the lethal concentration of a substance administered by inhalation.

Other acute toxicity tests included the concentration of a substance lethal to aquatic life or causing a decrease in aquatic flora. One chronic toxicity criterion used in several systems was the Threshold Limit Value (TLV) of a substance, (see discussion of TRW Systems Group).

The acute toxicity levels and flammability criteria utilized by different systems for classifying substances as hazardous are depicted in Table B. The Federal Hazardous Substances Labeling Act, (Title 15, U.S. Code, Sec.

1261), provided the definition of "highly toxic" substances that was subsequently incorporated into many classification systems.

This Act's definitions of "flammable" and "extremely flammable" substances were also incorporated into other classification systems, although these criteria were not as widely adopted as the "highly toxic" criterion. (The "combustible" classification shown in Table B, did not appear in the original act.)

Other classification criteria are not as similar among systems as the toxicological or flammability criteria. The source of each system's classification criteria is noted, if available. The "sufficient quantity" criterion is often an exemption for very small amounts or packages of a substance, i.e., too small for a label. Such exemptions are included in Title 15, U.S. Code, Sec. 1261, and CPSC and DOT regulations. In other cases, the quantity of a substance produced or discharged is a criterion considered when classifying substances (e.g., TRW Systems Group and Booz-Allen Applied Research, Inc.), or designating a substance as hazardous for purposes of regulation (e.g., Sec. 307 (a) and Sec. 311 (b) (2) (A) of the FWPCA).

Consumer Product Safety Commission

Title 16, (Commercial Practices), CFR, Part 1500

Section 30 (a) of the Consumer Product Safety Act (P.L. 92-573) vests authority granted under the Federal Hazardous Substances Labeling Act (P.L. 86-613, enacted July 12, 1960) in the Consumer Product Safety Commission. The Commission is therefore responsible for enforcement of labeling requirements governing hazardous substances intended or packaged in a form suitable for use in the household. (Exemptions to such requirements are allowed for food, drugs, cosmetics, fuels, small packages, minor hazards, and special circumstances.)

The Federal Hazardous Substances Labeling Act (referred to as "the Act") defines a "hazardous substance" as:

> any substance or mixture of substances which (i) is toxic, (ii) is corrosive, (iii) is an irritant, (iv) is a strong sensitizer, (v) is flammable, or (vi) generates pressure through decomposition, heat, or other means, if such substance or mixture of substances may cause substantial personal injury or substantial illness during or as proximate result of any customary or reasonably foreseeable ingestion by children.

---

* A Summary of Hazardous Substance Classification Systems, U.S. Environmental Protection Agency, December 1975.

TABLE A—CLASSIFICATION CRITERIA

Criteria

System	Toxicological	Flammability	Explosive	Corrosive	Reactivity	Oxidizing material	Radioactive	Irritant	Strong sensitizer	Bioconcentration	Carcinogenic, mutagenic, teratogenic	Sufficient quantity
Title 15, U.S. Code, Sec. 1261	X	X		X			X	X	X			X
CPSC-Title 16, CFR, Part 1500	X	X		X			X	X	X			X
Food, Drug, and Cosmetic Act	X									X	X	X
DOT-Title 49, CFR, Parts 100-199	X	X	X	X		X	X	X				X
Pesticides-Title 40, CFR, Part 162	X	X								X	X	
Ocean Dumping-Title 40, CFR, Part 227	X									X	X	X
NIOSH-Toxic Substances list	X										X	
Drinking Water Standards	X									X	X	
FWPCA Sec. 304 (a) (1)	X									X	X	
Sec. 307 (a)	X									X	X	X
Sec. 311 (b) (2) (A)	X											X
Clean Air Act-Sec. 112	X						X				X	
California State List	X	X	X	X			X	X				
National Academy of Sciences	X	X			X							
TRW Systems Group	X	X	X	X			X			X		X
Batelle Memorial Institute	X	X	X		X	X	X	X		X	X	
Booz-Allen Applied Research, Inc.	X	X	X		X							X
Dept. of the Army	X											
Dept. of the Navy	X	X	X	X	X	X	X					
National Cancer Institute	X									X	X	X

The term "toxic" is applied to any substance (other than a radioactive substance) which has the capacity to produce personal injury or illness to man through ingestion, inhalation, or absorption through any body surface.

The term "highly toxic" is defined as:

any substance which falls within any of the following categories: (a) Produces death within fourteen days in half or more than half of a group of ten or more laboratory white rats each weighing between two hundred and three hundred grams, at a single dose of fifty milligrams or less per kilogram of body weight when orally administered, (oral LD$_{50}$); or (b) produces death within fourteen days in half or more than half of a group of ten or more laboratory white rats each weighing between two hundred grams, when inhaled continuously for a period of one hour or less at an atmospheric concentration of two hundred parts per million, (ppm), by volume or less of gas or vapor or two milligrams per liter by volume or less of mist or dust, provided such concentration is likely to be encountered by man when the substance is used in any reasonably foreseeable manner, (inhalation LC$_{50}$); or (c) produces death

TABLE B—ACUTE TOXICITY AND FLAMMABILITY CRITERIA

SYSTEM	HIGHLY TOXIC SUBSTANCES †						FLAMMABILITY ‡			
	ORAL LD_{50} mg/kg	INHALATION LC_{50} (dust or mist) micrograms per liter	(gas or vapor) ppm	DERMAL LD_{50} mg/kg	AQUATIC LIFE LC_{50} mg/l	AQUATIC FLORA IL_m ppm	EXTREMELY FLAMMABLE	HIGHLY FLAMMABLE	FLAMMABLE	COMBUSTIBLE
Title 15, USC., Sec. 1261	50	200	2,000	200			$F_p \leq 20°F$		$20°F < F_p \leq 80°F$	$80°F < F_p \leq 150°F$
CPSC–Title 16, CFR, Part 1500	50	200	2,000	200			$F_p \leq 20°F$		$20°F < F_p \leq 80°F$	$80°F < F_p \leq 150°F$
DOT Title 49, CFR, Parts 100–199	50	200	2,000	200					$F_p < 100°F$	$100°F < F_p \leq 200°F$
Pesticides–Title 40, CFR, Part 162	50	200	2,000	200			$F_p \leq 20°F*$		$20°F < * F_p \leq 80°F$	
NIOSH–Toxic Substances List	5,000**		2,000**	2,800**						
FWPCA–Sec. 307(a)***	50	200	2,000	200	10					
Sec. 311(b)(2)(A)	50	200	2,000	200	500	100	$F_p < 20°F$		$20°F < F_p \leq 80°F$	
California State List	50	200	2,000	200						
National Academy of Sciences	5 to 50	50 to 200		20 to 200	1		$F_p \leq 100°F$ $B_p \leq 100°F$	$F_p \leq 100°F$ $B_p > 100°F$		
Battelle Memorial Institute	50		2,000	200	1,000	1,000	$F_p < 73°F$ $B_p \leq 100°F$			
Booz-Allen Applied Research, Inc.	50		5,000(man) 2,000(ecology)		100		$F_p < 100°F$ $B_p < 100°F$			

† Oral LD_{50}, Inhalation LC_{50}, Dermal LD_{50} are defined in Federal Hazardous Substances Labeling Act section; aquatic life LC_{50} and aquatic flora IL_m are defined in sections dealing with Sec. 307(a) and Sec. 311(b)(2)(A) of FWPCA, respectively. Unless otherwise indicated, substances having toxicities less than or equal to the amount indicated are classified as highly toxic

F_p: flash point; B_p: boiling point

*Proposed standard

**LCLo, LC50, LDLo, or LD50: no time limit for tests given, see NIOSH section.

***Currently under revision.

within fourteen days in half or more than half of a group of ten or more rabbits tested in a dosage of two hundred milligrams or less per kilogram of body weight, when administered by continuous contact with the bare skin for twenty-four hours or less, (dermal LD$_{50}$).

The term "corrosive" means any substance which in contact with living tissue will cause destruction of tissue by chemical action; such term shall not refer to action on inanimate surfaces. The term "irritant" means any substance not corrosive which in immediate, prolonged, or repeated contact with normal living tissue will induce a local inflammatory reaction.

The term "strong sensitizer" means:

a substance which will cause on normal living tissue through an allergic or photodynamic process a hypersensitivity which becomes evident on reapplication of the same substance, and which is designated as such by the Commission. Before designating any substance as a strong sensitizer, the Commission upon consideration of the frequency of occurrence and severity of the reaction, shall find that the substance has a significant potential for causing hypersensitivity.

The term "extremely flammable" shall apply to any substance which has a flash point at or below 20 F as determined by the Tagliabue Open Cup Tester, and the term "flammable" shall apply to any substance which has a flash point above 20 F, to and including 80 F, as determined by the same test.

These hazardous substance definitions are incorporated into Sec. 1261 of Title 15 (Commerce and Trade) of the U.S. Code. The only difference noted between the hazardous substance definitions found in the code and in the Act was the inclusion in the former of a "combustible" category for those substances with flash points greater than 80 F, but less than or equal to 150 F.

Title 16 (Commercial Practices), CFR, Part 1500.3 contains the following items which interpret, supplement, or provide alternatives to the aforementioned hazardous substance definitions, (which are also contained in this Part):

"Toxic substances" are defined as those substances falling within any of the following categories:

(1) Any substance having an oral LD$_{50}$ of more than 50 mg/kg but not more than 5 g/kg of body weight. Substances falling in the toxicity range between 500 mg and 5 grams per kilogram of body weight will be considered for exemption from some or all of the labeling requirements;

(2) Any substance having an inhalation LC$_{50}$ of more than 200 ppm but not more than 20,000 ppm by volume of gas or a vapor or more than 2 mg but not more than 200 mg per liter by volume of mist or dust; or

(3) Any substance having a dermal LD$_{50}$ of more than 200 mg/kg of body weight but not more than 2 g/kg of body weight.

"Irritants" are subdivided into "primary irritants to the skin" and eye irritants. Primary irritants are substances that are not corrosive and result in an empirical score of five or more when tested by the method described in Sec. 1500.41. Eye irritants are defined as those substances giving a positive test result when the test method described in Sec. 1500.42 is used.

"Extremely flammable solid" means a solid substance that ignites and burns at an ambient temperature of 80 F or less when subjected to friction, percussion or electrical spark. "Flammable solid" means a solid substance that when tested by the method described in Sec. 1500.44, ignites and burns with a self-sustained flame at a rate greater than one-tenth of an inch per second along its major axis.

A susbtance is considered "corrosive" to the skin, if when tested by the technique described in Sec. 1500.41, the structure of the tissue at the site of contact is destroyed or changed irreversibly in 24 hours or less. Criteria for determining whether a substance is hazardous because it generates pressure through decomposition, heat, or other means are also listed.

Department of Transportation

Title 49, (Transportation), CFR, Parts 100-199

Title 49, CFR, Parts 100-199 contain regulations covering the preparation of hazardous materials, (including their loading and storage), for transportation by common carriers, rail freight, rail express, rail baggage, highway or water. These regulations also cover construction containers, packaging, weight, marking, labeling, billing, and shipper's certificate of compliance. (Sec. 171.1(a)).

Section 173.2 lists the priority of hazards in the following order (from highest to lowest priority):

(1) Radioactive material
(2) Flammable or nonflammable gas
(3) Extremely toxic liquid or solid
(4) Flammable liquid
(5) Oxidizing material
(6) Flammable solid
(7) Corrosive material, (liquid)
(8) Highly toxic liquid or solid
(9) Irritating material

(10) Corrosive material, (solid)

(11) Combustible liquid

Sec. 173.326 lists thirteen poisonous gases that qualify as extremely dangerous poisons. The definition for less dangerous, liquid or solid poisons is contained in Sec. 173.343. The oral, inhalation, and dermal toxicities of a substance are used as criteria in determining this class of poisons. The toxicity levels chosen were identical to the "highly toxic" category of Title 15, U.S. Code, Sec. 1261.

Flammable and combustible liquids are defined in Sec. 173.115. Flammable liquids are those liquids having a flash point below 100 F. Combustible liquids are those liquids having a flash point above 100 F and below 200 F. Flammable compressed gases are defined in Sec. 173.300 (b).

Explosives forbidden from transportation by common carriers by rail freight, rail express or highway, or water are defined in Sec. 173.51. Sec. 173.52 categorizes explosives acceptable for transportation into three classes: (1) Class A explosives; detonating or otherwise of maximum hazard, (defined in Sec. 173.53); (2) Class B explosives; flammable hazard, (defined in Sec. 173.88); and (3) Class C explosives, minimum hazard (defined in Sec. 173.100).

Oxidizing materials are defined (Sec. 173.151) as those substances that yield oxygen readily, thereby stimulating the combustion of organic matter. Corrosives are defined as in Title 21 (Food and Drugs), CFR, Sec. 191.1 and tested by the method described in Sec. 191.11. A liquid is considered to have a severe corrosion rate if its corrosion rate exceeds 0.250 inches per year on steel (SAE 1020) or aluminum (nonclad 1075-T6) at a test temperature of 130 F, (Sec. 173.240 of Title 49). An acceptable test is described in NACE Standard TM-01-69.

Irritants, (defined in Sec. 173.381) are defined as those liquid or solid substances which upon contact with fire or when exposed to air give off dangerous or intensely irritating fumes, but not including any Class A, extremely poisonous materials.

An "etiologic agent" (defined in Sec. 173.386) means a microorganism, or its toxin which causes or may cause human disease and is limited to those agents listed in Title 42 (Public Health), CFR, Sec. 72.25(c). Radioactive materials are defined by fissile class and transport group in Sec. 173.389.

U.S. Coast Guard

Chapter I of Title 46 (Shipping), CFR, contains U.S. Coast Guard regulations governing shipment of materials. Most of the hazardous materials regulations are similar to those described in Title 49, Parts 100-199. One difference, however, is that Sec. 146.27-1 describes a hazardous article, (other than an explosive, inflammable liquid or solid, oxidizing material, corrosive liquid, compressed gas, poisonous article or combustible liquid) as one which:

(a) Is liable to spontaneous heating in excess of 10 F when subjected to a test of three continuous hours in a Mackey apparatus at or below a temperature of 212 F; or

(b) Liberates vapor susceptible to ignition by spark or open flame at or below a temperature of 300 F.

National Academy of Sciences[1] Systems for Evaluation of the Hazards of Bulk Water Transportation of Industrial Chemicals

This study classifies the hazards of bulk water transportation industrial chemicals in each of nine categories (see Table C—Summary of Hazard Evaluation Criteria)[1]. The hazard rating of an industrial chemical is evaluated independently for each category. No ranking of chemical hazard by combining the category ratings into one overall figure can be justified, however.

The fire hazard rating, Category 1, classifies those chemicals with a flash point of less than 100 F and a boiling point of less than 100 F as "extremely hazardous," while those chemicals with a flash point less than 100 F and a boiling point greater than 100 F are classified as "highly hazardous."

Categories II thru V are occupational health hazard ratings. The hazard of liquid contact with skin and eyes is classified in Category II. Those chemicals having a dermal LD_{50} of less than 20 mg/kg are classified as "extremely hazardous," while those chemicals with a LD_{50} of greater than 20 mg/kg, but not more than 200 mg/kg, are classified as "highly hazardous."

The hazard rating for the short-term inhalation of gas or vapor, Categories III and IV, classifies those chemicals with an inhalation LC_{50} of less than 50 ppm or less than 0.5 mg/1 as "extremely hazardous," while those chemicals with a LC_{50} of greater than 0.5 mg/1, but not more than 2 mg/1, are classified as "highly hazardous."

The hazard rating for repeated inhalation of gases and vapors, Category V, is based on Title 29 (Labor), CFR, Sec. 1910.93. Those substances listed in specified tables or regulated elsewhere in this section, having an 8-hour time weighted average exposure limit of less than 1 ppm are classified as "extremely hazardous," while those substances having an exposure limit of more than 1 ppm but not more than 10 ppm are classified as "highly hazardous."

TABLE C—SUMMARY OF HAZARD EVALUATION CRITERIA

GRADE	I FIRE	II Skin and Eyes	III Vapor Inhalation	IV Gas Inhalation	V Repeated Inhalation	VI Human Toxicity	VII Aquatic Toxicity	VIII Water Reaction	XI Self-Reaction
		HEALTH				WATER POLLUTION		REACTION	
0	Insignif. Hazard Non-combust	Insignif. Hazard All not described below	Insignif. Hazard All not described below	Not Applicable	Insignif. Hazard OSHA $\geq$ 1000 ppm	Insignif. Hazard LD_{50} > 5000 mg/kg	Insignif. Hazard TL_m >1000mg/l	Insignif. Hazard	Insignif. Hazard No appreciable self-reaction
1	Slightly Hazardous FPcc >140°F (60°C)	Slightly Hazardous Corrosive to eyes-	Slightly Hazardous Depressants, Asphyxiants	All those not described below	Slightly Hazardous OSHA 100-1000 ppm	Slightly Hazardous LD_{50} 500-5000mg/kg	Practically Nontoxic TL_m 100-1000 mg/l	Slightly Hazardous e.g. Cl_2	Slightly Hazardous May polymerize with low heat evolution
2	Hazardous FPcc 100°F-140°F (37.8°-60°C)	Moderately Hazardous Corrosive to skin	Moderately Hazardous LC50 200-2000 ppm	Moderately Hazardous LC50 200-2000ppm	Moderately Hazardous OSHA 10-100 ppm	Moderately Hazardous LD_{50} 50-500/mg/kg	Slightly Toxic TL_m 10-100 mg/l	Hazardous e.g. NH_3	Hazardous Contamination may cause polymerization, no inhibitor required
3	Highly Hazardous (37.8°C) FPcc <100°F BP >100°F (37.8°C)	Highly Hazardous LD_{50} 20-200mg/kg 24 hr. skin contact	Highly Hazardous LC_{50} 50-200ppm or $\leq$ 0.5-2 mg/l	Highly Hazardous LC_{50} 50-200 ppm	Highly Hazardous OSHA 1-10 ppm	Highly Hazardous LD_{50} 5-50 mg/kg	Moderately. Toxic TL_m 1-10 mg/l	Highly Hazardous e.g. Oleum	Highly Hazardous May polymerize, requires stabilizer
4	Extremely Hazardous (37.8°C) FPcc<100°F BP <100°F (37.8°C)	Extremely Hazardous $LD_{50} \leq$ 20mg 24 hr. skin contact	Extremely Hazardous $LC_{50} \leq$ 50ppm or $\leq$ 0.5 mg/l	Extremely Hazardous $LC_{50} \leq$ 50ppm	Extremely Hazardous OSHA <1ppm	Extremely Hazardous LD_{50} < 5 mg/kg	Highly Toxic TL_m < 1 mg/l	Extremely Hazardous e.g. SO_3	Extremely Hazardous Self-reaction may cause explosion or detonation

Water pollution hazard ratings, Categories VI, and VII, are concerned with the toxicological dangers, to humans and to aquatic life, associated with a chemical spilled or dumped into waterways. Those chemicals with an oral LD_{50} of less than 5 mg/kg are classified as "extremely hazardous," while those chemicals with oral LD_{50} of more than 5 mg/kg, but not more than 50 mg/kg, are classified as "highly hazardous." Substances having a 96-hour aquatic life LC_{50} (or median threshold limit, TLm) of less than 1 mg/1 are classified as "highly toxic." These ratings do not address the problem of bioaccumulation.

Reactivity hazard ratings, Categories VIII and IX, are concerned with the likelihood of chemical reaction with water, or chemical self-reaction, usually via polymerization. The "extremely hazardous" water reaction hazard rating is classified as a violent reaction of a chemical with an equal weight of water in a tank at ambient temperature. The "highly hazardous" rating is assigned to those chemicals that undergo a vigorous reaction with water under similar conditions. Chemicals that can undergo self-oxidation, and/or polymerization, possibly causing explosions or detonations, are classified as "extremely hazardous." Chemicals that may undergo self-reaction, and require special handling, such as incorporation of a stabilizer or polymerization inhibitor, are classified as "highly hazardous."

Department of Health, Education, and Welfare

Federal Food, Drug, and Cosmetic Act

The Federal Food, Drug, and Cosmetic Act mandates that any poisonous or deleterious substance added to any food shall be deemed to be unsafe in any quantity exceeding the limits fixed by the Secretary (of Health, Education, and Welfare) as necessary for the protection of public health (Sec. 406). Furthermore, under Sec. 408, those poisonous or deleterious pesticide chemicals not generally recognized as safe for use shall not exceed tolerance levels established by the Secretary with respect to use in or on raw agricultural commodities.

Tolerance levels for poisonous or deleterious substances are established on the basis of the best evidence currently available. Determination of a "safe" no-effect level in food is made considering toxicological studies, evidence of bioaccumulative properties, likelihood of body retention of the substance, amount of food likely to be consumed, and any other relevant factors.

Sec. 409 (c) (3) (A) stipulates that no food additive shall be deemed to be safe if it is found to induce cancer when ingested by man or animal. Thus, there are no "safe" no-effect levels for carcinogens. This provision does not apply with respect to the use of a substance as an

ingredient of feed for animals which are raised for food production if the Secretary of Health, Education, and Welfare finds that under specified conditions of use and feeding the additive will not adversely affect the animals for which the food is intended and that no residue of the additive will be found in any edible portion of such animal after slaughter or in any food yielded by or derived from the living animal.

National Institute of Occupational Safety and Health

Toxic Substances List.—The National Institute of Occupational Safety and Health (NIOSH) in compliance with Sec. 20(a)(6) of the Occupational Safety and Health Act of 1970, (P.L. 91-596), periodically publishes a toxic substances list. Dosages differentiating relatively toxic and nontoxic substances according to route of administration to experimental animals are listed in Table D. The limiting dosages were chosen on the basis of available data and toxicological judgment.

The oral route of administration is the preferred testing method. This method is the most widely used and documented. The inhalation and skin absorption routes of administration are used in that order when data for oral administration are not available.

The order of preferrence of test data by species is:

(1) Man	(6) Cat
(2) Rat	(7) Rabbit
(3) Mouse	(8) Pig
(4) Dog	(9) Guinea pig
(5) Monkey	(10) Hamster

The limiting dosages shown in Table D for substances administered by any route other than inhalation may be:

(1) LCLo: Lethal Dose Low—the lowest dose of a substance (other than LD_{50}) introduced over any given period of time, and reported to have caused death in man, or the lowest single dose introduced in one or more divided portions and reported to have caused death in animals; or

(2) LD_{50}: Lethal Dose Fifty—the dose of a substance expected to cause death of 50% of an entire population of an experimental animal species, as determined from the exposure to the substance of a significant number from that population. Other lethal dosages (LD_{10}, LD_{30}), are used in absence of an LD_{50}, if available.

Limiting dosages for substance administered by inhalation may be:

(1) LCLo: Lethal Concentration Low—the lowest concentration of a substance, other than an LC_{50} in air, which has been reported to have

TABLE D—LIMITING DOSAGES DIFFERENTIATING RELATIVELY TOXIC AND NONTOXIC SUBSTANCES ACCORDING TO ROUTE OF ADMINISTRATION TO EXPERIMENTAL ANIMALS OF A MAXIMUM TOTAL (ACUTE)* DOSE CAUSING DEATH**

ROUTES OF ADMINISTRATION (with abbreviations)

SPECIES (with abbreviations)	Oral (orl) Rectal (rec) Intraduodenum (idu) Intracervix (icv)	Inhalation (ihl) 24-Hour Maximum		Skin (skn)	Intraperitoneal (ipr) Intrapleural (ipl)	Parenteral — Subcutaneous (scu) Intradermal (idr) Implant (imp)	Parenteral — Intravenous (ivn) Intramuscular (ims) Ocular (ocu) Intracerebral (ice) Intratracheal (itr) Intraplacental (ipc) Intravaginal (ivg) Intrarenal (irn)	Other (par)	Unreported (unk)
	mg/kg	ppm	mg/m³	mg/kg	mg/kg	mg/kg	mg/kg	mg/kg	mg/kg
Hamster (ham), Frog (frg), Gerbil (grb)	2,500	5,000 (0.5%)	1,000	1,400	1,000	5,000	750	1,000	2,500
Rat (rat), Mouse (mus), Mammal, unspecified (mam)	5,000***	10,000 (1%)	2,000	2,800	2,000	10,000***	1,500	2,000	5,000
Rabbit (rbt), Guinea Pig (gpg), Chicken (ckn), Pigeon (pgn), Quail (qal), Duck (dck), Turkey (trk), Bird (brd)	10,000	20,000 (2%)	4,000	2,800***	4,000	20,000	3,000	4,000	10,000
Dog (dog), Monkey (mky), Cat (cat), Pig (pig), Cattle (ctl), Domestic Animals: sheep, goat, horse (dom)	10,000	20,000 (2%)	4,000	5,600	4,000	20,000	3,000	4,000	10,000

*Applies to those substances for which acute or short term toxicity characterizes the response, e.g., fast-acting substances, irritants, narcoses-producing substances and most drugs. Does not apply to substances whose characteristic response results from prolonged exposures, e.g., silica, lead, benzene, carbon disulfide, carcinogens. Concentrations more appropriately characterizing the toxicity of long- or slow-acting substances are derived from non-acute toxicity studies.

**Calculated from experimental data (Stokinger).

***From Hine and Jacobson, AIHAAP 15, 141, 54.

caused death in animals when they have been exposed for 24 hours or less; or

(2) LC$_{50}$: Lethal Concentration Fifty—the concentration of a substance, in air, exposure to which for 24 hours or less would cause the death of 50% of an entire population of an experimental animal species, as determined from exposure to the substance of a significant number from that population.

In addition to toxic substances reported to have caused death in experimental animals, the toxic substances list includes those substances reported to have produced any toxic effect in man. There is no qualifying limitation to the duration of exposure, nor to the quantity or concentration of the substance, nor to the circumstances that resulted in the exposure. Also, the list includes those substances reported to have produced any carcinogenic, teratogenic, mutagenic, or neoplastigenic effects in humans or animals. Qualifying toxic effects for animal data include the production of tumors (neoplastigenesis), whether benign or malignant (carcinogenesis), and the production of changes in the offspring, whether transmissible (mutagenesis) or not (teratogenesis). The reported effects may have been generated by exposure to the substance over any period of time and in any quantity or concentration of dose.

Priority List.—NIOSH prepared another list, known as the priority list, for the purpose of ranking substances that would subsequently have occupational standards developed and recommended to the Occupational Safety and Health Administration (OSHA). The ranking of substances on the priority list was done by calculation of an overall rating. The overall rating was derived by multiplying the number of workers exposed to a substance by a severity rating. The severity rating was developed by a Delphi technique in which a list of compounds was sent out to some 50 industrial hygienists who rated the substances using a scale for relative severity. Criteria documents were then developed for substances ranked highest on the priority list.

Criteria documents present background information on chemical exposure and morbidity. Evidence of the substance acting as a carcinogen, teratogen or mutagen, or inducing respiratory, cardiac, or nerve disease in humans or animals is used in developing the recommended standard. Though usually there is extensive data on chemical exposure and morbidity, in most cases few good correlations between exposure and morbidity can be inferred.

OSHA is responsible for the promulgation of occupational safety and health standards. To date, criteria document development by NIOSH has resulted in the promulgation of standards for only two substances—vinyl chloride and asbestos. Many chemical substances, however, are currently listed in Tables G-1, G-2, or G-3 of Sec. 1910.93 of Title 29 (Labor), CFR, and subject to regulations described therein. Also, 14 substances are designated as carcinogens and subject to regulations found in Sec. 1910.93c.

An employee's exposure to any material in Tables G-1 or G-3, the name of which is not preceded by "C", cannot exceed the 8-hour time weighted average, the acceptable ceiling concentration, or the acceptable maximum peak above the ceiling concentration for the given material. The acceptable maximum peak concentration may never be exceeded, and exposure to concentrations greater than the acceptable ceiling concentration cannot exceed the "maximum duration" time period listed in the table. The source of standards found in this table is the American National Standards Institute, 237 series.

Occupational health standards promulgated by OSHA for a given substance supercede those restrictions found in the previously referenced tables.

National Cancer Institute

The National Cancer Institute (NCI) is sponsoring a program to collect, analyze, and systematize information in the chemical description, production, distribution, and human exposure to carcinogenic chemicals which the public may come into contact with in significant amounts. The program is designated "A Research Program to Acquire and Analyze Information on Chemicals that Impact on Man and His Environment." Criteria used by NCI's Chemical Selection Committee for selecting chemicals for carcinogenic testing are:

1. The degree of overall exposure
2. Projected new or increased human exposure
3. Exposure of subpopulations important to society
4. Epidemiological clues (high cancer incidence subpopulations)
5. Relation to known carcinogens
6. Gaps in knowledge

Collected data include such information as the quantity of chemicals available for exposure, and the routes of exposure, i.e., oral, dermal, respiratory, and peritoneal.

Environmental Protection Agency

Office of Pesticide Programs

Title 40 (Protection of Environment), CFR, Part 162 contains regulations for the enforcement of the Federal

Insecticide, Fungicide, and Rodenticide Act (FIFRA), as amended. Sec. 162.8 lists the criteria for economic poisons "highly toxic" to man. The oral, dermal, and inhalation toxicity levels used to classify "highly toxic" poisons are identical to those listed in Table 15, U.S. Code, 1261. The classification of economic poisons is done for labeling purposes.

Proposed rulemaking to revise current procedures for the registration of pesticides and establish procedures for the registration and classification of pesticides to conform with the provisions of FIFRA, as amended, was published in the *Federal Register* of October 16, 1974. Proposed Sec. 162.10 describes labeling requirements. Four toxicity categories are listed, with Category I corresponding to the "highly toxic" classification under the current definition of economic poisons. In addition to prescribed oral, inhalation, and dermal toxicity levels, Toxicity Category I includes eye effects and skin irritation criteria. If a substance causes irreversible corneal capacity at seven days or severe skin irritation or damage at 72 hours, then it would be classified under Toxicity Category I.

The proposed rulemaking includes flammability labeling requirements. "Extremely flammable" and "flammable" categories correspond to those found in Title 15, U.S. Code, Sec. 1261. Also, a "Caution" label is required for non-pressurized containers having a flash point above 80 F and not over 150 F.

Acting under authority of Sec. 3(c)(5) of FIFRA, as amended, the Administrator proposes indices of presumptive refusal to register pesticides (Sec. 162.11). These indices include both acute and chronic toxicity criteria. "Use" classification criteria for newly and previously registered pesticide products are proposed (also in Sec. 162.11) under authority of Sec. 3(d) of FIFRA, as amended. Criteria for domestic, nondomestic, and outdoor applications of pesticides are delineated. Classification criteria for new pesticide registrations are generally less stringent than those for previously registered products.

Ocean Dumping

Sec. 102(a) of the Marine Protection, Research, and Sanctuaries Act of 1972 (P. L. 92-532) authorizes the Administrator to establish criteria for evaluating ocean dumping permit applications. Title 40, CFR, Part 227 is concerned with criteria for the evaluation of permit applications. Sec. 227.22(f) specifies that wastes with trace contaminants may be dumped under special permit when the following limits are not exceeded:

(1) Mercury and its compounds are not present in any solid phase of a waste in concentrations greater than .75 mg/kg, and the total con-

centration of mercury in the liquid phase of a waste does not exceed 1.5 mg/kg;

(2) Cadmium and its compounds are not present in any solid phase of a waste in concentrations greater than .6 mg/kg, and the total concentration of cadmium in the liquid phase of a waste does not exceed 3.0 mg/kg;

(3) The total concentrations of organohalogens do not exceed the limiting permissible concentrations of pollutants as defined in Sec. 227.71;

(4) The total amounts of oil and greases do not produce a visible surface sheen in an undisturbed water sample when added at a rate of one part waste material to 100 parts of water.

Sec. 227.71 defines limiting permissible concentration as:

(a) that concentration of waste material or chemical constituent in the receiving water which, after reasonable allowance for initial mixing in the mixing zone will not exceed .01 of a concentration shown to be toxic to appropriate sensitive marine organisms in a bioassay carried out in accordance with approved EPA procedures; or

(b) .01 of a concentration of a waste material or chemical constitutent otherwise shown to be detrimental to the marine environment.

The 96-hour median threshold value for aquatic life of a waste material or chemical constituent is used in computation of the limiting permissbile concentrations. Subsequent sections define the release and mixing zones.

The level of trace contaminants allowed in waste dumped under special permit and the definition of limiting permissible concentrations were derived from a working group's consensus judgment.

Drinking Water Standards

The current drinking water standards were promulgated by the U.S. Public Health Service in 1962. Proposed interim primary drinking water standards were published in the *Federal Register* on March 14, 1975, pursuant to Sec. 1412(a) of the Safe Drinking Water Act (P.L. 93-523). These standards will take effect 18 months after the date of promulgation. The current and proposed drinking water standards for inorganic and organic chemicals, and the proposed drinking water standards for pesticides are listed in Table E. (Note: Interim standards for aldrin, dieldrin, and DDT will be proposed pending the completion of a survey of selected water supplies. The survey is intended to estimate the current level of pesticides in U.S. drinking water supplies.) These standards are only concerned with those chemical substances regulated for health reasons. Other drinking water standards are concerned with such aesthetic qualities as odor, foaming and taste.

The rationale used in writing the current drinking water standards was that substances which may have deleterious physiological effects, or for which physiological effects were not known, should not be introduced into the water supply system in a manner which would permit them to reach the consumer. The total lifetime environmental exposure of man to a specific toxicant was considered when arriving at specific limits for the proposed interim primary drinking water standards. Choice of the concentration of a given substance permitted in drinking water was made after allowance for the amount of toxicant contributed by food (FDA guidelines for maximum allowable concentrations), or other environmental factors (such as the amount of lead asbestos through inhalation by people living in urban areas).

In formulating the proposed interim primary drinking water standards, consideration was given to the toxicities of a given substance and the likelihood of bioaccumulation, carcinogenicity, mutagenicity, or teratogenicity of that substance. Studies investigating the characteristics of a substance offered evidence that was often unclear, and sometimes conflicting, thus making the choice of a "safe no-effect" level in drinking water quite difficult. A safety factor, varying with the reliability of the evidence, was therefore used in derivation of the interim primary drinking water standards. The level of control chosen minimized the amount of a toxicant contributed by water, and this level was generally attainable by good quality control.

Federal Water Pollution Control Act, as Amended (FWPCA)

Sec. 304 (a)(1)-Water Quality Criteria.—The Administrator is obligated by this section to publish criteria for water quality taking into account identifiable effects on the propagation of fish and wildlife, recreational activities, public water supplies, and agricultrual, industrial and other activities. Inorganic and organic chemicals for which water quality criteria have been written for health reasons appear in Table F.

Considerations similar to those mentioned for formulation of the proposed interim primary drinking water standards were used in derivation of the water quality criteria. Also, allowance for the amount of toxicant contributed by other environmental factors (e.g., food, lead inhalation) was made when determining recommended concentrations of a given substance for each of the purposes mentioned above.

Sec. 307 (a)-Toxic Pollutant List.—This section of the FWPCA mandates the Administrator to publish a list

TABLE E—DRINKING WATER STANDARDS

	Current (mg/1)	Proposed Interim Primary (mg/1)
Arsenic	0.05	0.05
Barium	1.0	1.0
Cadmium	0.01	0.01
Chromium (Cr+6)	0.05	0.05
Cyanide	0.2	0.2
Fluoride		
Daily air temperatures (°F)		
50.0-53.7	1.7	2.4
53.8-58.3	1.5	2.2
58.4-63.8	1.3	2.0
63.9-70.6	1.2	1.8
70.7-79.2	1.0	1.6
79.3-90.5	0.8	1.4
Lead	0.05	0.05
Mercury	none	0.002
Nitrate	45. (undesirable)	10.
Organics-Carbon Absorbable	0.2	0.7
Selenium	0.01	0.01
Silver	0.05	0.05

Pesticides	Proposed Interim Primary (mg/1)
(i) Chlorinated Hydrocarbons	
Chlordane	0.003
Endrin	0.0002
Heptachlor	0.0001
Heptachlor Epoxide	0.0001
Lindane	0.004
Methoxychlor	0.1
Toxaphene	0.005
(ii) Chlorophenoxys	
2,4-D	0.1
2,4,5-TP Silvex	0.01

of toxic pollutants for which effluent standards will be promulgated. The criteria used for selection of the toxic pollutants were published in the *Federal Register,* Vol. 38., No. 173, Sept. 7, 1973. The proposed toxic pollutant effluent standards, *Federal Register,* Vol. 38, No. 247, Dec. 27, 1973, were recently invalidated in court. Consequently, the criteria used for selection of these pollutants are currently under revision.

TABLE F—WATER QUALITY CRITERIA

Substance	Domestic Water Supplies (mg/l)	Freshwater Aquatic life (mg/l)	Marine Aquatic Life (mg/l)	Other (mg/l)
Arsenic	0.05	none	none	0.1 - irrigation of crops
Barium	1.0	none	none	none
Beryllium*	none	0.011 - soft freshwater / 1.1 - hard freshwater	none	0.1 - continuous use on all soils / 0.5 - short-term irrigation on neutral to fine-textured soils
Cadmium *	0.01	soft water / hard water: 0.0004 / 0.003 - for cladocerans and salmonid fishes; 0.004 / 0.012 - for other, less sensitive, aquatic life	0.005	none
Chlorine	none	0.003 - salmonid fish / 0.01 - other freshwater organisms	0.01	none
Chromium	0.05	0.3	none	none
Copper	1.0 (aesthetic)	+	+	none
Cyanide	0.2	0.005	0.005	0.005 - wildlife
Iron	0.3 (aesthetic)	1.0	none	none
Lead	0.05	**	none	none
Manganese	0.05 (aesthetic)	none	0.1 - for protection of consumers of marine mollusks	none
Mercury	0.002	0.00005	0.0001	0.00005 - wildlife
Nitrate	10.0	none	none	none
Pthalate ester	none	0.003	none	none
Polychlorinated biphenyls	none	0.000001	0.000001	none
Selenium	0.01	***	***	none
Silver	0.05	***	***	none
Zinc	(aesthetic)	***	none	none

* Soft water is defined as having less than 100 mg/1 hardness as CaCO$_3$. Hard water has 100 mg/l or more of CaCO$_3$.

** The recommended limit is 0.01 of a 96-hour LC$_{50}$ value, using the receiving or comparable water as the diluent and soluble lead measurements (non-filtrable lead using a 0.45 micron filter) for sensitive freshwater fish species.

*** The recommended limit is 0.01 of the 96-hour LC$_{50}$, as determined by a bioassay of a sensitive resident fish species.

\+ The recommended limit is 0.1 of the 96-hour LC$_{50}$, as determined by a nonaerated bioassay of a sensitive aquatic resident species.

TABLE F—WATER QUALITY CRITERIA (continued)

Pesticides	Domestic Water Supplies (Micrograms/Liter)	Freshwater Aquatic Life (Micrograms/Liter)	Marine Aquatic Life (Micrograms/Liter)
Chlorinated Hydrocarbon Insecticides			
Aldrin – dieldrin*			
Chlordane	none	0.01	0.004
DDT*			
Endrin	0.2	0.01	0.01
Endosulfan	0.003	0.01	none
Heptachlor	0.01	0.01	0.1
Lindane	4.0	0.01	0.05
Methoxychlor	100.0	0.02	0.03
Toxaphene	5.0	0.01	0.01
Organophosphate Insecticides			
Demeton	none	0.1	0.1
Guthion	none	0.01	0.01
Malathion	none	0.1	0.1
Parathion	none	0.1	0.1
Chlorophenoxy Herbicides			
2, 4 – D	100.0	none	none
2, 4, 5 – TP	10.0	none	none
Organochloro Insecticide			
Mirex	none	0.001	0.001

* The persistence, bioaccumulation potential and carcinogenicity of these substances make avoidable human exposure or release into waterways unreasonably hazardous.

The criteria published in the *Federal Register* are briefly discussed as an indication of what the revised criteria might be. One criterion was concerned with a pollutant's toxicological properties. Evidence that a pollutant could be "highly toxic" to aquatic or non-aquatic life was reviewed. The term "highly toxic to aquatic life" was defined as a substance having a 96-hour LC_{50} of 10 mg/1 or less; this classification was taken from the National Academy of Sciences water pollution ranking system. The term "LC_{50}" means that concentration of a substance in water which is lethal to one-half of the test population in the specified time period. The term "highly toxic to non-aquatic life" was defined as in Title 15, U.S. Code, Sec. 1261. Less standardized data, including data from sub-acute and chronic testing, were reviewed as well.

Toxicological data were reviewed for evidence of bioaccumulation through aquatic mechanisms to the extent of reproductive impairment or contamination of foods. Also, evidence of a substance having carcinogenic, mutagenic, or teratogenic properties and being stable in the aquatic environment was reviewed.

Another criterion was concerned with the seriousness of discharges or potential discharges from point sources. The production, distribution, and use pattern of the pollutant were determined. Identification of toxic pollutant discharges should demonstrate that substantial quantities of the pollutant, both in absolute terms and relative to non-point sources, have been deposited in the aquatic environment. Also, these discharges must have caused substantial environmental harm, as supported by documented damage to important organisms.

Sec. 311(b)(2)(A)-Hazardous Substance Spills.—This section of the FWPCA authorizes the Administrator to promulgate regulations designating as hazardous substances those elements and compounds which, when discharged in any quantity into or upon the navigable waters, adjoining shorelines or the waters of the contiguous zone, present an imminent and substantial danger to the public health or welfare. The constraint in designation of hazardous substances of an "imminent danger" precludes use of criteria characterized by long-term effects, i.e., bioaccumulation, carcinogenesis, mutagenesis, or teratogenesis. Consequently, hazardous substance designation is based on acute effects only.

Criteria to be used for hazardous substance designation under this section were proposed in the *Federal Register,* Vol, 39, No. 164, August 22, 1974, ("Designation and Removability of Hazardous Substances from Water"). The oral, dermal, and inhalation toxicity criteria are identical to the "highly toxic" criteria found in Title 15, U.S. Code, Sec. 1261. Also, substances having a median inhibitory limit (ILm) of 100 ppm or less are considered as

having the potential for presenting an imminent and substantial danger to the environment when spilled. This criterion measures the decrease in aquatic flora, as indicated by a 50 percent decrease in cell count, biomass, or photosynthetic ability over a 14-day period. This criterion, while not specifically mentioned in Sec. 311, is included because aquatic plant life is of critical importance to aquatic fauna.

The last toxicological criterion used in designation of hazardous substances is a 96-hour aquatic LC_{50} of 500 ppm. The basis for the chosen limit is a calculation determining the achievable pollutant concentration in the average spill situation. Assumptions are made concerning the volume of water flowing in a stream, the size of a water body, uniform mixing within a given discharge time, and concentration of hazardous substances likely to result from typical capacity sizes of various transportation modes. The calculation then predicts that the concentration of a hazardous substance would rarely exceed 500 ppm.

Another criterion mandates that a reasonable potential for spillage of the material be established. Factors considered in making this evaluation include the production quantities, modes of transportation, handling and storing practices, past spill experience, and physical-chemical properties of each substance.

Additionally, EPA is soliciting comments on the economic impact of the proposed regulation. The cost of spill prevention or the impact of penalties borne by affected parties will be considered.

Clean Air Act-Sec. 112

This section of the Clean Air Act authorizes the Administrator to publish a list of hazardous air pollutants for which emission standards will subsequently be promulgated. "Hazardous air pollutants" are those which may cause, or contribute to, an increase in mortality or an increas in serious irreversible or incapacitating reversible illness. This definition is concerned only with the safeguarding of human health.

A preferred standards path analysis is performed for a pollutant being considered for hazardous designation. Ambient air data, health effects information, the technical feasibility of controlling emissions, and the economic implications of regulating the substance are assembled in an analytic document. Considering all this information and anticipating the practicality of emission standards implementation, decision-makers subjectively pass judgment as to whether or not a substance should be designated as hazardous. Asbestos, beryllium, and mercury, having been designated as hazardous, are currently regulated by emission standards.

EPA Sewage Sludge Working Group

This Working Group is presently formulating guidelines for the heavy metal content of sewage sludge applied to cropland or land used for cattle grazing. Recommended levels have yet to be determined.

TRW Systems Group Recommended Methods of Reduction, Neutralization, Recovery, or Disposal of Hazardous Waste

The TRW report initially classifies waste stream constituents into three categories: Category 1, candidates for National Disposal Sites; Category 2, candidates for Industrial Disposal; and Category 3, candidates for Municipal Disposal. Classification of waste stream constituents requires consideration of both the hazards associated with the constituents and the evaluation of the adequacy of the currently practiced treatment and disposal practices. The criteria used in classifying waste stream constituents are as follows:

> Category 1: Materials are very difficult to handle, have Threshold Limit Values less than or equal to 1 ppm, and/or are very highly flammable, and/or are currently regulated or considered for regulation by the Environmental Protection Agency.

> Category 2: Material can be handled by normal industrial procedures, can be incinerated with proper scrubbing equipment, buried without treatment in a Class 1 landfill, or broken down by biological processes as utilized by industry.

> Category 3: Materials are relatively easy to handle, can be incinerated without scrubbing, buried in a Class 2 landfill, or treated by municipal sewage treatment processes.

The landfill classifications are taken from the California State Department of Public Health, the Department of Water Resources. Class 1 landfills are those sites located over non-water bearing sediments or with only unusable ground water underlying them. The site location must provide complete protection from flooding, surface runoff or drainage, and waste materials and all internal drainage must be restricted to the site. Class 2 landfills are sites underlain by usable ground water, and they may be located adjacent to streams. A distance of separation must be maintained between the bottom of the fill and the water table. Surface water must be restricted from the site, to preclude water contact with the wastes.

A Profile Report is prepared for each of the waste stream constituents studied. The first three sections of the report present information characterizing the waste stream constituent. An introductory section contains such information as manufacturing techniques, production rates, use patterns, and types, quantities, sources, and distribution of wastes containing the constituent. Pertinent physical/chemical properties of the material are also summarized here. A toxicology section generally documents recommended Threshold Limit Values (TLV) for man, the TLm for fish, acute oral and dermal LD_{50} values for various forms of animal life, plant reactions following exposure, and exposure symptoms in man. (The TLV is that concentration of an airborne constituent to which workers may be exposed repeatedly, day after day, without adverse effect.) A section describing other than toxicological (or radiation) hazards completed the characterization of the constituent. The types of hazards included in this section are flammability, explosiveness, corrosiveness, and detectability.

The next section of the Profile Report discusses the overall hazard associated with the waste management sequence between waste constituent generation and waste disposal. Adequate waste management is defined in terms of maximum acceptable levels of occurrence in air, water, and soil, based mainly but not solely on the toxicological effects of the waste constituent. Maximum acceptable levels of occurrence are designated as Recommended Provisional Limits.

The Recommended Provisional Limits in air are equal in value to one-hundredth of the established TLV. For constituents without an established TLV, that of a structurally related compound is used. The Recommended Provisional Limits in water and soil are equal in value, based on the worst case assumption that contaminated soil is completely non-retentive, and that the contaminant eventually percolates to the ground water table, and eventually becomes potable water. These concentration limits are equal in value to current drinking water standards, or one-hundredth of the lowest reported drinking water study level in cases where no drinking water standard exists.

In the absence of drinking water standards or study limit values, provisional limits are calculated on the basis of the Stokinger and Woodward method. By this method, limiting concentrations for water are based on the TLV's of hazardous materials. The assumption underlying this method is that the noninjurious amount taken into the blood stream by inhalation is equal to that which can be taken orally in water. When this calculation is applied and no established TLV exists, then the TLV of a structurally related compound is used.

TRW reports that in a reasonable number of instances drinking water standards are roughly approximated by using a 100-fold safety factor to the lowest drinking water study levels or to the limits obtained by the Stokinger and Woodward method. Limiting concentrations for organic compounds present the greatest difficulty, however. The

TABLE G—CRITERIA FOR THE IDENTIFICATION OF CANDIDATES FOR NATIONAL DISPOSAL SITES

QUANTITY CRITERION

(1) Material is present in sizable quantities as a waste.

HAZARD CRITERIA

(1) Waste material is highly toxic.

(2) Waste material is toxic and not degraded, oxidized, reduced or combined to a nontoxic form by air, water, or soil organisms.

(3) Waste material is radioactive with long half-life and/or high level radiation.

(4) Waste material is spontaneously combustible or is an explosive sensitive to heat or mild shock.

TREATMENT CRITERIA

(1) No disposal method other than long term or permanent storage is considered adequate for the material.

(2) Adequate disposal techniques for the material are too specialized or complex for general application.

(3) Adequate disposal methods for the material are under development but not yet available, requiring short term storage.

Stokinger and Woodward limits often exceed the minimum concentration producing detrimental effects in fish. Also, the calculated value may greatly exceed known odor and taste thresholds for man. In the absence of additional toxicity or annoyance data, a 100-fold safety factor is applied to the calculated values to give the provisional limiting concentrations.

As an additional safety feature, these limiting concentrations are compared with "safe" concentrations for fish. The rule-of-thumb method for determining the "safe" concentration, as reported by Sprague, is to multiply the 48-hour TLm by one-tenth. Generally, recommendations for maximum levels are 0.1 to 0.05 toxic units for non-persistent pollutants, and 0.1 to 0.01 toxic units for persistent chemicals and pesticides, mostly of the lower figure. In the absence of TLV's for man, fish limits serve as tentative limiting concentrations for man.

The ratio of the five-day biological oxygen demand to the ultimate oxygen demand is used as a measure of persistence of a substance in the environment. Any substance with a value less than 20 percent is considered to be persistent. Available data on the accumulation of toxic substances in the ecologic cycle is also used in evaluating the persistence of a substance.

Department of Defense

Department of the Army, Materiel Command, Edgewood Arsenal

A ranked list of industrially supplied chemicals by acute toxicity, (mammalian oral consumption), is maintained by the Department of the Army, Materiel Command, Edgewood Arsenal. The oral LD50 of each chemical is classified on a scale ranging from an insignificant toxicity category, to progressively more toxic categories. The classification used at this facility is as follows:

	Classification	Oral LD50 (mg/kg)
0	Insignificant toxicity	above 5,000
1	Insignificant toxicity	500-5,000
2	Insignificant toxicity	50-500
3	Insignificant toxicity	5-50
4	Most Toxic	5

Department of the Navy Consolidated Hazardous Item List

The potential danger of supply materials is evaluated in regards to health hazard, flammability, reactivity, corrosivity, oxidation capacity, radioactivity, magnetic properties, or fire-fighting hazard. The degree of hazard classifications relative to health, flammability, and reactivity are those contained in NFPA Protection Guide No. 704M. This system utilizes five hazard levels, from four (4), indicating a severe hazard, to zero (0), indicating no special hazard. The degrees of hazard relative to flammability are outlined in this publication as follows:

Hazard Level	Criteria
4	Fp<73°F and Bp<100°F
3	Fp<73°F and Bp>100°F, or 73°F<Fp<100°F
2	100°F<Fp<200°F
1	Fp>200°F
0	Material will not burn.

CHAPTER TWO

TOXIC AND HAZARDOUS
SUBSTANCES

PRELIMINARY LIST OF CHEMICAL SUBSTANCES FOR FURTHER EVALUATION*

The Preliminary List which follows, beginning on the next page, contains in alphabetical order the names of chemical substances, mixtures and categories which are candidates for inclusion in the priority list under development by the Toxic Substances Control Act Interagency Testing Committee. The presence of a substance on the list indicates that the substance:

(a) Was on one of a number of previously developed lists of substances which may pose adverse health and/or environmental effects; and

(b) Was judged by the Committee to warrant further evaluation because significant human exposures and/or environmental release can be expected.

On the other hand, the presence of a substance on the Preliminary List does not, by itself, indicate that the Committee is making any statement on whether the substance should be regulated or on the need or priority for further testing of that substance. The Committee has not yet completely evaluated the substance's potential for adversely affecting human health or the environment. Nor has the Committee yet evaluated the adequacy of existing test data bearing on such effects. Only after the Committee has evaluated these factors will recommendations be developed. Moreover, the Committee does not exclude the possibility of recommending the testing of additional substances or categories, even through they do not appear on the Preliminary List. In general, categories of substances appearing on the Preliminary List reflect category or group entries on one or more of the source lists used by the Committee. The Committee will consider the rationale for those categories as well as the desirability of forming other categories in developing its recommendations.

---

* The Toxic Substances Control Act Interagency Testing Committee, Office of Toxic Substances, U.S. Environmental Protection Agency, July 1977.

Chemical or Category List

CHEMICAL OR CATEGORY NAME	NIOSH No.[1]	CAS No.[2]
Acetaldehyde, Chloro-	AB24500	000107200
Acetamide	AB40250	000060355
Acetamide, Thio-	AC89250	000062555
Acetic acid, Benzyl ester	AF50750	000140114
Acetic acid, Chloro-	AF85750	000079118
Acetic acid, Diazo-, ethyl ester	AG57750	
Acetic acid, (Ethylenedinitrilo)tetra-, tetrasodium salt	AH50750	000064028
Acetic acid, Iminodi-	AI29750	000142734
Acetic acid, Nitrilotri-	AJ01750	000139139
Acetic acid, Trichloro-	AJ78750	000076039
Acetonitrile	AL77000	000075058
Acetophenone, Chloro-	AM61250	001341248
Acetylene	AO96000	000074862
Acetyl peroxide	AP85000	000110225
Acrolein	AS10500	000107028
Acrylamide	AS33250	000079061
Acrylic acid	AS43750	000079107
Acrylic acid, 2-Cyano-, methyl ester	AS70000	000137053

(1) Identification number as given in the Registry of Toxic Effects of Chemical Substances, 1976 edition, National Institute for Occupational Safety and Health. In cases where the substance or mixture of substances was not included in the NIOSH Registry, an identification number was created for the substance in the same format as the NIOSH number. Such numbers are indicated by an asterisk (*).

(2) Chemical Abstracts Service (CAS) number as given in the NIOSH Registry.

Acrylic acid esters

e.g.	Acrylic acid, ethyl ester	AT07000	000140885
	Acrylic acid, 2-ethylhexyl ester	AT08550	000103117
	Acrylic acid, methyl ester	AT28000	000096333

Alkoxy alkanols

e.g.	Ethanol, 2-Butoxy-	KJ85750	000111762
	Ethanol, 2-(2-Butoxyethoxy)-	KJ91000	000112345
	Ethanol, 2-Ethoxy-	KK80500	
	Ethanol, 2-(2-Ethoxyethoxy)-	KK87500	000111900
	Ethanol, 2-Methoxy-	KL57750	000109864
	Ethanol, 2-(2-Methoxyethoxy)-	KL61250	000111773
	Ethanol, 2-(2-(2-Methoxyethoxy)ethoxy)-	KL63900	000112356
	2-Propanol, 1,1'-Oxydi-	UB87850	000110985

Alkyl adipates

e.g.	Adipic acid, Bis(2-ethylhexyl) ester	AU97000	000103231
	Adipic acid, n-octyl n-decyl ester	*ZZ02084	

Alkyl amines

e.g.	Cyclohexylamine	GX07000	000108918
	Diethylamine	HZ87500	000109897
	Dimethylamine	IP87500	000124403
	Dodecylamine	JR64750	000124221
	Ethanamine	KH21000	000075047
	1,2-Ethanediamine	KH85750	000107153
	Isopropylamine	NT84000	000075310
	Methanamine, N,N-Dimethyl-	PA03500	000075503
	Methylamine	PF63000	000074895
	Triethylamine	YE01750	000121448

Alkyl epoxides

e.g.	Butane, 1,2:3,4-Diepoxy-	EJ82250	001464535
	Butane, (+-)-1,2:3,4-Diepoxy-	EJ84000	000298180
	Butane, 1,2:3,4-Diepoxy-, meso-	EJ87500	000564001
	Ethylene oxide	KX24500	000075218
	Butylene oxide	EK36750	000106887

Alkyl phthalates (short chain)

e.g.	Dibutyl phthalate	TI08750	000084742
	Diethyl phthalate	TI10500	000084662
	Dimethyl phthalate	TI15750	000131113
	Dimethyl terephthalate	WZ12250	000120616

Alkyl phthalates (long chain)

e.g.	Bis(2-ethylhexyl) phthalate	TI03500	000117817
	Dicyclohexyl phthalate	*ZZ02069	
	Diisodecyl phthalate	*ZZ02104	
	Diisooctyl phthalate	*ZZ02051	
	Dioctyl phthalate	TI19250	000117840
	Ditridecyl phthalate	*ZZ02051	
	n-Octyl n-decyl phthalate	*ZZ02110	

Alkyl sulfates and sulfonates, linear

e.g.	Dodecyl sulfate, triethanolamine salt	*ZZ02105	
	Monododecyl sulfate, sodium salt	WT10500	000151213
	Octyl sulfate, sodium salt	*ZZ02056	
	Tridecyl sulfate, sodium salt	*ZZ02057	

Allylamine	BA54250	000107119
Aluminum distearate	*ZZ02128	
Ammonium, Alkyl(C8-C18)dimethyl 3,4-dichlorobenzyl-, chloride	BO32000	MX8023538
Aniline	BW66500	000062533
Aniline, o-Chloro-	BX05250	000095512
Aniline, 3,4-Dichloro-	BX26250	000095761
Aniline, N,N-Diethyl-	*ZZ02065	
Aniline, N,N-Dimethyl	BX47250	000121697
Aniline, N,N-Dimethyl-p-nitroso-	BX71750	000138896
Aniline, 4,4'-Methylenebis(N,N'-dimethyl)-	BY52500	000101611
Aniline, 4,4'-Methylenedi-	BY54250	000101779
Aniline, N-Methyl-N,2,4,6-tetranitro-	BY63000	000479458
Aniline, p-Nitro-	BY70000	000100016
Aniline, p-(Phenylazo)-	BY82250	000060093
Aniline, 2,4,5-Trimethyl-	*BZ08750500	

p-Anisidine	BZ54500	0C0104949
Anthranilic acid	CB24500	000118923
Antimony and Antimony compounds		
e.g. Antimony	CC40250	007440360
Antimony (III) chloride	CC49000	
Antimony trioxide		
Antimony trisulfide	CC94500	001345046
Arsine	CG64750	007784421
Aryl phosphates		
e.g. Cresyl diphenyl phosphate		
Triphenyl phosphate	TC84000	000115866
Tris(2-ethylhexyl) phosphate		
Tris(isopropylphenyl) phosphate		
Tritolyl phosphate	TD01750	001330785
Aryl sulfonic acids and salts		
e.g. Benzenesulfonic acid, Dodecyl-	DB64750	027176870
Xylenesulfonic acid, sodium salt	ZE51000	
Azelate, Di(2-ethylhexyl)-	CM20000	000103242
1-Aziridineethanol	CM70000	001072522
Aziridine, 2-Methyl-	CM80500	000075558
Azoxybenzene	CO40250	000495487
Benzaldehyde	CU43750	000100527
Benzene, Chloro-	CZ01750	000108907
Benzene, 1-Chloro-2-nitro-	CZ08750	000088733
Benzene, 1-Chloro-3-nitro-	CZ09400	000121733
Benzene, 1-Chloro-4-nitro-	CZ10500	000100005
Benzene, Dichloro-		
e.g. Benzene, p-Dichloro-	CZ45500	000106467
Benzene, o-Dichloro-	CZ45000	000095501
Benzene, Dinitroso-		
Benzene, Divinyl-	*ZZ02060	
Benzene, (Epoxyethyl)-	CZ96250	000096093
Benzene, Ethyl-	DA07000	000100414

Benzene, Hexachloro-	DA29750	000118741
Benzene, 1,2-(Methylenedioxy)-4-propenyl-	DA59500	000120581
Benzene, Nitro-	DA64750	000098953
Benzene, Pentachloro-	DA66400	
Benzidine, 3,3'-Dimethoxy-	DD08750	000119904
Benzimidazole, 6-Nitro-	DD98000	000094520
Benzoic acid, 2-((4-Dimethylamino)phenylazo)-	DG89600	000493527
Benzophenone, 4,4'-Bis(dimethylamino)-	DJ02500	000090948
p-Benzoquinone dioxime	DK49000	000105113
Benzothiazole, 2,2'-Dithiobis-	DL45500	000120785
Benzothiazole, 2-(Morpholino-thio)-	DL59500	000102772
2-Benzothiazolesulfenamide, N-Cyclohexyl-	DL61250	000095330
Benzoyl chloride	DM66000	000098884
Benzoyl peroxide	DM85750	000094360
Benzyl alcohol	DN31500	000100516
Beryl	DS14000	001302529
Beryllium oxide	DS40250	001304569
Biphenyl	*DU80500500	
Biphenylamines		
e.g. 2-Biphenylamine	DU88500	000090415
4-Biphenylamine	DU89250	000092671
2,4'-Biphenyldiamine	DV21000	000492171
Bismuth and Bismuth compounds		
e.g. Bismuth		
Bismuth, Tris(dimethyldithiocarbamato)-	EB34000	021260468
1,3-Butadiene	EI92750	000106990
Butadiene and Butylene fractions	*EI92750500	
1,3-Butadiene, 2-Chloro-	EI96250	000126998
1,3-Butadiene, Hexachloro-	EJ07000	000087683

1-Butene	*EM28900250	
2-Butene (cis and trans)	*EM28900500	
2-Butene, 1,4-Dichloro-, (e)-	EM49030	000110576
t-Butyl peroxide	ER24500	000110054
Carbon black		
Carbon disulfide	FF66500	000075150
Carbon tetrabromide	FG47250	000558134
Carbon tetrafluoride	FG49200	000075730
Cellulose tetranitrate	FJ60000	PM9004700
Chloral hydrate	FM87500	000302170
Chloramine		
Chlorinated paraffins, 35-64% chlorine	*RV03500500	
Chromium compounds	*GB42000500	
e.g. Chromic acid, calcium salt (1:1), dihydrate	GB28000	010060089
Chromic acid, dipotassium salt	GB29400	007789006
Chromium (III) oxide (2:3)	GB64750	001308389
Chromium (VI) oxide (1:3)	GB66500	001333820
Cobalt	GF87500	007440484
Cobalt (II) nitrate (1:2)	GG11090	010141056
Cobalt (II) sulfide	GG33250	001317426
Copper and Copper compounds	*GL53250500	
e.g. Copper (metal)	GL53250	007440508
Cresol		
e.g. Cresol	GO59500	001319773
o-Cresol	GO63000	000095487
m-Cresol, 4,4'-Butylidenebis(6-tert-butyl)-	GO70500	000085609
p-Cresol, 2,6,-Dinitro-	GO98000	
m-Cresol, 4,4'-Thiobis(6-tert-butyl)-	GP31500	000096695
Crotonaldehyde, (e)-	GP96250	000123739
Cumene	GR85750	000098828

Cyanamide, calcium salt	GS60000	000156627
Cyclohexanol	GV78750	000108930
Cyclohexanol, Methyl-	GW01750	
Cyclohexanone	GW10500	000108941
Cyclohexene	GW25000	000110838
1-Cyclohexene, 4-Vinyl-	GW66500	000100403
1,3-Cyclopentadiene	GY10000	000542927
Cyclopentadiene, Hexachloro-	GY12250	
Cyclopentane	GY23900	000287923
Cyclopentane, Methyl-	GY46400	000096377
Decaborane(14)	HD14000	017702419
Dibenzofuran	*HP45500500	
Diethylamine, 2,2'-Dichloro-N-methyl-	IA17500	000051752
Dimethyl sulfoxide	PV62100	000067685
p-Dioxane	JG82250	000123911
Diphenylamine	JJ78000	000122394
Diphenylamine, 2,2',4,4',6,6'-Hexanitro-	JJ92750	000131737
Diphenylamine, 4-Isopropoxy-	JJ95000	000101735
Diphenylamine, N-Nitroso-	JJ98000	000086306
Diphenylamine, 4-Nitroso-	JK01750	000156105
Ethane, Bromo-	KH64750	000074964
Ethane, Chloro-	KH75250	000075003
Ethane, 1,2-Dichloro-	KI05250	000107062
Ethane, 1,1,2,2-Tetrabromo-	KI82250	000079276
Ethane, 1,1,1-Trichloro-	KJ29750	000071556
Ethane, 1,1,2-Trichloro-	KJ31500	000079005
Ethanol, 2-Amino-	KJ57750	000141435
Ethanol, 2-Chloro-	KK08750	000107073

Ethanol, 2-Dimethylamino-	KK61250	000108010
Ethanol, 2,2'-Iminodi-	KL29750	000111422
Ethanol, 2,2',2''-Nitrilotri-	KL92750	000102716
Ether, 2-Chloroethyl vinyl	KN63000	000110758
Ethylene	KU53400	000074851
Ethylene, Bromo-	KU84000	000593602
Ethylenediamine, N-(1-Naphthyl)-, dihydrochloride	KV53300	001465254
Ethylene, 1,1-Dichloro-	KV92750	000075354

Ethylene, 1,2-Dichloro

   e.g. Ethylene, 1,2-Dichloro-   KV93600 000540590
         Ethylene, 1,2-Dichloro-,(e)- KV94000

Ethylene, Tetrachloro-	KX38500	000127184
Ethylene, Trichloro-	KX45500	000079016
Ferrocene	IK07000	000010254

Flame retardants (brominated alcohols)

   e.g. Dibromobutenediol
         Dibromoneopentyl glycol
         2,3-Dibromopropanol   UB01750 000096139
         Tribromoneopentyl alcohol

Flame retardants (brominated aromatic compounds)

   e.g. Decabromobiphenyl
         Decabromobiphenyl ether

         Hexabromobenzene
         Hexabromobiphenyl
         Hexabromocyclododecane

Flame retardants (halogenated phosphates and phosphonates)

   e.g. Bis(2-chloroethyl)vinyl phosphonate
         Chlorinated polyphosphates
         Diethyl 2-bromoethylphosphonate
         Tris(4-bromophenyl) phosphate
         Tris(2-chloroethyl) phosphate
         Tris(2,3-dibromopropyl) phosphate
         Tris(2,3-dichloropropyl) phosphate
         Tris(2,4,6-tribromophenyl) phosphate

Flame retardants (hexachlorocyclopentadiene derivatives)

    e.g. Bis(chlorendo) bicyclopentadiene
            Bis(chlorendo) cyclooctadiene
            Bis(chlorendo) furan
            Bishexachlorocyclopentadiene
            Chlorendic acid
            Chlorendic anhydride
            Chlorendic salts
            Chlorendocyclooctadiene
            Bromochlorendocyclooctadiene
            2,3,4,5-Tetrabromophenyl-2,2a,2a,3,4,5-hexachloro-bicycloheptadiene

Flame retardants (miscellaneous halogenated compounds)

    e.g. Ammonium bromide
            Tetrabromobisphenol A, Bis(2,3-dibromopropyl ether)
            Tetrabromophthalic anhydride
            2,2',6,6'-Tetrabromo-3,3',5,5'-tetramethyl-4,4'-dihydroxybiphenyl
            Tetrachlorobisphenol A
            Tetrachlorophthalic anhydride

Flame retardants (phosphonium compounds)

    e.g. Tetrakis(hydroxymethyl) phosphonium bromide
            Tetrakis(hydroxymethyl) phosphonium chloride
            Tetrakis(hydroxymethyl) phosphonium hydroxide
            Tetrakis(hydroxymethyl) phosphonium sulfate

Fluorescent brightening agents	*LM59500500	
e.g. 4,4'-Diamino-2,2'-stilbenedisulfonic acid	*ZZ02081	
Fluoroacetamide	AC12250	000640197
Fluorocarbons (excluding fully halogenated chlorofluoro-alkanes)		
e.g. Ethane, 1-Chloro-1,1-difluoro-	KH76500	000075683
Methane, Chlorodifluoro-	PA63900	000075456
Methane, Dichlorofluoro-	PA84000	000075434
Formaldehyde	LP89250	000050000
Formamide	LQ05250	000075127
Formamide, N,N-Dimethyl-	LQ21000	000068122
Formic acid	LQ49000	000064186
Fumaric acid·	LS96250	000110178
2-Furaldehyde	LT70000	000098011
Furan, Tetrahydro-	LU59500	000109999

Glycols (low molecular weight)

e.g. Diethylene glycol	ID59500	000111466
Ethylene glycol	KW29750	000107211
1,2-Propanediol	TY20000	000057556
Tetraethylene glycol	XC21000	000112607
Triethylene glycol	YE45500	000112276

Heptane	MI77000	000142825
Heptene (mixed isomers)	MJ88500	
Hexamethylenetetramine	MN47250	000100970
1,6-Hexanediamine	MO11800	000124094
Hexanes and other C6 hydrocarbons	*MN92750500	

e.g. Cyclohexane	GU63000	000110827
Hexane	MN92750	000110543
Pentane, 2-Methyl-	SA29950	000107835

1-Hexanol, 2-Ethyl-	MP03500	000104767

Hydrazine, methyl hydrazines, and their derivatives

e.g. Hydrazine	MU71750	000302012
Hydrazine, 1,1-Dimethyl-	MV24500	000057147
Hydrazine, Methyl-	MV56000	000060344
Hydrazine, monohydrate	MV80500	007803578
Hydrazine, 1,1-Diphenyl-	*MW26250500	

Hydrazobenzene	MW26250	000122667
Hydrocyanic acid	MW68250	000074908
Hydrogen selenide	MX10500	007783075
Hydrogen sulfide	MX12250	007783064
Hydroperoxide, alpha, alpha-Dimethylbenzyl-	MX24500	000080159
Hydroquinone	MX35000	000123319
Hydroxylamine	NC29750	007803498
Hydroxylamine, O-Methyl-	NC38500	000067629
Hydroxylamine, N-Phenyl-	NC49000	000100652
Isocyanic acid, p-chlorophenyl ester	NQ85750	000104121
Isophthalic acid	NT20000	000121915
Isoprene	NT40370	000078795

Ketones, asymmetric

e.g.	2H-Azepin-2-one, Hexahydro-	CM36750	000105602
	2-Butanone	EL64750	000078933
	Cyclohexanone, 2-Methyl-	GW17500	
	2-Heptanone	MJ50750	000110430
	3-Heptanone, 5-Methyl-	MJ73500	000541855
	2-Hexanone	MP14000	000591786
	2-Hexanone, 5-Methyl-	MP38500	000110123
	2-Pentanone, 4-Methyl-	SA92750	

Lauroyl peroxide	OF26250	000105748
Lead, Bis(dimethyldithiocarbamato)-	OF88500	019010663
Ligninsulfonic acid, calcium salt	*ZZ02119	
Ligninsulfonic acid, ferrochrome salt	*OI31500500	
Lithium hydride	OJ63000	007580678
Maleic acid, dibutyl ester	ON08750	000105760
Maleic anhydride	ON36750	000108316
Manganese	OO92750	007439965
Manganese, Tricarbonyl 2-methylcyclopentadienyl	OP14700	
Melamine	OS07000	000108781
p-Menthane-8-hydroperoxide	OS94500	
Mercaptans	*OU22750500	

e.g. Dodecyl mercaptan	*JR80500500	

Methacrylic acid esters

e.g.	Methacrylic acid, butyl ester	OZ36750	000097881
	Methacrylic acid, ethyl ester	OZ45500	000097632
	Methacrylic acid, methyl ester	OZ50750	000080626

Methane, Bis(2-chloroethoxy)-	PA36750	000111911
Methane, Bromo-	PA49000	000074839
Methane, Bromochloro-	PA52500	000074975
Methane, Bromotrifluoro-	PA54250	000075638
Methane, Chloro-	PA63000	000074873
Methane, Dibromo-	PA73500	000074953

Methane, Dibromodifluoro-	PA75250	000075616
Methane, Dichloro-	PA80500	000075092
Methane, Dimethoxy-	PA87500	000109875
Methane, Iodo-	PA94500	000074884
Methane, Tribromo-	PB56000	000075252
Morpholine	QD64750	000110918
Naphthalene	QJ05250	000091203
Naphthalene, Decahydro-	QJ31500	000091178
Naphthalene, 1-Nitro	QJ97200	000086577

Naphthalenes, chlorinated

e.g. Naphthalene, Pentachloro-	QK03000	
Naphthalene, Tetrachloro-	QK37000	
Naphthalene, Trichloro-	QK40250	001321659
"Chlorinated naphthalenes"	*QJ21000500	

Naphthenic acid, copper salt	QK91000	001338029
Naphthenic acid, lead salt	OG20250	
2-Naphthylamine, N,N-Bis(2-chlorethyl)-	QM24500	000494031
2-Naphthylamine, N-Phenyl-	QM45500	000135886

Nickel and Nickel compounds

e.g. Nickel (metal)	QR59500	007440020
Nickel (II) acetate(1:2)	QR61250	000373024
Nickel, compd with pi-cyclopentadienyl (1:2)	QR65000	001271289
Nickel (II) oxide (1:1)	QR84000	001313991

Nitrophenols

e.g. Phenol, m-Nitro-	SM19250	000554847
Phenol, o-Nitro-	SM21000	000088755
Phenol, p-Nitro-	SM22750	000100027

Nonene (mixed isomers)	RA85500	
Octadecanoic acid, 9,10-Epoxy-, butyl ester	RG15750	000106832
Octane	RG84000	
Oxalic acid	RO24500	000144627
2H-1,3,2-oxyazaphosphorine, 2-(Bis(2-chloroethyl)amino) tetrahydro-, 2-oxide	RP59500	000050180

Pentane	RZ94500	000109660
1,3-Pentanediol, 2,2,4-Trimethyl-	SA14000	000144194
1-Pentanol, 2-Methyl-	SA71750	000105306
2-Pentanol, 4-Methyl-	SA73500	000108112
Peroxide, Bis(alpha,alpha-dimethylbenzyl)	SD81500	000080433
Peroxide, Bis(dimethylethyl)	*SD78850500	
Peroxyacetic acid	SD87500	000079210
Peroxybenzoic acid, t-butyl ester	SD94500	000614459

Petroleum distillates (boiling point 35-130°C)

e.g. Petroleum spirits (ligroin, solvent naphtha)	SE75550	
Benzin	DE30300	008030306
Phenol, 2,4-Dichloro-	SK85750	000120832
Phenol, Dodecyl-	SL36750	001331573
Phenol, 4,4'-Isopropylidenedi-	SL63000	000080057
Phenol, Nonyl-	SM56000	025154523
Phenol, Tetrachloro-	SM91000	025167833
Phenol, 3,4,5-Trichloro-	SN16500	000609198
o-Phenylenediamine	SS78750	000095545
p-Phenylenediamine	SS80500	000106503
p-Phenylenediamine, dihydrochloride	ST03500	000624180
p-Phenylenediamine, N,N'-Diphenyl-	ST22750	000074317
o-Phenylenediamine, 4-Nitro-	ST29750	000099569
Phosphine oxide, Tris(1-aziridinyl)-	SZ17500	000545551
Phosphine	SY75250	007803512

Phosphines (PR$_3$)

i.e.   R = alkyl, aryl and alkoxy (mixed)

Phosphonic acid, bis(2-chloroethyl)(1-hydroxyethyl) ester

Phosphorane, Pentachloro-	TB61250	010026138

Phosphoric triamide, Hexamethyl-	TD08750	000680319
Phosphorotrithioic acid, S,S,S,-tributyl ester	TG54250	000078488
Phthalic acid	TH96250	000088993
Phthalic anhydride	TI31500	000085449
Picric acid		
e.g. Picric acid (dry)	TJ78780	
Picric acid (wet)	TJ88500	
Pigment blue 15, alpha and beta forms	*ZZ02123	
Pigment green 7	*ZZ02062	
Pigment yellow 12	*ZZ02076	
Pine oil	TK51000	MX8006880
Polyacrylonitrile (fibers)	*TQ03500500	
Polychlorinated diphenyl ethers	*KN89700500	
Polychlorinated triphenyls	TQ13800	
Potassium pyrophosphates	TT49000	
Propane, 1-Chloro-2,3,-epoxy-	TX49000	000106898
Propane, 1,2-Dichloro-	TX96250	000078875
Propane, 1,2-Epoxy-	TZ29750	000075569
Propane, 1-Nitro-	TZ50750	000108032
Propane, 2-Nitro-	TZ52500	000079469
Propane, 2,2'-Oxybis-	TZ54250	000108203
Propane, 1,2,3-Trichloro-	TZ92750	000096184
2-Propanol, 1-Chloro-	UA87500	000127004
1-Propanol, 2,3-Epoxy-	UB43750	000556525
2-Propanone, 1-Chloro-	UC07000	000078955
2-Propanone, 1,1,1,3,3,3-Hexafluoro-	UC24500	000684162
Propene	UC67400	000115071
Propene, 3-Chloro-	UC73500	000107051
Propene, 1-Chloro-2-methyl-	UC80450	

Propene, 3-Chloro-2-methyl-	UC80500	000563473
Propene, 2-Methyl-	UD08900	000115117
2-Propenoic acid, butyl ester	UD31500	000141322
Propionitrile, 3-Amino-	UG03500	000151188
Propyne, mixed with propadiene	UK49200	
Pyridine	UR84000	000110861
Quinoline, 1,2-Dihydro-2,2,4-trimethyl-	VB49000	000147477
8-Quinolinol	VC42000	000148243
Sebacic acid, Bis(2-ethylhexyl) ester	VS10000	000122623
Selenium dimethyldithiocarbamate		
Selenium, Tetrakis(diethyldithiocarbamato)-	VT07000	017156831
Silver iodide		
Soaps (fatty acid salts)		
Sodium dibutyldithiocarbamate	EZ38800	000136301
Sodium thiosulfate, pentahydrate	WE66600	010102177
Stearic acid, methyl ester	WI44600	000112618
Stibene	WJ07000	007803523
Styrene	WL36750	000100425
Styrene, alpha-Methyl-	WL52500	000098839
Styrenes, chlorinated		
e.g. Chlorostyrene	WL41500	001331288
Sulfide, Bis(dimethylthiocarbamoyl)	WQ17500	000097745
Terephthalic acid	WZ08750	000100210
Thiophene, 2,5-Dihydro-, 1,1-dioxide	XM91000	000077792
Thiophene, Tetrahydro-, 1,1-dioxide	XN07000	000126330
Titanium dioxide	XR22750	013463677
Toluene	XS52500	000108883
Toluene, alpha-Chloro-	XS89250	000100447

Toluene, p-Chloro-	XS90100	000106434
Toluene-2,4-diamine	XS96250	000095807
Toluene-2,4- and -2,6-diisocyanate (80/20 mixture)	*CZ63000500	
Toluene, 2,4, (and 2,6)-Dinitro-	*XT15750500	
Toluene, alpha,alpha,alpha-Trichloro-	XT92750	000098077
Toluene, Vinyl (mixed isomers)	XU03500	
Toluidines		

e.g.	m-Toluidine	XU28000	000108441
	o-Toluidine	XU29750	000095534
	p-Toluidine	XU31500	000106490

Triallylamine	XX59500	000102705
s-Triazine, Hexahydro-1,3,5-trinitro-	XY94500	000121824
s-Triazine, 2,4,6-Trichloro-	XZ14000	000108770
s-Triazine-2,4,6(1H,3H,5H)-trione, 1,3,5,-Trichloro-	XZ19250	000087901
Trichlorobenzenes, mixed		
Triethylenetetramine	YE66500	000112243
Tungsten	YO71750	007440337
Tungsten carbide		
Turpentine	YO84000	MX8006642
Vanadium compounds	*YW15750500	

e.g.	Vanadium pentoxide (dust and fume)	YW24500	001314621
		YW24600	

Vat blue 6	*ZZ02058	
Xylenes		

e.g.	Xylene (mixed isomers)	ZE21000	001330207
	"Mixed Xylene"	ZE21900	
	m-Xylene	ZE22750	000108383
	o-Xylene	ZE24500	000095476
	p-Xylene	ZE26250	000106423

Xylenols

e.g.	2,5-Xylenol	ZE57750	000095874
	3,4-Xylenol	ZE63000	000095658
	3,5-Xylenol	ZE64750	000108689

Xylidine	ZE85750	001300738
Zinc (metal)	ZG86000	007440666
Zinc, Bis(dibutyldithiocarbamato)-	ZH01750	000136232
Zinc, Bis(diethyldithiocarbamato)-	ZH03500	000136947

Basic Approach Adopted by the Committee

Committee is required to evaluate a number of factors for each chemical substance or mixture included in its list of priority recommendations to the EPA Administrator. In considering possible approaches to meeting these responsibilities, certain limitations became evident. There was no data system of consolidated chemical information which permitted retrieval of all of the required data. In addition, the many existing data systems were not formatted in such a manner to easily permit merging of data files, e.g., CAS numbers were not always available and chemical names were not always designated in a uniform manner. Certain information required by the Committee was often unavailable. This was particularly critical for chemical uses and occurrences thus limiting the Committee's knowledge of overall exposure to many chemicals.

In the light of these limitations as well as the time constraints of its charge, the Committee chose to use existing lists of prioritized potentially hazardous substances developed by other agencies and organizations as a primary starting point in its review process. Since the criteria used in the development of these lists were similar to those of the Committee, the lists were used in their original form. The sources of these starting lists are discussed in the following section.

When these lists were combined, an initial listing emerged containing approximately 3500 substances and categories. Included in this was a number of drugs, food additives or pesticides not subject to the authority of the Toxic Substances Control Act. These substances were dropped from the list unless they were judged likely to have another use subject to TSCA regulation. In addition, those substances judged not to be in commercial production were removed from the list on the assumption that they had low potential for human exposure or environmental contamination. After these deletions, the remaining list consisted of approximately 2100 substances which is hereafter referred to as the "Master File".

The Master File will be screened in a 3-stage process. The first stage has been completed and involved screening the substances on the basis of potential for human or environmental exposure. From this stage, approximately 300 substances and categories have been designated as the "Preliminary List" which is presented here for information and comment. In the second stage, the substances and categories on the Preliminary List will be further evaluated on the basis of their potential for adverse effects to humans and the environment. It is anticipated that perhaps 100 substances or categories, will emerge from the second phase of screening for a most intensive and detailed third stage review by the Committee.

Subsequent to the development of the Master File, additional information has been made available by the Consumer Product Safety Commission consisting of 1288 chemicals occurring most often in over 15,000 consumer products surveyed in 1976. From this new data source as well as comments received from the public regarding the Preliminary List, it is possible that additional chemicals will be considered in detail by the Committee.

Request for Comments on Preliminary List

The Preliminary List and this Background Document are being made available at this time to allow public comment on the procedures used by the Committee in developing the Preliminary List and on the specific chemical substances, mixtures and categories which should be further evaluated by the Committee for possible inclusion in its testing priority recommendations to be made to the EPA Administrator by October 1, 1977.

Only the Preliminary List will be distributed to the public. The Master File and the list of other chemicals or categories considered by the Committee will also be available for inspection at the Council on Environmental Quality and Headquarters and Regional Offices of the Environmental Protection Agency.

Methodology Used to Develop the Preliminary List

Overview

The statute imposes a deadline of October 1, 1977 for the Committee to make its initial recommendations to the

EPA Administrator. As discussed in previous sections, a method was chosen to maximize the retrieval and consolidation of available data in view of this imposed time constraint. The methodology possesses three main features: 1) it relies heavily on previous efforts to identify chemicals which may pose a hazard to man or the environment; 2) it makes maximum use of readily available chemical data; and 3) it incorporates subjective judgment in those areas where data are absent or not easily accessible.

Accordingly, the Committee adopted a methodology whereby an initial listing of chemicals was successively reduced to smaller lists through the application of screening criteria. The initial compilation of the various lists resulted in a compendium consisting of approximately 3500 different substances and categories that were previously identified either as potentially hazardous to man or the environment or in annual production in quantities of over one million pounds per year.

The Initial Listing

The initial listing included a number of substances that had pesticide, food additive, or drug uses, all of which are regulated under other Federal statutes and exempted from regulation by TSCA. Therefore, the initial listing was purged of substances with such uses by screening it against lists of pesticides prepared by the EPA and lists of food additives and drugs prepared by the Food and Drug Administration. The basis for comparing chemicals on the three listings was the Chemical Abstracts Service (CAS) Registry Number. The attempted purge of these regulated substances was incomplete, since some entries on source lists did not include CAS numbers. To compensate for this, a further manual purging was required. Consideration was also given to the fact that a chemical with a pesticide, food additive or drug use may have other uses that are subject to the authority of TSCA. Thus chemicals with over 10 million pounds of annual production were retained in the truncated list for further review of possible uses, as were chemicals which had known uses that were within the TSCA regulation.

The resulting file was further reduced by the elimination of chemicals which were judged not likely to be in commercial production. This was accomplished by comparing the file against EPA's Candidate List of Chemical Substances, prepared by the Office of Toxic Substances (dated April 1977). Again, the basis of comparison for this purge was an assigned CAS number. Consequently, this purge did not affect those chemicals on source lists for which no CAS number was given. In an attempt to eliminate substances which are not in commercial production, the following rule was adopted: any substance not identified by a CAS number which appeared on the NIOSH Registry and on none of the other source files, was judged not likely to be in commercial production. This decision was based on the fact that the NIOSH Registry lists any chemical for which toxic effects have been reported, including research chemicals. A study of the chemicals eliminated by the application of this rule upholds its validity: few of the purged chemicals were recognized to be in commercial production.

The Master File

As a result of the purges described above, a Master File of approximately 2100 substances emerged. The Committee reduced this list by the further application of a set of screening criteria. This screen was designed to truncate the Master File on the basis of each chemical's relative potential for entering the environment in appreciable quantities with consequent exposure to humans and other sensitive species. The screening factors selected for this purpose correspond to the first four factors set forth in TSCA Sec. 4(e) as ones which the Committee should consider when making its recommendations to the EPA Administrator. They are:

(i) quantity produced annually
(ii) amount released into the environment
(iii) number of individuals who are occupationally exposed
(iv) extent to which the general population will be exposed.

Using a combination of published data and judgment, an attempt was made to score each substance on the Master File on each of these four factors. A score was assigned to a substance only if information was available giving the use or uses of the substance. For approximately 1400 substances, inadequate data on uses prevented their scoring and further evaluation. The Committee will attempt to obtain the needed information on these substances to permit their consideration for subsequent Committee revisions.

The Preliminary List

In the first stage (exposure) screen, approximately 700 chemicals were assigned scores for the four exposure factors. Chemicals were ranked by summing the individual scores, assigning equal weighting to each factor. This ranking was the principal basis for selection of substances for the Preliminary List. In addition, the Committee exercised professional judgment in eliminating from current consideration many chemicals which are: a) currently under regulation or being considered for regulation, e.g., vinyl chloride and benzene; b) reasonably well-characterized as hazardous, e.g., mercury; c) considered essentially inert materials, such as certain polymers; and d) natural products which would be difficult to characterize for testing purposes, e.g., wood or gasoline.

The first stage of screening resulted in the identification of approximately 300 substances, mixtures or categories, which are designated as the Preliminary List. These chemicals will be further evaluated in the second stage for their potential for adverse effects to humans and the environment.

In several cases the Committee has grouped chemicals, fully cognizant of the difficulties in identifying appropriate groupings for testing purposes. Among the methods of grouping under consideration are: primary use, structural similarities, predicted toxic effect, etc. While some categories are presented on the Preliminary List, such groupings will be further considered in the review process.

The chemicals on the Preliminary List will be further screened on the basis of potential for carcinogenic, mutagenic and teratogenic activity, other human effects, ecological potential hazards and the need for further testing. This screening process will result in the Committee recommending to the EPA Administrator up to 50 substances or categories that require priority consideration for testing.

Types of Comments Sought and Recommended Response Formats

General Information

Comments are specifically sought by the Committee on the methodology used in developing the Preliminary List, on the content of the Preliminary List, and on the specific types of test data which are needed on substances appearing on the Preliminary List.

Because of the October 1, 1977 statutory deadline for the Committee's initial recommendations to the EPA Administrator, it is necessary that comments on the Preliminary List be timely and concise, but that they provide adequate information to allow the Committee to evaluate the recommendations made by commentors. This is particularly important in the case of comments on the content of the Preliminary List (i.e., recommendations that the Committee consider a substance not appearing on the Preliminary List or drop from consideration a substance which does appear on the Preliminary List).

As an aid to commentors, this section of the Background Document discusses the specific types of information which the Committee considers desirable to permit adequate evaluation of comments and provides recommended formats for the submission of information by commentors on the content of the Preliminary List. While it is not mandatory that commentors follow the recommended formats, failure to provide the types of information requested in the formats and marked deviations from the suggested formats may make it difficult for the Committee to take such comments into account in establishing its initial list of recommendations to the EPA Administrator.

While submission of adequate supporting data is important to the effective and timely consideration by the Committee of commentors' recommendations, submission of voluminous studies and large volumes of raw data may impede, rather than assist, the Committee's consideration of the commentor's views. Therefore, it is requested that commentors summarize information where possible in their submissions providing references to primary data sources on which the summarized comments are based, and where applicable identify an individual as a contact point.

It is also requested that commentors not submit with their comments data or copies of studies for which the commentor wishes to claim confidential treatment. Copies of all comments on the Preliminary List submitted to the Committee will be subject to public inspection. Where information which the commentor considers confidential is critical to adequately support the recommendations made in the comments, it is requested that the existence and general nature of the information be noted in the comments in a way which would not reveal any trade-secret data and that an individual be identified for the Committee to contact should it wish to subsequently request submission of the information.

Comments on Methodology Used in Developing the Preliminary List

Comments are sought by the Committee on the methodology it has used in developing the Preliminary List. Comments are specifically sought in the following areas:

a. The general approach used by the Committee in screening substances appearing on a number of existing lists of chemicals of potential hazard to health or the environment to identify those with the greatest potential for human exposure and/or environmental release. Suggestions of alternative approaches which might be considered by the Committee in preparing subsequent revisions of its recommendations are sought.

b. The various lists used by the Committee in constructing its initial listing. Commentors are requested to suggest additional sources for identification of potentially hazardous chemicals for review by the Committee.

c. Data sources for identifying drugs, food additives, pesticides, and non-commercial chemicals appearing on the initial listing.

d. Additional sources of data on production volumes, uses, environmental releases of, and exposure to chemicals.

e. The scoring and weighting factors used by the Committee to evaluate the relative levels of human exposure to and environmental release of chemicals for purposes of selecting items to be included on the Preliminary List.

Comments on the methodology used in developing the Preliminary List will, to the extent possible, be considered by the Committee in its further development of its initial recommendations of testing priorities. However, because of the statutory deadline for transmitting those recommendations, extensive modifications in approach will be considered principally by the Committee in developing procedures to be used for subsequent revisions of its recommendations.

Comments on the Content of the Preliminary List

Commentors may recommend specific modifications to the list, including additions or deletions of specific substances or exceptions of substances from inclusion in a category. In order to evaluate recommended additions, the Committee desires pertinent information about the substance, its production, use, environmental release, human exposure, and its health and environmental effects. A rationale for deletions or category modifications should also be presented.

In view of the Committee's statutory deadline for developing its initial testing priority recommendations, it is important that the needed information be submitted by the commentor in a form which will simplify the Committee's consideration. While commentors are encouraged to provide all the requested items, failure to do so will not jeopardize consideration of the comments. Commentors may supply additional data to support their recommendations.

Comments on Test Data Needs

The Committee also seeks comments on the needs for further testing of substances appearing on the Preliminary List. Commentors are requested to identify the substance on which they are commenting using the name as it appears on the Preliminary List and to specify the type of testing recommended by the commentor and the reasons for that recommendation. If a commentor wishes to recommend that further testing of a substance appearing on the Preliminary List is not needed, the substance should again be identified using the name as it appears on the Preliminary List and the reasons provided for the recommendation together with appropriate supporting data (such as a summary of existing test data with literature references).

Those wishing to submit comments on the relative priority for further testing of substances appearing on the Preliminary List (and other substances proposed for consideration by the commentor) are requested to provide a discussion of the reasons for their priority recommendations.

FIFTEEN TOXIC CHEMICALS SUBJECT TO IMMINENT REGULATION*

Introduction

This Document includes summary characterizations of fifteen chemicals of near-term interest to the U.S. Environmental Protection Agency. Chemicals of near-term interest are regarded as toxic and subject to imminent regulation. There are many other chemicals of interest, and in the future similar characterizations of other chemicals will be prepared.

The Document was prepared by the Office of Toxic Substances of EPA drawing on information provided by a number of authoritative sources. The characterizations are based on information available at the time of going to press. As additional information and interpretations of data become available, appropriate updating of the characterizations will be undertaken.

Arsenic

Why Should the Chemical Be of Concern at This Time?

In 1975, OSHA proposed a strict standard for workplace air exposure limits to inorganic arsenic. Earlier EPA sampling had found that atmospheric concentrations near two copper smelters exceeded the proposed limit (Anaconda, Montana; and Tacoma, Washington) and closely approached it at three other smelter sites. Preliminary results of an EPA-sponsored epidemiology study near an arsenical pesticide plant in Baltimore reveal lung cancer rates several times the national average. Congress has proposed that explicit attention be given to establishing an air standard by 1977 in an amendment to the Clean Air Act. A number of arsenical compounds are being considered for rebuttable presumption proceedings under FIFRA/FEPCA.

What Are the Health and Ecological Effects, and Environmental Behavior?

Liver, skin, lung, and lymphatic cancers, and adverse effects on the thyroid gland have been reported in epidemiological studies of occupationally exposed individuals. The main threat of arsenic as a carcinogen is in-

halation of the inorganic forms. A preliminary mortality study of the population surrounding Allied Chemical Company's arsenical pesticide plant in Baltimore revealed a lung cancer rate sixteen times the national average. A previous study had shown that retired workers from this plant suffer from lung cancer at a rate seventeen times the national average. A Dow Chemical Company study indicated an excess of lung and lymphatic cancers among their workers who had been exposed to arsenical compounds. Arsenic occurs in two forms: trivalent and pentavalent. Trivalent arsenic is much more toxic than pentavalent, both acutely and chronically. Pentavalent arsenic is often found in metallo-arsenicals, and is of concern because it can degrade into the trivalent form.

A 1972 outbreak of arsenic poisoning in Getchell, Nevada, is attributed to stack effluent from a gold smelter. Studies made abroad have suggested that arsenic may be a skin carcinogen when ingested in drinking water at levels as low as 0.3 mg/l. The debate over the carcinogenicity of arsenic is largely due to the fact that the animal studies conducted to date have not shown a relationship between ingested arsenic and cancer. Organic arsenical compounds may be more hazardous than previously believed. Carbarsone has been reported to produce liver cancer in trout through ingestion (480 mg/100g diet).

Arsenic is particularly toxic to legumes and other crop plants. Depending on the soil type, 6 ppm arsenic can cause a 50 percent growth reduction. Phytotoxic levels of arsenic have been found as far as two miles from the Tacoma smelter. Once combined in soil, arsenic is extremely persistent.

What Are the Sources, Environmental Levels, and Exposed Populations?

Inorganic arsenic is emitted to the air from several sources, including copper, lead, and zinc smelters, glass production plants, coal-burning facilities, cotton gins, arsenical-compound (including pesticides) production plants, and pesticide application. Organic arsenic discharges are associated with the manufacture and use of pesticides. Trivalent arsenic occurs naturally, is a common contaminant of ores, and is the major component of arsenic emissions from smelters. Based on EPA estimates, the 15 copper smelters contribute most heavily to air emissions of inorganic arsenic. The Anaconda copper smelter in Montana, and the ASARCO copper smelter

---

*   Summary Characterizations of Selected Chemicals of Near-Term Interest, Office of Toxic Substances, U.S. Environmental Protection Agency, April 1976.

and arsenic plant in Tacoma, Washington, have been identified as having the highest arsenic emissions. Other industrial sources generally emit less arsenic than copper smelters. Air levels in most urban areas for 1973 and 1974 were at or below the level of detection (0.001 ug/m^3). Levels in areas near smelters ranged from 0.003 to 4.86 ug/m^3.

The land disposal of arsenical wastes can become a long-range public health hazard. A good example is Perham, Minnesota, where eleven people were poisoned by contaminated well water in 1972.

A 1975 survey of drinking water supplies showed that about one percent exceeded the interim drinking water standard of 0.05 mg/l. Trivalent arsenic is found at high levels in some ground water. Underground injection of arsenical pesticide wastes in Philadelphia has contaminated a nearby stream which is being considered for use as a drinking water supply.

Three new technologies for energy production have important arsenic implications. Early data on coal gasification indicate that two-thirds of the arsenic present is volatilized. Oil shale exploitation and geothermal energy development may also release large quantities of arsenic.

What Are the Technologic and Economic Aspects?

In general, particulate control measures (multicyclones, balloon flues, and electrostatic precipitators) are used to reduce arsenic emissions. Baghouses offer the greatest potential for control, but have not been widely adopted by the smelting industry because of high capital and maintenance costs. Costs and feasibility of emission controls will vary from plant to plant. Significant control efforts are being planned at the ASARCO smelter in Tacoma, and are underway at Anaconda. Conventional water treatment technology has been shown to be effective in meeting the arsenic drinking water standard. Arsenic concentrations of 0.1 and 1.6 mg/l in wastewater can inhibit waste treatment by activated sludge and anaerobic digestion respectively. Thus, concentrations exceeding these levels can present an additional hazard in waste waters subjected to these treatment methods. Air and water pollution control efforts normally result in a solid waste or sludge. At present, these materials are being stored, pending development of acceptable disposal technologies.

What Steps Have Been Taken, and What Is Being Done?

EPA is locating and monitoring arsenical discharges, and is conducting several studies to determine the toxicity of various arsenical compounds. Limited epidemiological studies are planned to help determine effect levels. Studies have been initiated to determine control technologies and costs for arsenic reduction, and an Air Pollution Assessment Report on Arsenic has been prepared. EPA is considering the development of standards under Section 112 of the Clean Air Act. A review of the use of arsenical pesticides has recently been completed, and research into disposal techniques for arsenical wastes is planned. A Scientific and Technical Assessment Report is planned upon receipt of the NAS study of health effects.

In November 1975, OSHA proposed a workplace exposure limit for inorganic arsenic at 4 ug/m^3 (8 hour, TWA). The previous standard of 500 ug/m^3 for all forms of arsenic would remain in effect only for organic forms.

References

Air Pollution Assessment Report on Arsenic; EPA, Office of Air Quality Planning and Standards (December 1975).

Burruss, R. P. and Sargent, D. H., *Technical and Microeconomic Analysis of Arsenic and Its Compounds;* EPA, Office of Toxic Substances (performed under contract no. 68-01-2926) (September 1975).

Criteria for a Recommended Standard . . . Occupational Exposure to Inorganic Arsenic, New Criteria — 1975 (incorporating results of the Allied and Dow Chemical worker studies); HEW, National Institute for Occupational Safety and Health (publication no. NIOSH 75-149, 1975).

Faust, S. D., and Clement, W. H., *Investigation of the Arsenic Condition at the Blue Marsh Lake Project Site, Pennsylvania;* U.S. Army Corps of Engineers performed under contract no. DACW 67-71-C-0288 (1973).

Hazardous Waste Disposal Damage Reports (at page 1); EPA, Office of Solid Waste Management Programs (publication no. EPA 530/SW-157, June 1975).

Helver, J. E., "Progress on Studies on Contaminated Trout Rations and Trout Hepatoma;" *NIH Report* (April 12, 1962).

A Pilot Study on the Community Effects of Arsenic Exposure in Baltimore; EPA, Office of Toxic Substances (performed under contract no. 68-01-2490, in draft).

Tseng, W. P., et. al., "Prevalence of Skin Cancer in an Endemic Area of Chronic Arsenicism in Taiwan;" *J. National Cancer Inst.,* 40:453 (1968).

Wands, Ralph C., *Letter to APHA Panel on Arsenic Studies;* National Research Council (February 17, 1976).

Asbestos

Why Should the Chemical Be of Concern at This Time?

OSHA has proposed lowering its workplace standard by a factor of ten on the basis of recent epidemiological data suggesting wider spread health effects than previously suspected. A number of major commercial sources of airborne asbestos are limited by EPA regulations. The Agency is investigating taconite and other hard-rock mining operations, where asbestos is a major ore contaminant. There is renewed interest in hazards possibly presented by dust from asbestos brake linings and interior sources. Meanwhile, EPA continues to underscore the hazard of asbestos fibers in water supplies resulting from Reserve Mining Company's operations. EPA's nationwide sampling program is showing levels of asbestos fibers in water supplies, natural runoff, and discharges from manufacturing and mining sites.

What Are the Health and Ecological Effects, and Environmental Behavior?

Airborne asbestos fibers have been known to cause asbestosis, lung cancer, and pleural and peritoneal mesothelioma. OSHA cites a number of studies showing gastrointestinal (GI) cancer in workers exposed to asbestos. In one study of insulation workers in the United States, seven percent of deaths could be attributed to asbestosis, which generally appeared about 20 years after first exposure — the same latency period as for most cancers. Available epidemiological data show that lung cancer is responsible for as much as 20 percent of all deaths among certain asbestos workers; mesothelioma, 11 percent; and GI cancer, eight percent.

There are few if any data on the dose-response relationships of asbestos fibers in either air or water. Effects of airborne asbestos are far better documented than those of waterborne. OSHA cited workers who had developed mesothelioma at exposure levels below the previous standard of 5 fibers per cubic centimeter in its recent proposal to reduce the level by a factor of ten. There is some evidence that asbestos diseases, including mesothelioma, occur in families of workers exposed to asbestos at levels presumed to be much lower than direct occupational exposure.

Asbestos fibers are extremely resistant to degradation in the environment. Thus far, it has been impossible to demonstrate adverse effects on plants. Some adverse effects on animals have recently been reported.

What Are the Sources, Environmental Levels, and Exposed Populations?

The United States utilized approximately 800,000 tons of asbestos fiber in 1974. Asbestos products are widely used in the construction industry (asbestos-cement pipe, building and other construction products, and floor tile). Other products include friction materials (such as brake linings), felt and paper, packings and gaskets, and fireproof textiles. It has been estimated that 85 percent of the asbestos is tightly bound in products and is therefore not as available to the environment as are airborne and waterborne asbestos fibers generated in the mining and milling of asbestos ore, manufacture and fabrication of asbestos products, and disposal of solid wastes from these processes. Asbestos was used in spray insulation in buildings between 1950 and 1972. This may become a major source of environmental discharge as buildings constructed during this period are demolished.

Asbestos minerals are found throughout the United States. Significant quantities of asbestos fibers appear in rivers and streams draining from areas where asbestos-rock outcroppings are found. Some of these outcroppings are being mined. Asbestos fibers have been found in a number of drinking water supplies, but the health implications of ingesting asbestos are not fully documented. Emissions of asbestos fibers into water and air are known to result from mining and processing of some minerals. Asbestiform fibers in the drinking water of Duluth and nearby communities at levels of 12 million fibers per liter have been attributed to the discharge of 67,000 tons of taconite tailing per day into Lake Michigan by Reserve Mining.

Exposure to asbestos fibers may occur throughout urban environments. A recent study of street dust in Washington, D.C., showed approximately 50,000 fibers per gram, much of which appeared to come from brake linings. Autopsies of New York City residents with no known occupational exposure showed 24 of 28 lung samples to contain asbestos fibers, perhaps resulting from asbestos from brake linings and the flaking of sprayed asbestos insulation material.

What Are the Technologic and Economic Aspects?

Coagulation treatment and filtration are necessary to remove contaminant asbestos from water. Filtration technologies for air, while meeting the no-visible-emission standard, permit large quantities of asbestos fibers to escape. Fibrous glass has frequently been substituted for applications requiring insulative properties, but there is some debate over its safety. For some other applications, such as brake linings, economically feasible substitutes may not be available.

There is no inexpensive, standardized analytical method for measuring asbestos, and monitoring costs are very high.

What Steps Have Been Taken, and What Is Being Done?

An air standard has been promulgated for a number of major commercial sources of asbestos fibers. Hard-rock mining and taconite beneficiation, where asbestos is an ore contaminant, are being investigated. Effluent guidelines have been promulgated under the Federal Water Pollution Control Act which, together with the NPDES permit program, should reduce asbestos discharges.

EPA is sponsoring an extensive national asbestos monitoring program. Preliminary findings indicate that asbestos is a widespread contaminant of drinking water. NAS is reviewing the implications of these preliminary findings. EPA's Reserve Mining Task Force is monitoring efforts to halt the discharge of taconite tailings into Lake Superior. Standard analytical methods are being developed for both research and monitoring purposes. A number of epidemiology studies to further clarify the health risks of asbestos are being sponsored by EPA.

In 1972, OSHA established a workplace exposure standard. Last October, OSHA proposed a further reduction in the level. The National Institute of Environmental Health Sciences (NIEHS) is conducting ingestion experiments to clarify health hazards of this route of exposure; EPA is partially sponsoring these studies.

References

Asbestos, Its Sources, Uses, Associated Environmental Exposure and Health Effects; EPA, Office of Toxic Substances (September 1975).

Asbestos, The Need for and Feasibility of Air Pollution Controls; National Academy of Sciences (1971).

Biological Effects of Asbestos; HEW, National Institute of Health (February 1973).

Castleman, B. I., and Fritsch, A. J., *Asbestos and You;* Washington, Center for Science in the Public Interest (February 1973).

Effluent Guidelines for the Asbestos Manufacturing Industry: Building Materials and Paper; 40 Federal Register 1874 (January 9, 1975).

Effluent Guidelines for the Asbestos Manufacturing Industry: Friction Materials and Textiles; 40 Federal Register 18172 (April 25, 1975).

Haley, T. J., "Asbestosis — A Reassessment of the Overall Problem," *J. Pharm. Science* 64(9):1435 (1975).

"1975 — Review," *Asbestos* 57(7):12 (January 1976).

National Emission Standards for Hazardous Air Pollutants: Asbestos and Mercury; 39 Federal Register 38064 (October 25, 1974).

National Emission Standards for Hazardous Air Pollutants: Asbestos and Mercury (Amendment); 40 Federal Register 48292 (October 14, 1975).

Occupational Exposure to Asbestos: Proposed Rules; 40 Federal Register 47652 (October 9, 1975).

Occupational Safety and Health Standards: Subpart Z — Toxic and Hazardous Materials; 40 Federal Register 73072 (May 28, 1975).

Benzene

Why Should the Chemical Be of Concern at This Time?

Benzene, a component in gasoline and an important feedstock for the chemical industry, has been the subject of numerous published reports linking leukemia with worker exposure to it. Large quantities of benzene are discharged into the environment from automobiles, and probably from stationary sources. The Environmental Defense Fund has been particularly concerned about the identification of benzene in drinking water in the parts per billion range. NIOSH has recently recommended a reduction in workplace levels.

What Are the Health and Ecological Effects, and Environmental Behavior?

Numerous fatalities from occupational benzene poisoning have been reported since the early 1900's. After inhalation or ingestion, benzene is absorbed rapidly by the blood. At non-lethal concentrations, a variety of human central nervous system disorders are observed, depending upon the extent of exposure. These include euphoria followed by giddiness, headache, nausea and staggering gait, as well as fatigue, insomnia, dizziness, and unconsciousness. Observed human blood-forming system damage includes anemia, reduction in platelet numbers, and depression of the white blood cell count.

Chronic benzene exposure has also resulted in chromosome aberrations in human lymphocytes. As early as the 1930's, benzene was suspected in cases of leukemia. Available epidemiological data indicate that the compound does induce leukemia, although the data cannot be considered to constitute unequivocal evidence that benzene acting alone is leukemogenic. Attempts by NCI and others to induce leukemia in animals with benzene have not been successful. However, the results of inhalation experiments with mice, the species most susceptible to leukemia, are not yet available.

Based on its physical properties, benzene is expected to be quite mobile and probably persistent. Adverse effects on ecological resources have not been reported.

What Are the Sources, Environmental Levels, and Exposed Populations?

In 1973 over 10 billion pounds of benzene were produced from petroleum and coal in the United States. This volatile, colorless, flammable liquid is used mostly for synthesis of organic chemicals. It has been estimated that at least 80 million pounds of benzene may be lost to the environment during benzene production, storage, and transport, while an upper limit of 650 million pounds may be released during its use to produce other organics. The latter figure was calculated from the difference between 100% yield and the reported yield in these reactions. Therefore, this is only a crude measure of the worst-case benzene emissions during usage. The emissions would be concentrated in the Texas Gulf area and the Northeast.

It has been calculated that approximately one billion pounds of benzene were released with hydrocarbon emissions from motor vehicles in 1971 in a geographical pattern similar to population distribution. Another 22-24 million pounds of benzene may be released into the environment each year with spilled oil. Hydrocarbon emissions from non-transportation sources, such as coke ovens and power plants, may also contain considerable amounts of benzene. Additionally, benzene is an active ingredient in a number of insecticides and miticides, although the amount of release to the environment from this source has not been calculated.

In an EPA study of organic compounds in the drinking water of 10 cities, benzene was detected in water from four cities at concentrations ranging from 0.1-0.3 ug/l. Previous studies reported levels up to 10 ug/l. Average levels of benzene detected in air in a limited number of studies are in the low ppb range with one high reading of 23 ppm reported in the vicinity of a solvent reclamation plant. No data have been found on levels of benzene in soil, wildlife, and fish. Benzene is widely enough distributed that most people are probably exposed to very low levels; the health implications of this type of exposure are not known.

What Are the Technologic and Economic Aspects?

Reduction in organic compound emissions to achieve the National Ambient Air Quality Standard for oxidants should also result in some reduction in benzene emissions. As a result of lead removal from gasoline, the average content of aromatics, including benzene, in gasoline is likely to increase slowly. However, hydrocarbon emission controls on motor vehicles should result in a net reduction in benzene emissions.

What Steps Have Been Taken, and What Is Being Done?

In 1974 NIOSH published a criteria document for occupational exposure to benzene which recommended adherence to the existing Federal standard of 10 ppm as a time-weighted average with a ceiling of 25 ppm. OSHA is now in the final stages of reviewing its current standard. NIOSH is conducting a retrospective study of benzene mortality and a study of airborne benzene levels in service stations.

EPA has also initiated an air monitoring program which will determine benzene levels in selected areas. Qualitative results obtained to date indicate widespread low-level benzene contamination. Air regulations are not being considered at this time; however, further studies are planned. EPA has conducted a limited survey of drinking water supplies in which benzene was identified in some samples, and has begun a more extensive survey which will seek out benzene as well as a number of other pollutants. EPA has proposed to designate benzene a hazardous substance under section 311 of FWPCA, and ocean dumping is already strictly regulated. A National Academy of Sciences review of the health aspects of benzene being done for EPA should be completed soon. CPSC is also awaiting the results of the NAS study on the health effects of benzene, and will determine if action is appropriate when the results have been received.

References

Browning, Ethel, *Toxicity and Metabolism of Industrial Solvents;* Amsterdam, Elsevier (1965).
Criteria for a Recommended Standard . . . Occupational Exposure to Benzene; HEW National Institute for Occupational Safety and Health (NIOSH), (Publication No. NIOSH 74-137, 1974).
Deutsche Forschungsgemeinschaft, *Benzene in the Work Environment;* Weinheim, Verlag Chemie GmbH (1974).
Harris, Robert, *The Implications of Cancer Causing Substances in Mississippi River Water;* Washington, Environmental Defense Fund (November 6, 1974).
Occupational Safety and Health Standards: Subpart Z — Toxic and Hazardous Substances 29 CFR 1910.1000, Table Z-2.

IARC Monographs on the Evaluation of the Carcinogenic Risk of Chemicals to Man 7:203; Lyon, International Agency for Research on Cancer (1974).

Moran, John B., *Lead in Gasoline: Impact on Current and Future Automotive Emissions;* presented to Air Pollution Control Association meeting (June 1974).

Preliminary Assessment of Carcinogens in Drinking Water: Report to Congress; EPA, Office of Toxic Substances (December 1975).

Development of Analytical Techniques for Measuring Ambient Atmospheric Carcinogenic Vapors; EPA, Office of Air Quality Planning and Standards (Publication No. EPA-600/2-75-076, November 1975).

Sources of Contamination, Ambient Levels, and Fate of Benzene in the Environment, EPA, Office of Toxic Substances (Publication No. PB 244139, December 1974).

Benzidine

Why Should the Chemical Be of Concern at This Time?

Benzidine, a human carcinogen, is used as an intermediate in the manufacture of a number of azo dyes which color textile, leather, and paper products. In addition to the Agency's long-standing concern over liquid effluent discharges containing benzidine, recent research results suggest that some of the benzidine-derived azo dyes may reconvert to benzidine in man, or under certain environmental conditions. The AFL/CIO has expressed strong interest in any action taken on benzidine.

What Are the Health and Ecological Effects, and Environmental Behavior?

For a number of years, the manufacture and use of benzidine have been associated with a high risk of bladder cancer among exposed workers. Many scientists believe that tumors can result from ingestion, inhalation, or skin absorption. A number of animal studies have demonstrated the carcinogenic effects of benzidine. Mice, rats, and hamsters develop liver tumors, and dogs develop bladder cancer. Such studies have many deficiencies for estimating the risk associated with the levels of exposure to carcinogens likely to be encountered in the environment.

Free benzidine has been detected in the urine of monkeys fed benzidine-derived azo dyes, establishing a potential for reconversion of azo dye to benzidine. Metabolism of benzidine-derived azo dyes may be similar in humans. Japanese silk painters reportedly have a high incidence of bladder cancer, possibly resulting from licking brushes and spatulas coated with benzidine-derived azo dyes. However, the carcinogenicity of such dyes has not been specifically determined.

Industrial data indicate that benzidine entering a waterway dissipates and may be degraded by naturally occurring processes. Confirmatory investigations have not been conducted. Other aspects of environmental behavior have not been addressed. It has been hypothesized that azo dyes can reconvert to benzidine under certain undefined environmental conditions.

What Are the Sources, Environmental Levels, and Exposed Populations?

The three identified manufacturers (Allied, GAF, and Fabricolor) estimate that they produce 45 million pounds of azo dyes annually from benzidine. The dyes are used by about 300 major manufacturers of textile, paper, and leather products. The largest manufacturer (Allied) recently announced its intention to phase out benzidine production.

The principal environmental concern at benzidine production facilities has been the amount of benzidine in the waste effluents discharged to publicly owned waste water treatment works (POTWs). However, the only discharge measurements to date have been made by industry, which has contended that discharges at any facility usually do not exceed one pound per day. Benzidine is believed to be present in the sludge removed from industrial pretreatment plants. The environmental adequacy of land disposal of these sludges is unknown. According to industry data, discharges from the POTW are usually below the limit of detection. However, there are occasionally significant accidental releases to POTWs. Levels of benzidine exceeding 5 mg/l can inhibit anaerobic digestion wastewater treatment processes. Thus, concentrations above this level at the POTW present a problem to POTWs using this process, and a possible hazard to the receiving waters.

Free benzidine is present in the benzidine-derived azo dyes. According to industry, quality control specifications require that the level not exceed 20 ppm, and in practice the level is usually below 10 ppm. Industry has estimated a total environmental discharge at the 300 user facility site of 450 pounds per year or about 1.5 pounds per year per facility, assuming all of the free benzidine is discharged in the liquid effluent.

No measurements for benzidine in ambient air, surface water, or drinking water have been reported. Further, no measurements for free benzidine in finished products containing azo dyes have been reported.

What Are the Technologic and Economic Aspects?

The principal liquid effluent control technology currently being used is the reaction of benzidine with nitrous acid. While effective in destroying benzidine, hazardous decomposition by-products may be formed. Industry thus far has rejected carbon adsorption as uneconomical. The costs of treatment at the benzidine manufacturing plants are of far less concern than at the user plants. Thus, there is a continuing industrial emphasis on reducing the levels of free benzidine in dyes, which result from more complete reactions and release less benzidine into the environment.

If limitations were imposed on benzidine production or use, the vacuum would probably be filled by imported benzidine-derived dyes and substitute dyes. However, some of the possible substitutes, such as o-toluidine, are also of environmental and occupational health concern. Industry estimates that adequate substitutes would be three to five times more expensive. In some highly specialized uses, particularly for the halogenated benzidine dyes, a technically adequate substitute may not be available.

What Steps Have Been Taken, and What Is Being Done?

The stringent work place standards required by OSHA because of the carcinogenic nature of benzidine reduce environmental discharges resulting from inadequate housekeeping procedures at benzidine manufacturing sites.

EPA proposed a toxic pollutant effluent standard in December 1973 and is planning to repropose such a standard and a pretreatment requirement during the next few months. The results of current animal experiments at the National Center for Toxicological Research, addressing carcinogenicity and metabolic behavior, should be available within one year. Benzidine is also being examined in the expanded EPA drinking water survey.

References

Benzidine: Wastewater Treatment Technology; EPA, Office of Water and Hazardous Materials (June 1974).

Cranmer, Morris, Dr., *General Considerations in Setting a Benzidine Standard;* Testimony to EPA, Office of Water and Hazardous Materials.

EPA/SOCMA*, *Stipulation of Fact;* EPA, Office of Water and Hazardous Materials (Hearings, May 30, 1974).

*Synthetic Organic Chemical Manufacturers Association

Haley, Thomas J., "Benzidine Revisited: A Review of the Literature and Problems Associated with the Use of Benzidine and Its Congeners," *Clinical Toxicology,* 8:13 (1975).

Hazard Review of Benzidine; HEW, National Institute for Occupational Safety and Health (July 1973).

Occupational Safety and Health Standards: Subpart Z — Toxic and Hazardous Materials (Benzidine); 39 Federal Register 20 (January 29, 1974).

Rinde, Esther and Troll, Walter, "Metabolic Reduction of Benzidine Azo Dyes to Benzidine in the Rhesus Monkey"; *J. National Cancer Inst.,* 55:181 (1975).

SOCMA, *Comments on Production and Use of Benzidine;* submitted to EPA, Office of Water and Hazardous Materials (March 9, 1973).

SOCMA, *Affidavit of Mr. Kelvin H. Ferber;* submitted to EPA, Office of Water and Hazardous Materials (May 13, 1974).

SOCMA, *Benzidine Effluent Data;* submitted to EPA, Office of Water and Hazardous Materials (June 6, 1974).

SOCMA, *Industrial Hygiene and Environmental Control;* submitted to EPA, Office of Water and Hazardous Materials (May 30, 1975).

SOCMA, *Submission;* submitted to EPA, Office of Water and Hazardous Materials (June 5, 1975).

SOCMA, *Second Submission;* submitted to EPA, Office of Water and Hazardous Materials (August 5, 1975).

Cadmium (Cd)

Why Should the Chemical Be of Concern at This Time?

As evidence that cadmium levels in the environment may be increasing emerges, concern mounts over this substance's ability to accumulate in the body at low level exposures. Cadmium, which is used in a variety of commercial and consumer products, is believed to reach man through a number of routes, particularly as a contaminant of fish and other foods. There is recent concern over the presence of Cd in sludge which might reach the food chain as a result of leaching from disposal sites or its use as a soil conditioner. A proposed amendment to the Clean Air Act calls for explicit EPA attention by 1977 to a possible air standard for cadmium.

What Are the Health and Ecological Effects, and Environmental Behavior?

Cadmium accumulates in the kidney cortex, where it can cause damage to the renal tubules at levels on the order of 200 ppm. The results of autopsy studies show current levels of 15-50 ppm in the kidneys of people over the age of 50 who were not occupationally exposed; the higher levels generally reflect those found in individuals who had been smokers. Autopsy data on the occupationally exposed are inconclusive because samples have been too small.

At high levels of Cd exposure, other effects, such as bone brittling, have been observed, mainly in Japan, where widespread occurrence of Itai-Itai disease caused nearly 100 deaths. These effects resulted from an estimated intake of 600 ug/day. The average American diet contains 50-75 ug/day. Heavy fish eaters receive a higher dose, but well below the levels observed in Japan. About five percent of ingested Cd is retained in the body, and its biological half-life in humans is estimated to be at least 15 years.

Prolonged exposure to cadmium dust can cause emphysema. Recent epidemiological studies indicate abnormally high rates of several forms of cancer due to occupational exposure. Hypertension has been developed in laboratory animals after prolonged exposure to low levels. The presence of Cd in human fetal tissues during prenatal life shows that the metal traverses the placenta. Experimental studies in laboratory animals have confirmed this observation and have also shown that Cd is a potent teratogen.

Cadmium particulate in air falls out into water and soils. Plants take it up from the soil, and people and animals ingest Cd from these sources. Uptake from contaminated water has not been so well documented; it is suspected that this is the significant route of exposure for fish.

What Are the Sources, Environmental Levels, and Exposed Populations?

Cadmium is produced in conjunction with zinc refining. In 1974, the total U.S. consumption of cadmium was about 6300 metric tons, at a cost of about $8500 per metric ton. About one-third was imported. By 1985, demand is expected to reach 9600 metric tons. Of the total use in 1975, about 55 percent went to electroplating, 21 percent to plastic stabilization, 12 percent to pigments, 5 percent to batteries, and 7 percent to a variety of other uses. Major growth is expected in the nickel-cadmium battery industry.

An EPA-sponsored study estimated that a total of 1800 metric tons of Cd were released to the environment in 1974. Of this, about 20 percent was from zinc mining and smelting, via air, water, and tailings; fifty percent was from such indirect sources as fossil fuel combustion, fertilizer use, and disposal of sewage sludge; and thirty percent was from industrial uses, such as remelting of cadmium-plated scrap, incineration of plastics containing Cd, and electroplating.

The major sources of human exposure are food and tobacco contamination, while direct water and air intake appear to be very minor contributors. Groundwater contamination as a result of waste disposal, however, is common. FDA's marketbasket survey has been identifying low levels of Cd in most composite class samples. Thus, virtually everyone is exposed to trace levels of Cd. Recent studies indicate that, in Sweden, Cd concentrations in wheat may be increasing at a rate roughly proportional to levels of industrial use. Increasing soil levels may result from airborne fallout, fertilizer use, and cadmium in irrigation waters. Cadmium has been identified in soils at several locations at levels of 0.55 to 2.45 ppm.

Cadmium levels of 1 to 10 ug/l have been found in 42 percent of available ambient water samples, with more than 10 ug/l in four percent. Fifty-four percent of the samples did not contain measurable amounts. The annual release of Cd to the air at one copper smelter was estimated to be 250 tons per year. Ambient air levels averaged .031 ug/m^3. Soil levels of Cd were about 1.6 ppm, between one and five miles from the smelter, and were reflected by average findings of 4.7 ppm in leafy vegetables.

What Are the Technologic and Economic Aspects?

Substitutes are or will soon be available for most but not all electroplating uses and for plastic stabilizer use at comparable cost and efficacy.

A Cd level of 0.02 mg/l, has been shown to inhibit wastewater treatment by anaerobic digestion. Should Cd concentrations exceeding that level reach a wastewater treatment plant using anaerobes, a hazard may be presented to the receiving water. Trace contamination of air and water by Cd is common. Removal of such components is usually extremely costly.

What Steps Have Been Taken, and What Is Being Done?

NIOSH is expected to submit a criteria document to OSHA this year, at which time the existing workplace standard will be reviewed. FDA has banned certain uses of Cd pigments and cadmium-containing materials.

Epidemiological studies are being conducted by the World Health Organization to determine whether cadmium may be a factor in hypertension and cardiac disease in humans. NCI is sponsoring studies to investigate the carcinogenic potential of Cd metal, Cd oxide, and Cd sulfide.

EPA has prohibited the ocean dumping of cadmium, except as trace contamination. The effluent guidelines for the electroplating industry address Cd released from this segment of the economy, and hazardous spill regulations include some cadmium compounds among the substances for which spill penalties have been established. An Interim Primary Drinking Water Standard has been issued and pesticides containing Cd are being reviewed for possible Rebuttable Presumption Against Registration proceedings. A Scientific and Technical Assessment Report on Cd has documented health and technological concerns for EPA.

References

Chauve, S., et. al., "Zinc and Cadmium in Normal Human Embryos and Fetuses;" *Archiv Environmental Health* 26:237 (1973).

Cadmium and the Environment: Toxicity, Economy, Control; Paris, Environment Directorate, Organization for Economic Cooperation and Development (1975).

Ferm, V. H., "Developmental Malformations Induced by Cadmium;" *Biol Neonate* 19:101 (1971).

Friberg, Lars, et al, *Cadmium in the Environment, 2nd ed.;* Cleveland, Ohio, CRC Press (1974).

Fulkerson, William et al, *Cadmium: The Dissipated Element;* Oak Ridge National Laboratory, (report no. ORNL NSF-EP-21, January 1973).

National Inventory of Sources and Emissions: Cadmium-1968; EPA, Office of Air and Water Programs (report no. APTD-68, February 1970).

Regulations for Rebuttable Presumption Against Registration, 40 CFR 162.11.

Scientific and Technical Assessment Report on Cadmium; EPA, Office of Research and Development (publication no. EPA 600/6-75-003, March 1975).

Technical and Microeconomic Analysis of Cadmium and Its Compounds, EPA, Office of Toxic Substances (Report No. EPA 560/3-75-005, June 1975).

Webb, M., "Cadmium;" *Br. Med. Bull,* 31(3):246-250 (1975).

Ethylene Dibromide (EDB)

Why Should the Chemical Be of Concern at This Time?

Ethylene dibromide, which is used primarily as an additive in leaded gasoline, has been identified by the National Cancer Institute (NCI) as an extremely potent and fast-acting carcinogen when administered at high dose levels to animals. Recent monitoring data reveal very low concentrations in air collected from urban and rural areas, and near production and storage facilities. EDF has petitioned EPA to cancel all registered pesticides containing this chemical.

What Are the Health and Ecological Effects, and Environmental Behavior?

EDB is an extremely strong irritant. Chronic exposure can result in liver and kidney damage. High levels of exposure cause immediate depression of the central nervous system, usually resulting in death of laboratory animals. Laboratory animals survive only a few hours of exposure to 200 ppm EDB in air; adverse effects have been noted at exposures as low as 30 ppm. The exposure of domestic animals to EDB has revealed severe reproductive effects. In 1974, NCI issued a memorandum of alert citing preliminary findings of strong carcinogenesis in rats and mice. This document cites a high incidence of squamous cell carcinoma of the stomach in both rats and mice, with tumors observed after only six weeks of exposure. EDB is also mutagenic and teratogenic in animals. In humans, weakness and rapid pulse have been associated with EDB exposure, and, less commonly, cardiac failure leading to death.

The freshwater toxicity of EDB is indicated by the 48-hour Median Tolerance Limit for pan and game fish (15 to 18 ppm). The environmental behavior of EDB is poorly understood. EDB is reportedly short lived in the atmosphere. In soil, EDB is persistent for at least two weeks; however, within a two-month period, it is converted to other compounds. Ethylene, one of these, can reduce yields of fruits, vegetables, and flowering plants.

What Are the Sources, Environmental Levels, and Exposed Populations?

In 1973, approximately 175,000 tons of EDB were produced in the United States. Since sales appear higher than production figures, a small amount of EDB is probably imported. The major producers of EDB are: Houston Chemicals, Ethyl, Dow, PPG, Northwestern, and Great Lakes Chemical. Four of the manufacturing sites are located in Arkansas, one in Texas, and another in Michigan. Eighty to ninety percent of EDB produced in the United States is used as a lead scavenger in gasoline, and it is also registered for use as an insect fumigant and a soil nematocide. The pesticidal use of EDB is less than 1,000 tons per year. EDB is also used as an intermediate in the synthesis of dyes and pharmaceuticals, and as a solvent for resins, gums, and waxes. A small amount of EDB is used in the production of vinyl bromide.

The limited amount of monitoring data available indicates the presence of EDB residues throughout the environment. A Dutch study examining EDB residues in wheat, flour, and bread showed that flour made from wheat treated 13 weeks before analysis contained from 2 - 3 ppm of EDB. Bread made from this flour was found to contain about 0.02 - 0.12 ppm of EDB. The fumigation of apples reportedly results in detectable residues on the skin and in the outer pulp for from 4 - 28 days. Low levels of EDB have also been found in the ambient atmosphere of urban and rural areas. Ambient air collected in the vicinity of gasoline stations along highly trafficked arteries in Phoenix, Los Angeles, and Seattle has shown EDB concentrations of 0.07 - 0.11 ug/m^3. EDB also escapes into the environnent during the production and storage/transfer of gasoline products at oil refineries. Concentrations of EDB in the vicinity of manufacturing plants located in Arkansas ranged from 90 - 115 ug/m^3, while concentrations around a bulk transfer site in Kansas City were 0.2 - 1.7 ug/m^3. Since fully validated monitoring techniques are not yet available, these results should be considered qualitative rather than quantitative indicators.

Although few people live immediately adjacent to production plants where the ambient levels of EDB are the highest thus far recorded, very large populations frequent or live near highly trafficked areas of the type where EDB has been detected.

What Are the Technologic and Economic Aspects?

The elimination of most pesticidal uses for EDB will not result in significant reduction in media exposure, since less than 10 percent of production goes to this use. However, residues in food, which may pose a serious problem, have not been well studied. USDA used EDB to fumigate certain types of exported and imported grains and produce. Effective substitutes are probably not readily available for some uses, and may be more costly for others. Strict control of EDB emissions from production plants should pose few problems since production is basically a closed system. Major EDB emissions into the environment occur during packaging, and the recycling or capturing of vapors would seem practical. At present, the elimination of EDB from leaded gasoline is not being undertaken by manufacturers. Ethylene dichloride (EDC), which is being used in conjunction with EDB in leaded gasoline could be increased in concentration, and EDB eliminated; however, the environmental impacts of EDC are not clear.

What Steps Have Been Taken, and What Is Being Done?

Actions have been taken to phase out the use of leaded gasoline and are expected to result in a concurrent reduction of exposure to EDB. Gasoline transfer vapor recovery and vehicle evaporative emission controls should further reduce EDB emissions. Monitoring efforts have been initiated to define the zones of impact associated with various stationary and mobile sources of EDB. EPA will gather information on the quantities of EDB, EDC, and other gasoline additives produced, as a result of recently promulgated fuel and fuel additive regulations. Pesticides containing EDB are undergoing special review to determine whether they should be considered candidates for Rebuttable Presumption Against Registration procedures.

OSHA has established a standard for ethylene dibromide to guard against irritation and cumulative hepatic injury; however, the existing standard does not consider the new evidence on carcinogenicity. NIOSH and OSHA are currently reviewing the new evidence to determine whether more rapid action should be taken to revise the standard.

References

Alumot, (Olomucki), E. and Hardof, Z., "Impaired Uptake of Labelled Protein by the Ovarian Follicles of Hens Treated with Ethylene Dibromide;" Comp. Biochem. Physiol. 39B:6168 (1970).

Amir, D., "Sites of Spermicidal Action of Ethylene Dibromide in Bulls;" J. Reprod. Fert. 35(3):519-525 (1973).

Olson, et. al., "Induction of Stomach Cancer in Rats and Mice with Halogenated Aliphatic Fumigants;" J. National Cancer Inst. 51(6) 1993-1995 (1973).

Occupational Safety and Health Standards: Subpart Z — Toxic and Hazardous Substances. 20 CFR 1910.100, Table Z-2.

Memorandum of Alert: Ethylene Dibromide, National Cancer Institute (October 4, 1974).

Regulations for Fuels and Fuel Additives; 40 Federal Register 51995-52337 (November 7, 1975).

Regulations for Rebuttable Presumption Against Registration; 40 CFR 162.11.

Review of Selected Literature on Ethylene Dibromide; EPA, Office of Toxic Substances (November 1974).

Sampling and Analysis of Selected Toxic Substances, Task II — Ethylene Dibromide; EPA, Office of Toxic Substances (performed under Contract No. 68-01-2646) (Publication No. EPA-560/6-75-001, September 1975).

Toxicological Studies of Selected Chemicals, Task II: Ethylene Dibromide; EPA, Office of Toxic Substances (performed under Contract No. 68-01-2646, Preliminary findings, personal communication by C. C. Lee, Project Director) (March 1976).

Hexachlorobenzene (HCB)

Why Should the Chemical Be of Concern at This Time?

Despite recent steps by several States and several companies to reduce environmental discharges of HCB, environmental contamination persists. Recent reports of the occurrence of HCB in human adipose tissues (95 percent of those sampled), the food supply, effluents, drinking water, and pesticides (in addition to registered pesticidal use) add to earlier concerns of EPA, USDA, FDA, and other organizations. In 1973, EPA made a public commitment in response to a petition from USDA to set an HCB food tolerance in 1976.

What Are the Health and Ecological Effects, and Environmental Behavior?

The death of breast-fed infants and a epidemic of skin sores and skin discoloration were associated with accidental consumption of HCB-contaminated seed grain in Turkey in the mid-1950's. Doses were estimated at 50 to 200 mg/day for several months to two years. Clinical manifestations included weight loss, enlargement of the thyroid and lymph nodes, skin photosensitization, and abnormal growth of body hair. HCB levels of up to 23 ppb in blood are believed to have contributed to enzyme disruptions in the population of a small community in southern Louisiana in 1973.

Long-term (up to 3 years) animal ingestion studies show a detectable increase in deaths at 32 ppm, cellular alteration at 1 ppm, biochemical effects at .5 ppm, and behavioral alteration between .5 and 5 ppm. Apparently, the effective dosage to offspring is increased by exposure to the parent. A 12 percent reduction in offspring survival resulted when exposure to very low levels had been continuous for three generations. Teratogenic effects appear minimal.

While HCB appears to have little effect on aquatic organisms, a bioaccumulation factor of 15,000 has been demonstrated in catfish. HCB is toxic to some birds. Eighty ppm caused death, and 5 ppm caused liver enlargement and other effects in quail. The half life of HCB in cattle and sheep is almost 90 days. HCB is very stable. It readily vaporizes from soil into the air; emissions to air in turn contaminate the soil.

What Are the Sources, Environmental Levels, and Exposed Populations?

About 90 percent of the estimated 8 million pounds of HCB produced annually in the United States is as a by-product at 10 perchloroethylene, 5 trichloroethylene, and 11 carbon tetrachloride manufacturing plants. HCB is commonly detected in solid wastes and liquid effluents. Most of the remaining production is as a by-product at more than 70 other sites producing chlorine and certain pesticides. About 45,000 pounds per year are released into the environment during pesticide use. HCB has also been found in the waste tars from vinyl chloride and other chlorine-product plants.

In 1975, forty-six percent of the soil samples collected at 26 locations along a 150-mile transect in Louisiana were contaminated with HCB at levels from 20 to 440 ppb. Parallel sampling of aquatic sediments revealed concentrations of 40 to 850 ppb. Although water samples were generally below 3 ppb, one sample below an industrial discharge contained 90 ppb. Air immediately adjacent to production facilities has shown concentrations from 1.0 to 23.6 ug/m^3. Most of the HCB appeared to be associated with particulate, but low levels were found in the gaseous phase as well, which might result from volatization from solid wastes. Samples collected from pastureland near a known HCB production site revealed concentrations in the vegetation from 0.01-630 ppm and in the soil from 0.01-300 ppb.

HCB residues have been found in soil, wildlife, fish, and food samples collected from all over the world. In the United States, HCB residues have been reported in birds and bird eggs collected from Maine to Florida, duck tissue collected from across the country, and fish and fish eggs from the East Coast and Oregon. Animal foods, including chicken feed, fish food, and general laboratory feeds, have been found to contain HCB residues. The frequency of detection of HCB residues in domestic meats has been steadily increasing since 1972, in part because of closer scrutiny. HCB has been detected in trace amounts in only two drinking water supplies.

EPA's monitoring of human adipose tissues collected from across the United States reveals that about 95 percent of the population has trace HCB residues.

What Are the Technologic and Economic Aspects?

If a food tolerance is established by EPA at about .5 ppm (the interim tolerance), there is no reason to believe that substantial numbers of animals or crops will be held off the market. However, a level of .3 ppm or lower would probably prevent the marketing of some products. The feasibility and costs of air emission and water effluent controls, particularly the effectiveness of particulate reduction and of better housekeeping practices, have not been estimated. Effective incineration of wastes has been demonstrated. Proper landfill practice may serve this purpose; however, studies indicate that soil and other covers only delay volatilization.

What Steps Have Been Taken, and What Is Being Done?

In the wake of widespread HCB contamination of cattle in Louisiana in 1973, and concern over possible contamination of sheep in California, EPA established an interim tolerance of .5 ppm. Concurrently, the State of Louisiana and several companies took immediate steps to tighten up solid waste practices from manufacturing through disposal. Also, supplies of Dacthal containing 10 percent HCB as an inert ingredient were voluntarily withdrawn from the California market. Several toxicological, monitoring, economics, and related projects were initiated by EPA to provide a better basis for further actions, including the establishment of a tolerance. Also, additional toxicological efforts were undertaken by USDA.

As soon as the needed toxicological data are available, a food tolerance will be established. Also, all pesticidal uses of HCB, including pesticides which contain HCB as a contaminant, will be reviewed. Studies of land and other disposal methods have been completed. Ocean dumping of HCB-laden tars is prohibited. Although not directly addressed by the NPDES permit program, provisions relating to suspended solids, and oil and grease may provide some degree of control if HCB enters the effluent stream.

References

Assessing Potential Ocean Pollutants (p 188-208); National Academy of Sciences (1975).

Burns, J. E. and Miller, F. M. "Hexachlorobenzene Contamination: Its Effects on a Lousiana Population"; *Arch. Environ. Health,* 30:44-48 (1975).

Cam. C. and Nygogosyan, G. "Acquired Prophyria Cutanea Tarda Due to Hexachlorobenzene"; *J. Am. Med. Assoc.,* 183:88-91 (1963).

An Ecological Study of Hexachlorobenzene and Hexachlorobutadiene; Final Report under EPA Contract No. 68-01-2689 (to be published at NTIS April 1976).

Environmental Contamination from Hexachlorobenzene; EPA, Office of Toxic Substances, (July 1973), (to be published at NTIS April 1976).

Industrial Pollution of the Lower Mississippi River in Louisiana; EPA, Office of Water and Hazardous Materials (April 1972).

Sampling and Analysis for Selected Toxic Substances — Task 1, HCB and HCBD; EPA, Office of Toxic Substances (performed under Contract No. 68-01-2646) (to be published at NTIS, April 1976).

Survey of Industrial Processing Data, Task 1 — Hexachlorobenzene and Hexachlorobutadiene Pollution from Chlorocarbon Processes; (EPA, Office of Toxic Substances, Report No. 560/3-75-003, June 1975) (NTIS Publication No. PB 2436441/AS).

Survey of Methods Used to Control Wastes Containing Hexachlorobenzene; EPA, Office of Solid Waste Management (performed under Contract No. 68-01-2956) (November 1975).

Vos, J. G., *et al.* "Toxicity of Hexachlorobenzene in Japanese Quail with Special Reference to Porphyria, Liver Damage, Reproduction and Tissue Residues"; *Toxicol Appl. Pharmacol.* 18:944-957 (1971).

Zitko, V., and Choi, P.M.K., "HCB and P,P — DDE in Eggs of Cormorants, Gulls and Ducks from the Bay of Fundy, Canada"; *Bull. Env. Contam. and Toxicology,* 7(51):63-64 (1972).

Hydrogen Sulfide (H₂S)

Why Should the Chemical Be of Concern at This Time?

Hydrogen sulfide, a colorless gas characterized by its rotten-egg odor, is an emission product from a large number of industrial processes such as kraft paper mills, as well as a naturally occurring chemical. Recently, H_2S has been identified in the exhaust of improperly adjusted catalyst-equipped motor vehicles. This has resulted in considerable public concern because of its strong odor. It has also been identified in emissions from prototype stationary source NO_x reduction catalyst systems. As more and more vehicles become equipped with catalysts, and as the Agency begins to regulate NO_x emissions from stationary sources, H_2S may become a pollutant of greater concern because of its acute toxicity.

What Are the Health and Ecological Effects, and Environmental Behavior?

H_2S is readily absorbed into the blood, with the chief exposure route being inhalation. Test animals exposed to concentrations over 700 ppm exhibited no H_2S in exhaled breath. Systemic poisoning is characterized by respiratory paralysis, occurring when H_2S concentrations in the blood exceed the oxidation capacity. Despite the substance's characteristic odor, which is detectable at levels as low as 0.025 ppm, high concentrations producing toxic effects can be reached almost without warning, because of olfactory fatigue at levels above 50 ppm. Exposures in excess of 400 ppm are considered dangerous, and over 700 ppm, life threatening.

The subacute effects of H_2S exposure are manifested as irritation of the respiratory tract and eyes. Pulmonary edema or pneumonia may result from prolonged exposure to concentrations over 250 ppm. Such exposures over a shorter term may produce temporary symptoms, such as headache, excitement, nausea, dizziness, and painful sensations in the nose, throat, and chest. Chronic exposure to levels of 20 to 30 ppm can lead to conjunctivitis, and 50 to 300 ppm may result in corneal clouding or blurred vision.

What Are the Sources, Environmental Levels, and Exposed Populations?

The primary natural source of H_2S is anaerobic microbial action on organic materials using naturally occurring sulfates. It is also encountered in natural gases and geothermal exhausts, sewers and sewage treatment plants, waters of some natural springs, volcanic gases, and certain mining operations. Because H_2S is soluble in both water and petroleum, it can be transported considerable distances before it is released.

A variety of industrial activities result in the release of H_2S. It is a well-known pollutant at kraft paper mills, oil refineries, coke ovens, natural gas plants, chemical plants, rayon production facilities, rubber production plants, and sugar beet refineries. Fatal and near-fatal H_2S concentrations have been reported at industrial landfills. Environmental levels outside the workplace have not been adequately documented, although there have been a number of reports of odor intensity near plants where reduced sulfur compounds are emitted.

Recent studies show that H_2S is emitted from maladjusted vehicles with catalytic converters. With adequate oxygen, these catalysts normally convert some of the sulfur dioxide emitted from the system into sulfuric acid; these same catalysts reduce SO_2 into H_2S in the absence of adequate oxygen, usually caused by a malfunctioning air injection pump or a maladjusted carburetor. As more sophisticated catalytic control devices are employed to achieve automotive emission standards, particularly those for NO_x, the potential for H_2S emissions may increase.

Several catalytic reduction processes have been proposed to limit No_x emissions at stationary sources to achieve the Ambient Air Quality Standard for NO_2. Laboratory tests of prototype systems demonstrate increased production of H_2S. Thus, an increased potential for widespread human exposure to low levels of this substances exists, but the levels cannot be predicted at this time.

What Are the Technologic and Economic Aspects?

Current evidence is that H_2S is emitted in significant quantities from catalyst-equipped automobiles only when such vehicles are defective or maladjusted. Conditions leading to H_2S formation will also result in high levels of carbon monoxide and hydrocarbon emissions. Thus, efforts to ensure proper maintenance of vehicle emission control systems (such as state inspection/maintenance programs) should control automotive H_2S emissions as well, at no additional cost, at least for the 1975/77 model-year vehicles.

Desulfurization of fuels or stack gas feeds could largely eliminate the H_2S problem, but at very high cost and resulting in the generation of large quantities of troublesome sludge.

What Steps Have Been Taken, and What Is Being Done?

H_2S has been, and continues to be, assessed in exhausts from catalyst-equipped automobiles. A fact sheet on current findings has been released to the public. New Source Performance Standards for total reduced sulfur from pulp mills are being developed. These should result in lowered H_2S emissions.

EPA recently contracted with NRC's Committee on Medical and Biologic Effects of Environmental Pollutants to assess the potential environmental problems of H_2S. This study will focus upon effects on man, plants, and animals, and, in addition, will define control technology and areas needing additional research.

OSHA has a health standard of 20 ppm as a ceiling level with a maximum of 50 ppm for 10 minutes once per day only if no other measurable exposure occurs. NIOSH has scheduled H_2S for FY 77 Criteria Document development. The current American Conference of Governmental Industrial Hygienists has recommended an 8-hour TWA of 10 ppm to guard only against conjunctivitis.

References

Annual Catalyst Research Program Report; EPA, Office of Research and Development (Publication No. EPA-600/3-75-010, September 1975).
Assessment of Catalysts for Control of NO_x from Stationary Power Plants, Phase I, Volume 1 — Final Report;

EPA, Office of Research and Development (Publication No. EPA-650/2-75-001a, January 1975).

R. L. Klimisch, and J. G. Larson. *The Catalytic Chemistry of Nitrogen Oxides;* New York, Plenum Press, (1975).

News Report; National Academy of Sciences, Vol. XXVI (March 1976).

Objectionable Odors from Catalyst-Equipped Vehicles; EPA, Office of Mobile Source Air Pollution Control (Publication MSAPC fact sheet, FS-35, February 1976).

Mercury (Hg)

Why Should the Chemical Be of Concern at This Time?

Despite recent action by EPA to limit mercury discharges during sludge incineration and through pesticidal use, and earlier Agency efforts to control air emissions and liquid effluent discharges, mercury continues to enter the environment. While more stringent enforcement of existing regulations should be helpful, discharges of mercury from fossil fuel plants, especially those that have shifted to coal from other less contaminated fuels, leaching of mercury from land-disposal sites, particularly into ground water, and urban runoff, are among the currently uncontrolled problems.

What Are the Health and Ecological Effects, and Environmental Behavior?

Hg in many forms is highly toxic to man and other living things. In terms of toxicity, mercury and its compounds can be divided into three categories: 1) alkyl mercury compounds; 2) elemental Hg; 3) inorganic Hg salts and phenyl and methoxy ethyl compounds. Alkyl compounds, particularly methyl mercury, are the most toxic. Over 90 percent of ingested methyl mercury is absorbed in the gastrointestinal tract, and its whole-body biological half life is 70-90 days. Methyl mercury is transported in blood cells to, and concentrates in, brain and other central nervous system tissues where it can cause irreversible damage. In addition, it can cross the placental barrier and cause abnormalities in fetal tissues and irreversible damage to the fetus at levels that appear to cause no symptoms in the mother. Elemental mercury, phenyl and methoxy ethyl compounds, and inorganic mercury salts are far less dangerous than methyl mercury, because less are ingested and the rates of excretion are higher.

The FDA action level of 0.5 ppm of Hg for fish and shellfish, both raw and processed, is based on a 30 ug/day maximum intake of methyl mercury. This is one tenth of the 300 ug/day average intake resulting in a blood level of 0.2 ppb in adults, the lowest level at which neurological symptoms have been observed.

Hg is readily transported to water by leaching from soil and fallout from air; most forms of Hg soil and water can be biologically or chemically transformed to methyl mercury.

What Are the Sources, Environmental Levels, and Exposed Populations?

In 1973, United States use of mercury was slightly less than 1900 metric tons, at an estimated cost of $8800 per metric ton. The chief uses were for battery manufacture (29.9 percent) and chlor-alkali production (24.1 percent). Use in 1965 had been approximately 2700 metric tons; the reduction of use resulted largely from a recognition of the hazards of the substance. In the period 1965-1973, several uses (particularly as preservatives and in gold recovery) were eliminated. Hg is also used to make paints and industrial instruments.

NIOSH estimates that 150,000 workers are exposed to mercury. Because the vapor is colorless and odorless, overexposures can easily go unnoticed until symptoms appear. Of a total of 1900 metric tons used, it has been estimated that as much as 80 percent is discharged into the environment. Distribution of Hg discharges from man-related sources to the environment is about 31 percent to the air, 6 percent to water, and 36 percent to land. Concentrations in the various media are measured in terms of total Hg, rather than the more hazardous methyl forms; thus, the data collected do not represent the true hazard.

Mercury is also a contaminant of coal, and may be a slag runoff problem. In addition, landfills are a source of leaching Hg; this problem may be particularly severe in areas where drinking water supplies are drawn from ground water. Exposure to Hg is widespread, but inadequate documentation of levels of methyl mercury makes estimates of risk difficult.

What Are the Technologic and Economic Aspects?

Because most mercury losses occur in use and disposal of products, recycling may provide the best method for reducing environmental discharges from batteries and instruments. Mercury emissions from the chlor-alkali industry would still be significant, even if state-of-the-art controls are applied to the production stream. Diaphragm-cell technology could eliminate mercury emissions, but might add to problems associated with asbestos and lead. New developments in this technology are reducing use of lead and eliminating asbestos; thus increased future reliance on new diaphragm cells may offer the desired reduction in mercury emissions, without the added environmental burden.

What Steps Have Been Taken, and What Is Being Done?

FDA has proposed an action level for Hg in fish and shellfish. As a result of NIOSH recommendations, OSHA is considering revised workplace standards for inorganic and alkyl mercury.

EPA has set a hazardous air pollutant standard for mercury under section 112 of the Clean Air Act, and is considering New Source Performance Standards to require zero emissions of Hg from new chlor-alkali plants.

EPA has addressed the problem through effluent guidelines for a few industrial categories and may expand this coverage in the future. The National Interim Primary Drinking Water Standard for Hg is 2 ppb. Ocean dumping is tightly controlled.

The EPA Administrator recently ordered an end to the registrations of most pesticides containing Hg, and particularly those used in paints, although the decision has been stayed pending completion of judicial review.

References

Clarkson, T. W., "The Transport of Elemental Mercury into Fetal Tissue"; *Biolheoncke* 21:239 (1972).

Decision of the Administrator on the Cancellation of Pesticides Containing Mercury (FIFRA Dockets No. 256 *et al.*) (February 17, 1976).

"Materials Balance and Technology Assessment of Mercury and Its Compounds on National and Regional Bases"; EPA, Office of Toxic Substances (Publication No. EPA 560/3-75-007, October 1975).

National Emission Standards for Hazardous Air Pollutants, Asbestos and Mercury; 39 Federal Register, 38064, (October 25, 1974).

National Emission Standards for Hazardous Air Pollutants: Asbestos and Mercury (amendments); 40 Federal Register, 48292 (October 14, 1975).

News Item (untitled), *Environmental News,* February 18, 1976.

News Item (untitled), *Toxic Materials News,* January 14, 1976.

Notice of Suspension of Mercurial Pesticides for Use on Rice Seeds, in Laundry, and in Marine Antifouling paints; Federal Register, April 6, 1973.

Proposed designations and effluent standards under Sec. 307(a), FWPCA; Federal Register, December 1973.

Pathological, Chemical and Epidemiological Research About Minamata Disease, Ten Years After (2nd Year) TR-509-75.

Platinum (Pt)

Why Should the Chemical Be of Concern at This Time?

EPA research efforts nearing completion indicate that platinum is more active biologically and toxicologically than previously believed. It methylates in aqueous media, establishing a previously unrecognized biotransformation and distribution mechanism. Because platinum complexes are used as antitumor agents, the potential for carcinogenic activity is present; tests to clarify this aspect should be completed within several months. While low levels of emissions of platinum particulate have been observed from some catalyst-equipped automobiles, the major potential source of Pt is from the disposal of spent catalysts.

What Are the Health and Ecological Effects, and Environmental Behavior?

Prior to EPA's research efforts, the literature cited platinosis as the only known adverse health effect. Platinosis is an allergenic respiratory sensitization to the substance, resulting in a severe asthmalike reaction to low concentrations by sensitized individuals. It is of particular concern to industrial users of Pt, as 50 percent of the workers have shown this syndrome; there is no way to predetermine if individuals are susceptible to it.

Other literature sources indicated that metallic and insoluble forms of Pt were toxicologically inert. Organic complexes of Pt have been used on a limited scale as antitumor treatments, sometimes resulting in toxic effects to the liver.

The health research program has determined the following:

— Metallic Pt and insoluble Pt compounds accumulate in many animal organ tissues, causing various abnormalities.

— Pt methylates in aqueous systems in much the same fashion as does mercury, suggesting a similar, but heretofore unknown, ecosystem transport mechanism for Pt.

— Although Pt is not found in urban or rural air, water, or soil, autopsies have found it in human fat at low levels. It has been hypothesized that this phenomenon may derive from Pt used in dental fillings, thus establishing that Pt may be soluble in human tissue, primarily lipid, and possibly by methylation.

— Pt in automotive exhaust is principally associated with large particulates coming off the catalyst.

What Are the Sources, Environmental Levels, and Exposed Populations?

Sources of Pt in the United States have been extremely limited, being principally associated with two Pt refining plants and Pt metal fabricators for electrical, chemical, refining, dental, jewelry, and glass industries. Total production and use was about 1.4 million troy ounces (52 tons) in 1971; the projection for 1981 is 2.7 million troy ounces (92.5 tons). Recent introduction of the automotive catalytic converter may result in nationwide exposure to Pt should Pt leach from discarded catalysts. Each catalyst contains 2-5 grams of Pt. Monitoring data to date have revealed virtually no Pt in air, water, or soil.

What Are the Technologic and Economic Aspects?

Should the results of current EPA research efforts to document health hazards of Pt suggest a need to control these exposures, disposal controls would be the most promising, since catalyst disposal is expected to be the largest contributor of Pt to the environment. In addition, the value of the metal would help to offset the cost of reclaiming the Pt from discarded catalysts. If direct vehicular emissions of Pt are found to be significant, particulate traps, which are available at reasonable cost, may provide a technological solution. Other noble metals have been suggested as Pt substitutes; however, even less is known about their potential for hazard.

What Steps Have Been Taken, and What Is Being Done?

An extensive health-effects research program is in progress in EPA, and efforts to characterize platinum in vehicular exhausts are also underway.

OSHA has established a standard for soluble salts of Pt as 0.002 mg/m³ total Pt, 8-hour TWA. This standard was designed to prevent development of platinosis in workers exposed to airborne Pt.

References

Annual Catalyst Research Program Report; EPA, Office of Research and Development (Publication No. EPA-650/3-75-010, September 1975).

Brubaker, P. E., et al., "Noble Metals: A Toxicological Appraisal of Potential New Environmental Contaminants;" Environmental Health Perspective Vol. 10 (1975).

Duffield, F. V. P., et al., "Determination of Human Body Burden Baseline Data of Platinum through Autopsy Tissue Analysis;" EPA, Office of Research and Development (accepted for publication in Environmental Health Perspectives, March 1976).

A Literature Search and Analysis of Information Regarding Sources, Uses, Production, Consumption, Reported Medical Cases, and Toxicology of Platinum and Palladium; EPA, Office of Research and Development (publication no. EPA 650/1-74-008, April 1974).

Occupational Safety and Health Standards: Subpart Z — Toxic and Hazardous Substances. 29 CFR 1910.1000, Table Z-1.

Polybrominated Biphenyls (PBB's)

Why Should the Chemical Be of Concern at This Time?

In 1973, one to two tons of PBB's, a highly toxic flame retardant, were accidentally mixed into an animal feed supplement and fed to cattle in Michigan. Contamination also resulted from traces of PBB's being discharged into the environment at the manufacturing site and at other facilities involved in handling PBB's. Approximately 250 dairy and 500 cattle farms have been quarantined, tens of thousands of swine and cattle and more than one million chickens have been destroyed, and law suits involving hundreds of million of dollars have been instituted. Before the nature of the contamination was recognized, many of the contaminated animals had been slaughtered, marketed, and eaten, and eggs and milk of the contaminated animals also consumed. Thus, large numbers of people have been exposed to PBB's. While commercial manufacture and distribution of PBB's have currently ceased, the full extent of the problem has not yet been assessed.

What Are the Health and Ecological Effects, and Environmental Behavior?

Among the 10,000 people who have been identified as having consumed PBB-contaminated meat, milk products, poultry, and eggs, no overt symptoms have been reported to date. Health effects can only be extrapolated from animal data. Based on experimental data, PBB's may be much more toxic than PCB's.

Although no long term toxicity data are available, short term rat, mice, and cattle studies have shown that PBB's may interfere with reproduction and liver functions, promote nervous disorders, and react as a teratogenic agent in tissues. PBB's have produced pathological changes in the livers of rats, mice, guinea pigs, cows, and rabbits. In an experiment with guinea pigs, the chemical was demonstrated to be an immuno-suppressant agent. About 400 cows in herds fed contaminated feed for about 16 days exhibited anorexia, decreased milk production, increased frequency of urination, some lameness, abnormal hoof growth, and shrinking of the udder. Later signs of toxic effects included bloody blebs, malformed or dead fetuses, abscesses, weight loss, and high susceptibility to stress. Non-lactating cows died within six months while the lactating animals survived and gradually improved. Massive liver abscesses were found in dead animals.

Fish taken from streams known to have been contaminated by PBB's have demonstrated that PBB's can bioaccumulate 20,000 to 30,000 times the ambient levels. PBB's are believed to be quite persistent in the environment, with perhaps one-half the lifetime of PCB's. PBB's readily vaporize.

What Are the Sources, Environmental Levels, and Exposed Populations?

PBB's have been used commercially as flame retardant additives in synthetic fibers and molded thermoplastic parts. PBB's have been incorporated into the plastic housings of many commercial products, such as typewriters, calculators, and microfilm readers, and consumer products, such as radio and television parts, thermostats, shavers, and hand tools.

Michigan Chemical Corporation produced approximately 11 million pounds of PBB's from 1970 to 1974. The White Chemical Corporation produced approximately 100,000 pounds of the closely related compounds, octabromobiphenyl and decabromobiphenyl, from 1970 through 1973. In addition, nine companies have been suppliers of laboratory quantities of PBB's, each producing about five pounds per year. There is no indication of importation of the material.

Monitoring in the Pine River near the facility where PBB's were produced indicated that levels diminished from 3.2 ppb in the ambient stream near the effluent discharge to .01 ppb eight miles downstream. Fish obtained in this eight miles stretch had levels of .09 to 1.33 ppm.

Detailed data are available on PBB levels found in cattle and hogs, with the highest level detected being 2.27 ppm. Data are not yet available on the levels found in any of the 10,000 or more exposed persons.

What Are the Technological and Economic Aspects?

Michigan Chemical Corporation reportedly has paid $20 million in settling a $270 million suit, with claims of $500 million still outstanding. However, the financial dimensions of the incident are still not known.

Among the substitutes for PBB's are the more expensive decabromobiphenyl oxide and several halogenated aliphatic compounds. However, the environmental acceptability of these compounds has not been assessed.

Monitoring methods for PBB's are in the developmental stage. Air monitoring has not yet been attempted.

What Steps Have Been Taken and What Is Being Done?

The State of Michigan has been the focal point for responding to the contamination incident. In addition, USDA, HEW (including FDA, NCI, and CDC), EPA, Michigan State University, and the University of Michigan support a wide array of epidemiological, toxicological, analytical, and related projects to clarify the effects of PBB's on humans and animals and to assess the extent of contamination. The HEW Toxicology Coordinating Committee is preparing a synthesis of available health effects information and will issue a report by June 1975. EPA is providing assistance in monitoring environmental levels of PBB's.

FDA has set temporary action levels for PBB's in contaminated foods and in animal feed. The State of Michigan has issued warnings to sport fishermen along the Pine River.

References

Survey of Industrial Processing Data, Task II — Pollution Potential of Polybrominated Biphenyls; EPA 560/3-75-004, NTIS PB 243-690, Prepared by Midwest Research Institute, June 1975.

Hesse, John L., *Water Pollution Aspects of Polybrominated Biphenyls Production: Results of Initial Surveys in the Pine River in Vicinity of St. Louis, Michigan,* Presentation to the Governor's Great Lakes Regional Interdisciplinary Pesticide Council, October 17, 1974.

"Michigan's PBB incident: Chemical Mix-up Leads to Disaster," *Science,* April 16, 1976.

The Contamination Crisis in Michigan, Polybrominated Biphenyls. A report from the Michigan State Senate Special Investigating Committee, July 1975.

Head, James D., *A Case Study: Polybrominated Biphenyls,* National Academy of Sciences, 1975.

Polynuclear Aromatic Hydrocarbons (PNA'S)

Why Should the Chemicals Be of Concern at This Time?

Increased exposure to polynuclear aromatic hydrocarbons (PNA's) and other air pollutants has been implicated by some researchers in increased rates of cancer, especially of the lung. Over 30 PNA's have been identified as urban air pollutants, including several carcinogens. PNA's are emitted during fossil fuel combustion, in natural combustion processes, and as a result of a variety of human activities. A proposed amendment to the Clean Air Act calls for explicit attention by 1977 to a possible air standard for PNA. PNA's have been found at low levels in liquid effluents, some drinking water supplies, and food.

What Are the Health and Ecological Effects, and Environmental Behavior?

Certain PNA's which have been demonstrated as carcinogenic in test animals at relatively high exposure levels are being found in urban air at very low levels. Various environmental fate tests suggest that PNA's are photo-oxidized, and react with oxidants and oxides of sulfur. Because PNA's are adsorbed on particulate matter, chemical half-lives may vary greatly, from a matter of a few hours to several days. One researcher reports that photo-oxidized PNA fractions of air extracts also appear to be carcinogenic. Environmental behavior/fate data have not been developed for the class as a whole.

It has been observed that PNA's are highly soluble in adipose tissue and lipids. Most of the PNA's taken in by mammals are oxidized and the metabolites excreted. Effects of that portion remaining in the body at low levels have not been documented.

Benzo[a]pyrene (BaP), one of the most commonly found and hazardous of the PNA's has been the subject of a variety of toxicological tests, which have been summarized by the International Agency for Research on Cancer. 50-100 ppm administered in the diet for 122-197 days produced stomach tumors in 70 percent of the mice studied. 250 ppm produced tumors in the forestomach of 100 percent of the mice after 30 days. A single oral administration of 100 mg to nine rats produced mammary tumors in eight of them. Skin cancers have been induced in a variety of animals at very low levels, and using a variety of solvents (length of application was not specified). Lung cancer developed in 2 of 21 rats exposed to 10 mg/m^3 BaP and 3.5 ppm SO$_2$ for 1 hour per/day, five days a week, for more than one year. Five of 21 rats receiving 10 ppm SO$_2$ for 6 hr/day, in addition to the foregoing dosage, developed similar carcinomas. No carcinomas were noted in rats receiving only SO$_2$. No animals were exposed only to BaP. Transplacental migration of BaP has been demonstrated in mice. Most other PNA's have not been subjected to such testing.

What Are the Sources, Environmental Levels, and Exposed Population?

PNA's can be formed in any hydrocarbon combustion process and may be released from oil spills. The less efficient the combustion process, the higher the PNA emission factor is likely to be. The major sources are stationary sources, such as heat and power generation, refuse burning, industrial activity, such as coke ovens, and coal refuse heaps. While PNA's can be formed naturally (lightning-ignited forest fires), impact of these sources appears to be minimal. It should be noted, however, that while transportation sources account for only about one percent of emitted PNA's on a national inventory basis, transportation-generated PNA's may approach 50 percent of the urban resident exposures.

Diesel powered vehicles produce more particulate emissions than gasoline powered; the nature of the fuel is such that the emissions would be expected to contain greater amounts of PNA's, and limited studies have confirmed this. EPA has tested gasoline-powered passenger vehicles to determine the amount of PNA's in the exhaust. However, this characterization is of particulate-associated PNA's; little is known of vaporous components.

PNA's have been detected in urban water supplies at low levels. PNA's in water and soils are adsorbed on minerals or organic particulate matter. Algae and invertebrates contain concentrations as high as 200 times those of the surrounding waters. Levels detected in plants, on the other hand, are slightly lower than soil levels. Sludge samples taken near a steel refinery showed combined benzo[e]- and benzo[a]pyrene levels of 0.91-19.0 mg/kg (dried weight). Liquid effluents did not appear to contain these substances.

Although a variety of PNA's have been observed in particulates from urban air samples, these are not now routinely monitored. Atmospheric concentrations of PNA's are generally represented by measurements of benzo[a]pyrene (BaP) concentrations. In heavily industrialized areas, BaP levels have been as high as 20 nanograms (ng)/m^3. Urban BaP levels are generally 2-7 ng/m^3; and rural, 0.3 ng/m^3. In 1971-73, nationwide annual emissions of BaP were estimated at 900 tons. It has been estimated that BaP represents 2-5 percent of the total PNA's emitted from automobiles; a similar and as yet undetermined relationship may exist for stationary source emissions.

Because of the large number of sources, most people are exposed to very low levels of PNA's. BaP has been detected in a variety of foods throughout the world. A possible source is mineral oils and petroleum waxes used in food containers and as release agents for food containers. FDA's studies have indicated no health hazard from these sources.

What Are the Technologic and Economic Aspects?

Good particulate emission controls can substantially reduce PNA emissions. However, the costs that would be incurred in further limiting PNA emissions from stationary and vehicular combustion sources are not known. The application of oxidation catalyst exhaust treatment has been effective in dramatically reducing PNA emissions from automobiles when such systems are operating properly. Similar controls at stationary sources may have a similar effect.

What Steps Have Been Taken, and What Is Being Done?

Limitation of carbon monoxide and hydrocarbon emissions from motor vehicles have simultaneously and dramatically reduced PNA emissions. A 1974 analysis of stationary source problems concluded that control regulations designed specifically for BaP or PNA were not warranted or practical, but noted that compliance with existing regulations for incinerators, open burning, coal combustion, and coking operations could significantly reduce PNA emissions. Additional efforts to document stationary source emissions, atmospheric chemistry, and human exposure have been initiated on a limited scale. EPA's STAR document and an NAS report for EPA have detailed much of the hazard and technologic aspects of this class of compounds.

References

Begeman, C. R., and Colucci, J. M., *Polynuclear Aromatic Hydrocarbon Emissions from Automotive Engines;* Warrendale, PA, Society of Automotive Engineers, Inc. (Publication No. 700469, 1971).

Gross, Herbert, *Hexane Extractables and PAH in the Black River (Ohio);* EPA, Office of Enforcement and General Counsel (October 1974).

IARC Monographs on the Evaluation of Carcinogenic Risk of the Chemical to Man: Certain Polycyclic Aromatic Hydrocarbons and Heterocyclic Compounds (Volume 3); Lyon, International Agency for Research on Cancer (1973).

Moran, J. B., *Assuring Public Health Protection as a Result of the Mobile Source Emission Control Program;* Warrendale, PA, Society of Automotive Engineers, Inc. (Publication No. 740285, March 1974).

Moran, J. B., *Lead in Gasoline: Impact of Removal on Current and Future Automotive Emissions;* for the Air Pollution Control Association (APCA) (June 1974).

Moran, J. B., Colucci, A., and Finklea, J. F., *Projected Changes in Polynuclear Aromatic Hydrocarbon Exposures from Exhaust and Tire Wear Debris of Light Duty Motor Vehicles;* Internal Report, ERC - RTP (May 1974).

Particulate Polycyclic Organic Matter; National Academy of Sciences (1972).

Preferred Standards Path Report for Polycyclic Organic Matter; EPA, Office of Air Quality Planning and Standards (October 1974).

Scientific and Technical Assessment Review of Particulate Polycyclic Organic Matter; EPA, Office of Research and Development, (1975).

Trichloroethylene

Why Should the Chemical Be of Concern at This Time?

Trichloroethylene (commonly referred to as tri), has been identified by NCI as a carcinogen in laboratory animals. It is widely used for vapor degreasing of fabricated metals and, to a lesser extent, in cleaning fluids. In addition to extensive worker exposure, tri has been detected in ambient air and water in industrial areas, in food, and in human tissues.

What Are the Health and Ecological Effects, and Environmental Behavior?

Tri induces tumors in mice at high dose levels, predominantly liver cancer with some metastases (transfer) to the lungs, according to NCI. Tri is absorbed rapidly by the lungs; only a small amount is eliminated by exhalation, 58-70% being retained. This is slowly eliminated in the urine as trichloroacetic acid (TCA) or trichloroethanol. The first major review of tri poisoning studied 284 cases, including 26 fatalities, in European plants where tri vapors were inhaled. Results indicated that toxic action involves the central nervous system. A number of short-term studies indicate that exposure to a concentration of 100 ppm in air may interfere with psychophysiological efficiency. In one study, six students exposed to 110 ppm for two four-hour periods separated by

1-1/2 hours showed significantly lower levels of performance in perception, memory, and manual dexterity tests. A confirmatory test using six tri workers produced almost identical results. There is a reported case of a man operating a metal degreaser who lost his sense of taste after one month's exposure to concentrations of tri which "occasionally escaped in sufficient quantities to be visible". Two months later he lost facial mobility and sensation, and developed EEG changes which did not clear up during the following two years.

Tri has been frequently detected in the environment; however, the behavior and transport have not been documented. Adverse ecological effects have not been reported.

Because of its low solubility, high vapor pressure, and high photodegradation rate at sea level (half life in air is about eight hours), tri is not expected to accumulate in the atmosphere. Its half life in water is on the order of months.

What Are the Sources, Environmental Levels, and Exposed Population?

Domestic tri production in 1974 was about 215,000 tons by five producers. Over 90% is used for vapor degreasing of fabricated metals. Ambient concentrations in the atmosphere of industrialized areas have been estimated by industry to be 2-16 ppt. Water concentrations are about 0.1 ppt. The character of the water was not defined, but trace amounts of tri have been identified in drinking water by EPA.

Over 200,000 workers are exposed to tri. The general public is exposed via inhalation of cleaning fluids and ingestion of foods, spices, and medicines from which undesirable components have been removed by extraction with tri. Residues ranging from 0.02 to 22 ppt have been detected in foodstuffs, and concentrations of up to 32 ppt have been detected in human tissue.

There is only one reference to degreasing equipment which concerns control of vapor emissions. In those cases where both an exhaust system and a degreaser vapor condensation system were in use, the average concentration of tri in air near the operation dropped from 105 ppm to 30 ppm.

What Are the Technologic and Economic Aspects?

Both methyl chloroform and perchloroethylene should be considered as possible substitutes for tri as degreasing agents. Both appear less damaging to air quality. However, methyl chloroform may adversely impact the ozone layer. All three are comparably priced and, since many users have already made this change, the economic impact should be minor. Closed loop systems could permit recovery of tri; however, this may represent a higher cost factor than use of substitutes. Preliminary NCI studies indicate that perchloroethylene may also present health hazards.

What Steps Have Been Taken, and What Is Being Done?

In October 1975, OSHA proposed a reduction in the workplace standard, and is currently reviewing the proposal, together with possible standards for methyl chloroform and perchloroethylene. CPSC is preparing a monograph on consumer exposures and possible hazards.

Tri producers are undertaking epidemiological studies, long-term animal studies, long-term animal inhalation studies, and an in-depth literature survey which have recently been started for the Manufacturing Chemists Association.

Since trichloroethylene contributes to photochemical smog, State Implementation Plans provide a mechanism for limiting emissions. Detailed health, environmental, and economic analyses are planned. NPDES permits limiting BOD_5, COD, and suspended solids also provide some control over effluent discharges of tri.

References

Air Pollution Assessment of Trichloroethylene; EPA, Office of Air Quality Planning and Standards (September, 1975, Draft).

Criteria Document: Occupational Exposure to Trichloroethylene; HEW, National Institute for Occupational Safety and Health (June 1973).

Memorandum of Alert: Trichloroethylene; HEW, National Cancer Institute (March 21, 1975).

Mitchell, A. B. S., and Parsons-Smith, B. G., "Trichloroethylene Neuropathy"; *Br. Med. J.,* 1:422-23 (1969).

News item (untitled), *Chemical and Engineering News* (May 19, 1975).

Preliminary Study of Selected Potential Environmental Contaminants — Optical Brighteners, Methyl Chloroform, Trichloroethylene, Tetrachloroethylene, Ion Exchange Resins; EPA, Office of Toxic Substances (Publication No. EPA-560/2-75-002, July 1975).

Stuber, K., "Injuries to Health in the Industrial Use of Trichloroethylene and the Possibility of Their Prevention," *Arch Gewerbepathol Gewerbehyg* (Ger.) 2:398-456 (1932).

Trichloroethylene (Background Report); HEW, National Institute for Occupational Safety and Health (June 6, 1975).

Trichloroethylene (data sheet 389) (revised); Chicago, National Safety Council (1964).

Tris (2,3-Dibromopropyl) Phosphate (TBPP)

Why Should the Chemical Be of Concern at This Time?

TBPP is presently the most popular flame retardant additive for acetate and polyester fabrics which are widely used in children's sleepwear. Recent EPA-funded experiments have shown that TBPP is a mutagen in a microbial assay that is also being considered for use as a screen for carcinogenic potential. Crude experiments have suggested that TBPP is present in the waste water from home laundry of such sleepwear. The Environmental Defense Fund recently requested the Consumer Product Safety Commission (CPSC) to take steps to reduce the hazards associated with TBPP and Ralph Nader has asked the Senate Commerce Committee to hold hearings. Both CBS and NBC have raised the issue of continuing use of TBPP in their National news coverage.

What Are the Health and Ecological Effects, and Environmental Behavior?

TBPP is mutagenic in the "Ames bioassay" which is regarded by some toxicologists as a screen for carcinogens. This has been confirmed by two independent investigators.

In feeding experiments, dose-related accumulations of TBPP, or its metabolites, were found in rats given 100 to 1000 ppm for 28 days. Six weeks after withdrawal, no residues were found and no histopathology was detected. A degradation product (2,3-dibromopropanol) was found in the urine of rats after oral administration or dermal exposure, indicating that TBPP reaches a number of organs.

Laboratory experiments have shown that aqueous extracts of fabrics containing TBPP at extremely low levels are lethal to fish even after the fabric had been laundered. Erratic behavior, possibly resulting from central nervous system involvement, was noted before death. Full confirmation of these findings has not yet been performed. Environmental biodegradation occurs, but the rate is not known.

TBPP is also a mild sensitizing agent in humans. However, no allergic responses have been reported from consumer or occupational exposure.

What Are the Sources, Environmental Levels, and Exposed Populations?

About 65% of the 10 million pounds of TBPP produced annually in the United States by six manufacturers are applied to fabrics used for children's clothing, with the remainder used as a flame retardant in other materials, such as urethane foams. A significant portion of the total, perhaps ten percent, reaches the environment from textile finishing plants and laundries. Most of the rest will eventually find its way into solid wastes (manufacturing waste and used clothing). Environmental levels of TBPP near manufacturing plants, dumps, mills, and laundries have not been measured.

TBPP is added to fabrics used for children's garments to the extent of 5-10 percent by weight. A child wearing such a garment and chewing on a sleeve or collar could easily ingest some TBPP, particularly if the garment had not been laundered before use. The effects of saliva, urine, or feces on the extractability of TBPP or on its absorption through the skin have not been measured.

What Are the Technologic and Economic Aspects?

At current production levels, TBPP gross sales are about 10 million dollars per year. Substitute materials or methods for meeting the flammability requirements for children's sleepwear are probably available. However, quantitative data on the cost, performance, and safety of substitute materials and methods are not available. The major costs would be in the product development and application areas.

What Steps Have Been Taken, and What Is Being Done?

The mutagenicity of TBPP was originally discovered in a small EPA screening program. The limited available data on the environmental effects of TBPP have been compiled and transmitted to the Consumer Product Safety Comission, the National Institute for Occupational Safety and Health, the Toxicology Coordinating Committee of the Department of Health, Education, and Welfare, the NAS committee on the Fire Safety Aspects of Polymeric Materials, and several other groups concerned with flame retardants, including some TBPP manufacturers and users. The National Cancer Institute is currently conducting carcinogenicity tests on TBPP in rats and mice; results are expected in early 1977.

References

Harris, Robert H., and Manser, Philip J., *Petition Pursuant to 15 USC 2059 to the Consumer Product Safety Commission to Commence a Proceeding for the Issuance of a Consumer Product Safety Rule;* Washington, Environmental Defense Fund (March 24, 1976).

Nader, Ralph, *Letter* to Chairman, Senate Commerce Committee (undated) (Copy received by Consumer Product Safety Commission March 25, 1976).

Prival, Michael J., *Information Available to Date Relevant to the Mutagenicity of Tris (2,3-dibromopropyl) Phosphate;* EPA, Office of Toxic Substances (internal memorandum December 2, 1975).

A Study of Flame Retardants for Textiles; EPA, Office of Toxic Substances (Report No. EPA-560/1-76-004, December 31, 1975).

Vinylidene Chloride (VDC)

Why Should the Chemical Be of Concern at This Time?

Vinylidene chloride (VDC), an important monomer in the manufacture of methyl chloroform and of Saran and other plastics, is of particular concern because the manner in which the problem is emerging is similar to earlier developments concerning vinyl chloride. In January 1976, NIOSH reported that about 60 percent of examined workers in a New Jersey plant using VDC had developed liver disorders, and announced its intention to follow up. Previous laboratory animal studies had suggested that VDC might be a liver carcinogen, as well as produce a number of other adverse health effects. A substantial amount of VDC is vented to the atmosphere during production, polymerization, and fabrication.

What Are the Health and Ecological Effects, and Environmental Behavior?

Vinylidene chloride has recently been reported to cause liver impairment. Twenty-seven of forty-six workers examined at the BASF Wyandotte VDC polymerization plant in South Kearny, New Jersey, showed 50 percent or greater loss in liver function. Other examinations indicate that VDC is biochemically altered in the body and may form intermediates similar to the cancer-producing metabolites of vinyl chloride.

Inhaled VDC is reported to produce liver tumors in rats at 200 ppm. Inhalation experiments with animals showed that VDC causes liver and kidney damage. When rats were pre-exposed to vinyl chloride and then tested with VDC, the acute toxicity of VDC was greatly enhanced.

Concurrent exposure reduces the acute effects and may potentiate the carcinogenic effects. This is important because a significant part of polymer production involves the use of both chemicals.

As yet, the ecological effects and environmental behavior of VDC in either air or water have not been studied. Its highly reactive nature would seem to support a thesis that it is relatively short-lived in air, possibly on the order of several hours.

What Are the Sources, Environmental Levels, and Exposed Populations?

Dow Chemical and PPG Industries annually produce 270 million pounds of VDC monomer in three Gulf Coast plants. About 50 percent is used in the production of methyl chloroform by PPG. The remainder is polymerized to plastic resins at 12 facilities owned by a number of companies throughout the country. The resin is then fabricated into plastics at 60 to 75 plants. It has been estimated that about four million pounds of VDC were lost to the air in 1974. One EPA-funded report estimates that as much as 25 percent of the VDC used in any given Saran production run is disposed of in landfill, primarily in polymerized form, but there are no estimates of the levels of unreacted monomer.

In the past, worker exposure has generally not been monitored. Tests demonstrate that 20,000 ppm can easily be attained in the immediate vicinity of a spill. In some cases, past worker exposures to VDC may have exceeded those to vinyl chloride (which were measured at 300-1000 ppm before OSHA limits were imposed). The odor threshold of VDC is 500 ppm.

What Are the Technologic and Economic Aspects?

The primary requirement for reduction of exposure to VDC would be to limit emissions through improved housekeeping procedures in the industry. The type of control technology used to control vinyl chloride should be applicable to VDC production.

What Steps Have Been Taken, and What Is Being Done?

The American Conference of Governmental Industrial Hygienists has established a threshold limit value of 10 ppm.

NIOSH is planning to monitor the follow-up studies on workers at the South Kearny BASF plant and will survey other VDC production sites in 1977 to determine if a workplace standard should be recommended.

EPA is preparing an assessment of the air pollution problems associated with VDC production and use. Fetotoxicity and embryotoxicity have been demonstrated under EPA-funded contracts. Data on environmental effects of VDC are also being obtained.

References

Air Pollution Assessment Report: Vinylidene Chloride; EPA, Office of Air Quality Planning and Standards (Prepared under contract) (Report to be published June 1976).

Jaeger, R. J., Trabulus, M. J., and Murphy, S. D., "Biochemical Effects of 1,1-Dichloroethylene in Rats: Dissociation of its Hepatotoxicity from a Liperoxidative Mechanism;" *Toxicology and Applied Pharmacology,* 24:457-567 (1973).

The Merck Index, 8th edition; Rahway, NJ, Merck and Co., Inc. (1968).

Prendergast, J. A., Jones, R. A., Jenkins, L. J., Jr., and Siegel, J., "Effects on Experimental Animals of Long-Term Inhalation of Trichloroethylene, Carbon Tetrachloride, 1,1,-Trichloroethane, Dichlorofluoromethane and 1,1-Dichloroethylene"; *Toxicology and Applied Pharmacology,* 10:270-289 (1967).

Vinylidene Chloride Monomer Emissions from the Monomer, Polymer, and Polymer Processing Industries; EPA, Office of Air Quality Planning and Standards (Prepared under contract No. 68-02-1332, Task 13) (Report to be published April 1976).

Wessling, R., and Edwards, F. G., "Poly (Vinylidene Chloride)," *Kirk-Othmer Encyclopedia of Chemical Technology, 2nd edition,* (21:275-303); New York, Interscience Publishers (1967).

HEALTH AND FIRE HAZARDS FROM COMMON INDUSTRIAL CHEMICALS*

Explanation of Table Headings

Column headings of the table are explained as follows.

Listing of Substances

The "Table of Chemical Hazards" lists those physical properties associated with the fire and health hazards of common industrial chemicals.

Data in the Table have been taken from the following principal references: Standard 49, *Hazardous Chemicals Data,* and Standard 325M, *Flammable Liquids, Gases, and Volatile Solids,* both published by the National Fire Protection Association, 60 Batterymarch St., Boston, Mass. 02110, and *Handbook of Organic Industrial Solvents,* published by the American Mutual Insurance Alliance, 20 N. Wacker Dr., Chicago, Ill. 60606.

Column 1

FLASH POINT is the lowest temperature at which the liquid gives off sufficient vapor to form an ignitible mixture with air and produce a flame when an ignition source is brought near the surface of the liquid. The flash point is used by the NFPA to define and classify the fire hazard of liquids.

A standard closed container is used to determine the closed-cup flash point of a liquid and a standard open-surface dish is used for the open-cup flash point determination, as specified by the standards of the American Society for Testing and Materials (ASTM).

Unless otherwise specified, the flash point values in the Table are closed-cup determinations.

Column 2

FLAMMABLE OR EXPLOSIVE LIMITS are those concentrations of a vapor or gas in air below or above which propagation of a flame does not occur on contact with a source of ignition. The lower limit is the minimum concentration below which the vapor-air mixture is too "lean" to burn or explode. The upper limit is the maximum concentration above which the vapor-air mixture is too "rich" to burn or explode.

Flammable or explosive limits are given in the Table in terms of percentage by volume of gas or vapor in air, and, unless otherwise noted, at normal atmospheric pressures and temperatures. Increasing the temperature or pressure lowers the lower limit and raises the upper limit. Decreasing the temperature or pressure has the opposite effect.

Column 3

VAPOR VOLUME is the number of cubic feet of solvent vapor formed by the evaporation of one gallon of a liquid at 75 F.

Column 4

SEVERITY AND TYPE OF HAZARD. Each hazardous substance presents a distinct problem and must be treated individually in the light of its own characteristics. Conclusions regarding the hazards of a product cannot safely be drawn either from the properties of the materials from which it is formed or by analogies based upon chemical structure. Mixtures of two or more chemicals may have properties that vary in kind or degree from those of the individual components. Impurities may contribute hazardous properties and should not be overlooked.

Signal word—This word is intended to draw attention to the presence of hazard, and to indicate the degree of severity. The Signal Words used, in the order of diminishing severity of hazard are:

D DANGER—Serious, severe, hazardous.

W WARNING—Moderate, intermediate, harmful.

C CAUTION—Minor, mild, irritating.

The degree of severity can be expressed only in relative terms. DANGER! is the strongest of the three words and is used for those products presenting the most serious hazards. CAUTION! is for those compounds presenting the least serious hazards. WARNING! is intermediate between DANGER! and CAUTION!

Type of hazard—This statement gives notice of the hazards that are present in connection with the customary

* From *Fundamentals of Industrial Hygiene,* by the National Safety Council. Reprinted by permission.

or reasonably anticipated handling or use of the product. Many chemical products will present more than one type of hazard; in which case, appropriate statements for each significant type are included. In general, the most serious hazard is stated first.

Severity of toxic or fire hazard from the undiluted material as normally used. Impurities, mixtures, and conditions of use may influence this.

Combine degree of severity (signal word) with description statement below.

1. Flammable material.

2. Oxidizing material—Contact with other combustible material may cause fire.

3. Gas or vapor rapidly toxic or extremely irritating on exposure for a short time or to low concentration.

4. Gas or vapor harmful or irritating on prolonged or repeated exposure, or exposure to high concentrations.

5. Gas or vapor physiologically inert, but displaces oxygen available for breathing.

6. Dust hazardous when inhaled or touched.

7. Irritant, sensitizer, corrosive—Causes skin irritation or burn.

8. Toxic through skin absorption.

Column 5

PRECAUTIONS TO TAKE. These instructions are intended to supplement the statement on "Severity and Type of Hazard" by briefly setting forth measures to be taken to avoid injury or damage from stated hazards.

To minimize hazards, take precaution indicated by signal number listed below.

1. Keep away from heat, sparks, open flame.

2. Avoid spilling, contacting skin, eyes, clothing.

3. Use adequate ventilation or personal protection. Avoid breathing dust, fumes, mists, gases, vapors.

4. Avoid contact with acids, moisture, combustibles.

5. Do not handle or use until safety precautions outlined in data sheets, or recommended by consultant or manufacturer are understood.

Many products present no hazard in normal handling and storage. For these products, no precautionary statements are necessary. The development of new chemical products, and the introduction of chemical processes into ever-widening fields, have accentuated the need for obtaining appropriate information in those cases where there are hazards requiring special precautions. Precautionary information should, so far as is practicable, reach every person using, handling, or storing hazardous substances.

Column 6

ORAL TOXICITY RATING. The numerical toxicity rating of column 6 is largely explained in Table C-1. To use toxicity ratings effectively, their many implications and limitations must be appreciated, as noted below.

- The rating is based on mortality, not morbidity; that is, it is really a lethality rating.

- Unless otherwise noted, each rating is based on the acute toxicity of a single dose when taken by mouth. Other dose regimens and other routes of administration are not represented by the rating.

- The toxicity rating reflects an estimate of the probable or mean lethal dose, not the minimal fatal dose. Perhaps because of personal idiosyncrasy or hypersensitivity or predisposing disease, minimal lethal doses recorded in the clinical literature are usually considerably lower than those implied by the current ratings.

- With only a few compounds are clinical data adequate to establish a toxicity rating. Most of the values here are based on laboratory determinations of mean lethal doses (LD_{50}) in small laboratory mammals (rat, mouse, guinea pig, rabbit; sometimes cat, dog, and monkey). Implicit in the use of such data is the conventional assumption that the mean lethal dose in man lies in the same class as does the LD_{50} for the test animals.

- For most corrosive agents (such as mineral acids, alkalis, and bleaches), no toxicity rating is suggested. In these cases, death is usually the result of severe local tissue injury, with secondary complications such as toxemia, shock, perforation, infection, hemorrhage, and obstruction. The intensity of the local lesion and of the results that follow is often determined by the concentration of the corrosive substance, whereas the volume and "dose" are secondary considerations. For such agents, no single toxicity rating is an appropriate measure of lethality, unless the concentration is also specified. Actually, no simple parameter can describe this relation in a way which is thought to be clinically useful.

In Table C-1, common units of measure are used to describe lethal doses for an adult of average size (body weight of 150 lb or 70 kg). For patients who are heavier or lighter, probable lethal doses are proportionately larger or smaller. It is assumed that lethal doses are proportional to body weight, irrespective of age. The reader is urged to consult the reference given above for more complete information.

Toxicity rating—Amount to produce death when swallowed by an average (150-lb) man.

TABLE—C-1

Toxicity Rating or Class	Probable LETHAL Dose (human)	
	mg/kg	For 70 kg man (150 lb)
6 Super toxic	Less than 5	A taste (less than 7 drops)
5 Extremely toxic	5–50	Between 7 drops and 1 teaspoonful
4 Very toxic	50–500	Between 1 tsp. and 1 ounce
3 Moderately toxic	500–5 gm./kg.	Between 1 oz. and 1 pint (or 1 lb.)
2 Slightly toxic	5–15 gm./kg.	Between 1 pt. and 1 quart
1 Practically nontoxic	Above 15 gm./kg.	More than 1 quart

1. Practically nontoxic—Takes more than one quart (2 lb).

2. Slightly toxic—1 pint to 1 quart or

3. Moderately toxic—1 oz. or to 1 pint.

4. Very toxic—1 teaspoonful to 1 oz.

5. Extremely toxic—7 drops to 1 teaspoonful.

6. Super toxic—A taste (<7 drops).

Column 7

ACTION ON SKIN. Local action on normal skin of the undiluted material.

A Relatively harmless.
B Sensitizer—Can cause allergic reactions.
C Primary skin irritant—Brief contact can cause inflamation or burns.
D Can cause solvent irritation-type dermatitis.

Column 8

NFPA HAZARD CLASSIFICATIONS. Fires and other emergency situations may involve chemicals that have varying degrees of toxicity, flammability, and reactivity (instability and water reactivity). The National Fire Protection Association grading of these relative hazards (under fire conditions) is given in the columns marked "NFPA Health" "NFPA Flammability," and "NFPA Reactivity."

For a full description of the NFPA classifications, see NFPA Standard No. 704M, *Identification of the Fire Hazards of Materials.* A complete listing is given in NFPA Standard No. 49, *Hazardous Chemicals Data,* and in Standard No. 325M, *Fire Hazard Properties of Flammable Liquids, Gases, and Volatile Solids.*

An explanation of the degrees of hazard follows.

NFPA Health Hazards—In general the health hazard in fire fighting is that of a single exposure which may vary from a few seconds up to an hour. The physical exertion caused by fire fighting or other emergency may intensify the effects of any exposure.

Health hazards arise from two sources: (a) the inherent properties of the material, and (b) from the toxic products of combustion or decomposition of the material. (Common hazards from burning of ordinary combustible materials are not included.)

The degree of hazard should indicate (a) that people can work safely only with specialized protective equipment, (b) that they can work safely with suitable respiratory protective equipment, or (c) that they can work safely in the area with ordinary clothing.

A health hazard, as defined by the NFPA, is any property of a material which either directly or indirectly can cause injury or incapacitation, either temporary or permanent, for exposure by contact, inhalation, or ingestion.

The degrees of hazard under fire conditions are ranked according to the probable severity of hazard to personnel, as follows:

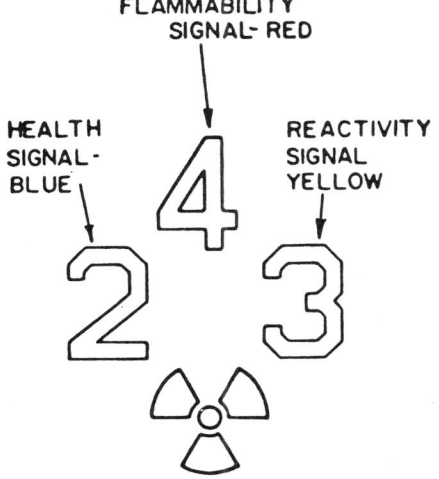

FIG. C–1.—Correct arrangement and order of signals used on equipment for quick identification of material hazards. For details, see NFPA Standard 704M.

4 Materials which on very short exposure could cause death or major residual injury even though prompt medical treatment were given, including those which are too dangerous to be approached without specialized protective equipment. This degree should include:

● Materials which can penetrate ordinary rubber or synthetic protective clothing;

● Materials which under normal conditions or under fire conditions give off gases which are extremely hazardous (i.e., toxic or corrosive) through inhalation or through contact with or absorption through the skin.

3 Materials which on short exposure could cause serious temporary or residual injury even though prompt medical treatment were given, including those requiring protection from all bodily contact. This degree should include:

● Materials which give off highly toxic combustion products;

● Materials corrosive to living tissue or toxic by skin absorption.

2 Materials which on intense or continued exposure could cause temporary incapacitation or possible residual injury unless prompt medical treatment is given, including those requiring use of respiratory protective equipment with independent air supply. This degree should include:

● Materials giving off toxic combustion products;

● Materials giving off highly irritating combustion products;

● Materials which either under normal conditions or under fire conditions give off toxic vapors lacking warning properties.

1 Materials which on exposure would cause irritation but only minor residual injury even if no treatment is given, including those which require use of an approved canister type gas mask. This degree should include:

● Materials which under fire conditions would give off irritating combustion products;

● Materials which on the skin could cause irritation without destruction of tissue.

0 Materials which on exposure under fire conditions would offer no hazard beyond that of ordinary combustible material.

NFPA Flammability Hazards—deal with the degree of susceptibility of materials to burning, even though some materials that burn under one set of conditions will not burn under others. The form or condition of material, as well as its properties, affects the hazard.

The degrees of hazard are ranked according to the susceptibility of materials to burning, as follows:

4 Materials which will rapidly or completely vaporize at atmospheric pressure and normal ambient temperature or which are readily dispersed in air, and which will burn readily. This degree should include:

● Gaseous materials;

● Cryogenic materials;

● Any liquid or gaseous material which is a liquid while under pressure and having a flash point below 73 F (22.8 C) and having a boiling point below 100 F (37.8 C). (Class IA flammable liquids.)

● Materials which on account of their physical form or environmental conditions can form explosive mixtures with air and which are readily dispersed in air, such as dusts of combustible solids and mists of flammable or combustible liquid droplets.

3 Liquids and solids that can be ignited under almost all ambient temperature conditions. Materials in this degree produce hazardous atmospheres with air under almost all ambient temperatures or, though unaffected by ambient temperatures, are readily ignited under almost all conditions. This degree should include:

● Liquids having a flash point below 73 F (22.8 C) and having a boiling point at or above 100 F (37.8 C) and those liquids having a flash point at or above 73 F (22.8 C) and below 100 F (37.8 C). (Class IB and Class IC flammable liquids);

● Solid materials in the form of coarse dusts which may burn rapidly but which generally do not form explosive atmospheres with air;

● Solid materials in a fibrous or shredded form which may burn rapidly and create flash fire hazards, such as cotton, sisal and hemp;

● Solids which burn with extreme rapidity, usually by reason of self-contained oxygen (e.g., dry nitrocellulose);

● Materials which ignite spontaneously when exposed to air.

2 Materials that must be moderately heated or exposed to relatively high ambient temperatures before ignition can occur. Materials in this degree would not under normal conditions form hazardous atmospheres with air, but under high ambient temperatures or under moderate heating may release vapor in sufficient quantities to produce hazardous atmospheres with air. This degree should include:

● Liquids having a flash point above 100 F, but not exceeding 200 F;

● Solids and semisolids which readily give off flammable vapors.

1 Materials that must be preheated before ignition can occur. Materials in this degree require considerable preheating, under all ambient temperature conditions, before ignition and combustion can occur. This degree should include:

- Materials which will burn in air when exposed to a temperature of 1500 F for a period of 5 minutes or less;
- Liquids, solids and semisolids having a flash point above 200 F;
- This degree includes most ordinary combustible materials.

0 Materials that will not burn. This degree should include any material which will not burn in air when exposed to a temperature of 1500 F for a period of 5 minutes.

NFPA Reactivity (Instability) Hazards—deal with the degree or susceptibility of materials to release energy. Some materials are capable of rapid release of energy by themselves (as by self-reaction or polymerization), or they can undergo violent eruptive or explosive reaction if contacted with water or other extinguishing agents or with certain other materials.

The violence of reaction or decomposition of materials may be increased by heat or pressure, by mixture with certain other materials to form fuel-oxidizer combinations, or by contact with incompatible substances, sensitizing contaminants, or catalysts.

Because of the wide variations of accidental combinations possible in fire emergencies, these extraneous hazard factors (except for the effect of water) cannot be applied in a general numerical scaling of hazards. Such extraneous factors must be considered individually in order to establish appropriate safety factors such as separation or segregation. Such individual consideration is particularly important where significant amounts of materials are to be stored or handled. Guidance for this consideration is provided in NFPA Standard No. 49, *Hazardous Chemicals Data.*

The degree of hazard should indicate to fire fighting personnel that the area should be evacuated, that the fire may be fought from a protected location, that caution must be used in approaching the fire and applying extinguishing agents, or that the fire may be fought using normal procedures.

The relative reactivity of a material is defined as follows.

REACTIVE MATERIALS are those which can enter into a chemical reaction with other stable or unstable materials. For purposes of this guide, the other material to be considered is water and only if its reaction releases energy. Reactions with common materials, other than water, may release energy violently. Such reactions must be considered in individual cases, but are beyond the scope of this identification system.

UNSTABLE MATERIALS are those which in the pure state or as commercially produced will vigorously polymerize, decompose or condense, or become self-reactive and undergo other violent chemical changes.

STABLE MATERIALS are those that normally have the capacity to resist changes in their chemical composition, despite exposure to air, water, and heat as encountered in fire emergencies.

The degrees of hazard are ranked according to ease, rate and quantity of energy release as follows:

4 Materials which are readily capable of detonation or of explosive decomposition or explosive reaction at normal temperatures and pressures. This degree should include materials which are sensitive to mechanical or localized thermal shock at normal temperatures and pressures.

3 Materials which are capable of detonation or of explosive decomposition or explosive reaction but which require a strong initiating source or which must be heated under confinement before initiation. This degree should include materials which are sensitive to thermal or mechanical shock at elevated temperatures and pressures or which react explosively with water without requiring heat or confinement.

2 Materials which are normally unstable and readily undergo violent chemical change but do not detonate. This degree should include materials which can undergo chemical change with rapid release of energy at normal temperatures and pressures or which can undergo violent chemical change at elevated temperatures and pressures. It should also include those materials which may react violently with water or which may form potentially explosive mixtures with water.

1 Materials which are normally stable, but which can become unstable at elevated temperatures and pressures or which may react with water with some release of energy but not violently.

0 Materials which are normally stable, even under fire exposure conditions, and which are not reactive with water.

Column 9

REFERENCES. References to publications on chemical safety have been included in the Table as an aid to the safety professional who is seeking more information on safe handling of a particular chemical. A brief description of the format and content of each of these publications follows. Copies of the publications and further information can be obtained from the organizations listed.

- *N.S.C. Data Sheet*—National Safety Council, 425 North Michigan Ave., Chicago, Ill. 60611.

The "Industrial Data Sheets, Chemical Series," cover the major hazards associated with a single substance or family of substances, discuss methods of controlling or eliminating the hazards, and concisely outline safe and efficient procedures. They are written for the safety professional, foreman, and other management representatives.

A typical "Chemical Series Data Sheet" is composed of the following topics:

1. Properties
2. Uses
3. Containers
4. Shipping regulations
5. Storage
6. Personnel hazards
7. Handling
8. Personal protective equipment
9. Ventilation
10. Fire and explosion hazards
11. Electrical equipment
12. Symptoms of poisoning
13. First aid
14. Treatment of burns
15. Toxicity
16. Threshold limit values
17. Medical examinations
18. Waste disposal

- *MCA Data Sheet*—Manufacturing Chemists' Association, 1825 Connecticut Ave. N.W., Washington, D.C. 20009

Of interest to both producers and consumers, the Chemical Safety Data Sheets are based on the recommendations of chemical engineers. Topics covered are:

1. Properties
2. Hazards
3. Engineering control of hazards
4. Employee safety
5. Fire fighting
6. Handling; storage
7. Tank and equipment cleaning and repairs
8. Waste disposal
9. Medical mgt.
10. First aid

- *AIA Bulletin*—American Insurance Association, Engineering and Safety Department, 85 John St., New York, N.Y. 10038.

The "Chemical Hazards Bulletins" present data interpreted in terms of insurance coverage and therefore are of value in inspection and underwriting problems.

A typical topic outline includes:

1. Properties
2. Identification
3. Health hazards
4. Public liability hazards
5. Product liability hazards
6. Boiler and machinery hazards
7. Fire hazards
8. Control measures
9. Storage and shipping
10. Hazardous procedures and operations
11. Detection and determination
12. First aid
13. Waste disposal

These bulletins are available only through those insurance companies that subscribe to the Engineering and Safety Service of the AIA.

- *AIHA Hygienic Guide*—American Industrial Hygiene Association, 66 S. Miller Rd., Akron, Ohio 44313.

The "Hygienic Guides," which appear in the bimonthly *AIHA Journal*, summarize data from experimental work and industrial experience to establish the degree of chemical hazards and recommend control procedures.

A typical Guide covers:
1. Hygienic standards
   a. Recommended maximum atmospheric concentration
   b. Severity of fire and health hazards.
   c. Short exposure tolerance
   d. Atmospheric concentration immediately hazardous to life
2. Toxic properties
   a. Inhalation
   b. Skin contact
   c. Eye contact
   d. Ingestion
3. Industrial hygiene practice
   a. Recognition
   b. Evaluation
   c. Recommended control procedures
4. Medical information
   a. First aid
   b. Special medical procedures
5. References

TABLE OF CHEMICAL HAZARDS

Substance	1 Flash Point (deg F)	2 Flammable or Explosive Limits (% by volume) Lower	2 Upper	3 Vapor Volume (cu ft per gal)	4 Severity and Type of Hazard	5 Precautions to Take	6 Oral Toxicity	7 Action on Skin	8 NFPA Health Flammability Reactivity	9 N.S.C. Data Sheet	9 MCA Data Sheet	9 AIA Bulletin	9 AIHA Hygienic Guide†
Acetaldehyde	−36	4.1	55	58	D-1,4,7	1,3,2	3	B,C	2 4 2		SD-43		9–55
Acetic acid (glacial)	109	5.4	16.0 (212 F)	57	D-7,4,1	2,3	Cor.	B	2 4 2	410	SD-41	C-52	9–72
Acetic anhydride	129	2.7	10.0	35	D-7,3,1	2,3	Cor.	B	2 2 2		SD-15		1–71
Acetone	0	2.6	12.8	44	D-1,4	1,3	3	D	1 3 0	398	SD-87	C-46	3–57
Acetonitrile	42	–	–	62	D-1,3	1,2,3	3	D	3 3 3				6–60
Acetylene	gas	2.5	81	–	D-1,5	1,3	–	A	1 4 3	494	SD-7		4–67
Acetylene Dichloride				–	D-1,3	1,3,2	3	–	3 0 1				
Acids, mixed				–	D-7,3,2	2,3,4	Cor.	C			SD-65	C-66	
Acrolein	<0	2.8	31	–	D-1,3,7	1,2,3	Cor.	C	3 3 3	436	SD-85	C-66	6–63
Acrylonitrile	32*	3.0	17.0	–	D-3,8,1	3,2,1	4	C	4 3 2		SD-31	C-65	3–57
Adipic acid				–	C-1,4	1,3	–	–	1 1 –	438		C-68	
Allyl alcohol	70	2.5	18.0	48	D-3,8,1	5,3,1	4	C	3 3 1				12–63
Allyl chloride	−25	3.3	11.1	40	D-1,3,7	1,3,2	Cor.	C	3 3 1			C-42	12–63
Aluminum (powder)		0.035 oz/cu ft	–	–	C-1	1	–	A	0 1 1				6–63
Aluminum chloride				–	W-7	2,4	3	B		435	SD-62		
Ammonia, anhydrous	gas	16	25	–	W-3,1,7	3,1,2	Cor.	C	3 1 0	251	SD-8		2–71
Ammonia, aqua				–	W-7,3	2,3	Cor.	C			SD-13		
Ammonium dichromate				–	W-1,2,6	1,2,3	Cor.	C	– – –		SD-45		
Ammonium nitrate				–	W-2	1,4	3	C	2 1 3	536, 604	A-10		
n-Amyl acetate	77	1.1	7.5	22	C-1,4,7	1,2,3	3	D	1 3 0	208			4–65
iso-Amyl alcohol	109	1.2	9.0 (212 F)	30	D-1,3	1,2,3	3	D	1 2 0				
n-Amyl alcohol	91	1.2	10.0 (212 F)	30	D-1,3	1,2,3	3	D	1 3 0				
Aniline	158	1.3	–	36	D-8,3	2,3,5	4	B	3 2 0	409	SD-17		12–55
Antimony		metal and sulfides flammable		–	C-7,6	3,2	–	A		408	SD-66		12–59
Antimony trichloride				–	D-7,4	2,3	5	B			SD-66		
Arsenic trioxide				–	D-4,7	3,2,5	5	B,C		499	SD-60		12–64
Arsine	gas			–	D-3	3	–	–		499			8–65

TABLE OF CHEMICAL HAZARDS (Continued)

Substance	Flash Point (deg F)	Flammable or Explosive Limits (% by volume) Lower	Upper	Vapor Volume (cu ft per gal)	Severity and Type of Hazard	Precautions to Take	Oral Toxicity	Action on Skin	NFPA Health	Flammability	Reactivity	N.S.C. Data Sheet	MCA Data Sheet	AIA Bulletin	AIHA Hygienic Guide†
	1	2		3	4	5	6	7	8					9	
Asphalt, cutback	<50			–	C-7	2	–	B	0	3	0	215, 582			
Benzene	12	1.3	7.1	37	D-1,4,8	1,3,5	4	C	2	3	0	308	SD-2	C-77	6–70
Benzoyl peroxide				–	D-2,1,7	5,1,4	3	C	1	4	4		SD-81		
Benzyl chloride	153	1.1	–	28	W-7,4	2,3	Cor.	C	2	2	0		SD-69		
Beryllium				–	D-6,7	5,3	–	B,C	4	1	1	562		C-1	12–64
Bleaching compounds				–	C-7,2	2,4	Cor.	C	2	1	0	343			
Boron compounds				–	D-1,3	5,1,3	–	C				508	SD-84	C-80	
Bromine		nonflammable		–	D-7,4,2	5,3	Cor.	C	4	0	1	313	SD-49	C-49	8–58
1,3-Butadiene	gas	2.0	11.5	–	D-1,3	1,3	–	C	2	4	2		SD-55		2–63
n-Butane	gas	1.9	8.5	–	D-1,4	1,3	–	–	1	4	0				
l-Butanol	84	1.4	11.2	36	W-7,4,1	3,2,1	3	C	1	3	0				9–55
l-Butene	gas	1.6	9.3	–	D-1,4	1,3	–	–	1	4	0				
n-Butyl acetate	72	1.7	7.6	25	C-1,4,7	123	3	D	1	3	0				12–62
n-Butylamine	10	1.7	9.8	33	W-1,4,8	1,2,3	Cor.	C	3	3	0				12–60
Butyllithium				–	D-1	5,1,2	–	–					SD-91		
Butyraldehyde	20	2.5	–	37	D-1,4,7	1,2,3	1	D	2	3	–		SD-78		
Cadmium				–	D-6	3	–	B				312			12–62
Calcium carbide				–	D-1,7,6	4,1,2	Cor.	C	1	4	2		SD-23		
Calcium oxide				–	D-7,6	2,3	Cor.	C	1	0	1	241			
Carbon dioxide		nonflammable		–	W-5,7	2,3	–	C				397			10–64
Carbon disulfide	−22	1.3	44	54	D-1,4,8	1,3,2	3	C	2	3	0	341	SD-12	C-40	12–56
Carbon monoxide	gas	12.5	74	–	D-4,1	3,1	–	A	2	4	0	415			8–65
Carbon tetrachloride		nonflammable		34	D-3,8	3,2,5	4	D				331	SD-3	C-6	12–61
Chlorates				–	W-2,7,6	4,2,1	4	C							
Chlorine				–	D-3,2,7	5,3,4	–	C	3	0	1	207	SD-80	C-43	
Chlorine dioxide		>10		–	D-2,3,1	5,4,3	–	C				525			6–58
Chlorobenzene	84	1.3	7.1	32	D-1,3	1,3,2	4	–	2	3	0			C-20	2–64
Chloroform		nonflammable		41	W-4,7	3,2	3	D					SD-89	C-16	12–65
Chlorosulfonic acid				–	D-2,7,3	5,4,3	–	C	3	0	2		SD-33		
Chromic acid				–	D-2,6,7	4,2,3	4	C	1	0	1		SD-44		6–56

TABLE OF CHEMICAL HAZARDS (Continued)

	1	2		3	4	5	6	7	8 NFPA			9 References			
Substance	Flash Point (deg F)	Flammable or Explosive Limits (% by volume) Lower	Upper	Vapor Volume (cu ft per gal)	Severity and Type of Hazard	Precautions to Take	Oral Toxicity	Action on Skin	Health	Flammability	Reactivity	N.S.C. Data Sheet	MCA Data Sheet	AIA Bulletin	AIHA Hygienic Guide†
o-Cresol	178	1.4 (300 F)	–	32	D-8,7,4	2,3	4	C	2	2	0		SD-48		10–58
p-Cresol	202	1.1 (300 F)	–	32	D-8,7,4	2,3	4	C	2	1	0		SD-48		10–58
Cumene	111	0.9	6.5	23	W-1,4	1,3	4	D	0	2	0				12–61
Cyanides				–	D-6,8	5,3,2	6	–	3	2	–		SD-30	C-5	
Cyanogen	gas	6	32	–	D-1,3	1,3,4	–	–	4	4	3				
Cyclohexane	–4	1.3	8.0	30	D-1,4,7	1,3	3	D	1	3	0		SD-68		10–63
Cyclohexanol	154			31	C-4,1,7	1,2,3	3	D	1	2	0				
Cyclohexanone	111	1.1	–	32	C-4,1,7	1,2,3	3	D	1	2	0				12–65
Cyclopropane	gas	2.4	10.4	–	D-1,3	1,3	–	--	1	4	0				
DDT				–	C-6,8	3,2	4	–				303			10–59
Decaborane	176	0.8 (approx.)	98 (approx.)		D-1,6	5,13			3	2	1	508	SD-84		
Diacetone alcohol	148	1.8	6.9	26	C-4,1,7	1,2,3	3	D	1	2	0				
1,2-Dibromoethane		nonflammable		38	W-3	2,3	3	D	3	–	–				
o-Dichlorobenzene	151	2.2	9.2	29	C-3,7,1	3,2,1	3	C	2	2	0		SD-54	C-21	6–64
p-Dichlorobenzene	150			29	C-3,7,1	3,2,1	3	C	2	2	0		SD-54		6–64
1,2-Dichloroethane	56	6.2	16.0	42	W-1,3	1,3	3	D	2	3	0	350	SD-18	C-15	8–65
1,1-Dichloroethylene	5*	5.6	11.4	42	W-1,3	1,3	3	D	2	4	2				
1,2-Dichloroethylene	43	9.7	12.8	43	D-1,3	1,3	3	D	2	3	2			C-14	
Dichloroethyl ether	131			28	W-1,3	1,3	3	D	2	2	0			C-19	
1,1-Dichloro-1-nitro-ethane	168*			32	W-1,3	1,3	4	D	–	2	3				
1,1-Dichloro-1-nitro-propane	151*			27	W-1,3	1,3	4	D	–	2	3				
1,2-Dichloropropane	60	3.4	14.5	33	W-1,3	1,3	4	D	2	3	0			C-18	
Diethylamine	<0	1.8	10.1	32	D-1,3,7	1,2,3	Cor.	C	3	3	0				6–60
Diethylene glycol	255			35	C-3,1	2,1	3	D	1	1	0				
Diethylene glycol mono-ethyl ether	201			24	C-3,1	2,1	3	D	1	1	0				

TABLE OF CHEMICAL HAZARDS (Continued)

Substance	Flash Point (deg F)	Flammable or Explosive Limits (% by volume) Lower	Upper	Vapor Volume (cu ft per gal)	Severity and Type of Hazard	Precautions to Take	Oral Toxicity	Action on Skin	Health Flammability Reactivity (NFPA)	N.S.C. Data Sheet	MCA Data Sheet	AIA Bulletin	AIHA Hygienic Guide†
Diethylene-triamine	215*			–	D-7,4	2,3,5	Cor.	C	3 1 0		SD-76		6–60
Diethyl sulfate	220			–	D-7,4,8	5,2,3	Cor.	C	3 1 1				
Diisobutyl ketone	140	0.8 (212 F)	6.2 (212 F)	–	W-3,1	3,1,2	–	D	1 2 0				12–62
Diisopropyl-amine	30*			–	D-1,4,7	1,2,3	Cor.		3 3 0				
Dimethylaniline	145			26	D-4,8,1	2,3	4	C	3 2 0			C-67	
Dimethyl ether	gas	3.4	18.0	–	D-1,4	1,3	–	D	– 4 0				
Dimethyl-formamide	136	2.2 (212 F)	15.2	–	D-3,1,8	3,1,2	Cor.	C	1 2 0				9–57
1,1-Dimethyl-hydrazine (unsym.)	5	2	95	–	D-1,3,8	5,1,3	Cor.	C	– – –				4–63
Dimethylsulfate	182			33	D-7,4,8	5,2,3	Cor.	C	4 2 0		SD-19	C-53	
1,2-Dinitro-benzene	302			30	D-8,6,1	3,2,1	4	B	3 1 4				
Dinitrotoluene				–	D-8,6,1	3,2,1	4	B	3 1 3		SD-93		
Dioxane (diethylene dioxide)	54	2.0	22	39	W-1,4,8	1,2,3	3	D	2 3 1				12–60
Epichlorohydrin	105			42	D-3,7,8	2,3,5	4	C	3 2 0				12–61
Epoxy resin systems				–	W-7,4	2,3,5	–	B,C		533		C-64	10–58
Ethanolamine	185			33	W-4,1	1,2,3	Cor.	C	2 2 0				6–68
Ethyl acetate	24	2.5	9.0	33	W-1,4,7	1,2,3	3	D	1 3 0		SD-51		4–64
Ethyl acrylate	60			–	W-1,4,8	1,2,3	–	C					
Ethyl alcohol	55	4.3	19.0	56	W-1,4	1,3	2	D	0 3 0	391			3–56
Ethylamine	<0*	3.5	14.0	50	D-1,4,7	1,2,3	Cor.	C	3 4 0				
Ethylbenzene	59	1.0	–	27	W-1,4	1,3	–	C	2 3 0				6–69
Ethyl bromide		6.75	11.3	–	W-1,4	1,3	–	C	2 3 0				4–65
Ethyl butyl ketone	115*			–	W-1,4,7	1,2,3	–	D	1 2 0				
Ethyl chloride	−58	3.8	15.4	46	D-1,4	1,3	–	D	2 4 0		SD-50		10–63
Ethyl ether	−49	1.9	48.0	31	D-1,3	1,3	3	D	2 4 –	396	SD-29	C-36	2–66
Ethyl formate	−4	2.7	13.5	41	D-1,3	1,2,3	3	C	2 3 0				
2-Ethyl hexanol	185*			21	W-1,4	1,2,3	–	D	2 2 0				
Ethyl mercaptan	<80	2.8	18.0	–	D-3,7,1	1,3,2	–	C	2 4 0				

TABLE OF CHEMICAL HAZARDS (Continued)

	1	2		3	4	5	6	7	8 NFPA		9 References		
Substance	Flash Point (deg F)	Flammable or Explosive Limits (% by volume) Lower	Upper	Vapor Volume (cu ft per gal)	Severity and Type of Hazard	Precautions to Take	Oral Toxicity	Action on Skin	Health Flammability Reactivity	N.S.C. Data Sheet	MCA Data Sheet	AIA Bulletin	AIHA Hygienic Guide†
Ethyl silicate	125*			–	W-1,4	1,2,3	–	–	2 2 0				12–68
Ethylene	gas	3.1	32	–	D-1	1,3	–	C	1 4 2				
Ethylene chlorohydrin	140*	4.9	15.9	49	D-3,8,1	3,1,5	4	D	3 2 0			C-22	12–61
Ethylene-diamine	110*			40	D-7,4,1	2,3,1	Cor.	C	3 2 0				2–70
Ethylene dichloride	56	6.2	16.0	42	W-1,4,7	1,2,3	3	D	2 3 0	350	SD-18	C-15	
Ethylene glycol	232	3.2	–	59	C-4	3	3	D	1 1 0				8–70
Ethylene imine		3.6	46		D-1,3,8	5,1,3	Cor.	–	3 3 3				2–65
Ethylene oxide	<0	3	100	–	D-1,3,7	1,3,2	–	C	2 4 3		SD-38	C-55	12–58
Ferrosilicon										352			
Fluorides				–	W-6,7	3,2,5	4	C		442			8–65
Fluorine				–	D-2,3	5,4,3	Cor.	C	4 0 3			C-61	12–65
Formaldehyde	gas	7.0	73	–	W-7,4	3,2	3	B,C	– 4 –	342	SD-1		4–65
Formic acid	156			–	W-7,4	3,2	Cor.	C	3 2 0				
Fuel oil No. 2	100			–	C-1,3	1,2	3	D	0 2 0				
Fuel oil No. 5	130			–	C-1,3	1,2	3	D	0 2 0				
Fuel oil No. 6	150			–	C-1,3	1,2	3	D	0 2 0				
Furfural	140	2.1	–	39	D-8,4,1	3,2,1	3	C	1 2 1				4–65
Furfuryl alcohol	167*	1.8	16.3	–	D-8,4,1	3,2,1	3	C	1 2 1				
Gasoline	–45	1.4	7.6	24 to 32	D-1,3	1,3	3	D	1 3 0				
Glycerine	320			45	C-4	3	1	A	1 1 0				
n-Heptane	25	1.2	6.7	22	D-1,4	1,3	–	D	1 3 0				2–59
n-Hexane	–7.0	1.1	7.5	25	D-1,4	1,3	–	D	1 3 0				2–59
sec-Hexyl acetate	113			–	C-1,4,7	1,2,3	3	D	1 2 0				
Hexylene glycol	215*			31	C-4	3	3	D	1 1 0				
Hydrazine	100	4.7	100	–	D-1,3,8	5,3,1	–	B,C	3 3 2			C-57	12–56
Hydrochloric acid				–	W-7,4	2,3	Cor.	C	3 0 0		SD-39	C-30	8–58
Hydrocyanic acid	0	6	41	–	D-3,1,8	3,1,2	6	C	4 4 2		SD-67		2–70
Hydrofluoric acid				–	D-3,7	5,2	Cor.	C	4 0 0	459	SD-25	C-59	3–56
Hydrogen	gas	4.0	75	–	D-1,5	1,3	–	A	0 4 –				

TABLE OF CHEMICAL HAZARDS (Continued)

	1	2		3	4	5	6	7	8	9			
		Flammable or Explosive Limits (% by volume)		Vapor Volume (cu ft per gal)					NFPA	References			
Substance	Flash Point (deg F)	Lower	Upper		Severity and Type of Hazard	Precautions to Take	Oral Toxicity	Action on Skin	Health Flammability Reactivity	N.S.C. Data Sheet	MCA Data Sheet	AIA Bulletin	AIHA Hygienic Guide†
Hydrogen peroxide (52%)	nonflammable			–	C-2,7	4,2	Cor.	C	2 0 3		SD-53		9–57
Hydrogen sulfide	gas	4.3	45	–	D-3,1	5,3,1	–	–	3 4 0	284	SD-36	C-38	2–63
Iodine				–	D-3,6,7	3,2	5	C		457			8–65
Isocyanates (TDI and MDI)	–	0.9	9.5	–	D-3,7	3,2,5	–	D	2 1 0	489	SD-73	C-70	2–67
Isophorone	205	0.8	3.8	22	W-3,7,1	3,2,1	–	C	2 1 0				
Isoprene	–65			–	D-1,3	5,1,3	–	–	2 4 1				
Kerosene	100	0.7	5	–	C-1,3	1,2,3	3	D	0 2 0				
Lead				–	W-6	3,5	4	A		443	SD-64		4–58
Lime (quick)				–	D-7,6	2,3	Cor.	C	1 0 1	241			
Lithium				–	C-1,7,6	5,1,3	3	A	1 1 2	566		C-81	
Lithium hydride				–	D-1,3	5,4,3	Cor.	C	1 4 2				8–64
LP-gas				–	D-1,4	1,3	–	–	1 4 0	479			
Magnesium		0.02 oz/cu ft		–	C-1,7	1,5	3	A	0 1 2	426		C-72	2–60
Maleic anhydride	215	1.4	7.1	–	W-7,4	2,3	–	C	3 1 0		SD-88	C-71	6–70
Manganese				–	C-6,7	3,2,5	3	A		306			6–63
Mercury				–	W-4,8	3,2,5	–	C		203		C-60	6–66
Mesityl oxide	87			28	D-1,3	1,2,3	–	–	3 3 0				10–69
Methane	gas	5.3	14.0	–	D-1,5	1,3	–	–	1 4 0				
Methyl acetate	14	3.1	16	41	D-1,4,7	1,2,3	–	D	1 3 0				6–64
Methyl acrylate	27	2.8	2.5	–	W-1,4,8	1,3,2	–	D	2 3 2		SD-79		
Methylal	0*			37	D-1,4	1,3	–	D	2 3 2				
Methyl alcohol	52	7.3	36	80	D-1,4,7	1,3,5	3	D	1 3 0	407	SD-22	C-24	12–57
Methylamine	gas	4.9	20.7	–	D-1,7,4	1,3,2	Cor.	C	3 4 0		SD-57		
Methyl amyl alcohol	106	1.0	5.5	–	D-1,4,7	1,3,5	–	D	2 2 0				
Methyl aniline	185			31	D-3,8	3,2,5	4	B	3 2 0		SD-82		
Methyl bromide	gas	10	16	–	D-3,7,8	3,2	4	D	3 1 0		SD-35	C-41	4–58
Methyl butyl ketone	95	1.2	8	27	D-1,3	1,2,3	3	D	2 3 0				
Methyl isobutyl ketone	73	1.4	7.5	–	D-1,3	1,2,3	3	D	2 3 0				
Methyl chloride	gas	10.7	17.4	–	W-1,4,7	1,3,2	–	D	2 4 0		SD-40		12–61

TABLE OF CHEMICAL HAZARDS (Continued)

Substance	1 Flash Point (deg F)	2 Flammable or Explosive Limits (% by volume) Lower	Upper	3 Vapor Volume (cu ft per gal)	4 Severity and Type of Hazard	5 Precautions to Take	6 Oral Toxicity	7 Action on Skin	8 NFPA Health	Flammability	Reactivity	9 References N.S.C. Data Sheet	MCA Data Sheet	AIA Bulletin	AIHA Hygienic Guide†
Methyl chloroform				32	W-4,7	3,2	3	D				456	SD-90	C-10	12–61
Methylcyclohexane		1.2	–	26	C-4,1,7	1,2,3	3	D	–	3	0				
o-Methylcyclohexanol				27	C-4,1,7	1,2,3	3	D	–	2	0				
Methylcyclohexanone				27	C-4,1,7	1,2,3	3	D	–	2	0				
Methylene chloride	non-flammable	15.5 (in oxygen)	66	51	C-4,7	3,5	3	D	2	0	0	474	SD-86	C-17	12–65
Methyl ethyl ketone	21	1.8	10	36	W-1,4,7	1,2,3	3	D	1	3	0		SD-83	C-51	3–57
Methyl formate	−2	5.9	20	53	D-1,3	1,2,3	3	–	2	4	0				
Methyl hydrazine	<80			–	D-1,3	1,2,3	Cor.	C	–	–	–				
Methyl mercaptan		3.9	21.8	–	D-1,3	1,3	–	–	2	4	0				
Methyl propyl ketone	45	1.5	8.0	31	D-1,3	1,2,3	3	D	2	3	0				
Methyl styrene	134*	0.7	–	–	C-4,1,7	3,2,1	–	D	2	2	0				
Mixed acid				–	D-7,3,2	2,3,4	Cor.	C					SD-65		
Molybdenum															2–60
Morpholine	100*			9	D-4,1	1,2,3	4	C	2	3	0				
Naphtha (coal tar)	100–110–			37	D-1,4	1,2,3	–	D	2	2	0				
Naphtha (petroleum)	<0	1.1	5.9	20	D-1,4	1,2,3	–	D	1	4	0				8–63
Naphthalene	174	0.9	5.9	–	W-4,7	3,2	–	B,C	2	2	0	370	SD-58		10–67
Naphthylamine				–	W-3,1	2,3	4	–	2	1	0		SD-32		
Nickel carbonyl				–	D-3,1	5,3,1	6	–							6–68
Nicotine		0.7	4.0	–	D-4,8	5,2,3	6	C	4	1	0				
Nitrate-nitrite salt baths				–	W-6,7	2,3	4	–				270		C-35	
Nitric acid				–	D-2,3,7	2,4,3	Cor.	C	2	0	1		SD-5	C-13	8–64
Nitric oxide				–	–							206			
p-Nitroaniline	390			–	D-8,3,1	2,3,1	5	B,C	3	1	1		SD-94		
Nitrobenzene	190	1.8 (200 F)	–	32	D-8,3,1	2,3,1	5	B,C	3	2	0		SD-21	C-33	2–59
o-Nitrochlorobenzene	261			–	D-3,1	2,3,1	5	–	3	1	–				
Nitroethane	82	3.4	–	46	W-1,4,7	5,1,3	3	C	1	3	3				12–61
Nitrogen dioxide				–	D-2,3,7	5,4,3	–	C	–	–	–	206		C-34	6–56

TABLE OF CHEMICAL HAZARDS (Continued)

Substance	Flash Point (deg F)	Flammable or Explosive Limits (% by volume) Lower	Upper	Vapor Volume (cu ft per gal)	Severity and Type of Hazard	Precautions to Take	Oral Toxicity	Action on Skin	NFPA Health	Flammability	Reactivity	N.S.C. Data Sheet	MCA Data Sheet	AIA Bulletin	AIHA Hygienic Guide†
Nitrogen (liquid)				–	D-5,7	3,2	–	C	–	–	–				
Nitromethane	95	7.3	–	61	W-1,4,7	5,1,3	3	C	1	3	4				12–61
1-Nitropropane	120*	2.6	–	37	W-1,4,7	5,1,3	3	C	1	2	3				6–60
2-Nitropropane	103*	2.6	–	36	W-1,4,7	5,1,3	3	C	1	2	3				6–60
Nitrotoluene	223			–	D-3,1	2,3,1	5	C	–	1	3				
Octane	56	1.0	3.2	20	D-1,3	1,3	3	D	0	3	0				
Oils (cutting), emulsions, and drawing compounds				–	–	2	3	D				501			
Oleum				–	D-7,3	2,3	Cor.	C	3	0	1	210			
Oxalic acid				–	W-6,7	2,3	4	C				406			
Oxygen (gas)	gas			–	D-2	4	–	A				472, 360			
Oxygen (liquid)				–	D-2,7	5,4	–	C				283			
Ozone				–	D-3,2	3,4	Cor.	–						C-63	4–66
Paraformaldehyde	158			–	W-7,4	2,3	4	C	2	2	1	342	SD-6		
Parathion				–	D-3,1	5,3,2	6	–	4	1	0				6–69
Pentaborane	spontaneous	0.8 (approx.)	98 (approx.)	–	D-1,3	5,1,3	–	–	–	–	–	508	SD-84		6–66
Pentachlorophenol				–	W-6,7	3,2	4	C						C-32	8–70
n-Pentane	<−40	1.5	7.8	29	D-1,4	1,3	–	D	1	4	0				4–66
Perchloric acid				–	D-2,7	4,2,5	3	C	3	0	3	311	SD-11	C-44	
Phenol	175			–	D-8,7,4	2,3,5	4	C	3	2	0	405	SD-4		12–57
Phenylhydrazine	192			–	D-3,7,1	5,3,1	–	B,C	3	2	0				
Phosgene				–	D-3,7	3,2,5	–	C					SD-95		6–68
Phosphoric acid					C-7,4	2,3	Cor.	C					SD-70		6–57
Phosphoric anhydride					C-7,6	2,3	Cor.	C					SD-28		6–58
Phosphorus (red)					C-1	1	–	A	0	1	1		SD-16		
Phosphorus (white)					D-6,1,7	5,2,3	6	C	3	3	1	282	SD-16		
Phosphorus oxychloride					D-7,3	2,3,5	Cor.	C					SD-26		
Phosphorus pentasulfide					W-6,7,2	5,4,3	–	C	3	1	2		SD-71		
Phosphorus trichloride		nonflammable			D-7,3,1	5,4,3	Cor.	C	3	0	2		SD-27		

TABLE OF CHEMICAL HAZARDS (Continued)

	1	2		3	4	5	6	7	8			9		
		Flammable or Explosive Limits (% by volume)		Vapor Volume (cu ft per gal)	Severity and Type of Hazard	Precautions to Take	Oral Toxicity	Action on Skin	NFPA Health Flammability Reactivity		N.S.C. Data Sheet	MCA Data Sheet	AIA Bulletin	AIHA Hygienic Guide†
Substance	Flash Point (deg F)	Lower	Upper											
Phthalic anhydride	305	1.7	10.5	–	C-7,6	2,3	Cor.	B	2 1 0			SD-61		8–67
Picric acid	explodes				D-1,2,6	1,4,3	5	B,C	2 4 4		351			
Potassium hydroxide				–	D-7,6	2,3	Cor.	C	3 0 1			SD-10		
Propionaldehyde	15–19*	3.7	16.1	–	D-1,3	1,3,2	3	D	2 3 1					
Propionic acid	130			–	C-1,3	1,2,3	Cor.	C	2 2 0					
n-Propyl acetate	58	2.0	8	28	W-1,3	1,3	3	D	1 3 0					
iso-Propyl acetate	40	1.8	8	28	W-1,3	1,3	3	D	1 3 0					
n-Propyl alcohol	77	2.1	13.5	44	W-1,3	1,3	3	D	1 3 0					12–61
iso-Propyl alcohol	53	2.0	12	43	W-1,3	1,3	3	D	1 3 0					12–61
Propylene	gas	2.0	11.1	–	D-1,4	1,3	–	A	1 4 1			SD-59		
Propylene dichloride	60	3.4	14.5	33	W-1,3	1,3	4	D	2 3 0				C-18	6–67
Propylene glycol	210	2.6	12.5	45	C-4	3	3	D	0 1 0					
Propylene oxide	−35	2.1	21.5	–	D-1,3	1,3	Cor.	C	2 4 2					6–59
Propyne	gas	1.7	–	–	D-1,5	1,3	–	A	2 4 2					
Pyridine	68	1.8	12.4	41	W-1,3,7	1,2,3	3	B,D	2 3 0		310			8–63
Selenium				–	W-7,6	3,2,5	–	C			578		C-56	6–59
Silica or silicates				–	C-6	3,5	1	A			531			4–58, 9–57
Silver nitrate				–	C-2,3	4,2,3	4	C	1 0 1					9–57
Sodium				–	D-1,7	5,4,2	Cor.	C	3 1 0		231	SD-47		
Sodium chlorate				–	W-2,7	5,4,2	4	C	1 0 2		371	SD-42	C-62	2–58
Sodium cyanide				–	D-6,7	5,3,2	6	–				SD-30		
Sodium dichromate				–	W-6,7,2	3,2,5	4	C				SD-46		
Sodium hydroxide				–	D-7,6	2,3	Cor.	C	3 0 1		214, 373	SD-9		
Stoddard solvent	105	0.8	5.0	20	C-1,4	1,3	3	D	0 2 0					
Styrene monomer	90	1.1	6.1	28	C-4,1,7	3,2,1	–	D	2 3 –			SD-37	C-48	10–68
Sulfur				–	C-6,7,1	3,2,1	3	A	2 1 0		612	SD-74		
Sulfur chlorides	245*			–	W-7,3	3,4,5	–	C	2 1 0			SD-77		
Sulfur dioxide				–	D-3	3,5	–	C	3 0 0			SD-52	C-39	12–55
Sulfuric acid				–	D-7,3,2	3,4	Cor.	C	3 0 1		325	SD-20	C-12	9–57

TABLE OF CHEMICAL HAZARDS (Concluded)

	1	2		3	4	5	6	7	8			9			
		Flammable or Explosive Limits (% by volume)		Vapor Volume (cu ft per gal)					NFPA			References			
Substance	Flash Point (deg F)	Lower	Upper		Severity and Type of Hazard	Precautions to Take	Oral Toxicity	Action on Skin	Health	Flammability	Reactivity	N.S.C. Data Sheet	MCA Data Sheet	AIA Bulletin	AIHA Hygienic Guide†
1,1,2,2,-Tetrachloroethane		nonflammable		31	D-8,3	5,3,2	4	D					SD-34	C-9	
Tetrachloroethylene	–	nonflammable		31	W-4,7	3,2	3	D					SD-24	C-11	12–65
Tetraethyl lead	185*			–	D-3,1,8	5,3,1	6	–	3	2	3				8–63
Tetrahydrofuran	6	2	11.8	40	D-1,3	1,2,3	4	D	2	3	1				6–59
Tetralin	160	0.8 (212 F)	5.0 (212 F)	24	D-3,1	3,2,1	3	D	1	2	0				
Tetramethyl lead	100*			–	D-3,1,8	5,3,1	6	–	3	3	3				
Titanium dust		0.045 oz/cu ft		–	C-6	3	1	A				485		C-75	6–59
Toluene	40	1.2	7.1	31	W-1,4,7	1,2,3	4	C	2	3	0	204	SD-63	C-77	6–57
Toluidine	185			–	D-3,8	3,2,5	4	B	3	2	0		SD-82		
Trichloroethylene	99‡ (practically	12.5‡ nonflammable)	90‡	36	W-4,7	3,2	–	D	–	–	–	389	SD-14	C-8	2–64
Trinitrotoluene				–	D-1,3,8	5,1,3	5	–	2	4	4	314			10–64
Turpentine	95	0.8	–	18	W-4,1,7	3,1,2	3	B,C	1	3	0	367		C-54	6–67
Vinyl acetate	18	2.6	13.4	35	D-1,4,7	1,3,2	–	C	2	3	2		SD-75		
Vinyl chloride	gas	4	22	–	D-1,4	1,3	–	C	2	4	2				8–64
Vinylidene chloride monomer	5*	5.6	11.4	–	W-1,4,7	1,3,2	–	C	2	4	2			C-14	
o-Xylene	90	1.0	6.0	27	W-1,4,7	1,3,2	4	D	2	3	0	204		C-77	10–71
m-Xylene	84	1.1	7.0	27	W-1,4,7	1,3,2	4	D	2	3	0	204		C-77	10–71
p-Xylene	81	1.1	7.0	27	W-1,4,7	1,3,2	4	D	2	3	0	204		C-77	10–71
Zinc		0.48 oz/cu ft	–	–	W-1,6	1,3,5	3	A	0	1	1	267			8–69 (oxide)
Zirconium		0.19 oz/cu ft	–	–	D-1,6	1,3,5	–	–	1	4	1	382	SD-92	C-74	10–58

TOXICITY TESTING PROCEDURES*

I. Introduction

In order to provide the employee with a safe work environment, the consumer with safe products, and satisfy certain federal regulations designed to promote these concepts, it is paramount for industry to recognize and appreciate the degree of toxicity associated with its starting materials, its intermediates, and its products. Essentially all things can be toxic and it is basically the amount of material (absorbed) by an individual which determines the toxicity of the substance.

For example, everyone is exposed to lead and takes in small amounts of this element daily without detectable injury. However, individuals in circumstances where there is a much higher than normal exposure, and consequently intake, may develop signs and symptoms of lead poisoning.

It is the function of the toxicologist and his staff to describe qualitatively and quantitatively the toxicity of substances and recommend procedures necessary for the safe use of those substances.

The purpose of this compendium is to briefly describe the various types of toxicity testing available at Haskell Laboratory for Toxicology and Industrial Medicine and provide basic information concerning toxicology which may be of assistance in selecting a useful approach for recognizing and handling potential toxicological problems. In essence, it is hoped that this will serve as an everyman's guide to toxicity testing at Haskell.

Before selecting an approach, familiarity with some of the basic concepts of industrial toxicology, described below, is essential.

Toxicity—The ability of a chemical to produce injury in a living organism following exposure to that substance. It is the intent of toxicity testing to determine the type of injury produced, the level(s) of exposure which produce the injury and the biological malfunction(s) responsible for that injury.

Hazard—The likelihood that a toxic effect will occur under the conditions of use for a particular substance. This concept implies that the degree of toxicity of a material is already known, so that meaningful procedures can be recommended and implemented for the safe handling of the material in the industrial and consumer environments.

Specific toxicity tests are needed to critically evaluate the toxicity and potential hazard of a substance. Unfortunately, no single test or set of standardized tests will provide all necessary toxicity data. Toxicity testing for each substance requires careful planning, so that the data produced can be transplanted into meaningful results, which then form the basis of guidelines for handling, labeling, use and disposal of the substance.

In designing a test, there are two major considerations: (1) the probable routes of exposure, and (2) the duration of exposure.

Routes of exposure include: skin or eye contact, ingestion, inhalation and occasionally injection (parenteral) or any combination of these. The toxicity of many substances is known to vary, depending upon how the material enters the body. For example, inhalation of silica dust can product a disabling lung disease, but skin contact or ingestion poses little or no problem.

In the industrial environment, skin and eye conditions constitute approximately 70% of all instances of occupational illness. Skin contact with chemicals may elicit primary irritation, contact sensitization, ulceration, photosensitization acne, pigmentary changes, nodules, vesicles, and tumors. In addition, the large surface area of the skin provides an opportunity for a substance to be absorbed and thus produce systemic toxicity.

Eye contact often results from an accidental spill or splash. Eye exposure may result in simple reversible irritation or irreversible scarring, cataracts, as well as systemic toxicity from ocular absorption.

A single exposure by the oral route is of major importance in the consumer environment as illustrated by the incidence of accidental poisonings. Repeated low-level exposure is the prime mode in industry and often results from inadvertent food contamination or the swallowing of inhaled material trapped in the upper respiratory tract. Finally, the oral route is very important in the food industry, particularly with food additives and pesticide residues.

Exposure to toxic substances by inhalation constitutes both an important type of industrial exposure for the worker and an important type of community air exposure for the general public. Toxicity of inhaled materials depends upon a number of important factors: airborne concentration, particle size of airborne material, length and frequency of exposure and the physical-chemical nature of

---

* Toxicity Testing at Haskell Laboratory, E. P. DuPont de Nemours & Co. A presentation to the House of Representatives Committee on Interstate and Foreign Commerce, July 1975.

the test substance. The physical properties are the most crucial because they determine how much of the airborne material is deposited in and absorbed by the body.

Materials may exist in the air as gases (vapors), as aerosols i.e., (dusts, mists, fogs or fumes) or as a combination of these. Gases usually are rapidly absorbed and often produce direct toxic effects upon the tissues. Aerosols may behave differently; the expression of their toxicity depends primarily upon the size of the inhaled particles. Particles greater than approximately 5 μm in diameter fail to reach the alveoli, because they are efficiently removed by the filtering action of the upper respiratory tract. Deposited material is transported to the pharynx by the muco-ciliary escalator system where it is either expectorated or swallowed. Particles between approximately 5 to 0.01 μm in diameter are capable of depositing deep within the lung and thus can exert systematic toxicity by absorption or produce localized pulmonary effects if they remain in the alveoli.

The duration of exposure is the second major consideration in designing an experiment. For convenience, studies are labeled as acute, prolonged or chronic.

Acute studies signify a single exposure to the test substance. It is the first determination of a substance's toxicity. Results usually are expressed as a simple all or none response—death or no death over a broad range of dose levels and recorded as mg substance per kg body weight. In addition, any signs and symptoms of toxicity are recorded and evaluated. Knowledge of acute toxicity is valuable in cases of accidental poisoning and in planning for long-term toxicity studies.

Prolonged studies consist of repeated daily exposure to a substance and last up to and including 90 days. They are often referred to as subacute studies because the dosage used is generally a fraction of that found to produce lethality (from an acute study). These tests simulate a common type of human exposure. A prolonged study is intended to determine cumulative toxic effects from low level exposure. They are often used to detect the teratogenic potential of a substance. Results are expressed as graded changes of biological parameters, rather than the all or none response of acutes.

Chronic studies consist of repeated daily exposure to a substance for more than 90 days and can last the lifetime of the experimental animal. They often are one or two years in length. These studies simulate human lifetime or work-lifetime exposures to a substance. Chronic studies are designed to obtain a nondetectable toxic effect level, a detectable toxic effect level and a safe level of exposure. They are often used for detecting the carcinogenic potential of a substance. Results are expressed similarly as those for a prolonged study but often include additional biological measurements.

Finally, additional variables such as animal species, number of animals, dose levels, biological parameters to be monitored and other factors not mentioned here are discussed within the appropriate tests.

II. Acute Studies

This section outlines the acute toxicity tests performed at Haskell. Acute tests are designed as the initial study for gathering information on the toxicity of a substance. Results from acute studies form the basis for determining dose levels for prolonged and chronic studies. Furthermore, these investigations simulate single human exposures to substances by the dermal, oral, or inhalation route.

The acute studies constitute the largest variety of tests available, so that in selecting a test or tests the use of each test should be known. In particular, there are only slight differences in Federal Hazardous Substances Act (FHSA) tests and Class B poison tests, but these differences are important and should be appreciated.

FHSA tests are special toxicity tests required by the federal government to determine whether a substance is a hazardous substance—"Any substance or mixture of substances which is toxic, corrosive, an irritant, a strong sensitizer, flammable or combustible, or generates pressure through decomposition, heat, or other means, if such substance or mixture of substances may cause substantial personal injury or substantial illness during or as a proximate result of any customary or reasonably foreseeable handling or use, including reasonably foreseeable ingestion by children." Code of Federal Regulations, Title 16, Commercial Practices, Part 1500.3, January 1, 1974, U.S.G.P.O., Washington, D.C. Pesticides, foods, drugs, cosmetics, and radioactive material subject to other federal regulations are not included under this Act.

Class B poison tests are special toxicity tests required by the federal government for determining proper labeling and packaging for the shipping of chemicals in interstate commerce. R. M. Graziano's Tariff No. 25, April 24, 1972, in the Federal Register, Vol. 38, No. 28, Section 173.343, February 12, 1973.

Skin and Eye—Primary Irritation

The intent of these tests is to determine whether a material will cause injury to the skin when placed in direct contact with it. Primary irritation is a localized reaction of the skin to the test substance and ranges from minor redness (erythema) through swelling (edema) to corrosive damage indicated by cell death (necrosis). There are currently five types of primary irritation tests available; two are of Haskell design and the other three are procedures from federal regulations: the Federal Hazardous Substance Act test (FHSA test) and the Department of Transportation Class B poison and Skin Corrosion (DOT) tests. Each of the above tests are designed for a specific purpose which is described along with the appropriate test.

Skin Sensitization

Another type of skin response commonly experienced is of immunological origin and is termed contact or dermal hypersensitivity, allergic dermatitis or skin sensitization. This reaction is similar to that of any irritation response but generally requires only minute amounts of material to trigger a response in the sensitized individual. Usually repeated exposures to a substance are necessary to produce sensitization. Sensitizers may or may not be primary irritants so that testing for both responses with one test allows these important differences to be determined.

A Skin & Eye

1. Primary Skin Irritation and Sensitization (PI+S) Haskell—This is somewhat of an all inclusive type of test which consists of two parts—the first, irritation; the second, sensitization. Guinea pigs are the best available animal for this test, because they closely parallel the human response.

Part 1—Primary Irritation.—The test material is prepared as a solution in a suitable solvent or as a solution/suspension if not completely soluble. The substance is applied topically at various concentrations to the shaved intact skin of albino guinea pigs and lightly rubbed to effect intimate contact. No covering is applied. The reaction site is evaluated for evidence of erythema, edema and necrosis at 24 and 48 hours after contact. In addition, the minimal concentration required to produce irritation is noted.

Part II—Sensitization.—The guinea pigs from Part I are given a total of four intradermal injections of a 1% solution of the test substance over a three-week period in order to induce dermal sensitivity. Two weeks after the final injection these guinea pigs, as well as controls of similar age, are treated topically as in Part I. Sensitization reactions are determined and graded by visual inspection of the challenged site 24 and 48 hours post treatment.

On occasion two weeks after these guinea pigs have demonstrated contact sensitization, a second challenge may be applied to determine whether a structurally similar chemical will cross-react, or whether the same chemical, but now incorporated into the composition of some substrate or product, is capable of producing a sensitization reaction. Thus, although a chemical itself may cause sensitization, it may be bound within the substrate or product and therefore be unavailable to further react or be at such a low concentration it will not initiate a sensitization response when tested.

2. Human Studies—Primary Skin Irritation and Sensitization—After being assured that a material does not produce notable skin irritation and/or sensitization through animal experiments, tests on human skin may be instituted. These tests are mainly used to test fabrics or other materials which are likely to be in repeated contact with human skin during normal use. A panel of human subjects representing a wide variety of adult human skin of both sexes is tested as described below to estimate the maximum percentage of humans responding with irritation and/or sensitization to the test substance. Tests may be run on a 20 subject and a 200 subject volunteer panel. The 20 subject panel is used as a guide and pilot for the 200 subject panel or for nonapparel industrial use. The 200 subject panel is used as the estimate to measure the potential response expected in a large population.

Both tests utilize the same procedure. The test material is placed in contact with the skin of the upper arm or thigh and secured under a taped occluded patch for 6 days. This tests for irritation. Two weeks later, a second patch is applied for 48 hours. This tests for irritation. Two weeks later, a second patch is applied for 48 hours. This tests for sensitization. The test site is evaluated at 2 and 6 days post-contact and 48 hours after applying the challenge patch. The guidelines for human experimentation set forth by the Helsinki code are followed throughout these studies.

3. Human Patch Test—Special—This special test is used to test for sensitization or cross-reactivity of materials in persons already known to be sensitized to the particular substance. The number of individuals on these panels is small, generally one to 15 persons. The test procedure consists of placing the test material on the skin under an occluded patch and evaluating the test site 48 hours later when the patch is removed and again 24 hours after removal of the patch.

Eye—The danger of eye injury from an inadvertent spill or splash is ever present in the industrial and consumer environment, especially when there is improper handling of materials and/or failure to wear adequate eye protection. Liquids are most often involved in these instances followed to a lesser extent by dust, mists, vapors, and gases. Besides producing a direct toxic action upon the eye itself, the eye remains a potential portal of entry into the body for a substance, so that systemic poisoning from eye exposure does occur and requires monitoring.

Haskell currently has two types of eye tests available, the Haskell Eye Irritation Test and the FHSA Eye Irritation Test.

4. Haskell Eye Irritation Test—This is a general type of eye toxicity test. The test material is administered, as supplied, to the test animal. Usually, 0.1 ml of a liquid or 100 milligrams of the solid, finely divided, is placed into the lower conjunctival sac (eye lid) of one eye of a rabbit; the other eye serves as an untreated control. A second rabbit is treated similarly, but after 20 seconds of exposure the test eye is washed with tap water to stimulate actual first aid conditions. In the case of cosmetic ingredients the material remains in eye contact for 4 seconds before it is washed out.

Treated and control eyes are examined at one and four hours and at 1, 2, 3, 7, and 14 days post-exposure. A thorough ophthalmologic examination is performed using an ophthalmoscope as well as a biomicroscope aided by examination with fluorescein dye which stains for corneal and conjunctival lesions. Injury to the cornea, iris and conjunctiva is graded according to a standard scoring technique.

5. FHSA—Eye Irritation Test—This study is a shortened, slightly modified form of the Haskell test and is mainly intended for evaluating materials destined for household rather than industrial use. The complete test procedure and scoring system requirements are found in Code of Federal Regulations, Title 16, Commerical Practices, Part 1500.42, Jan. 1, 1974, U.S.G.P.O., Washington, D.C.

The eyes are treated exactly the same as in the Haskell protocol, but evaluation is made at 24, 48, and 72 hours after exposure with an ophthalmoscope and injury is scored according to a defined system. Substances rated as irritants must be properly labeled. Six rabbits are used.

6. FHSA—Skin Irritation Test—This test is designed for evaluating primary irritation of substances intended for human use and is used for labeling information of these materials. The test is listed in the Code of Federal Regulations, Title 16, Commercial Practices, Part 1500.42, Jan. 1, 1974, U.S.G.P.O., Washington, D.C.

The test material, 0.5 gms of solid or 0.5 ml liquid, is applied to a 1 inch square area of both intact and abraded skin of the clipped back of rabbits for 24 hours under semiocclusion. The reaction sites are graded at 24 and 72 hours post-exposure according to a defined scoring system.

A material is adjudged either irritating or non-irritating on the basis of the score.

7. Department of Transportation Skin Corrosion Test (DOT Test)—This test is designed to determine whether a substance produces a corrosive effect upon the skin under the condition listed below. It is primarily aimed at the shipping requirements for classifying, labeling and packaging these corrosive materials. A complete description of the test is found in R. M. Graziano's Tariff No. 25,

April, 1972 and in the Federal Register, Vol. 38, No. 28, Section 173.245, Feb. 12, 1973. Corrosive action is defined in this procedure as tissue destruction or irreversible alteration, i.e., ulceration or necrosis of the skin.

The test substance is applied to two sites on intact shaved skin of albino rabbits for four hours under an impervious wrap. After exposure the wrap is removed and the reaction sites are evaluated. The site is washed free of test material and subsequently readings are made at 24 and 48 hours.

Besides producing localized irritation, chemicals may penetrate the skin, become absorbed, and thus have the potential of producing systemic toxicity. Since skin contact is one of the most common types of industrial exposure, knowledge concerning the toxicity of a substance, which in normal use includes skin contact, is vital for safe handling instructions for the material.

Acute skin absorption testing provides the necessary means for determining the toxicity of a material mediated by skin absorption. Two measures of acute toxicity by skin absorption are commonly employed: (1) approximate lethal dose (ALD), a range finding study prerequisite for determining the more meaningful and accurate (2) median lethal dose (LD_{50}). Rabbits are the species of choice for these two studies. In addition, there are federal protocols for FHSA and Class B poison classification. Animals may be sacrificed for microscopic tissue examination in any of these studies in order to evaluate the toxic effects in detail.

8. Approximate Lethal Dose—Skin Absorption (ALD-Skin Absorption)—One rabbit is used for each of six dose levels, milligrams of substance per kilogram of body weight (mg/kg), and each dose level differs in concentration by a constant factor from the preceding one. The test material is placed on the shaved back skin of the animal under an occlusive wrap for 24 hours. After exposure the wrap is removed from the surviving animals and the material is washed off. Mortality and body weight are recorded for 14 days post-exposure.

Animals are dosed at various levels until such dose levels are found that at two consecutive levels the animals survive, but at the next two higher levels the the animals die.

9. Median Lethal Dose (LD_{50}) Skin Absorption—A preliminary study often precedes the LC_{50} in order to determine the range of dosing to be used in the LD_{50}. In the LD_{50}, there are usually at least six, but not more than 10 rabbits per dose level and 3 or 4 dose levels. The material is applied as in the ALD. Fourteen day post-exposure mortality and body weight of survivors is recorded. Surviving animals may be sacrificed for histopathologic examination in order to locate any specific systemic effects. LD_{50} is calculated from the mortality data.

10. FHSA—Acute Dermal Toxicity—This study is designed to test acute dermal toxicity of substances which are destined for human use and is described fully in the Code of Federal Regulation, Title 16, Commercial Practices, Parts 1500.3 and 1500.40, Jan. 1, 1974, U.S.G.P.O., Washington, D.C.

Ten rabbits weighing between 2.3 and 3.0 kilograms are prepared for testing by clipping the skin free of hair around the trunk; and half of these animals are further prepared by abrading the skin. The test materials are placed in continuous contact with the skin by use of an impervious sleeve for 24 hours at dose levels of 200 mg/kg or 2000 mg/kg. At the end of 24 hours the sleeves are removed, and the excess material wiped off. Fourteen-day post-exposure mortality is recorded. If the amount of material produced death in half or more of the treated animals within this period, the substance is classified as either highly toxic or toxic respectively under the FHSA. If lethal in less than half of the 2000 mg/kg dose group, the substance is classified as non-toxic.

11. Class B Poison Test—This test from the Department of Transportation, R. M. Graziano's Tariff No. 25, April 24, 1972 in the Federal Register, Vol. 28, Section 173.343, Feb. 12, 1973, is designed for determining whether a substance is a class B poison by skin absorption. Class B poisons are defined as those substances which produce death by skin absorption within 48 hours after exposure in half or more of a group of 10 or more rabbits when listed according to the following procedure: The undiluted test material is applied under an impervious wrap to the clipped skin of albino rabbits at a dose level of 200 mg/kg or less for 24 hours, after which time the wrap is removed and mortality is noted for another 14 days.

B. Oral

Acute oral toxicity is of major importance in accidental ingestion of chemicals, and in the development of basic toxicity data for designing prolonged and chronic exposure studies. The generation of acute oral toxicity data is obtained by performing the Approximate Lethal Dose test (ALD) and/or the oral LD_{50} test, the FHSA test and the Class B poison test. These tests may be performed with fasted or non-fasted animals depending upon the nature of the study.

1. Approximate Lethal Dose (ALD)— Oral—This is a single dose exposure designed to determine the ALD which is an end point in itself or is used as a range finding study for determining the more accurate oral LD_{50}. Animals, generally albino rats, are dosed at various levels until a series of consecutive doses, differing from one another by a factor of 1.5, is arrived at wherein death oc-

curs at least at three higher levels and survival at least at three lower levels.

The test material is administered as a suspension or a solution via a catheter to insure that the whole amount of the dose is introduced into stomach. After dosing, the animals are checked periodically for the development of toxic signs, weighed daily and mortality is recorded up to 14 days post-exposure. Histopathologic examination of tissues may be elected in order to determine the cause of death or observe any tissue changes.

2. Median Lethal Dose (LD_{50})—Oral—This is a statistically derived value obtained from experimental data. It is the amount of orally administered material, in mg/kg, which is calculated to produce death in 50% of a given animal population. Two techniques can be used to determine the oral LD_{50}. In one, 10 rats per dose level are used with a minimum of three dose levels that produce fractional mortalities. The LD_{50} is determined from the 14-day post-exposure mortality data and is generally stated with its 95% confidence limits.

In the other, five rats per dose level are used with the dose level differing from one another by a factor of two. The LD_{50} is again determined from the 14-day post-exposure mortality data.

3. FHSA—Oral—The test originates from the federal protocol found in the Code of Federal Regulation, Title 16, Commercial Practices, Part 1500.3, Jan. 1, 1974, U.S.G.P.O., Washington, D.C. Procedurally, 10 albino rats, weighing between 200 to 300 grams are given the test material orally at a dose level of 50 mg/kg, 500 mg/kg or 5000 mg/kg. Fourteen-day post-exposure mortality is recorded. If the amount of material produces death in half or more of the treated animals within this period, the substance is classified as either highly toxic, toxic or non-toxic, respectively, by FHSA.

4. Class B Poison Test—Oral—This test originates from R. M. Graziano's Tariff No. 25, April 24, 1972 in the Federal Register, Vol. 38, No. 28, Section 173,343, Feb. 12, 1973. Class B poisons are those chemicals which when administered orally, at doses of 50 mg/kg or less produce death within 48 hours post-exposure in half or more than half of a group of 10 or more albino rats weighing between 200-300 gms.

C. Inhalation

Acute inhalation studies are designed to determine the effects of a single, short-term relatively high concentration exposure. This toxicity data is used for selecting atmospheric levels to be used in prolonged or chronic studies and for estimating acute toxicities relative to known substances. Methods for generating aerosols must be tailored to fit each test material, because a simple standard technique does not exist which is applicable for all materials.

Currently there are five acute inhalation toxicity tests available at Haskell and they are described below. A discussion of methodologies for producing inhalable substances is beyond the scope of this report and therefore is not included.

1. Approximate Lethal Concentration (ALC)— The ALC is the lowest atmospheric concentration in mg/m^3 for solids and liquids or ppm (vol/vol) for gases or vapors which produces death in the animals exposed for a specified time. Generally four or five groups, six albino rats per group, are exposed to each of four or five graded atmospheric concentrations of the material for a single 4-hour period. Atmospheric concentrations may be measured. Mortality is recorded up to 14 days post-exposure. This test may be used as an end point or as a range finding test for the LC$_{50}$ determination.

2. Lethal Concentration—50% Mortality (LC$_{50}$)—This is exactly like the LD$_{50}$ except that the LC$_{50}$ is expressed in terms of an atmospheric concentration of the material which is calculated to be lethal to 50% of the test animals, rather than in a dose per animal. In this study five groups, 6-10 albino rats per group, are exposed to five graded airborne concentrations of test material. Mortality is recorded up to 14 days post-exposure and the LC$_{50}$ is statistically computed from these results.

*3. FHSA—Inhalation—*This test is from the Code of Federal Regulations, Title 16, Commercial Practices, Part 1500.3, Jan. 1, 1974. Experimentally, 10 albino rats are exposed to an atmospheric concentration of 200 or 20,000 ppm by volume of gas or vapor, or 2 mg/liter or 200 mg/liter of dust or mist for a single 1-hour exposure period. If half or more of the exposed animals die within 14 days post-exposure, the substance is classified as highly toxic or toxic respectively by FHSA. If less than half die at the higher concentration, it is classified as non-toxic.

*4. Class B Poison Test—Inhalation—*This test, from the Department of Transportation, R. M. Graziano's Tariff No. 25, April 24, 1972, in the Federal Register Vol. 38, No. 28, Section 173.343, Feb. 12, 1973 is designed to provide guidance in labeling and interstate shipping of chemicals. Class B poisons are classified as those substances which produce death by inhalation within 48 hours after exposure in half or more of a group of 10 or more albino rats when tested according to the following procedure. Rats, weighing between 200 to 300 gms, are exposed to an atmospheric concentration of 2.0 mg/liter or less of an aerosol or dust or 200 ppm of a gas or vapor for a 1-hour or less exposure period.

5. Approximate Lethal Temperature (ALT)— This test is used to determine the temperature (range 150-450°C) to which a material must be heated in order to produce death in one or more albino rats exposed to an atmosphere containing the original volatilized material and/or its pyrolysis products. This information is interpreted in terms of a thermal safety margin between the normal operating temperature and the temperature at which acutely important concentrations of toxic materials are produced. In this study the test substance is heated in a furnace and the gaseous products produced are passed into the exposure chamber containing the animals. Analysis of the exposure chamber atmosphere is useful in identifying the pyrolysis products which may be responsible for the observed toxicity. Generally, six animals per group are used, and each group is exposed for 4 hours to an atmosphere at the selected temperature. Fourteen-day post-exposure mortality is recorded.

*6. Pyrolysis Products—*A material that is involved in a fire or other high temperature situation may undergo pyrolytic decomposition and/or combustion producing degradation products of significant toxicity (which may be quite different than that of the original substance).

To determine the nature and toxicity of these decomposition products, the following procedure is used:

Rats (six animals per group, from six up to 20 groups per study) are exposed for anywhere between 15 minutes and 4 hours, at ambient temperatures to various atmospheres of thermal degradation products produced at selected temperatures (200°-1200°C). During exposure, signs of toxicity and mortality are recorded. The levels of common degradation products in the test atmosphere are monitored by spectrophotometry, gas chromatography or ion specific electrodes. Surviving animals are kept for 14 days post-exposure so that any delayed effects and mortality may be determined.

III. Prolonged Studies

This section contains the prolonged or subacute toxicity tests performed at Haskell. Prolonged tests are designed for determining cumulative toxic effects and to simulate the human experience whereby an individual is exposed daily for extended periods of time. In general, prolonged tests consist of repeated exposure to a substance at a fractional part of the ALD (or ALC) or LD$_{50}$ (or LC$_{50}$) dose and range from daily exposures for a week, five (weekends excluded) or seven exposures to daily exposures up to and including 90 days (five exposures per week for 13 weeks). Prior to implementing a prolonged study the following information needs to be known: The proposed use of the test substance, its physical and chemical properties, its acute toxicity and its pharmacokinetic activity.

Rather than using mortality as the primary end point, as in acute testing, prolonged studies are concerned with the more subtle cumulative toxic effects found by monitoring biochemical, hematological and histopathological parameters. To meet these demands, prolonged studies are usually custom designed so that the unique considerations concerning the properties of the materials or exposure conditions may be fulfilled.

A. Dermal

There are two basic prolonged dermal procedures available at Haskell. As in all prolonged tests, cumulative toxic effects are the objective.

1. Prolonged (subacute) 10 days—Skin Absorption—This is the most often employed prolonged dermal test. Rabbits are exposed for 10 consecutive days or 10 consecutive work days, 6 hours per day, to a fractional part of the ALD or LD_{50} under occlusion on shaved skin. Following each exposure, the material is washed off and the animals are returned to their cages until the next exposure. Throughout the study the general health status of the animals is observed and any deviations are recorded and evaluated. In addition, clinical biochemistry may be performed. During and/or at the completion of the study, the animals are sacrificed for histological evaluation of tissues. These results provide insight into the possible mechanism of action and are valuable in uncovering subtle toxic effects not seen in an acute study.

2. Prolonged (subacute) 21 days—Skin Absorption—This is a prolonged study in which the exposure simulates a type of human industrial exposure, repeated contact for 5 days per week for a total of 3 weeks (excluding weekends). Protocol details are the same as those in the above 10-day skin absorption study.

3. Acnegenic Test—This is a special purpose test designed to investigate whether a substance produces an acne-like reaction in the skin of treated animals. Certain substances, mainly halogenated aromatics i.e., chloronaphthalenes and polychlorinated biphenyls have produced an acne-like lesion in workers exposed to them. The usual PI+S test or skin absorption toxicity tests fail to detect this type of response. However, by using the following procedure this effect can be detected. The test material is applied topically to the external ear canal of albino rabbits for 10 days over a 2-week period. Daily observations are made to check for the development of this lesion. Further confirmation is obtained from a histopathological examination of the test site at the conclusion of the study.

B. Oral

There are two basic forms of prolonged oral toxicity tests available at Haskell. Each should be considered as a framework in which modifications can be made in order to accommodate special requirements particular to the test substance or its intended use. As in all prolonged tests, cumulative toxic effects are the objective.

1. Prolonged (subacute)—10 days—Oral—This test requires that the ALD or LD_{50} of the substance be known. Procedurally, one-fifth of the ALD or LD_{50} is administered orally to a group of six albino rats daily for a total of ten doses over a 2-week period (excluding the weekend). In this way each animal receives an amount of material equal to twice the ALD. A matched control group which receives only vehicle is also included. Daily records are made of body weight changes and any clinical signs that appear.

Four hours after the administration of the final dose, one-half of the animals are sacrificed and their tissues are examined for gross and histopathologic changes. The remaining animals are allowed a 14-day recovery period, after which time they are sacrificed and examined as previously described.

2. Prolonged—(subacute)—90 days—Oral—Similar in concept to the previous test, this exposure lasts for 90 days. In addition, a wider variety of biological parameters are monitored to provide insight into the mechanism of action of the substance. Rats and/or dogs are the species of choice for these studies.

For dogs, the test is divided into two parts, a preparatory phase and the exposure phase.

The preparatory phase requires 7 to 10 days during which time body weight changes, biochemical, urinary, and hematological measurements can be made to establish base line values.

In the test there are usually four groups of animals, with one group serving as a control. Each group is fed a different dietary level of the test substance for 90 days. The controls consume the same diet but without the test material. Intubation is also used. Twenty rats or eight dogs, equal in males and females are used. Diet consumption by each dog or each group of rats is recorded weekly.

Randomly selected rats and all dogs from each group are subjected to biochemical and hematological examination at 30, 60, and 90 days of exposure. At the completion of the test, the surviving animals are sacrificed and submitted to gross pathological evaluation. Select tissue from various animals are subjected to detailed histopathologic study.

C. Inhalation

There are three basic types of prolonged inhalation tests available at Haskell. Each should be considered as a framework in which modifications can be made in order to accommodate special requirements particular to the test substance or its intended use. As in all prolonged test, cumulative toxicity is the primary concern.

1. Prolonged (subacute)—10 days—Inhalation— The ALC or LC_{50} of the test substance is a prerequisite. In this study a group of six rats or other species is exposed continuously to one-fifth of the ALC or LC_{50} for 4 hours per day; 5 days per week over a period of 2 weeks (excluding weekends) for a total of 10 exposures. At the conclusion of the test, a portion of the animals are sacrificed for histopathologic examination. Those remaining are observed during a 14-day recovery period, after which time they are sacrificed and tissues are examined grossly and microscopically.

2. Prolonged—(subacute)—90 days—Inhalation— The exposure is for a period of 90 days. Prerequisite toxicity data is required. Generally a minimum of two animal species is used; a rodent (rat), and non-rodent (dog). Three atmospheric concentrations of the test substance and a zero level (control) usually are used. The animals are exposed for 6 hours per day; 5 days per week (excluding weekends) for 13 weeks. A randomly selected portion of animals usually are sacrificed for histopathological examination at designated times during and at the conclusion of the experiment.

The remaining animals are sacrificed and examined as before during a recovery period, which varies according to the study, but generally does not exceed 90 days.

*3. Insufflation—*This test is used to determine the pulmonary tissue response following a single intratracheal injection of the test substance directly into the lung. The quantity of material is designed to be of sufficient size so that it usually will remain in the lung for a number of weeks and therefore is considered a continuous prolonged exposure. Generally rats are used. The animals are sacrificed at 2, 7, 14, 28, and 56 days, and at 6, 12, 18, and 24 months post-insult for histopathologic evaluation.

IV. Chronic Studies

This section describes the chronic toxicity tests performed at Haskell. Chronic studies are intended to simulate human working lifetime or lifetime exposures. Chronic studies require careful planning so that the animal exposed is comparable to human exposure in terms of time, routes of administration, daily dose, etc. In addition, background information from previous studies is necessary and should include physical and chemical properties of the test material, acute toxicity data and pharmacokinetic data in an animal species which is similar to man in these respects. The selection of exposure levels should range from one where toxic effects are produced to one where no toxic signs are manifested. The number of

animals employed should be sufficiently large so that at the conclusion of the study there are enough survivors left so that meaningful statistical analysis of biological parameters can be performed. The selections of indices to be monitored requires careful consideration. Currently there are no chronic dermal tests.

A. Oral

A variety of basic chronic studies is available. As in all long-term tests experimental design is much more flexible than in acute studies and this flexibility should be considered when planning a study.

*1. Two-year Feeding—Rats—*The purpose of this test is to determine long term oral toxicity of a substance. Approximately 250 albino rats are used in the study. In food additive studies, first generation (F_a those whose parents were exposed to the substance) are used.

During the study, all animals are weighed once per week in the first 6 months, every 2 weeks during the second 6 months, and once every month during the second year. Food consumption is recorded following the same time schedule. Data from the above are combined to determine food efficiency and average daily dose. Throughout the test, animals are examined for abnormal behavior and toxic signs. Mortality is recorded and tissues from animals that die or are sacrificed are preserved for histopathological evaluation.

At 1, 2, 3, 6, 9, 12, 15, and 18 months, routine hematology, urinalysis, and biochemical measurements are performed on 20 randomly selected animals each time.

At the end of the first 12 months, approximately 10-20% (depending upon mortality during this time) of the animals are sacrificed and subjected to detailed histopathologic appraisal. At 18 months, animals also may be sacrificed for histopathology depending on the 12-month results. At 24 months or when the group size is reduced to a critical level (20% of the initial group) through deaths, all remaining animals are sacrificed and the high dose level and control rats undergo histopathologic evaluation. If necessary, rats from the lower dose levels are also examined histopathologically to determine a no effect level for the target organs.

*2. Two-Year Feeding—Dogs—*The beagle dog is commonly employed as the non-rodent species for chronic oral studies. Eight beagles, between 1 to 2 years of age and equal in males and females, are assigned to each of four groups: three dose levels and one control. The test is divided into two parts: a preparatory phase and the exposure phase. The preparatory phases requires seven to 10 days during which time body weight changes, biochemical and hematological measurements can be made to establish base line values. Throughout the study, diet consumption, and weight changes are recorded on a weekly basis. The dogs are observed daily for toxic signs. Routine hematological, biochemical and urine examinations are performed on each dog at 2 weeks, 1 month, and alternate months thereafter for the first year, and quarterly during the second year.

After the 2 years of continuous feeding, both control and test animals are sacrificed and subjected to histopathological appraisal. All tissues from animals at the highest dose level are evaluated microscopically and then if necessary, select tissue from the other animals are examined.

B. Inhalation

Specially designed studies similar in scope to the 90-day prolonged investigation are available. The details of a study are designed to fit the particular requirements for each test material.

V. Special Tests

This section describes special toxicity tests which are available at Haskell. These studies are designed to evaluate substances for particular biological effects not usually tested for in routine investigations. Tests cataloged as miscellaneous are also included in this section.

A. Carcinogenesis

A carcinogen is defined as a substance which has the ability to induce tumors in humans. Testing for the carcinogenic potential of a substance can be done using a variety of procedures. However, only the long-term daily exposure of animals to the test substance is recognized as the accepted technique. In general, these tests are modifications of chronic studies which include one or more species of animals exposed to the substances by dermal, oral or inhalation routes. Exposure periods last a lifetime or significant fraction, e.g., from 1 to 2 years with rats and 5 to 7 years with dogs. Gross and microscopic examination of tissues is the prime method for detecting onset, development, and morphology of tumors.

B. Teratology

Teratology may be defined as the study of the effects of substances or environmental factors which will produce permanent structural or functional aberrations, i.e., malformations and embryo mortality, during embryogenesis. A teratogen is a substance which can either in-

duce or increase the incidence of congenital malformations and generally does not produce any adverse effects on the mother. Thalidomide is the classic example of such a material. Protocols for teratogenic testing of each substance require individual consideration and are derived from the basic plan which is described below. The test substance usually is administered to one or more species of pregnant laboratory animals during organogenesis (the mid-gestational period) when the embryo is most sensitive to chemical effects. Examination is made of near-term fetuses and consists of recognition and enumeration of external visceral and skeletal gross morphologic malformations and grossly recognizable function defects. Microscopy may also be included. The uterus is examined for evidence of embryo resorption.

Teratogenic information can be obtained from a three-generation reproduction study (see below), but in that study both the male and female are dosed so it generally cannot be determined whether teratogenic effects were due to sperm or egg defects (mutations) caused by the chemical or due to an *in utero* effect of the chemical. The three-generation reproduction study also is usually started sometime before mating, so enzymatic adaption with reduced toxicity is possible. Also, because of the shorter duration of the above teratogenic test, higher doses often are used.

C. Genetic Toxicology

The primary function of this discipline is testing for mutagens, substances which produce mutations, when incorporated in a living organism. Current interest stems from the fact that man is being exposed to large numbers of new chemicals each year and some of these substances may be capable of producing genetic alterations. In addition, studies have shown that there is a high correlation between substances which are mutagenic and those which are carcinogenic. The value of mutagen testing, in this context, is that results are obtained sooner and the costs are more economical than long term animal carcinogen study. For example, a standard mutagen test, the host-mediated assay, costs less than 1/30th of an animal carcinogen study and is completed in 1/20th of the time. Therefore, mutagen testing can be used as a screening program for selecting substances which should be tested in animal systems for carcinogenicity.

Presently, Haskell offers three tests for evaluating a potential mutagen: dominant lethal test, cytogenetic study (karyotyping) and bacterial mutagensis.

1. Dominant Lethal Test—Dominant lethal mutations are due to chromosomal aberrations caused by a test substance and result in death of the individual carrying these aberrations. Both males and females may be used.

Present studies are only involved with dominant lethal mutations in male germ cells.

At least 10 adult male rats per dose level, at three dose levels and one control, are exposed to the test substance by either the oral, parenteral or inhalation route. Immediately following, or concurrently with exposure, these males are mated with unexposed virgin females. The females are sacrificed 9-10 days (midgestation), after mating to detect pregnancy and to score the number of viable corpora lutea, implantation sites, deciduomata (dead implants), and embryos. Calculations based upon these data are used to determine whether a substance causes a dominant lethal mutation.

2. Cytogenetics—Cytogenetics is defined as the study of chromosomes aberrations using light microscopy. Changes are categorized as either structural (rearrangements and breaks) or numerical (increased or decreased number of chromosomes, hyper- or hypoploidy, respectively).

A. Metaphase Study

Young rats are administered the test material by an appropriate route in a single or several daily doses. The animals are sacrificed 6 to 48 hours after treatment depending upon experimental protocol. Colcemid, an inhibitor of cell division at the metaphase, is given 3 to 4 hours prior to sacrifice. After sacrifice, bone marrow cells from test and control animals are examined for chromosomal aberrations.

B. Micronucleus Test

Similar to the above, except the Colcemid treatment is not used. Bone marrow cells from test and control animals are examined for micronuclear bodies.

3. Bacterial Mutagenesis—The test substance and various strains of bacterium *Salmonella typhimurium* are incubated together with and without metabolic activation (in combination with mammalian tissue extracts). The number of mutants induced in the treated cultures is compared to those of the controls. From these results mutagenicity is determined.

C. Reproduction

There are two reproduction studies, one is an offshoot from a 90-day rat feeding study and the other is from a 2-year rat feeding study. The former generally ends with the first generation (F_1), the latter with the third (F_3). They are described below.

The offshoot from the 2-year rat feeding study includes a variety of parameters designed for detecting toxic manifestations in the reproduction process of the exposed animals. After 12 weeks in the feeding study, an equal number of male and female animals are randomly selected for this test and they constitute the parental (F_0) generation. These animals are mated.

Approximately 10 days after the first litter (F_{1a}) has been weaned, the F_0 animals are again mated to produce the F_{1b} generation. After weaning of the F_{1b} generation, the F_0 are returned to the original study, F_{1b} animals are placed on the same diet as their parents F_0. At the age of about 110 days these F_{1b} animals are mated to produce F_{2a} animals and F_{2b} animals. The same procedure is followed to produce a F_{3b} generation. F_{3b} animals are sacrificed at the age of 21 days and subjected to gross and histopathologic evaluation. From the above data the following are calculated: fertility index (% of matings resulting in pregnancy); gestation index (% of pregnancies resulting in birth of live litters); viability index (% of rats born that survive 4 days or longer); and lactation index (% of rats alive at 4 days that survive to be weaned).

D. Antidote

The ability to reverse or stop the development of toxic effects produced by a substance is a useful technique and may be life saving under certain conditions. Essentially this type of study may be instituted in cases where substances produce rapid development of toxicity, such as inadvertent exposure to large amounts of organic phosphate pesticides. The experiment must be specially designed for each substance.

E. Cardiac Sensitization

The purpose of this test is to determine whether a test substance has the ability to sensitize the heart to the effects of epinephrine. This test is primarily used to investigate aerosol can propellants, refrigerants and other volatile low molecular weight halocarbons which could be inhaled.

Beagle dogs are utilized in the study and are given an injection of epinephrine while breathing air and then given another epinephrine injection while being exposed to an admixture of air and the test substance. Cardiac activity is monitored with an electrocardiograph throughout the study in order to detect any abnormal effects produced by the test substance.

F. Wildlife

1. Aquatic Toxicology—These studies are aimed at determining the toxic effects of substances on aquatic life forms such as fish, mollusks and plants. These investigations include effects of effluents from manufacturing plants to pesticide runoffs from crop treatment. The experimental design must be individually tailored for each study. Currently, Haskell offers static bioassays (TLm) on freshwater fishes and macro-invertebrates and brine shrimp bioassays. An enlarged aquatic facility is under construction and when completed will have the capacity for undertaking comprehensive aquatic toxicity studies. Meanwhile, consultation for securing outside contractors is available.

2. Terrestrial Wildlife Toxicology—These are similar in scope to that of aquatic toxicity testing and are designed to study possible toxic effects of substances on terrestrial wildlife forms such as rabbits, mallard ducks, pheasants, or Japanese quail. Experimental design is custom tailored for each study.

G. Human Exposure

The use of large controlled climate chambers at Haskell provides an environment for investigating human physiologic and psychologic activity under various combinations of temperature, humidity and exercise. These types of studies are useful in evaluating human performance in various work environment conditions. These chambers can also be used for human inhalation studies. Tests are designed to fit individual requirements. As in all human testing performed by Haskell, the Helsinki code of ethics for human experimentation is followed.

VI.   Research

In addition to performing routine toxicology evaluation of substances, Haskell invites investigations into the research frontiers of toxicology. This includes improved methods for detecting toxic effects, development of cell culture techniques for toxicity testing, improved animal models, new analytical procedure for measuring low level of airborne toxins, pharmacokinetic evaluations, radiotracer studies, computer technology for data collection and toxicologic information storage and retrieval.

In areas closely related to toxicology such as biochemistry and industrial hygiene, Haskell offers routine and non-routine analytical services. They include trace metal analysis of air, water, or soil, monitoring the environment for organic compounds, and biochemical assays for biological monitoring of human exposures. Instrumentation capabilities include atomic absorption, ultraviolet, visible and infrared spectrascopy, gas and liquid chromatography, electrophoresis, an automatic clinical analyzer, a liquid scintillation counter and a Coulter Counter.

VII. Submission of Test Materials

In the submission of test materials to Haskell for study it is imperative that as much physical and chemical information as possible be included so that testing can be started as soon as the sample is received. This knowledge should be incorporated in the form which accompanies the material; formula, lot number, purity, source of sample, solubility and proposed use are most valuable. The amount of material shipped should be of sufficient quantity so that the same material can be used for a complete toxicity evaluation. Further information concerning these requirements are found in the cost list or can be directly obtained from Haskell.

Sample containers should be labeled with contents and be leakproof. Handling materials of unknown toxicity requires utmost caution. A transmittal letter accompanying the sample or sent separately is required.

CHAPTER THREE

IMPACT ON INDUSTRY

COST ESTIMATES FOR COMPLIANCE*

Purpose

The purpose of this paper is to review three studies that estimate the costs to industry of the proposed Toxic Substances Control Act. The three studies we have reviewed are:

1. *Draft Economic Impact Assessment for the Proposed Toxic Substances Control Act, S.776,* U.S. Environmental Protection Agency, June 1975 ("EPA study");

2. *Study of the Potential Economic Impacts of the Proposed Toxic Substances Control Act as Illustrated by Senate Bill S.776 (February 20, 1975),* Manufacturing Chemists Association, June 1975 ("MCA study");

3. *Statement on S.776 and the Toxic Substances Legislative Issue,* Dow Chemical U.S.A., April 1975 ("Dow study").

The three estimates vary considerably. In this paper, the component costs of each estimate are set against each other in Table 1, so the reader can see the reasons for the differences. We then examine the economic rationale behind the estimates and attempt to determine which is most reasonable. Our analysis is confined almost entirely to the information contained in the three studies. We did not attempt to verify the accuracy of the basic technical data, such as the cost of performing a certain test for toxicity. Our criticism deals mainly with the economic methodology of the studies—whether they use the data properly to estimate costs.

Toxic substances are potentially dangerous to people and the environment during each stage of their existence—manufacture, use, and disposal. The main goal of the proposed Toxic Substances Control Act (TSCA) is to insure that information is provided on the toxicity of new and existing chemical substances so that appropriate steps can be taken to guard against their dangers. Unquestionably, there are certain costs of producing such information. The economic question is whether these costs are justified by the benefits. The objective of the three studies is to estimate the magnitude of the costs that the TSCA would impose upon the technical industry. The MCA study also considers the impact upon related industries. Benefits are scarcely discussed, and costs to the rest of society are not mentioned. Because these studies are confined to only part of the problem, none of them is suffi-

cient to determine whether the TSCA should be passed or not.

Going beyond the simple question of benefits versus costs, one could ask the more difficult question of how strict the testing requirements should be, i.e., at what level of strictness of testing requirements do marginal testing costs start to exceed the marginal benefits of the information generated. The MCA study does not address this question directly, but it does give some insight into the manner in which testing costs increase as the thoroughness of testing and product coverage is increased. This is an important point because while it is obviously true that *some* testing should be required, it is also possible to require testing far in excess of what is necessary.

Overview of the Three Studies

The EPA Report

The EPA study is said to be based upon two years of intensive staff work, four years of legislative history, experience with other environmental laws, and consultations with experts in the field. On this basis, it would appear to be realistic in specifying the number of chemicals that would have to be screened and tested under the Act. This, in turn, would appear to make the estimates relatively reliable. However, very few sources are cited in support of specific assumptions and statements. In other words, the study is not documented well enough to allow the reader to check the basis for all of the assumptions.

For each component cost, it gives a range of figures; the totals then add up to "low" and "high" figures of $79 million and $141.5 million. In order to keep the comparisons simple, we will generally refer to the "high" figures, even though this may somewhat misrepresent the main thrust of the EPA report. Both "high" and "low" figures are reported in Table 1.

The Dow Study

The Dow Study gives the highest cost figure—$2 billion per year—and uses a relatively unsophisticated approach. Its estimate of costs for the industry is based solely upon its estimate of what Dow Chemical's costs would be. The method is to multiply Dow costs by the ratio of industry sales to Dow sales, which is about 24 to 1. In other words, the study makes projections from a 4 percent *non-random* sample. If Dow Chemical is unrepresentative of the industry in such crucial (and varying) factors as new products

---

* A Comparison of Three Estimates of Costs of the Proposed Toxic Substances Control Act, U.S. General Accounting Office, December, 1975.

TABLE 1—COMPARISON OF TSCA COST ESTIMATES

	EPA (low)	EPA (high)	MCA #4 (lowest)	MCA #1 (highest)	Dow
1. Screening and Testing					
a. Screening					
# Chemicals	1,000	1,000			
Cost per chemical ($)	3,500	5,500	Included in Testing		
Total (million $)	3.5	5.5			
b. Testing new chemicals					
# Chemicals	150	150	1,230	7,900	916
Cost per chemical ($)	20,000	40,000	51,900	33,700	382,000
Total (million $)	3	6	63.8	266.5	350
c. Testing existing Chemicals					
# Chemicals	200	200	65	100	482
Cost per chemical ($)	22,500	42,500	465,000	411,000	1,000,000
Total (million $)	4.5	8.5	30.2	41.1	482
TOTAL (million $)	11	20	79.0[2]	292.6[2]	712[1]
2. Administration and Reporting (million $)	19.5	41.5	64	82	133
3. Delays (million $)	10.0	19.5	24	86	145
4. Bans, Restrictions (million $)	37.5	60	165	195	965
5. All Other (million $)	0.5	0.5	26	669.4	--
TOTAL COST (million $)	78.5	141.5	358	1,325	2,000

[1] Net of $120 million for current testing.
[2] Net of $15. million for current testing.

per dollar of sales, or in fraction of products that have to be tested, then the total figure is biased. The smallness of the "sample" assures that this bias can be a large dollar amount.

This study's high cost estimate derives from assuming that it will cost $1 million per test of existing chemicals. This figure far exceeds that of the other studies. The number of existing chemicals assumed to be tested also exceeds that of the other studies by a significant amount.

The MCA Study

This study is by far the most ambitious in data collection. It attempts to estimate costs from data collected in a survey of chemical firms. Some of these data are not available elsewhere and give valuable insight into the composition of output of the chemical industry. Unfortunately, the study has several important defects:

1. It makes what seems to be an excessively broad interpretation of the screening and testing requirements of the Act.

2. Its "maintenance of innovation scenario", which gives large cost estimates, purports to be based upon sound economic theory and upon the results of the survey. In fact, it is based upon neither. In fairness to the study, the apparently bogus figure of $600 million for "maintenance of innovation" is used in only one scenario. It is this scenario, however, that seems to find its way into various references to the report. This point is discussed later in this study.

3. Too much is made of the supposed "economic impact" on the rest of the economy. As is discussed below, the MCA study uses questionable methodology to arrive at the large negative impacts.

The cost estimates range between $360 million and $1.3 billion per year, plus over $100 million "start-up" costs.

Thus, we see how widely the estimates vary. The EPA study gives a range of cost estimates that differ, from high to low, by a ratio of almost two to one. The Manufacturing Chemists' report shows a range, from high to low, of more than three to one. The ratio of the lowest EPA estimate to the Dow Chemical estimate is 25 to one, though it must be said that the high Dow estimate is probably the least reliable of all the figures.

Component Costs of TSCA

In the simplest terms, what could the TSCA require? Firms would have to give advance notice and brief descriptions of new chemical substances that they plan to produce. Some of these, as well as some existing chemicals, would have to be tested. Depending upon the results of the tests, some chemicals would be banned or restricted in use.

The costs of this process would be as follows:

1. Direct costs of screening and testing,

2. Associated costs of administration and reporting,

3. Delays in the introduction of beneficial products,

4. Costs associated with developing products that are later shown to be dangerous and are therefore banned or restricted,

5. All other indirect costs to the firm associated with testing requirements.

The studies' estimates of these costs are shown in Table 1. Each of these costs is discussed below, and each study's method of estimation is compared. Each number is taken as it appears in the studies, even though some rounding is in order, so that the reader can more easily find each number in the original study.

Direct Costs of Screening and Testing

These costs depend upon the testing cost per chemical and the number of chemicals that have to be tested. Here the three studies vary considerably, not only in these two aspects but also in their interpretation of what the TSCA would require in the way of reporting.

The EPA views the process as screening (or simply reporting upon) about 1,000 chemicals per year at low cost, and testing 150 of them plus 200 existing chemicals at a cost of about $40,000 each. The MCA study, however, assumes that all new chemical substances (defined so as to number up to 7,900) will be subject to some fairly costly testing in the early stages. The Dow study assumes that about 900 new chemicals per year will be tested.

The MCA study shows how widely costs of testing can vary depending upon how stringent they are; the most thorough testing would cost $800,000 per test. The averages presented in Table 1 do not reflect this variation, but merely show the general magnitudes involved. MCA Scenario No. 4 assumes that fewer tests are required, with the least costly ones dropping out. Thus, costs per test are higher.

The EPA and MCA studies are in near agreement about the cost per test of new substances; it is the number of tests that accounts for the large differences in total cost. The EPA figure, based on extensive low-cost reporting and selective testing, seems most consistent with the TSCA requirements.

Significant differences appear among the studies in their treatment of testing existing chemicals. The EPA uses a cost near $40,000, the MCA figure is ten times higher, and the Dow figure is more than double *that* figure—$1 million per test. The Dow study assumes 482 tests per year, far more than the other studies.

It would appear that the two industry reports have a valid point here. Testing costs of other substances thought to be dangerous have been quite high. The MCA report lists the costs of a whole battery of tests, the (high) costs of which are fairly well known. If testing is required for a given product, then there is a presumption that the product could be dangerous in some uses, even though it had passed some of the preliminary, less costly, tests. The industry would then be in a position of having to prove the absence of danger, which naturally would require thorough testing. Firms would be more likely to have a vested interest in an existing product than in a new one, and they would therefore, go to greater expense to exonerate an existing product before deciding not to produce it. Perhaps a figure of $30-40 million would be appropriate here; this is the MCA estimate. (Note, however, that the EPA assumes 200 existing products per year would be tested, while the MCA figures are 65 to 100.)

Another possibility for reconciling the two studies would be to use EPA's number of tests and MCA's cost per test. Using the lowest figures in each case would give a cost for testing existing chemicals of $82.2 million. This, however, would probably be an overestimate; the MCA figures imply a decreasing marginal cost of testing as the number of chemicals tested increases. (The reasons for this are logical but need not be elaborated here.) Taking decreasing costs into account would give a cost figure of around $60-65 million, depending upon the assumptions made. Given the contents of the studies, a case could be made for any of these figures. Our point here is to show the wide range of plausible estimates that exists.

Administration and Reporting

The EPA figures range between $19 million and $41.5 million, which includes annual reports, record keeping, and preparation of a bibliography of health and safety studies. Pre-market screening is called an "administrative cost" but we have included it in Table 1 under Screening and Testing. MCA's figures range from $64 million to $82 million. The two studies appear to be referring to costs for the same administrative activities, so there is no way to account for the difference except to realize that this type of cost is impossible to forecast accurately. As might have been expected, the government estimate of administrative costs is lower than the industry's estimate; unfortunately, this does not tell us whether one (or both, or neither) is biased.

The Dow study estimate of administrative and reporting costs is $133 million, far above the other two. According to the study, this figure is the low side, since it does not figure in the possibility that progress reports on each test-mouse and test-rat might have to be sent to Washington. The Dow study estimate appears to be little more than a guess, especially when extrapolated to the entire industry; the other two studies appear more plausible.

Delays

The TSCA will cause delays in marketing chemicals. The 90-day waiting period could be relatively minor, especially if the required reporting is done early enough so that the 90 days are spent on other necessary product development. Much more important costs would be incurred for chemicals for which extensive testing is required. Assuming that the chemical ultimately reaches the market, the firm will have incurred the cost of having to wait longer for the beginning of the stream of income from selling the product.

Example: A product costs $500,000 to develop and, starting in one year, it will produce a net income of $50,000 per year forever. This is a rate of return of 10% and is the same as depositing $500,000 in a bank that pays 10% compounded annually. If the income stream does not start until two years from now, then the firm has lost next year's $50,000. The discounted value of this loss is $50,000 divided by (1 + the interest rate), or $45,454.

This method is used by both the EPA and MCA studies, except that in the above example they would simply multiply $500,000 by 10%, for a "cost" of $50,000. This is a slight overestimate, but it pales in comparison to the other uncertainties and can therefore be overlooked.

Using the same method the two studies come up with somewhat different figures for the cost of delay: $19.5 million for the EPA, and $24 million to $82 million for the MCA. The MCA study uses lower discount rates and assumed capital investment per product. If it had used the EPA figures on discount rates and investment, its "high" estimate would have been over $280 million, instead of $82 million. The higher MCA estimate, therefore, is attributable to its assumption that *thousands* of products will be delayed, not just 750 as EPA assumes.

Because, as was noted earlier, the EPA study seems to take a more plausible view of the number of chemicals that would have to be tested, it would appear that its estimates of delay costs are more believable.

The Dow study gives $133 million as the cost of delays, but does not present much support for this figure. Our previous comment on the Dow study's extrapolation method still applies.

Because all of the studies deal only with cost to the chemical industry, they ignore a very important aspect of delay costs — the costs to the consumer. Just as the firm's profits are postponed, so too are the benefits to the consumer, i.e., the difference between the value to the consumer and what he pays for the product. The cost of delay to the consumer could very well equal the cost of delay to the industry.

This concept requires further explanation. "Consumer surplus" is the benefit that a consumer receives from being able to buy a certain good. It is the difference between the value of the good to the consumer and the amount of money that he has to pay for it. A textbook example: A thirsty man buys a drink for $1 but would have been willing to pay as much as $5. His consumer surplus is $4. Depriving him of the opportunity to buy the drink would impose a cost of $4 in lost benefits.

Economists have attempted to measure consumer surplus in many specific cases, such as the introduction of a new good. More to the present point, there have been attempts to measure the cost of delays in introducing new goods. Sam Pelzman's study of the 1962 Kefauver-Harris Amendments to the Food, Drug, and Cosmetics Act (*Journal of Political Economy*, Sept./Oct. 1973, pp. 1049-91) found that benefits foregone on *effective* new drugs were substantial. Although Pelzman's actual cost estimates, as well as his measures of cost avoided by preventing the use of harmful drugs, may be questioned, his study does indicate that delay can be a significant cost to the consumer.

Bans and Restrictions

As with the costs of delay, the cost estimates differ considerably. Here, however, the EPA and MCA studies are roughly in agreement over the number of chemicals expected to be banned or limited in use, but they differ on the cost per banned chemical. The same methodology is used as with delays: Not being allowed to produce a chemical eliminates the return to invested capital. The difference between the two estimates appears to be due mainly to a "multiplier" of 2.5 by which the MCA study multiplies the cost of banning a major substance. Though not fully explained, it appears to be an estimate of the cost to industries that formerly used the banned product but which no longer can do so. As such, it appears to be an attempt to capture the lost "consumer surplus" that was discussed in the previous section. In our view, this is a legitimate procedure, though the accuracy of this particular estimate is in doubt.

The EPA study is no clearer than the other studies as to the costs of bans and restrictions. Although the EPA study presents two pages of numbers with detailed assumptions about such quantities as "R & D investment for each chemical at time of ban" and "percentage of R & D investment recoverable", not a single number is documented. The reader is given no information as to where

the numbers came from or why they are what they are rather than something else. We therefore have no great confidence that the range of cost figures is likely to include the actual cost that the Act would incur due to bans and restrictions.

In a larger sense, it is possible to take the view that this type of cost need not be estimated very accurately for the purpose of deciding whether or not to enact the proposed legislation. If it is assumed that chemicals will be banned or restricted only if it is determined that the benefits of the ban exceed the costs, then we could be sure that this part of the Act's costs would indeed be canceled out by benefits.

Other Costs

Maintenance of Innovation—The MCA Scenario No. 1 ascribes a cost of $600 million to what it refers to as "maintenance of innovation." It appears to be an estimate of the extra cost required to maintain the same rate of innovation as before the imposition of testing requirements. As will be discussed, it is not clear why the firm would want to do this. Neither is it clear how the study arrived at the number, because the innovation questions on the survey do not seem to get at this at all.

From the point of view of economic theory, there are two ways in which the TSCA might affect the rate of spending on research and development:

1. Testing costs would tend to reduce the rate of return to the discovery of new products, and this would reduce the optimal level of R & D spending. Assuming that firms would want to maintain the rate of innovation implies that firms do not pay attention to the rate of return on R & D spending and, therefore, do not try to maximize profits.

2. It is possible that some R & D would be undertaken to find new ways of reducing the cost of testing, either by discovering cheaper tests or new products that would not have to be tested. If successful, such spending would reduce testing costs by an amount at least equal to its cost. Therefore, if such increases are counted as costs, the cost of testing should be reduced. Or, if testing costs are estimated in full, then it is double-counting to add in this type of R & D cost.

It follows from this that the $600 million figure should not be counted as a cost. MCA omitted it from Scenario No. 4.

Economic Impacts—The MCA study considers the effects of the Act on GNP, employment, and international trade. The value of this analysis is questionable. In the first place, no information is given about the models that were used, so we don't know the magnitudes of such crucial parameters as the elasticity of demand for chemical products or the elasticity of demand for exports. The values of these parameters are subject to some difference of opinion among economists and differ from model to model.

Second, the economic process that is assumed to take place is not realistic. Costs of testing would result in higher prices which decrease the demand for chemical products. So far, so good (except that the original estimates of testing costs may be too high, which would overstate the consequent price increases). The next and final step, however, appears to be that this decrease in output by the chemical industry reduces sales by other industries to the chemical industry, and so affect GNP by means of a reverse multiplier. The estimate in the study has GNP decreasing by as much as $1.6 billion. This reasoning ignores the possibility that if less money is spent on chemicals more will be spent in other sectors, perhaps enough to cancel out the decrease in demand for chemicals. The testing industry will certainly expand by the amount of the testing costs. Consumers who previously bought chemical products will divert their spending to other goods. We therefore conclude that these estimates should not be considered in the evaluation. Although there will be an economic impact on other industries, we do not believe that the studies so far have managed to quantify that impact very successfully. The crucial question at this point is the estimation of direct costs to the chemical industry.

Conclusions

Cost estimates for the TSCA entail many analytical steps and many assumptions about how the Act would work in practice. It is hazardous to generalize about the studies because so many separate elements are involved. Nevertheless, on the basis of our discussion, some general impressions emerge:

1. Despite the efforts that have been made so far, the costs to industry due to the Toxic Substances Control Act are, at this stage, quite uncertain.

2. The Dow study is the least plausible of the three.

3. The differences between the EPA study and the MCA study exist mainly because the MCA study assumes many more tests would be required. Only with the cost of delays does the EPA study assume a higher cost per delay, but that is partly because the MCA study assumes many more small delays, which bring down the average.

4. If it is assumed that the EPA study is based upon more accurate knowledge of the scope of the TSCA's requirements, then one could conclude that the EPA figures are more accurate, with the following exceptions:

   a. For tests of existing chemicals, the MCA figures seem more plausible. Substituting the MCA No. 1 figure of $41.1 million for the EPA (high) figure of $8.5 million would add $32.6 million to the EPA cost estimate.

   b. Because all of the studies consider costs to industry, but no cost to consumers, no figures are given for the costs of delays to consumers. These costs could be substantial.

ECONOMIC IMPACT OF TOXIC SUBSTANCES CONTROL LEGISLATION*

Introduction

Because of conflicting and broadly differing estimates of the annual cost to the chemical industry of a toxic substances law by government, at a low of $45 million, and by a chemical company, at a high of $2 billion, the Manufacturing Chemists Association (MCA) engaged Foster D. Snell, Inc., to conduct an economic impact study.

The comments in respect to the chemical industry and the total U.S. economy are based on the Snell analysis of 45 companies with annual sales ranging from $100 million to more than $1 billion. These represent about 23 percent of total industry sales of chemicals and allied products, which last year came to $86.8 billion. Results were projected to encompass the entire industry.

General Analysis

Using the provisions of the Senate Bill 776 and 1975 cost figures, the survey documents annual economic dollar impacts on the industry ranging from a conservative $358.0 million for the "selective" model to $1.3 billion for the "broad" model. (In this context, "selective" means the exercise of thoroughly justified and careful discrimination by EPA; "broad" means very little selectivity in EPA's activities however within the limits of the ultimate law.) This does not include one-time, non-recurring preparation expenses estimated at from $78.2 to $114.5 million. Even to an industry as large as the chemical industry, these figures are not insignificant, but more significant is the impact of these increased costs on the nation's economy in terms of inflationary effects, jobs and the balance of trade.

The models cover requirements in the proposed legislation, the broader the restraints, the higher the costs. The overall economic effect of the House of Representatives Bill 7229 could be even greater than for the S. 776 broad model. The reason is the sweeping, discretionary powers given (the Administrator including his mandate to list "not less than 300 chemical substances (other than new chemical substances)" with the highest priority for testing.

The June 6, 1975, Staff Working Draft of S. 776 also could increase costs substantially. This would depend on the number of mixtures to be tested and the proportion of longer-term, more costly tests required to prove or disprove preliminary indications of the "potential to induce cancer."

Some likely economic impacts remain unquantified. For example: company failures and/or increased industry concentration through acquisition or merger of smaller firms; higher risk and more conservative return-on-investment criteria; the "wasting" of patient time during testing, and various Toxic Substances Control Act compliance phases which would become evident only as implementation developed.

Results by Category

If EPA were to impose relatively unselective H.R. 7229 demands, the yearly cost to the chemical industry would be $1.3 billion, not including start-up, non-recurring costs. Broken down, costs would be:

(1) Start-up—$114.5 million

These non-recurring preparation expenses involve administrative organization, establishment of record-keeping and reporting systems and submission of inprogress health and safety studies. The latter encompass new and existing substance reports including 40-year bibliographies of 50,000 or more chemical compounds.

(2) Institutional—$1.125 billion

Included are:

(a) $82 million to administer the program, to interpret regulatory developments and to consult with EPA on testing protocols and all other responsibilities including record maintenance for reporting adverse effects.

(b) $600 million in additional research and development funds to overcome a projected 30 percent reduction in the introduction of new products to the marketplace due to restraints and thus maintain the status quo. The industry's current level of R/D funding is $2 billion a year.

(c) $41 million for formulating protocols and testing existing substances plus reporting results to EPA. This evaluation excludes testing existing products which might be on the proposed list of 300 substances. If they were added, costs would range from $90 million to $120 million

---

* Manufacturing Chemists Association. A presentation on the potential economic impacts of the proposed Toxic Substances Control Act made to the House of Representatives Committee on Interstate and Foreign Commerce, July 1975.

as the Administrator would be required to perpetuate the list. Short- and long-term environmental tests, as well as sub-acute and chronic animal testing, would be included since the compounds would have suspected significant hazard-potential or would be classified as having unreasonable risks.

(d) $225.6 million for premarket screening of 6,500 new and modified products that reach the marketplace each year.

(e) $35.9 million for premarket screening of significant new applications of some 450 older products.

(f) $86.0 million annually in lost income due to delays of four or more years in the introduction of new products.

(g) $50.5 million in added manufacturing costs as a result of tighter quality control tests, new test protocols, labeling and other in-house record keeping.

(h) $19.4 million for all other incidental costs such as fees to EPA for submitting test data, judicial action, if required, and defense of third-party civil suits.

(3) Extraordinary—$195 million

These costs cover those products which could be banned from or restricted in the marketplace, without adequate justification and thus preventing maximum utilization of productive capacity.

Interpretation

Cost levels for each factor and totals may be visualized readily in this table for the "broad" and most "selective" models:

To better understand the magnitude of difference between the models, let us examine one factor, premarket screening. If the Administrator's authority were limited to selectively testing only those substances he finds pose an unreasonable risk to health and environment, the estimated annual cost would be $63.4 million, the sum of screening new substances and new applications.

If his authority covered the broad model and he exercised more speculation than firm justification in requiring testing of as many as 6,500 new and modified products marketed yearly, then the cost would be $261.5 million.

This scaling process between the "broad" and "most selective" models is applicable to any of the proposed bills.

Obviously, total testing costs would depend on the number of substances to be tested and the testing costs per substance. In the four calculated models shown in the full report, mixtures are omitted. Provision for such testing, however, clearly is included in H.R. 7229. According to the Snell survey, there are some 10,000 significant new mixtures and more than 350,000 minor formulation changes each year. Conservatively, inclusion of mixtures readily could double the number of products to be tested.

Of even greater economic impact significance is the current language of S. 7776 that would require tests which could show "...the potential to induce cancer ..." The costs would be far higher because of the need for vastly more extensive validating tests. Clearly the Administrator's discretion bears heavily on economic impact, and in this respect, the Snell figures again are most conservative.

Toxicological Testing Costs

The Snell study also surveyed toxicological testing laboratories to determine current test costs on mam-

ECONOMIC IMPACT
[**In millions of dollars**]

Cost factor	Selective	Broad
Nonrecurring preparation expenses	78. 2	114. 5
Recurring annual institutional expenses:		
Administration	64. 0	82. 0
Maintenance of innovation		€00. 0
Testing of existing substances	30. 2	41. 1
Premarket screening (new substances)	€0. 3	225. 6
Premarket screening (new application)	3. 5	40. 9
Less testing without act	(15. 0)	(15. 0)
Delays (time costs of completed R. & D., testing)	24. 0	86. 0
Product labeling, QC, monitoring	21. 7	50. 5
Application legal fees	4. 3	18. 9
Subtotal	193. 0	1, 130. 0
Extraordinary costs, limiting and banning of production	1€5. 0	195. 0
Total, continuing cost	358. 0	1, 324. 0

malian, wildlife and aquatic species. To adequately evaluate the chronic, metabolic and reproductive effects of a single chemical substance, costs already run $800,000 and may escalate further. Much less costly techniques are being developed to indicate mutagenic and teratogenic effects, with some foundation for extrapolating to carcinogenicity. These, however, are still screening tests, not always applicable to broad classes of substances, and at the present state of the art inevitably require validation by much longer-term and far more costly testing protocols. It follows that the Administrator's action based on screening level or preliminary test data could be unjustified and could, in fact, be contraindicated by results of the ultimately necessary longer-term tests.

These factors are at the heart of the Manufacturing Chemists Association's emphasis that the Administrator explicitly determine the economic impacts of his final rule-making activities.

Other Related Factors

In addition to these direct costs, there were other limiting factors examined in the Snell study.

One of these is the ability of the toxicology testing industry to respond to the obviously increased demands to which it would be subjected. The level of demand will depend upon the criteria finally expressed in the law. As stated earlier, we believe the level of action in H.R. 7229 would be even greater than stringent interpretation and enforcement of the original S. 776. Our figures forecast doubling of the present $350 million testing business in five years. The number of toxicologists could not be doubled in this time. Consequently, major changes would be required in the administration and work force of testing facilities, probably with some adverse effects on quality and/or quantity of output.

These facilities, especially the larger laboratories, are strained to capacity now. This reflects industry's continual shift to a more responsible level of testing employing more sophisticated protocols. At stake is the setting of a precise standard for a more responsible level of testing in good balance with the real need.

A Forward-Looking Analysis of Production Limitation

To substantiate forecasts on costs of limiting production, an analysis was made of the possible prohibition of markets for polychlorinated biphenyls, "PCB." These are highly stable synthetic liquids now used soley as dielectric fluids in highly restricted applications. For this one closely related group of compounds, costs are estimated at a one-time expense of $13.7 million and annually at $110 million for 20 to 30 years, based on the life cycle of involved end-use products. These figures confirm the very conservative basis of the forecasted levels for this type of restrictive action, which seems highly probable.

Chemical Process Industries

An area of this legislation which has received little attention but which warrants serious consideration, is its effect of the chemical process industries ("CPI"), as opposed to the primary chemical producers. The CPI includes elements from fourteen major SIC groups such as paper and allied products, petroleum refining, etc. all of which could be affected by reporting requirements and testing provisions. Because these effects are so diffused, no attempt was made to quantify them. Their exclusion adds to the conservative nature of this report's estimates.

Macroeconomic Effects

To this point, we have noted only those economic effects relating to the chemical industry. However, costs in economic sectors are interrelated and affect others. Obviously, the more basic the industry and the larger its product contribution to base-line industry, the greater its impact as incremental costs flow through the total economic matrix.

The U.S. chemical industry is one of the economy's most basic. What happens elsewhere when it is adversely affected can best be determined by employing widely accepted principles through computerized economic models. This technique is generally recognized as valid and is used by EPA in many of its macroeconomic effect analyses.

While such analysis is designed to determine various effects, most significant are the impacts on:

Inflation
Unemployment
Foreign trade
Economic growth

These have been analyzed in our study employing the macroeconomic input/output model developed by the University of Maryland—popularly known as the Maryland Model. The results can be examined quickly in tabular form:

MACROECONOMIC EFFECTS
[Dollar amounts in billions]

Factor (per year)	Base forecast	Selective model	Effect	Broad model	Effect
Gross national product...	$1,603.33	$1,603.07	−$0.26 (1972)...........	$1,601.76	−$1.6 (1972).
Inflationary impact: Wholesale price index (1967 = 100).	201.50	201.50	No effect...............	202.42	+0.5 percent or $11.89 fewer (1985 dollars).
Unemployment (percent).	4.45	4.52	80,000 fewer jobs not created.	4.47	20,000 fewer jobs because: 54,000 jobs will be created.
Favorable Export-Import balance for 1972.	$4.26	$4.48	Trade balance appears to grow since exports are unaffected.	$3.12	−25 percent trade balance.

Summary of Economic Impact of Proposed Toxic Substances Control Legislation

Capital Investment

While fixed assets would remain, there would be a reduction in capital investment in the chemical industry.

For example:

Unanticipated additional R/D expenditures of $600 million per year to maintain present rate of innovation will reduce available capital for investment.

Expansion or improvement of existing facilities and construction of new facilities would be hampered and restricted.

The birth of new chemical companies would be greatly reduced because little if any risk or venture capital would be allocated to so severely regulated an industry plagued with product marketing uncertainties.

Employment

The U.S. chemical industry directly employed about 1,000,000 people in 1972. Indirect employment, in the chemical process and regulated and dependent industries, was estimated at 12 million in 1970. Unemployment effects, therefore, must be considered on both direct and indirect bases. Elements which would contribute to unemployment include:

(1) Direct:

(a) Curtailed expansion and limited formation of new companies due to scarce capital and high risk would have a negative effect on new employment.

(b) The unjustified banning or restricting of existing products also would have a negative effect.

(c) Plant closings, a particularly logical conclusion as related to small and medium-sized companies, would result in short and long-term unemployment.

(2) Indirect:

The relationship to dependent industries, integrated both forward and backward, would create a "ripple" effect resulting in two types of unemployment: direct as a consequence of immediate forestallment or curtailment of activities, and indirect through no new hiring. Should legislation such as H.R. 7229 be enacted in 1975, the number of jobs not created could reach 80,000 in 1985, according to the selective model. In the chemical industry, there are losses of about 20,000 and 9,000 jobs in the productive sector under the "selective" and "broad" models, respectively. Under the latter model, this loss is more than offset by an increase in service-sector jobs attributed to the chemical industry.

Contribution to GNP

An analysis of GNP often is considered of major economic significance. The inflated value of goods produced, due to added costs, would appear to be a positive contribution to the GNP. However, these are nonproductive costs and make no real addition to the BNP. In fact, the survey's broad model forecasts a decrease in the GNP of $1.6 billion a year, based on the 1972 dollar.

Imports/Exports

The chemical industry is a massive contributor to a favorable balance of trade. This would change as increased costs would place us at an economic disadvantage to foreign producers subject to less or no regulation. Furthermore, there would be a strong tendency toward even greater U.S. chemical production abroad.

Both factors would combine to affect adversely the export of U.S. chemical production and negatively influence the industry in two ways: (1) a direct loss of sales and volume, and (2) a fostering of development of the foreign chemical industry.

While the selective model indicates a slightly favorable effect, the net economic effect of the broad model projects a reduction in the trade balance of 25 percent, or $1.14 billion, an amount this nation can ill afford.

Research and Development

While a world leader, the U.S. chemical industry has no corner on scientific breakthroughs. Even the position now held would be threatened because there undoubtedly would be a reduction in the industry's research and development investment and activity. The entire complex of R&D-related questions would require the most serious evaluation.

Inflation

While the selective model indicates no significant effect on the wholesale price index, the broad model indicates that a substantial inflationary impact would flow from this legislation. Our figures show 0.5 percent increase in the wholesale price index or close to $12 billion by 1985. This would be felt in direct additional costs as well as in "additive" costs incurred as products flow through the manufacturing, processing, distribution and marketing channels.

Another factor, difficult to quantify but which must be considered, would be the effect on discretionary spending. This is a very sensitive economic indicator. Additional expenses incurred in the purchase of any given entity adversely affect the availability of money for discretionary spending, thus helping to disrupt and distort the economy. The consumer ultimately pays all these costs.

In another area, excessive costs could force small producers to close and lead to further chemical industry consolidation. Larger companies, able to absorb such costs, might grow even larger, generating undesirable effects on normal marketing competition.

Effects on Dependent Industries

While the input/output analysis looks to quantification of direct economic effects, there would be many indirect effects on dependent industries. Dependent industries are those that produce and sell goods to the chemical, chemical process and allied industries, as well as those that are dependent upon goods produced by the chemical process and allied industries. Logic dictates that any negative forces affecting the chemical industry would "ripple" through dependent industries. This would further affect employment in all sectors of the economy.

Industrial Transfer

There would be considerable emphasis toward the redirection and growth of chemical process industries abroad. Multinational firms would seriously reconsider their positions with a strong emphasis on foreign investment. Foreign producers, taking advantage of the U.S. situation, undoubtedly would increase capital, R&D and related investment in their home countries thus effecting a further transfer of chemical industry growth as well as dominance from the U.S.

Benefits of Toxic Substances Control Legislation

Preventive and corrective benefits for the chemical industry and the total economy are possible in a toxic substances control law designed to provide effectiveness in acceptable balance with cost.

Premarket screening cannot be an entirely reliable indicator of long-term effects which may require decades to be manifested. It would, however, reduce risk to workers, consumers and to the environment. The data base provided by this and other requirements ultimately should provide helpful analytical comparisons with later epidemiological studies and so improve our understanding of the true relationship between chemical substances and health and the environment.

This information—continually augmented and improved by reporting, inspection and adulteration provisions—also should help reduce the frequency of voluntary product discontinuations by industry and regulatory actions by EPA. Further, through the Inter Agency Coordinating Committee and other communication routes among all concerned federal agencies, the information could provide a sound basis for regulations related to any substance.

Many sectors of the chemical industry already are exercising risk-management actions of the types mandated by H.R. 7229, but a law based on this bill could make performance more uniform and reduce the probability of extreme aberrations.

It has been impossible to quantify the potential benefits of this legislation in terms of "dollars saved" principally because benefits most likely would be diffused throughout the economy and no analytical techniques have been established to relate them to gross health and environmental expenditures.

Conclusion and Recommendations

In closing, there is an additional subject area to consider. Our comments have been confined to toxic substances control legislation. To our knowledge, no one has addressed the problem of multiple economic effects—micro and macro—of all major legislation and the totality of the cumulative effect on industry, the public and U.S. and world economics. In addition, there are the ever-increasing, non-legislated cost burdens under which industry and the public must function.

We could not undertake the enormous project needed to quantify and inter-relate all pertinent material. We do, however, call attention to this matter. Of particular importance is the need to appreciate the direct and indirect increasing economic cost factors under which industry, the public and the economy now and in the future will operate in respect to:

Energy
Crude oil feedstock
Pollution control
OSHA
Labor
Social benefits programs

From the foregoing, it is not to be surmised that the chemical industry does not have a profound concern for human life and safety. Where decisions involving human life and safety are at issue, it is obvious that numerous factors come into play requiring carefully reasoned judgments and cost is only one of them.

Examined objectively and on balance, however, it always is necessary to evaluate consequences as well as benefits associated with any action. Therefore, the impact of the proposed legislation must be understood and carefully assessed. Specifically we recommended that:

(1) The economic effects of toxic substances control legislation—not on the chemical industry alone but on the U.S. and world economies—be carefully analyzed and weighed;

(2) Steps be taken to more clearly delineate toxic substances control legislation needs and the Administrator's authority to act under any such legislation, and

(3) Provision be made for a specific requirement that the Administrator include an economic impact analysis in any of his final rule-making procedures.

THE LABORATORY COSTS FOR TOXICITY TESTING*

The following tabular presentation gives a detailed breakdown of laboratory cost estimates for toxicity testing and studies. The estimates were developed by the Haskill Laboratory of the DuPont Company. For detailed descriptions of actual toxicity testing and study procedures see page 121 of Chapter Two.

	Estimated cost [1]	Time for completing experimental phase		Amount of test material required
		Weeks	Days	
DERMAL				
Acute skin absorption (ALD): 2–6 rabbits, 4 dose levels:				
(a) With pathology	$1,800	[2] 3		70 g.
(b) Without pathology	980			
(c) If biochemistry is required, cost estimate must be increased according to number and type of examinations requested.	300			
Subacute skin absorption: 18 rabbits, 10 doses:				
(a) With pathology	5,375	4		Depends on dose level, rabbits weigh 2–3 kg.
(b) Without pathology	2,500			
(c) If biochemistry is required, cost estimate must be increased according to number and type of examinations required.	1,000			
Acute skin absorption (LD50): 18–30 rabbits:				
(a) With pathology on 10 rabbits	3,000	[2] 3–4		100 g.
(b) Without pathology	1,500			
Skin irritation and sensitization: (guinea pigs):	900	6		10 g.
(a) Regular test not requiring pathology.				
Human patch test:				
20 subjects	740	3		440 in[2] fabric or 300 g fiber.[3]
200 subjects	4,500	3		500 in[2] fabric or 300 g fiber.[3]
1 subject (sensitized)	320	1		1 in[2] or 1 g.
Skin irritation: 6 rabbits	370	2		10 g.
FHSA skin irritation: 6 rabbits	450	1		5 g.
FHSA skin absorption: 6–10 rabbits	585	3		70 g.
Acnegenesis, rabbit ear	450	1		
DOT skin corrosion: 6 rabbits	350		1	50 g.
Class B poison, skin absorption: 10 rabbits	540		3	100 g.
EYE				
Irritation (Haskell procedure; 2 rabbits)	475	2		3 g.
Irritation (AG, chemical; 2 rabbits)	475	2		3 g.
Irritation (FHSA procedure; 6–18 rabbits)	650	2		5 g.
INHALATION				
Acute: LC50–ALC 36 rats—6 exposures	4,350	[4] 3–4		150–4,500 g depending on toxicity, m.wt and whether gas, liquid or dust.
Subacute inhalation—12 rats (10 4-hr exposure):				
(a) With pathology	4,880	4		40–4,000 g depending on factors mentioned above.
(b) If biochemistry is required, cost estimate must be increased according to number and type of examinations required.				
Inhalation:				
Long-term, 2-yr	300,000			Depends on design.
Long-term, 90 days	80,000			
Human exposure, 8 subjects	30,000			Do.
Approximate lethal temperature (ALT)—36 rats, 6 exposures, with pathology.	2,550	[4] 3–4		180 g.
Intratracheal insufflation: 40 rats, 1 exposure, with pathology.	8,300	[5] 9–104		5 g.
Class B poison—10 rats	1,230	1		100 g.
FHSA—10 rats (includes pathology)	785	[6] 3		60–100 g
Aspiration—Gerarde method	3,250	[6] 3		5 g.
Inhalation—pyrolysis	1,000	2		

* Haskell Laboratory, Cost Estimate for Studies, E.I. DuPont DeNemours & Co. A presentation to the House of Representatives Committee on Interstate and Foreign Commerce, July 1975.

HASKELL LABORATORY, COST ESTIMATES FOR STUDIES—JANUARY 1975—Continued

	Esti-mated cost [1]	Time for completing experimental phase Weeks	Days	Amount of test material required

ORAL

Approximate lethal dose (ALD): 6–10 rats, 6–10 dose levels:

(a) With pathology	$1,085	[6] 3		25 g.
(b) Without pathology	335			
Oral LD on fasted male and female Rats—36 Rats: (a) Without pathology	455	[6] 3		50 g.
Subacute—12 male rats, 1 does level: (a) With pathology	1,850	4		100 g 5,000 mg/kg.
(b) Without pathology	650			
(c) If biochemistry is required, cost estimate must be increased according to number and type of examinations requested.				
90-day feeding study: 110 rats, 3 dose levels (includes 1 generation reproduction): (a) Cost will depend on scope of test.	18,000		90	500 g at 5,000 p/m in diet.
90-day feeding study: 32 dogs, 3 dose levels: (a) Cost will depend on scope of test.	29,500		90	1,500 g. at 5,000 p/m in diet.
90-day feeding studies: Rats and dogs: (a) Cost will depend on scope of test, combination of above 2 estimates.	47,500		90	2,000 g. at 5,000 p/m in diet.
Reproduction study: 128 rats, 3 dose levels (3 generation study).	30,500	78		12 kg (total for both test 42 and 36) at 5,000 p/m in diet.
2-year feeding study: 400 rats, 32 dogs including reproductive studies.	231,000	104		
FHSA, 10 rats	335	[2] 3		25 g.
Class B poison, 10 rats	390	1		5 g.
Antidote, oral	8,000			
Carcinogenic, rats only, 3 levels, 2 yr	87,000	104		Depends on design.

PATHOLOGY

Mutagenicity Screening	1,200			
Dominant lethal	10,000			
Karyotyping	10,000			
Teratogenic (oral administration)				
Exact cost depends on test design				
Standard test on rats with 3 dose levels	14,750		15	

PHYSIOLOGY

Literature search	3,000	[7]	[7]	
Industrial hygiene survey	1,500		[8]	
Cardiac sensitization	6,200	[9]	[9]	
Pulmonary function studies	90,000			

AQUATIC

Fish toxicity (all bioassays in static system):

Brine shrimp bioassay (48 hr TLm)	220	1		50 g single compound, 1 liter effluents.
Freshwater fish (96 hr TLm)	275	1		100 g single compound, 1 gal effluents.
Freshwater macroinvertebrate (96 hr TLm)	275	1		Do.
Marine fish (96 hr TLm)	320	1		Do.
Dynamic, LC_{50}	750		4	Do.
Marine macroinvertebrate (96 hr TLm)	320	1		Do.
Bioaccumulation study	18,000		180	Do.
Residue study	10,000		42	Do.
Stream monitoring study	25,000	26		Do.

HASKELL LABORATORY, COST ESTIMATES FOR STUDIES—JANUARY 1975—Continued

	Esti-mated cost [1]	Time for completing experimental phase		Amount of test material required
		Weeks	Days	

SPECIAL STUDIES

Following studies are also carried out, but test cost depends on the test design, which is variable:

(a) Intravenous toxicity	$1,000			
(b) Intraperitonal toxicity, acute	1,000			
(c) Subcutaneous toxicity	1,000			
(d) Studies on rabbits, quail and other wildlife (cage or field conditions).	2,000			
(e) Teratogenicity and mutagenicity by skin absorption.	15,000			
(f) Carcinogenicity by skin absorption	95,000			
(g) Enzyme assay	5,000			
(h) Trace metals	5,000			
(i) Metabolites	1,000			
(j) Metabolism	25,000			
(k) Biodegradation	25,000			
(l) Blood level versus dose	7,500			
(m) Blood and/or urine levels of materials	5,000			
(n) Trace analyses of organic compounds	5,000			
(o) Demyelination, chickens	2,500			

[1] In many cases the estimate is in the middle of a wide range, with the cost of a particular study depending on its design.
[2] Dosing usually complete within a week. There is a two-week post-week post-treatment period.
[3] Some lot of material must be used for 20 and 200 subject tests.
[4] These studies require a two-week post-exposure observation period. Dosing is usually complete in one two-week period.
[5] This study has an eight-104 week post-treatment observation period. Dosing is usually complete within a week.
[6] Dosing usually complete within a week.
[7] For 100 hr of technical time (considered average).
[8] Average.
[9] 2 concentration levels, 12 dogs per level. If literature search or new method development is required, cost would increase according to design.

REPORTING REQUIREMENTS UNDER THE TOXIC SUBSTANCES CONTROL ACT

SECTION ONE

THE CHEMICAL SUBSTANCES INVENTORY REPORTING REGULATIONS

These regulations shall take effect on January 1, 1978. In accordance with 5 U.S.C. 553 (d)(3), the Administrator finds for good cause that the effective date of these regulations will not be postponed until 30 days after publication in the *Federal Register*. TSCA section 8(b) provides that a chemical substance may be included on the inventory only if it was manufactured or processed within three years before the effective date of these regulations. If these regulations are effective on January 1, 1978, any chemical substance manufactured or processed for a commercial purpose since January 1, 1975 may be reported for the inventory. The January 1, 1975 date has been relied on by the industry in preparing for reporting under these regulations. Any greater delay in the effectiveness of these regulations would interfere with orderly and timely reporting for the inventory.

Dated: December 12, 1977

Douglas M. Costle
Administrator
Environmental Protection Agency

Part 710—Inventory Reporting

Sec.
710.1   Scope and Compliance
710.2   Definitions
710.3   Applicability: Reporting for the Initial and Revised Inventory
710.4   Scope of the Inventory
710.5   How to Report for the Inventory
710.6   When to Report
710.7   Confidentiality
710.8   Effective Date

AUTHORITY: Subsection 8(a), Toxic Substances Control Act (TSCA) (90 Stat. 2003, 15 U.S.C. 2607 (a))

710.1 SCOPE AND COMPLIANCE

(a) This Part establishes regulations governing reporting by certain persons who manufacture, import, or process chemical substances for commercial purposes under section 8(a) of the Toxic Substances Control Act (15 U.S.C. 2607 (a)). Section 8(a) authorizes the Administrator to require reporting of information necessary for administration of the Act and requires EPA to issue regulations for the purpose of compiling an inventory of chemical substances manufactured or processed for a commercial purpose, as required by section 8(b) of the Act. Following an initial reporting period, EPA will publish an initial inventory of chemical substances manufactured or imported for commercial purposes. After a supplemental reporting period, EPA will publish a revised inventory including those additional chemical substances processed or used for commercial purposes or imported for commercial purposes as a part of a mixture or article. Further, in accordance with section 8(b), EPA periodically will amend the inventory to include new chemical substances which are manufactured or imported for a commercial purpose and reported under section 5(a)(1) of the Act. EPA also will revise the categories of chemical substances and make other amendments as appropriate.

(b) Section 15(3) of TSCA makes it unlawful for any person to fail or refuse to submit information required under these reporting regulations. In addition, section 15(3) makes it unlawful for any person to fail to keep, and permit access to, records required by these regulations. Section 16 provides that any person who violates a provision of section 15 is liable to the United States for a civil penalty and may be criminally prosecuted. Pursuant to section 17, the Government may seek judicial relief to compel submission of section 8(a) information and to otherwise restrain any violation of section 15. (NOTE: As a matter of traditional Agency policy, EPA does not intend to concentrate its enforcement efforts on insignificant clerical errors in reporting.)

(c) Each person who reports under these regulations shall maintain records that document information reported under these regulations and, in accordance with the Act, permit access to, and the copying of such records by EPA officials.

710.2 DEFINITIONS

For the purposes of this Part:

(a) The following terms shall have the meaning contained in the Federal Food, Drug, and Cosmetic Act, 21 U.S.C. 321 *et seq.,* and the regulations issued under such Act: "cosmetic," "device," "drug," "food," and "food additive." In addition, the term "food" includes poultry products, as defined in the Poultry Products Inspection Act, 21 U.S.C. 453 *et seq.;* meats and meat food products, as defined in the Federal Meat Inspection Act, 21 U.S.C. 60 *et seq.;* and eggs and egg products, as defined in the Egg Products Inspection Act, 21 U.S.C. 1033 *et seq.*

(b) The term "pesticide" shall have the meaning contained in the Federal Insecticide, Fungicide, and Rodenticide Act, 7 U.S.C. 136 *et seq.,* and the regulations issued thereunder.

(c) The following terms shall have the meaning contained in the Atomic Energy Act of 1954, 42 U.S.C. 2014 *et seq.,* and the regulations issued thereunder: "byproduct material," "source material," and "special nuclear material."

(d) "Act" means the Toxic Substances Control Act, 15 U.S.C. 2601 *et seq.*

(e) "Administrator" means the Administrator of the U.S. Environmental Protection Agency, any employee or authorized representative of the Agency to whom the Administrator may either herein or by order delegate his authority to carry out his functions, or any other person who shall by operation of law be authorized to carry out such functions.

(f) An "article" is a manufactured item (1) which is formed to a specific shape or design during manufacture, (2) which has end use function(s) dependent in whole or in part upon its shape or design during end use, and (3) which has either no change of chemical composition during its end use or only those changes of composition which have no commercial purpose separate from that of the article and that may occur as described in section 710.4(d)(5); except that fluids and particles are not considered articles regardless of shape or design.

(g) "Byproduct" means a chemical substance produced without separate commercial intent during the manufacture or processing of another chemical substance(s) or mixture(s).

(h) "Chemical substance" means any organic or inorganic substance of a particular molecular identity, in-

cluding any combination of such substances occurring in whole or in part as a result of a chemical reaction or occurring in nature, and any chemical element or uncombined radical; except that "chemical substance" does *not* include:

(1) any mixture,

(2) any pesticide when manufactured, processed, or distributed in commerce for use as a pesticide,

(3) tobacco or any tobacco product, but not including any derivative products,

(4) any source material, special nuclear material, or byproduct material,

(5) any pistol, firearm, revolver, shells, and cartridges, and

(6) any food, food additive, drug, cosmetic, or device, when manufactured, processed, or distributed in commerce for use as a food, food additive, drug, cosmetic, or device.

(i) "Commerce" means trade, traffic, transportation, or other commerce (1) between a place in a State and any place outside of such State, or (2) which affects trade, traffic, transportation, or commerce described in clause (1).

(j) "Distribute in commerce" and "distribution in commerce" when used to describe an action taken with respect to a chemical substance or mixture or article containing a substance or mixture, mean to sell or the sale of, the substance, mixture, or article in commerce; to introduce or deliver for introduction into commerce, or the introduction or delivery for introduction into commerce of, the substance, mixture, or article; or to hold, or the holding of, the substance, mixture, or article after its introduction into commerce.

(k) "EPA" means the U.S. Environmental Protection Agency.

(l) "Importer" means any person who imports any chemical substance or any chemical substance as part of a mixture or article into the customs territory of the U.S. and includes: (1) the person primarily liable for the payment of any duties on the merchandise, or (2) an authorized agent acting on his behalf (as defined in 19 CFR 1.11).

(m) "Impurity" means a chemical substance which is unintentionally present with another chemical substance.

(n) "Intermediate" means any chemical substance (1) which is intentionally removed from the equipment in which it is manufactured, and (2) which either is consumed in whole or in part in chemical reaction(s) used for the intentional manufacture of other chemical substance(s) or mixture(s), or is intentionally present for the purpose of altering the rate of such chemical reaction(s). (NOTE: The "equipment in which it was manufactured" includes the reaction vessel in which the chemical substance was manufactured and other equipment which is strictly ancillary to the reaction vessel, and any other equipment through which the chemical substance may flow during a continuous flow process, but does not include tanks or other vessels in which the chemical substance is stored after its manufacture.)

(o) ''Manufacture'' means to produce or manufacture in the United States or import into the customs territory of the United States.

(p) ''Manufacture or import *'for commercial purposes'* '' means to manufacture or import

(1) for distribution in commerce, including for test marketing purposes, or

(2) for use by the manufacturer, including for use as an intermediate.

(q) ''Mixture'' means any combination of two or more chemical substances if the combination does not occur in nature and is not, in whole or in part, the result of a chemical reaction; except that ''mixture'' does include (1) any combination which occurs, in whole or in part, as a result of a chemical reaction if the combination could have been manufactured for commercial purposes without a chemical reaction at the time the chemical substances comprising the combination were combined and if, after the effective date or premanufacture notification requirements, none of the chemical substances comprising the combination is a new chemical substance, and (2) hydrates of a chemical substance or hydrated ions formed by association of a chemical substance with water.

(r) ''New chemical substance'' means any chemical substance which is not included in the inventory compiled and published under subsection 8(b) of the Act.

(s) ''Person'' means any natural or juridical person including any individual, corporation, partnership, or association, any State or political subdivision thereof, or any municipality, any interstate body and any department, agency, or instrumentality of the Federal government.

(t) ''Process'' means the preparation of a chemical substance or mixture, after its manufacture, for distribution in commerce (1) in the same form or physical state as, or in a different form or physical state from, that in which it was received by the person so preparing such substance or mixture, or (2) as part of a mixture or article containing the chemical substance or mixture.

(u) ''Process for 'commercial purposes' '' means to process (1) for distribution in commerce, including for test marketing purposes, or (2) for use as an intermediate.

(v) ''Processor'' means any person who processes a chemical substance or mixture.

(w) ''Site'' means a contiguous property unit. Property divided only by a public right-of-way shall be considered one site. There may be more than one manufacturing plant on a single site. For the purposes of imported chemical substances, the site shall be the business address of the importer.

(x) ''Small manufacturer or importer'' means a manufacturer or importer whose total annual sales are less than $5,000,000, based upon the manufacturer's or importer's latest complete fiscal year as of January 1, 1978, except that no manufacturer or importer is a ''small manufacturer or importer'' with respect to any chemical substance which such person manufactured or imported in quantities greater than 100,000 pounds during calendar year 1977. In the case of a company which is owned or controlled by another company, total annual sales shall be based on the total annual sales of the owned or controlled company, the parent company, and all companies owned or controlled by the parent company taken together. (NOTE: The purpose of the exception to the definition is to ensure that manufacturers and importers report production volumes for all chemical substances which they manufactured or imported in quantities equal to or greater than 100,000 pounds during calendar year 1977.)

(y) ''Small quantities for purposes of scientific experimentation or analysis or chemical research on, or analysis of, such substance or another substance, including any such research or analysis for the development of a product'' (hereinafter sometimes shortened to ''small quantities for research and development'') means quantities of a chemical substance manufactured, imported, or processed or proposed to be manufactured, imported, or processed that (1) are no greater than reasonably necessary for such purposes and (2) after the publication of the revised inventory, are used by, or directly under the supervision of, a technically qualified individual(s). (NOTE: Any chemical substance manufactured, imported or processed in quantities of less than 1,000 pounds annually shall be presumed to be manufactured, imported, or processed for research and development purposes. No person may report for the inventory any chemical substance in such quantities unless that person can certify that the substance was not manufactured, imported, or processed solely in small quantities for research and development, as defined in this section.)

(z) ''State'' means any State of the United States, the District of Columbia, the Commonwealth of Puerto Rico, the Virgin Islands, Guam, the Canal Zone, American Samoa, the Northern Mariana Islands, or any other territory or possession of the United States.

(aa) ''Technically qualified individual'' means a person (1) who because of his education, training, or experience, or a combination of these factors, is capable of appreciating the health and environmental risks associated with the chemical substance which is used under his supervision, (2) who is responsible for enforcing appropriate methods of conducting scientific experimentation, analysis, or chemical research in order to minimize such risks, and (3) who is responsible for the safety assessments and clearances related to the procurement, storage, use, and disposal of the chemical substance as may be appropriate or required within the scope of conducting the research and development activity. The responsibilities in clause (3) of this paragraph may be delegated to another individual, or other individuals, as long as each meets the criteria in clause (1) of this paragraph.

(bb) "Test marketing" means the distribution in commerce of no more than a predetermined amount of a chemical substance, mixture, or article containing that chemical substance or mixture, by a manufacturer or processor to no more than a defined number of potential customers to explore market capability in a competitive situation during a predetermined testing period prior to the broader distribution of that chemical substance, mixture or article in commerce.

(cc) "United States," when used in the geographic sense, means all of the States, territories, and possessions of the United States.

710.3 APPLICABILITY: REPORTING FOR THE INITIAL INVENTORY AND REVISED INVENTORY: WHO MUST REPORT; WHO SHOULD REPORT

Based on reports from manufacturers and some importers of chemical substances, EPA will compile an initial inventory of chemical substances manufactured for commercial purposes. Paragraph (a) of this section identifies who must report for this initial inventory and who should report.

After publication of the initial inventory, EPA will compile a revised inventory of chemical substances manufactured or processed for a commercial purpose based on reports from processors of chemical substances, and from importers of chemical substances as a part of mixtures or articles. Paragraph (b) of this section identifies who may report for this revised inventory.

Paragraph (c) of this section identifies the persons not subject to the initial inventory.

(a) *The Initial Inventory—* (1) *Domestic Manufacturers Who Must Report Concerning Chehical Substances*

Any person who manufactured a chemical substance(s) in the United States for a commercial purpose during calendar year 1977 must report concerning:

(i) All chemical substances which that person manufactured in the United States during calendar year 1977 at each site for which:

(A) Thirty percent or more of the weight of the products distributed from that site consists of products of the types described under Standard Industrial Classification (SIC) Group 28 or 2911, or

(B) The total pounds of reportable chemical substances manufactured at that site equals one million pounds or more; and

(ii) Any chemical substance not reported under paragraph (a)(1)(i) of this section that was manufactured at a site during calendar year 1977 in quantities equal to or greater than 100,000 pounds.

(NOTE: Any person who is a "small manufacturer," as defined in section 710.2, and who has more than one site, is exempt from separately reporting the chemical substances manufactured at each site.)

(2) *Importers Who Must Report Concerning Chemical Substances*

Any person who imported a chemical substance into the United States for a commercial purpose during calendar year 1977 must report concerning:

(i) All chemical substances which that person imported into the United States during calendar year 1977 if:

(A) Thirty percent or more of the weight of the products imported consists of products of the types described under Standard Industrial Classification (SIC) Group 28 or 2911, or

(B) The total pounds of reportable chemical substances imported equals one million pounds or more; and

(ii) Any chemical substance not reported under paragraph (a)(2)(i) of this section that was imported during calendar year 1977 in quantities equal to or greater than 100,000 pounds.

(NOTE: These reporting requirements include all chemical substances imported in bulk form, including in cans, bottles, drums, barrels, packages, tanks, bags and other containers, but do not include chemical substances imported as part of mixtures or articles.)

(3) *Other Manufacturers and Importers Who Should Report Chemical Substances*

(i) In order to ensure that a chemical substance is included in the initial inventory, any person who manufactures or imports, or who has manufactured or imported a chemical substance (including the importation of a chemical substance as part of a mixture or an article) for a commercial purpose since January 1, 1975, may report concerning that chemical substance.

(ii) Any person permitted to report under paragraph (a)(3) of this section may either report individually or, in accordance with section 710.5(f), authorize a trade association or other agent to report on his behalf.

(b) *Revised Inventory* (1) During the reporting period for the revised inventory (section 710.6(c)), a person may report concerning a chemical substance which was not included in the initial inventory if:

(i) the person has processed or used the chemical substance (including use in the manufacture of a mixture or article containing that chemical substance) for a commercial purpose since January 1, 1975; or

(ii) the person has imported the chemical substance as part of a mixture or article for a commercial purpose since January 1, 1975.

(2) Any person permitted to report under paragraph (b) of this section either may report individually or, in accordance with section 710.5(f), may authorize a trade association or other agent to report on his behalf. (NOTE: The premanufacture notification requirements of section 5(a)(1)(A) of the Act for manufacturers of new chemical substances and importers of new chemical substances in bulk will begin 30 days after the publication of the initial inventory and will apply to all chemical substances not included in the initial inventory. The premanufacture notification requirements of section 5(a)(1)(A) will not be applied to importers of chemical substances as part of a mixture until 30 days after publication of the revised inventory. In addition, section 15(2) of the Act as it relates to section 5(a)(1)(A) will not be applied to persons who process or use for a commercial purpose chemical substances not on the inventory until after publication of the revised inventory.)

(c) *Persons Not Subject to The Initial Inventory* Persons who have only processed or used a chemical substance for a commercial purpose are not subject to the initial inventory requirements.

710.4 SCOPE OF THE INVENTORY

(a) *Chemical Substances Subject to These Regulations*—Only chemical substances which are manufactured, imported, or processed "for a commercial purpose," as defined in section 710.2, are subject to these regulations.

(b) *Naturally Occurring Chemical Substances Automatically Included*—Any chemical substance which is naturally occurring and

(1) which is (i) unprocessed or (ii) processed only by manual, mechanical, or gravitational means; by dissolution in water; by flotation; or by heating solely to remove water; or

(2) which is extracted from air by any means, shall automatically be included in the inventory under the category "Naturally Occurring Chemical Substances." Examples of such substances are: raw agricultural commodities; water, air, natural gas, and crude oil; and rocks, ores, and minerals.

(c) *Substances Excluded by Definition or Section 8(b) of TSCA*—The following substances are excluded from the inventory:

(1) Any substance which is not considered a "chemical substance" as provided in subsection 3(2)(B) of the Act and in the definition of "chemical substance" in section 710.2(h);

(2) Any mixture as defined in section 710.2(q);

(NOTE: A chemical substance that is manufactured as part of a mixture is subject to these reporting regulations. This exclusion applies only to the mixture and not to the chemical substances of which the mixture is comprised. The term "mixture" includes alloys, inorganic glasses, ceramics, frits, and cements, including Portland cement.)

(3) Any chemical substance which is manufactured, imported, or processed solely in small quantities for research and development, as defined in section 710.2(y); and

(4) Any chemical substance not manufactured, processed or imported for a commercial purpose since January 1, 1975.

(d) *Chemical Substances Excluded from the Inventory* The following chemical substances are excluded from the inventory. Although they are considered to be manufactured or processed for a commercial purpose for the purpose of section 8 of the Act, they are not manufactured or processed for distribution in commerce as chemical substances *per se* and have no commercial purpose separate from the substance, mixture, or article of which they may be a part. (NOTE: In addition, chemical substances excluded here will not be subject to premanufacture notification under section 5 of the Act.)

(1) Any impurity.

(2) Any byproduct which has no commercial purpose.

(NOTE: A byproduct which has commercial value only to municipal or private organizations who (i) burn it as a fuel, (ii) dispose of it as a waste, including in a landfill or for enriching soil, or (iii) extract component chemical substances which have commercial value, may be reported for the inventory, but will not be subject to premanufacturing notification under section 5 of the Act if not included.)

(3) Any chemical substance which results from a chemical reaction that occurs incidental to exposure of another chemical substance, mixture, or article to environmental factors such as air, moisture, microbial organisms, or sunlight.

(4) Any chemical substance which results from a chemical reaction that occurs incidental to storage of another chemical substance, mixture, or article.

(5) Any chemical substance which results from a chemical reaction that occurs upon end use of other chemical substances, mixtures, or articles such as adhesives, paints, miscellaneous cleansers or other house-keeping products, fuels and fuel additives, water softening and treatment agents, photographic films, batteries, matches, and safety flares, and which is not itself manufactured for distribution in commerce or for use as an intermediate.

(6) Any chemical substance which results from a chemical reaction that occurs upon use of curable plastic or rubber molding compounds, inks, drying oils, metal finishing compounds, adhesives, paints, or other chemical substances formed during manufacture of an article destined for the marketplace without further chemical change of the chemical substance except for those chemical changes that may occur as described elsewhere in this section 710.4(d).

(7) Any chemical substance which results from a chemical reaction that occurs when (i) a stabilizer, colorant, odorant, antioxidant, filler, solvent, carrier, surfactant, plasticizer, corrosion inhibitor, antifoamer or defoamer, dispersant, precipitation inhibitor, binder, emulsifier, de-emulsifier, dewatering agent, agglomerating agent, adhesion promoter, flow modifier, pH neutralizer, sequesterant, coagulant, flocculant, fire retardant, lubricant, chelating agent, or quality control reagent functions as intended or (ii) a chemical substance, solely intended to impart a specific physico-chemical characteristic, functions as intended.

(8) Chemical substances which are not intentionally removed from the equipment in which they were manufactured. (NOTE: See note to definition of "intermediate" at section 710.2(n) for explanation of "equipment in which it was manufactured.")

710.5 HOW TO REPORT

(a) *General Instructions* (1) Except for small manufacturers or small importers, any preson who is required to report under section 710.3(a)(1) or (2) shall follow the reporting procedures of paragraphs (b), (c), and (d) of this section.

(2) Any person who reports under section 710.3(a)(3) shall follow the reporting procedures of paragraphs (b), (c), (d)(1) and (d)(3) of this section. In addition, the Agency encourages these persons to report in accordance with paragraphs (d)(2) and (d)(4) of this section. A trade association or other agent may report aggregated production data under paragraph (d)(4) of this section.

(3) Any person who is required to report under section 710.3(a)(1) or (2) and who is a small manufacturer or small importer as defined in section 710.2 shall follow the reporting procedures of paragraphs (b), (c), and (d)(1) and (3) of this section except that such person is exempt from reporting production volume (for quantities less than 100,000 pounds) and site information.

(4) Any person who reports under section 710.3(b) shall follow the reporting procedures of paragraphs (b), (c), and (d)(1) of this section.

(b) *Reporting the Identity of a Chemical Substance*—(1) Any person reporting under these regulations should first read and carefully follow the reporting instructions, "Reporting for the Chemical Substance Inventory," published by and available through EPA.

(2) To report a chemical substance, a person should first consult the TSCA Candidate List of Chemical Substances and any amendment to the Candidate List. For assistance in using the Candidate List, consult the "Guide to the Use of the TSCA Candidate List of Chemical Substances."

(3) All persons except "small manufacturers and importers" must use a separate Form A, B, or C to report chemical substances for each site. Small manufacturers and importers may report several chemical substances manufactured at different sites on one form, as appropriate.

(4) To report a chemical substance found in the Candidate List, or in an amendment to the list, a person must complete, sign, and submit EPA inventory report Form A (EPA Form No. 7710-3A). All forms, A through D, have OMB No. 1585 77011.

(5) To report a chemical substance not found in the Candidate List, or in an amendment to the list, but for which there is a Chemical Abstracts Service (CAS) Registry Number, a person must complete, sign and submit EPA inventory report Form B (EPA Form No. 7710-3B).

(6) To report a chemical substance which is not found in the Candidate List, or in an amendment to the list, and for which there is no known CAS Registry Number, a person must complete, sign, and submit EPA inventory report Form C (EPA Form No. 7710-3C). Persons must describe chemical substances on Form C as specifically as possible, in accordance with the instructions published by EPA, "Reporting for the Chemical Substance Inventory."

(7) To report a chemical substance whose chemical identity is claimed to be confidential, a person must complete, sign, and submit EPA inventory report Form C (EPA Form No. 7710-3C). In addition, he must substantiate the claim that the chemical identity is confidential at the time he submits the form to EPA, in accordance with

instructions published in "Reporting for the TSCA Inventory" and section 710.7.

(NOTE: The reporting instructions also describe a reporting Form D (EPA From No.7710-3D). This is for additional voluntary reports which may be submitted by any person who manufactures trademarked products comprised of chemical substances and is not a substitute for any of the reports required by these regulations.)

(c) *Reporting Polymers* (1) To report a polymer a person must list in the description of the polymer composition at least those monomers used at greater than two percent (by weight) in the manufacture of the polymer.

(2) Those monomers used at two percent (by weight) or less in the manufacture of the polymer may be included as part of the description of the polymer composition.

(NOTE: The "percent (by weight)" of a monomer is the weight of the monomer expressed as a percentage of the weight of the polymeric chemical substance manufactured.)

(d) *Reporting Other Information Concerning a Chemical Substance*— (1) For purposes of the initial inventory, designate whether the person manufactures and/or imports the chemical substance. For purposes of the revised inventory, designate whether the person processes and/or imports the chemical substance.

(2) Report the site(s) at which the person manufactures and/or imports the chemical substance. The site, as defined in section 710.2(w), for importers is their business address.

(3) Designate whether the person manufactures and processes the chemical substances only within a site *and* does not distribute the chemical substance, or any mixture or article containing that substance, for commercial purposes outside that site. (NOTE: This requirement does not apply to importers.)

(4) Report the amount of the chemical substance which the person manufactured at each site and/or imported during calendar year 1977. For each substance, report the digit (e.g., 0 through 9) which corresponds to the appropriate volume range, according to the following table. Enter "N" in the space provided for production amounts if the person did not manufacture or import the substance during calendar year 1977. Small manufacturers or importers, as defined in paragraph 710.2(x), should enter "X" in the space provided for production amounts of less than 100,000 pounds (45,400 kilograms). If a small manufacturer or importer reports these production amounts, that person shall enter both "X" and the appropriate digits (e.g., SO, S1, or S2). For other production ranges, do not include an "X" (e.g., 3 through 9). Trade associations or other agents should enter "A" in the space provided for production amounts. If trade associations or agents report production volumes,

they should enter both "A" and the appropriate digits (e.g., A2 or A6).

(0) Less than 1,000 pounds;
Less than 454 kilograms

(1) 1,000 to 10,000 pounds;
454 to 4,540 kilograms

(2) 10,000 to 100,000 pounds;
4,540 to 45,400 kilograms

(3) 100,000 to 1 million pounds;
45,400 to 454,000 kilograms

(4) 1 million to 10 million pounds;
454,000 to 4.54 million kilograms

(5) 10 million to 50 million pounds;
4.54 million to 22.7 million kilograms

(6) 50 million to 100 million pounds;
22.7 million to 45.4 million kilograms

(7) 100 million to 500 million pounds;
45.4 million to 227 million kilograms

(8) 500 million to 1 billion pounds;
227 million to 454 million kilograms

(9) over 1 billion pounds;
over 454 million kilograms

(A) Trade associations or other agents.

(e) *Importers*— (1) Any importer who reports a chemical substance for the inventory may authorize the foreign supplier of the imported chemical substance(s) to report to EPA on his behalf, if both the foreign supplier and the importer sign the declarations provided on the reporting forms. A foreign supplier may authorize an agent to act in his behalf.

(2) The importer has the ultimate responsibility for reporting all information required by this Part and for the completeness and truthfulness of such information. If certain information is not or cannot be provided by the foreign supplier or his duly authorized agent, it must be provided by the importer.

(f) *Trade Associations or Other Agents*— (1) A trade association or other agent may report on behalf of any person who is not required to report for the initial inventory under section 710.3(a)(1) and (a)(2). Accordingly, a trade association or other agent may report on behalf of a manufacturer or importer of a chemical substance who chooses to report under section 710.3(a)(3), or any processor or user of a chemical substance, or any importer of a chemical substance as part of a mixture or an article who chooses to report under section 710.3(b).

(2) For every chemical substance reported by a trade association or other agent under this section, at least one manufacturer, importer or processor must have certified to that agent, and be able to document to EPA, in accordance with section 710.1(c), that the chemical substance

was manufactured, imported, or processed for a commercial purpose since January 1, 1975.

710.6 WHEN TO REPORT

(a) A manufacturer, importer, or processor may claim ted by May 1, 1978.

(b) All reports concerning chemical substances which are manufactured or imported for a commercial purpose for the first time during the period from May 1, 1978 to the effective date of premanufacture notification requirements shall be submitted when such manufacturing or importation begins.

(c) All reports for the revised inventory shall be submitted within 210 days after publication of the initial inventory.

710.7 CONFIDENTIALITY

(a) A manufacturer, importer, or processor may claim that for a particular chemical substance any or all of the following items of information submitted under this Part are entitled to confidential treatment:

(1) Company name.

(2) Site.

(3) The specific chemical identity.

(4) Whether the chemical substance is manufactured, imported, or processed.

(5) Whether the chemical substance is manufactured and processed only within one site and not distributed for commercial purposes outside that site.

(6) The quantity manufactured, imported, or processed.

(b) Any claims of confidentiality must accompany the information at the time it is submitted to EPA. The claims must appear on the form on which the information is submitted to EPA and in the manner prescribed on the form. In addition, any claims of confidentiality must be substantiated at the time the information is submitted to EPA in the manner specified in the form instructions.

(c) Any information that is covered by a claim made as specified will be disclosed by EPA only to the extent permitted by, and by means of, the procedures set forth in this section and in Part 2 of this Title (41 FR 36902).

(d) If no claim accompanies information at the time it is submitted to EPA, the information may be made public by EPA without further notice to the submitter. Failure to provide substantiation of any claim asserted on the forms will be considered a waiver of the claim and will result in a

determination that the information is not entitled to confidential treatment.

(e) (1) A claim of confidentiality may be asserted concerning the specific chemical identity of a particular chemical substance. This claim may be asserted by any submitter who believes that inclusion of the specific chemical identity on the inventory would reveal the trade secret fact that the particular chemical substance is manufactured or processed for commercial purposes.

(2) If a submitter asserts such a claim the submitter must

(i) report the specific chemical identity,

(ii) propose a generic name which is only as generic as necessary to protect the confidential identity of the particular chemical substance,

(iii) provide a detailed, written substantiation of the claim as specified in the reporting instructions,

(iv) agree that EPA may disclose to a person with a *bona fide* intent to manufacture the substance (as defined in paragraph (g) of this section) the fact that the particular chemical substance is included in the inventory for purposes of TSCA section 5(a)(1)(A) premanufacture notification, and

(v) have available, and agree to furnish to EPA upon request, for the particular chemical substance, either an x-ray diffraction pattern (in the case of inorganic substances) or a mass spectrum for the particular chemical substance (in the case of most other substances), a sample of the substance in its purest form, an elemental or structural analysis, any additional or alternative spectra, or other data that may be required to resolve uncertainties with respect to the identity of the substance.

Failure to meet any of these five requirements will be considered a waiver of the claim and will result in inclusion of the particular chemical identity on the inventory.

(f) (1) If a submitter asserts that the identity of a particular chemical substance should not be included on the inventory, the submitter has met the five requirements specified in paragraph (e) of this section, and the EPA General Counsel has made a determination, in accordance with Part 2 of this Title, that the particular chemical identity should not appear on the inventory because inclusion would disclose a trade secret, EPA will publish a generic name in an appendix to the inventory rather than place the specific chemical identity on the inventory. Publication of a generic name in the appendix does *not* create a category for purposes of the inventory. Any person proposing to manufacture a substance included in the apprendix under a generic name must submit notice under section 5(a)(1)(A) of the Act unless specifically exempted by EPA (see paragraph (g) of this section).

(2) EPA will examine the generic name proposed by the submitter claiming confidentiality.

(i) If EPA determines that the generic name proposed by the submitter asserting the claim is only as generic as necessary to protect the confidential identity of the particular chemical substance, EPA will place that generic name on the inventory.

(ii) If EPA determines that the generic name proposed by the submitter asserting the claim is more generic than necessary to protect the confidential identity, EPA will ask the submitter to submit further proposed generic names.

(iii) If EPA does not agree with the further proposed generic names, EPA will choose a generic name that EPA determines is only as generic as necessary to protect the confidential identity. EPA will give 30 days notice of this choice to the submitter asserting the claim. After the end of the 30-day period EPA will place the chosen generic name on the inventory.

(g) (1) If the particular chemical substance a person is proposing to manufacture is not included on the inventory by specific name but does fall within one of the generic names in the appendix entitled "Confidential Identities," the person may ask EPA whether the specific substance is included on the inventory. EPA will answer such an inquiry only if EPA determines that the person has a *bona fide* intent to manufacture the substance.

(2) In order to establish a *bona fide* intent to manufacture the specific chemical substance the person proposing to manufacture the chemical substance must submit to EPA:

(i) a signed statement that that person intends to manufacture the substance for commercial purposes,

(ii) a description of the research and development activities he has conducted to date and the purposes for which the substance will be manufactured,

(iii) an elemental or structural analysis,

(iv) either an x-ray diffraction pattern (in the case of inorganic substances) or a mass spectrum (in the case of most other substances) of the particular chemical substance,

(v) a sample of the substance in its purest form, if requested, and

(vi) any additional or alternative spectra, or other data that may be required to resolve uncertainties with respect to the identity of the chemical substance.

(3) (i) Upon receipt of the information specified in paragraph (g)(2) of this section, EPA may require the submitter who asserted the confidentiality claim for a specific chemical substance within the generic name to submit to EPA:

(A) either an x-ray diffraction pattern or a mass or alternative spectrum for the substance,

(B) an elemental or structural analysis of the substance,

(C) a sample of the substance in its purest form, if requested, and

(D) any additional spectral or other data that may be required to resolve uncertainties with respect to the identity of the substance.

(ii) Failure to submit any of the information required by EPA under this paragraph (g)(3) will be construed as a waiver of the submitter's confidentiality claim, and EPA will place the specific chemical identity on the inventory without further notice to the submitter.

(4) EPA will compare the information submitted by the proposed manufacturer under paragraph (g)(2) of this section with the information submitted under paragraph (g)(3) of this section.

(5) If (a) the comparison of the elemental analyses and either the x-ray diffraction patterns or mass or alternative spectra is sufficiently similar to be consistent with a presumption that the chemical substances are the same, and (b) comparison of any of the other submitted information affirms this presumption, EPA will tell the person proposing to manufacture the particular chemical substance that the particular chemical substance is included on the inventory and, therefore, that premanufacture notification is not required.

(6) If (a) the comparison of either the x-ray diffraction patterns or the mass or alternative spectra is not sufficiently similar to be consistent with a presumption that the chemical substances are the same, and (b) comparison of the other information affirms this conclusion, EPA will tell the person proposing to manufacture the particular substance that the information submitted does not support a conclusion that the substance is included on the inventory, and, therefore, that premanufacture notification is required.

(7) A disclosure to a person with a *bona fide* intent to manufacture a particular chemical substance will not be considered a public disclosure.

710.8 EFFECTIVE DATE

These regulations shall take effect on January 1, 1978.

A Summary and Analysis of the Regulations

The inventory reporting regulations were initially proposed on March 9 and August 2, 1977 in the FEDERAL REGISTER and supplemented thereafter. The final inventory reporting regulations were signed on December 12, 1977. Specifically, these regulations require some persons who manufacture of import chemical substances:

(1) to report the identity of each chemical substance manufactured at each site of manufacture, or imported into the United States, for a commercial purpose;

(2) to estimate the amount of each chemical substance manufactured at each site, or imported during calendar year 1977; and

(3) to indicate whether each such chemical substance is manufactured and used only within one site.

Based upon the reports of manufacturers and importers, the U.S. Environmental Protection Agency (EPA) will publish an initial inventory of chemical substances. After publication of the initial inventory, these regulations authorize reporting by processors of additional chemical substances. EPA will publish a revised inventory including these substances in 1979.

DATES: These regulations are effective January 1, 1978. Reporting for the initial inventory by manufacturers and importers of chemical substances will begin January 1, 1978 and end May 1, 1978. During the 210 days after publication of the initial inventory in late 1978, processors of chemical substances, and importers of chemical substances as part of mixtures or articles, may report additional chemical substances for a revised inventory. (NOTE: Many of the terms used in these regulations, such as "manufacturer" and "processor" have a special meaning for purposes of these regulations. Persons should read the regulations, especially the definition section carefully, and be sure they understand the special meanings of these terms.)

AUTHORITY, PURPOSE AND SCOPE OF THE REGULATIONS

These regulations are promulgated pursuant to the authority of section 8(a) of the Toxic Substances Control Act (TSCA). They accomplish two of the purposes contained in that section of the Act. In the first place, in accordance with section 8(a)(1) of the Act, they require reporting for compilation of the inventory of chemical substances manufactured or processed for a commercial purpose in the United States. The Administrator is required to compile and publish an inventory of chemical substances under section 8(b) of the Act. In the second place, under the authority of TSCA section 8(a)(1)(A), these regulations require reporting of production and site information on chemical substances, which is reasonably necessary for establishing a profile of the chemical industry, monitoring chemical substances in the environment, and setting Agency priorities for implementing other provisions of TSCA.

In the interest of accomplishing these objectives, section 710.3(a) requires any person who manufactured or imported chemical substances during calendar year 1977 to report concerning all such substances if (a) thirty percent or more of the weight of the products consists of products of the types described under Standard Industrial Classification (SIC) groups 28 or 2911, or (b) if the total pounds of reportable chemical substances manufactured or imported equals one million pounds or more. In addition, any person who manufactured or imported a chemical substance in 100,000 pound quantities or greater during calendar year 1977 must report concerning that chemical substance.

By directing the reporting requirements to those persons who are significantly engaged in manufacturing chemical substances, EPA will create a profile of the chemical industry useful in future implementation of TSCA. As a minimum, the Agency will know the site of manufacture of all chemicals manufactured in quantities greater than 100,000 pounds. Moreover, for every plant site substantially engaged in producing chemical substances or chemical products for commercial purposes, EPA will know the identities of the substances manufactured there and the relative quantities in which they are produced.

EPA desires to minimize duplicative reporting to the extent consistent with its needs in implementing TSCA. Manufacturers and importers who do not meet the criteria for required reporting do not need to report individually. Instead, they may report through a trade association or rely upon another manufacturer or an importer to report the substances for inclusion on the inventory. Further, in the interest of minimizing duplicative reporting, persons who are processors of a chemical substance they neither manufacture nor import are not subject to the initial reporting requirements. Persons who process or use chemical substances for commercial purposes may report those chemical substances not included in the initial inventory during a special reporting period. In addition, persons who import chemical substances as a part of a mixture or article may report during the special reporting period. EPA expects to publish a revised inventory based on these additional reports sometime in fall 1979.

SMALL MANUFACTURER

As provided in TSCA section 8(a) EPA may require "small manufacturers" to submit only information necessary for compilation of the inventory or concerning a chemical substance which is subject to a proposed rule or order under TSCA section 4, 5 or 6 or court action under section 5 or 7. At this time, any person who is required to report and who is a small manufacturer is only required to submit information required for compilation of the inventory. Accordingly, any "small manufacturer" need only identify chemical substances and report certain limited information required for purposes of the inventory. Small manufacturers are exempt from reporting production volumes and, if a small manufacturer has more than one plant site, he need not separately report for each site. Since these regulations define "manufacture" to include "import," the provision applies equally to "small importers."

The definition of "small manufacturer or importer" proposed in August has been revised to exempt from these additional reporting requirements those manufacturers and importers with total sales of less than $5 million. This exemption, however, does not apply with respect to any chemical substance produced by a manufacturer or imported in quantities equal to or greater than 100,000 pounds during calendar year 1977. Accordingly, no manufacturer will be considered a "small manufacturer or importer" with respect to any chemical substance manufactured or imported in quantities over 100,000 pounds.

In deciding how to define "small manufacturer or importer," EPA considered both the relative burden to manufacturers and importers to submit additional information and the value of that information to EPA and other federal agencies. Under this definition, the total cost of reporting for the inventory for the marginal small firm, in most cases, will be about 1 percent of profits. While the firms exempted under this definition represent nearly 80 percent of chemical firms, they account for only 4 percent of sales of chemical substances and only 6 percent of employment in SIC groups 28 and 2911.

In the interest of creating a data base that is at least complete with respect to the volumes of those chemical substances produced in substantial quantities, this definition does not exempt any manufacturer or importer from reporting the volumes of individual substances produced in quantities of over 100,000 pounds during calendar year 1977. Reporting production volume will involve simply providing the digit that is associated with a broad range (e.g., "4" is for production between 1 million and 10 million pounds), as provided in the table at section 710.5(d)(4). Since the manufacturer or importer will already be reporting the names of substances, the additional burden of supplying production information in terms of large ranges should be minimal, especially for substances produced in significant quantities.

REPORTING SCHEDULE AND ENFORCEMENT

These regulations provide for publication of an initial inventory based on reporting by manufacturers and importers of chemical substances, followed by publication of a revised inventory based on reporting by processors of chemical substances and importers of chemical substances as part of mixtures and articles. Reporting for the initial inventory will begin on January 1, 1978, and end May 1, 1978. Manufacturers and importers of chemical substances must report, as provided in section 710.3(a), during this first reporting period. Processors are only subject to the second reporting period which will begin after publication of the initial inventory and end 210 days later.

Thirty days after publication of the initial inventory, premanufacture notification will begin. After that date, any person who intends to manufacture or import (in bulk) a chemical substance not included on the inventory must submit premanufacture notice under section 5(a)(1)(A). Processors, and users of a chemical substance for a commercial purpose and importers of a chemical substance as a part of a mixture or article will be able to supplement the initial inventory during the second reporting period, as provided in section 710.3(b).

TSCA section 15(1) makes it unlawful for any person to fail or refuse to comply with the premanufacture notification requirements of section 5. TSCA section 15(2) makes it unlawful for a person to use for a commercial purpose any substance which he had reason to know was manufactured in violation of section 5. Sections 15(1) and 15(2) as they relate to section 5(a)(1)(A) will not be applied to persons who process or use for a commercial purpose chemical substances not on the inventory or who import chemical substances as a part of a mixture until after publication of the revised inventory. By reporting any chemical substance not included on the initial inventory during the second reporting period, these persons will be able to protect themselves from prosecution under section 15(2) with respect to the requirements of section 5(a)(1)(A) of TSCA. Importers of a chemical substance as part of a mixture or article will not be subject to premanufacture notification requirements for "new" chemical substances until 30 days after publication of the *revised* inventory. Under these reporting requirements, persons who import chemical substances as part of articles do not have to report concerning those chemical substances for the initial inventory. EPA is still considering whether importers of certain chemical substances as part of articles will be subject to premanufacture notification requirements under section 5(a)(1)(A). EPA will address this issue prior to publication of the initial inventory.

Finally, EPA recognizes that it is inevitable, considering the large volume of information to be compiled and transmitted, that there may be some unintentional clerical errors in reporting. Accordingly, the note in section 710.1(b) of these regulations provides that EPA does not intend to focus its enforcement efforts on reporting violations that are clerical in nature. Instead, EPA will give priority to bringing actions against persons who (1) report false information, (2) report for inclusion on the inventory chemical substances which are excluded under section 710.4(c) of these regulations, (3) fail to report, or (4) fail to maintain records documenting reported information.

CONFIDENTIALITY

There is an apparent conflict between section 14 and sections 8(b) and 5(a) of TSCA with respect to the inclusion of the identities of certain chemical substances on the inventory. Section 8(b) requires EPA to publish a list of *each* chemical substance which is manufactured or processed in the United States. Such list shall include *each* chemical substance which *any* person reports, under section 5 or subsection (a) of this section, is manufactured or processed in the United States.'' (emphasis added) The list has two purposes, to inform the public concerning which chemical substances are manufactured or processed for a commercial purpose and to which the public may be exposed, and to define what constitutes a "new chemical substance" for purposes of premanufacture notification requirements under section 5(a). However, section 14 states that any information reported to EPA under TSCA that is exempt from disclosure under the Freedom of Information Act fourth exemption (5 U.S.C. 552(b)(4)) may not be disclosed except in specific circumstances set out in section 14(a) and (b).

In the absence of the requirements of section 8(b) and 5(a), EPA would publish an inventory that would not include the identities of specific chemical substances for which the fact that the particular substance is manufactured or processed for commercial purposes is confidential. In the absence of the requirements of section 14, EPA would publish a list of *all* chemical substances manufactured or processed for commercial purposes. Since the term manufacture includes "to import," this discussion and the regulations apply equally to imported chemical substances.

Having no explicit statutory guidance about how to resolve this conflict, EPA has attempted to balance the concerns of section 14 with those of sections 8(b) and 5(a). EPA believes that Congress did not intend manufacturers to be required to furnish EPA premanufacture notification on existing chemical substances whose identity for purposes of the inventory has been claimed as trade secret. EPA believes that Congress did intend EPA to preserve confidentiality to the maximum extent practicable without impairing administration of TSCA. Accordingly, EPA has developed the approach set forth in section 710.7, and explained in greater detail in Appendix A, Response to Significant Comments. The approach balances confidentiality under section 14 with the regulatory scheme of sections 8(b) and 5(a). This approach will be used for the submission of confidential identities for the inventory now under section 8(a) and subsequently under section 5(a)(1)(A). Should the approach fail to achieve its stated purposes, EPA will re-examine the approach and consider alternatives.

EPA will allow manufacturers, importers and processors to claim as confidential the fact that a particular chemical substance is manufactured or processed in the United States for commercial purposes. The manufacturer or processor making such a claim must provide certain information and agree to certain provisions specified in section 710.7(e) of these regulations. EPA will make a final determination concerning entitlement to confidentiality, in accordance with EPA's procedures for handling confidentiality of business information in 40 CFR Part 2, Subpart B (41 FR 36906, September 1, 1976). If EPA determines that the fact the particular chemical substance is manufactured or processed in the United States for commercial purposes is confidential, EPA will not place the specific chemical identity on the published inventory. Instead, EPA will publish a generic name in an appendix to the inventory.

The generic name will inform the public of at least the generic types of confidential chemical substances manufactured or processed for a commercial purpose in the United States. Further, the generic name will be helpful to manufacturers who consult the appendix to the inventory to determine whether they must submit premanufacture notification under section 5(a)(1)(A) for a proposed "new" chemical substance. The generic name will alert them to the possibility that the proposed "new" chemical substance may be included on the inventory under that name. The generic name will *not* establish a category of chemical substances for purposes of the inventory and premanufacture notification requirements.

EPA wants to avoid the anti-competitive impacts which may arise if new entrants into an existing market were required to give premanufacture notification while the existing manufacturer who claimed the identity of the chemical substance as confidential was able to continue to manufacture it. If a manufacturer is required to give premanufacture notification on a chemical substance, he cannot manufacture the substance for at least the 90-day notice period. This delay may be considerably longer if a testing rule under section 4 requires the manufacturer to develop and submit certain test data. EPA is also interested in distinguishing a "fishing expedition" by a competitor from a *bona fide* inquiry concerning the identities of confidential chemical substances on the inventory.

Accordingly, section 710.7(g) permits an inquiring manufacturer to submit certain information to establish his *bona fide* intent to manufacture the chemical substance. A manufacturer is not required to establish this intent; he can simply submit a premanufacture notification. If a manufacturer establishes *bona fide* intent to manufacture a chemical substance, EPA will tell the inquiring manufacturer whether the chemical substance is included on the inventory as a confidential identity, and therefore, whether he must submit premanufacture notification under TSCA section 5(a)(1)(A). The submitter who claimed that the specific chemical identity should not appear on the inventory will be required by section 710.7(e)(2) to agree to have available and furnish to EPA upon request certain identifying information on the chemical substance and agree that EPA may disclose to a person with a *bona fide* intent to manufacture the substance whether the particular chemical substance is included on the inventory. Failure to furnish this information to EPA upon request will be construed as a waiver of the claim of confidentiality, and the specific identity will be placed on the inventory. The Agency believes that this resolution of the conflict between sections 5(a), 8(b), and 14 of the statute balances the equities and interests of all parties.

DEFINITIONS

EPA wishes to emphasize that the terms used in these regulations may not be wholly consistent with the ordinary usage of such terms. For example, the term "manufacturer" includes importers. As used in tthese regulations, the terms "manufacturer" and "processor" may both apply to a person who normally would consider himself one or the other. "Intermediate" refers only to those intermediates which are isolated or removed from the equipment in which they are manufactured. Persons should be sure they understand the special meanings of the terms used for purposes of these regulations.

SECTION TWO

INVENTORY REPORTING FORMS

The EPA has developed four forms for reporting under the chemical substances inventory regulations, as follows:

- Form A
- Form B
- Form C
- Form D

Form A is used for reporting chemical substances which are included on the Toxic Substances Control Act (TSCA) Candidate List of Chemical Substances, a list of approximately 33,000 chemical substances. Form A provides space for the Chemical Abstracts Service (CAS) registry number and the EPA code designation according to the Candidate List. For any listed chemical, a manufacturer merely has to report these numbers on Form A rather than the chemical name or description of the chemical substance.

Form B is used for reporting chemical substances which have chemical Abstracts Service Registry Numbers but are not included in the Candidate List.

Form C is used for reporting either chemical substances which do not have a Chemical Abstracts Service Registry Number or whose identities are confidential.

Form D may be used by manufacturers and importers of trademarked products to report their product trademarks. If such products contain chemical substances which are required to be reported for the Chemical Substance Inventory, it must be certified that these substances have been reported.

On the following pages you will find samples of all these forms with accompanying instructions for completion.

Approved OMB Form No.
158 S 77011

U. S. ENVIRONMENTAL PROTECTION AGENCY
CHEMICAL SUBSTANCE INVENTORY REPORT
(Section 8(a) and (b) Toxic Substances Control Act 15 USC 2607)

FORM

A

I. CERTIFICATION STATEMENT: I hereby certify that, to the best of my knowledge and belief:(1) the chemical substances identified below have been manufactured or imported for a commercial purpose since January 1, 1975, and can be reported for the inventory (40 CFR 710); (2) all information entered on this form is complete and accurate; and (3) the confidentiality statements on the back of this form are true as to that information for which I have asserted a confidentiality claim. I agree to permit access to, and the copying of, records by a duly authorized representative of the EPA Administrator, in accordance with the Toxic Substances Control Act, to document any information reported here.

SIGNATURE   DATE   NAME/TITLE (TYPE OR PRINT)

EPA USE ONLY	II. CORPORATION
MID	

III. PLANT SITE NAME/ADDRESS

NAME

ADDRESS

CITY   STATE

COUNTY   ZIP

DUN & BRADSTREET NO.

IV. PRINCIPAL TECHNICAL CONTACT(S)

FORM NO.

V. TSCA CANDIDATE LIST CHEMICAL SUBSTANCES (LIST ADDITIONAL SUBSTANCES ON SEPARATE FORMS)

NUMBER	CAS REGISTRY NUMBER (INCLUDE HYPHENS)	EPA CODE DESIGNATION (INCLUDE HYPHEN)	PRODUCTION RANGE	ACTIVITY		SITE LIMITED	CONFIDENTIALITY CLAIMS						EPA USE ONLY	NUMBER
				MANUFACTURE	IMPORT		(a) MANUFACTURE	(b) IMPORT	(c) SITE LIMITED	(d) PRODUCTION	(e) CORPORATION	(f) PLANT SITE		
1														1
2														2
3														3
4														4
5														5
6														6
7														7
8														8
9														9
10														10
11														11
12														12
13														13
14														14
15														15
16														16
17														17
18														18
19														19
20														20
21														21
22														22
23														23
24														24
25														25
26														26

EPA FORM NO. 7710-3A (11-77)

EPA USE ONLY

Instructions

Form A
TSCA Candidate List Chemical Substances

U.S. Environmental Protection Agency
Chemical Substance Inventory Report
Section 3(a) and (b)—Toxic Substances Control Act
(15 U.S.C. 2607)

Form A may only be used to report, for the Toxic Substances Control Act (TSCA) Section 8(a) and Section 8(b) Inventory, chemical substances that are identified in the EPA publication "Toxic Substances Control Act (TSCA), PL 94-469, Candidate List of Chemical Substances," April 1977, GPO Stock Number 055-007-00001-2, or in any addendum to that list published by EPA in the FEDERAL REGISTER. Chemical substances with known Chemical Abstracts Service (CAS) Registry Numbers but which do not appear in the TSCA Candidate List of Chemical Substances should be reported using Form B. A chemical substance which has no known CAS Registry Number and/or whose identity as a commercial chemical substance is claimed confidential, must be reported on Form C. Before completing this form, carefully read the inventory reporting regulations.

U.S. Environmental Protection Agency
Office of Toxic Substances
Columbus, Ohio 43210

EPA will acknowledge receipt of the forms to the addressee identified in block III of the form.

TYPE OR USE A BLACK BALL POINT PEN (Press Firmly).

BLOCK I. CERTIFICATION STATEMENT AND SIGNATURE:

The certification statement must be signed by a person authorized by the company to sign official documents for the company. If a trade association reports on behalf of one or more persons, a duly authorized official of the trade association must sign the form. If an importer elects to have his foreign supplier/manufacturer complete block V of this form, the importer must, nevertheless, sign the form, and a duly authorized official of the foreign supplier/manufacturer (identified in block IV) must sign in the space below the importer's signature.

DATE: Enter the month, day, and year that the form was signed.

NAME and TITLE: Enter the name and title of the person who signed the form.

BLOCK II. CORPORATION:

Enter the complete name of the domestic corporation of which the plant site identified in block III is a part or, if that corporation is controlled by another domestic cor-

poration, enter the complete name of the controlling corporation. If the plant site is owned by an unincorporated entity, enter the company name. A trade association should enter its complete name.

BLOCK III. COMPANY NAME AND PLANT SITE ADDRESS

Enter the company name and address of the plant where the chemical substances identified in block V are manufactured or processed. An importer should enter his company name and business address. A trade association should enter its name and headquarters address.

BLOCK IV. PRINCIPAL TECHNICAL CONTACT(S)

Enter the name, address, and telephone number (including area code) of the person(s) whom EPA may contact for clarification of information submitted on this form. An importer who elects to have his foreign supplier/manufacturer complete block V should enter the name and address of his foreign supplier/manufacturer.

BLOCK V. CANDIDATE LIST CHEMICAL SUBSTANCES:

CAUTION: The TSCA Candidate List of Chemical Substances inappropriately includes some mixtures and certain chemical substances which, as explained in the inventory reporting regulations, are excluded from the inventory. Do *not* report mixtures or excluded chemical substances. Furthermore, the Candidate List includes some trademarks. Do *not* use Candidate List entries which are trademarks to identify and report chemical substances. Trademarks will not be included on the inventory.

Up to 27 Candidate List chemical substances may be reported on this form. Manufacturers and processors should report on this form only TSCA Candidate List chemical substances which are manufactured or processed at the plant site identified in block III.

For each chemical substance entered in block V:

1. Enter in the column labeled "CAS Registry Number" the Chemical Abstracts Service (CAS) Registry Number as it appears in the Candidate List. Include hyphens.

2. Enter in the column labeled "EPA Code Designation" the code number (including hyphen) which accompanies the CAS Registry Number in the Candidate List.

3. As specified below, make the appropriate entry in the box under "Production Volume". Quantities should be entered in pounds and be expressed accurate to two (2) significant figures (for example, report 175,411 as 180,000; or 2,550 as 2,600)

a) Manufacturers and Importers: Enter the quantity manufactured and/or imported during calendar year 1976; except that (i) if there was no manufacture or importation during 1976, enter the quantity projected for manufacture and/or for importation during calendar year 1977, or (ii) if there was no manufacture during 1976 or 1977, enter the quantity manufactured or imported during calendar year 1975, or (iii) if you did not manufacture or import the chemical substance since January 1, 1975, enter the average annual quantity distributed in commerce since that date.

b) Processors: If you only processed the chemical substance since Januuary 1, 1975, make no entry.

c) Trade Associations: The estimated aggregate quantity manufactured by your member companies during calendar year 1976 may be entered.

4. Enter a check in the appropriate box(es) under the general heading "Activity" to indicate whether you manufacture, process, and/or import the chemical substance. Check as many boxes as applicable.

5. Enter a check in the box under "Site Limited" if you manufacture the chemical substance within the plant site identified in block III and do not distribute the chemical substance, or any mixture or article containing that substance, for commercial purposes outside that site.

6. *Confidentiality Claims:*

Enter checks in the appropriate boxes to indicate which information is claimed confidential. Trade associations are not permitted to make any confidentiality claims.

(a) By checking the box under "Manufacture" for a particular chemical substance, you assert that the fact that you manufacture the chemical substance at the plant site identified in block III for commercial purposes is confidential.

(b) By checking the box under "Process" for a particular chemical substance, you assert that the fact that you process the chemical substance at the plant site identified in block III for commercial purposes is confidential.

(c) By checking the box under "Import" for a particular chemical substance, you assert that the fact that you import the chemical substance for commercial purposes is confidential.

(d) By checking the box under "Site-Limited" for a particular chemical substance, you assert that the the fact that the chemical substances is not distributed for commercial purposes outside of the manufacturing site identified in block III is confidential.

(e) By checking the box under "Production Volume" for a particular chemical substance, you assert that the production volume of the chemical substance for the plant site identified in block III is confidential.

(f) By checking the box under "Corporation" for a particular chemical substance, you assert that the link of this particular chemical substance to the corporation identified in block II is confidential because the corporation is not known to the public as a manufacturer, importer, or processor of this particular chemical substance for commercial purposes.

(g) By checking the box under "Plant Site" for a particular chemical substance, you assert that the link of this chemical substance to the plant site identified in block III is confidential because it is not known to the public that the particular chemical substance is manufactured, imported, or processed for commercial purposes at this particular plant site.

U. S. ENVIRONMENTAL PROTECTION AGENCY
CHEMICAL SUBSTANCE INVENTORY REPORT
(Section 8(a) and (b) Toxic Substances Control Act 15 USC 2607)

FORM **B**

Instructions
Form B
Chemical Substances with CAS
Registry Numbers

Form B may only be used to report, for the Toxic Substances Control Act (TSCA), Section 8(a) and Section 8(b) Inventory, chemical substances with known Chemical Abstracts Service (CAS) Registry Numbers. Chemical substances which appear in the TSCA Candidate List of Chemical Substances, should be reported using Form A. A chemical substance which has no known CAS Registry Number and/or whose identity as a commercial chemical substance is claimed confidential, must be reported on Form C.

Before completing this form, carefully read the inventory reporting regulations.

U.S. Environmental Protection Agency
Office of Toxic Substances
Columbus, Ohio 43210

EPA will acknowledge receipt of the forms to the addressee identified in block III. TYPE, OR USE A BLACK BALL POINT PEN (Press Firmly).

BLOCK I. CERTIFICATION STATEMENT AND SIGNATURE:

The certification statement must be signed by a person authorized by the company to sign official documents for the company. If a trade association reports on behalf of one or more persons, a duly authorized official of the trade association must sign the form. If an importer elects to have his foreign supplier/manufacturer complete Block V of this form, the importer must, nevertheless, sign the form and a duly authorized official of the foreign supplier/manufacturer (identified in block IV) must sign in the space below the importer's signature.

DATE: Enter the month, day, and year that the form was signed.

NAME and TITLE: Enter the name and title of the person who signed the form.

BLOCK II. CORPORATION:

Enter the complete name of the domestic corporation of which the plant site identified in block III is a part or, if that corporation is controlled by another domestic corporation, enter the complete name of the controlling corporation. If the plant site is owned by an unincorporated entity, enter the company name. A trade association should enter its complete name.

BLOCK III. COMPANY NAME AND PLANT SITE ADDRESS:

Enter the company name and address of the plant where the chemical substances identified in block V are manufactured or processed. An importer should enter his company name and business address. A trade association should enter its name and headquarters address.

BLOCK IV. PRINCIPAL TECHNICAL CONTACT(S)

Enter the name, address, and telephone number (including area code) of the person(s) whom EPA may contact for clarification of information submitted on this form. An importer electing to have his foreign supplier/manufacturer complete block V, should enter the name and address of the foreign supplier/manufacturer.

BLOCK V. CHEMICAL SUBSTANCES WITH CAS REGISTRY NUMBERS

Up to 10 chemical substances may be reported on this form. Manufacturers and processors should report on this form only chemical substances with CAS Registry Numbers which are manufactured or processed at the plant site identified in block III.

For each chemical substance entered in Block V:

1. Enter in the column "CAS Registry Number" the Chemical Abstracts Service (CAS) Registry Number. Include hyphens.

2. Enter in the column labeled "Specific Chemical Name" the systematically derived or other specific chemical name. Enter only *nonconfidential* chemical names. All names reported in this column will be published in the inventory with the CAS Registry Number.

3. As specified below, make the appropriate entry in the box under "Production Volume". Quantities should be entered in pounds and be expressed accurate to two (2) significant figures (for example, report 175,411 as 180,000; or 2,550 as 2,600).

    a) Manufacturers and Importers: Enter the quantity manufactured and/or imported during calendar year 1976; except that (i) if there was no manufacture or importation during 1976, enter the quantity projected for manufacture and/or for importation during calendar year 1977, or (ii) if there was no manufacture during 1976 or 1977, enter the quantity manufactured or imported during calendar year 1975, or (iii) if you did not manufacture or import the chemical substance since January 1, 1975, enter the average annual quantity distributed in commerce since that date.

    b) Processors: If you only processed the chemical substance since January 1, 1975 make no entry.

    c) Trade Associations: The estimated aggregate quantity manufactured by your member companies during calendar year 1976 may be entered.

4. Enter a check in the appropriate box(es) under the general heading "Activity" to indicate whether you

manufacture, process, and/or import the chemical substance. Check as many boxes as applicable.

5. Enter a check in the box under "Site Limited" if you manufacture and process the chemical substance only within the plant site identified in block III and do not distribute the chemical substance, or any mixture or article containing that substance, for commercial purposes outside that site.

6. CONFIDENTIALITY CLAIMS

Enter checks in the appropriate blocks to indicate which information is claimed confidential. Trade associations are not permitted to make any confidentiality claims.

(a) By checking the box under "Manufacture" for a particular chemical substance, you assert that the fact that you manufacture the chemical substance at the plant site identified in block III for commercial purposes is confidential.

(b) By checking the box under "Process" for a particular chemical substance, you assert that the fact that you process the chemical substance at the plant site identified in block III for commercial purposes is confidential.

(c) By checking the box under "Import" for a particular chemical substance, you assert that the fact that you import the chemical substance for commercial purposes is confidential.

(d) By checking the box under "Site-Limited" for a particular chemical substance, you assert that the fact that the chemical substance is not distributed for commercial purposes outside of the manufacturing site identified in block III is confidential.

(e) By checking the box under "Production Volume" for a particular chemical substance, you assert that the production volume of the chemical substance for the plant site identified in block III is confidential.

(f) By checking the box under "Corporation" for a particular chemical substance, you assert that the link of this particular chemical substance to the corporation identified in block III is confidential because the corporation is not known to the public as a manufacturer, importer, or processor of this particular chemical substance for commercial purposes.

(g) By checking the box under "Plant Site" for a particular chemical substance, you assert that the link of this chemical substance to the plant site identified in block III is confidential because it is not known to the public that the particular chemical substance is manufactured, imported, or processes, for commercial purposes at this particular plant site.

CONFIDENTIALITY STATEMENTS
(For Chemical Substance Inventory Report Forms A and B)

By signing the statement appearing in block I of this form, the person signing the form certifies that the following statements are true for all information on this form that has been claimed as confidential by checking one or more of the boxes under the heading "Confidentiality Claims."

a. By checking the box under "Manufacture" for a particular chemical substance, I assert that the fact that we manufacture the chemical substance at the plant site identified in block III for commercial purposes is confidential.

b. By checking the box under "Process" for a particular chemical substance, I assert that the fact that we process the chemical substance at the plant site identified in block III for commercial purposes is confidential.

c. By checking the box under "Import" for a particular chemical substance, I assert that the fact that we import the chemical substance for commercial purposes is confidential.

d. By checking the box under "Site-Limited" for a particular chemical substance, I assert that the fact that the chemical substance is not distributed for commercial purposes outside of the manufacturing site identified in block III is confidential.

e. By checking the box under "Production Volume" for a particular chemical substance, I assert that the production volume of the chemical substance for the plant site identified in block III is confidential.

f. By checking the box under "Corporation" for a particular chemical substance, I assert that the link of this chemical substance to the corporation identified in block II is confidential because the corporation is not known to the public as a manufacturer, importer, or processor of this particular chemical substance for commercial purposes.

g. By checking the box under "Plant Site" for a particular chemical substance, I assert that the link of this chemical substance to the plant site identified in block III is confidential because it is not known to the public that the particular chemical substance is manufactured, imported, or processed for commercial purposes at this particular plant site.

General Statement

For ALL of the claims I have asserted by checking any of the boxes under "Confidentiality Claims" the following statements are true:

1. We have taken reasonable measures to protect the confidentiality of the information, and we intend to continue to take such measures.

2. The information is not, and has not been, reasonably obtainable without our consent by other persons (other than governmental bodies) by use of legitimate means (other than discovery based on a showing of special need in a judicial or quasi-judicial proceeding).

3. The information is not publicly available elsewhere.

4. Disclosure of the information would cause substantial harm to our competitive position.

U. S. ENVIRONMENTAL PROTECTION AGENCY

CHEMICAL SUBSTANCE INVENTORY REPORT

(Section 8(a) and (b) Toxic Substances Control Act 15 USC 2607)

FORM

C

I. CERTIFICATION STATEMENT: I hereby certify that, to the best of my knowledge and belief: (1) the chemical substance identified below has been manufactured or imported for a commercial purpose since January 1, 1975, and can be reported for the inventory (40 CFR 710); (2) all information entered on this form is complete and accurate; and (3) the confidentiality statements on the back of this form are true as to that information for which I have asserted a confidentiality claim. I agree to permit access to, and the copying of, records by a duly authorized representative of the EPA Administrator, in accordance with the Toxic Substances Control Act, to document any information reported here.

SIGNATURE   DATE   NAME/TITLE (TYPE OR PRINT)

Foreign Supplier Signature _____ Date _____

EPA USE ONLY   **II. CORPORATION**

MID

III. PLANT SITE NAME/ADDRESS   **IV. PRINCIPAL TECHNICAL CONTACT(S)**

NAME

ADDRESS

CITY   STATE

COUNTY   ZIP

DUN & BRADSTREET NO.

FORM NO.

NUMBER

EPA USE ONLY

EPA USE ONLY

(f) PLANT SITE

(e) CORPORATION

(d) PRODUCTION

(c) SITE LIMITED

(b) IMPORT

(a) MANUFACTURE

CONFIDENTIALITY CLAIMS

SITE LIMITED

IMPORT

MANUFACTURE

ACTIVITY

PRODUCTION RANGE

CLASS 1

CLASS 2

V. CHEMICAL SUBSTANCE WHERE THE IDENTITY IS CONFIDENTIAL (AND/OR) THE CAS REGISTRY NUMBER IS UNKNOWN.

SPECIFIC CHEMICAL NAME

(SEPARATE MULTIPLE NAMES WITH A SEMI-COLON)

CAS REGISTRY NUMBER (IF KNOWN)

IN THE SPACE PROVIDED BELOW, PROVIDE STRUCTURAL INFORMATION, MOLECULAR FORMULA, AND OTHER SUPPLEMENTAL INFORMATION TO AID IN THE SPECIFIC IDENTIFICATION OF THE CHEMICAL SUBSTANCE:

SEE ATTACHED SHEETS (WRITE FORM NO. ON ALL ATTACHMENTS)

NO OF SHEETS

MOLECULAR FORMULA

NUMBER

CHEMICAL SUBSTANCE IDENTITY IS CONFIDENTIAL

(1) SUBSTANTIATION: No. of sheets attached (write form number on all substantiation sheets).

Proposed Generic Name:

(3) I agree to the terms of CONFIDENTIAL CHEMICAL SUBSTANCE IDENTITY STATEMENT on the back of this form.

SAMPLE Do Not Use

Instructions
Form C
Chemical Substances Whose Identity Is Claimed Confidential or Whose CAS Registry Number Is Unknown

Form C may only be used to report, for the Toxic Substances Control Act (TSCA). Section 8(a) and Section 8(b) Inventory, a chemical substance whose identity as commercial chemical substance is claimed confidential or whose Chemical Abstract Service (CAS) Registry Number is unknown. Chemical substances which appear in the TSCA Candidate List of Chemical Substances should be reported using Form A. Chemical substances with known CAS Registry Numbers but which do not appear in the TSCA Candidate List of Chemical Substances should be reported using Form B.

Before completing this form, carefully read the inventory reporting regulations.

U.S. Environmental Protection Agency
Office of Toxic Substances
Columbus, Ohio 43210

EPA will acknowledge receipt of the forms to the addressee identified in block III.

TYPE, OR USE BLACK BALL POINT PEN (Press Firmly).

BLOCK I. CERTIFICATION STATEMENT AND SIGNATURE:

The certification statement should be signed by a person authorized by the company to sign official documents for the company. If a trade association reports on behalf of one or more persons, a duly authorized official of the trade association should sign the form. If an importer elects to have his foreign supplier/manufacturer complete block V of this form, the importer must, nevertheless, sign the form, and a duly authorized official of the foreign supplier/manufacturer (identified in block IV) must sign in the space below the importer's signature.

DATE: Enter the month, day, and year that the form was signed.

NAME and TITLE: Enter the name and title of the person who signed the form.

BLOCK II. CORPORATION:

Enter the complete name of the domestic corporation of which the plant site identified in block III is a part or, if that corporation is controlled by another domestic corporation, enter the complete name of the controlling corporation. If the plant site is owned by an unincorporated entity, enter the company name. A trade association should enter its complete name.

BLOCK III. COMPANY NAME AND PLANT SITE ADDRESS:

Enter the company name and address of the plant where the chemical substances identified in block V are manufactured or processed. An importer should enter his company name and business address. A trade association should enter its name and headquarters address.

BLOCK IV. PRINCIPAL TECHNICAL CONTACT(S):

Enter the name, address, and telephone number (including area code), of the person(s) whom EPA may contact for clarification of information submitted on this form. An importer electing to have his foreign supplier/manufacturer complete block V, should enter the name and address of the foreign supplier/manufacturer.

BLOCK V. CHEMICAL SUBSTANCE WHOSE IDENTITY IS CONFIDENTIAL and/or CAS REGISTRY NUMBER IS UNKNOWN

A. SPECIFIC CHEMICAL NAME: Indicate whether the chemical substance proposed for inclusion in the inventory falls within class 1 or 2, described as follows:

Class 1 chemical substances are distinct chemicals which can be represented by definite structural diagrams. For a class 1 chemical substance, propose, if possible, a systematically derived name that uniquely defines the chemical species. Also provide synonyms known to you, other than trademarks, by which the chemical substance is commonly known.

Class 2 chemical substances are those which can not be named by a class 1 description. For a class 2 chemical substance, propose a name which is as descriptive of the substance as possible. Also provide synonyms known.

B. ACTIVITY: Check the appropriate box(es) to indicate whether you import, manufacture, or process the chemical substance at the plant site identified in block III. Check as many boxes as applicable.

C. PRODUCTION VOLUME: As specified below, make the appropriate entry in the blank provided for "Production Volume". Quantities should be entered in pounds and be expressed accurate to two (2) significant figures (for example, report 175,411 as 180,000; or 2,550 as 2,600).

1. Manufacturers and Importers: Enter the quantity manufactured and/or imported during calendar year 1976; except that (i) if there was no manufacture or importation during 1976, enter the quantity projected for manufacture and/or for importation during calendar year 1977, or (ii) if there was no manufacture during 1976 or 1977, enter the quantity manufactured or imported during calendar year 1975, or (iii) if you did not manufacture or import the chemical substance since January 1, 1975, enter the average annual quantity distributed in commerce since that date.

2. Processors: If you only processed the chemical substance since January 1, 1975, make no entry.

3. Trade Associations: The estimated aggregate quantity manufactured by your member companies during calendar year 1976 may be entered.

D. CONFIDENTIALITY CLAIMS:

Enter checks in the appropriate boxes to indicate which information is claimed confidential. Trade associations are not permitted to make any confidentiality claims.

1. By checking the box under "Chemical Identity" for the substance reported, you assert that the chemical identity of the particular substance on the TSCA inventory is confidential. Enter the CAS Registry Number (including hyphens), if known. Check one or more of the justification statement boxes which refer to statements appearing on the back of this form under Item 1. These statements explain the reasons for asserting the identity to be confidential. EPA must know the reason for asserting the identity to be confidential.

    In addition, provide a generic name for inclusion on the Inventory which is only as generic as necessary to protect the confidential identity of the particular chemical substance.

2. By checking the box under "Manufacture" for a particular chemical substance, you assert that the fact that you manufacture the chemical substance at the plant site identified in block III for commercial purposes is confidential.

3. By checking the box under "Process" for a particular chemical substance, you assert that the fact that you process the chemical substance at the plant site identified in block III for commercial purposes is confidential.

4. By checking the box under "Import" for a particular chemical substance, you assert that the fact that you import the chemical substance for commercial purposes is confidential.

5. By checking the box under "Site-Limited" for a particular chemical substance, you assert that the fact that the chemical substance is not distributed for commercial purposes outside of the manufacturing site identified in block III is confidential.

6. By checking the box under "Production Volume" for a particular chemical substance, you assert that the production volume of the chemical substance for the plant site identified in block III is confidential.

7. By checking the box under "Corporation" for a particular chemical substance, you assert that the link of this particular chemical substance to the corporation identified in block II is confidential because the corporation is not known to the public as a manufacturer, importer, or processor of this particular chemical substance for commercial purposes.

8. By checking the box under "Plant Site" for a particular chemical substance, you assert that the link of this chemical substance to the plant site identified in block III is confidential because it is not known to the public that the particular chemical substance is manufactured, imported, or processed for commercial purposes at this particular plant site.

E. STRUCTURAL AND SUPPLEMENTAL INFORMATION

For Class 1 chemical substances, provide a structure diagram indicating the atoms and the nature of the bonds joining the atoms. Stereochemistry, if known, and ionic charges should be shown. In addition, provide a molecular formula which is an inventory of the kinds and numbers of atoms present in the molecule without regard to how the atoms are bonded.

For Class 2 chemical substances, describe, in the form of a reaction scheme, the final reaction sequence used to produce the reported chemical substance. Such description should identify all immediate precursor substance(s) and the nature of the reaction. All reactants should be identified by their CAS Registry Numbers, if known. In addition, provide, to the extent possible, a partial structural diagram.

Supplemental instructions for the proper identification of chemical substances is provided in Appendix A of the inventory reporting regulations (40 CFR 710).

CONFIDENTIALITY STATEMENTS
(For Chemical Substance Inventory Report Form C)

By signing the statement appearing in block I of this form, the person signing the form certifies that the following statements are true for all information on this form that has been claimed as confidential by checking one or more of the boxes under the heading "Confidentiality Claims" or the box entitled "Chemical Identity."

1. By checking the box entitled "Chemical Identity," I assert that the chemical identity of this chemical substance is confidential for one or more of the following reasons (as indicated by a check by the appropriate statement or statements):

A. ☐ This chemical substance is known to exist; however, no one knows that this chemical substance is being manufactured, imported, or processed for commercial purposes. If our competitors knew that this chemical substance is being manufactured, imported, or processed for commercial purposes, it would show them that the chemical substance has commercial potential and might lead them into research concerning its use. No one knows that this chemical substance has commercial possibilities except us to the best of our knowledge.

B.☐ This chemical substance is known to exist; however, no one knows that this chemical substance is being manufactured, imported, or processed for commercial purposes. If our competitors knew that this substance is being manufactured, imported, or processed for commercial purposes, they would immediately conclude that we had reported it. The fact that we manufacture, import, or process this chemical substance for commercial purposes is confidential.

C.☐ This chemical substance is not known to exist. If our competitors knew that this chemical substance does exist and that it is manufactured, imported, or processed for commercial purposes, it would show them that the chemical substance has commercial potential and might lead them into research concerning its use. No one knows that this chemical substance has commercial possibilities except us to the best of our knowledge.

2. By checking the box under "Manufacture" for a particular chemical substance, I assert that the fact that we manufacture the chemical substance at the plant site identified in block III site for commercial purposes is confidential.

3. By checking the box under "Process" for a particular chemical substance, I assert that the fact that we process the chemical substance at the plant site identified in block III for commercial purposes is confidential.

4. By checking the box under "Import" for a particular chemical substance, I assert that the fact that we import the chemical substance for commercial purposes is confidential.

5. By checking the box under "Site-Limited" for a particular chemical substance, I assert that the fact that the chemical substance is not distributed for commercial purposes outside of the manufacturing site identified in block III is confidential.

6. By checking the box under "Production Volume" for a particular chemical substance, I assert that the production volume of the chemical substance for the plant site identified in block III is confidential.

7. By checking the box under "Corporation" for a particular chemical substance, I assert that the link of this chemical substance to the corporation identified in block III is confidential because the corporation is not known to the public as a manufacturer, importer, or processor of this particular chemical substance for commercial purposes.

8. By checking box under "Plant Site" for a particular chemical substance, I assert that the link of this chemical substance to the plant site identified in block III is confidential because it is not known to the public that the particular chemical substance is manufactured, imported, or processed for commercial purposes at this particular plant site.

General Statement

For ALL of the claims I have asserted by checking any of the boxes under "Confidentiality Claims" or the box entitled "Chemical Identity" the following statements are true:

1. We have taken reasonable measures to protect the confidentiality of the information, and we intend to continue to take such measures.

2. The information is not, and has not been, reasonably obtainable without our consent by other persons (other than governmental bodies) by use of legitimate means (other than discovery based on a showing of special need in a judicial or quasi-judicial proceeding).

3. The information is not publicly available elsewhere.

4. Disclosure of the information would cause substantial harm to our competitive position.

U. S. ENVIRONMENTAL PROTECTION AGENCY
VOLUNTARY PRODUCT TRADEMARK REPORT
(IN CONJUNCTION WITH THE TOXIC SUBSTANCES CONTROL ACT INVENTORY REPORTING)

FORM
D

I. CERTIFICATION STATEMENT: I hereby certify that, to the best of my knowledge and belief, each trademark listed below identifies a product which I manufacture or import and that all component chemical substances that are permitted to be reported for the inventory (40 CFR 710) have been reported either by me or by others. I agree to permit access to, and the copying of records, by a duly authorized representative of the EPA Administrator, in accordance with the Toxic Substances Control Act, to document any information reported here.

EPA USE ONLY — MID — SIGNATURE — DATE — NAME/TITLE (TYPE OR PRINT)

II. CORPORATE NAME/ADDRESS — NAME, ADDRESS, CITY, STATE, COUNTY, ZIP, CORPORATE DUN & BRADSTREET NO.

III. PRINCIPAL TECHNICAL CONTACT(S)

IV. LIST OF PRODUCT TRADEMARKS

NO.	PRODUCT TRADEMARKS (NAMES)	NO.	NO.	PRODUCT TRADEMARKS (NAMES)	NO.
1		1	29		29
2		2	30		30
3		3	31		31
4		4	32		32
5		5	33		33
6		6	34		34
7		7	35		35
8		8	36		36
9		9	37		37
10		10	38		38
11		11	39		39
12		12	40		40
13		13	41		41
14		14	42		42
15		15	43		43
16		16	44		44
17		17	45		45
18		18	46		46
19		19	47		47
20		20	48		48
21		21	49		49
22		22	50		50
23		23	51		51
24		24	52		52
25		25	53		53
26		26	54		54
27		27	55		55
28		28	56		56

SAMPLE Do Not Use

EPA FORM NO. 7710-3D (11-77)

Instructions
Form D
Product Trademarks

Form D may be used by manufacturers and importers of trademarked products to report their product trademarks. If such products contain chemical substances which are permitted to be reported for the Toxic Substances Control Act (TSCA), Section 8(a) and Section 8(b) Chemical Substance Inventory by the inventory reporting regulations (40 CFR 710), the manufacturer or importer must certify that those chemical substances have been reported. Form D may *not* be used to report chemical substances. Chemical substances must be reported using Chemical Substance Inventory Report Forms A, B, or C, whichever is applicable.

From reports voluntarily submitted using Form D, EPA will compile and publish a Product Trademark List in conjunction with the TSCA Chemical Substance Inventory. The list will serve primarily two purposes. First, it will allow manufacturers and importers of trademarked products to assure customers that all reportable chemical substances contained in their products appear in the TSCA Chemical Substance Inventory. Second, processors and users, who may add chemical substances to the Inventory during a special 120-day reporting period following its publication, will be able to consult both the Inventory and the Product Trademark List to determine if the chemical substances they process or use have been reported.

Before completing this form, carefully read the inventory reporting regulations.

U.S. Environmental Protection Agency
Office of Toxic Substances
Columbus, Ohio 43210

EPA will acknowledge receipt of the form to the addressee identified in block II of the form.

TYPE OR USE A BLACK BALL POINT PEN (Press Firmly).

BLOCK I. CERTIFICATION STATEMENT AND SIGNATURE:

The certification statement must be signed by a person authorized by the company to sign official documents for the company. By signing the statement, you certify for each product trademark listed in block IV that all chemical substances permitted to be reported under the Toxic Substances Control Act Section 8(b) inventory reporting regulation (40 CFR 710) which comprise that trademarked product have been reported by someone for inclusion on the Chemical Substance Inventory.

BLOCK II. CORPORATE NAME AND ADDRESS:

Enter the complete name and address of the domestic corporation which manufactures or imports the trademarked products. For unincorporated entities, enter the company name and address.

BLOCK III. PRINCIPAL TECHNICAL CONTACT(S):

Enter the name, address, and telephone number (including area code) of the person(s) whom EPA may contact for clarification of information submitted on this form.

BLOCK IV. PRODUCT TRADEMARKS:

List the trademarks for products which you manufacture or import. Trademarks which cover a line of products may be listed in aggregated form if the certification statement is true for all products within that line.

REREGISTRATION REQUIREMENTS UNDER THE AMENDED PESTICIDES ACT

SECTION ONE

INSTRUCTIONS FOR REREGISTERING PESTICIDE PRODUCTS*

The Federal Insecticide, Fungicide, Rodenticide Act as amended, (FIFRA) requires that all *presently registered pesticide products shall be reregistered* under new requirements by *October 21, 1977.* In an effort to complete this task within the time frame established by Congress, the Registration Division has developed a special procedure, which, with your cooperation and effort, will expedite the necessary processing.

Because of the volume of products which must be reregistered—about 32,500—it is imperative that reregistration proceed on an orderly basis. Consequently the Registration Division intends to *call in applications* for reregistration according to a set schedule over the time frame allowed for reregistration. This schedule of call-in for reregistration was published in the Federal Register.

Do not submit applications for reregistration until you are specifically requested to do so.

We cannot over emphasize this point: to allow applications to be submitted without an orderly schedule would only cause confusion and delay in processing *all* applications, a situation neither you nor we can afford. Therefore, any application submitted prior to being called in *will be returned without action.*

It is extremely important that you understand thoroughly the reregistration process and the general instructions *before* the call-in process begins. Only a limited time can be allowed for the submission of applications once they are called in. When you are notified to submit an application for reregistration, *you must be prepared to respond within the time frame specified.* The Federal Register Notice of call-in was intended to give advance notice for purposes of planning and preparation.

This package contains general information concerning all products subject to reregistration. At the time of the actual call-in of your application, you will be sent a Specific Guidance Package covering the products being called in.

Together these two packages have been designed to provide you with all the information you will need to fulfill the requirements of the law. Options will be discussed where applicable. If you follow the instructions explicitly, you can avoid needless and time-consuming rejections.

If you have any questions about the reregistration process in general, you may contact the Registration Division, by calling the appropriate number below:

Insecticide-Rodenticide Branch (202) 755-9315
Herbicide-Fungicide Branch (202) 755-1806
Disinfectants Branch (202) 426-2635

General Instructions

The "Batch" Process

The Registration Division will use a so-called "batch" process to accomplish reregistration. Basically, the "batch" concept involves the following steps:

1. Grouping of products by active ingredient into defined "batches" having the same general toxicity or other characteristics.

2. Determination of reregistration requirements for the "batch" as a whole.

3. Development of a specific guidance package for the batch, which will clearly tell registrants what they must do to meet the requirements of 2 above for each of their products in the "batch".

4. Call-in and simultaneous processing of all products in the batch.

The "batch" concept will assure that all products of the same type (i.e., in the same batch) are handled consistently in reregistration and classification. The standards and requirements developed will apply across the board to all products in the batch.

Let's look at these four steps in more detail.

* U.S. Environmental Protection Agency, February 1976.

Product Grouping—All products presently registered have been or will be grouped into batches for reregistration call-in purposes. For example, a batch might consist of all iodine products, or all elemental copper products below a given percentage. Some active ingredients will be split into several batches, depending on broad use patterns. For products with a mixture of active ingredients, the determining factor as to batch will generally be the ingredient of highest toxicity.

Determination of Reregistration Requirements— For each batch, the Registration Division will determine (1) whether sufficient data have been submitted previously to support reregistration, and (2) whether such data are adequate to classify the batch products as GENERAL or RESTRICTED.

1. If all necessary data are present, the products in the batch are subject to call-in for reregistration and classification.

2. If data obtainable on a short-term basis are lacking such as acute toxicity data or fish and wildlife data, products in the batch *cannot be classified,* and therefore, cannot be reregistered until the missing data are supplied for the Administrator's consideration. All registrants of products in these batches will be notified as soon as possible, to allow the data gaps to be filled prior to the 1977 deadline.

3. If long-term studies, such as chronic toxicity studies, are not available, registrants of affected products will be notified as early as possible, so that the data can be developed. However, since long-term studies cannot be expected to be completed by October, 1977, the products falling into such batches will be reregistered on an *interim* basis, for a stated period of less than five years, while the studies are being conducted. Interim reregistrations are non-renewable.

Guidance Package—Once the data requirements for the particular batch of products have been established, the Registration Division will assemble and distribute to each affected registrant of the products included in the batch, a Specific Guidance Package related to the batch in question. The guidance package will include detailed instructions, which must be followed explicitly, on how to submit your application. The following instructions will be included:

1. A list of product registrations which are in the "BATCH".

2. The necessary forms for submission of applications.

3. A deadline.

4. Options for the specific Method of Support statements which will be acceptable, consistent with Section 3(c)(1)(D) of the Act as amended by HR 8841 (November 28, 1975) and the modified Statement of Interim Policy (41 FR 3339, January 22, 1976).

5. The proposed classification of the products in the batch together with any criteria which if satisfied, may make a candidate for Restriction acceptable for General Classification. Such criteria might include adding restrictive labeling statements, increasing a pre-harvest interval, revising the directions for use to be more specific, changing the method of application, or packaging in child-resistant containers.

If, in your opinion, the proposed classification is incorrect, and there are reasons why your product should be classified differently, you may propose such reasons in your reregistration application, together with data supporting your conclusions, and they will be considered.

6. The exact wording of the precautionary statements, the fish and wildlife cautions, the physical and chemical hazard statements, and the re-entry statement (where necessary), and other prescribed labeling statements.

7. Guidance concerning data requirements for reregistration, including citations of data which have been determined to satisfy the requirements, and indication of any unsatisfied data requirements. Names and addresses of all registrants of products in the "BATCH" covered by the Guidance Package will be included to benefit those registrants who may desire to cooperate with other registrants in developing any required data needed for reregistration.

8. Instructions regarding the directions for use. In those cases where the agency is aware of problems or inconsistencies which may exist in use directions of products in the "BATCH" special instructions will be given to modify certain directions for use. Examples of such instructions are:

   a. FOR AGRICULTURAL USES—Mode of application—the use directions must clearly state the method of application in order that the user has no confusion regarding ground and aerial application. If use directions for aerial application do not appear on the label, then aerial application will be deemed a use inconsistent with the label of that product.

   b. DOMESTIC VERSUS NONDOMESTIC USES— In some cases, labels include directions for use on sites which do not specify whether the site is domestic (in, on or around all structures, vehicles or areas associated with the household or home life . . .) or nondomestic (areas other than domestic). This distinction may be crucial in determining the classification of a use and the registrant may be required to clearly distinguish the type of site. An example of such a use pattern might be a use on apple trees. In this case, the registrant might be required to specify whether the apple trees are lawn trees around a home or trees in a commercial orchard.

Call-in of Reregistration Applications— Reregistration application batches will be called in a defined sequence. Each registrant of a product included in a particular batch will be notified directly that his application(s) for reregistration must be submitted. The notification will be accompanied by the Specific Guidance Package for the batch into which his product falls. A specified time period will be allowed for the submission of applications. PRODUCTS FOR WHICH APPLICATIONS ARE NOT RECEIVED BY THE DATE SPECIFIED IN THE NOTIFICATION WILL BE SUBJECT TO CANCELLATION.

Disposition of Applications—

1) If the application for reregistration is acceptable and in compliance with the Act and Regulations, you will be notified by a letter which will request five copies of the final printed labeling with necessary revisions.

2) If your draft labeling is acceptable without revisions, you may be notified by telephone to submit the required final printed labeling. In this case, your final printed labeling should be accompanied by a letter certifying that you have complied with our telephone request and that your final printed labeling is identical to the draft originally submitted.

In both cases, upon receipt of acceptable final printing labeling, your product will be officially reregistered and you will be so notified.

Status of Other Registration Activities During the Reregistration Period—

Amendments to the registration *may* be submitted concurrently with reregistration applications, or at any time during the reregistration period. We strongly encourage however that applications for amendment *NOT* be submitted concurrently with reregistration applications. Such amendments must be reviewed independently, delaying both the amendment and the reregistration response.

Only those amendments which can be processed without disruption of the reregistration operation will be considered. This decision will be made on a case by-case basis. The more complex the revisions proposed, the less likely that your application for amendment will be processed immediately. Applications which will jeopardize reregistration will be delayed.

You should adjust your registration activities accordingly.

New applications will continue to be processed; however, because of EPA's constrained resources during the period of reregistration they may also be delayed somewhat.

New applications for supplemental registration of distributors will continue to be processed as usual, except that new forms will be used. These are not affected by the reregistration process. Distributor products *previously accepted* need not be submitted again and do not require reregistration action. However, note that once a basic product is reregistered, all distributor product labels must be brought into compliance.

During the reregistration period, the five-year renewal program has been suspended. No renewal actions will be taken until after reregistration is complete.

Preparation of Application Packages

All applications for reregistration should be submitted to the following address:

Registration Division (WH-567)
ATTN: (Product Manager _____) (Batch No. _____)
Office of Pesticide Programs
Environmental Protection Agency
401 M Street, S.W.
Washington, D.C. 20460

All applications for reregistration must include the following:

1. Application for New Pesticide Registration (EPA Form 8570-1)
2. Confidential Statement of (EPA Form 8570-4)
3. Label Technical Data Sheet (EPA Form 8570-10)
4. Two compies of draft labeling.

Do *not* submit samples of the pesticide product or packaging, unless specifically requested to do so.

Forms to be Used

New forms have been developed by the Registration Division for use reregistration. You should familiarize yourself with them and the information they require, since they will be used not only for reregistration, but for all future applications for registration. For purposes of reregistration, on the Application for New Registration, the registration number should be put in block No. 3 (Company Product No.)

Application for New Registration (EPA Form 8570-1)

The old application forms have been substantially revised, to comply with the new provisions of the FIFRA, and to make the format more responsive to the needs of registrants and the Agency. Several new features have been incorporated into the form:

1. Provision for a proposed classification for each product. The Guidance Package contains instructions on the proposed classification of products in the batch. If you feel you have justification for a different classification you may propose it together with your reasons.

2. Space for separate listing of the various types of data physically submitted, with the application. If only references to data in the guidance package are used, do not list them there.

3. Because the requirements for data compensation have changed, an amended offer to Pay Statement Sheet will be attached to the form. Do not use the offer to Pay Statement on the form itself; it has been marked out. Check the appropriate block under No. 13, method of support. Note that the 2c option is no longer a viable option.

4. A block for you to tell us who in your firm or organization should be contacted if questions arise about your application. This is not necessarily the person who signed the form, but the person who is most knowledgable about the actual information in the application.

Confidential Statement of Formula (EPA Form 8570-4)

In an effort to update our files, we are requiring that each application for reregistration include this form. Some changes and additions have been made, including the following:

1. Additional information on the properties of the chemical. Such items are solubility rate, pH, and flash point, which previously were submitted separately, are now included on the Confidential Statement Form.

2. The purpose in the formulation of each component listed. Simple terms such as emulsifier, stabilizer, perfume, solvent, etc. will suffice. This is in accordance with the Regulations, which state that an ingredient for which no purpose can be determined may be required to be deleted from the formulation, or the application for registration may be denied.

3. A change in the procedure for dealing with supplier's statements of formula. *You must submit for each active ingredient a letter from your supplier authorizing us to refer to his confidential formula information in our files.* Suppliers should not submit a copy of the Confidential Statement of Formula as in the past. The authorization letter must accompany your application for reregistration.

Label Technical Data Sheet (EPA Form 8570-10)

This entirely new form contains simple check-offs describing the basic characteristics of your product for the purpose of developing a profile of all registered products. One copy is required with each application for reregistration.

Labeling—

1. Draft Labeling

Labeling submitted with applications for reregistration should be in draft form unless you have been specifically requested to submit final printed labels. The draft labeling should be either type-written text on 8 1/2 x 11 inch paper or a mock-up of the label prepared in such a way as to facilitate storage in 8 1/2 x 11 inch files. The draft label must indicate the intended colors of the final label, clear indication of the front and side panels and the intended type sizes of the text.

2. *Final Printed Labeling*

Final Printed Labeling is defined as that labeling which will accompany the pesticide product to market, and includes not only the label on the container, but also *all accompanying technical information, brochures, etc.*

Final printed labeling is to be prepared for storage in 8 1/2 by 11 inch files. Labels may be mounted or photoreduced to meet the size requirements provided the printing is legible and is of microfilm reproduction quality. Should photoreduction make any of the text illegible, the text must be typed out on an accompanying sheet of paper.

1. *Paste-on Labeling.* This should be submitted as is, unless it requires photoreduction.

II. *Screen Print Labeling.* Labels for submission can be obtained by taping paper on the container as it goes through the printing process.

III. *Embossed Labeling.* Submit photo copies.

IV. *Unusual Size Labeling.* Labeling for large bags or boxes must be photoreduced. Submit either the entire label on one reduction or in sections of 8 1/2 by 11 inches in size.

In those cases where it may not be feasible to complete a press run to produce actual copies of the finished label, copies of the printer's proof or what is commonly called "blue line" or "brown line" copy, may be submitted. However, at such time as final labels are printed, copies must be submitted to Registration Division for inclusion in the products' registration file.

In the case of photo reductions, reproductions, or printers' proofs, all type sizes and colors which will appear on the final label must be clearly indicated at the appropriate locations on the label copy.

In no case should containers or packages be submitted, nor labels too large to reasonably fit into 8 1/2 by 11 inch files.

Note that supplemental literature accompanying the product at any time during distribution, sale, or use is considered labeling and must be submitted. This is particularly important in those cases when use directions appear in attached booklets or other such literature. *Only those uses which you actually propose for reregistration will be considered for reregistration.*

Classification of Products

The law requires that during reregistration (and during registration, for new products) all products be classified, according to their use, as GENERAL or RESTRICTED. Products classified GENERAL may be used, *in accordance with label directions and precautions,* by members of the general public. Products classified RESTRICTED may be used only by a Certified Applicator who has been trained in the use of such pesticides.

THE SPECIFIC GUIDANCE PACKAGE WILL CONTAIN INFORMATION ON THE PROPOSED CLASSIFICATION OF YOUR PRODUCTS.

If you believe there are reasons why your product should not be classified as the Agency proposes, you may propose another classification, enclosing with your application an explanation of the reasons and documentary evidence in support of your arguments. We will review your arguments and data and will notify you of our decision. Because this review requires additional consideration independent of the batch processing, you should be aware that your reregistration will be delayed beyond that normally necessary for batch processing.

Should the Agency's decision remain unchanged after review, you may request a hearing under Section 6(b) of the Act.

A product's classification is determined by the classification of the uses appearing on the label. If product labeling contains both RESTRICTED and GENERAL uses, and no factors exist which would further delineate the proposed classification, several courses of action, given below, are available to you:

1. You may delete all RESTRICTED uses and reregister your product containing only those uses classified GENERAL.

2. GENERAL uses may appear on a RESTRICTED label, but not vice versa. Therefore, you have the option of using a RESTRICTED label bearing all of your directions for use. If you choose this option, you may *not* distinguish those uses that are GENERAL from those that are RESTRICTED.

3. You may register two separate products with identical formulations, one containing only GENERAL uses and the other RESTRICTED uses. To do so, submit two applications for reregistration, each containing all forms and necessary labels. Both applications should be submitted simultaneously in response to the Guidance package and both will be processed simultaneously.

Note that the products must have different product names and will be assigned separate registration numbers.

Addition of Classification Statement to the Label

When registration of a product is approved, the letter of approval will include instructions to label products with the new label within a specified timeframe. These instructions will differ for general use and restricted use classification as follows:

1. General Use Classification — All products packaged and labeled on or after October 21, 1977 must bear the reregistered label. However, the registrant has the option of using the reregistered label at any time in advance of October 21, 1977.

2. Restricted Use Classification — *All* products reregistered with restricted uses must bear the reregistered label, complete with restricted use statement by October 21, 1977. This includes *All* products being offered for distribution or sale. All products found in channels of trade without the *complete* reregistered label after December 21, 1977 will be subject to enforcement action under the FIFRA, as amended. In order to allow registrants to proceed with relabeling in advance of this date, products may be labeled using the reregistered label with a blank space in the area at the top of the front panel where the restricted use statement appears. Stick-on labeling may be utilized to add the restricted use statement on October 21, 1977.

The Label

The pesticide label is the final result of the registration and/or reregistration process and describes the risks and benefits of a given pesticide to the user. The label is not only the primary source of information to the user, it is also the primary tool of pesticide regulation. The label, in a sense, is a legal document. Improvement of pesticide labels benefits both the industry and the public.

The reregistration process provides for overall improvement in labels. Certain of the changes in label are required by the Regulations; other changes are recommendations. The charts that follow will clearly indicate what is required and what recommended. It is hoped that avoidance of all-inclusive requirements will allow flexibility in those labels which may not fit the standard situations.

The Section 3 Regulations require that certain statements must appear at certain locations on the label. The designation of specific areas of the label for specific information is known as *format labeling*. This is not a new concept, i.e., signal words have been required to appear on the front panel. The application of the concept to the entire label is new. If pesticide users know where to look on labels for certain kinds of information, we should be able to better train and educate people, thus improving the understanding of the proper use of pesticides. We ask your cooperation in making the format label a viable concept. We recognize the potential conflict between marketing concepts of product identity and the standardization which results from format labeling. However, the primary purpose of the pesticide lable is to communicate information and regulate use. Marketing is a secondary purpose, although a very real purpose. Leeway for product individuality is provided in format labels, but only with the recognition that the label is a legal document to instruct the user on use and safety. As you might expect, there is less leeway in the case of RESTRICTED use labels since these products are not for use by the general public.

The tables at the end outline the basic elements of the pesticide label and are keyed to the sample labels which follow the tables. These sample labels represent a typical three panel label for a RESTRICTED USE product of highest toxicity and a three panel label for a GENERAL Classification product. In the tables, Column 2 identifies each label element. Column 3 outlines the applicability of the requirement. Columns 4 and 5 describe the location of the element on the label and indicate what is required and what is recommended. Column 6 contains any additional comments or information on format. This chart is intended as a general guide. Registrants should refer to §162.10 of the Section 3 regulations for the exact labeling requirements.

Type Size Requirements

The table below shows the minimum type size requirements on various sized labels, as set forth in the Regulations.

Size of label on front panel in square inches	"Restricted Use Pesticide" (if required) and Signal Word All Capitals	Heading "Storage and Disposal" "Keep Out of Reach of Children"
5 and under	6 point	6 point
above 5 to 10	10 point	6 point
above 10 to 15	12 point	8 point
above 15 to 30	14 point	10 point
over 30	18 point	12 point

NOTE: All required label text must be set in 6-point or larger type.

Physical and Chemical Hazard Statements

Flammability

The chart on the next page illustrates in flow form the precautionary statements relating to flammability of pesticide products. Select the proper statement based on the product's characteristics. Note that the statement is not preceded by a signal word.

Other physical and chemical hazards statements may be specified in the Specific Guidance Package.

Storage and Disposal Statements

The regulations under Section 3 which provide for registrations and reregistration require that a storage and disposal statement appear on the label of *every pesticide product*.

Present storage and disposal statements are inadequate because:

1. Storage and disposal have been determined to be conditions of use of a pesticide product, and improper storage and disposal can thus be a violation of the misuse provisions of the law.

2. Regulations have been proposed which prohibit certain conditions of pesticide disposal and storage.

In order to avoid conflict with the proposed Regulations, and at the same time adopt a set of disposal options to direct the label reader, the EPA has developed a set of interim storage and disposal statements for use until date on disposal of pesticides and pesticide containers have been submitted. As data are developed on the disposal of various pesticides, the statements may be modified.

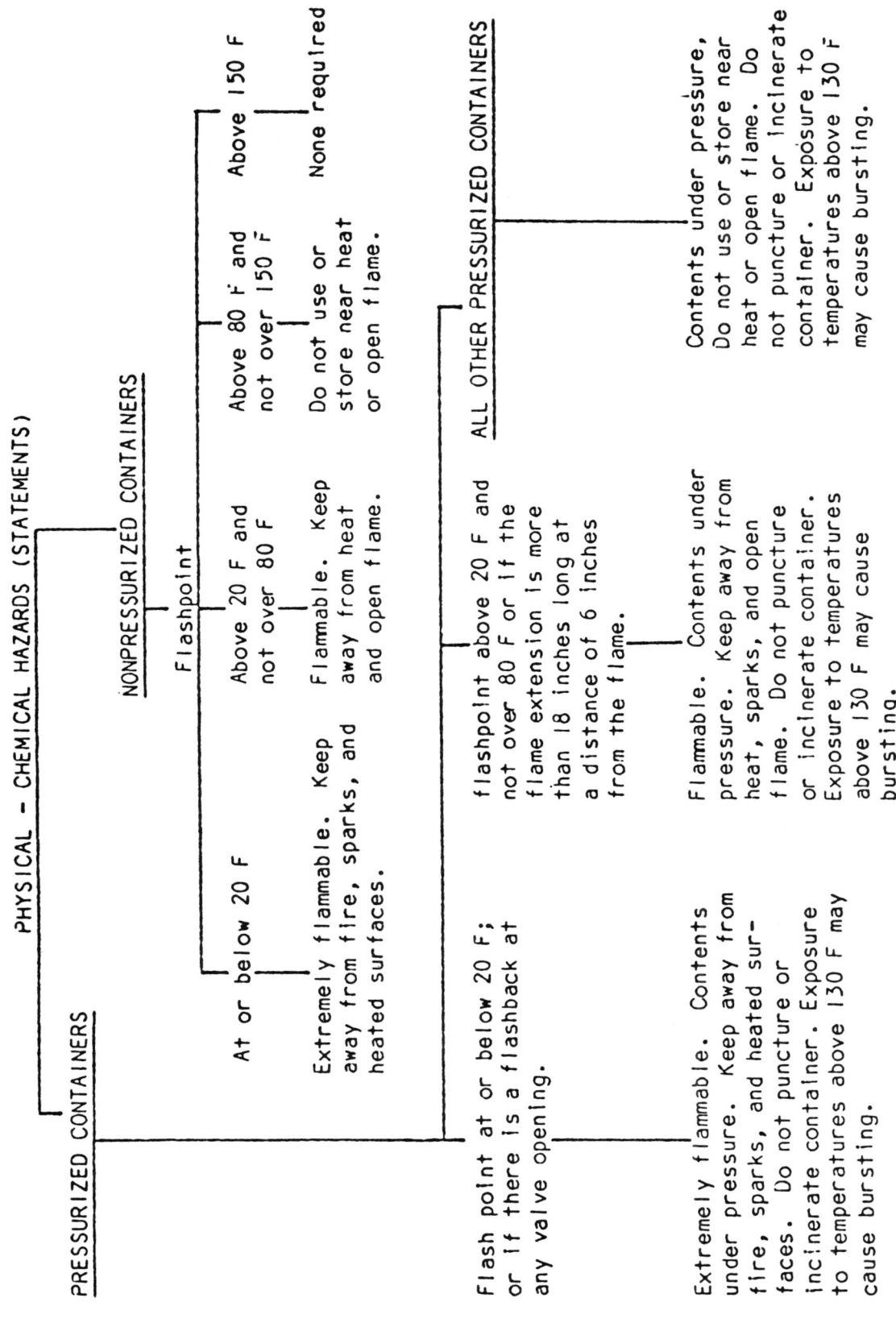

LABELING REQUIREMENTS OF THE FIFRA, AS AMENDED
[Refer to the sample labels following]

ITEM	LABEL ELEMENT	APPLICABILITY OF REQUIREMENT	PLACEMENT ON LABEL REQUIRED	PREFERRED	COMMENTS
1	Product Name	All Products	Front panel	Center front panel	
2	Company Name and Address	All Products	None	Bottom front panel or end of label text	If registrant is not the producer, must be qualified by "Packed for**" "Distributed by ***", etc.
3	Net Contents	All Products	None	Bottom front panel or end of label text	May be in metric units in addition to U.S. units .
4	EPA Reg. No.	All Products	None	Front panel	Must be similar type size and run parallel to other type
5	EPA Est. No.	All Products	None	Front panel, immediately before or following Reg. No.	May appear on the immediate container instead of the label. Immediate container does not include the lid.
6a	Ingredients Statement	All Products	Front panel	Immediately following product name	Text must run parallel with other text on panel
6b	Pounds/Gallon Statement	Liquid products where dosage given as lbs. al/ unit area	Front panel	Directly below the main Ingredients statement	

LABELING REQUIREMENTS OF THE FIFRA, AS AMENDED
[Refer to the sample labels following]

ITEM	LABEL ELEMENT	APPLICABILITY OF REQUIREMENT	PLACEMENT ON LABEL REQUIRED	PREFERRED	COMMENTS
7	FRONT PANEL PRECAUTIONARY STATEMENTS	All Products	Front panel		All front panel precautionary statements must be grouped together preferably outlined in black
7a	Keep Out of Reach of Children (Child Hazard Warning)	All Products	Front panel	Above signal word	Note type size requirements
7b	Signal Word	All Products	Front panel	Immediately below Child Hazard Warning	Note type size requirements.
7c	Skull & Crossbones and word "POISON" (in red)	All products which are Category I based on oral, dermal or inhalation toxicity	Front panel	Both in close proximity to signal word	
7d	Statement of Practical Treatment	All products in Categories I, II, and III	Category I: Front panel unless referral statement is used. Others: None	Front panel for all	

LABELING REQUIREMENTS OF THE FIFRA, AS AMENDED
[Refer to the sample labels following]

ITEM	LABEL ELEMENT	APPLICABILITY OF REQUIREMENT	PLACEMENT ON LABEL		COMMENTS
			REQUIRED	PREFERRED	
7e	Referral Statement	All products where precautionary labeling appears on other than front panel	Front Panel		
8	SIDE/BACK PANEL PRE-CAUTIONARY STATEMENTS	All Products	None	Top of side or back panel preceding Directions for Use	Must be grouped under the headings given in 8a, 8b, and 8c; Preferably outlined in black
8a	Hazard to Humans and Domestic Animals	All Products in Categories I, II, and III	None	Same as above	Must be preceded by appropriate signal word
8b	Environmental Hazards	All Products	None	Same as above	Environmental hazards include the bee caution where applicable
8c	Physical or Chemical Hazards	All pressurized products; others with flash points under 150 F	None	Same as above	
9a	Restricted Block	All RESTRICTED products	Top center of front panel	Preferably blocked	Includes a statement of the terms of restriction. the words "RE-STRICTED USE PESTICIDE" must be same type size as signal word.

LABELING REQUIREMENTS OF THE FIFRA, AS AMENDED
[Refer to the sample labels following]

ITEM	LABEL ELEMENT	APPLICABILITY OF REQUIREMENT	PLACEMENT ON LABEL		COMMENTS
			REQUIRED	PREFERRED	
9b	Statement of Classification	All products classified GENERAL	Immediately following heading of Directions for Use		
9c	Misuse Statement	All Products	GENERAL: Immediately following Statement of Classification RESTRICTED: Immediately following heading directions for Use		
10a	Re-Entry Statement	Certain Cholinesterase inhibitors; see CFR 40, Part 170	In the Directions for Use	Immediately after Misuse Statement	
10b	Category of Applicator	All RESTRICTED products which are limited to certain categories of applicator	In the Directions for Use	Immediately after Re-entry Statement (when used)	
10c	Storage and Disposal Block	All products	In the Directions for Use	Immediately before specific directions for use or at the end of directions for use	Must be grouped together, and preferably outlined in black. Heading must be same type size as Child Hazard Warning
10d	Directions for Use	All products, except for exemptions provided in the Regulations	None	None	May be in metric units as well as U.S. units.

CROP

CROP

CROP

CROP

CROP

WARRANTY STATEMENT

9A

1

6A

6B

7A

7C

7D

7E

3

RESTRICTED USE PESTICIDE

FOR RETAIL SALE TO AND APPLICATION ONLY BY CERTIFIED APPLICATORS OR PERSONS UNDER THEIR DIRECT SUPERVISION

PRODUCT NAME

ACTIVE INGREDIENT _____ %
INERT INGREDIENTS _____ %
TOTAL   100.00 %

7

THIS PRODUCT CONTAINS _____ LBS OF _____ PER GALLON

KEEP OUT OF REACH OF CHILDREN

7B DANGER—POISON

STATEMENT OF PRACTICAL TREATMENT

IF SWALLOWED _____
IF INHALED _____
IF ON SKIN _____
IF IN EYES _____

SEE SIDE PANEL FOR ADDITIONAL PRECAUTIONARY STATEMENTS

2

4

5 MFG BY _____
TOWN STATE _____
ESTABLISHMENT NO _____
EPA REGISTRATION NO _____
NET CONTENTS _____

10A

10B

10C

8

PRECAUTIONARY STATEMENTS

HAZARDS TO HUMANS
(& DOMESTIC ANIMALS)

DANGER

ENVIRONMENTAL HAZARDS

PHYSICAL OR CHEMICAL HAZARDS

DIRECTIONS FOR USE

It is a violation of Federal law to use this product in a manner inconsistent with its labeling

RE-ENTRY STATEMENT
(If Applicable)

CATEGORY OF APPLICATOR

STORAGE AND DISPOSAL

STORAGE _____
DISPOSAL _____

CROP

8A

8B

8C

9C

10C

PRODUCT NAME

ACTIVE INGREDIENT
INERT INGREDIENTS
TOTAL 100.00 %

THIS PRODUCT CONTAINS _____ LBS OF _____ PER GALLON

KEEP OUT OF REACH OF CHILDREN

CAUTION

STATEMENT OF PRACTICAL TREATMENT
IF SWALLOWED
IF INHALED
IF ON SKIN
IF IN EYES

SEE SIDE PANEL FOR ADDITIONAL PRECAUTIONARY STATEMENTS

MFG BY
TOWN STATE
ESTABLISHMENT NO
EPA REGISTRATION NO

NET CONTENTS

CROP
CROP
CROP
CROP
CROP
WARRANTY STATEMENT

PRECAUTIONARY STATEMENTS
HAZARDS TO HUMANS
(& DOMESTIC ANIMALS)
CAUTION

ENVIRONMENTAL HAZARDS

PHYSICAL OR CHEMICAL HAZARDS

DIRECTIONS FOR USE

GENERAL CLASSIFICATION

It is a violation of Federal law to use this product in a manner inconsistent with its labeling

RE ENTRY STATEMENT
(if Applicable)

STORAGE AND DISPOSAL

STORAGE
DISPOSAL

CROP

SECTION TWO

REREGISTRATION FORMS FOR PESTICIDE PRODUCTS*

Under the provisions of Section Three of the Federal Insecticide, Fungicide, and Rodenticide Act, all pesticide products must be registered with the Environmental Protection Agency prior to their use. This requirement applies to all persons who distribute, sell, ship, deliver, or receive pesticide products.

In order to facilitate these requirements, the following forms must be completed and filed with the EPA:

Form 8570-1 — This form will provide basic registration information to the EPA for all who distribute, sell, ship, deliver or receive pesticides. Reregistration is also provided through this form, as EPA information guidelines are revised periodically.

Form 3540-8 — This form is required of all pesticide producing establishments. Approval of the application is required prior to the manufacturing of pesticides.

Form E — This form is also required of all persons subject to the initial registration requirements; it is used by the EPA for determining the classification of pesticides and whether they will be for general or restricted use.

Form F — This form, required as part of the registration process, provides the labeling information that will appear on pesticide containers. If the labeling of a pesticide is changed the registration must be amended accordingly.

* U.S. Environmental Protection Agency, February 1976.

Form Approved
OMB No. 158-R0066

| U.S. ENVIRONMENTAL PROTECTION AGENCY OFFICE OF PESTICIDES PROGRAM *(WH-567)* WASHINGTON, D.C. 20460 | 1. REFERENCE CODE | 2. EPA USE ONLY |

APPLICATION FOR PESTICIDE: ☐ REGISTRATION ☐ REREGISTRATION

(Please read instructions on reverse before completing) **A**

3. COMPANY/PRODUCT NO.	4. PROPOSED CLASSIFICATION
	☐ GENERAL
	☐ RESTRICTED

5. NAME AND ADDRESS OF APPLICANT *(Include ZIP Code)*

☐ CHECK IF THIS IS A NEW ADDRESS

6. TYPE OF CONTAINER
☐ METAL
☐ PLASTIC
☐ GLASS
☐ PAPER
☐ OTHER *(Specify)*

7. WILL CHILD RESISTANT PACKAGING BE USED?
☐ YES   ☐ NO

8. PRODUCT NAME

9. EXPERIMENTAL PERMIT NO.

10. LOCATION OF LABEL DIRECTIONS
☐ ON LABEL

☐ ON MATERIAL ACCOMPANYING PRODUCT

11. MANNER IN WHICH LABEL IS AFFIXED TO PRODUCT
☐ LITHOGRAPH   ☐ OTHER *(Specify)*
☐ PAPER GLUED
☐ STENCILED

12. TYPES OF DATA SUBMITTED		FOR EPA USE ONLY			
01. NONE	1201				
02. PRODUCT CHEMISTRY	1202				
03. RESIDUE CHEMISTRY	1203				
04. ENVIRONMENTAL CHEMISTRY	1204				
05. EFFICACY	1205				
06. PHYTOTOXICITY	1206				
07. HUMAN SAFETY	1207				
08. DOMESTIC ANIMAL SAFETY	1208				
09. FISH AND WILDLIFE SAFETY	1209				
10. BENEFICIAL INSECT SAFETY	1210				
11. ACCIDENT EXPOSURE EXPERIENCE	1211				
12. OTHER *(Specify)*	1212				
13. OTHER *(Specify)*	1213				

13. METHOD OF SUPPORT *(See instructions)*
☐ Required Supporting Data Attached. *(2A)*
☐ Required Supporting Data is Submitted by Reference. *(2B)*

OFFER TO PAY STATEMENT
I hereby offer to pay reasonable compensation to the extent provided under Section 3(c)(1)(D) of the Federal Insecticide, Fungicide, and Rodenticide Act, as amended, and in accordance with the Regulations and Guidlines published thereunder for use of any test data which has been submitted to the U.S. Environmental Protection Agency in support of an application for the registration of a pesticide for the first time on or after January 1, 1970 and which may be used in support of the registration application for the subject pesticide.

14. CONTACT POINT

Complete items directly below for identification of individual to be contacted, if necessary, to process this application.

NAME

TITLE

TELEPHONE NO. *(Include Area Code)*

15. DATE APPLICATION RECEIVED *(Stamped)*

16. SIGNATURE	17. TITLE
18. TYPED NAME	19. DATE SIGNED

INSTRUCTIONS
GENERAL

This form is to be used for all applications for new registration of a pesticide product. In addition, it will be used for all applications for reregistration, and at such other times as specifically directed by the Agency.

In order to process an application for new registration submitted on this form, the following material must accompany the application:

1. Label Technical Data *(EPA 8570-10)*
2. Confidential Statement of Formula *(EPA 8570-4)*
3. Two copies of draft labeling
4. Two copies of any data submitted
5. Any required authorization letters regarding ingredients

Submission of Labeling: Labeling should first be submitted in the form of draft labels with all applications for new registration. Such draft labels may be in the form of typed label text on 8½ x 11" paper or as a mock-up of the proposed label. If prepared as a mock-up it should be constructed in such a way as to facilitate storage in an 8½ x 11" file. Mock-up labels significantly smaller than 8½ x 11" should be mounted on 8½ x 11" paper for submission.

Submission of Data: Data submitted in support of this application must be submitted in two copies. In order to facilitate review, each type of data identified in Item 12 must be submitted, bound separately, and identified on the front cover to include the information in boxes 3, 5, 8 and 12.

A copy of the application form and a copy of the label should be bound in each separate volume of the data.

SPECIFIC

Box Number

1. Reference Code. For use only when specifically directed by the Agency. Not for use on routine applications for new registration.
2. For EPA Use Only.
3. Company/Product Number. Insert your assigned company number, if one has been assigned. This number may have been assigned to you as a basic registrant, a distributor, or as an establishment.
4. Proposed Classification. Specify the proposed classification for this product.

5. Name and Address of Applicant. The name of the firm or person and address shown in your application is the person or firm to whom registration will be issued. If you are acting in behalf of another party, you must submit authorization from that party to act for them in registration matters.

> An applicant **NOT** residing in the United States must have an authorized agent residing in the United States to act for them in all registration matters. The name and complete mailing address of such agent must accompany this application.

6. Type of Container. Specify all the types of containers in which this product will be packaged.
7. Will Child Resistant Packaging Be Used? If this product is to be marketed in special child resistant packaging, mark "yes" in the box provided and submit a sample of the child resistant packaging with the application for registration.
8. Product Name. The name in this box must match the product as shown on the labeling submitted. Do not list the active ingredients.
9. Experimental Permit Number. If this application was preceded by an experimental permit issued to permit development of data to support this application, specify the experimental permit number in this box.
10-11. Self-explanatory.
12. Types of Data Submitted. Specify each type of data submitted or specify none. See "Submission of Data" above.
13. Method of Support. This box is mandatory for all applications for new registration or reregistration. The Federal Insecticide, Fungicide, and Rodenticide Act, as amended, requires that the applicant prove the efficacy and safety of a proposed product. This box identifies the method by which you are attempting to provide such proof.
    2A is used when all the data required to support the proposed product is actually attached to the application.
    2B is used when all of the data required to support the proposed product is either attached or submitted by specific references
14. Self-explanatory.
15. EPA Use Only.
16-19. Self-explanatory. Note that the signature in Box 16 certifies the offer to pay statement printed above as well as certifying the accuracy of the entire application package.

THE THIRD COPY OF THIS FORM MAY BE RETAINED BY THE APPLICANT.

U.S. ENVIRONMENTAL PROTECTION AGENCY

APPLICATION FOR REGISTRATION OF PESTICIDE-PRODUCING ESTABLISHMENTS

(Section 7, Federal Insecticide, Fungicide, and Rodenticide Act,
as amended, 86 Stat. 973-999)

Form Approved
OMB No. 158—R0105

NOTE: Read instructions on reverse of last page before completing.

1. EPA REGISTRANT NUMBER

2. COMPANY NAME AND ADDRESS

3A. ENTER CODE WHICH BEST DE-SCRIBES THE TYPE OWNERSHIP
1 - INDIVIDUAL
2 - PARTNERSHIP
3 - COOPERATIVE ASSOCIATION
4 - CORPORATION
5 - OTHER (Specify)_____

COMPLETE IF ITEM 3A IS CORPORATION

3B. DATE OF INCORPORATION (Mo, Day, Yr)

3C. STATE OF INCORPORATION

4. ENTER THE CODE WHICH DESCRIBES THE ACTIVITY OF THE COMPANY IDENTIFIED IN ITEM 2
1 - COMPANY IS A REGISTRANT AND IS ENGAGED IN PRODUCTION OF PESTICIDES.
2 - COMPANY IS A REGISTRANT BUT IS NOT ENGAGED IN PRODUCTION OF PESTICIDES. (Skip Item 5)
3 - COMPANY IS NOT CURRENTLY A REGISTRANT BUT IS ENGAGED IN PRODUCTION OF PESTICIDES.

5. NAMES AND ADDRESSES OF ESTABLISHMENTS	EPA USE ONLY
	EPA ESTABLISHMENT NO.

A
NAME
STREET
CITY   STATE   ZIP CODE

B
NAME
STREET
CITY   STATE   ZIP CODE

C
NAME
STREET
CITY   STATE   ZIP CODE

D
NAME
STREET
CITY   STATE   ZIP CODE

E
NAME
STREET
CITY   STATE   ZIP CODE

☐ CONTINUED ON ATTACHED SHEET

6A. SIGNATURE OF COMPANY OFFICER

6B. NAME AND TITLE (Typed)

6C. DATE (Mo, Day, Yr)

6D. TELEPHONE NUMBER

EPA USE ONLY

7. DATE APPLICATION RECEIVED (Mo, Day, Yr)

The establishments assigned numbers above have been registered in accordance with Section, 7, FIFRA, as amended.

8A. SIGNATURE

8B. EPA OFFICE

8C. DATE (Mo, Day Yr)

EPA Form 3540-8 (Rev. 4-76)   PREVIOUS EDITION TO BE USED UNTIL SUPPLY IS EXHAUSTED.

INSTRUCTIONS

Section 7 of the Federal Insecticide, Fungicide, and Rodenticide Act, as amended *(86 Stat. 987, 7 U.S.C. 136e)*, provides that no person shall produce any pesticide subject to this Act or any device subject to this section in any State unless the establishment in which it is produced is registered with the Administrator. In order to comply with this requirement, this application is to be completed, signed by a responsible officer of the company and the first page returned to the U.S. Environmental Protection Agency Office having jurisdiction over the State in which the company is located. Addresses of the Regional offices are found below.

ITEM(S)

1 EPA REGISTRANT NUMBER. Enter the registrant *(company)* number assigned by EPA, i.e., the numbers preceding the hyphen in the product registration number. If the company has no products registered, leave the item blank.

2 COMPANY NAME AND ADDRESS. Enter the complete name and address of the home office or headquarters of the company submitting the application. If item is already completed, check for accuracy and make any necessary changes.

3 Self-explanatory. For State of incorporation, use abbreviation as shown below in STATES COVERED.

4 Self-explanatory.

5 NAME AND ADDRESS OF ESTABLISHMENT. Enter the names and addresses of all establishments *(production sites)* whose production is controlled solely by the company listed in Item 2. Use the State abbreviations as shown below in STATES COVERED. Do not list those establishments not under the control or ownership of the company listed in Item 2. If the company's production occurs at the address contained in Item 2, repeat complete name and address in Item 5.

EPA ESTABLISHMENT NUMBER. Do not complete this item. Establishment Numbers will be assigned by EPA.

Check the box indicating continuation on attached sheet if applicable. Continuation sheets should repeat Items 1 and 2 and provide the information required in Item 5. Continue lettering F, G, H, etc.

6 SIGNATURE OF COMPANY OFFICER. An individual authorized to sign official documents for the company should enter his signature, title, telephone number and the date.

7 and 8 For use by the U.S. Environmental Protection Agency only.

ADDRESSES OF REGIONAL OFFICES

ADDRESSES OF REGIONAL OFFICES	STATES COVERED
U.S. Environmental Protection Agency Region I, Pesticides Branch John F. Kennedy Federal Building Boston, Massachusetts 02203	Massachusetts (MA), Vermont (VT), Connecticut (CT), Maine (ME), New Hampshire (NH), Rhode Island (RI)
U.S. Environmental Protection Agency Region II, Pesticides Branch 26 Federal Plaza New York, New York 10007	New York (NY), New Jersey (NJ), Puerto Rico (PR), Virgin Islands (VI)
U.S. Environmental Protection Agency Region III, Pesticides Branch Curtis Building 6th and Walnut Streets Philadelphia, Pennsylvania 19106	Pennsylvania (PA), District of Columbia (DC), Delaware (DE), Maryland (MD), Virginia (VA), West Virginia (WV)
U.S. Environmental Protection Agency Region IV, Pesticides Branch 1421 Peachtree Street, N.E. Atlanta, Georgia 30309	Georgia (GA), North Carolina (NC), South Carolina (SC), Kentucky (KY), Alabama (AL), Florida (FL), Mississippi (MS), Tennessee (TN)
U.S. Environmental Protection Agency Region V, Pesticides Branch 230 South Dearborn Street Chicago, Illinois 60604	Illinois (IL), Indiana (IN), Michigan (MI), Ohio (OH), Minnesota (MN), Wisconsin (WI)
U.S. Environmental Protection Agency Region VI, Pesticides Branch 1600 Patterson Street Dallas, Texas 75201	Texas (TX), Louisiana (LA), New Mexico (NM), Oklahoma (OK), Arkansas (AR)
U.S. Environmental Protection Agency Region VII, Pesticides Branch 1735 Baltimore Avenue Kansas City, Missouri 64108	Missouri (MO), Kansas (KS), Iowa (IA), Nebraska (NB)
U.S. Environmental Protection Agency Region VIII, Pesticides Branch 1860 Lincoln Street Denver, Colorado 80203	Colorado (CO), Montana (MT), Utah (UT), Wyoming (WY), North Dakota (ND), South Dakota (SD)
U.S. Environmental Protection Agency Region IX, Pesticides Branch 100 California Street San Francisco, California 94111	California (CA), Hawaii (HI), Arizona (AZ), Nevada (NV), Guam (GU), American Samoa (AS), Trust Territories of Pacific Islands (PI), Wake Island (WK)
U.S. Environmental Protection Agency Region X, Pesticides Branch 1200 Sixth Avenue Seattle, Washington 98101	Washington (WA), Oregon (OR), Idaho (ID), Alaska (AK)

Applications from companies located in foreign countries should be submitted to

U.S. Environmental Protection Agency
Pesticides Enforcement Division
Scientific Support Branch
401 M Street, S.W.
Washington, D.C. 20460
U.S.A.

EPA Form 3540—8 (Rev. 4—76)   PREVIOUS EDITION TO BE USED UNTIL SUPPLY IS EXHAUSTED.

Form Approved
OMB No. 158-R0066

U.S. ENVIRONMENTAL PROTECTION AGENCY OFFICE OF PESTICIDE PROGRAMS (WH-567) WASHINGTON, D.C. 20460 **CONFIDENTIAL STATEMENT OF FORMULA** (See instructions on back of last page) **E**	1. PAGE ____ OF ____	2. REGISTRATION/FILE SYMBOL

3. NAME AND ADDRESS OF APPLICANT/REGISTRANT (Include Zip Code)	4. CONTAINER SIZE(S)	5. LOCATION OF NET CONTENTS ☐ CONTAINER ☐ LABEL

	6. COUNTRY WHERE FORMULATED	FOR EPA USE ONLY

7. NAME AND ADDRESS OF PRODUCER (Include Zip Code)	8. WEIGHT/GALLON DENSITY	9. WEIGHT/UNIT TABLET	10. SOLUBILITY RATE
	11. pH	12. PERCENT FREE ALKALI	13. DRUM TEST

14. PRODUCT NAME	15. FLASH POINT/FLAME EXT	16. DIELECTRICAL BREAK-DOWN VOLTAGE

FLAG	17. COMMERCIAL COMPONENT (List each as actually introduced into this formulation. Give Trade Name (if any), Name of Supplier, and EPA Registration Number if applicable)	18. AMOUNT OF EACH COMPONENT*	19. PERCENT BY WEIGHT	20. PURPOSE IN FORMULATION

*If liquid measure, give specific gravity or pounds per gallon.	TOTAL WEIGHT OF BATCH	100.00%	
21. SIGNATURE AND TITLE	22. NAME OF FIRM		23. DATE SIGNED

EPA Form 8570-4 (Rev. 12-74)

Form Approved
OMB No. 158 - R0066

U.S. ENVIRONMENTAL PROTECTION AGENCY OFFICE OF PESTICIDE PROGRAMS (WH-567) WASHINGTON, D.C. 20460 **LABEL TECHNICAL DATA** (See *INSTRUCTIONS* on back of last part) F	1. COMPANY/REGISTRATION NO.	2. EPA USE ONLY — —
	3. PRODUCT NAME	

4. APPLICATION SITES (Check all that apply)	5. PEST TYPE (Check all that apply)	7. USER TYPE (Check all that apply)
01 CROPS (*Fruit*)	01 ALGAE	01 UNSPECIFIED GENERAL USE
02 CROPS (*Vegetable*)	02 AMPHIBIAN/REPTILE	02 UNSPECIFIED RESTRICTED USE
03 CROPS (*Field*)	03 BACTERIA	03 HOMEOWNER USE
04 CROPS (*Spice*)	04 BIRDS	04 JANITORIAL USE
05 CROPS (*Nut*)	05 FISH	05 PEST CONTROL OPERATOR USE
09 CROPS (*Other*)	06 FOULING ORGANISMS	06 COMMERCIAL APPLICATOR USE
10 SOIL TREATMENT (*No crop specified*)	07 FUNGI	07 FARMER USE
20 FOREST	08 INSECTS AND MITES	08 MEDICAL USE
30 ORNAMENTALS	09 MAMMALS	09 VETERINARY USE
40 TURF	10 NEMATODES	10 GOVERNMENT AGENCY USE
50 STORED PRODUCTS TREATMENT	11 PLANTS	11 MANUFACTURING USE
61 ANIMALS (*Livestock*)	12 RODENTS	8. FORMULATION
62 ANIMALS (*Dairy*)	13 SLIME	(Check one only)
63 ANIMALS (*Pet*)	14 SLUGS AND SNAILS	01 TECHNICAL CHEMICAL
64 ANIMALS (*Laboratory*)	15 VIRUS	02 FORMULATION INTERMEDIATE
69 ANIMALS (*Other*)	16 OTHER (*Specify*)	03 DUST
71 OUTDOOR (*Nocrop Agricultural*)		04 GRANULAR
72 OUTDOOR (*Resident/Commercial*)		05 PELLETED/TABLETTED
73 OUTDOOR (*Non agricultural*)		06 WETTABLE POWDER
81 BUILDINGS (*Agricultural*)	6. MODE OF ACTION	07 WETTABLE POWDER/DUST
82 BUILDINGS (*Commercial*)	(Check all that apply)	08 CRYSTALLINE
83 BUILDINGS (*Food Processing*)		09 MICROENCAPSULATED
84 BUILDINGS (*Medical*)	01 ATTRACTANT	10 IMPREGNATED MATERIALS
85 BUILDINGS (*Residential*)	02 BIOLOGICAL CONTROL	11 SELF-GENERATING SMOKE
91 EQUIPMENT (*Commercial*)	03 CHEMOSTERILANT	12 EMULSIFIABLE CONCENTRATE
92 EQUIPMENT (*Food*)	04 DEFOLIANT	13 INVERT EMULSION
93 EQUIPMENT (*Agricultural*)	05 DESICCANT	14 FLOWABLE CONCENTRATE
94 EQUIPMENT (*Medical*)	06 FEEDING DEPRESSANT	15 SOLUBLE CONCENTRATE
95 EQUIPMENT (*Transportation*)	07 GROWTH INHIBITOR	16 SOLUTION (*Ready to Use*)
96 LAUNDRY AND DRY CLEANING	08 GROWTH REGULATOR	17 OILS (*No added pesticide*)
97 INDUSTRIAL PRESERVATIVES	09 POISON (*Single dose*)	18 PRESSURIZED (*Gas*)
98 PESTICIDE (*Manufacturing only*)	10 POISON (*Multiple Dose*)	19 PRESSURIZED (*Liquid*)
99 OTHER (*Specify*)	11 PRESERVATIVE	20 PRESSURIZED (*Dust*)
	12 REPELLENT	21 OTHER (*Specify*)
	13 OTHER (*Specify*)	

REMARKS

EPA Form 8570-10 (11-74)

CHAPTER FOUR

MANAGEMENT TECHNIQUES
FOR TOXIC SUBSTANCES

CONTROL PROCEDURES IN TEN MAJOR INDUSTRIES*

SECTION ONE

ANALYSIS OF TECHNIQUES

Abstract

By analysis of use/exposure data compiled for toxic substances under federal regulations (29 CFR 1910.1000), ten industries and twelve common industrial operations were selected for study. Each industry/operation was investigated to determine whether effective engineering controls exist to prevent excessive employee exposures to toxic agents utilized in the equipment and processes involved. Recommendations for engineering control research were developed which addressed identified needs. These recommendations were then prioritized according to their perceived capability to result in beneficial technology.

The industries studied included pharmaceutical manufacturing, plastic and foam manufacturing, paint and coatings manufacturing and application, rubber production, pesticide manufacturing and application, adhesive manufacturing and application, dye and ink manufacturing, soap and detergent manufacturing, perfume manufacturing, and printing. The common industrial operations were abrasive machining; chemical processes; cleaning and maintenance; drying and curing oven use; grinding, crushing, and screening; laboratory operations; non-spray application of volatile substances; open-surface tank use; spray-finishing; coating and dyeing; and welding.

* Engineering Control Research Recommendations. National Institute for Occupational Safety and Health. February, 1976.

1. Introduction

1.1 Background

The National Institute for Occupational Safety and Health (NIOSH) and The Occupational Safety and Health Administration (OSHA) have initiated a joint program, the Standards Completion Program (SCP), with the objective of providing specific standards for each of the over 400 toxic substances identified by 29 CFR 1910.1000. As part of this program, Arthur D. Little, Inc., (ADL) has been developing engineering, personal protective, and administrative control standards under Contract No. CDC-99-74-44 and has developed recommended industrial ventilation guidelines for selected industries or operations under Contract No. CDC-99-74-33. The recommended ventilation guidelines will be published in report form to demonstrate the feasibility of engineering control use in the specific process and operation categories investigated.

It is imperative for NIOSH to systematically and selectively identify those problem areas where research can have the greatest impact on the health and safety of the American working population. Consequently, NIOSH identified a need for prioritized research recommendations pertaining to the provision of adequate engineering controls for prevention of worker exposure to toxic substances utilized in industrial processes and operations.

ADL conducted a program to identify industrial processes which have significant potential for resulting in excessive worker exposures and to determine whether engineering controls of proven adequacy exist for them. Additionally, it developed and prioritized recommendations for engineering control design and performance research where needs for such research were identified.

1.2 Program Objectives

It was the purpose of this study to assist NIOSH by taking advantage of the data accumulated by ADL in the performance of aforementioned contracts. Specifically, the objectives of the program were:

1. To identify industries or specific operation and process categories in industry which have considerable potential for exposing large numbers of workers to toxic substances;

2. To examine and evaluate these problem areas with the objective of determining the availability and/or adequacy of engineering controls;

3. To justify, recommend, and specify the research activities necessary to provide the technology for effective engineering controls, where needs are identified; and

4. To prioritize the recommendations and/or the problems they represent in a reasonable and logial manner.

2. Methodology

2.1 Introduction

To most effectively apply the resources allocated to this study, it was necessary to develop an approach which utilized the available data to best advantage. In the following, the sources and compilations of data which were available are described, and we present the rationale for the work approach utilized in this program.

2.2 Literature Sources of Information

The primary data base for this study consisted of a large compilation of literature collected during our previous work for NIOSH. To this data base, other information was to be added as needed on a case-by-case basis. The data base consisted of a library and a system of chemical-specific, industry, and engineering control data files.

The carefully selected library served as the core of the data base. Included were well-known texts on toxicology, industrial hygiene, and ventilation, and texts which specifically described the operations and processes conducted in various industries.

In the initial stages of the program to develop engineering, personal protective and administrative control standards for the individual chemicals regulated by 29 CFR 1910.1000, an intensive literature search was conducted of reports and periodicals to compile data on the uses of individual chemicals, and their physical, chemical, and toxicological properties. These data were available in a set of "chemical-specific" files, one for each chemical of concern. Each file also contained a computer printout of abstracts of papers, etc. which the NIOSH Technical Information Service had provided for each chemical of interest.

This initial, intensive literature search also generated a data base which was termed as an "industry file." Here were collected, by individual industry, various papers, booklets, reports, etc. which described the processes and operations conducted therein and the problems with exposure to hazardous materials. Since these files often contained data of interest for specific chemicals, they were cross-referenced with the "chemical-specific" files.

Prior to and during the study to develop recommended industrial ventilation guidelines, yet another literature search was conducted. This search was directed at gathering available information on the application of general and local exhaust ventilation for controlling exposures, and resulted in "engineering control" files for various industries and operation categories.

2.3 Use/Exposure Documents

To provide information essential to defining where toxic substances are primarily employed in industry and what their impact is upon overall worker health, ADL prepared for NIOSH what is termed a "use/exposure" document for each of the chemicals treated to date in the SCP. Each of these documents identifies for its chemical of concern those industrial operations or processes which cause significant and common employee exposures to occur and presents its findings in a list which is rank ordered by a methodology which was developed. The operations appear in order of their relative hazard potential for a specific chemical and any given ranking is not directly comparable to the ranking of an operation, similar or otherwise, conducted with another chemical.

Table 1 shows a typical example of one such document. The left column lists the individual operations or processes identified from literature and other sources of information in an order indicated by the aforementioned methodology. The center column notes the primary routes of exposure by which the substance acts, and the final column reports which personal protective and engineering controls have been cited as being applicable for preventing harmful exposures. At the point in time when this program was initiated, 210 of the approximately 400 such documents to be prepared had been completed.

TABLE 1—USE/EXPOSURE AND CONTROL DOCUMENT

ISOBUTYL ACETATE

Use/Exposure	Principal Route of Entry	Currently Used Control Methods
1. Inhalation of vapor during spray application of varnishes and nitrocellulose lacquers as protective and finish coatings for wood (especially in the furniture industry), plastic, metal, leather and other surfaces	A	Local exhuast ventilation; personal protective equipment (respiratory protective devices)
2. Inhalation of vapor during application of varnishes and nitrocellulose lacquers as protective and finish coatings for wood, paper (printing ink vehicle), metal, leather, and other surfaces by dipping, roller coating, tumbling, knifing or brushing	A	Local exhaust ventilation; general dilution ventilation
3. Inhalation of vapor during oven baking of phenolic and epoxy coatings	A	General dilution ventilation
4. Inhalation of vapor during air or oven drying of varnishes and lacquers	A	General dilution ventilation
5. Inhalation of vapor during application of nitrocellulose, cellulose acetate, cellulose acetate butyrate and cyclized rubber adhesives by machine spraying, dipping, roller coating, tumbling, knifing, or brushing. Most common industrial applications are in shoe manufacturing, book binding, packaging, leather processing, photographic film manufacturing, and paper processing.	A	Local exhaust ventilation; general dilution ventilation
6. Inhalation of vapor and skin contact with liquid during manual application of nitrocellulose (household cements), cellulose acetate, cellulose acetate butyrate and cyclized rubber adhesives	A,B	Local exhaust ventilation; general dilution ventilation; personal protective equipment (gloves and aprons)
7. Inhalation of vapor and skin contact with liquid during cleaning and maintenance of acetate processing equipment such as kettles distillation columns and storage vessels	A,B	Personal protective equipment (respiratory protective devices, gloves, aprons, barrier creams, eye goggles)
8. Inhalation of vapor during the manufacture of some perfumes, cosmetics, and flavoring agents	A	General dilution ventilation
9. Inhalation of vapor during spray application of vinyl based primers, maintenance paints and other industrial coatings	A	Local exhaust ventilation; personal protective equipment (respiratory protective devices)

A — Inhalation
B — Skin contact resulting in localized irritation
C — Ingestion
D — Skin contact resulting in absorption and subsequent systemic poisoning

2.4 Selection of Operations Categories for Study

The primary source of data specified for use in selecting operation categories for study was the set of use/exposure documents prepared for the chemicals of interest to the SCP. As previously noted, these documents listed and ranked the most significant uses of each chemical in a "relative" manner, i.e., though the ranking procedure used resulted in use/exposures "relatively" ranked within each document, the ranking was not "absolute" in the sense that the significance of the uses and exposures of one chemical could be compared to those of another.

To assemble these data into a form which would allow overall analysis, the use/exposure documents for the first 210 chemicals addressed in the SCP were employed to develop a set of matrices indicating the name of each compound, and the significant operations, processes, or industries it is utilized in. The set of these matrices, given in order of the "sets" of chemicals in which the SCP was divided, is presented in Appendix A. To be noted is that the operation categories described across the top of these matrices are somewhat generalized to accommodate the multitudinous specific uses listed in the actual documents. Indeed, many of the checked blocks represent more than one specific use/exposure.

The determination of which of these categories were most significant overall required that they be somehow ranked by order of their potential to result in exposures. Attempts to do so, utilizing various procedures, resulted in the conclusion that the only practical basis for doing so was by simply determining the number of chemicals utilized in any individual category and by listing the categories according to this ranking. Table 2 presents the operation categories in the order determined by this procedure. While it is realized that the mere fact that more toxic substances are used in one process than another does not prove that it is more hazardous than another, no other procedure which a priori indicated the degree of hazards involved could be developed from these data. Furthermore, there did appear to be considerable merit in an approach which makes no a priori assumptions that a particular type of operation can be effectively controlled and which results in the investigation of processes and operations in industry which involve the use of the largest numbers of toxic substances.

Examination of Table 2 indicated that some categories listed are inherently included within others, and that some were breakdowns of a more general category. For example, it could be seen that the categories of fuel production, lubricant manufacture, heat transfer fluid use, petroleum refining, and refining and dewaxing of mineral oil could be considered as being inherently included in closed-system "chemical processes". Additionally, such categories as "textile coating processess" and "textile dyeing"

could be envisioned as the single category of "textile coating and dyeing processess," while "ink manufacturing" and "dye manufacturing" could be reasonably combined into "ink and dye manufacturing" because of perceived similarities in processes utilized and/or hazards involved. Before proceeding further, therefore, the list in Table 2 was condensed, using the reasoning exampled, into the list presented in Table 3. While this table was very similar to Table 2, its differences served to better organize the operations into more identifiable categories and to eliminate redundancy.

It was then observed that Table 3 contained two types of categories; those which represent entire industries, and those which represent some set of similar equipment or activities which may be utilized or conducted in a number of industries. Pharmaceutical, plastic and foam, and paint and coating manufacturing are examples of the former type while chemical processes, cleaning and maintenance, and open-surface tank use are examples of the latter. Also observed was that some significant common operation categories were missing from Table 3; material handling and laboratory operations being prime examples.

With the realization that any investigation of engineering control use in a specific industry would involve consideration of common operations, it was perceived to be desirable to somehow delineate these common operations. This would allow their separate and complete study and reduce the possibility of redundant efforts. Furthermore, since an objective of the program was to develop research recommendations having the greatest potential benefit to the working population, it was decided that study of these common operations must be given priority over study of any industry-specific problem areas. To accomplish these objectives, a "two-tiered" study approach, as described in the following paragraphs, was instituted.

TABLE 2—OPERATION CATEGORIES IDENTIFIED FROM USE/EXPOSURE DOCUMENTS

Rank	Operation Category	No. of Chemicals
1.	Chemical processes	170
2.	Cleaning and maintenance operations	103
3.	Pharmaceutical manufacturing	74
4.	Extractant or solvent use	73
5.	Plastic and foam manufacturing	73
6.	Paint manufacturing	71
7.	Rubber production	53
8.	Pesticide manufacturing	53
9.	Manufacture and application of adhesives	50
10.	Dye manufacturing	49
11.	Non-spray application of paints, etc.	41
12.	Spray-finishing	45
13.	Textile coating processes	43
14.	Pesticide application	37

15.	Cleaning and degreasing	37
16.	Textile dyeing	34
17.	Photographic film and chemical manufacturing	34
18.	Fuel production	32
19.	Ink manufacturing	28
20.	Detergent manufacturing	28
21.	Perfume manufacturing	26
22.	Ink application	21
23.	Electric equipment manufacturing	20
24.	Leather tanning and treatment	20
25.	Lubricant manufacture and use	19
26.	Synthetic fiber manufacturing	19
27.	Dry cleaning operations	17
28.	Manufacture of polishing compounds	16
29.	Heat transfer fluid use	15
30.	Use of paint or varnish removers	15
31.	Aerosol manufacturing and use	15
32.	Mining operations	15
33.	Manufacture and use of pyrotechnics and explosives	13
34.	Refrigerant use	13
35.	Petroleum refining	12
36.	Machining, grinding, buffing, and polishing	11
37.	Food additive use	10
38.	Manufacture and use of corrosion inhibiters	10
39.	Ore refining and metal processing	10
40.	Ceramics and glass production	9
41.	Welding	8
42.	Drying of paints, varnishes, etc.	8
43.	Manufacture and use of fire extinguishers	8
44.	Refining and dewaxing of mineral oil	7
45.	Foundry processes	7
46.	Cosmetic manufacturing	6
47.	Anesthetic use	5
48.	Artificial leather manufacturing	4
49.	Water purification	4
50.	Safety glass manufacturing	3
51.	Food processing	3
52.	Smokeless powder manufacturing	3
53.	Pulping of wood	3
54.	Paper impregnation	2
55.	Straw hat manufacturing	2
56.	Synthetic pine oil manufacturing	1

TABLE 3—CANDIDATE INDUSTRIES AND OPERATIONS FOR INITIAL STUDY

1. Chemical processes
2. Cleaning and maintenance
3. Pharmaceutical manufacturing
4. Plastic and foam manufacturing
5. Paint and coatings manufacturing
6. Rubber production
7. Pesticide manufacturing and application
8. Adhesive manufacturing and application
9. Dye and ink manufacturing
10. Non-spray application of volatile substances
11. Spray-finishing
12. Textile coating and dyeing processes
13. Open-surface Tank Use

14. Soap and Detergent Manufacturing
15. Perfume manufacturing
16. Printing
17. Electric equipment manufacturing
18. Leather tanning and treatment
19. Synthetic fiber manufacturing
20. Dry cleaning
21. Manufacture of polishing compounds
22. Aerosol manufacturing and use
23. Mining operations
24. Pyrotechnics and explosives manufacturing
25. Abrasive machining
26. Food processing
27. Manufacture and use of corrosion inhibiters
28. Ore refining and metal processing
29. Ceramics and glass production
30. Welding
31. Dryer and oven use
32. Manufacture and use of fire extinguishers
33. Foundry processes
34. Cosmetic manufacturing
35. Anesthetic use
36. Artificial leather manufacturing
37. Water purification
38. Pulping of wood
39. Paper impregnation
40. Straw hat manufacturing
41. Synthetic pine oil manufacturing

In Section 3 of this report, the first 10 industries identified in Table 3 are individually studied and reported upon. These "industry evaluations" attempt to describe the activities conducted in each industry, determine how, when, and where exposures to toxic substances may occur, and assess the availability and adequacy of engineering controls. For this latter task, they concentrate upon those activities which can be considered somewhat unique to the industry. Those operations common to many industries are simply noted as being such and listed for further consideration.

Section 4 of the report is concerned with an evaluation of the availability and adequacy of engineering controls for the common operation categories either identified in Section 3 or appearing in Table 3. Thus, this "two-tiered" approach results in detailed study of not only a set of industries in which large numbers of toxic substances are utilized, but also in the study of common operations conducted throughout industry from which exposures are likely to occur.

2.5 Prioritization of Recommendations

Section 5 of the report summarizes the research recommendations developed in Sections 3 and 4 and prioritizes them according to their perceived capability to result in beneficial technology. The approach taken in this endeavor is presented and discussed fully in that section.

2.6 Limitations of the Approach

There are a number of limiting factors to the approach utilized vis-a-vis its appropriateness in fulfilling the objectives of this study. These are discussed in the following:

The first area of concern involves the data from which Tables 2 and 3 were derived. As previously reported, only 210 of the 400 use/exposure documents to be prepared had been completed at the point in time their use was necessary for this study. This suggests the possibility that a similar analysis procedure conducted at the completion of the SCP program might lead to listings of industries and operations of somewhat different ordering. Since we only studied those selected from the top portion of Table 3, however, we can be confident that they would be present at or near the top portion of any future listing prepared.

The number of industries investigated in Section 3 represents a second limitation. It was originally intended that all industries on the list would receive attention. However, it was determined that the amount of effort required per industry evaluation was beyond the resources allocated to the study. While considerable information existed which described the activities of any given industry, that which described the exposure problems and the availability and adequacy of engineering controls were not so readily available. Thus, data collection efforts would have been necessary beyond compilation of the information already inhouse and/or readily available.

Restriction of the number of industries selected from Table 3 to a total of 10 indicates that any unique problem areas in the industries not studied could not be identified. The "two-tiered" approach of study, however, leads us to the conclusion that problems common to many industries were sufficiently addressed. While it cannot be guaranteed that all such common problem areas were identified, it can be said with confidence that very few, if any, others exist which might be of a significance to warrant research activities.

Finally, some comment is warranted concerning the validity of the numerous conclusions drawn throughout the report. Wherever possible, such conclusions can be seen to logically evolve from the data presented. However, in some cases, it was necessary to base conclusions upon our own judgment and a limited amount of information available in the public domain. Though we have considerable confidence in the validity of these judgments, one can never ensure that such a conclusion is entirely a correct one.

2.7 Other Topics

In Section 6 of this report, the authors present a discussion of some problem areas and topics not previously

noted in preceding sections. Included are a number of ideas and concepts which evolved during performance of this work and which do not properly belong in preceding sections.

3. Industry Evaluations

3.1 Introduction

For each of the first 10 industries listed in Table 3, this section contains an individual evaluation. These evaluations describe the operations and processes conducted in the industry, indicate which production stages or areas have been noted in the literature as presenting exposure hazards to workers, and discuss the availability and adequacy of engineering controls where needs for them are evident.

At the end of the evaluations for the first six industries addressed, a summary list of operations common to many industries is presented. The operations so listed, together with others found in Table 3, constitute the total of those investigated in Section 4 of this report. Such lists are not presented in the remaining four evaluations because of the redundancy of doing so. No new common operations were identified in these latter evaluations.

3.2 Pharmaceutical Manufacturing

Introduction—The pharmaceutical industry has grown from predominantly individual endeavors into an industry characterized by the drugstore pharmacist and the complex organizations of modern corporations. Advancements in life sciences have resulted in the production of numerous varieties of medicines from chemical, animal, and vegetable sources.

According to Kirk-Othmer[1],* pharmaceuticals are designed to perform one or more of a variety of functions— (1) destroy disease producing organisms by interference with their cell processes, (2) produce a physiological effect by interference with the body's cell chemistry, (3) stimulate the body to produce protective substances, (4) replenish or supplement the body's natural supply of a substance and/or (5) produce some sort of functional physical effect.

When these substances are properly administered, they can provide benefits to victims of disease or trauma. Alternatively, where workers are excessively and unintentionally exposed to them, they can result in undesirable health effects. In the following, we describe the operations involved in the production of pharmaceuticals and discuss the availability and adequacy of engineering controls where needs for them can be identified. The major subject

---

* References to footnote numbers can be found beginning on page 289.

areas of concern are process development, manufacturing, and packaging.

Process Development—Process (and product) development involves the typical steps of development and testing of products in the laboratory, design and evaluation of production processes in small-scale pilot plant operations, and the subsequent scale-up, design, and start-up of a full-scale plant. The processes and operations involved are by no means unique to this industry. Laboratory operations are usually typical of those conducted in chemical and biology labs. The operations conducted in the pilot plant phase of development and in the start-up of full-scale operations are characteristic of the chemical industry and the processes conducted therein. Included are evaporation, distillation, absorption, esterification, nitration, hydrogenation, and others.

Manufacturing—It is in the manufacturing of pharmaceuticals that one finds those operations or types of equipment which are primarily unique to the industry. In this characterization again cannot be included those operations or processes utilized to synthetize many basic ingredients, processes which must be considered as common to many industries and usually typical chemical unit operations.

Like so many other industries, the passing of time and the identification and appreciation of hazards to workers arising from physical and toxic agents have resulted in reductions, and even eliminations, of such hazards in the pharmaceutical industry. Indeed, one might consider that the present, somewhat dual regulation of the industry by the Food and Drug Administration and OSHA has accelerated the "desire" to control operations to a significant degree. This is not meant to imply, however, that protection of employees is always provided by engineering controls.

In 1947, Watrous[2] described the health hazards of the pharmaceutical industry. He noted, "In recent years the pharmaceutical industry has come to occupy a unique position in regard to problems of industrial hygiene and toxicology; practically no other single commercial enterprise presents such a wide variety of potentially toxic exposures or such a rapidly-changing advent of new chemical substances." His paper continues by noting the hazards of many of the substances used at that time and repeatedly notes the need for sufficient knowledge of the hazards of toxic substances, proper instructions and warnings to those handling them, and proper control measures.

Bresler[3], in a 1949 paper concerning nitroglycerin reactions among pharmaceutical workers, reported "the reaction from exposure to nitroglycerin represents an unpleasant and undesirable complication, causing the worker considerable discomfort and illness, and resulting in a production problem of some magnitude. The phar-

maceutical industry has been faced with this problem for many years. Many control procedures have been instituted, none of which have proved entirely satisfactory." After describing the toxic effects of the substance and its method of manufacture, he then discusses methods of control. Of interest are his statements "Dust and vapors should be removed at their source by suitable ventilation measures. This should include a hood for the powder mixer. The tablet machine and the drying units should be enclosed and exhausted."

Speaking of his experiences "decades ago," Chilson[4] reflects "in those days dust collection was inadequate, if not totally missing in most plants. Very dusty products were isolated within partitions. Cleaning up meant hosing down once a week. And . . . conditions were truly appalling . . ." In another article[5], he states "all factories were, by current standards, dirty. Dust collection was virtually unknown, even when the products being made were hazardous. When such products were exceptionally dirty, irritating, or hazardous, the areas where they were made were enclosed within partitions. These areas usually had windows equipped with exhaust fans so that some of the dust could be spread all over the neighborhood. Men working in these areas were provided with dust masks, and, in the fine chemical industry at least, when poisonous chemicals were dumped into hoppers, the hoppers were designed to provide a downdraft. But such hoppers were used only when there was a danger that the dust might get through the masks and kill the operators."

The above describes typical conditions in the industry "decades ago" and represent one extreme of the conditions under which employees worked. Since then, of course, conditions have improved to the extent that the literature contains references to manufacturing systems which approach the opposite extreme. A description[6] of a process for manufacturing aspirin tablets in 1967 fully illustrates this point.

In comparison to other substances manufactured in the pharmaceutical industry, aspirin must be ranked with those being the least hazardous. Nevertheless, at least one manufacturer has been capable of designing a system which virtually eliminates exposures. Indeed, once ingredients are manually fed to hoppers at the very beginning of the process, no other manual contact with the product is necessary. By use of tubular conveyors, the ingredients are fed at controlled rates to blenders, sifters, millers, compactors, granulators, and finally, to tableting presses —all, or most, of which are totally enclosed and/or controlled by other means to reduce contaminant releases. The author reports that "strict sanitary requirements are met by the totally enclosed design of the conveyors and that no storage area is required as all material is either in the pipes, being conveyed, or in the system's compact storage hoppers."

The description of this system suggests it would be a simple matter to design all product lines in this industry in a similar, totally enclosed manner and thereby virtually eliminate toxic chemical exposures. Unfortunately, for this particular industry, this generally is neither a feasible or practical concept because a high volume, continuously-in-demand product such as aspirin is more an exception than a rule. If one could consider there is a "rule" of any kind, he would have to accept the concept that production equipment in the pharmaceutical industry is designed to be flexible and usable in various combinations to allow "changeovers" from one product line to another and the rapid development of methods to produce new products with existing equipment.

The 1971 design of Gerhart et al.[7] for a potent drug production area can be considered to represent more realistically the layout of manufacturing areas in this industry. Essentially, the authors illustrate a layout consisting of a number of rooms, one for each of the major operations conducted; weighing, blending, drying and grinding, tablet compressing, and packaging. Batches of materials are manually handled at almost every step of the operation and manually transferred to the next area in line.

Where products other than tablets are manufactured, other types of operations may also be conducted. In the following, we therefore generally describe the important and currently utilized equipment and operations conducted in this industry for the three major forms of products — tablets, capsules, and liquids.

Tablet Manufacturing[1,8]

Generally speaking, there are three types of tablets-compressed, coated, and effervescent. The machinery utilized to produce the tablets from granulations are compressing machines of the single punch or rotary multiple-punch variety. These machines are peculiar to the pharmaceutical industry, and may produce thousands of tablets per minute.

A major problem at this stage of the operation is the dependence on a steady supply of free-flowing, uniform granulations. Each tablet is measured by volume instead of weight so that granulations must possess nearly identical physical properties, even though they may be prepared from numerous blended materials. Three methods are usually employed to ensure uniformity.

In the wet method, the active ingredients are milled and mixed, fillers and coloring materials are added, the mass is mixed again, and a binding agent is added. The resulting wet mass is screened, dried, blended with quantities of other ingredients, and then charged into a compressing machine. The dry method for granules is used where the nature of the ingredients does not allow exposure to heat or moisture without decomposition. A heavy-duty compressing machine is therefore used to compress powders into large "slugs" which are then broken up into the desired granulations, blended with other ingredients, and charged into a compressing machine. The direct method is employed when materials possess the physical properties desired without additional treatment.

Coated tablets are produced from compressed tablets with rounded edges (for even coating). Previous to coating, the tablets are screened to remove dust and broken tablets. Some compressing machines have screen chutes or vacuum dust removing devices for these purposes. Equipment used in actually coating the tablets consists of metal rotating pans, canvas-lined polishing pans, or jacketed kettles (for syrups).

Effervescent tablets are commonly made by the addition of an alkali bicarbonate with citric or tartaric acid to the formulation. The method followed is to warm the preparation in a rotating pan, rapidly dry under a vacuum and compress in a room with low humidity. Special precautions are necessary against moisture so that decomposition does not occur.

Capsule Manufacturing[1]

Capsules are of two types, a hard type, made of gelatin and water and used to contain powder type medications, and a soft type, of gelatin containing glycerol, usually used for containing oily medications. Hard capsules are made leakproof by a very close fit and the processes for making them are usually fully automatic. The capsule body is formed, filled with powder from a hopper, united with its "mate" and blown dust free. Soft capsules are produced by a special rotating die process. This process makes and fills capsules simultaneously, and the capsule is made leak-proof by pressure sealing.

Liquid Processing[1]

Liquids produced are of three types — aqueous, hydroalcoholic or oily. The formation of any one of these types generally involves mixing, filtration, emulsification and/or homogenization.

Mixing of ingredients is accomplished in chemically resistant tanks of various types. Some may be jacketed to allow heating, cooling, or sterilization. Some are designed to withstand pressures to 50 psi and/or a moderate vacuum. Storage tanks, sometimes fitted with agitators, are used to store batches of product until needed. The former of these may be exposed at the top for easy access.

Once a medication has been formulated, it may be filtered to ensure a clear, particle free appearance The filter equipment used, such as the plate and frame type, is common to that generally used in industry. Homogenizing and emulsification are conducted on some preparations to give them a more uniform consistency using colloid mills, homogenizers, or a combination of both.

Ointments, fatty preparations of semisolid consistency which melt when applied to skin, are manufactured in the same manner as liquids except that they may be passed through an ointment mill of the roller or rotor-and-stator type. The former type consists of parallel rollers set to close tolerances whereas the latter type consists essentially of a high-speed rotor and stator enclosed in a head which may be jacketed for heating and cooling.

Suppositories can also be considered as a semi-solid product. Classified according to their vehicle or base, the common types are cocoa butter, glycerinated gelatin, glycerin, and polyethylene glycols. Some of these types are made by mixing active ingredients with the grated base and then compressing in manual or automatic machines. Others may be formed by mixing active ingredients with a molten base and pouring the resulting mixture into molds.

Where liquid injectable products are manufactured, there is an extreme need for maintaining aseptic conditions. Kirk-Othmer reports that "Areas reserved for compounding and filling are usually small, enclosed, and constructed and furnished in such a manner as to facilitate cleaning. In some installations where disinfectant sprays are employed, the ceiling is sloped toward one end to obviate dripping of condensed vapors. Control of air purity, temperature, and humidity is a necessary adjunct to other measures designed to support aseptic operations, and rooms are generally maintained under slight positive pressure to present influx in air from unconditioned spaces." These and other measures are taken to ensure that bacteria in the area are kept to a minimum and "In most manufacturing laboratories a periodic check is made of the bacterial count of the air at different stations."

Though more a chemical unit operation than a procedure unique to the industry, extraction must be considered an important operation in manufacturing pharmacy. Essentially, it involves the grinding, possibly drying, and the charging into a closed vessel of animal or plant tissue. Solvents are introduced which are capable of dissolving desired active ingredients contained in the tissue. The solution is then recovered from waste material by filtration, centrifugation, heat vacuum evaporation, or a combination of these methods.

Packaging [1]

The pharmaceutical industry uses about every form of packaging available to protect products against damage, contamination, pilferage, and decomposition. Depending upon the particular nature of the medication prepared, these operations may involve bag filling with powders, bottle filling with pills or liquids, blister sealing, the filling of aerosols, and others. The procedures and equipment used for most products do not have unusual characteristics which make them unique to this industry. Some unique methods are used, however, for the filling of ampuls and vials.

Filling of ampuls or vials with liquids is done with either single fill or multiple fill equipment. The hypodermic syringe is the basis for single-fill equipment. Multiple-fill apparata are of various designs but all have the purpose of placing a precise quantity of medication into a receiving container. Prior to sealing ampuls, a jet of steam may be used to remove droplets remaining on the lip, or a warming flame may be used to evaporate residual moisture. The ampuls then pass a cross-fire of higher temperature sealing flames, and finally pass through annealing flames which relieve stresses in the glass.

Filling of dry powders into ampuls or vials is often accomplished in a small hood with sleeved outlets into which the operator inserts his gloved hands (i.e. a glove box). The operation may be done manually or by means of an automatic feeder but in any case may be completely enclosed.

Specific Operations, Exposures, and Controls

Compressing Machines

In general, compressing machines for tablet production, whether of single punch or rotary head construction, are designed to produce a maximum number of tablets in a minimum amount of time. These machines operate in the following sequence of events: (1) the powdered ingredients are gravity-fed from a hopper into a hole in a flat surface, (2) a scrapper removes excess material from the flat surface, leaving the hole full of the preparation, (3) a plunger possibly subjects the material to precompression prior to final compaction, (4) applications of pressure form hardened particles and promote bonding, and (5) the finished tablet is ejected down a short chute to a collection container. Excess powder scraped from the surface is sent down another chute to a separate container.

These actions of the machine, conducted at high velocity, can create dust problems, problems which could appear to be fully appreciated. By reviewing periodicals of 1968[9], it was found that advertisements for such machinery clearly showed series of flexible hose type suction devices incorporated into their design for dust control. The same periodicals, published in 1972-1973, contained advertisements for compressing machines almost identical to those shown in 1968. In these latter ads, however, the machines were seen to not only have the suction tubes, but also to have full plexiglass enclosures over all parts from which dust may be generated. These parts not only included those on or by which the tablets were actually made, but also the various chutes for finished tablets and excess powders.

The fact that commercially available machinery for such operations inherently contain full enclosures and local exhaust ventilation systems in their design indicates that engineering controls are fully available and utilized. The nature of the operation and the controls installed suggest the controls must be considered adequate when used.

There is one feature of the use of these machines which is worthy of note, and this concerns the fact that, since they are often used for batch operations, there is sometimes a requirement that the operator manually fill the bin above a machine[10]. It is not unusual to consider that the pouring of a fine powder from one container to another may release some of the product to the atmosphere. The fact that such exposures would be infrequent, might be controlled by general ventilation or an extension of the local exhaust controls installed nearby, and would easily be observed if they were significant, leads to the conclusion, however, that there is no need for any sort of research into the matter.

Rotating Pan Use

Rotating pans are used for tablet coating purposes. Such equipment primarily consists of a metal "pan" resembling a small cement mixer tilted at some angle, or a canvas-lined polishing pan similarly mounted. To minimize dust releases, dry coatings applied, and dry the pan after wet cleaning methods, such units are provided exhaust systems to pull air through the pan opening and the entire bed of tablets. The ACGIH ventilation manual[11] recommends a fact velocity of 100-150 fpm through the pan opening and a minimum transport velocity of 3000 fpm. Where heated air is supplied to the pan, it suggests "add volume of heated air to exhaust." Considering that the substances added to the pans include talc (to prevent sticking) and usually innocuous syrups (e.g., sugar, chocolate, and the like), and considering the nature of the operation, it can be concluded that controls for this operation are available, adequate, and indeed necessary to remove fines which would flaw the finish of tablets.

Capsule Filling Machines

Kirk-Othmer generally reports of hard-capsule manufacturing machines that "The precision of measurement with which the modern capsule machine produces capsules complements the remarkable accuracy of the modern capsule-filling machine. Here precise dimensions are essential." This comment, together with the comment that the process is completely automatic, suggests that excesses of ingredients are unlikely to be released from the operation. Nevertheless, it is further noted that capsule bodies are filled from a drug hopper and are often "blown dust-free" after filling. The fact that they need be cleaned in this or some other manner suggests that some dust is evolved from the operation.

In soft capsule making a rotary die process is used in which two continuous gelatin ribbons are fed between dies that revolve toward each other to form capsules at their points of convergence. An injection wedge at the last instant injects a measured quantity of the product into the closing cpasule, and the capsule is then pressure-sealed.

By reviewing advertisements in trade magazines[9] for precise descriptions of the equipment utilized, it was found that at least one manufacturer of automatic capsule filling machines factured that their units were "totally enclosed, ultra-hygienic, (and) anti-dust." An inspection of an included picture demonstrated that, as reported, all operating parts of the machine were surrounded with transparent, sliding panels. This indicates that control of airborne contaminants has been given consideration and can most likely be adequately achieved for these types of machines in general with applications of basic engineering control principles.

Mixers, Blenders, and Other Common Operations and Equipment

The literature contains numerous references to the use of mixers and blenders for preparing solid or semi-solid formulations of proper and uniform consistency. The types of machinery used are similar to those commonly used throughout industry and do not warrant any special discussion concerning use in the pharmaceutical industry. That controls may be necessary for such units is evidenced by the statement of Gerhart et al.[7] concerning a production line for oral contraceptives. They note, "we are currently investigating the use of closed mixers, or processors, as opposed to our present open mixers. These would process larger formulas, and require less labor and material handling than the present system. A reduction of required labor and *exposure* to the product would carry important personnel considerations."

Numerous other types of equipment are used and operations conducted which are similar to those commonly conducted in other industries and which also do not appear to require any special consideration when used in this industry. Among these are laboratory operations, materials handling operations, open-surface tank use, the use of dryers for wet granulations, chemical synthesis in "closed" systems, cleaning and maintenance, etc.

Overall Discussion and Conclusions—This general overview of the industrial hygiene problems of the pharmaceutical industry leads to conclusions that are not in any way surprising. The most significant of these is clearly stated by the ILO[12], i.e. "The production of pharmaceutical preparations in large modern chemical factories does not involve any special dangers for employees working there. The experiences of the chemical laboratory and the results of the toxicological and pharmacological tests have supplied detailed knowledge of the substances handled."

Whether engineering controls are available and/or adequate for controlling exposures is a separate issue however. The paper of Gerhart et al. indicates that respirators are often primarily relied upon to reduce vapor or dust exposures, and protective clothing and devices relied upon to reduce skin and eye contact. Indeed, besides the notation that air conditioning "air exchange rates were chosen to minimize the concentration of airborne drug particles throughout the (production) area, and to facilitate solvent removal," there are no indications that engineering controls were used. It is specifically stated, furthermore, that all employees were required to wear dust masks in the production areas except for those working with solvents. These latter personnel were provided self-contained breathing apparatus.

Taking into consideration the fact that no significant exposure problems were noted from operations being unique to the industry, and that respirators are nevertheless necessary to control exposures, we are left with the overall conclusions that the equipment and operations which require further study are those which are common to other industries.

Summary of Operations Requiring Study—The equipment and operations identified in this evaluation as requiring further study are:

Chemical Processes
Cleaning and Maintenance
Dryer Use
Grinding, Crushing, and Screening
Laboratory Operations
Materials Handling
Open-Surface Tank Use
Spraying

3.3 Plastic and Foam Manufacturing

Introduction—A plastic essentially consists of an organic substance of large molecular weight which is solid in its finished state but which at some stage of production can be shaped by flow. The terms plastic and resin are used in overlapping senses, but resin applies more specifically to chemically homogenous polymers used as starting materials while plastics signifies the final solid product which may contain numerous additives.

The ingredients added to basic chemical resins include plasticizers, fillers, dyes and pigments, stabilizers, and mold lubricants. Lefaux[13] devotes an entire book to a discussion of the toxicology of plastics and the substances used to manufacture them. Literally hundreds of toxic substances are described and a discussion given of how and why they are used and the hazards they present to workers. A review of this text quickly confirms the validity of placing the plastics industry high on a list of industries which utilize large numbers and varieties of hazardous substances. The presence of literally hundreds of reports and papers on the subject in the literature further confirms this conclusion.

Processes and Operations—Lefaux and others[14-17] describe the principles and methods of processing resins

and plastics in a similar manner, and it appears that the operations conducted within the industry can be generally characterized by those conducted in two major stages of production. The first of these involves the synthesis of monomers from basic elements and/or petroleum feedstock, and the polymerization or condensation of the monomers into resins. The specific operations and processes involved can be considered typical closed-system chemical processes as conducted by the chemical process industries in general.

The second major stage of manufacturing must be considered to be of primary concern in any discussion of exposure hazards. It is here that polymeric compounds are converted into finished products using methods which require a greater degree of active participation by employees. Specifically, the operations involved may include drying, to drive off solvents used in solution polymerization; grinding, to powder granules of resins or additives; measuring and blending of various fillers, plasticizers, antioxidants, colorants, flame retardants, and stabilizers; mixing and blending of additives into the plastic; and molding and other processing methods for forming finished articles.

The "conversion" methods which are somewhat unique to the industry are those by which finished articles are prepared. These methods include blow molding, injection molding, compression molding, extrusion, calendering, casting, coating, forming, foaming, laminating, and thermofusion. These are described in the following.

Blow molding involves the extrusion of a section of tubing into an open mold, the closing of the mold, and subsequent pressurization of the interior of the tubing with air or steam. This sequence of events causes the tubing to conform to the configuration of the mold and is often used for manufacturing bottles and similar hollow articles.

Injection-molding involves the charging of the plastic into a cylindrical chamber and the application of heat and pressure to force the fluid mass into a mold. In the past, the method was used primarily for the manufacture of small articles but presently is used for large objects also.

Compression molding involves application of heat and/or pressure to force plastic to flow into, fill, and conform to the shape of a cavity. For some thermosetting substances molded with pressure only, the product is subsequently cured in an oven.

For continuous manufacturing of tubes, rods, sheets, or films, extrusion or calendering may be used. In extrusion, the polymer is propelled continuously along a screw through regions of high temperature and pressure where it is melted, compacted, and finally forced through a shaped die. Calendering involves the passing of a granular resin or a thick plastic sheet through pairs of highly polished and heated rolls under high pressure.

A variety of casting methods are used. The simplest, as used in the casting of metal articles, involves pouring a liquid material into a mold and solidifying it by physical or chemical means. Casting of thermosetting resins and thermofusion techniques for thermoplastic resins involves the filling of a mold and its subsequent placement in an oven for curing or fusion. Cylindrical articles such as drums and pipes may be made by rotating a cylindrical metal mold charged with granular polymer. In this process, the granules are fused by applying heat to the outside of the rotating mold.

Films, such as photographic film and cellophane, may be formed by flowing or spraying a solution of the polymer onto an extremely smooth surface in the form of a large polished wheel or, occasionally, a metal belt or band. After the solvent has evaporated (or, in the case of cellophane, the polymer has coagulated) the film is stripped from the casting surface.

Coating processes include "dipping" forms into open-surface tanks and "slush molding." In this latter process, a viscous latex or "slush" of partly plasticized material is poured into a hollow mold. The excess is then poured out, leaving a film which is heat treated and removed.

Often, paper and board used in packaging and box-making respectively are treated in a process having attributes of film forming and coating. This application involves the extrusion of a molten film of plastic onto the substrate under conditions in which the plastic adheres to it.

Lamination and low-pressure molding involve the impregnation of sheets with a dissolved thermosetting resin, the assembly of the individual sheets, and curing. A bag inflated with steam may be used for curing assemblies in the low-pressure molding process. Hot-press methods are used for lamination. The makeup of these assemblies may involve automatic machinery or the "wet lay-up" of work by hand. In this latter method, workers manually apply coatings of uncured resins and reinforcement materials to parts of a mold.

Forming of thermosetting resins involves heating a laminated sheet, shaping it quickly in a mold or around a form, and holding it in place with light pressure until it sets up. Vacuum forming is widely used for thermoplastic sheets. In this procedure, the sheet is warmed and laid across a hollow mold cavity. A vacuum drawn on the cavity causes the sheet to conform to the mold.

The production of plastic foams is accomplished by generating a gas in a fluid polymer. theromoplastics can be foamed by the addition of a blowing agent, which decomposes to a gas at elevated temperatures or by the addition of an inert gas. The basic foam liquid generating machine consists of supply vessels, supply and recirculating lines, metering pumps, an integral valve block, a mixing head, nozzle, and solvent flush system. The resulting liquid is either poured into molds for subsequent curing, or is sprayed onto surfaces to be coated.

As reported by the Modern Plastics Encyclopedia[14], other operations conducted in the industry which are of interest, but by no means unique to the industry, include mixing and compounding, parts removal, size reduction, slitting and winding, electroplating, flocking, dyeing, hot stamping, painting, printing, vacuum metalizing, materials handling, abrasive or laser machining, and others. Of interest is their comment that "virtually all materials handling systems used in plastics processing plants are of the pneumatic type, in which the material is conveyed in a stream of air under either negative (vacuum) or positive pressure or both."

Literature Review—Though numerous sources[12,18,19,20] indicate that the majority of cured plastics have a low order of toxicity, they also note that hazards may be encountered from unreacted starting materials, the various additives, and products of thermal decomposition. An exception to the rule of low toxicity is demonstrated by the isocyanates and other additives used to generate foams; toluene diisocyanate being of the highest concern from a hazard viewpoint. In the following, various papers which describe specific health hazards to workers and the means by which they have been controlled are reviewed to point out which specific operations have been noted as being hazardous.

Kingsley[21] describes controls installed to reduce the high incidence of sensitization occurring in a plant processing epoxy resins. A principal design consideration was the separation of operations in which fumes were evolved from those in which dry plastics only were handled. "Uncured material could (thus) be confined to areas where special exhaust and handling equipment was available." The operations considered to be hazardous were "wet layup" and curing.

The area where wet plastics were handled was designed in the manner of a large water-wash booth. As he notes, "this area is 15 feet deep from the water tank to the outer edge of the hood. Exhaust capacity is 24,000 cubic feet per minute. All mixing and wet layup are done in this area." To ensure cleanliness, "not only is butcher paper used under the mold but the floor covering, a plastic film, is scraped off and replaced once a month."

Curing in this special area was sometimes conducted in a 50-ton press which also required an exhaust system. Thus, flexible ductwork was attached to the movable top of the press and sheet-metal panels were installed "to restrict the flow of air to the immediate area around the plastic."

For some work, a downdraft table with a rotating top was used. A feature of this unit was that air could be drawn across the work when a panel was removed from the front of a plenum to one side of the table. Baffles were used so that the down-draft flow was not stopped. Thus, the unit simultaneously worked as a side-draft hood and a down-draft table.

Key[18] reports that enclosure of processes used in manufacturing is seldom practical but that they usually can be isolated "as in a resin-mixing booth. Exhaust ventilation can be provided at the booth or at imbedding and encapsulating machines. Push-pull ventilation is helpful in controlling fibrous glass reinforced plastics. Alternately, a housing and a vacuum cleaner line can be attached to sanders and routers."

Harris[19] describes the case history of a worker who developed a dermatitis from powdered polymethyl methacrylate. The worker's job involved the sieving of the powder after its transfer from a drum to the sieve by means of a small hand scoop. Of interest is the observation that "this has also occurred as a result of cutting sheets of the polymer with a band-saw when a spray of fine powder or 'swarf' may be ejected over the man's hands . . .''

Among the operations involving health hazards to workers, Eckardt and Hindin[20] include grinding, sanding and polishing epoxy resins; and sawing or grinding of asbestos-containing plastics.

Bourne and Milner[22] discuss the hazards from forming, molding, curing and finishing fiberglass articles containing polyester resins. The operations specifically described include spray-finishing, manual mixing of resin systems, compression molding, hand-sanding of articles, sanding of a continuous belt energy sander, polishing on soft textile buffing wheels, and dry-polishing on a buffing wheel. They note the use of spray-booths for spray operations, and "efficient exhaust ventilation apparatus" for the belt sander. The comment is made that "in the sanding rooms a fine dust of hardened polyester resin is given off, and it is almost impossible to obtain a completely dust-free atmosphere."

Lefaux[23] reports for compression molding machines that "the mixtures of resins and various added substances may give off gases or vapours above the presses, and it is important to install in the workshops extraction systems over the presses capable of removing these gases and vapours as rapidly as they are produced. Simply airing or ventilating the workshops is not enough, and early symptoms of poisoning have been found in places without suction hoods." Other operations noted by Lefaux to require controls include spray-finishing, changing of filters, machining of plastics, and materials handling. For the latter he states "operations involving weighing powders or emptying them from containers involve risks depending on the toxic nature of the materials involved. . ." For control of such operations, he notes "a vertical air current between the worker and the materials being handled ena-

bles him to reach these while keeping his body the other side of the air curtain, and affords excellent protection."

Referring to the health hazards of "conversion processes" the ILO reports "in normal processing, chemical hazards are for all practical purposes non-existent. One would, however, advise provision of adequate ventilation in shops where moulding and extrusion machinery is operating. However, in the fabrication of polytetrafluoroethylene, there is a danger of overheated polymer decomposing and releasing highly toxic decomposition products that may cause polymer fume fever in exposed workers. In practice, (however) almost every case of polymer fume fever has been associated with smoking tobacco contaminated with traces of polytetrafluoroethylene."

Concerning polytetrafluoroethylene, Harris[23] writes "the monomer is of a low order of toxicity but its purification results in the production of highly toxic residues; the process is therefore fully enclosed and handling is by remote control for the protection of the worker."

The Michigan Department of Public Health,[24] writing of the hazards of polyurethane processing, reports "operations requiring ventilation are the pre-polymer preparation kettles, mixing equipment, mold filling stations, curing ovens, crush rolls, and waste disposal points." Though not included in the above list, it also addresses controls for spray-finishing operations. Additionally, it pictures exhaust hoods used for mold release spraying and crush rolls, and a pouring station in front of a ventilated mold tunnel. The crush rolls appear similar to open-roll mill and calendering equipment in design.

Problems in controlling emissions of vinyl chloride monomer (VCM) in polyvinyl chloride (PVC) operations are the subject of a recent paper by Mack.[25] He notes that a typical PVC processing operation involves "first the high speed mixing and cooling of the dry blend, followed by transfer to a compounding process, and then final extrusion or molding of the product. At all stages of the process, PVC is heated and VCM is evolved."

Concerning the operations conducted in mixing and compounding areas, Mack notes that published documents show VCM levels are highest in these areas and that recent investigations have shown that atmospheric concentrations of VCM can be reduced "without significant engineering alterations to the plant, and with only minor modifications to the existing mixing process." Such systems are noted as being commercially available and adaptable to all existing PVC dry blending operations. From his description and an illustration, it is seen that the system consists of a typical mixer or cooler to which a cover is attached. To one side of the cover, a filtered air inlet is positioned. Near the center of the bowl, an exhaust port, connected to an aspirator unit, draws contaminated air away from the work area. As a refinement to the system, it is noted that air can be injected at specific points in the bowl to pass air upwards through the material and enhance stripping of VCM from the mix. To reduce operator attendance, and achieve maximum processing efficiency, the aspirator system can be integrated into automatic filling, mixing, and discharge cycles of the equipment.

Also reported is that "in most extrusion and molding operations with proper ventilation, the atmospheric monomer levels are generally lower (than in mixing and compounding areas), usually in the order of about 1 ppm. Nevertheless, the risk remains that unexpectedly high monomer concentrations can occur in any plant using PVC dry blend compounds containing high residual monomer levels." To be noted is that PVC compounds are extruded at relatively low temperatures, below 350°F, for reasons of thermal stability. Citing an advantage of stripping VCM from compounds before extrusion, he notes "ventilation of the extrusion area is usually more difficult because of the physical limitations presented by the process." Nevertheless, he reports "that some advanced design compounding extruders can also contribute to a general reduction of residual monomer levels in the finished compound. This is accomplished through vacuum venting ports located along the extruder barrel and by employing large quantities of air for pellet conveying and cooling."

Discussion and Conclusions—As previously noted, the operations and processes utilized for the synthesis of polymers from raw material feedstocks must be considered to involve what is termed "chemical processes" in this report, i.e. operations conducted within "closed-systems." Exposures which may occur from such operations would primarily result from "normal" leakage of system components, venting practices, cleaning and maintenance of equipment and similar conditions or situations. These operations are common to many industries and will therefore not be discussed separately for the plastics industry. That they may need some special attention in regards to their use in this particular industry is, however, evidenced by the toxicity of some of the typical raw materials used.

Most resins do not appear to present any significant exposure hazards per se. However, since some portion of the product may contain unreacted monomer, subsequent handling of resins may result in excessive exposures, as may operations which by heat or other physical means promote some degree of decomposition. Directly after polymerization, the concentrations of such unreacted materials can be considered to be higher than they will be in subsequent processing steps. Filtering, drying, grinding, and materials handling operations at this point therefore appear to require consideration for the necessity of engineering controls. If the resin is to be packaged and shipped before conversion processes are conducted, the filling and packaging operation must be included in the problems of materials handling.

The measuring, adding, mixing, and blending operations conducted in mixing and compounding areas are similar to "masterbatching" operations conducted in the rubber industry. In this, and that industry, these operations are considered to involve considerable potential for over-exposures. The specific operations conducted can be characterized as materials handling operations, and operations utilizing mixers and blenders for both solids and liquids. Again, these operations are common to numerous industries, but would appear to already have been given special attention in the plastics industry where such monomers as vinyl chloride may be evolved.

Some of the subsequent processing steps in "conversion workshops" are also similar to those conducted in the rubber industry. Specifically, these include compression molding, curing oven use, extrusion, and calendering. Calender roll exhaust ventilation is a subject addressed by the Industrial Ventilation Manual[11] for rubber processing. Whether the same exhaust system performance criteria are necessary and/or adequate for the processing of various plastics is uncertain. The use of ovens and dryers is common in industry and does not appear to necessitate special consideration in this industry.

Whether extruders and compression molding machines (primarily those which heat the substance) require controls any more sophisticated than general mechanical ventilation is a subject of contention. Lefaux and Kingsley indicate that local exhaust ventilation is necessary. The ILO states that general ventilation is sufficient. Mack's comments indicate that the ILO may be right. He cites VCM concentrations of only 1 ppm as typical where general ventilation is used in the vicinity of extruders. While the temperatures applied are relatively low, the volatility of vinyl chloride is relatively high.

There is no indication in the literature that blow molding, injection molding, casting, forming, or thermofusion operations release amounts of airborne contaminants. However, since the plastic is heated in most of these, one must consider the possibility that some vapors or fumes may be released when heated chambers are opened or molten plastics or plastics dissolved in solvents are used in dip tanks, poured into molds, sprayed onto surfaces, etc.

The high-toxicity of the various isocyanates used in the manufacture of foam plastics obviously requires adequate control measures. In view of the above descriptions of operations, the comments of Reference 24 seem most appropriate and accurate. To reiterate, the statement of interest was that "operations requiring ventilation are the pre-polymer preparation kettles, mixing equipment, mold filling stations, curing ovens, crush rolls, and waste disposal points."

"Wet-layup" operations conducted manually with such materials as epoxy resins definitely require installation of special controls. Though Kingsley generally describes a production area which provides adequate control, specific ventilation system design criteria would not appear to be available in the literature of such operations.

A number of sources discuss the hazards of abrasive machining operations conducted upon polymeric substances. Controls for much of the equipment used in such operations have been described in various sources where metals are handled, but not for non-metallic substances.

In general, a review of the preceding information and the sources it was derived from does not allow any absolute conclusions to be drawn whether engineering controls are commonly available and/or adequate for all the numerous combinations of toxic chemicals and operations conducted in this industry. Concerning those operations conducted which are somewhat unique to the industry, the strong impression is, however, that serious problems do not exist, that various control measures have been adapted for use where necessary, and that real problems, such as the control of VCM, are receiving considerable attention, attention which is resulting in significant reductions in exposures.

A limited research effort to confirm or disprove this supposition may be in order. Of interest would be the determination of whether uncured plastics evolve excessive quantitites of airborne contaminants at various stages of the production process which are somewhat unique to the industry, and whether adequate engineering controls exist for such operations. Particular operations to be studied would include those involving molding, curing, extrusion, calendering, casting, foaming, wet lay-up, etc. A level of effort of 9 man-months and a time-frame of 6 months would probably be sufficient if personnel highly knowledgeable of the industry were utilized. The results of the survey would lead to the firm determination of whether specific engineering control research is necessary.

Other operations conducted are similar to those in other industries. As such, they are included in a following list of operations requiring further study. Of note is that the chemical processes and the materials handling operations conducted in the initial stage of production may be ideal candidates for specific study to develop control technology applicable to many industries. This, by virtue of the toxicity of some of the substances utilized or produced (e.g., vinyl chloride, epoxy resins, etc.).

Summary of Operations for Further Study—The list of operations and equipment which require further investigation in this report is as follows:

Abrasive Machining (non-metals)
Chemical Processes
Cleaning and Maintenance
Dryer and Oven Use

Grinding, Crushing, and Screening
Laboratory Operations
Material Handling
Non-Spray Application of Volatile Substances
Open-Surface Tank Use
Spray-finishing

3.4 Paint and Coatings Manufacturing and Application

Introduction—Paints, varnishes, and lacquers are widely and commonly employed throughout industry for coating objects with protective and/or decorative finishes. For 1975, it was estimated that over one billion gallons of such coatings would be manufactured with a dollar value approaching $4 billion. The paint and coatings industry is comprised of some 1,200 companies having 1,600 plant locations. A large number of these companies are very small since the industry requires modest capital investment for equipment and relies on relatively simple and readily available technology. The four largest companies in the field account for about 22% of sales while the largest 50 account for 61% of sales. [26,27]

In general the formulae for paints consist of three basic components. These include binders, solvents, and pigments to which various ingredients are added to impart special qualities. The binder forms the film, the solvent keeps the mixture fluid for ease of application, and the pigment provides color and improves film properties. Binders consist of natural resins, drying oils, or synthetic polymers. The solvent may be either water, as in water-based (latex) paints, or may be an organic solvent. Pigments consist of inorganic oxides, sulfates, carbonates, and chromates. [12,26]

Materials in Paint and Their Toxicity—It is estimated that approximately 2,000 different raw materials are used by the paint industry and that any given plant might keep as many as 600 of these in stock. Included among this vast number of raw materials are many which can and have been shown to cause occupationally related illness and disease. The ILO[12] specifically describes some of the substances commonly used which are of concern.

Among binders it lists the use of shellac, linseed oil, alkyds, formaldehydes, polyesters, acrylics, epoxies, polyurethanes, vinyls, rubber derivatives and bitumens. Pigments include various metallic oxides, lithopone, calcium carbonate, magnesium silicate, barium sulphate, cadmium selenide and sulphide, ferric ferrocyanide, and carbon and lamp black. A wide range of hydrocarbons, alcohols, esters, ketones, and glycol esters are noted as being employed for solvents (toxic solvents such as benzene, tetrachloroethane and carbon tetrachloride have largely been eliminated from the industry though trichloroethylene is still used). Among special additives, mercury aryl compounds, pentachlorophenol or its sodium salt, and organic lead compounds are listed as fungicides. Metallic salts of organic acids, tricresyl-o-phosphate, zinc, aluminum, silica, asbestos fiber, antimony oxide and cadmium red are listed for other special purposes.

ILO reports lead pigments as being seldom used although a few yellows and greens may contain lead chromate. Metal primers, however, are noted as usually containing red lead. The EPA[26] reports that the use of red lead has been slowly increasing while the use of white lead has been rapidly decreasing. Estimated figures for 1975 show that 15.8 million pounds of red lead and 2.6 million pounds of white lead will be used this year.

Piper,[28] listed the various raw materials utilized in paints and discussed the toxicological effects of exposure. Thus, it is well documented that a large number of known toxic materials are used daily in the paint and coatings industry. Sources describing these materials and their toxicity are readily available and should be consulted for additional information.

Processes and Equipment—Bidlack and Fasig[29] provide a detailed description of the operations involved in manufacturing paint. In order, they discuss the receiving and storage of raw materials, the assembly of batch ingredients, mixing, grinding, thinning, tinting, filling, labeling, packing, storage, and shipping.

The operations involved in receiving and storage of raw materials are similar to those in numerous other industries. Essentially, they consist of materials handling operations, and the possible sampling and analysis of substances in quality control laboratories. "Assembly of batch ingredients" includes the weighing or measuring out of the various ingredients and their placement in position adjacent to the manufacturing unit.

Assembly is followed by the mixing of pigments and other additives with a sufficient amount of the vehicle (i.e., liquid binder and/or solvent) to wet the dry materials and produce a paste for grinding. When the paste is to be ground on roller mills, such mixing may be accomplished in low vats or tanks which are equipped with heavy stirrers. Lead mixers, dough mixers, and portable change-can or pony mixers are ordinarily used for this purpose. Pebble and ball mills or functionally similar devices may alternatively be used to combine mixing and grinding into one operation. It is generally recommended that the mixers be located so that pastes can flow by gravity to the grinding units and that piping facilities be provided to permit direct feeding of commonly used vehicles from metering or weighing devices to the mixing equipment.

Various types of grinders are used to reduce or disperse the paste to a fine and uniform consistency. The afore-

mentioned roller mills consist of a number of hollow steel rollers which operate in a manner similar to that of all multi-roll mills; i.e., the substance is passed through closely spaced rollers which revolve at different speeds and in opposite directions to effect some degree of dispersion and/or size reduction. Ground paste is taken off the last roll by means of a scraper knife and either collected into a portable change can or pumped to thinning tanks.[29,30,31]

Pebble and ball mills depend on the contact between stone or porcelain pebbles and steel balls within revolving, horizontal cylinders. Water jackets are sometimes provided around the revolving mill for cooling purposes. The operating procedure for such units involves placing all pigments and a predetermined portion of vehicle into the mill and bolting down the loading hatch cover prior to revolution. The mill may be stopped periodically to allow for the insertion of additional vehicle. Access to the unit is through the aforementioned loading hatch and at a pouring spout on the opposite side of the cylinder.[29,31,32,33]

Other types of dispersion equipment used include high-speed stone mills, dough mixers, sand mills, high-speed dispersers, Uniroll mills, Kady mills, attritors, resonant mills, vibratory mills, and stator-rotor mills.

High-speed dispersion units basically consist of a tank in which is placed a high-speed stirrer with a speical type of impeller.[31,34,35] High-speed stone mills contain a stationary stone and a rotating stone between which the pigment is subjected to a shearing action.[31] Dough mixers have two roughly S-shaped blades which overlap and rotate in opposite directions for dispersing very heavy pastes.[31,36] A sand mill is a cylinder containing sand and rotating impeller disks through which pigment slurrys can be passed continuously (a screen retains sand in the mill).[31] Grinding in a stator-rotor mill is carried out in a specially designed head at the end of a shaft.[36] An attritor mill is mounted vertically and contains balls which are agitated by means of a series of arms attached to a vertical shaft. In such a mill, the mixture can be passed through either on a batch or continuous basis.[36]

A Uniroll mill is single, horizontal, water-cooled roller in an enclosure. A feeding hopper directs the mixture over the roll, and a vane bar, forced against the side of the roller, provides the grinding action. A scraper removes the treated paste from the roller face and directs it to an outlet spout.[32] Resonant mills are a somewhat recent development which consist of racks of tubes which contain a grinding or dispersing medium and which are vibrated at their resonant frequency.[37] A vibratory mill is similar to a high-speed disperser but has the added feature that the grinding chamber is vibrated.[38] A Kady mill is also very similar to the high speed dispersion unit except for details of design and the fact the batch is generally smaller.[39]

Thinning operations involve the further addition of vehicle to the paint formulation and follow the mixing and grinding processes. These operations may be performed either in stationary tanks or in portable change-cans. The stationary tanks have attached agitator units to mix the ground paste and vehicle thoroughly and uniformly. The portable tanks are placed under a change-can mixer unit for this purpose.

Tinting of batches involves the addition of tinting color to white bases or the adjustment of solid color bases which have been formulated as complete products.

The first step in the filling operation is the straining or filtering of the paint to remove any foreign material. This can be accomplished by the use of fabric or screen filters, or where high-grade finishes are desired, by the use of variable-speed centrifugal clarifiers. After such treatment, the paint is filled into cans which are subsequently sealed, labeled, packed, stored and/or shipped. Filling stations are usually small tanks which support hand filling operations or which are connected to filling machines by lengths of hose.

The paint industry is also involved with the manufacture of lacquers and varnishes, coatings which can be included in the generic term "paint" but which are produced by employing somewhat different equipment and procedures. For example, since lacquers differ from most other coatings in that the film dries entirely by evaporation, it is often impractical to manufacture them in the customary manner because the volitilization of the fast-evaporating solvents results in the deposition of a dry, solid film on equipment. Thus, the tanks, mixers, and other equipment used are usually of the completely closed variety to reduce solvent loss, ensure ease of clean-up, and provide fire safety.

"Varnish" consists of processed resins and oils, solvents, and metallic driers or catalysts. While some are produced without the application of heat, others require the heating of the oils and resins in kettles at temperatures ranging from 450-625°F. The heating process results in the evolution of resin and oil decomposition products.

The methods of application of paints and coatings include brushing, dipping, roller coating, flow coating, and spraying. Production lines often use driers and ovens for "curing" purposes. The specific details of these operations are well-understood and common knowledge and, therefore, do not require further elaboration.

Literature Review-Exposures and Controls—
Stern and Horowitz[40] report that "in the paint factory, dust exposures occur primarily at the mixers, ball or pebble mills and varnish kettles, and in the handling of bags of pigments, extenders, and resins for use in the mixers. A lesser exposure occurs in the handling and mixing of tinting pigments." Concerning vapor exposures, they state "the principal vapor exposures in paint manufacture occur at the reducing (thinning) and tinting operations, in can filling, in cutting nitrocellulose and resins and in the cleaning of equipment with a solvent. Vapor exposures also result from leakage from pump lines, meters, pumps, filters, etc. A small steady drip can allow a whole workroom's atmosphere to become charged with solvent vapor."

Stern and Horowitz discuss various exposures and their controls in detail and make some observations and recommendations apropos for many industries. Dumping bags, discarding empty bags and adding less than full bag quantities by means of a scoop are noted as sources of dust. Recommended controls include specially designed hoods over the top of mixers and ball mill loading spouts, covers with hinged sections for access to open-top mixers (with a 200 fpm downdraft through the opening), and/or lateral draft hoods near loading openings and at floor level chutes. Exhaust hoods over mills from which solvents are evaporated; general mechanical ventilation for can filling, reducing, and "churning" rooms; local exhaust of open-surface tanks and can filling operations involving lacquers and varnishes; and use of toxic solids in pellet form are recommended for other vapor and dust sources.

Bidlack and Fasig agree with the above and state "some provision for dust removal is desirable when dry pigments or powders are being handled. This can be accomplished by having exhaust fans so placed that the air is drawn over the mixing equipment and away from the operator and carried to the outside atmosphere, or by means of a direct-suction system which has openings mounted over each mixing unit so that the dust is directly exhausted therefrom." They also discuss fume recovery systems in varnish plants. From their descriptions and illustrations of typical plants, it is evident that adequate fume control technology and equipment for these processes are available. They not only address the control and recovery of vapors from heated kettles but also those from thinning tanks.

Patty[4] indicates that "the mixing of dry pigments with oils and varnishes constitutes the principal health hazard of this industry." He notes that various pigments are purchased in paper sacks and "a dangerous dust exposure occurs when the pigments are added to the mixers unless exhaust ventilation is used."

Freriks[39] addresses the hazards associated with the use of flammable solvents in the paint industry. Since lower flammable limits are generally well above toxic concentrations in air, his comments are applicable. He reports "it is commonly accepted today that a positive air movement equal to about one cubic foot per minute per square foot of floor space is adequate. Since solvent vapors are heavier than air the suction should be within a few inches of the floor. Exhaust and make-up air vents should be located to sweep the area of vapors."

Concerning dispersion equipment, Freriks reports that the too costly and too slow ball mill is being replaced by the high speed dispersion mill, the sand grinder, the attritor, and to some extent, by the Kady mill. For the high speed dispersion mill, it is said that unless local ventilation is used, the mill can release flammable vapors. This latter statement is also made for sand mills and continuous attritors. For batch attritors and Kady mills, it is noted that they are covered during operation and present no hazard greater than a covered tank of paint.

Arnstein[42] describes several factors to be considered when installing and maintaining a pumping system in a paint plant. Of interest is that the toxicity of the substance being handled and the leak-resistance of the system components are not mentioned.

Martens[43] reports that water-base paints are growing in popularity in industrial applications because of the principal reasons of "reduction of the fire hazard, toxic solvents, and air pollution." For example, in the area of ambient air quality, Los Angeles County officials[26] estimate that the application of paint contributes 360 tons of hydrocarbons per day to total organic emissions.

The hazards associated with application techniques are well appreciated and, again, do not require further elaboration other than to say that non-spray application techniques and drier use may expose workers to solvent vapors, and spray techniques expose them to the entire paint formulation.

Discussion and Conclusions—The paint and coatings industry relies upon relatively simple technology and utilizes equipment and procedures which generally require modest capital investment. These equipment and procedures do not include any which are unique to the industry, but rather, consist of those which are very commonly utilized or conducted in a wide variety of other industries. Consequently, there does not appear, at this time, to be any reason to investigate this industry separately and specifically to determine the availability and adequacy of engineering controls.

The large number of potentially toxic substances used in the industry, and the lack of specific data to identify the exposures which occur, does however suggest that some sort of exploratory study is warranted to determine the nature and extent of occupational exposure to recognized hazardous substances and of any observable health effects that can be associated with exposure to occupational environments in the industry. Such a study would provide sufficient information to permit responsible evaluation of the need for more thorough and rigorous investigations and might concentrate on all toxic substances, not only those regulated by OSHA.

Since the objective of this report is to investigate the availability and adequacy of engineering controls where a need for their use can be identified, the recommendation of such a study is beyond the scope of this work. Therefore, the following list simply identifies those operation categories which are pertinent to the industry, but which, by reason of their commonness, require study on a broader scale.

Summary of Operations for Further Study—The list of operations and equipment which are to be further investigated in this report is as follows:

Chemical Processes
Cleaning and Maintenance
Dryer and Oven Use
Grinding, Crushing, and Screening
Laboratory Operations
Material Handling
Non-Spray Application of Volatile Substances
Open-Surface Tank Use
Spray-Finishing Operations

3.5 Rubber Production

Introduction—The rubber industry is large and heterogeneous, involving a wide variety of processes, operations, and substances. Many of the toxic substances encountered play a role at any number of points in the compounding of rubber, either natural or synthetic.

The rubber industry, according to McCormick[44], involves "not only the conversion of the natural and synthetic polymers into usable articles, but the manufacture of chemicals, plastics, and numerous other materials."

Morris[45] writes, "The rubber industry is one of America's largest industries, in which great numbers of workers are employed in compounding, mixing, fabricating, and sheeting the various rubber compounds."

Processes, Equipment, and Exposures[12,46]—The processes and operations employed in rubber compounding and fabrication include those which are common to many industries and those which are found primarily in the rubber industry. Among the former can be included closed-system chemical processes, materials handling, abrasive machining, drying, and others. The latter include vulcanization, mastication/mixing, and formation. The specific processes and operations involved are discussed and described in the following.

Stephenson[47] relates that natural "rubber is a natural exudation from various plants when they are cut or injured. The ordinary rubber comes from the tree 'Heavea braziliensis', a native of South America which is now grown commercially chiefly in the Far East. Two other forms of natural rubber are also known, namely, gutta-percha and balata. All three of these natural rubbers are polymers of isoprene (and) 2-methyl-1,3-butadiene" and constitute the basic feedstock for production of natural rubber articles.

Stephenson also gives an account of the processes by which various types of rubbers are synthetized. The emulsion polymerization process for butadiene-styrene copolymer (former GR-S, now usually called SBR), the most commonly used rubber, is described as consisting of a series of reactors, a blowdown tank, a pressure flash tank, a vacuum flash tank, and stripping columns. Nitrile rubbers are said to be produced in a similar process and the process trains for other rubbers are generally noted as utilizing similar and typical chemical processing equipment. Exceptions to this generalization, however, are found to include the use of tunnel driers during the production of solid butyl rubber and ethylene-propylene elastomers.

Mallette[48] writes, "The points of potential exposure in these processes are in the handling of raw materials, coagulation, centrifugation, and drying." With regard to controls in the synthetic rubber plant, he says, "Adequate local and general exhaust ventilation must be provided for coagulating tanks and centrifuges. The escape of vapor from the dryers can best be prevented by maintaining a slight negative pressure within them. It may be necessary to provide hoods over the dryer outlets to control vapors of styrene and other substances driven off at this point."

McCormick[49] gives a comprehensive review of the hydrocarbons encountered in the synthetic rubber industry, specifically styrene-butadiene rubbers, referred to in his paper as GR-S (Government Rubber-Styrene). Both butadiene and styrene are said to be produced, stored, transported, and handled in closed systems, and neither substance is considered to be more than midly hazardous.* However, some of the minor constituents, specifically dinitrochlorobenzene, sodium sulfide, and triphenylphosphite are noted to pose potential dangers. The author states that "precautions must be taken to prevent skin contact (with dinitrochlorobenzene)," and recommends process ventilation to control the sodium sulfide hazard. (The danger here is from the production of hydrogen sulfide in the presence of sulfide). Process ventilation is also recommended for the triphenyl phosphate which produces phenol upon hydrolysis.

Open-roll mills, enclosed Banbury mixers or Gordan Plasticators are employed to "breakdown" rubber elastomers prior to compounding, the Banbury being the most commonly used for this purpose. The open-roll mill is usually ventilated with a semi-enclosing hood with a specific face velocity. Ventilatory control for Banbury mixers is a subject addressed by the ACGIH ventilation manual.[11] Gordon Plasticators operate on an extrusion principle, with the raw material being forced through screens, around disks, and through a die by a rotating screw.

---

*These substances may be hazardous according to more recent information.

After breakdown, resins and pigments are added to the rubber in a process termed master-batching. In this operation, some of the more hazardous materials are introduced into the rubber, including solvents, carbon-black, sulfur, and organic accelerators. To provide for homogeneity of the additives in the rubber, the components are mixed on a mill or in a Banbury. If a mill is used, a considerable amount of manual handling may be required to help pass the rubber through the rollers. The adequacy of any ventilation system is more important at this point due to the increased number of potentially toxic materials added in this procedure. Since the rubber is handled in batches, the exhaust systems must be capable of handling sudden, intense dust and vapor loadings.

Master-batching is initiated in a weighing room where the accelators, pigments, anti-oxidants, and other components are stored, weighed, and added to the rubber. These steps are simply materials handling operations but may result in significant dust and vapor levels. In writing on the use of rubber additives, Bourne[50] notes that "The main problem areas discovered were: mixing or milling of rubber stocks with the additives, curing or vulcanizing, and dusting of semifinished and finished products." Almost 600 different additives were encountered in a survey of 140 plants, involving 19,400 workers.

Schwartz[51] reports that "Dermatitis has been known to exist in the rubber industry for a long time, the terms 'rubber itch' and 'rubber poisoning' being commonly applied to this condition." He refers to chlorides, stearic acid, lead oxide, litharge, yellow pigment, arsenic, carbon disulphide, and many others. In reference to the National Safety Council Transactions, Schwartz reports that "compounding is rated as one of the most hazardous processes in the industry, and that insurance companies class compounders among those whose mortality is expected to run from 50 percent to 100 percent above the average. Mixing room workers are thought to be exposed to similar hazardous compounds."

Once the rubber has been properly mixed with additives, it is either stored for future use, or employed directly as stock for calenders or extruders. A calender is generally a multi-roll mill which can be adjusted to produce various thicknesses of rubber sheeting, impregnate fabric with rubber, or skin rubber to one side of a fabric. Again, a moderate degree of rubber handling is required by the mill operator, who may be exposed to solvent fumes. The ACGIH manual contains hood design criteria to control such exposures.

Extruders, which function by forcing rubber through a die by a rotating steel screw, are used in stock preparation, as in the production of tire threads, or to form the basic product immediately, as in rubber band production. In either situation, the process is the same and will generally require handling by the operator. If extrusion takes place at high temperatures, fumes may be evolved in the operation. Excess talc, applied at this point to eliminate the stickiness of the rubber may result in excessive dust exposures.

Depending upon the article being manufactured, any number of machines and tasks may be involved to convert stock into its final, or nearly final, form before vulcanizing or curing. Tire-building operations, for example, are complex and generally require a high degree of manual handling done in the presence of petroleum spirits used to keep the rubber fresh and workable.

The next step is curing, i.e. vulcanization, of the assembled product. This operation is achieved on several types of machines, including presses, autoclaves, pressure vulcanizers, hot water tanks, or Rotocures. Some of these are enclosed units, releasing gases and vapors solely when opened, or when improperly maintained, while others, such as the Rotocure, operate in a semi-open mode.

The mold cure is the most elementary operation. Here, the fabricated article is inserted between two halves of a mold which is then closed and heated to the vulcanizing temperature. After a specified time, the mold is opened and the article removed. It is precisely at this stage that vapors and fumes may be released to the environment. Of note is that toxic substances which were not originally present may be evolved because of the elevated temperatures.

In steam curing, the article is placed in an autoclave and exposed to steam at a specific pressure. For such operations the article must be preformed since no shaping occurs as when a mold is used. When an autoclave is opened after curing, a cloud of steam and toxic contaminants are released similar to the mold curing process. Not all curing operations are conducted in a batch. Where conveyor belting or packing is treated, units such as the Rotocure are utilized in a continuous operating mode.

Final finish operations involve an assortment of buffing, trimming, and inspection tasks highly specific to the product involved. Buffing may be considered as an abrasive machining operation from which an airborne dust may be produced.

Engineering Controls—A review of the literature indicates that specific ventilation system design criteria for operations unique to the industry exist solely, in the general literature at least, for Banbury mixers and calenders. Other recommendations found were definitely of a more general nature. For example, where solvents are used, Greenburg and Moskiwitz[52] simply recommend the three standard methods of approach, i.e., (1) substitution, (2) general ventilation, and (3) local exhaust ventilation and give the substitution of toluene for benzene as an example of the first stratagem. When the work area is large, they report that general ventilation can be employed. If

the work is localized and fixed, local ventilation is noted as probably being most effective. The ILO[12] recommends "good general ventilation of areas where volatile materials are used," effective ventilation of process vessels prior to entrance for maintenance, and standard controls, such as gloves, aprons, and other protective garments. Particular problem areas within the rubber industry appear to be the formation and vulcanization stages of manufacturing. Given the controls recommended by the ACGIH for the calender mill, it would appear that similar controls may be necessary for open-roll mills, extruders, and other machines in this category. The nature of curing operations indicates that controls are also desirable for these processes.

Discussion and Conclusions—As is evident, a wide variety of operations are conducted in this industry, and industrial hygiene problems, both specific and non-specific to the materials and processes employed therein, must be considered to exist. Indeed, the nature of many methods used would regularly involve direct insult to the worker from an assortment of physical and toxic agents if proper control measures were not instituted.

Whether practical engineering control measures exist or are adequate for all the various operations unique to the industry involves questions which cannot be simply answered from this investigation. That it is likely that they do not, however, is evidenced by the fact that as a result of an agreement between the International Union of Rubber Workers (IURW) and the major tire manufacturers, Harvard University and the University of North Carolina are involved in a multi-million dollar effort to investigate the industry's problem areas. Though this investigation is limited to the tire industry, it can be assumed that many of the problems and solutions that are discovered and developed will be applicable to the industry as a whole.

The fact that such a large-scale effort is being conducted by the private sector leads to the conclusion that it is not at this time necessary for NIOSH to participate in research activities which may parallel or repeat work currently being conducted. It is therefore recommended that NIOSH attempt to keep abreast of the findings and results of these activities, encourage publication and dissemination of worthwhile findings to the entire industry, and if then deemed necessary, support research to fill any data gaps which become evident.

Summary of Operations Requiring Study—In the above, the opinion has been expressed that current activities of the rubber industry itself suggest a "wait and see" attitude is appropriate on the part of NIOSH. The operations and processes which at present warrant further study, therefore, are those which are common to many industries; these being:

Abrasive Machining
Chemical Processes

Cleaning and Maintenance*
Dryer and Oven Use
Grinding, Crushing, and Screening
Laboratory Operations*
Material Handling
Open-Surface Tank Use*
Spray-Finishing*

3.6 Pesticide Manufacturing and Application

Introduction—Pesticides are agents used to destroy the plant and animal pests of man. These compounds encompass many different classes of chemicals and act upon numerous target pests; hence, the more specific terms insecticide, herbicide, rodenticide, fungicide, nematocide, molluscicide, etc. used to describe them. In addition to agricultural uses, these substances are employed in a wide variety of other pursuits such as structural pest control, wood treatment, mothproofing, and maintenance of utility and transportation rights-of-way. Thus, potential occupational exposures to pesticides are numerous and varied.[12]

For the purposes of this study, three major aspects of pesticide manufacture and use are of concern: manufacture of the active ingredient; formulation of the pesticide product; and application.

Pesticide Manufacturing—The first stage of the pesticide industry involves the synthesis of the active ingredients. The specific operations and processes involved are generally closed-system chemical syntheses as practiced by the chemical process industries. Most of the producers of basic pesticides, indeed, are multi-product chemical companies that produce a variety of products[53]. The technical grade pesticide compound produced is usually transported in bulk in liquid form (tank cars or barrels) or as a powder or dust (in barrels or bags) to be mixed with other ingredients into a finished pesticide product.

Formulation of Product—In addition to one or more active ingredients or toxicants, a trade-name pesticide product contains synergists, surface-active agents, diluents, and other substances. The mixing of these ingredients is the formulation stage.

The compounding of solid insecticides involves the addition of liquid or solid toxicant to a dust carrier such as clay dust, talc, lime, or silica. In the case of the former, liquid toxicant is spray during blending, and the wetted mixture passed through several mixing stages (usually through ribbon blenders) and/or ground in a pulverizer. (It should be noted that the wetted mixture, although containing as much as 50% liquid by weight, acts as a solid because of the porosity of the solid carrier.) The wetted dust is then conveyed to a holding bin for packaging into bags, boxes or other containers. The insecticide Toxaphene is produced in this manner. The active ingredient, melted and mixed together with kerosene, is sprayed into clay dust and blended thoroughly.[54]

Active ingredients in solid form are ground and mixed with a carrier in several passes through standard grinding and size-reduction equipment. An example of this process is the compounding of DDT dust.[54]** Emptied from sacks into a hopper, DDT flakes are passed through a crusher and a pulverizer where finely ground silica (a stabilizing agent) is added. After thorough mixing in a ribbon blender, the DDT is conveyed to a barrel-filling unit where it is packed for aging. The aged material is blended further and finely ground in several stages (utilizing high-speed grinding mills and high-pressure air mills and cyclone separators). Finally, the product is packed into barrels for shipment.

Liquid products can be formulated as a solution, emulsion or suspension. DDT, for example, was often formulated as a liquid containing the active ingredient, two emulsifying agents, and an organic solvent.[54] Formulation of such products generally simply involves addition of ingredients to an agitated tank and subsequent packaging operations.

Pesticide Application—Application operations involve the preparation of the pesticide product into a final form for distribution, the transfer of the product to the application equipment, and the actual application of the pesticide. The specific nature of these operations is determined by the properties of the formulation, the nature of the target pest, and the site of application.

The EPA recognizes 15 distinct physical forms of pesticide products ranging from crystals, granules, and wettable powders to emulsions, liquids, and pressurized gases and aerosols.[55] Many of these forms require some pre-application preparation in accordance with manufacturers' instructions. Typically, this involves the mixing of the concentrated product with a diluent or vehicle such as food (use as bait), water, kerosene, oils, or organic solvents.[12] Since most pesticides are still handled in 5 to 50 pound bags and cartons or 1 to 5 gallon containers, these mixing operations, and subsequent transfer operations, may involve considerable manual handling, although the use of closed systems for transfer to application equipment must be noted as increasing.

* The industry evaluation did not specifically demonstrate that these operation categories are involved in this industry. Nevertheless, they are included either by reason of the logicality of doing so when the nature of the industry is given consideration, or because of personal experience of the authors in noting their use.

** Although DDT is no longer in use in the U.S. and is produced by only one pesticide company, the formulation process is felt to be representative of the procedures for other pesticide products formulated from solid toxicants.

Application equipment for pesticides include mist blowers, hydraulic ground sprayers, backpack and handheld sprayers, specially outfitted aircraft, dusters, soil injectors and incorporators, and aerosol bombs.

Spraying is the most common method of pesticide application. For organic insecticides, concentrated sprays (toxicant 10-95% of mixture) are being widely used. The development of air blast nozzles for pesticide spray equipment has made this technique possible since the droplet size produced by these nozzles is 30-80 μ compared with the 200-500 μ droplets produced by hydraulic nozzles[54] Concentrated sprays are also conducive to aerial application, the most economical way to treat large tracts of land with spray, dust, or granules.[12]

Increasingly, farmers and others have been employing specially trained pest control applicators because of the cost of specialized equipment (such as airplanes) and the hazards of pesticide use. Only large-scale pest control concerns are likely to have specialized facilities for handling and mixing of pesticides, facilities equipped with suitable engineering controls. Smaller volume pesticide users, such as most pest-control operators and individual farmers, usually lack such facilities and must rely on respirators and protective clothing for exposure control.

Other Operations—Other occupational groups exposed to pesticides include harvesters of sprayed crops,[56] other agricultural workers that reenter fields too soon after pesticide application,[12] and employees in food and other agricultural product processing plants.[12] The disposal of unwanted pesticides and pesticide containers may also result in occupational exposures. For example, workers in pesticide container recycling operations (common in California) may be exposed.[57]

Literature Search—According to Hamilton and Hardy[58], "it is the continuing objective of the chemical and agricultural industries to develop materials which control specific categories of pests and have minimal potential for detrimental effects in man and other desirable animal and plant species. Unfortunately this goal has not yet been achieved, and examples of clinical intoxication continue to appear, usually as the result of accidents or incorrect methods of distribution and application." In many cases, these accidents result from disregard of instructions pertaining to the handling or application of the active ingredients.

A very recent closing of a chemical plant in Virginia demonstrates not only the continuing occurrence of occupational exposures during synthesis of pesticide active ingredients, but also the possible serious effects of these exposures.[59] Following reports of illness among workers, state health officials visited a small plant manufacturing the active ingredients for a chlorinated hydrocarbon related to DDT, used in controlling banana pests, ants, and roaches. As reported by one of the visitors, "There was Kepone dust all over and piled up an inch deep in some places. Nobody was wearing proper safety gear, and I could tell just by looking at them that seven of the 10 workers there that day were sick. One could hardly stand up." Equipment in use was termed "kind of a Rube Goldberg contraption." Ex-employees reported numerous spills and leaks. One commented that, "the plant was filthy and there was Kepone all over the place, even on the picnic table in the gas station part where I'd eat." Health effects found in 110 workers tested ranged from high levels of the chemical in their blood (50% of the workers tested) to severe memory loss and liver damage (requiring hospitialization of 13). "A small group tested for effects on their reproductive systems was found to be sterile."

The complete lack of engineering controls and disregard of protective equipment in the plant reported upon in this incident must be noted as representing a rare situation in the industry. The report indicates that the serious occupational and environmental problems observed might have been avoided if "a more established chemical concern" had been manufacturing the chemical.[59] Nevertheless, the very fact that such a situation could be allowed to develop indicates a lack of understanding on the part of employers and employees alike of the hazards of toxic substances.

Wolf and Armstrong[60] studied the potential maximum exposures to DDT of workers in two plants formulating 50% water wettable DDT powder. Dermal and respiratory exposures (assuming no protective gear) were measured for four work situations: mixing, bagging, sewing of bags, and packing of bags into cartons. The potential maximum exposure was found to be greatest in the bagging operations in both plants. During bagging operations, powder was blown into the air during removal of filled bags from the spout and closure of the filler hole by a folded flap. Inadequate ventilation and periodic malfunction of the filler spout mechanism in one plant sometimes resulted in gross contamination of the operator. ("The authors have observed that these two problems are not uncommon in pesticide formulating plants.") The main exposure in the packing stations was powder blown from inside the carton as the worker pressed the bags into place. Of the four work situations monitored, mixing of the various dry ingredients resulted in the lowest total exposure. In all operations, higher potential exposures were found in the plant observed to have the poorer ventilatory controls and housekeeping.

Calculated exposures when "recommended protective gear" was worn were determined to be 2.5 to 10 times less than the calculated potential maximum exposures (i.e., no respirator or protective clothing). Low levels of a DDT metabolite in urine of workers in these plants indicated that the protective gear actually worn* was effective. However, the authors reported the wearing of clothing which provides only minimum protection and non-use of a respirator are not particularly uncommon in formulating plants. . . Our observations in formulating plants indicate that, when workers are at the bagging station and not wearing respirators, it is almost impossible for them to avoid inhaling large quantities of the compound which is sometimes obviously present in high concentrations near the breathing zone.''

Danielson[61] addresses the air pollution aspects of insecticide formulation procedures. He comments that some of the sources emitting dust are bag packers, barrel fillers, hoppers, crushers, conveyors, blenders, mixing tanks, and grinding mills. He considers the crushing and grinding operations most hazardous and recommends maximum enclosure with indraft velocities of 400 fpm or higher. For the other operations, where dust is released at low velocities, velocities of 200 to 300 fpm are suggested through areas in enclosures.

Inhalation and dermal exposures of 52 farmers engaged in seasonal spraying of apple orchards and vegetable and grain fields were studied by Jegier.[62] With respect to personal protective measures, 48% of the operators had no protection at all; 39% had head protection, 20% used cloth overalls, 27% used gloves, and only 11% used respirators. Bare hands were used during mixing by 21% of the operators, and 85% were observed to exercise "insufficient care" (e.g., disregarding wind direction during tank preparation). The total exposures of operators were nevertheless calculated to be less than 1% of the toxic dose for each insecticide used. "While the investigation did not indicate that spray operators were subjected to doses approaching acute toxic levels, field observations during the course of spray operations revealed a laxness in the attitude of the majority of spray operators towards cautious spraying practices."

California is the largest agricultural state in the U.S. and one of the largest state users of pesticides. Reports of occupational disease in California, including that resulting from agricultural chemicals, must be filed by the attending physician with the State Department of Industrial Relations. In 1957[63], 710 cases of occupational disease (including one fatality) were attributed to pesticides. Of these, more than one-half involved cases of dermatitis, 251 cases involved systemic poisoning, and 52 cases involved respiratory illness. Workers in agriculture (both on farms and in services, including commercial ground and air spraying) had 60% of the reported occupational disease, 65% of the reported cases of systemic poisoning, and 53% of all reported dermatitis. Statistics[64] of disabling

work injuries for 1970 in this state (i.e., those resulting in loss of time from work) include 216 (no fatalities) attributed to insecticides, sprays, and fumigants. Of these, 113 occurred on field crop, vegetable, orchard, and general farms, and 69 involved agricultural service workers (including commercial pest control operators.) Since the number of disabling injuries in 1957 was 340, it appears that advances in control of occupational exposures to pesticides in the 13 year period resulted in a net 36% reduction in disabling injuries. While this reduction is significant, it does not overshadow the fact that a substantial problem yet remains.

Discussion and Conclusions—The operations and processes of the pesticide industry are by no means unique; chemical processing, wet and dry mixing, bagging and materials handling, and spray application are common operations throughout industry. However, unlike many other industries, the pesticide industry predominantly produces or utilizes compounds of significant toxicity, many of which are not only hazardous when inhaled but also when in contact with the skin. The nature of end use is very widespread, occurring for the most part out-of-doors or in places (such as barns, grain storage bins, etc.) where traditional engineering controls such as process enclosure and exhaust ventilation are generally considered impracticable.

In its roles as regulator, the EPA has sought to protect the pesticide worker, as well as the environment. Occupational safety is a consideration during registration of both the active compound and the formulation. Restricted-use registrations issued for some pesticides are designed in large part to protect the applicator (only those with proper protective equipment and training are licensed to apply these products).

No unique problem areas requiring study by NIOSH were identified for this industry. Thus, the following list of operation categories simply encompasses those which are common to many industries.

Summary of Operations Requiring Study—The operation categories which are to be further studied in this report are:

Chemical Processes
Cleaning and Maintenance
Grinding, Crushing and Screening
Material Handling
Open-Surface Tank Use
Spray Application

3.7 Adhesive Manufacturing and Application

Introduction—In the United States, the adhesive manufacturing industry is highly fragmented with over 800 companies operating approximately 1200 plants.[65,66]

---

*Rubber boots, cap, respirator, cloth overalls with long sleeves, and rubber gauntlet gloves.

Industrial end-users of adhesive products number in the hundreds of thousands. In both manufacturing and end-use applications, there are numerous opportunities for worker exposure to the toxic substances incorporated in many adhesive formulations.

Adhesive manufacturers can be classified by their position in the "raw materials to end-use" sequence. The three primary classifications are:

- Companies that produce raw materials for adhesives, but also manufacture limited quantities of finished products for outside distribution;

- Companies that purchase raw materials and specialize in the manufacture of finished adhesives for commercial distribution; and,

- Companies that purchase raw materials and manufacture finished adhesives for captive use in other products.

The first of these classifications encompasses the producers of natural and synthetic elastomers and resins. The manufacture of finished adhesives is not a major part of the plastics industry; however, the plastics industry is a primary contributor of substances for manufacturing adhesives.

The manufacturers of commercial adhesives comprise the most important classifications. A typical adhesive plant is very small compared to most other plants of the chemical process industries. The largest adhesive plants usually have work forces of about 100 people; while small 2- or 3-man operations are frequently encountered.

The third classification—captive manufacturing—also represents a major part of adhesive manufacturing. Most of these companies are of moderate to large size. The poor economics associated with small scale manufacturing is often incentive for small companies to purchase their finished adhesives.

The 1980 projection for domestic end-use of all types of adhesives has been set at nearly 5 billion pounds.[67] Applications cover a wide range of areas which relate, either directly or indirectly, to nearly every aspect of the manufacturing industry in the U. S. The primary applications are:

1. In the conversion of forest products:
   - Plywood, veneer and particle board used in residential construction and durable goods.
   - Furniture, prefabricated structures and other wood products.
   - Paper and packaging such as boxes, bags and envelopes.

2. In the conversion of non-forest products:
   - Non-durable goods such as plastic laminates, tires, sealants, shoes, and apparel.

- Durable goods such as glass insulation, automobiles and aircraft.

3. As replacements for conventional fasteners:
   - In interior and exterior construction.
   - In musical instruments and other consumer goods.
   - In sealing metal cans and other food industry packaging.

Therefore, both adhesive manufacturing and end-use applications have the potential for affecting large numbers of workers.

The chemical and physical processes associated with adhesives are best described for both manufacture and end-use applications separately.

Manufacturing Processes—Batch processing, rather than continuous processing, is the most common method of adhesive manufacture. Batches of 1000 gallons are considered large by industry standards. The single unit operation that is common throughout is mixing. In fact, various aspects of mixing comprise 80-90% of operations used in adhesive manufacture.[66]

The processing necessary in adhesive manufacture is best delineated by a discussion of each primary adhesive category.[68-71]

A. Animal Glues

Although manufactured by only 4 companies at the present time, glues derived from animal materials have been the foundation of the adhesive industry for several hundred years. Such adhesives are usually one-part systems supplied to industrial users as water-soluble dry flake. The manufacture of these glues from animal hide and bone involves several preparatory steps. Bones must be ground and hides must be cut, solvent-degreased and washed. The protein products from each must be extracted, filtered and further processed to final flake form.

B. Aqueous Adhesives

This category includes most one-part adhesives based on water soluble natural and synthetic materials.

(1) Natural Adhesives

- Starches are water-dispersible, natural polymers derived from grains and roots. Preparation of the raw materials for use in glues requires extraction, drying and milling. Moreover, additional starch-based glues—such as oxidized starches and dextrins—are formed by hydrolysis and repolymerization of the original starches. Many of these preparatory steps are carried out by the grain processing industry. The final glues vary from dry powders to liquid solutions.

- Soybean glues are manufactured by wetting, shearing, defoaming, dispersing, denaturing and finally crosslinking the original soybean materials. The grain processing industry is involved in this area also.

• Cellulosics are derived from the structural elements of plants such as cotton and wood pulp. Most require dissolution in organic solvents, but some are water soluble. Processing is similar to starches and soybean glues. The final form can be a solvent solution, powder or thermoplastic (hot melt).

• Milk, as the starting material for casein glue, must go through extraction and drying operations to produce the final powdered products.

(2) Synthetic Adhesives

Synthetic materials will account for most of the anticipated growth in adhesive manufacturing during the next 5 years.[72] However, this growth estimate includes all synthetic adhesives, not just the aqueous-based adhesives discussed here.

• Phenol resin adhesives are predominately the phenolformaldehyde glues as used in plywood manufacture. The phenol and formaldehyde are combined in batches to produce dry resins tailored to the needs of the user. Water dispersions are formed at the time of application.

• Polyvinyl acetate is polymerized and used as a dispersion of solid resin in water. This is the most widely used form of PVAC.

C. Solvent Adhesives

This category also includes adhesives based on natural and synthetic elastomers and resins. The use of organic solvents substantially increases exposures to toxic substances.

(1) Natural Adhesives

• Cellulosics requiring dissolution in organic solvents are also derived from the structural elements of plants such as cotton and wood pulp. As with water soluble cellulosics, these adhesives can be formed as solvent solutions, powders or thermoplastics.

• The rubber industry supplies raw materials to adhesives manufacturing. These include vulcanized and unvulcanized rubbers in latex solution and bulk form. Adhesives are prepared by mastication (milling) and mixing of rubbers with various additives such as tackifiers, resins, fillers, plasticizers and curing agents. Products are available in latex form, as solvent solutions and as mastics.

• Other natural hydrocarbon materials—asphalt, pitches and organic resins—are included in this adhesive category. Very little modification other than blending with various additives is carried out on these materials after initial petrochemical production.

(2) Synthetic Adhesives

• Elastomeric adhesives and cements are manufactured by mastication (milling) of the raw elastomers and blending with solvents and other additives. The most frequently used elastomers are butyl rubber, polyisobutylene, nitrile rubber, SBR and neoprene.

• In-use polymerization of polysulfides forms the basis for some adhesives and sealants. Such products are usually two-part systems which are mixed at the point of application. However, monomers, curing agents, reaction retarders, reinforcing agents and fillers are premixed in one or more of the parts.

• Thermoplastic adhesives are prepolymerized materials that become fluid and are applied upon heating to 175°F and above. They do not require solvents, but rather rely on the melted state for proper viscosity. The manufacture of thermoplastic adhesives consists mostly of polymerization reactions followed by blending with plasticizers, fillers and reinforcing agents. Products are packaged as powders, sheets, cubes and pellets. Included are polyethylene, isobutylene, polyamides and some cellulosics.

• Thermosetting adhesives are essentially reaction type adhesives. These products are most often multipart systems and must be mixed at the point of application. Monomers, curing agents, reaction retarders, reinforcing agents and fillers are premixed in one or more of the parts. Some common thermoset adhesives are epoxies, urethanes, phenolformaldehyde, polyvinyl butyral, and cyanoacrylates.

• Progress in adhesives development has paralleled progress in polymer development over the past 20 years. As more exotic or stable materials have become available, adhesives capable of functioning in difficult circumstances—such as in high temperatures or with difficult to bond surfaces—have been developed. The organosilicones, s-Triazine and modified phenolics are examples. These adhesives may be solvent-based, thermoplastic or thermosetting, but all require some type of component mixing in manufacture.

Manufacturing and Application Processes—Most adhesives are applied as fluids at their point of use. This ensures full coverage and allows wetting of the surfaces to be bonded. Therefore, in addition to the mixing required for multipart systems, the primary processes in adhesives application are based on spreading adhesives as liquids—either cold or hot. This is usually followed by drying and possibly compression of the bonded joint. The equipment most often used in the adhesives industry can be described in general terms for manufacturing, but must be related to adhesive types for application.

Manufacturing Equipment

The formulation of most one-part and multipart liquid and powder adhesive systems requires the mixing of resins, curing agents, catalysts, reaction retardants, reinforcing agents, fillers, plasticizers, tackifiers, wetting agents, and water or organic solvents. This is accomplished in open or closed mixing tanks with single or double mixing propellers or blades. Heavy duty mixing tanks, sigma mixers, Banbury mixers, and vertical and horizontal churns are used with high viscosity liquids and elastomers.

Some adhesives require special processing and therefore special equipment prior to final formulation. For example, animal and vegetable glue formulations require the use of grinding and cutting equipment, solvent-degreasing and washing tanks, protein extraction apparatus, filters and dryers. Starches must be milled prior to use in starch-based glues.

Perhaps the largest use of special equipment occurs in the manufacture of those organic adhesives requiring chemical modification of the original raw materials. Reactors must be used in oxidizing or hydrolizing starches, in cross-linking cellulose and soybean materials, and in partially or fully polymerizing resin based thermoplastic and thermosetting adhesives. Many of these reactions can be carried out in mixing tanks; however, others must be carried out in enclosed, jacketed reactors.

Application Equipment

Adhesives application equipment is best described in terms of both one-part adhesives and multipart adhesives.[68,73]

One-part adhesives are applied by hand when used in piece-by-piece operations—as in furniture, automobile, aircraft, plywood and shoe manufacture. The adhesives may be applied by brush, dip, trowel, knife, roller, spray gun or other special applicator. When continuous adhesive application is practical—as in coating paper, textiles and plastic sheeting—continuous knife applicators; spray applicators; roll, drum and wheel applicators; extruders and curtain applicators are used. Each of these devices usually is associated with peripheral equipment such as pumps or other feed apparatus, backing rolls or drums, and adhesive reservoirs.

Thermoplastic adhesives, which are applied as hot-melts, require specialized equipment. Either a heated reservoir is used to supply adhesive as a fluid to an applicator or the applicator itself is heated to bring adhesive to its melting point. Once fluidized, hot-melts are applied by heated roller, knife, pressure gun or wheel.[74]

Most emphasis in the proper use of multipart adhesives is placed on mixing of the parts. Because of the limited lifetime of these adhesives, mixing is most often done in batches, either by hand or with laboratory scale mixing equipment. Adhesive application on a piece-by-piece basis makes use of the same manual equipment described under one-part systems. Continuous application requires complex metering, mixing and dispensing equipment. Usually only enough adhesive is mixed as is needed by the

dispensing equipment. Such equipment may consist of pumps, sprays or extruders.

The curing of many adhesives is enchanced by forced air and heat. The equipment used for this purpose includes various types of driers and ovens. Batch pressing or continuous roller pressing of bonded materials is often conducted while adhesives are setting up.

Exposures and Controls—The operations involved in the manufacture of synthetic adhesives are identical to those conducted in the plastics and rubber industries to convert resins into finished products. The only differences appear to involve the final form into which the product is prepared. Whereas in the plastics and rubber industries the product may be in the form of a solid object after molding and curing, the adhesive industry supplies solutions or dry powders which can be applied to various surfaces before curing. Since these "masterbatching" operations are identical, the evaluations for these other industries can be considered adequate in describing the operations and their associated exposure hazards. Indeed, it appears that the plastic and rubber industries themselves include prime examples of companies which manufacture limited quantities of finished adhesive products for outside distribution or captive use.

The manufacture of natural adhesives generally involves some sort of extraction or reaction process followed by filtering, drying, milling, and/or dissolution in solvents. Preparatory steps for animal glues also include grinding and solvent-degreasing. Various aspects of these operations, especially where toxic solvents are used, can be expected to have the potential for resulting in excessive exposures, just as they have been noted to do so in other industries investigated.

Overall, the conclusions can be made that there is considerable potential in the manufacturing operations conducted within this industry for exposures to occur, that these operations are by no means unique to this industry, and that the problems which would occur and the controls appropriate for them are covered elsewhere in this report.

The application of adhesives for the bonding of materials might be conducted in numerous industries ranging from the manufacture of military aircraft to that of wooden furniture. Wherever solvent based glues are used, one can expect that solvent vapors will be evolved. Where adhesives are applied at elevated temperatures, the ILO[12] reports that there may be "high evolution of fume and vapor."

Knife-over-roll, air knife and kiss roll coating techniques are used for coating paper, fabric or other flexible materials. Spraying may also be used. Houwink and Salomen[75] report that cooled rollers facilitate the application of adhesives containing solvents that evaporate quickly, "although generally speaking such adhesive solutions are best applied by spraying."

The spray techniques utilized include the well-known air and airless spray techniques and the use of flame spray guns. A feature of these latter devices is that they handle a fusible powder instead of a liquid. The powder is melted as it passes through a flame and is deposited in molten form on the surface of the adherend. An advantage of their use is that the need for a solvent is eliminated. The other spray techniques obviously require the use of solvents to keep the adhesive in a fluid state.

Numerous other types of complex and specialized machinery are used for applying adhesives in paper bag making, book binding, furniture, cabinet and door manufacturing, boat and airplane manufacturing, boot and shoe making, etc. At the other extreme, a variety of simple hand-held glue spreaders, rollers, brushes, and other applicators exist for manual applications.

There appears to be little information in the literature concerning whether occupational exposures to toxic substances result from the application of adhesives. One may infer, however, from comments made in the literature, that the potential is there; again, especially where solvents are used as a vehicle. For example, it is noted that a large weight fraction of solvent-based adhesives may consist of ketones, lower alcohols (including methanol), esters, hydrocarbons, ethers, and other solvents. Concerning urea formaldehyde adhesives, Houwink and Salomen report "the pungent smell of formaldehyde is not only objectionable during the use of the glue, but may persist in the glued material . . . for months or years." For adhesives containing toluene diisocyanate, they state "the substance is toxic and has a relatively high vapor pressure, making it unpleasant to handle." The possible sensitivity reactions of workers to fumes from epoxy adhesives are, of course, well-appreciated and were reported upon in the evaluation for the plastics industry.

Summary of Operations for Further Study—The operations and equipment concluded to require further investigation in this study are:

Abrasive-Machining (for removing excess glue)
Chemical Processes
Dryer and Curing Oven Use
Grinding, Crushing and Screening
Laboratory Operations
Materials Handling
Non-Spray Application of Volatile Substances
Open-Surface Tank Use
Spray-Finishing
Textile (and paper) Coating Operations

3.8 Dye and Ink Manufacturing

Introduction—Dye sales in 1969 amounted to an estimated $380 million. About 50% of this could be accounted for by four American-owned firms, 30 to 40% by four foreign-owned companies, and the remaining 10 to 20% by 41 other U.S.- and foreign-owned companies.[76] Two thirds of these dyes are used in textile processing.

Inks are available for printing on virtually every type of paper, film, foil, paperboard, etc. A 1972 census[77] of the printing ink industry reported that the industry includes 398 manufacturing establishments (belonging to 200 odd companies), 140 of which have 20 employees or more. The number of production workers at that time was 5,000. The total value of product shipments amounted to $484 million.

Process Description

Dyes

Beginning with benzene, naphthalene, and anthraquinone, the steps to a dye or pigment through benzenoid intermediate compounds are many and difficult, and essentially involve a series of chemical reactions in typical equipment used in the chemical industry. Because of the exacting nature of dye manufacture, the chemical reactions are run in batches and the processes are labor intensive. According to Stinson[76] "reactors must be loaded from drums by hand. Solutions with precipitated product are filtered in huge presses. The filter cake is taken from the press by hand and carried to ovens for drying." The reactors referred to are closed-top, agitated tanks with loading hatches, the presses are typical filter presses, and the dryers are usually of the tray type. The filter cake is removed from the plates of the press by shoveling into the dryer trays.

Dye may be sold as dry powder, which the dyer has to disperse himself, or in liquid or paste forms. Liquid dyes are usually water-based. However, considerable efforts are taking place to perfect methods of solvent dyeing textiles. The companies involved feel that solvent dyeing with chlorinated solvents might eventually solve the water pollution effluent problems of the industry. Since solvents could be redistilled after use, aqueous effluents would be eliminated. Low heats of vaporization would cut down fuel costs, though a solvent recovery system would possibly balance off against the cost of alternative waste treatment plants.

Inks

Various sources[1,12,78,79] agree in their descriptions of how inks are manufactured. The operations and equipment involved are shown to be the same as those utilized in the paint industry. The similarities are so striking, indeed, that no further description appears necessary.

Discussion—The production processes for dyes contain no procedures which are unique. For the purposes of this study, one can consider that the operations which may result in exposures include chemical processing, the weighing out of ingredients to be added to tanks, dumping of ingredients into the tank, the emptying of filters, the use of drying ovens, and the packaging of products into bags, drums, or cans. Also to be included must be laboratory operations required for analyzing samples throughout the production process and cleaning and maintenance operations. Considering the spillage which can take place in the transfer of filter cakes to drying trays and the transport of the trays to the oven, one might expect that a considerable amount of dry powder would be present after any spilled, wet, filter cake dries in the workplace.

The review of the production process for inks indicated that it is essentially identical to that used for making paints. The hazards and control of these operations were described in the industrial evaluation for the paint and coatings industry and do not warrant further discussion here.

Whether any particular health problems currently exist in either industry could not be determined from the available information. That considerable potential exists for them can however be appreciated. Many of the intermediate compounds in dyes are derivatives of benzene, naphthalene, and anthraquinone. Aniline and its compounds are also widely used. Inorganic pigments used include salts or oxides of lead, cadmium, antimony and titanium. Some dyes are suspected as being carcinogenic.[12] Substances used in inks are also toxic. In another section of this report, it is noted that printing pressmen have been found to have abnormally high incidences of various serious diseases. If it is considered that the use of various inks can lead to serious health problems, then it must also be considered that workers manufacturing these substances might be subjected to similar hazards.

Conclusions and Recommendations—For both of these industries, there appears to be a need for some sort of exploratory study to determine the nature and extent of occupational exposure to recognized hazardous substances, and the nature and extent of any observable health effects that can be associated with occupational environments in the industry. Such a survey would be especially appropriate for the ink industry if the current study of the printing industry verifies that substances in common inks are more hazardous than once believed. Given the commonness of the various manufacturing equipment and procedures utilized, however, there does not appear, at this time, to be any need for engineering control research directed specifically towards these industries.

3.9 Soap and Detergent Manufacturing

Introduction—According to the ILO,[12] "detergent" is a general term presently applied to all synthetic washing compounds, although it was originally applied only to "soaps" made from natural fats and oils. For the purpose of this investigation, these terms will be utilized in their current context.

The yearly production of detergents for household use currently approaches 5 billion pounds per year. Soap sales have leveled off at a rate of 1 billion pounds per year. Overall, the companies which manufacture these products constitute a multi-billion dollar industry.[61]

Soap

Ingredients[61]

Soap is a product of the reaction of the combining of a fat or fatty acid with a saponifying alkali. Though more fatty acids are produced synthetically today than by the splitting of natural fats and oils, the soap industry uses fatty acids produced almost exclusively from natural products. The saponifying alkali most often used is sodium hydroxide, but potassium hydroxide is used to some degree because of the greater solubility of resulting products. Metallic soaps may be manufactured from alkaline earth, metal, or heavy-metal salts of fatty acids.

Finished products are not generally 99.4% "pure" soap. Minerals such as soda ash, caustic potash, sodium silicate, sodium bicarbonate and trisodium phosphate are used as builders or fillers. Other additives may include tetrasodium pyrophosphate, sodium tripolyphosphate, carboxymethyl cellolose, and various preservatives, pigments, dyes, perfumes and antioxidants or chelating compounds.

Fatty Acid and Glycerine Production[61]

Three methods are currently used in the soap industry for fatty acid production. The high-pressure hydrolysis process, the only continuous technique, is the one most often used. The equipment train consists of various flash, blending, settling, and holding tanks; a hydrolizer column; heat exchangers; a vacuum still, condensers; and associated pumping and piping systems. An autoclave operating at temperatures ranging from 300 to 450°F and pressures ranging from 75 to 450 psig is used for some batch operations. The simplest method, the Twitchell process, is conducted in unpressurized tanks at 212 to 220°F.

Soap and glycerol are the reaction products of a fat and a hydroxide. The production of commercial and pharmaceutical grades of glycerine is therefore a part of the soap industry. Typical glycerine purification operations involve equipment trains consisting of typical process plant equipment such as tanks, heat exchangers, condensers and evaporators. Centrifuges and filter presses may be used to separate precipitates.

Soap Manufacturing[1,61]

The soap-making processes can be batch or continuous in nature. Batch processes include the kettle or full-boiled process, the semiboiled process, the cold process, the autoclave process, the methyl ester process and the jet saponification process. Continuous processes are proprietary and include the Sharples and Mon Savon processes. Regardless of their specific details, all accomplish the same steps of soap manufacture.

The kettle or full-boiled process is one which follows the historical and traditional methods used since the beginning of the industry. It simply consists of a number of steps or operations conducted in a single kettle or a series of kettles. The semiboiled process involves mixing and heating of ingredients in a steam-jacketed crutcher. Externally applied heat is not required in the similar cold process. Since reactions between fats and oils are exothermic, mixtures are simply allowed to stand in insulated containers. In the autoclave process, reactants are pumped through externally heated coils and the hot mass is flashed into a vacuum chamber. Jet saponification involves the use of a special steam nozzle or jet through which hot fats and caustic soda solution are proportioned. Emulsions formed drop into conventional kettles where the reaction goes to completion.

The continuous processes mentioned are conducted in closed vessels. The Sharples process utilizes high speed centrifuges and reaction vessels. The Mon Savon process utilizes a high-speed homogenizer for mixing and then discharges the emulsion onto the hot inner walls of a jacketed reaction chamber.

Soap Finishing[1,61]

Finishing of soap involves its processing into liquid, powder, granule, chip, flake or bar form for consumer use. During these operations various additives mentioned above may be blended with the basic soap stock.

Liquid soaps are blended with other materials in tanks and then packaged in standard bottle filling equipment. They are rarely manufactured today except for some very specialized products.

The oldest and most seldom used method for forming bar soaps involves the pouring of a semiliquid paste into molds, the subsequent sawing of the resulting blocks into roughly bar size pieces, and the stamping of these into the final form. "Milling" processes are most commonly used for bar soaps. In one method, the soap is batched in a mixer called a "crutcher," is flowed onto chill rolls, and then flaked off and passed through a dryer. The flakes can be "plodded" (mixed in a screw or tubular type mixer), fed to a roll mill, plodded again, and extruded in a continuous shape for cutting and stamping. The flakes could also be packaged as is or ground into powders. Other methods, though differing in detail, involve the same types of equipment and operations.

Detergents[1,61]

Ingredients

The raw materials used to manufacture active ingredients of anionic detergents, those which by far represent the bulk of such substance produced, include substances to form the hydrophobic portion of the surfactant molecule, those which form the hydrophilic portion, and those used for purposes of neutralization. Among the first type can be included alkylates made from long straight-chain normal paraffins reacted with benzene, and normal fatty alcohols produced synthetically or from natural fats and oils. The second category includes sulfur trioxide or one of its hydrates, sulfuric acid or oleum. The last group includes sodium hydroxide, sodium bicarbonate, other sodium bases, ammonia, potassium, diethanolamine, and triethanolamine.

The additives incorporated into detergent formulations are numerous and varied in nature. For the sake of brevity, a summary list of them is presented in Table 4.

Surfactant Manufacturing

The processes for sulfonation or sulfation of organic bases to produce the most common detergents are separate and distinct. Variously employed are oleum, sulfur trioxide, sulfuric acid or chlorosulfonic acid in liquid or vapor form. The equipment trains utilized to synthesize the active surfactants are all typical of closed system chemical processes and do not warrant further description.

Detergent Product Manufacturing

The only production process for detergent formulations which is not similar to those used in soap production is that of spray-drying. Since the control of spray-dryer emissions is an air-cleaning problem not of interest to this study, no further elaboration appears to be necessary.

Exposures—The operations in soap manufacturing which can be regarded as dust sources in the workplace include: addition of powdered and fine crystalline materials to crutchers, mechanical sawing and cutting of cold frame soap, milling and plodding soap, air drying soap in steam heated dryers, milling, forming, and packaging. These operations and the grinding of soap chips are noted by Danielson[61] as causing internal plant hygiene problems which require control for worker comfort and safety.

The batching and mixing of fine dry ingredients to form slurries for the production of detergent granules can also cause dust problems. Particular problem areas reported by Danielson are scale hoppers, mixers, and crutchers. This author summarizes the problems encountered in detergent product manufacturing by stating "Conveying, mixing, packaging, and other equipment used for granules can cause dust emissions. The granule particles, which are hollow beads, are crushed during mixing and conveying and generate fine dusts. Dusts emitted from screens, mix-

TABLE 4—TYPICAL ADDITIVES USED IN DETERGENT PRODUCTION

Compounds	Purpose	Physical Form Used
Sodium tripolyphosphate	Sequestering agent	Powder, prill, or granule
Tetrasodium pyrophosphate	Sequestering agent	Powder, prill, or granule
Nitrilotriacetic acid and its sodium salts	Sequestering agent	Powder
Sodium sulfate	Filler	Powder
Sodium carbonate	Filler	Powder
Amides of various types	Supplementary sur-factants	Liquids
Trisodium phosphate	Alkali	Powder
Carboxy methylcellulose	Prevention of dirt redeposition	Powder or granule
Sodium silicate	Corrosion inhibition, others	Water solutions
Various	Optical brightners	Powder or liquid
Various	Perfume	Liquid
Sodium perborate and magnesium silicate	Bleaches	Powders
Enzymes	Stain removal	Powders
Others	Perservatives Antioxidants Foam-suppressors Scouring cleansers (including finely pulverized silica)	Varies

ers, bins, mechanical-conveying equipment, and air-conveying equipment are quite irritating to eyes and nostrils with continuous exposure. Some of the additive materials, such as enzymes, also cause serious health problems."

Houghton in Reference 80 reports that arsenic, in concentrations ranging up to 700 ppm, has been found in detergents. He also discusses the hazards of sodium nitrilotriacetic acid and of enzymes. Of the latter he reports that extreme precautions are needed to protect workers from the inhalation of enzyme dust and that the Soap and Detergent Association has published a detailed list of precautions to be taken.

The ILO[12] notes "although the production of detergents and cleaners is largely automated, diseases of the respiratory system may occasionally occur if dust control in the working area is not provided for." It also reiterates the problems associated with the handling of enzymes. Other than occasional skin allergies and dermatitis which may arise from manual handling of chemicals, no health problems are noted for soap manufacturing.

Kirk-Othmer[1], however, reporting on soap manufacturing, states that high dust concentrations may result from some flake and powder processing operations. The dust is noted as being irritating to the mucous surfaces of the nasal passages and throat but not as being particularly injurious. A method for dedusting flakes and powders is described which involves dropping them through a tower against a rising forced current of air.

Discussion and Conclusions—There is not a great deal of information available on the health problems of this industry. That which there is indicates that dust may be generated from a variety of operations and that this dust is simply irritating and not "particularly injurious." The majority of operations which appear to represent the problem areas are very common ones involving batching, mixing, materials handling, grinding, milling, and packaging. Dusts containing enzymes are noted to be a significant problem which presumably has been solved by the special handling procedures formulated by the industry.

3.10 Perfume Manufacturing

Introduction—A perfume can be defined as a mixture of substances which are incorporated into a vehicle to produce a pleasing olfactory sensation. Originally, perfumes were only produced from naturally occurring botanical or animal products. Recently, however, the tendency has been to market substances which have no counterpart in nature. These perfumes are made from synthetic materials produced by a variety of chemical processes. The perfumes considered the finest are blends of natural products and synthetic materials. The synthetic substance

serves to enhance the odor of the natural perfume, reduce the market price of the formulation, and produce an entirely unique fragrance.

Ingredients[1,12,69,81]

Vehicle/Solvent

A somewhat volatile solvent helps project the odor of a perfume away from its carrier. The most widely used vehicle in the industry is ethyl alcohol in various dilutions. Ethyl alcohol is fairly inert to the solutes and is relatively non-irritating to the skin.

Fixative

Fixatives are substances which have a lower volatility than that of perfume oil. Their purpose in a formulation is to retard and even up the rate of evaporation of various odorous constituents. Fixatives may or may not add to the odor of the product. Some main categories of fixatives are animal secretions, essential oils, resinous products and synthetic chemicals. The latter are usually odorless esters used to substitute from some imported animal fixatives.

Odorous Substances

The odorous substances used in "perfumery" can mostly be represented as essential oils, isolates, and synthetic or semisynthetic chemicals. The essential oils are of vegetable origin and can consist of mixtures of compounds representing most types of chemical structures (i.e., various esters, alcohols, aldehydes, acids, phenols, ketones, lactones, terpenes, and hydrocarbons). An isolate is a pure chemical compound, which as its name suggests, is isolated from some essential oil or other natural perfume material. Synthetics are obviously manufactured by chemical synthesis. Semisynthetics is a term utilized to describe constituents chemically synthesized from an isolate or other natural starting material. Included in these latter categories are such substances as vanillin, ionone, terpineols, coumarin, diphenyl oxide or ether, cinnamic aldehyde, benzyl benzoate, phenylethyl alcohol, benzaldehyde, and others.

Manufacturing Processes

Synthetic Constituents

Acetylene and aromatic coal tar constituents are usually the starting materials for the synthesis of perfumes and essences. The manufacturing processes and equipment used are typical of those used throughout the chemical industry and can be considered to be closed-system chemical processes.

Natural Constituents

Naturally occurring compounds are processed by steam distillation, expression, fat or solvent extraction, exudation, dry distillation, and even by ultrasonic treatment. The first three of these are of most interest.

Steam distillation is commonly conducted in stills of about 600 gallon capacity. The material to be subjected to the steam is first ground, chipped, or cut into small pieces before charging to the still. Although distillation is most often conducted at atmospheric pressure, the process may be run in a vacuum if the constituents of the oil are easily subject to hydrolysis. The steam is collected with condensors, and separators are used for dividing the resulting oily layer from the aqueous one. The crude oils retrieved may be further processed before use by a variety of physical or chemical treatments.

Expression involves machine pressing operations to squeeze out oils present in fruit. The enfleurage process is a cold-fat extraction process for some flowers. The fat usually consists of a mixture of tallow, lard, and a preservative. Once the fat has been exposed to a sufficient quantity of flowers, it in turn is subjected to extraction with an alcohol.

The most successful solvents for solvent extraction processing are petroleum ether and benzene. Others used include acetone, toluene, dichloromethane, and ethyl alcohol. Extraction equipment consists of stills for fractionating the solvent, batteries of extractors, holding tanks for solvents and solutions, and an evaporator. Two types of extractors are employed; the rotary drum type and the stationary vertical cylinder type. Use of either type of extractor involves the charging of the flowers or other materials, repeated washing with the solvent, and solvent recovery from the residue with either steam or vacuum.

Exposures—The ILO[12] simply recommends that where toxic solvents or other toxic substances are used (e.g. benzene, acetone, toluene, dichloromethane, etc.), their concentrations in equipment and in the workplace air should be maintained at levels which do not present explosion or health hazards. Addressing the toxicity of the ethereal oils, it reports that the results of research into the toxicity of these substances "indicates that little precise information is available on the toxicity of the main components and that nothing at all is known of the toxicity of the minor components or oxidation products." It further states that "the continuing use of perfumes and essences in food or for external application should provide sufficient reason for systematic studies on the pharmacology and chronic toxicity of all components to ensure that these substances do not contain certain harmful or even carcinogenic constituents."

Addressing the hazards of perfumes in general, Sax[82] lists them as being acute and chronic allergens of slight toxic hazard. The acute and chronic effects by inhalation are similarly rated while the acute systemic effects of ingestion are rated as being of moderate toxic hazard.

Other information concerning the health hazards of the industry was not found.

Discussion and Conclusions—From the descriptions of the various equipment used and operations conducted in the perfume industry, it appears that those in any way unique to the industry (i.e. expression, fat extraction, etc.) are not of a nature to result in exposures to known toxic substances. Whether they result in exposures to substances which are of yet unrecognized toxicity is another matter.

The other operations conducted appear to include those which are quite common to many industries and which have already been listed as such in preceding industry evaluations.

3.11 Printing

Introduction—The printing industry today is very complex, involving processes and operations ranging from art-and-copy preparation to presswork, binding and other finishing operations.

The products of the printing industry are ever-increasing in complexity, variety, and quantity. The ILO[12] points out that, in America, "printing is the third most important industry in the production of non-durable goods, preceded only by food and clothing," and that about 4% of the total work force of all manufacturing industries is involved in the printing industry. It also notes that "between 50% and 70% of all printing establishments employ fewer than 25 people."

There is recent evidence to indicate that respiratory diseases can be caused by agents in the workers' environment in printing plants. An investigation of members of the Printing Pressmen's Union has shown that their mortality rate from emphysema is much greater than the national average.[83] Thus, not only is this industry representative of one which utilizes a large number and variety of toxic agents, but one in which serious worker health problems currently exist.

In the following, we generally describe the basic processes and equipment used in the printing industry, the health hazards associated with these operations, and the controls which have been implemented.

Processes and Equipment—There are three basic methods of printing: letterpress, lithography, and gravure.

Letterpress

Marsailes[84] explains that letterpress printing involves the application of ink to a raised or relief surface by ink rollers, and then transfer of an image by applying the surface, under pressure, to paper. Strauss[78] notes that letterpress printing is actually only one of four basic relief printing methods. The others are: newspaper relief printing, flexographic printing, and indirect relief printing.

These methods differ widely in their products, printing inks, stocks, and presses.

The ILO points out that in small businesses most or all of the letterpress operations and processes are performed in one work area, whereas in the larger firms, the individual operations (i.e. composing, stereotyping, binding) are carried out in separate work areas.

Composing can be done either manually or automatically. When the composing is done by hand, the compositor holds a composing stick in one hand, picks up the type with the other hand, puts the type in the stick, and then spaces the words so they fit the line properly. For the setting of large quantities of type, a machine operator uses a keyboard (like a typewriter keyboard) to tap out copy. Linotype and Intertype machines are usually used for casting solid lines of type, whereas Monotype machines are used for casting lines of individual characters. Molten metal is used to cast the type.

After the type has been cast and placed in shallow metal trays, it is proofed and corrected as necessary. The material is then assembled into units, each representing a page. "Imposition" is the process of assembling pages for presswork so that once the printing is finished and the pages are trimmed and folded, the pages will be in the right order.

Duplicate plates are usually made to increase the speed of the printing process and prevent the original type or blocks from becoming damaged. Strauss points out that type forms cannot be curved, and therefore are not useful for the rotary presses, the fastest type of letterpress. Since duplicate plates can be either flat or curved, they can be suitable for rotary presses. There are four principal types of duplicate plates: stereotypes, electrotypes, rubber plates, and plastic plates.

Stereotyping and electrotyping are the two main duplicating processes by which plates are made. These operations may be performed in the foundry section of a printing firm, but usually are contracted out. In stereotyping, a mold is used to cast the molten metal, though rubber or plastic "stereos" are often used in rotary presses. In electrotyping, a thin layer of copper or nickel is plated out on a mold of wax, lead, or plastic which has been made from the original form.

After the type, block or duplicates have been prepared, the form goes to press. According to Strauss, there are three general categories of presses for letterpress printing: platen presses, flatbed cylinder presses, and rotary presses. Rotary presses are usually used for high-speed mass production printing, whereas platen and flatbed cylinder presses are used for short and medium-length runs.

Platen presses are the smallest and simplest of the letterpress presses and are often found in small printing firms. They may be sheetfed by hand ("open presses"),

by machine ("job automatics"), or roll-fed (webfed). The ILO points out that a platen machine "puts the whole of the printing surface under pressure in one operation."

Flatbed cylinder presses can print directly from type or duplicates. These presses can print larger areas than platen presses without requiring as much impression power. Instead of printing the whole area at one time as the platen presses do, they make impressions in small segments. There are three kinds of flatbed cylinder presses: single-color presses, two-color presses, and perfecting presses.

The third kind of press for letterpress printing is the rotary press. As noted, the main benefit of rotary presses is that they are more efficient than the others, and therefore more useful in mass production printing. Rotary letterpresses can either be roll-fed or sheetfed and incorporate either of two basic construction principles. For printing one side of paper, sheetfed presses are usually used, whereas for printing both sides of the sheet or web, roll-fed presses are used.

Lithography

In lithography, certain areas of a flat surface are chemically treated to be "ink-receptive" and other areas are treated to be "water-receptive" (or "ink-repellent"). The "ink-receptive" areas become the printing areas, and the "ink-repellent" areas become the non-printing areas.[78]

In the early days of printing the image to be printed was drawn in greasy, ink-receptive chalk on a porous stone and the remainder of the stone was wetted.[12] Next, a greasy-ink which only adhered to the greasy chalk was rolled over the stone. The ink was transferred to the paper by rolling a cylinder covered with paper over the treated surface. In later times zinc, aluminum, or stainless steel plates were substituted for the stone. Since the aluminum and zinc plates could be curved, rotary lithography presses came into being.

Platemaking is the last step in converting original images into image carriers for lithographic printing. There are four main stages in lithographic platemaking: (1) selection and preparation of the base metal for the plate, (2) photomechanics for lithographic platemaking, (3) formation and protection of image areas, and (4) formation and protection of non-image areas.

To prepare plates for coating, the plates must be "grained" and "counter-etched." Through graining, old plates can be prepared for reuse. There are several methods of graining. In one method, the plate is put in a "tub graining" machine which contains metal balls ("marbles") and an abrasive water mixture. In "brush graining" electrically rotated brushes prepare the surface for coating. Once the plate-surface has been grained, it is counter-etched if it is rough. In addition, the plate surface is pretreated (cleaned and deoxidized) so that the coating necessary for the photomechanical image formation can adhere properly to its surface[78]. Once coated and exposed, the light-sensitive coating is developed and a photomechanical stencil is made.

In order to form and protect the printing image, plates are treated with various chemicals and lacquers, the choice of which depends upon the type of plate being prepared (i.e. surface or deep-etch plates). The purpose of this step is to ensure maximum adhesion of the image carrier, to provide a high degree of water repellency, and to provide a surface which is not easily damaged.[78] The non-image areas are also treated with various chemicals to provide an extremely water-receptive surface to which ink cannot adhere. After proofing, the plates are then ready for the press.

In offset lithography, the ink is first transferred to a rubber blanket, and then from the blanket to the paper. Newer presses based on the offset rotary principle operate at high speeds and can be either sheetfed or web-fed.

Gravure[12]

Gravure printing (intaglio) utilizes printing-image carriers in which the printing areas are sunken below non-printing areas. Preparation of gravure plates or cylinders for the press first involves the exposure of a screen and then a positive onto a carbon tissue (a layer of light-sensitive gelatine attached to a sheet of paper). After fixing this to the cylinder and developing with warm water, an etching solution is applied in stages to the surface. The etching process can be performed manually, but automatic etching machines and electronic engraving machines are available.

In the fast-operating rotary gravure presses, a very volatile, solvent-based ink is used. This type of ink allows rapid refilling of the depressions in the cylinder and evaporates rapidly. Since the evaporation rate is significant, exhaust hoods are often present over these presses to minimize fire, safety, and health risks caused by solvent vapors. Special drying units are sometimes used to dry the deposited ink.

Photogravure is a special form of intaglio printing. Here, the design is etched into a copper-coated cylinder or a copper-plate. An ink of low viscosity is applied to the surface of the gravure cylinder or plate, and a "doctor blade" scrapes the surface to force ink into image areas and remove it from non-image areas. When the paper and the plate or cylinder meet, the ink leaves the recesses and transfers to the paper. Since this is a rotary method, it is also suited for mass-production printing.

Binding and Finishing

Additional work is usually necessary to finish the product after the paper has come off the press. Some printers enlist the services of private businesses for such work, but others have their own departments for these processes.

Most printers, however, have at least a "power-cutter" or some other binding and finishing machine.

According to Strauss, there are nine main stages of binding: "(1) planning for printing to be bound, (2) cutting of sheets, (3) folding of the printed sheets, (4) assembling of the folded material for binding, (5) binding proper, or fastening the assembled material in book form, (6) trimming, (7) preparing the book for covering, (8) cover-making, and finally (9) combining of cover and bound book. The ILO points out that automatic and semi-automatic machinery can be used to perform most of these operations.

Health Hazards

Fairley[85] reports there are numerous hazardous chemicals used in the printing industry but that "in most cases the risks from these substances are adequately controlled by ventilation, proper handling precautions, etc., or are comparatively slight, and most workers have little need to worry about their health during their working lifetime. Where harm may result, however, is from accidental breakage, spillage, leaks, from carelessness, undue sensitivity or perhaps because the dangers are unrecognized or unknown. The main health or safety hazards from these causes with which the printer is likely to be concerned are the dangers to the eyes and the skin, through ingestion and inhalation, or from the risk of fires and explosions."

Fairley[86] states that there have been no reportable cases of lead poisoning in the printing industry in England during recent years. He notes, however, that when lead poisoning does occur, it occurs through inhaling dusts or fumes of lead, not through the handling of type metal. Sources of lead fumes and dusts are listed as:

COMPOSING DEPARTMENT
Dust in type cases, not from handling type or type metal.
Wire brushing of metal pot plungers.
Metal dust from matrices after casting slugs.
Drossing of metal pots.
Careless dumping of dross on floor.
Sweeping up and bagging of dross.

STEREO AND FOUNDRY DEPARTMENTS
Drossing of metal.
Careless dumping of dross on floor.
Sweeping up and bagging of dross.

The author adds that since "the temperature of metal pits in printing shops is not high enough to give off lead vapors during type founding, . . . caster attendants, lino operators and stereotypes are not formally in any danger . . . Great care should be taken in the skimming of metal pots and in the handling of dross. The remelting and drossing of metal should not be carried out by unskilled workers who have no knowledge of the dangers involved." Also stated is that dust in type cases does not contain much lead but, nevertheless, should never be removed by blowing out.

Bronze powders have presented a significant exposure hazard in the past, since lead used to be (but is no longer) one of their main constituents. Now, however, the only hazard of bronze powders is from inhalation of the dust.[86] Solvents used in the printing industry in inks and for cleaning type and presses can obviously present health hazards. "There is a weak anaesthetic action from solvent vapours which may cause deterioration of health if breathed continuously over a period."[87]

Finally, it has been discovered that there is an alarmingly high prevalence of respiratory impairment among members of the International Printing Pressmen and Assistants' Union of North America (IPP & AU). Dr. Ruben Merliss of Los Angeles found that 90% of the web pressmen working in the L.A. Herald-Examiner pressroom have "lung problems ranging from lung cancer, emphysema, bronchitis to pneumonia. In testing inks he found possible cancer caustive factors in mineral oil and lampblack . . . both are used in ink formulas."[88] In addition, such toxic substances as "chrome and methyl mercury in color inks, and talc and asbestos in paper dust" have been found in pressroom air.[89]

A research group headed by Dr. Irving Selikoff at the Mount Sinai School of Medicine in New York City has undertaken an in-depth study of the respiratory diseases among printing pressmen, documenting "the degree to which members working at the trade have respiratory problems. Further study of causes and corrective measures will follow if, as expected, the research team substantiates the information developed in the preliminary investigation."[90]

Controls—Concerning controls in the printing industry, Fairley[85] states that "if chemicals cannot be eliminated or substituted for others in the process, then the necessity for physical contact with liquids, dusts, or fumes should be reduced or removed. Here, the best method of protection lies in total enclosure or complete mechanization of the process, wherever possible. To this end, the printing and the graphic reproduction industries now have available to them equipment for the automatic processing of films and plates. Automatic etching of rotogravure cylinders is also possible and this means that instead of the etching room being the messiest part of the plant it is the cleanest . . . Reduction of contact may also be achieved through the use of mechanical aids such as tongs, trays, dispensers, etc., or by the use of splash guards and screens. Local exhaust ventilation is also an important factor where dust or fumes are involved. Hooded exhaust systems are seen in many factories to overcome this problem and probably need no comment. Fume extraction tables, which provide a complete safeguard both for the health of the operator from toxic etch fumes and against

corrosion of nearby equipment, are also available from a number of suppliers."

Marsailes[84] points out that "the nature of the various printing processes necessitates special considerations in the design of air management systems of printing plants. For example: drying time of inks dictates allowable humidity tolerances, as does incidence of static electricity; evacuation of particulate and chemical pollutants is a special need; some presses operate at high temperatures, affecting ventilation system design; drafts can cause misalignment of paper entering the presses; and while accounting for the special needs of the machinery, human comfort cannot be neglected."

Greiner[91] discusses different techniques for attempting to achieve dust free cutting. He indicates that the increasing use of offset printing for long, high speed runs has intensified the concern for dust control in the printing industry since valuable time is lost if a printer has to shut down presses to clean off press blankets which have accumulated dust. He describes methods of cutting which do not produce dust, and suggests that these methods may be more economical for the control of dust rather than removing dust once it has been produced.

Gadomski et al[92] report data from several controlled and uncontrolled metal decorating processes. In defining this type of operation, they state that "metal decorating can best be described as the application of ink (printing process) and various types of coatings (coating operation) to a flat sheet rock or metal plate. The standard printing process utilized is sheet-fed (offset) lithography." The lithographic inks for metal decorators consist of alkyd resins, solvents (approximately 15%), and various sizings and coatings. Generally, coating materials consist of vinyls, acrylics, alkyds, and oleoresinous or phenolic lacquers. After the resinous material (and then perhaps a pigmented material) is applied to the metal, a coating of varnish is usually applied by a varnishing unit before the metal sheets go into the baking oven. In the oven, low velocity heated air is passed over the sheets to dry the metal. An exhaust outlet vents the solvent vapors from the oven to the atmosphere. The authors point out that "further studies are definitely needed to determine the actual extent that an oven's operational structural, and mechanical characteristics can affect the level of pollution ultimately emitted from the oven."

In describing the processes and controls for letterpress printing, Marsailes states that, since sheet-fed printing is a slow operation, the ink is virtually dry once the sheets come off the press. In order to prevent "offsetting or the transferrance of an image from one sheet to another," each sheet is usually dusted with a powder or a starch after printing to hold them apart until they dry properly. It is pointed out that, since the dispersed starch particles are large (30-40μ and tend to settle out on surfaces, they present a housekeeping problem. Apparently "an effort has been made to provide hoods to capture starch particles but results were inconclusive. An adequate capture velocity tends to pick up starch off paper sheets defeating the prime purpose of the starch." In order to eliminate the housekeeping problems, the possibility of substituting the starch with a silicon fluid of low toxicity has been considered.

Marsailes also discusses the ventilation of the driers for printing. He states that the driers for weboffset printing and for webfed letterpress are the same type. For weboffset lithography, however, drying is not as major a concern, since not as much ink is applied and the presses do not operate at as high speeds.

Both weboffset printing and web letterpress printing involve the use of heatset inks. These inks "dry on the printed surface by evaporation of organic solvents under the influence of heat and air current. Drying is achieved through either a direct flame hot air (d.f.h.a.) or high velocity hot air (h.v.h.a.) dryer. In the former, the web passes between burners in which flame impinges on the sheet surface and a series of air blasts are blown against the web to remove volatile gases and to help carry off solvents. In the latter type, preheated high velocity air impinges on the sheet surface to drive off solvents." In either method of drying, the atmosphere becomes contaminated with evaporated solvents.[93]

Marsailes mentions two other letterpress (and presumably offset lithography) driers: open flame gas cup and steam drums. The exhaust volumes required are noted to vary according to the type of drier used and the press speed, but in general to range between 7000 and 15,000 scfm, with exhaust temperatures ranging from 250 to 400°F.

Marsailes states that, in rotogravure, ink dries by the evaporation of solvent, leaving a "solid film of pigment (clay), and resin." The author points out that there are separate driers for each individual printing unit, and the airflow through each drier is approximately 4000-6000 scfm. "Exhaust of from 1000 to 1500 scfm is provided at the floor between each unit to prevent any concentrations from accumulating. This is primarily protection against any spillage which may accidentally occur."

Concerning the ventilation for platemaking processes, this author states that most operations have lateral or downdraft exhaust systems. Since the plates are usually lowered into the tanks from above for plating, etching, or other operations, the ductwork must be below or to the side of tank, thus making lateral exhaust the most commonly used ventilation technique for open surface tanks. He points out that "exhaust quantities could vary somewhat depending upon the nature of the solution but generally 100 scfm/sq ft of surface of solution is provided and/or a control velocity of 50 fpm at the side of the tank

opposite the exhaust intake. It is highly desirable (1) to provide tank covers if possible; and (2) to provide rear and side walls for tanks. Air quantities may then be minimized and/or a more efficient exhaust system may be obtained. Cross ventilation across open tanks which may be caused by drafts of supply air distribution should be avoided."

Marsailes also addresses the need for make-up air in printing establishments. He points out that it is difficult to distribute air properly and suggests that the main duct trunks in production areas be installed adjacent and parallel to the equipment and that, at frequent intervals, ductwork be installed which extends down to about 10 ft from the floor.

Discussions and Conclusions—As clearly shown, there are numerous processes and operations in the printing industry which can result in industrial hygiene and occupational health problems. Some of the exposure hazards are specific to the printing industry, and others are common to many industries.

As is true for the rubber industry, a full-scale effort has been underway for the printing industry to investigate the specific health hazards associated with this industry. NIOSH should not repeat the research which has been and is currently being conducted, but should obtain information concerning the results and activities of this study. Subsequently, it is recommended that NIOSH support research to investigate any specific problem areas identified which would appear to benefit from engineering control research.

4. Evaluation of Common Operations

4.1 Introduction

This section of the report contains the results of investigations conducted to determine the availability and adequacy of engineering controls for those operations categories which are common to a number of industries. Included are all such categories found in Table 3 and those which were identified from the industry evaluations in Section 3 of this report.

4.2 Abrasive Machining Operations

Introduction—The term "abrasive machining operations" is used here to include all operations involving the shaping or surfacing of solid workpiece materials using abrasive machine tools. The operations include grinding, polishing, sanding, abrasive sawing, and buffing, but do not include abrasive blasting.

An abrasive machining operation is characterized by the removal of material from a workpiece by the cutting action of abrasive particles contained in or on a machine tool. The workpiece material is removed in the form of small particles and, whenever the operation is performed dry, these particles are projected into the air in the vicinity of the operation. Ventilation in its various forms is used routinely with abrasive machining operations to control employee exposures to airborne particulates generated by the operations.

Existing Ventilation Criteria—Current recommended practice in the application of ventilation methods to various types of abrasive machining operations is summarized in the ACGIH Industrial Ventilation Manual.[11] Ventilation methods are described which have been found through industrial practice to be effective in controlling exposures of workers to airborne dusts generated by the operations. The methods described in the Industrial Ventilation Manual are recommended primarily for control of inert dusts resulting from machining non-toxic workpiece materials with abrasive tools. The methods include open and enclosing hoods as applied to metal and woodworking operations.

A second source of ventilation criteria for abrasive machining operations is a report on a NIOSH-sponsored investigation of ventilation requirements for grinding, buffing, and polishing of metals.[94] In this investigation, quantitative criteria were developed which relate breathing zone dust concentrations to machine and ventilation system parameters. These criteria allow the results of the investigation to be extrapoltated (with caution) to machine sizes and speeds outside the ranges investigated and to the control of moderately toxic contaminant materials. This investigation was directed primarily toward the performance of open and shaped exhaust hoods.

A third source of ventilation criteria is a report on experience with the control of beryllium dust.[95] In this report, methods of controlling beryllium exposures from grinding operations are described. The methods are based on the use of ventilated enclosures.

Many other publications are available which generally describe ventilation methods applicable to machining operations. A bibliography of such publications is contained in Reference 94. However, these publications do not provide sufficient information to guide an operator in the selection and operation of ventilation equipment.

Identification of Problem Areas—Problem areas in the ventilation of abrasive machining processes can be identified best by means of a matrix defining combinations of ventilation methods and contaminant materials. The matrix, shown in Table 5, defines 12 material-method combinations or problem areas. The matrix could be extended in a third dimension to include different classes of abrasive machining equipment. However, this added degree of complexity is not necessary for a discussion of research requirements. In the matrix, the problem areas are identified for which ventilation criteria exist. The spaces in the matrix without entries represent problem areas for which ventilation method performance data and design criteria are not available or do not exist. However, not all of the open spaces represent problems of equal significance. For example, the use of enclosures for the control of inert materials is not common except for complex machines such as automatic buffing equipment. In the next section, the problem areas requiring investigation are diecussed in an order which represents the priorities attached to the data requirements.

Discussion of Problem Areas

Non-Metals/Open and Shaped Hoods

Abrasive operations are widely used in the woodworking industry and in manufacturing many types of building materials. To a lesser extent, abrasive operations are used in the manufacture of ceramics, masonery materials, stone, glass, asbestos products, and other non-metal products. Most contaminant materials generated by these operations are inert. However, asbestos and silica are encountered in some operations.

Very few data are available on the contaminant exposures associated with abrasive machining of non-metals or on the performance of ventilation methods used to control these contaminants. Detailed guidelines for ventilation system design and operation are available only for woodworking operations,[11] and ventilation criteria for abrasive machining of metals may not be applicable because of differences in contaminant generation rates. Consequently, it is not possible with existing data to assess the significance of the problem of employee exposure to contaminants from abrasive machining of non-metals, nor is it possible to define ventilation requirements wherever such problems are observed.

To assess the significance of this problem and to provide a basis for ventilation guidelines, it is recommended that a program be initiated similar to the previous program sponsored by NIOSH on ventilation requirements for grinding, buffing, and polishing operations.[94] This new program, however, should be directed toward the control of non-metal contaminants from abrasive machining of non-metals, (2) to compile a data base on the performance of ventilation critria for these operations. The program should consist of three corresponding tasks.

The first task should consist of a survey of abrasive machining operations used with non-metals and should provide a classification of these operations and the contaminant materials which they generate. In this task, persons or organizations should be used who are familiar with these operations.

The second task should include field measurements of worker exposures with various classes of operations and

TABLE 5—MATRIX OF PROBLEM AREAS IN VENTILATION OF ABRASIVE MACHINING OPERATIONS

Materials Methods	Inert		Toxic	
	Metals	Non-Metals	Metals	Non-Metals
Open and Shaped Hoods	(Refs. 11 & 94)*	(Ref. 11 – Wood)	(Ref. 94)	
LVHV Hoods				
Enclosures	(Ref. 11)		(Ref. 95)	

*Notes refer to sources of ventilation method performance data.

contaminant materials. These measurements should be conducted as far as possible under actual industrial operating conditions.

The third task should involve the development of ventilation guidelines for the various classes of operations based on the exposure data compiled in the second task. The ventilation performance models described in Reference 94 can be used in the development of these guidelines.

Metals and Non-Metals/Low-Volume-High-Velocity Hoods

The low-volume-high-velocity (LVHV) hood is a relatively new approach to exhaust hood design. The approach appears to be a simple extension of the use of open hoods; however, the operating principle of LVHV hoods is different from that of open hoods. The LVHV hood is designed and operated to control air contaminants at the point at which they are generated by inducing exhaust velocities at the generation point higher than the release velocity of the contaminants. With this design approach, the contaminants are not allowed to travel beyond the generation point and, hence, a small hood is sufficient to achieve a high capture efficiency with attendant savings in energy and equipment requirements.

At the present time, there are no data in the public domain on the performance of LVHV hoods, and guides to their use are non-specific.[11] Consequently, the design and application of these hoods are trial-and-error processes. It is recommended that a program be initiated to investigate the performance characteristics of LVHV hoods used for the control of contaminants from abrasive machining processes.

The LVHV hood program should have three objectives and consist of three tasks analogous to those recommended for the previous program. These objectives and tasks should be directed toward (1) classification of operations and contaminants, (2) compilation of ventilation performance data, and (3) development of ventilation system design and performance criteria.

The classification task should identify operations with which LVHV hoods are used currently, as well as other operations which are potential applications of LVHV hoods. In this task, the experience of manufacturers of LVHV hoods should be utilized as far as possible.

The data compilation task will require a laboratory test program because of the limited use of LVHV hoods in industry at the present time. A laboratory program will be necessary in order to provide all of the hood applications which should be investigated.

The criteria development task will require formulation of ventilation performance models specific to LVHV hoods. These models can then be used to correlate the data compiled in the previous task.

Metals and Non-Metals/Enclosures

Enclosures are most often used for control of highly toxic contaminants or for controlling contaminants from complex machinery. The operational criterion for an enclosure generally is a simple prescription of a minimum face velocity.

Collection of performance data on enclosures is not regarded as a high-priority requirement. However, additional data on enclosure performance would provide a firmer base for the development of guidelines for the design and operation of enclosures used with abrasive machine operations. Such a data base would be most useful for operations involving toxic materials.

It is recommended that a program be considered wherein enclosure performance data would be collected from existing facilities. These data would be collected as a part of a simple field measurement program which might also include investigations of ventilation method performance for other industrial operation categories.

Summary of Program Recommendations—Three programs have been recommended for the development of ventilation criteria for abrasive machining operations. Suggested durations and levels of effort for these programs are listed below.

Program	Duration	Effort
Non-Metals/Open Hoods	12 months	2-3 Man-years
LVHV Hoods	12 months	2-3 Man-years
Enclosures	9 months	1-2 Man-years

4.3 Chemical Processes

Introduction—The processes and operations which can be termed as "chemical processes" are numerous and varied in nature. It is therefore not surprising that such a high percentage of the toxic substances with which we are concerned have been described in the literature as being utilized in chemical processes.

According to ILO[12], such processes in the chemical industry involve "chemical or physical changes in the nature of substances and particularly in the chemical structure and composition of the substances. This covers a much wider field than just 'chemicals' since it also includes such principal products as fertilizers, dyestuffs, pharmaceuticals and mecidinal products, explosives, plastics, resins, adhesives, cosmetics, synthetic rubber, photographic films, synthetic and artificial fibres, detergents, soap, paints, as well as the multiplicity of inorganic and organic substances which can be classified as 'miscellaneous' chemicals."

The ILO describes "chemical processes" as including "crushing, grinding, size separation, filtration, drying, heating, cooling, solvent extraction, absorption, distillation, fractionation, electrolysis, mixing, blending, analysis and process control, packaging and transport.'' As is evident, this list comprises most of the operations utilized in any of the manufacturing industries, and in some service industries. Indeed, it covers many of the more specific categories of uses/exposures which will be treated individually and in detail in this report. For this reason, we will here only consider those equipment types which are usually specific to the actual manufacture of chemical feedstocks, i.e. the ''raw'' materials from which industry manufactures the products which we utilize in our everyday lives. If in making this differentiation we realize that we have excluded materials handling and grinding, crushing, and screening equipment for solids, it becomes evident that we are primarily left with equipment or overall process systems which cause the contacting, reaction, or separation of gases and/or liquids. If we additionally impose the condition that chemical process companies involved with the manufacture of large volumes of chemicals utilize ''closed'' systems, i.e. where the chemicals are not blatantly open to the atmosphere, we can safely theorize from an engineering, economic, and practical viewpoint that the exposures to toxic substances in chemical process plants are primarily due to:

1. Releases (i.e. leaks) from joints, closures, etc. in system components normally and functionally considered to be a part of a ''closed system'' (i.e. fugitive emissions),

2. Intentional releases of contaminants from system component vents,

3. Stack gases from combustion processes, and,

4. Accidental or unintentional releases due to equipment malfunction or failure.

Literature Review—Ludwig[96], is a paper discussing exposures in chemical process plants, indicates that the specific equipment and practices which must be given ''considerably more attention'' include valves; conservation, tank and process vents; and relief valves and disks. He makes the point that ''many ordinary valves leak continuously even when supposedly shut'' and that such leaks ''can build up the exposure level to the limits set for air contaminants by OSHA.'' Concerning the practices of routinely releasing purges, minor overpressures and system breathing to the atmosphere (i.e., heretofore common venting practices), he notes that the contaminants should be subjected to recycle collection, scrubbing, reaction, or other measures to produce safe products or to eliminate their release completely. In doing so, he correctly points out that even if excessive employee exposures are not a problem, EPA emission requirements may still mandate such measures.

Another source of contaminant release discussed by Ludwig and mentioned above is the operation of safety and relief valves. It is noted that specifications for bubble-tight seating and reseating of valves ''are receiving considerable attention,'' that the composition and temperature of some gas streams can make some soft seat materials unacceptable, and that the hard-seat design has been improved to cope with the ever-increasing range of vapor compositions.

The prevention of releases of large volumes of contaminants into the environment from activated relief valves requires that ventings be somehow collected. Ludwig reports an increase in efforts to collect such ventings into headers, and to control fume releases by means of smokeless flare systems. He indicates that this is not a simple task because a header connecting a number of system components must be designed to handle a wide variety of operating conditions, including the possibility of simultaneous release of several streams.

Concerning rupture disks it is stated that they are capable of ensuring bubble-tightness until they relieve. An interesting application is illustrated in which a rupture disk is positioned in a line preceding a relief valve to protect the valve until an overpressure situation becomes imminent.

Wiley[97] supports the statement that venting practices contribute to contaminant concentrations by noting ''vapor losses from tank vents and process vents, and during chemical loading operations, are the primary sources of hydrocarbon emissions in the chemical process industries.'' His paper then summarizes and discusses the various techniques available for the collection, and recovery or disposal of the displaced vapors.

Discussing compression packings for pumps, Rothman[98] reports fibrous packing around the shaft, rod, or plunger in ''a typical stuffing-box design'' requires lubrication for proper operation. It is noted that this lubricant is provided by the fluid being handled as it flows through the slight clearance between the moving parts and the packing. He goes on to give a ''desired leakage rate'' of 8-10 drops per minute and to note that such visible leakage should not be misinterpreted as something more serious by inexperienced mechanics. Such leakage would obviously increase contaminant concentrations.

Danielson[61] discusses how the operation of pumps handling fluids in petroleum process units can result in the release of air contaminants. He describes the various means of sealing pumps and then presents the results of a study to measure losses from pumps with packing glands and those with mechanical seals. During the handling of highly volatile hydrocarbons, it is shown that pumps in continuous service lose an average of 18.3 and 7.9 pounds per day for packed and mechanical seals respectively. This indicates that mechanical seals are far more efficient when running continuously. On spare or standby service, the packed seals were shown to be more effective, losing 1.8 pounds per day versus 4.4 pounds for mechanical seals.

The amount of contents lost was obviously less when less volatile substances were conveyed. The largest leak encountered in the study was from a reciprocating pump on intermittent service handling LPG. Its leak rate was 266 pounds per day.

Other types of pumps described by Danielson include the canned, diaphragm, and electromagnetic types. All of these are devoid of seals and do not leak. A pressure-seal-type application is noted capable of reducing packing gland leakage.

Leakage from valves is another area of concern for Danielson. Of the valves of refineries which have been inspected in Los Angeles County, it was found that 16.5% of those in gaseous service leaked while 11.5% of those in liquid service leaked. Of those in the first category, 70% leaked less than 0.2 pounds per day. In the second category, 90% leaked less than 0.2 pounds per day. The remaining large leaks were of an average ratof 9.1 and 8.1 pounds per day respectively.

Browning[99], in a paper concerning the estimation of loss probabilities, gives some relative probability figures for failures of containing equipment under various inplant environmental conditions. Of interest is that the probability of a failure in a process system with generally screwed fittings is shown as being an order-of-magnitude greater than for systems with generally welded and flanged fittings. One might consider that this also gives some indication of the relative leak resistance of such systems.

Templeton[100] notes that screwed connections have many disadvantages. Discussing them, he notes that leaks are difficult to seal, threads are subject to crevice corrosion, easily galled materials are difficult to tighten, and that "leaks from screwed joints may produce conditions on the exterior that will rapidly attack piping and valves." He further makes the point that welded piping systems give the maximum assurance of leak tightness.

Russell[101] reports that spiral-wound gaskets with centering rings are normally specified for use with welding neck or lap joint flanges for "high reliability piping systems." He makes the point in his paper that the dimensional tolerances specified by industry standards can allow the outer portion of the spiral gasket to extend beyond the raised faces of a flange pair, a condition which can ultimately result in gasket failure or leakage.

A method of reducing gas leaks from flange closures is the subject of a paper by Fitzpatrick.[102] He reports on a particular problem with a reaction system in a polyethylene plant where rapid temperature changes resulted in undesirable leaks. The successful solution of the problem involved the rather novel application of Belleville springs on the closure bolting.

Baturin[103], discussing normal leakage rates from equipment and pipes in which gases under pressure are handled, states that "with acceptably sound equipment the leakage of a gas, depending on its molecular weight, is approximately 7-12 percent per hour of the internal volume of the equipment." He also presents a formula for more specifically estimating leakage rates, one which takes into account the pressure of the gas stream.

A recent news item[104] described the success of Sweden's largest chemical company, KemaNord, in achieving 8-hr workplace vinyl chloride concentrations of 1.5 ppm at two vinyl chloride plants. The article reports that typical previous performance was around 30 ppm. These "unusually low concentrations" were realized by the use of local exhaust ventilation at points in the process where vinyl chloride release was "particularly possible", and by mechanically tightening up the equipment train. The latter involved investment in new valves and replacement of numerous screwed couplings by flanged ones that could be tightened easier and more readily equipped with gaskets.

A more recent article[105] discussed the attempts of PVC manufacturers to reduce exposures by controlling fugitive emissions. As would be expected, the tightening of flanges, the use of welded or flanged couplings instead of "leak-prone" threaded couplings, investigation of better gasket materials and pump seals, the use of larger capacity equipment, process changes to reduce the frequency of cleaning operations, and other subjects are receiving attention. Of interest is that the EPA "plans to insist on double-sealed or canned pumps (or an equivalent); the agency also will call for rupture disks under relief valves, to avoid leaks if a valve reseats improperly."

Conclusions

The information given above clearly indicates that, within any given chemical process plant, there are numerous sources of air contaminants. The most significant of these, from the viewpoint of quantity, are those involving process venting practices and the flue gases from combustion processes. Ludwig and Wiley both support the fact that venting practices are significant factors. That uncontrolled flue gas emissions also can contribute to exposures, both inplant and in the surrounding environment, is so well appreciated that further elaboration is unnecessary.

Since these two sources are of such significance, they have become a primary concern of the Environmental Protection Agency (EPA) and state and local air pollution agencies, mostly as a result of the Clean Air Act. Consequently, not only have emission rates and the number and location of sources been regulated, but this regulation has encouraged considerable research and development effort for pollution control systems. For this reason, it can be concluded that participation of NIOSH in such efforts is not warranted, unless of course it becomes evident in the future that the control methods being utilized are inadequate to eliminate these sources as significant inplant contaminant sources.

In this report, we are primarily concerned with those potential contaminant sources which more directly affect the inplant working environment. We have thus elaborated in the above upon such leaks as may result from "closed" process system components, and have noted the possibility of accidental releases due to equipment malfunction or failure. Utilizing the data presented, a knowledge of process plant design practices, and an appreciation of the difficulties involved in controlling exposures to low limits, we are logically and intuitively led to the following conclusions:

1. Process plant equipment trains, even those which are normally considered as representing, "closed systems," inherently contain numerous small sources of airborne contaminants. Generally speaking, every pipe joint assembly, valve stem, closure flange, relief valve seat, and opening into a vessel for the insertion of measurement devices or mixer/agitator driveshafts, etc., is a potential source which contributes to overall exposures.

2. For most slightly and moderately toxic substances, the application of general mechanical ventilation in enclosed areas or the placement of systems in the open (i.e. the layout of a typical hydrocarbon refinery) has been adequate to maintain contaminant concentrations at or below exposure limits.

3. Until possibly very recently, process engineering design groups did not consciously consider the effect on ambient contaminant concentrations of their choice of specific fitting or seal types. Systems components were generally selected on the traditional bases of cost, reliability (in the length of service sense), and performance of function.

4. With the advent of regulations requiring extremely low average and ceiling contaminant concentrations for highly toxic substances, general ventilation methods can no longer be considered solely adequate to maintain acceptable exposure levels. As proof, witness the difficulty of industry in complying with vinyl chloride standards and the experience of KemaNord. Their equipment and plants were not originally designed to be leak-tight to the extent which now has become necessary.

5. There is little specific information in the literature which allows an *accurate* estimation of the gas or liquid leakage rates from system fittings in either a relative or absolute manner, or which suggests how best such sources can be controlled or eliminated.

Possible Course of Action—It is evident that the feasibility of reducing work place contaminant concentrations by the proper selection and use of process equipment

is being demonstrated by the efforts of PVC manufacturers. Their efforts are somewhat on an individual basis, however, and cannot be expected to result in design guidelines which are formulated and disseminated to other industries which utilize the same types of equipment. The potential benefits of eliminating such emissions at their source appear to be of substantial significance. Not only would worker exposures be reduced, but the possible additional costs involved might be overshadowed by reductions in product loss and ventilation system capacity.

There is considerable need at this time for some sort of consolidation of the available information on this basic control measure and the provision to industry of guidelines for the proper design of chemical process systems. Hence, it is recommended that a research program be conducted to compile all available information and to attempt to present it in a form which allows appreciation by industry of the potential benefits of choosing equipment of maximum leak-tightness. The results of such a program would serve to identify data gaps for further research consideration. A 2 man-year effort over a time period of 1 year is considered reasonable for this task.

Some time in the future, NIOSH and OSHA might consider the formulation from these data of design codes for industry use. These might allow the estimation of total leakage rates from systems consisting of known numbers and types of fittings and couplings, might include construction and/or performance specifications for various fittings, and might attempt restriction of the use of certain "leaky" types under particular circumstances.

The National Electrical Code[106] contains standards for the use of electrical equipment in hazardous locations, i.e. locations where flammable gas concentrations are, or may be, present. The code assigns classifications including a "Group" (A-D inclusive) based on the ease of ignition of the substance, and a "Class," (I-III inclusive), based upon the type of operating environment. In a somewhat general manner, construction specifications are given for equipment to be used in the various classified locations, equipment "approval" is required for the Class of location and the chemical Group involved, and it is specified that equipment be marked to show the Class, Group, and operating conditions for which it is approved.

Standards or codes for equipment handling toxic substances could benefit from a similar format. One can envision Group assignments based upon the vapor pressure and the Threshold Limit Value of a substance, and Classes based upon general ventilation rates, proximity of employees to system components, location of equipment (i.e., size of room, outdoors, etc.), number of fittings per unit area, etc. Loss probability data, such as those presented by Brown, could be utilized in the code development procedure to ensure adequate system integrity over the life of the plant and to thereby reduce the potential for massive equipment failures.

Development of such codes or standards requires that a firm and complete body of data be available on the present performance characteristics of the equipment of concern. Since the research program recommended may identify numerous areas for research by NIOSH, EPA or industry, no attempt will be made to estimate the extent of efforts required to develop such codes at this time. This might be more properly achieved as a part of the program recommended.

4.4 Cleaning and Maintenance

Introduction—In 1973, a "loom fixer" in a cotton mill wrote to the S.C. Department of Health. He noted that "The looms are blowed off with an air hose while everyone is working with the looms. This dust is terrible . . . I have pulmonary emphysema."[107]

Wheeler and Sutherland[108] report that a single pump repair can release 1 to 10 pounds of vinyl chloride monomer (VCM) to the atmosphere while a once full line scavenged with an inert gas can still contain several tenths of a pound of VCM. They add, "rarely was provision made to control emissions from maintenance work."

These are the types of comments which can be found in the literature regarding the hazards involved in the cleaning of work places and the maintaining of equipment. That the operations are of a type which are difficult to characterize and are widely varying in nature can be well-appreciated. In the following, a brief look is taken at these operations, and some ideas presented on what might be done to reduce exposures.

Cleaning Operations—The primary cleaning methods utilized in industry include:

- Manual wiping with a dry or solvent soaked rag;
- Use of hand or mechanical brushes;
- Use of a scraper;
- Wet mopping;
- Use of a wet sponge;
- Hydroblasting;
- Steam-cleaning;
- Use of compressed air; and the
- Use of vacuum devices (for solids *and* liquids)

The only common feature of all of these is that by some physical and/or chemical action a contaminant is dislodged from the surface it is adhering to, collected, and removed from the immediate environment for disposal. The locations and circumstances under which each method may be used vary considerably.

It is difficult to approach this subject in the manner in which other engineering control problem areas have been investigated. The appropriateness of various methods appears to have been given only cursory treatment in the literature. Any guidelines which have been stated can be summarized in but a few phrases. These are: use wet methods for dusty substances, or use vacuum devices instead of "blowing" techniques; limit cleaning operations to periods when the least number of employees are present; use "safe" solvents if any are needed; provide workers respiratory protective devices when toxic substances are involved; and finally, minimize releases of toxic agents into the work area by use of process change, local exhaust ventilation, enclosure, etc.

The latter of these recommendations is, of course, the most satisfactory. If a toxic substance is not "free" in a work area, then it cannot harm a worker attempting to clean the area. Unfortunately, however, this solution is not one which can be solely relied upon. As long as equipment trains leak and containers or packages of toxic substances are handled in the working environment, there will be need for periodic housekeeping activities to remove accumulations from the floor, from machinery, from walls, and from overhead fixtures.

Looking at the above list of cleaning methods, one can make some observations concerning how excessive exposures may occur. Where a rag is used with some solvent, the solvent vapors may be the problem. This might also occur if solvents were used in wet mopping or with a sponge.

The use of a brush to sweep up dusty substances, or a scraper to dislodge built-up cakes of dry substances, can disperse dust into the air, especially if work is being done at some elevation above the floor. The use of compressed air to "blow" dust from surfaces is probably the most hazardous technique of all.

Hydroblasting and steam cleaning are essentially wet methods which might, on first consideration, appear to be methods which would suppress the generation of contaminants. However, in at least one situation we are aware of, hydroblasting of equipment to remove a solid which crystallized out of leaked solutions resulted in short-term contaminant concentrations which were 2-3 times that normally found (and substantially above permissible exposures). Because of the temperatures and forces involved, steam-cleaning may be assumed to be capable of producing similar conditions under some circumstances.

It is the use of vacuum devices, i.e. vacuum cleaners, which appears to be the most satisfactory method of collecting contaminants. Since such devices utilize the same principles as local exhaust systems, i.e. they provide a sufficient air velocity at a "hood face" to dislodge and thereby "capture" contaminants, and a sufficient "duct velocity" to transport contaminants to an air cleaning device, one might consider their use of the utmost desirability. Such devices are available which not only can vacuum dusts but also pools of liquids.

Discussion and Conclusions—There are two observations which can be made about these methods. The first is that some are clearly more appropriate for use than others depending upon the physical and chemical characteristics of contaminants and the location from which they are to be removed. The second is that the use of vacuum devices has the potential to substantially reduce the generation of air contaminants where their use is practical.

Recommendations—Both observations made suggest that some worthwhile studies may be conducted by NIOSH in this subject area. For example, since some methods must be considered to be better than others under certain circumstances, NIOSH might attempt to define what these circumstances are based upon the cleaning problems commonly found in industry. This could be accomplished by a survey investigating the experience of plants in a variety of industries and the reductions in contaminant concentrations which might be achieved with alternative methods. The results of the survey, together with other "good practice" type information could then be utilized to produce some sort of booklet or manual to be provided industry. Topics would include discussions of the need for keeping a workplace clean, the types of cleaning problems typically encountered, the cleaning methods available, and the appropriateness of the various methods in various situations. Within these categories, discussions could be given of the relative safety of various solvents and cleaning fluids, the difficulty of achieving permissible exposure levels by any means if the workplace is significantly contaminated, and other pertinent topics. An 8 man-month effort over a time span of 6 months would probably be adequate for such a program.

One question was raised during consideration of the use of vacuum cleaners that also may deserve some attention. This involves the fact that the types of self-contained vacuum units commonly utilized are analagous in many respects to a local exhaust system which provides for cleaning and subsequent recirculation of exhaust air back to the working environment. Whenever the topic of recirculation of exhaust air is addressed, a primary factor given consideration is the need for the exhaust air to be sufficiently free of contaminants. The analogy of large-scale recirculation systems with vacuum cleaners lead to the observation that similar attention might be given to these devices to ensure that respirable toxic dust collected is not passed through filters and blown back into the workroom air. A small program to look at this perceived problem area would simply involve the collection of pertinent information from the manufacturer of such filters and possibly some limited experimentation to determine if the amounts of contaminants which pass through might be significant for some toxic substances. A 2 man-month effort over a time span of 2 months would probably be adequate. This recommendation is presented simply as an idea. We have no knowledge if this might be a realistic problem area.

Maintenance—A dictionary definition of the above word indicates we are concerned with the act of keeping implements used in an operation or activity in an existing state of repair, efficiency, or validity. From an occupational health viewpoint of the matter, we may restrict this definition to those operations which require a worker to open or approach a device containing or releasing toxic substances and to perform some activity in its immediate vicinity. Included would be such diverse operations as the changing of a filter media, the repair or replacement of a pump packing, the tightening of a leaking flange, and similar acts.

These operations, because of their characteristic of being possibly necessary at any point in an equipment train, would not appear to have been given specific consideration for the provision of engineering controls, except, of course, where they are repeatedly conducted in a specific location. Thus, one can safely make the statement that engineering controls are generally unavailable for them.

There is, however, one approach to provide effective control which can be envisioned, one which might at first be considered a bit far-fetched but which may be feasible with current technology. This entails the development of self-contained, portable, local exhaust systems with internal air cleaning devices for vapors and fumes.

Ideally, such devices would have the following features:

- Be not much larger than a 55-gallon drum,
- Be electrically powered,
- Be provided wheels for portability,
- Have a stiff but flexible hose (i.e. exhaust duct) which is self-supporting,
- Have lightweight open or partially enclosing exhaust hood attachments for the open end of the exhaust duct,
- Be provided with a limited number of air cleaning sections which are interchangeable within the basic unit, each being effective for a large number of chemicals of similar chemical structure (i.e. ketones, alcohols, aldehydes, etc.) and
- Be of reasonable cost.

One can imagine that such units might find widespread use in industry if they were practical and effective. Not only might they be used for "one-shot" maintenance work, but also, in small firms, as relatively low-cost, easily installed units for reducing overall contaminant releases for both EPA and OSHA compliance purposes. In all types of firms, they might find use as emergency control measures for controlling small but potentially hazardous leaks.

The development of such devices into actual working prototypes might require the expenditure of considerable resources, or alternatively, might be simple and require little effort depending on the type of contaminant to be treated. For flammable ketones and the like, the air cleaning section might simply include an incinerator followed by a catalytic converter, much like the controls on automobiles. For other substances, more sophisticated measures may be necessary.

Given these uncertainties, the first step in any such program would involve a user need and engineering feasibility study. Such a study would review the list of toxic substances regulated by OSHA, determine which are most hazardous from acute exposures which can be conceived of in industry, survey plants utilizing these substances to investigate how excessive exposures during maintenance work are now avoided, determine the ideal operating characteristics of the device, and review the state-of-the-art of air cleaning technology to suggest reasonable development approaches. This might require a one man-year effort over a time span of nine months.

4.5 Drying and Curing Oven Use

Introduction—In the industrial terminology, "drying" is defined as "the removal of a liquid from a solid by thermal means."[8] There are many situations in industry in which drying is beneficial. These include (1) facilitating handling in subsequent operations, (2) allowing satisfactory use of final product, (3) reduction of shipping costs, (4) increasing capacity of equipment used in following processes, (5) preservation of a product, and (6) treatment of waste products, among others.

Much of the equipment used for drying purposes are also used for "curing", i.e. the application of heat to bring about a physical or chemical change in a substance.

A logical classification system for this equipment is based on the method used to transfer heat to the material being dried and/or cured. The first major category includes direct dryers in which hot gases are in direct contact with the material, and carry away any vaporized substances to be exhausted. The second category includes indirect dryers in which the drying heat is transferred to the moist solid through a conducting wall. In this type of dryer, any vaporized substance is removed independently of the heated air. A third, limited-use class includes radiant-heat and dielectric-heat dryers. The operation of the former is based on the generation, transmission, and absorption of infra-red rays. The latter rely on heat generation within the solid when it is placed in a high frequency electric field.

Direct Dryers—Direct dryers can be sub-categorized with regard to their operating mode, i.e., continuous or batch. In the continuous type, operation is not interrupted as long as wet feed is supplied. Any continuous dryer can be operated intermittently or on a batch basis if the need arises. Batch dryers can only operate on a fixed amount of wet stock for a fixed time cycle. These dryers operate at atmospheric, or near-atmospheric pressures.

Continuous tray dryers function by circulating heated air across a wet material until sufficient drying has occurred. This can be accomplished in a vertical turbo-dryer, consisting of a series of stacked trays through which the material is successively dropped. In the upper stages, hot air is brought across the trays by a vertical series of exhaust turbines.

Tunnel drying involves the placement of material on small trucks and movement of the trucks through a hot gas tunnel in a semi-continuous fashion. Airflow in the chamber can be either parallel or perpendicular to the movement of the trucks. In the former, the heated air may flow with or against the material flow. Effective control of the humidity and temperature is more easily achieved, however, if perpendicular air flow is employed.

Continuous through-circulation dryers can be used when the material to be heated is permeable. Heated gases are blown through the bed of wet material as it passes continuously through the dryer on a moving screen. Drying rates are increased due to the increased effective surface area and the decreased travel distance for the internal moisture.

Spray drying is accomplished by establishing a highly dispersed liquid state in a high temperature gas zone. The material to be dried must therefore be amenable to atomization. The spray droplets formed fall through a chamber which has an upward flow of heated air. Because the surface area-to-volume ratio of the material in droplet form is quite large, drying is accomplished quite rapidly, usually in less than 30 seconds.

A direct rotary dryer consists of a cylinder revolving at a slight horizontal tilt. Hot gases pass through the cylinder, drying the material as it is introduced at the elevated end. The airflow may be either counter-current or parallel to the material flow. The inside of the drum is generally fitted with a series of baffles to lift and agitate the material. Some rotary dryers operate on an indirect-direct basis, with the gases heating the cylinder and contacting the material directly.

Continuous sheeting dryers convey a continuous length of material, such as a fabric, rug, paper, or textile through a chamber in which it is exposed to a heated air stream. The configuration in which the material is suspended is dependent upon the substance being dried. Some materials must be kept taut during drying, others can be draped in loops. In all cases, the material must be strong enough in its wet state to withstand any tension applied.

The final dryer in this category is the pneumatic conveying model. Often performed in conjuction with grinding and pulverizing, removal of moisture occurs when the material to be dried is dispersed in a hot gas zone and transported at high velocities. Cyclones, bag collectors or wet scrubbers are used to retrieve the dried product.

There are two types of dryers which operate in the direct batch mode. The first is a through-circulation type in which the material is positioned on stationary trays through which hot air is forced. The second type consists of tray and compartment dryers. These dryers support material on trays across which hot air is passed. The method of suspension is dependent on the physical characteristics of the solid. Proper operation of these units requires a constant, uniform flow into all parts of the chamber, since the material to be dried remains stationary.

Indirect Dryers—Cylinder dryers are employed to continuously treat paper, cellophane, and some textiles. The cylinder is steam heated, and drying generally occurs at atmospheric pressure, although a few designs incorporate a vacuum chamber to facilitate drying. The process may be enhanced by blowing air across the surface. Removal of any wet air is usually accomplished by the use of hoods and exhaust fans.

Drum dryers are similar to cylinder dryers. Here, a slurry, paste, or liquid solution is applied to the exterior of the heated drum. In less than one rotation, the material, now dried, is scraped off the drum by a fixed blade and falls onto a conveyor belt. Operations ensue at atmospheric pressure or under vacuum, and the vapor is removed by local ventilation.

Steam tube rotary dryers are similar to direct and indirect-direct rotary dryers. They consist of a rotating cylinder in which material is dried as it traverses the length of the unit. In this situation, however, drying is totally indirect, with the rotating shell being heated by tubes or cylinders placed within it. These devices can be used to continuously dry granular or powdery materials that must not be contaminated with combustion products.

There are four types of machines used for indirect batch drying; the agitated pan dryer, the vacuum rotary dryer, the vacuum tray dryer, and the freeze dryer. These are easily adapted to work in vacuum conditions, thus lowering operating temperatures.

The agitated pan dryer consists of a circular tray which is steam heated from underneath. The material in the pan is stirred by an agitator so that fresh material is constantly coming into contact with the bottom of the pan. Solvent recovery is attainable with this device and the device is particularly useful for small batches.

Vacuum rotary dryers consist of a large, stationary cylinder to which heat is applied through a jacket in the shell. A set of internal paddles agitates the substance. Vacuum is applied and maintained throughout the drying cycle. A less common type of unit has a rotating jacketed shell.

Vacuum tray dryers consist of a chamber containing shelves. The shelves are hollow and connected to heaters through which the heating medium is supplied. Conduction of heat occurs between the shelves and the metal trays in which the material is placed. Vacuum allows low temperature drying and solvent vapor recovery.

In a freeze drying process, frozen solvent is removed by sublimation. The machinery may be of the shelf drying, cylindrical vessel, or horizontal rotary vacuum design. An advantage to prefreezing the material is that the warming temperature can be lowered to the point where damage will not occur to the substance. The loss of volatile constituents is also minimized, thus making the process attractive for food processing.

Radiant-Heat and Dielectric-Heat Dryers—Radiant heat dryers transfer radiant energy to the material. This can be accomplished with infrared lamps, electric resistance elements, or incandescent, gas-heated refractories. The latter add the advantage of convective heating. Infrared drying is used primarily in baking or drying paint films or heating thin layers of material. It is employed infrequently in the chemical industries.

Dielectric-heating involves heat generation by use of a high-frequency electromagnetic field. Since these dryers operate on the principle of heat generation within the solid, the potential for drying large objects is being explored. A severe drawback is the increased power demand, which can be ten-fold the requirements of conventional machinery.

Literature Review—The ACGIH Industrial Ventilation Manual[11] contains criteria for the control of employee exposures to contaminants from drying ovens. The dryer illustrated is of a type used for continuous processing. The controls recommended include a slot type hood located around the top portion of the entrance or exit doors and a canopy type which also is installed over these doors. For the slot hood, an exhaust volume of 100 cfm per sq ft of door area plus 1/2 the products of combustion is recommended. For the canopy, the recommended rate is 200 cfm per sq. ft of hood face plus the same correction for combustion products. Notes on the illustration suggest a separate vent be added for products of combustion, and if flammable solvents are evolved, that the oven ventilation rate be adjusted to maintain solvent concentrations at or below 25% of the lower explosive limit (LEL).

The National Fire Codes[109] also address the need for ventilation of ovens and furnaces to control flammable or toxic vapors. However, they simply give a calculation procedure for determining the ventilation rate necessary to maintain concentrations at 25% of the LEL in continuous process ovens. A procedure with similar objectives is given for powder fusing or curing ovens. Batch process ovens are reported to usually require no more than 380 cfm per gallon of solvent present.

The Michigan Department of Health [110] notes that although oven ventilation system design involves basic application of well-known principles, it is often conducted poorly. The authors report that oven fumes, sometimes including carbon monoxide, are often released into the workroom, as are a variety of solvents and other substances found in the drying or curing products. Particular problem areas noted include inlet and outlet openings in continuous mechanized ovens and opened doors in batch ovens. While suitable exhaust from the oven interior which causes an indraft into all openings is noted as being "sometimes" satisfactory in conveyorized units, it is recommended that ventilation system designs include local exhaust hoods at inlet and outlet openings and at batch oven doors.

Feiner and Kingsley[111] somewhat disagree with other sources addressing the control of continuous ovens. They recommend that make-up air enter the enclosure through the work entrance opening, traverse the entire heating zone and be exhausted from the dry end of the oven. Of interest is their comment that "in a number of continuous ovens, particularly for cloth coating machines, the nature of the different products manufactured is such that conditions will vary due to different coating materials used, different concentrations of material in the solvent, varying thicknesses of coat applied, changing temperatures, and adjustable conveyor speeds. Under these circumstances, ventilation should be provided to satisfy the worst possible conditions."

Feiner and Kingsley also discuss the use of "air seals." These are described as vestibules attached to the ends of an oven where the work enters and leaves; their purpose being to act as an expansion chamber in which the air leaving the oven proper can cool and contract. It is noted that these are "usually provided at ovens processing large pieces of work where the entrance and exit openings are so large that the ventilation rates required for maintaining safe atmospheres within the oven are not high enough to maintain controlling velocities through the large areas of openings." These "air seals" are more clearly described in other sources which equate them to "air curtains" installed at oven openings.

Constance[112] simply recommends an exhaust volume of 100 scfm per sq ft of booth cross-section for drying, baking, and curing ovens. For the door opening of a continuous washer, dryer or oven, he gives canopy hood design criteria of 150 scfm per sq ft of door opening or 200 scfm per sq ft of hood face.

Hemeon[113] discusses exhaust system design considerations for hot processes. Because air streams from hot surfaces behave differently than those from cold surfaces, it is noted that the operating temperature of a dryer must be included as a design parameter. Hemeon states that when the air in an exhausted enclosure is hotter than the ambient air, the enclosure may permit the escape of contaminated air in some situations. He writes, "This phenomenon may sometimes be observed in drying ovens having small openings or cracks at the top, or which are open at the ends . . ." The thermostatic head due to the temperature difference can be determined by Hemeon's equations and counter-balanced by creation of a negative static pressure within the enclosure. If the pressure difference is equal to or greater than the termostatic head, no leakage will occur.

Boyle and Novack[114] have developed a set of equations for calculating the dilution ventilation requirement to maintain concentrations at or below 25% of a LEL in batch process ovens. They prefer their method of determination to the NFPA requirement of 380 cfm per gallon of solvent because the latter "does not always insure a ventilation rate at 25% of the LEL."

There are several variables, according to these authors, that must be considered in the determination of a safety factor value. These include: (1) elevated operating temperature, (2) thicker coating of material, (3) excessive oven loading, (4) alterations in coating material diluent ratios, (5) changes in diluent, (6) variations in air drying time between coats, and (7) poor oven circulation.

Danielson[61] reports that dust can be a problem in any dryer in which the material is agitated or stirred during drying. Those which are listed as "prolific" dust producers include direct-fired rotary dryers, flash dryers, and spray dryers. Types noted to produce less dust are indirect-heated rotary dryers, pan dryers, and cylinder dryers. Tray dryers, sheeting dryers, and dryers for massive solids are noted as possibly emitting no dust. In general, he states that emissions may include vapors, mists, odors, and smoke whenever an organic liquid is evaporated. Also generally addressed by this source is the necessity for controls for many specialized ovens used in industry. Among these are included printing system ovens, can lithograph ovens, paint baking ovens, and others.

Discussion—There are two distinct reasons to control the environment in and around a drying oven aside from the obvious requirement of providing effective heating and/or moisture removal. The first source of concern is the maintenance of a low concentration of flammable solvent vapor to eliminate the possibility of fire or explosion. The second is to ensure that toxic contaminants are not released into the breathing zones of workers.

As noted by the Michigan Department of Health, oven ventilation system design involves the application of basic principles. This statement must be agreed with. Review of the literature indicates that the factors which must be given consideration are many but can each be taken into account with relatively simple and available technology.

What is of concern is that the designing of an adequate exhaust system for such equipment requires that the designer be thoroughly familiar with all the various factors which must be given consideration and with the calculation procedures and techniques available to properly do so. It is clear that oven ventilation design is now a subject to be approached only by persons experienced in the field. Hence, comments in the literature that ovens are often poorly designed can be accepted as reasonable, especially since those most common literature sources which address the subject (e.g. NFPA and ACGIH) do not discuss many important factors. Discussions of these are only found in such sources as Feiner and Kingsley, sources which are not easily available and which are rather qualitative in nature.

Conclusions and Recommendations—From the above, the conclusion is reached that engineering controls for drying and curing ovens are available and adequate but that the factors which can lead to ineffective control are many and not sufficiently stressed. Consequently, it is recommended that NIOSH consider a program to review the state-of-the-art of oven ventilation design, to identify and fill identified data gaps (e.g. how to control emissions from types of dryers other than those which resemble tunnels or boxes), and to present the correct procedure for enclosure and ventilation system design of these units in some sort of ventilation guideline manual. Conducted by a firm which specializes in this subject area, such a program might be completed with a 6 man-month effort over a time span of 5 months.

4.6 Grinding, Crushing, and Screening

Introduction—Many industries utilize grinding, crushing, and screening equipment to effect size reduction or dispersion of solids or pastes. The intent of this discussion is to review the types of equipment commonly used for these purposes, and to examine the available information pertaining to the control of dust or vapor emissions resulting from their use.

Equipment[1,8]

Crushing and Grinding

Size reduction refers to the mechanical reduction in size of solid particulate material. Two of the principle methods of achieving size reduction are crushing and grinding, but the terms are not synonymous. Crushing generally refers to a relatively slow compressive action on individual pieces of coarse material ranging in size from several feet to under one inch. Grinding is performed on finer pieces and involves an attrition or rubbing action as well as interaction between individual pieces of material. Pulverizing and disintegrating are terms related to grinding. The former applies to an operation producing a fine powder; the latter indicates the breakdown of relatively weak interparticulate bonds, such as those present in caked powders.

Jaw crushers are employed for primary crushing of hard materials. The crushing cavity is formed by two wear plates, one rigid and one moving. The motion of this latter plate varies with the design, but the essence of the movement is to crush the material until pieces are of a size to drop through an opening between the plates.

A gyratory crusher consists of a cone-shaped pestle rotating within a cone-shaped bowl. The spacing between the two faces diminishes toward the bottom. Units can be used as either primary or secondary crushers.

Roll crushers either contain two rollers which rotate towards each other at a fixed, set distance, or simply one roller which revolves downward into a "nip" between the roller and a rigid breaker plate. To increase effectiveness, the roller usually has some type of protuberance to grab the material and force it into the gap.

Disk mills can be used for either crushing or grinding. The various forms include the rotary crusher, rotary grinder, stone mill and colloid mill. The rotary crusher simply consists of two faces revolving at different speeds within close proximity of one another. The rotary grinder operates on the same principle but provides much finer grinding.

Roller mills are used to process powders and pastes. In the most common roller mills, the substance is passed through closely spaced rollers which revolve at different speeds and in opposite directions to effect some degree of dispersion and/or size reduction. Removal is accomplished by means of a scraper on the final roller. Variations of roller mills include the pan type, the ring roller, and the ball roller.

Hammer crushers can be used for either crushing or pulverizing. The hammers are mounted to a rotor shaft which runs in a housing containing grinding plates or liners. The rotor is enclosed by a cylindrical screen or granting through which the product is removed. The grinding action results from attrition, while the crushing arises due to impact. To attain effective pulverization, hammer mills are employed which differ from hammer crushers in specifics, but not in basic design.

Cylinder mills or tumbling mills consist of a horizontally-mounted chamber containing a loose packing or grinding media which move about the grinding charge to provide the necessary impaction and attrition. The media can be balls, tubes, rods or pebbles.

Other machines used for grinding and crushing include jet mills, pan crushers, non-rotary ball or bead mills, dispersion mills, and others. A number of these were previously described in the industry evaluation for paint and coatings manufacturing.

Screening

Once the size of individual pieces of material has been reduced, it is often necessary to assure size uniformity, or at least a maximum cut-off. This can be accomplished through several techniques, including screening, centrifugal classification, pneumatic classification, and aqueous classification. Screening is the most commonly used method and is of most concern to this discussion.

Matthews[115] writes that "Screening is a unit operation that is an essential part of many different processes". It involves the mechanical separation of particles on basis of size, and is also known as sizing, sifting, sieving, or separation. There are three types of screen surfaces: (1) parallel bars, (2) punched plates, and (3) woven wire or fabric. In general, these are in order of increasing fineness. Screening surfaces are generally moved or vibrated to facilitate material flow.

Perry[8] divides screening machines into five categories: (1) grizzlies, (2) revolving screens, (3) shaking screens, (4) vibrating screens, and (5) oscillating screens. In general, grizzlies are used for the coarser materials and oscillating screens for the finest.

Grizzly screens contain a set of parallel bars held at a specified spacing. These can be stationary or vibrating and are often used prior to a primary crusher to remove fines. Stationary grizzlies are often used to retain occasional pieces too large for processing by following plant equipment.

Revolving screens (trommel screens) consist of a cylindrical frame surrounded by a mesh wire cloth or a punched plate. Mounted on an incline, the material to be screened enters the rotating cylinder on the upper end, the oversize exits at the bottom and the desired product falls through the openings. Low capacity and efficiency have limited the use of these screens.

Shaking screens are mounted on a slightly inclined rectangular frame which is suspended by cables or rods. The surface itself can be a punched plate or wire cloth. The raw material is deposited on the screen at the upper end and is advanced down the slope by the forward stroke of the screen.

Vibrating screens fall into two subcategories, (1) mechanically vibrated screens and (2) electrically vibrated screens. The former are similar to the shaking screens described above but have a much higher capacity and efficiency. Electrically vibrated screens are drive by electromagnets at intense, high-speed, low amplitude vibration. These are commonly used in the chemical industry for handling light, fine, dry materials and powders.

Oscillating screens operate by applying an oscillating motion in the plane of the screening surface. These are fine screens used for light, free-flowing materials. Reciprocating screens, gyratory screens, and gyratory riddles are also used in industry.

Controls—The fact that hazardous dust exposures may be produced in size reduction and separation processes is fairly obvious. These operations involve small particles which can easily be suspended in air and released to the plant environment if proper controls are not utilized. Where organic solvents are used in paste or ointment formulations, solvent vapors may also be evolved.

Both Hemeon[113] and the ACGIH[11] recognize the potential for dust exposures from classifying screens. Hemeon states, "Vibrating and rotary screens may produce tremendous quantities of dust which can be controlled only by complete enclosure. Indeed, it is possible to enclose them so well as to eliminate the need for any exhaust." Nevertheless, the ACGIH presents a combined enclosure-exhaust system for both flat-deck and cylindrical screens. Called for are a face velocity of 200 fpm for the flat deck screen and 100 cfm/sq ft circular cross section for the cylindrical screen. In the first case, the minimum given is 50 cfm/sq ft of screen area, while in the latter it is 400 cfm/sq ft of enclosure opening or a face velocity of 400 fpm.

Baturin[103] devotes a chapter in his book to the "Local Exhaust of Dust in Crushing, Pulverizing, and Transporting of Free-Flowing Materials". In it, he provides design criteria for the construction of local exhaust systems for many of the pieces of equipment discussed above, including jaw-type and roll-crushers as well as vibrating sieves. His recommended exhaust volumes for the crushers are presented as a function of the volume of the casing, with the values ranging from 300 to 700 cfm.

The ACGIH manual also contains design criteria for the ventilation of tumbling mills of the trunnion and stave types, and mixers and mullers used in foundry operations.

Danielson[61] provides brief notes on the basic requirements of a system ventilating this machinery. Grinders and crushers are noted to require enclosure with a 200 fpm face velocity, while mixers are reported to require 100 to 200 fpm. He also, somewhat confusingly, suggests 500-1000 fpm through enclosure openings.

When flammable solvents are used as the vehicle for pastes to be processed, all duct and hood velocities must be reconsidered, bearing in mind the allowable concentration of the contaminant in the work place environment and the lower explosive limit (LEL) of the solvent.

Discussion—Depending upon the equipment used, the materials handled, and operating conditions, many of these operations may result in exposures if uncontrolled. These may occur as a result of the actual operation or as a result of loading and unloading the equipment. The controls recommended throughout the literature are of two types; exhausted enclosures for equipment, and various open hoods placed near loading and unloading hatches. For certain types of equipment used in specific industries, specific design criteria are available. For others, only the need for controls is expressed together with general approach for providing them.

There is cause to consider whether it would be advantageous for NIOSH to catalog these equipment, determine their points of contaminant emissions, and the specific control design criteria applicable to each type. While much of the equipment of concern is commercially available with enclosures and exhaust duct take-offs already attached, some of it is not and requires the user to provide such controls. If a sufficient body of data is compiled, NIOSH may then consider the development of ventilation guidelines for these equipment. To be noted is that attempts to accomplish this latter task were aborted in a previous program to develop recommended industrial ventilation guidelines when sufficient data of confirmed validity could not be found.

Recommendations—It is recommended that NIOSH consider a research program with the objectives of cataloging the types of grinding, crushing, and screening equipment available; compiling available data on control of emissions from these equipment; identifying and filling data gaps; and developing engineering control guidelines for the equipment and its attendant material transfer points. It is estimated that an effort of 24 man-months over a time span of 12 months would be appropriate for such a program.

4.7 Laboratory Operations

Introduction—The term "laboratory operations" is used here to include small-scale experimental research and testing activities conducted by academic, research, and industrial operations. Operations of concern are those with the potential for creating excessive exposures of laboratory personnel to airborne contaminants.

Laboratory operations are highly variable and, to a large degree, unpredictable in their nature. The types of air contaminants which can be generated and their modes of generation are unlimited. Some operations will produce air contaminants in a "normal" operating mode such as with the evaporation of volatile liquids from storage or process vessels. With many operations, contaminants may be produced accidentally due to spillage, container failure, or explosion.

General practice in protecting laboratory personnel from air contaminants involves the use of general purpose

protective facilities. The most common protective measure is the use of ventilated enclosures such as laboratory hoods, glove boxes, and biological cabinets. These enclosures also can be designed to protect personnel from explosions, fire, and equipment breakage. To a lesser extent, free-standing, open exhaust hoods also are used in laboratories for air contaminant control.

Existing Ventilation Criteria—Ventilation criteria are contained in the Industrial Ventilation Manual[11] for three types of ventilation equipment used in laboratories: glove boxes, laboratory hoods, and open hoods. Ventilation criteria for laboratory hoods are also found in References 116 and 117. The criteria found in these sources are consistent in that, in all cases, operations are classified according to the toxicity levels of the contaminants generated, and minimum ventilation rates are prescribed for each contaminant class. However, there are differences in the contaminant class definitions and in the ventilation rates prescribed. None of these sources present supporting data for their recommended ventilation criteria.

Criteria for the use of open hoods are applicable to many laboratory operations. In addition to those found in Reference 11, additional criteria are presented in References 118 and 119. These criteria generally are based on the same principles of open hood operation, and the design and operational requirements are consistent.

Ventilation criteria also exist for laboratory operations involving specific types of air contaminants. Criteria for the control of radioisotopes and carcinogens have been published as Federal Regulations.[120,121]

The control of biological agents presents a problem which is different from other contaminant control problems in that two-way contamination control generally is required. In addition to protecting laboratory personnel from biological agents, it is necessary to isolate the biological materials from contaminants in the laboratory environment. Approaches to designing biological cabinets have been published,[122] but no comprehensive guidelines are available.

Identification of Problem Areas—Problem areas in the ventilation of laboratory operations can be identified as shown in Table 6 by defining combinations of contaminant types and ventilation methods. Entries in the table indicate the existence of ventilation guidelines for specific combinations.

From Table 6, we can see that ventilation guidelines exist for nearly all combinations of contaminant types and ventilation methods of practical interest. The only exception is for biological cabinets which, as discussed above, represent a unique problem of contaminant control. Thus, the need for ventilation guidelines for biological cabinets is the first problem area identified. The second problem area is a need to verify the effectiveness of the existing

TABLE 6—PROBLEM AREAS IN VENTILATION OF LABORATORY OPERATIONS

Contaminant Types	Ventilation Methods			
	Open Hood	Laboratory Hood	Glove Box	Biological Cabinet
Nuisance Materials	11, 118, 119	11, 116, 117	NA	NA
Low Toxicity Materials	11, 118, 119	11, 116, 117	NA	NA
High Toxicity Materials	NA	11, 116, 117	11	NA
Radioisotopes	NA	11, 120	11, 120	NA
Carcinogens, Mutagens & Teratogens	NA	121	11	NA
Biological Agents	NA	NA	NA	

Notes: Numerical entries indicate references to existing ventilation criteria.

NA = not applicable (not of practical interest).

guidelines for other ventilation methods. This latter problem stems from the lack of supporting data for the existing guidelines.

Discussion of Problem Areas—

Biological Cabinets

Experiments with biological agents often entail the dual requirements of protection of personnel from the agents and isolation of the agents from atmospheric contaminants. These requirements can be met by means of a biological cabinet which provides an internal, filtered, recirculatory airflow which protects the experiment from external contamination. Other features of the cabinet prevent escape of contaminated air from the cabinet.

Biological cabinets are in use in universities, research institutions, and commercial laboratories where experiments are conducted with biological agents. Such cabinets are manufactured and sold by a number of companies. However, at the present time, there are no generally ac-

cepted design or operational criteria for these cabinets. Consequently, there is no way of knowing whether laboratory personnel are adequately protected against exposures to the agents involved in the experiments.

We suggest that a study be initiated for the purpose of evaluating the present state of the technology of biological cabinet design. The objectives of the study should be:

1. To classify biological agents and to define exposure limits appropriate to each class of agents,

2. To classify biological cabinets in terms of their applications and characteristic design parameters,

3. To determine the performance characteristics of biological cabinets in terms of exposure control as a function of cabinet design and operational variables, and

4. To formulate guidelines for the design and operation of biological cabinets.

It would be logical to conduct this study in phases with the first phase devoted to the collection and analysis of existing information. This phase would serve to identify the information gaps and to develop a program plan for later phases wherein these data gaps would be eliminated.

Verification of Existing Guidelines

It is also suggested that a study be initiated with the purpose of verifying the effectiveness of existing ventilation guidelines for laboratory operations. The objectives of the study should be:

1. To develop a procedure for measuring the performance of laboratory hoods and glove boxes which is easier to use than existing measurement methods,

2. To characterize "typical" releases of air contaminants in laboratory operations and to develop test methods for simulating releases of various magnitudes,

3. To measure the performance of laboratory hoods and glove boxes using the improved performance measurement procedure and release simulation methods. Performance measurements should be made under broad ranges of conditions both within and external to the hoods and glove boxes, and

4. To formulate revised ventilation guidelines for laboratory operations based on the performance measurements. Such guidelines might be in the form of a manual discussing all aspects of health and safety in the laboratory. Topics might include hazards in general, a list of "preclassified" chemicals, ventilation and other contaminant control techniques, protective clothing, spill and accident response, safe operating procedures (i.e., various "do's and don'ts"), etc.

This study could be conducted most economically by a field test program conducted with existing ventilation equipment.

Summary of Program Recommendations—Two programs have been recommended for the development and validation of ventilation guidelines for laboratory operations. Suggested durations and levels of effort for these programs are listed below.

Program	Duration	Effort
Biological Cabinet Guidelines (Phase 1)	9 months	1/2 man-year
Existing Guidelines Validation	18 months	2/3 man-years

4.8 Material Handling Operations

4.8.1. Introduction—Within the broadest sense of the term, "material handling" can be interpreted to include every type of industrial operation in which a substance in any form is moved or processed. However, since the processing of substances involves equipment or procedures which are covered separately in this report, and/or are unique onto themselves, the phrase is limited here to the operations involved in simply transporting materials or filling or emptying their containers. In this definition is not included the use of such principally individual units as industrial trucks, tractors and trailers, cranes, hoists, etc., nor the use of piping systems to transfer liquid.

4.8.2 Bulk Handling Systems—Buffington[123] and others[8,124] describe the various mechanical conveyors and elevators used throughout industry. It is rather commonly noted that few bulk solid handling systems are complete without a dust control unit. Trouble spots noted include inlet and discharge points, and long chutes with free-falling material. Elevators are noted as requiring dust pick-up connections at both head and boot sections. Ventilation of bins into which materials are discharged is also suggested. In the following, the various controls suggested in the literature for these equipment are described and discussed more fully.

Belt Conveyor Transfer Points

Literature Review

Morrison[125] notes that "belt conveyors emit dust almost exclusively from two points—at the tail pully where material is received from prior equipment, and at the head pulley where material is discharged". Appropriately, he reports that the amount of dust released at a transfer point depends upon the physical characteristics of the bulk material and the manner in which the material is handled. Concerning the physical characteristics of the material, he notes that there is no convenient way for a designer to ascertain whether or not a material is dusty enough to warrant dust control measures. (The only guideline appears to be the "Standard Method of Test for 'Index of Dustiness' of Coal and Coke" approved by ASTM in 1941[126], a guideline which may or may not be applicable to various other substances.)

For enclosing dust sources, Morrison discusses four subject areas—the size of the enclosure, its ease of disassembly, access doors, and the installation of skirting and curtains at enclosure openings. He recommends that enclosures at transfer point be spacious to permit internal recirculation of dust-laden air and to suppress pressure surges caused by inflowing material and ingress of induced air. Specifically, it is said that enclosure heights be made not less than 2/3 the belt width and lengths from four to six times the belt width.

Designing the enclosure in removable sections for ease of maintenenace is considered to be worthwhile. A note of caution is given that some form of gasketing material should be employed at jointing to reduce the risk of dust escaping. Hinged access doors, preferably self-closing by gravity and attached with quick-opening clamps, are reported to be useful and convenient to facilitate routine inspection and maintenance. Rubber skirting, attached to the enclosure sides by nut and bolt, or clamps, is recommended for along the enclosure sides between the lower enclosure edges and belt surfaces. Strips 1/2 in. thick by 9 in. wide with a No. 40 Durometer reading are noted as working very well. Further suggested is that the skirting be adjustable in height to allow balancing and fine turning of airflows.

A rubber curtain is suggested for the open end of the enclosure from which the conveyor exits. "The bottom edge should be cut to conform to the cross-sectional profile of the material conveyed on the belt. In addition, vertical slits 2 in. long and approximately 1/2 in. apart along the bottom edge of the curtain will regulate the size of the opening and prevent it from being torn loose when periodic surge loadings occur on the belt."

Morrison states that these recommendations only address the utilitarian aspects of dust control, aspects which have been developed by trial-and-error operating experience and which can be considered as conveniences which contribute to effective dust control. Appropriately, he reports that without proper exhaust volumes effective control is impossible. Consequently, he reviews the various popular "rule-of-thumb" methods for determining exhaust volumes given in Steel Mill Ventilation,[127] the ACGIH Industrial Ventilation Manual,[11] ANSI,[118] and others.

This review reveals wide disparities in recommended exhaust volumes for various belt speeds and material drop heights and it is observed that some recommended air volumes are exceedingly conservative while others are questionably low. Morrison concludes, therefore, that consideration of only a few basic system parameters is insufficient to evaluate any given problem and that relationships must be established between all pertinent variables before reliable exhaust volumes can be calculated. Among the variables to be given consideration are included belt speed, material drop height, flow rate, lump size, density, and temperature, and enclosure opening dimensions.

Taking these various factors into consideration, the author attempts to develop a comprehensive approach. Using an equation by Hemeon,[113] and the results of his own study, he proposes an expression for determining the amount of induced (i.e., entrained) air by the action of the falling material. Similarly, he develops expressions for determining the amount of air displaced by the entering material and an expression for the amount of air required for proper control. The volume of air determined from addition of these three volumes, corrected for temperature differences by an empirical expression he developed, results in a total exhaust volume which he feels necessary for effective control.

Pring, Knudsen, and Dennis[128] support the statement that the criteria given most commonly in various sources may be considerably inaccurate. To illustrate their point, they note that the exhaust volume recommended for a 54-in. conveyor with a speed of 350 fpm is given as 2250 cfm by ANSI. For such a set-up, however, it was experimentally determined that 9000 cfm was necessary for effective control. "The reason for the discrepancy is, of course, that width and speed of conveyor have only indirect bearing on tonnage of material handled and no relationship to the height of fall onto the belt. Similar error is possible in the reference of exhaust volume to the area of openings in bins and enclosures."

Discussion

These authors make a very strong case that recommended control volumes in popular sources vary considerably and do not rigorously take into account important factors. The calculation methods Morrison suggests appear, on the surface, to be quite appropriate and much more apt to result in reliable and accurate exhaust volume determinations than those typically recommended. In many respects, however, what he has admirably accomplished must be considered to be more the qualitative conceptualization of a proper approach than the actual quantitative development of one which can be applied throughout industry. Reasons for this statement are given below.

While the paper repeatedly addresses the need for "effective control", there is no indication of what Morrison's definition is of this phrase. Since his paper was published in an issue of "Rock Products" and includes a picture of a belt conveyor transporting some type of crushed rock, one must question whether his method provides effective control for materials other than nuisance dusts.

For determining the amount of air entrained by falling material, he modifies the equation given by Hemeon for the turbulent-accelerating velocity case of falling spheres. His reason for doing so is that the "equation yielded exhaust volumes which were considerably higher than field experience indicated were necessary." No justification is given other than "field experience lends empirical support to making the change." There appears to be a need, therefore, to review this subject area using all available data to produce correlations of confirmed validity. Reference 103, references 128 to 135 inclusive, and reference 113 are some of the information sources available on the subject.

As previously noted, Morrison's calculation procedure also includes a method for correcting for the temperature of the falling material. "Based on actual temperature measurements in the field, this author has derived an empirical expression which related the conveyed material temperature and the ambient air temperature to the temperature of the exhaust air." Since there is no other description of how this expression was arrived at, and the possibility exists that it does not apply to all substances and particle size distributions, there appears to be justification for questioning its overall applicability.

Conclusions and Recommendations

A number of subject areas, therefore, appear to require basic research to better define and quantify the factors which influence the achievement of effective control of air contaminants generated from belt conveyor transfer points. The resulting technology might also lead to better understanding and control of other processes in which materials fall from one plane to another (e.g. bag or drum dumping into a tank, bin loading, etc.).

The first question which enters a designer's mind when approaching such a problem must be whether the operation requires control at all. As noted by Morrison, the subject has been given little attention and no convenient method exists for such an a priori determination. While his point is well-taken that "only in a few instances would a truly representative sample of material be available for test prior to system design," the fact that he presumably is referring only to rock products leads to the conclusion that this may not generally be the case throughout industry. Hence, it is suggested that a study be conducted to develop a method whereby a sample substance can be subjected to the forces present in the bulk transfer and falling of solid substances. Ideally, measures such as shape-rugosity, hardness, apparent surface, dispersibility, etc. could be correlated in a manner which would allow indication of how much, if any, respirable dust might be produced from a given sequence of events, and whether the operation requires engineering control. Hopefully, in time, as the method proves itself, designers would have sufficient data to perform cost-benefit analyses on the various control measures available for use.

Such a program could be conducted in phases to help ensure its success before additional resources were committed. The first phase would essentially be a feasibility study to confirm the need for such a program, to critically review the available literature on the subject, to describe where and when in industry such a method might be useful, and to define the approach to be taken to develop the method. Six man-months of effort over a time period of six months would probably be sufficient.

Actual development and trial of a test method could constitute the second and final stage of the program. Having chosen typical samples of materials, researchers would apply various test methods and correlate their results with actual test data. It is somewhat difficult to estimate the efforts required for such a program, but a guess would be 2 man-years over a 1 year period.

Determination of the amount of air entrained with a mass of falling material is, as noted, a subject which has been studied, but which seems to require a critical review to determine the merits of the various methods proposed in the literature. Such a review and analysis could most likely to accomplished with 3 man-months effort over a time span of 4 months and could be expected to point out which method is "best" and what further study, if any, appears warranted. A similar study in effort and scope could be accomplished for developing methods for temperature correction.

The volume of air which must be withdrawn for "control" purposes from an enclosure or container into which a toxic substance is falling might also be given consideration for study. While an exhaust volume of X cfm per sq ft of open area in an enclosure may be capable of reducing the amount of dust released, it may provide an unnecessary degree of control for nuisance dusts while providing far too little control for highly toxic substances.

Once these separate subject areas had been adequately studied, the final task would be to join the various calculation procedures together into one single procedure, and to confirm the validity of the method using actual field data. An 8 man-month effort conducted over a time span of 6 months would probably be sufficient to collect data from transfer points and to compare them to the results obtained from the 3 parts of the overall procedure studied. To determine the amount of "control" air necessary, the conveyor speed at one site might be varied and dust counts taken or a number of similar operations which involve different speeds might be investigated. These air sampling data could hopefully be correlated with the various belt widths, speeds, and types of materials involved and criteria developed for effective control of materials of varying physical and toxicological characteristics.

Belt Conveyor Straight Runs

Various sources contain design criteria for enclosing and ventilating the length of a belt conveyor and for a belt wiper to clean off the belt before its return run. The first of these is to control dusting off the top of the load; the second, to prevent dusting off the return belt.

The controls recommended by the ACGIH manual for the load include a full enclosure exhausted at least every 30 feet with a tapered exhaust connection drawing 350 cfm per ft of belt width. It is not immediately apparent how these figures were arrived at, but if the enclosure was well designed, one would expect that an adequate degree of control could be achieved with this set-up. Buffington notes that slider-bed felt conveyors, similar to other belt conveyors except that idlers are not used to support the belt, can be employed for handling light loads where a dust-tight enclosure is necessary. For handling asbestos on a belt conveyor, Goldfield and Brandt[136] describe a full and ventilated enclosure which goes well beyond the types of enclosures normally found. It is essentially a complete metal box about all parts of the machinery.

The mining industry and others utilize water and other sprays on material being transported to reduce dust generation. This control would also tend to reduce contaminants from most operations involving industry substances and "wetting" is often an adequate alternative to ventilation.

A belt wiper design, presented by the ACGIH manual, is essentially a scraper combined with a slot hood against the bottom of the belt. With a recommended exhaust volume of 200 cfm per ft of belt width and a slot velocity of 2000 fpm, it appears to be adequate to remove any dust which might adhere to the belt. Indeed, this combination of a scraper and a high velocity hood appears to be substantially better than just the provision of rotating brushes or a scraper. However, Goldfield and Brandt report that "in spite of all cleaning equipment on the market, the return belt often carries dust to the idlers. In extreme cases, it is necessary to enclose the return belt with a complete leak-tight enclosure and to clean that enclosure by means of scraper conveyors."

Other Types of Conveyors[123,137]

The belt conveyor is but one of many types available to industry. Others include the pan, drag chain, en masse, zipper, flight, screw, vibrating, pivoted bucket carrier and pneumatic types. Little information appears to be available in the literature concerning whether engineering controls are required to prevent exposures when these types are used. That information which could be found, and that which could be inferred, are summarized in the follwoing.

Drag Conveyors

Used frequently for handling hot clinkers, ashes, coal, sawdust and similar materials, this conveyor is made of a special wide chain which is dragged through a hard trough made of white iron, steel, or concrete. Illustrations in the literature indicate that the trough could probably be covered to restrict dusting problems during operation. Since the chain returns to its starting point over rollers above the trough, any dust sticking to it may be shaken loose. As with other conveying mechanisms, dust can also be assumed to be generated at loading and discharge points.

En Masse Conveyors

The en masse or continuous-flow conveyor moves material as a single mass through a tube by using stirrup-shaped or solid flights. It is especially useful with non-abrasive materials that must be handled gently and has numerous other advantages for use. The primary disasdvantage appears to be that it requires more power for a given load than some other types. Buffington reports that they can be sealed to prevent contamination and to contain fumes, "an important factor in solvent extraction processes." This type of system is noted as being one of the two mechanical types available which does not require special considerations when materials are handled which dust excessively, the other mechanical type being the screw conveyor.

Flight Conveyors

Working well with sticky or abrasive solids, flight conveyors are simple in design and slow moving. They consist of one or two strands of chain that drag wood, steel, or plastic flights through a trough. The returning flights travel above the trough in the opposite direction and material adhering to them may possibly be shaken loose. Loading and discharge points appear to be of concern as with other conveyors which are gravity-fed or discharged.

Pan Conveyors

A pan conveyor is often used to carry large, heavy lumps and is frequently found in metallurgical plants feeding raw ore to a crusher. Consisting of metal pans carried between two rows of chains, it may be used to carry molten glass while it cools. Examination of illustrations of this type show that the pans return in an upside-down fashion back to their starting point. Any material which sticks to the pans can thus be expected to be shaken off during the return trip. Furthermore, since they are similar to belt conveyors, loading and discharge points may need controls when dusty substances are handled.

Pneumatic Conveyors

Pneumatic conveyors are, in simple terms, a pipeline through which materials are moved by means of a vacuum, under positive pressure, or by a combination of both. Since they are completely enclosed, product contamination, material loss and dust emission are reduced or eliminated. Their prime use is to convey dry, granular or powdered materials via pipelines to remote plant areas that would be hard to reach economically with mechanical conveyors. An article in Food Engineering[138] demonstrates how the use of closed pneumatic conveyor systems has eliminated dust problems in the filling of hoppers above filling machines for food products.

Screw Conveyors

Screw conveyors can be used to convey, feed, mix, agitate, dig and blend. They essentially consist of a long screw which, by revolving in a trough or tube, propels the material forward. According to Buffington, they can be effectively sealed to keep in dust and fumes and keep out moisture and contaminants. An article previously cited[6] reported that the use of these conveyors in a pharmaceutical plant allowed strict sanitary requirements to be met.

Vibratory Conveyors

Vibratory or oscillating conveyors propel the load forward by imparting forces generated from vibrating the supporting system at its natural frequency. The equipment may be of the open trough design or fully enclosed. Standard units have capacities up to 500 tons/hr and can handle materials at temperatures up to 250°F. They are versatile in that they can be used with small sand-like particles or very large pieces such as casting.

Pivoted Bucket Carriers

This and the "Peck" carrier operate in the same manner, and the unit is much like a bucket elevator except that the buckets pivot so they can travel horizontally until a discharge point is reached and they are turned over. The points at which they are loaded or discharged appear to have the potential of generating dust. Each bucket is, after all, dumped from some height into a chute opening or bin.

Discussion

In our previous program to develop recommended ventilation guidelines, an attempt was made to extrapolate the control measures recommended for belt conveyors to conveyors of other types. At transfer points, where dusty material falls from one plane to another, it can be expected that the forces and airflows generated will produce airborne dust regardless of the specific type of conveying mechanism used. During straight runs of non-enclosing conveyors, it can be considered that dusting off the top of the load or dust falling from the return side would add to the problem. Thus, it appeared that providing examples of enclosures and recommended exhaust volumes, based on analogy of belt conveyor controls, would be useful. Reviewers of the resulting guideline, however, strongly objected to this extrapolation and it was therefore deleted from the final version. The reasoning was that the guidelines upon which it was based were developed solely for belt conveyors and were not proper for direct use with other conveyor types. Having pointed out the disadvantages of using the control measures given in popular sources, this conclusion can now be fully agreed with.

Conclusions and Recommendations

The first of any efforts directed towards normally unenclosed conveyors other than belt conveyors should be to determine if indeed the use of these mechanisms in some industries results in excessive, or hard to control, concentrations of airborne contaminants. A limited survey of this type could probably be accomplished with 5 man-months effort over a time span of 8 months. Its results would characterize the types of substances generally handled by each type of conveyor, the controls used to reduce exposures, and the airborne concentrations evident. If it were found that there was a problem requiring further investigation, the following suggested efforts would be warranted.

In the discussion above for belt conveyor transfer points a program was suggested for developing a comprehensive approach for calculating accurate exhaust volume requirements. If and when such a program were to be conducted,

it could be expected that modest additional effort would be required to modify the procedure for use with other types of mechanisms. These modifications would probably involve only the part of the procedure for estimating control volumes, and could probably be simply and accurately determined from theoretical considerations or from the data obtained from the belt conveyor program. An effort of 1 man-month over a time span of 2 months would probably suffice if field verification were not felt necessary.

The ACGIH Ventilation Manual, as noted for belt conveyors, also gives recommended criteria for ventilating straight runs of belts and for the design of belt wipers, the former of these to control dusting off the top of the load and the latter for dust from the return belt. To some extent, these phenomena may possibly also occur with other mechanisms depending upon the material handled and the transport speed. Another area of interest might be the determination of how analagous controls may be developed for these other types of conveyors.

Bin Filling

Literature Review

For mechanical loading of bins, the ACGIH suggests that the loading point be enclosed and that an exhaust connection drawing 200 cfm per sq ft of all open areas be installed either directly over the loading point or at a location on the top of the bin remote from the loading point. For hoppers which are manually loaded, presumably into bins, it recommends a three-sided booth over the hopper with an exhaust connection at the rear which draws 150 cfm per sq ft of face area. Specifically for "grain" elevators, feed mills, (and) flour mills", it recommends 550 cfm per bin.

ANSI and Steel Mill Ventilation recommend a capture or control velocity of 150-200 fpm through openings in the enclosure and an exhaust rate of 0.5 cfm per cubic foot of bin volume. The latter adds that these velocities are based on general velocities in material handling systems and that a more detailed analysis requires use of correlations which take into account material flow rate, size, the free fall trajectory distance, and the enclosure opening area.

Kruse and Bianconi,[135] Pring, Knudsen and Dennis[128] and others report that such recommendations as are in the ACGIH manual and ANSI are largely empirical and based on trial-and-error methods or on the personal experience of a designer.

Discussion

It has repeatedly been pointed out in the above that operations characterized by the falling of a material from one plane to a lower one involve effects which are not taken into consideration in many of the guidelines found in popular sources. This also appears to be entirely true for bin filling operations.

Conclusions and Recommendations

If the program suggested for conveyor transfer point controls is conducted, the resulting technology should be utilized to develop a method for accurate determination of required exhaust volumes for bin filling operations.

Bucket Elevators and Chutes

Many of the same sources cited above, and below, recommend controls for operations involving bucket elevators and chutes. As for belt conveyor transfer points and other operations discussed in this section, it appears that given recommendations are based on personal experience and trial-and-error research on a limited number of operations. Thus, detailed reasoning is not presented here for why the results of research conducted for such transfer points should be applied to these devices also, but it follows from the preceding that it should be.

4.8.3 Bag and Drum Filling

Bag Filling

The bagging of powdered materials such as plastic resins, paint pigments, pesticides, cement, and the like is generally accompanied by the generation of airborne dusts. This occurs due to the displacement of air from the bag, spillage, and motions of the bagging machine and the worker. When the dust is toxic, the potential exists for the operation to adversely affect the health of workers.

In a limited survey ADL personnel recently conducted in a chemical plant, attention was given to the actions involved in a simple bag filling operation. The steps of the operations were: (1) the operator secured a bag on the filling tube; (2) a measured amount of the powder was dumped into the bag from a hopper; (3) the bag was removed from the filling tube and placed onto a small scale; (4) a scoop and a nearby drum of the product were used to bring the weight of the bag into specification; (5) the operator closed the flaps of the paper bag; (6) he lifted it onto a pallet; and (7) he "patted" it down.

The filling of the bag was conducted within a three-sided enclosure with an exhaust takeoff at the top. Nevertheless, the operation produced excessive contaminant concentrations because of the necessity of steps (3) thru (7). The "plopping" down of the bag on the scale, the pouring in of a partial scoop of powder, and the placement of the bag on the pallet all appeared to release some amount of dust to the atmosphere at locations obviously uncontrolled by the enclosure. The floor was covered with the highly toxic substance and the bags were lightly covered with a layer of dust. Crossdrafts produced by a nearby open garage door served to stir up the spilled contaminant during gusts. These observations led to the conclusion that it is not the actual bag filling operation which

is often difficult to control as it is the ancillary actions required of the operator. Recognizing this fact, the management of the company themselves suggested the need for machinery which automatically and accurately fills and seals the bag and delivers it to a pallet loading station. To further reduce the amount of the chemical which may spill from the bag, and thereby cause exposure problems in subsequent warehousing and transport operations, it was decided to place a plastic liner within the paper bag, a liner which could be securely heat-sealed.

Literature Review

Hama[139] studied the control of dust from bagging operations in 1948, and determined "that bagging machine workers at uncontrolled operations are exposed to concentrations of dust in the magnitude of 100 million to 1000 million particles of dust per cubic foot of air." Discussing the various local exhaust hood configurations practical for use, he found that side draft hoods and slots and the double-canopy type (where the air is exhausted in the concentric space between the hood and the bag-filling tube of the machine) are not capable of completely controlling the dust. "From a number of dust counts taken on various types of local exhaust hoods," he therefore concluded that the best control is achieved by a simple three-sided booth enclosure. Features of his recommended design for such a hood included a plexiglass window to reduce the open face area and a hinged plate edging on the front edge of the floor to confine spilled material within the booth.

For bagging machines located where high-velocity crossdrafts are present, Hama recommends face velocities from 200 to 250 fpm. Under calmer conditions, he remarks that 150 to 200 fpm will be adequate. Since the area of the greatest dust release is near the top of the booth, he further recommends that the branch exhaust duct be connected to the top of the hood. As an example of the degree of control which can be achieved, he illustrates an operation which, before ventilation was installed, produced dust counts of 173 to 182 million particles per cubic foot of air, and after installation, produced breathing zone concentrations of the order of 3.5 mppcf. The airflow distribution in this booth was: top third of face, 250 fpm; center of face opening, 200 fpm; bottom third of face opening, 150 fpm.

The ACGIH manual contains two exhaust system designs for such operations, one for "bag filling" and another for a "bag tube packer." The first design shows a bin which feeds into a funnel to which the bag is securely attached. The open area between the top of the funnel and the bottom of the bin is pointed out as being the principal dust source and is partially enclosed with a hood. The exhaust volumes recommended are 400-500 cfm for "nontoxic dust" and 1000-1500 cfm for "toxic dust". The maximum recommended face velocity for the open bottom portion of the hood is given as 500 fpm.

The exhaust system design for the "bag tube packer" includes a number of exhaust hoods. One hood is installed on the feed hopper and has a recommended exhaust volume of 500 cfm. Another is installed at the side of a spill hopper under the machine and has a recommended exhaust volume of 950 cfm. Others appear to be installed over the individual filling tubes and have recommended exhaust volumes of 500 cfm each.

Constance[112] recommends 130 to 150 scfm/sq ft of hood face opening for a "hooded bag or drum filler from pulverizer", 450 to 500 scfm/sq ft of pouring slot for "bag filling", and 125 scfm/sq ft of face area for a "hood over bag packing from belt packer".

Kane[140] reports that New York Code 34 for silica dust in stone crushing operations recommends a booth or enclosure for bagging machines (with a spillage hopper) and exhaust requirements as follows: Paper bag - 100 cfm/sq ft open area; Cloth bags - 200 cfm/sq ft open area.

ANSI gives essentially the same recommendations reported by Kane. For pulverized sand, however, it recommends 400 fpm at the point of origin.

Burke[141] describes the various bag filling machinery available to the chemical industry, the accuracy of various systems for weighing out substances and the economics involved in choosing the correct setup for a particular operation. An interesting observation that results from a review of his paper is that the simpler the bag filling and weighing operations are, the more prone they are for resulting in bags of product which are further from a desired weight specification. A look at a graph he gives for savings per year vs. bags filled per hr indicates that the savings potential from using machinery of greater accuracy can in the course of a few years justify the initial cost of installation of better machinery. Thus, one gets the impression that a manual weight checking step and other actions can be eliminated in such operations with advantages both in reducing exposures and in increasing profits.

Discussion

Hama indicates he could control dust counts to 3.5 mppcf using the enclosure and exhaust takeoff configuration he recommends. Given the present exposure limit of 50 mppcf for total nuisance dust, this must be considered to be more than adequate control for such substances. For highly toxic substances, however, this degree of control must be considered inadequate. For the bag filling operation we observed, it was estimated that 100,000 particles of a fine dust (<2 microns) would approximate the TLV. The other guidelines, given only as examples of many which can be found, obviously vary widely in their recommendations. One must therefore wonder how and when

they are applicable, to what types of substances they apply, and whether they provide adequate control if installed.

Conclusions and Recommendations

The actions involved in bag filling operations are simple to comprehend and the sources of air contaminants easy to characterize. Nevertheless, there appears to be some degree of confusion in the literature as to how such operations should be controlled. It is therefore advised that a program be conducted to classify the types of bags and bagging machinery currently available and the actions required of workers, and to determine how best each of these operations may be controlled for substances of various physical and toxicological characteristics. The study should then devote some efforts to better quantify the manpower and product savings advantages and disadvantages of utilization of inherently "cleaner" machinery. This type of limited field survey and information collection and analysis effort could probably be accomplished with 8 man-months effort over a time span of 6 months.

Drum Filling

Just as the filling of bags can generate air contaminants, so can the filling of drums. The problem does not appear to be one of similar severity, however, since numerous references to the hazard of the operation do not appear in the literature as they do for bag filling.

Numerous recommendations exist in popular sources as to how these operations can be controlled. The ACGIH manual, for example, gives four possible hood configurations and attendant suggested exhaust volumes. Two of these appear in Hemeon. At least one of these, a full enclosure with a suggested exhaust volume of 150 cfm per sq ft of open area, can be considered to be adequate to control most toxic substances.

Thus, while some methods may be better than others depending upon the toxicity and physical characteristics of the substance, specific research for this operation does not appear to be warranted at this time. However, if the program suggested for belt conveyor transfer points is conducted, it would be expected that the technology would then exist to allow more accurate determinations of required exhaust volumes for effective control. As for that operation, it would be expected that the rate of flow, particle size, material density, material temperature, fall height, and other factors would influence the determination of the amount of air which must be removed for effective control.

4.8.4 Storage and Transportation of Packaged Materials

Package Ingrity

One source of toxic chemicals released into the industrial environment is from broken bags, a sight which can be seen in almost every warehouse, boxcar, loading platform, or internal plant area where large numbers of bags are stored or handled. For many substances, a broken bag doesn't present a serious hazard to workers. For some, however, those which are highly toxic, it may. When such bags are transported on fast-moving fork-lift trucks, it is likely that some powder released will be dispersed into the air. When they leave a ''pile'' of substance on the floor, drafts from open loading doors and airflows caused by mobile machinery may do the same.

Uncles[142] writes that some breakage and product loss must be expected in bag shipments, that the odd rip or tear or puncture or snag is inevitable, and a damage rate of 0.5% (1 out of 200) is not considered bad shipping experience. He notes that it may be good economics to attempt to reduce this figure by adding to the container cost, but that it frequently turns out that the cause of the damage would have the same effect regardless of the number of plies or weight of bag used. Also noted by this author is that the Department of Transportation limits the types of containers which may be used for hazardous materials to reduce the probability of unintentional release.

It is most certain that the Department of Transportation has expended considerable effort in ensuring that interstate transport of hazardous commodities does not take place using containers which can easily rupture and cause an immediate danger to the public. However, it is also certain that many highly toxic substances are transported intrastate, or are quite properly transported interstate in paper or plastic bags, probably the most fragile of the types of containers in which such chemicals are handled in moderate quantities.

There is some cause to consider whether NIOSH should review DOT regulations as to how they affect the probability of exposures occuring in the workplace. Such a study would be concerned with determining the physical and toxicological characteristics of toxic substances which might lead to hazards in the workplace, would compare these characteristics to DOT classifications for hazardous commodities, and would suggest what changes, if any, might be suggested to DOT. Since suggested changes would need to be fully justified and well-documented, such a program might require 2 man-years of effort over a time span of 1 year.

Emptying of Bags and Drums

For a number of the industries investigated in this report, there has been quoted at least one authority who flatly states or implies that the dumping of a fine powder into a tank or a hopper, or the scooping of an amount of a powder into some type of receptable, generates dust, sometimes much of it, into the breathing zones of workers. In the following, we therefore review and discuss various controls recommended for such operations in the literature.

Literature Review

Owen[143] reports on how fine, toxic dust in the ''difficult-top-capture'' 0.10 to 10 micron range was controlled. ''At two of the 200-gal (paint) mixing tanks, when 50-lb bags of pigment powder were added, bursts of dust resulted from dumping the bags. This problem was solved by designing the collection hoods for these tanks as 'flip tops', which raise to permit adding the sacks of pigment, and then close until the dust is taken into the (exhaust) system.''

Stern and Horowitz[40] graphically describe the hazards of these operations. Not only do they report that the dumping produces a ''puff'' of dust, but also, that excessive dust is generated when bags are shaken or struck against the side of a mixer or mill or an ''empty'' bag is tossed aside. They recommend covers which are hinged for access for open top mixers and specially designed hoods at mill loading spouts. To allow bag or drum dumping into the mixers, they suggest that a hole be cut in the front of the cover and that vertical side and rear sheets, extending up at least a foot above the opening, be installed around the opening. A downward air velocity of 200 fpm is recommended through the opening from an exhaust connection at the mixer cover. To avoid the possibility of sucking out any raw materials from the mixer, ''a bug-a-boo in the minds of many executives'', a long converging connection upward from the rear cover is suggested. A lateral draft hood over loading openings is recommended as an alternate to the above. The exhaust volume recommended is 1000-1500 cfm.

To avoid scooping of small amounts of highly toxic substances, Stern and Horowitz suggest that they either be weighted, mixed and dispersed in a vehicle in the laboratory (under a hood if necessary) or that the containers of the substances and the scale be enclosed in an open front booth.

Goss and Ross[144] investigated the hazards of operations involving the handling of lead compounds. Finding that the health hazard was significant, they instituted new procedures in the plant they were studying, procedures which were successful. To summarize, they required that all transfer of dusty materials be restricted to a single exhausted enclosure. Here, dusty materials were either processed into non-dusty pastes by mixing with a liquid or, when this was not practical, the materials were weighed into plastic bags which were hermetically sealed before being brought into the factory proper. Since they were in a plant processing polyvinyl chloride resin, the plastic bags were just thrown by the operator into mixers or blenders, where eventually the bag dissolved in the mix.

Hemeon reports that dust is generated from two sources when a scoop or shovel is used; at the point the scoop or shovel is filled, and at the point where it is discharged into a receptable. For containers and receptables not permanently placed at one location, he recommends the use of flexible metal hoses which can be conveniently placed over each container. For charging of open mixers, he illustrates a slot type exhaust hood along one top edge with an exhaust volume selected on the basis of 75 to 100 cfm per sq ft of top area. He notes that a tightly fitting cover can obviate the need for any exhaust during mixing.

The ACGIH manual addresses manual loading of hoppers. The design shown is essentially an open front booth with a rear exhaust connection. The exhaust volume recommended is 150 cfm per sq ft of face area.

Steel Mill Ventilation recommends a capture or control velocity of 150-200 fpm for openings in bin and hopper enclosures and notes that these velocities are based on general averages in material handling systems. ANSI simply recommends a 150 fpm control velocity through manholes or inspection openings of closed tanks.

An advertisement in Iron Age[145] illustrates how dust concentrations can be controlled or reduced in many operations by using substances in pellet form. Since phenolic dust was a problem for molders, General Electrics Plastic Department developed a new barrel replacement for molding machines which could handle the substances in this form. They note that "there's no need for separate molding areas or special maintenance and ventilation." Whether they realize it or not, they make a pun in stating their "marketplace looks almost exhaustless."

In one situation, one of our staff suggested the use of a flexible, tubular screw conveyor to transfer the contents of drums of a very fine powder into a tank. One end of the tube was to be placed through a hole in a typical drum cover. The other end was to be placed through an opening in the leaky, but essentially closed top of the tank and down to a point just above the liquid surface. This solution had many advantages.

The workers liked the idea because they would not have to lift the heavy drums up to dump them and were concerned about the excessive dust usually generated. Management expressed interest because clean-up of the fine powder was difficult but necessary on a frequent basis. Furthermore, the conveyor was relatively inexpensive and had a flowrate which would not slow production when all factors were given consideration. Unfortunately, it was never determined whether the idea was implemented. It can only be assumed that it would be useful since the drum would not have to be violently agitated, a tight drum cover on a drum from which material was being drawn would cause an induced airflow into the drum through any openings (e.g. the opening the tube passed through), the tube was obviously sealed to its other end, the flowrate was slower and more uniform, and the drop height of the substance from the discharge end would be greatly reduced.

The extreme toxicity of asbestos fibers has required that measures be developed for the opening and emptying of bags in a manner which does not result in excessive exposures. Goldfield and Brandt illustrate an enclosed and ventilated bag opening station which even has provision for pushing the empty bag into another clean bag for subsequent disposal. The discharged asbestos falls into a hopper which is connected to a screw feed, chute, belt conveyor, etc. Hills[146] describes a similar station and includes a picture of one in operation.

Conclusions and Recommendations

A small-scale effort to more fully review these material handling procedures and the controls available to industries which utilize packaged solids might be warranted. The researchers would survey such industries as the paint, plastics and rubber industry to observe the controls commonly utilized, to catalog the various methods found to be useful, and to produce a report which describes the work practices and engineering controls available for use (along with their individual advantages and disadvantages). If the research recommended in the "conveyor belt transfer" part of this section were conducted, the relationships developed for entrained air, etc. might be applied to these operations to improve upon the determination of proper exhaust volumes where ventilation is to be used as the control. A 6 man-months effort over a time span of 5-7 months would most likely be adequate.

4.8.5. Other Operations

Pouring Stations For Liquids

Wherever "volatile" substances are poured from a spout into a container, some release of contaminants can be expected. In the foundry, fumes are released when molten metal is poured into ladles or molds. In the paint and other coatings industries, solvents may be released when the contents of mills or mixers are poured into portable change cans. While ladles and change cans can be covered while being moved through a workplace, they must of necessity be open at the points they are filled or discharged.

Pouring stations for molten metals are a subject covered by the ACGIH and in the American Foundrymen's Society manual of "Foundry Environmental Control". The hood designs and recommended capture or control velocities of about 200 fpm in these sources appear to be fully adequate to capture any fumes generated. The various designs given for side-draft hoods, enclosures, and open hoods can also be expected to be adequate for controlling vapors evolved from other liquids being poured. Thus, while a study to develop a method to calculate accurate required exhaust volumes based on the projected area of the stream, its fall height, volatility, toxicity, etc. might be somewhat useful, the benefits of such research does not appear to warrant expenditures of time or funds.

4.8.6. Overall Discussion, Conclusions and Recommendations—We have identified and described the engineering controls commonly utilized within industry to reduce exposures to airborne contaminants generated during material handling operations and processes. The overall conclusion arrived at from this review is that, while numerous recommended ventilatory controls exist in the literature, many of these are contradictory, are not based on proven scientific principles, and may easily lead to inadequate control of operations. It is not to be implied that these guidelines are without merit. In the hands of knowledgeable practitioners, they are highly useful tools and may be modified as necessary to fit a particular situation. There is concern, however, that when these guidelines are rigorously applied by the uninitiated, they could unknowingly result in ineffective control.

There appears to be considerable merit in the suggestions to investigate, identify and confirm those principles and factors which influence operations involving falling toxic substances. Such programs would firmly base calculation procedures upon proven principles and would give them a much higher probability of resulting in adequate control. For the expert in the field, they would also be useful. When faced with a novel situation, estimates of what specific controls are necessary would tend to be more accurate. Control systems could be designed, constructed and installed simultaneously with plants themselves, and the need for trial-and-error "patchwork" modifications would be reduced or eliminated. A priori cost-benefit analyses could more accurately be performed to compare different methods known to be satisfactory.

For both bag filling and container emptying operations, specific programs were recommended to survey the state-of-the-art of controls and report on them. Given the number of literature sources which illustrate the hazards of these operations, the amount of attention they have received for controls seems to be insufficient. Indeed, there was rarely an author reviewing the health hazards of an industry which did not state that the bag filling and mixing and compounding areas were those of highest contaminant concentration.

The review of DOT shipping regulations recommended is of somewhat more uncertain merit. It is presented as an idea to be further discussed and investigated before action is taken. The idea essentially evolved from observations in two plants which handled large numbers of bags; one which filled, stored and shipped them, and the other which received, stored and used their contents. In both cases, spillage from broken or leaking bags was troublesome. Since bag shipments of these materials were industry practice and completely proper under current DOT regulations, it was rather impractical to suggest a change in container type to either management.

A review of the various descriptions and discussions above indicates that there are numerous ways to control each type of operation and it is evident that each has its own particular advantages and disadvantages. Popular sources of recommended control measures are usually very brief and described but one way to control an exposure. Thus, as a final recommendation for this subject, it is suggested that NIOSH consider the publication of a manual concerning the uses of engineering controls for materials handling operations. A manual of this sort could serve a useful function. It would describe the hazards of various operations, the types of equipment typically used, and the various control measures which are applicable. For each of the controls, for which specific design or other criteria would be given, it could discuss advantages and disadvantages, degree of control provided, operating costs involved, etc. Periodic updates of the manual would keep the publication "state-of-the-art". Its initial preparation would probably take 2 man-years effort over a time span of 1 year.

4.9 Non-Spray Application of Volatile Materials

Introduction—Whenever a substance containing volatile constituents is applied to a surface in an industrial environment, there is obviously potential for any vapors evolved to enter the breathing zones of workers. If the volatiles evaporate at a sufficient rate and/or the particular operation is such that workers must remain in the immediate vicinity of the "wet" coating, these vapors may result in excessive exposures.

The specific types of industrial operations and situations in which such exposures can occur are many and varied in nature. Off-hand, one might consider the plight of painters working in poorly ventilated and/or enclosed spaces, workers at benches manually applying adhesives to various materials, and even workers who might enter a room painted or coated with some substance sometime previously, a substance which is still evolving volatile constituents at some finite rate. In this latter case, it does not really matter whether the coating was applied by spray or other technique.

In practice, according to Riley,[148] the nature of the hazard and the greatest likely exposure may be determined in three ways. The first of these involves observation of an actual application in the field or in a test room. The second involves an assumption of an application rate and an estimation of the maximum concentration in air from the evaporation of all volatile materials. Finally, one may devise a scale model of the process to simulate severe plant conditions.

Of the first technique above, Riley points out that it presents some economic and logistic problems and develops information which lacks definition, since the conditions are hard to control or define. The second is noted as being relatively simple but as providing an unrealistic evaluation of the hazard. The third approach is noted as being open to some criticism but having the advantage of largely avoiding the limitations of the others.

The various approaches outlined by Riley appear to be most applicable for determining the rate of general ventilation necessary in an enclosed area to maintain average concentrations at or below exposure limits. However, where workers in some sort of production line continuously apply volatile coatings, localized concentrations of contaminants may greatly exceed average concentrations. In the following, therefore, various methods are reviewed for specifying not only general ventilation criteria but also local exhaust ones.

Literature Review—Riley reports upon a laboratory model designed to simulate a small, somewhat poorly ventilated room and to develop quantitative information concerning the rate of evaporation and the probable exposure resulting when the ceiling and four walls are painted rapidly with a coating under investigation. Use of the model is claimed to allow determination of whether the coating contains ingredients likely to create a health hazard from a proposed application, and of the nature of the hazard. It also is said to be capable of defining the greatest exposure likely to be encountered in actual use and of allowing comparison with other products or materials available.

The experimental apparatus essentially consisted of a 1-foot cube with a slot inlet and a sampling port which exhausted the enclosure at a fixed rate of 0.1 cfm. To approximate the conditions to be expected when a 10 x 10 x 10 feet room, ventilated at a rate of 100 cfm, was rapidly coated on 5 of its 6 sides, one-half of one side of the model cube was coated with the test material before the test commenced. Results of the test procedure included concentration in the enclosure versus time data for the various components in the coating material.

Riley acknowledges that, while the basic principles of his approach can be applied to many industrial processes, his model was designed for a specific purpose and, as constructed, provides information applicable only to his specific application. He notes that "care should be used in extrapolating from these conditions to different temperatures and different ventilation rates and patterns."

In considering the application of his experimental data to the estimation of the probable exposure, Riley comes to the conclusion that the rate of release of volatile components from the coating would be a more useful value than the concentration in the box. He notes that for plant conditions significantly different from the test model, these data could be used to compute probable concentrations. Using relationships between the various data available, he therefore presents a method for determining the specific evaporation rate (weight per unit area per unit time) from the coating from the available data.

Boyle and Novak[114] also address the subject of predicting ventilation requirements for coating materials. Though they are primarily concerned with determining the evolution rate of substances evolved in curing ovens, the test method they describe also results in curves of specific evolution rate versus time at some given temperature and air humidity.

The ACGIH Industrial Ventilation Manual[11] recommends the second type of approach discussed by Riley, i.e. the assumption of an application rate, an estimation of the maximum concentration in air from the evaporation of all volatile materials, and the calculation of the necessary amount of dilution air required from dilution formulae presented. It is noted that factual data are needed on the rate of vapor generation or on the rate of liquid evaporation, and that such data "can be obtained from the plant if they keep any type of adequate records on material consumption."

The ACGIH lists four limiting factors for the use of dilution ventilation. These are: (1) the quantity of contaminant generated must be not too great or air volume necessary for dilution will be impractical; (2) workers must be far enough away from contaminant evolution, or evolution of contaminant must be in sufficiently low concentrations so that workers will not have an exposure in excess of the established TLV value; (3) the toxicity of the contaminant must be low; and (4) the evolution of contaminants must be reasonably uniform. To ensure adequate control, a multi-purpose safety factor K is described for application to the calculated air volume. It is said that K should vary from 3 to 10 depending upon the toxicity of the material, the evolution rate of contaminant and the effectiveness of the ventilation. No attempt is made, not even a qualitative one, to describe the conditions under which various K values are appropriate.

Hemeon[113] contains a detailed chapter on the application of general ventilation. Equations are derived which allow calculations of the rate of concentration build-up when contaminant is injected into a ventilated space and the rate of decrease of concentration with time when some fixed rate of ventilation is applied to a space initially charged with contaminant. Addressing the subject of general ventilation for industrial solvents, Hemeon gives an equation for estimating the amount of dilution air necessary which is quite similar to the one given by the ACGIH. Instead of a K factor, however, he uses a contaminant concentration somewhat less than the TLV.

Hemeon explicitly states that it is necessary to explore the question of local concentrations caused by work requiring the worker to maintain a position close to a source of solvent vapor. It is noted that breathing zone samples during normal working operations would determine whether concentrations were excessive or not, "but in the planning of new operations it would be helpful to have a method for estimating the need for ventilation measures supplementary to the general ventilation." He then presents various calculation methods which allow estimation of breathing zone concentrations as a function of the nature of the vapor source, the working distance, and the rate of contaminant evolution.

To provide local control of operations which expose workers to excessive contaminant concentrations, Hemeon discusses the design and location of booths, slot hoods, and canopy hoods. Though use of a canopy hood in a configuration which causes vapors to be drawn past the breathing level is generally reported to be an undesirable control method, Hemeon demonstrates that if the exhaust volume is specified correctly, the breathing zone concentrations can be sufficiently diluted due to entrained fresh air. For operations involving the manual application of coatings to walls or floor, he illustrates the use of portable exhaust plenum chambers with slots. Both of these latter control measures are designed to collect vapors from their source, thereby reducing concentrations in breathing zones, and then to exhaust the vapors, together with entrained air, back into the work place where general ventilation can remove them.

ANSI,[118] in demonstrating the proper location for manufacturing hoods, generally illustrates a "cementing table" having lateral exhaust slots along two sides. Detailed design data for such a table are not presented however.

Kingsley[21] describes the hazards involved in manual "wet layup" operations involving uncured epoxy resins. To reduce the incidence of sensitization among workers, he designed an area for handling these substances which resembled a large water-wash booth. The area was 15 feet deep from the water tank to the outer edge of the hood. Total exhaust volume was 24,000 cfm. For some work, a downdraft table with a rotating top was used. A feature of the unit was that air would be drawn across the work when a panel was removed from the front of a plenum to one side of the table. Baffles ensured that the down-draft flow was not stopped.

Discussion and Conclusions—This entire subject area is a complicated one entailing numerous factors to be given consideration, especially when there is a desire to determine the controls necessary in the planning stage of new operations. From the above, it appears that Hemeon approaches the overall subject most completely. Not only does he provide the means for determining appropriate general ventilation rates, but he is also concerned with the means for dispersing any localized areas of high concentration.

Regardless of the specific design methods given in the literature, however, one of the basic parameters which appears to require better definition is the evaporation rate of volatile constituents out of an applied coating. While estimates based upon the total volatiles within the coating material lead to conservative estimates, there does appear to be a need for developing a more precise method, one which also provides data useful for other purposes, i.e. the comparison of the hazards of different coatings when applied under similar circumstances.

Overall, it appears that three aspects of the problem might be given closer attention. Such attention would result in the following:

1. A method of comparing the exposure hazards of various coatings for allowing selection of that coating which presents the least overall hazard,

2. A method of determining how long it will take under given environmental conditions for average concentrations of volatile substance evolved from coating materials to reach safe concentrations in freshly coated rooms, and

3. Designs for local exhaust ventilation systems for manual coating operations (for reducing localized "high" concentrations).

The first item provides a means for applying the engineering control of substitution. Its use would not only help reduce exposures to personnel directly involved in coating operations but also to those who may be exposed to substances being evolved long after the actual coating operation has been completed. Houwink and Salomen[75] report that the smell of formaldehyde from urea formaldehyde adhesives is not only objectionable during the application of the glue "but may persist in the glued material . . . for months or years." The papers of Riley and Boyle and Novak are both directly concerned with the rate of evolution of isocyanates from coating materials. The rate of evolution of these substances need not be appreciable for hazardous conditions to arise.

A priori estimation of the length of time necessary for average contaminant concentrations to "decay" to acceptable levels under given environmental conditions can prevent returning employees from being exposed to short-term, high concentrations. A sufficiently accurate technique would allow modification of environmental conditions as necessary to return the contaminated area to an acceptable state within the desired time interval (e.g. by increasing ventilation capacity or temperature).

Hemeon describes the application of various hood configurations for collecting and/or dispersing localized areas of high vapor concentration. The design of these devices involves the application of basic design principles once the evaporation rate and source area are somewhat defined. Nevertheless, some of the designs illustrated are quite different from those usually found in the popular literature. There appears to be merit in the conclusion, therefore, that some further efforts directed towards developing generalized designs for the sort of operation being considered would be desirable.

Recommendations—Riley, Boyle, and Novak have laid the groundwork for the development of a test method which allows the determination of which of a number of coatings available for use is the least hazardous for workers applying it to large areas. The nature of the test procedures also allows their use to determine the "safe entry" time after the coating has been applied. It would appear to be advantageous for NIOSH to consider a program to continue efforts for refinement of such a method. Such a program would develop a testing apparatus which could be easily duplicated in a laboratory. It would then proceed to refine a calculation method for estimating necessary general ventilation rates, or conversely, for estimating the "safe entry" time for a fixed ventilation rate, and to confirm the validity of the method with actual field data. An 8 man-month effort over a time span of 6 months would most likely be adequate for such a program.

Riley reported that, at various times, the use of urethane formulations have been discontinued (in the company he works for) when it was not practical to provide adequate control. Use of the data generated by his method has, however, allowed safe resumption of the use of materials which release low concentrations of diisocyanate. Of interest is his remark that this company would not accept for use any material which generated concentrations greater than one half the current threshold limit value for any constituent.

If a method is developed which allows accurate estimation of contaminant concentrations, NIOSH may sometime in the future have the basis for the development of a regulation which limits the use of certain coating formulations in certain interior, and even exterior, applications. If there is any sort of operation for which industry is going to claim that only respirator use is feasible for control, it can be envisioned as one which involves "one-shot" applications of large amounts of surface coatings.

Whether there is any real need to develop a set of specific ventilation system design criteria for "cementing tables," wall coating operations, and the like is open to question. However, to demonstrate the feasibility of controlling such operations, notably wall and floor coating ones, by combinations of local and general exhaust ventilation, some worthwhile research efforts may be directed towards this subject area. A small effort to develop design criteria for ventilating bench surfaces, for designing canopy hoods which entrain sufficient clean air volumes to reduce breathing zone concentrations, and for dissipating the vapors generated in wall or floor operations could also probably be accomplished with a 8 man-month effort over a time span of 6 months.

4.10 Open-Surface Tank Use

Process Description—Open-surface tanks are utilized by industry for numerous purposes. Among their applications can be included the common operations of degreasing, electroplating, metal stripping, fur and leather finishing, dyeing, and pickling.

An open-surface tank operation is defined as any operation involving the immersion of materials in liquids, or in the vapors of liquids, which are contained in pots, tanks, vats, or similar containers. Excluded from consideration in this definition, however, are certain similar operations such as surface-coating operations and operations involving molten metals for which different engineering control requirements exist.

A large number of toxic substances are utilized in open-surface tank operations. Degreasing is usually done with trichloroethylene, and this solvent is the principal air contaminant in metal degreasing operations. A variety of other solvents may be employed in open surface tanks used for the cleaning and finishing of cloth and other products. The finishing of animal hides may involve open surface tanks containing depilatory agents whereas electroplating and metal pickling and stripping may involve tanks containing acids. The contaminant released from an open-surface tank may be a vapor (as is typical around degreasers) or a mist (as found around pickling tanks).

Review of Controls Employed—Ventilation is commonly employed to control emissions from open-surface tanks and a standard prescribing specific ventilatory controls can be found in 29 CFR 1910.94.[121] A limited amount of research has been conducted to assess the adequacy of the design methods presented in the standard and to assess the effectiveness of other control measures for application to open-surface tanks.

The most recent work on engineering controls for open-surface tanks is a program conducted for NIOSH by Battelle Columbus Laboratories.[149] This program involved a series of experiments with an open-surface tank mock-up to assess the adequacy of ventilation recommendations contained in two design guides,[11,150] to quantify contaminant generation rates, and to examine the effectiveness of floating plastic balls in reducing the need for ventilation. A number of tests were also conducted on the mock-up with a push-pull ventilation system to review design criteria for this type of system. The report for the program concludes that the use of existing design guides frequently results in the overdesign of ventilation systems and presents an alternative method of calculating ventilation requirements. The method presented has shortcomings in that it is based on laboratory data collected with a simplified and idealized experimental mock-up and requires data on evaporation rate or gassing rate that is not currently available. Thus, Battelle's work must be considered as a limited first step toward the understanding of open-surface tank contaminant control. While a significant body of data was collected, the laboratory mock-up was limited in size and had a single contaminant in a given size room with controlled general ventilation and drafts. Furthermore, the limited resources allocated for the study limited the evaluation of mist generation phenomena.

Esman and Clearwater[151] in work sponsored by the US Army have shown that the use of a cover on vapor degreasers can result in a significant reduction in solvent losses and in airborne solvent vapor concentrations, and have suggested a possible cover design. The effectiveness of a surface-active agent in reducing chromic acid mist generation from a chromium plating tank was investigated by Hama, Frederick, Millage and Brown.[152] Their work demonstrates that surface-active agents can reduce air contaminants caused by gassing in actual industrial settings and presents the characteristics required of a surface-active agent for use in chromium plating. There is nothing in the literature, however, suggesting a generalized procedure for selecting a surface-active agent for other open-surface tank operations.

Areas Needing Further Research—The most significant weakness of current standards and design guides on open-surface tanks is the absence of a general guideline on the evaluation of contaminant generation rates. Evaporation rates for vapors are based on work by Doolittle[154] that cannot be readily applied to substances he did not study. Gassing rates are classified as high, medium, low or nil in the current OSHA standard with no guidance provided on the definitions of these terms. It is therefore recommended that a program be undertaken to evaluate contaminant generation rates both analytically and experimentally. The objective of this program should be the development and validation of a correlation of the generation rate with properties characterizing the contaminant and the operation. The correlation developed would be confirmed through an evaluation of actual open-surface tank operations and then incorporated into design guides.

A review article by Skinner[153] points out that the most important source of air contaminant in many degreasing operations may be "drag-out" of solvent when the article being degreased is removed from the solvent bath. It suggests that ventilation rates determined from evaporation rate estimates may be inadequate for such operations when the effects of "drag out" are not accounted for. Battelle's work included an evaluation of the cycling of a "parts barrel" on concentrations around the open-surface tank mock-up and confirmed that the operating temperature and cycling time can significantly influence the determination of adequate ventilation rates.

It is readily conceivable that the effect of "drag out" is determined primarily by the speed with which an item is removed from a liquid bath, the projected area of the item being moved, and its wetted surface area. It is also understandable that a combination of work practice and engineering controls are required to ensure effective contaminant control. Since sufficient information does not exist to provide guidance on how controls should be instituted to ensure their effectiveness where "drag out" effects can be significant, it is recommended that a study be performed to develop appropriate guidelines. The study should review the data available in Battelle's report and any other available data on the subject, and, by evaluating typical industrial open-surface tank operations, attempt to quantify the effects of "drag out" and the controls, both work practice and engineering, necessary to overcome them.

Ancillary control techniques such as surface-active agents, float plastic balls, and tank covers have been shown to be effective in reducing the generation of air contaminants from open-surface tanks. The extent to which these techniques can be effective, however, has never been determined in a generalized quantitative manner that would permit the reduction of ventilation requirements where ancillary controls are used. Since ancillary techniques are much less expensive and less energy intensive than ventilation, there would be advantages to the development of engineering control criteria that recognized their use. A program is therefore also recommended to identify criteria for the effective use of ancillary controls on open surface tanks. The objective of this program would be the development of guidelines for the use of tank covers, surface-active agents, and floating solids as means of reducing the generation rate of air contaminants. Specification of ventilation requirements appropriate when these controls are used would follow. The program should involve analytical prediction followed by validation on actual open-surface tank operations.

Summary of Program Recommendations—Three programs have been recommended for the development of engineering control criteria for open-surface tank operations. Suggested durations and levels of effort for these programs are listed below.

Program	Duration	Effort
Evaporation and Gassing rates	9 months	15-18 man-month
Drag-out	6 months	9 man-months
Ancillary Controls	9 months	1 man-year

4.11 Spray-Finishing Operations

Introduction—Spray-finishing operations involve methods by which organic or inorganic materials in mist or droplet form are projected and thereby deposited onto surfaces to be coated, treated, or solvent cleaned. As an operation category, it is not commonly considered to include metal spraying or metallizing operations, or spray washing and degreasing as conducted in self-contained washing and degreasing machines or systems.

We will initially consider engineering control of such methods for the spray application of paints, and then, will attempt to assess the adequacy of these and similar controls for spray application of other types of substances. This is done because paint spraying operations have received considerably more attention in the literature than similar operations involving other materials.

Control Measures Used and Available—Control of paint spray operations conducted indoors is usually accomplished by the use of ventilated spray booths. The ACGIH Industrial Ventilation Manual[11] contains specific design criteria for both small and large spray booths, for large booths in which vehicles are painted, and for ventilation of spray painting operations conducted within trailer interiors. Additionally, there is an ANSI standard[155] for the design, construction, and ventilation of spray-finishing operations. Both these sources were extensively reviewed and utilized in the development of recommended ventilation guidelines for spray-finishing operations.

As best as can be determined, ventilation has not been considered to be an adequate control measure where large structures such as bridges, ships, or storage tanks are painted. For such operations, and indeed, for many "one-shot" operations in confined spaces, the operator is usually required to wear an approved respirator. Though ventilation has been provided in such situations, its purpose has been primarily to ensure that explosive concentrations of solvents do not accumulate in enclosed spaces.

Reichenbach[156] agrees in this by stating that "in general it is seldom possible to furnish sufficient ventilation to enclosed areas of ships which are being spray-painted to eliminate the health hazard from solvents and diluents." He continues that "it is possible and mandatory to ventilate at a rate which maintains the area free from explosive hazards." Hama and Bonkowski[157] also agree and report "the painting of large structural steel parts in steel plants poses a rather difficult problem in providing practical ventilation. Spray painting booths that would accommodate all the parts would often have to be of tremendous physical size, and exhaust air volumes would need to be extremely large." They therefore note that a large part of spraying is done at night to reduce exposures to other workers and that spray paint operators are provided respirators.

The comments of the ILO[12] and other sources suggest that "substitution" is occasionally used as an engineering control when the substances involved are shown to be highly toxic.

Available Ventilation System Criteria and Data—The ACGIH manual, as previouely noted, gives specific design criteria for various types and sizes of spray booths. For large walk-in booths and air spray operations, it recommends an exhaust volume of 100 cfm/sq ft. of booth cross section except where the booth is very large and very deep. In this latter case, it suggests a rate of 75 cfm/sq ft. In general, it states that the operator may require an approved respirator. For large booths into which the operator need not enter, an exhaust volume of 100-150 cfm/sq ft is suggested. The exhaust volume for airless spray operations in walkin booths is given as 60 cfm/sq ft; for operator outside the booth, 60-100 cfm/sq ft.

Small spray booth exhaust system criteria are given separately for booths with a face area up to 4 sq ft and for booths with larger face areas. With air spray paint operations the volumes are respectively 125 and 100 cfm/sq ft.

Exhaust volumes of 100 and 60 cfm/sq ft of cross-section area are given for air and airless spray operations respectively in auto spray paint booths, and 50 cfm/sq ft is given for large drive-through booths and trailer interiors which are being coated. For this latter operation, it is noted that the operator must wear an air-supplied respirator.

In 1962, airless spray painting was a new method. At that time Brandt[158] wrote "studies were made independently by several industrial hygiene organizations and all arrived at the same conclusion, namely, that the atmospheric lead concentration in the vicinity of a painting operation with a lead-containing paint, when the pressure and the heat-with-pressure techniques were used, was not sufficiently different from the lead concentration when conventional spray painting was done to justify any change in the control measures. This was quite a blow to all concerned because the visual impression when observing the two types of spray painting was that much less mist was present when the new application techniques were used. This merely proves what all industrial health engineers know only too well, that is, conditions are not always what they appear to be."

Hama and Bonkowski investigated ventilation requirements for airless spray painting and presented their findings in a paper which is the source of recommendations in the ACGIH Industrial Ventilation Manual. Their conclusion, based upon comparative test data was "that contamination of air in a worker's breathing zone by paint mist and vapor appears considerably less with airless spray painting than with conventional compressed air spray painting." Consequently, they recommended air flow rates, for both general and local ventilation, of approximately 60 percent of conventional rates, and made the comment that "they (the rates) include some safety factor, and it is possible that they may safely be reduced further when more test data become available."

Though the aforementioned paper was primarily written with the purpose of demonstrating that airless spray operations require lower air flow rates for effective control than air spray operations, it contains data for the latter sort of operation which are of interest. For example, in a spray booth with a face velocity of 115 fpm, it is shown that the solvent vapor concentration ranged from 220 to 400 ppm. In a booth with a face velocity less than 75 fpm, the solvent vapor concentration ranged from 290-850 ppm. To be noted is that the table of recommended ventilation rates given clearly points out that "highly toxic materials such as high lead paints will require larger air volumes."

"Although comparisons are not completely valid," the authors report that, "where types of solvent were similar," airless spraying in a booth with a 120 fpm face velocity resulted in solvent vapor concentrations of 25-50 ppm. With the ventilation turned off, the concentrations were 25-230 ppm. Where such operations were conducted in an indoor area with only general ventilation, concentrations of 30 to 80 ppm were measured.

Also of interest in this paper is a table which gives the "minimum spray booth inflow air velocity to offset bounce-back" of spray from an object being coated by air paint spraying. The table, based upon a German study, lists an air velocity of 69 fpm for a distance between the spray gun and object of 59 inches, 98 fpm for 39 inches, 158 fpm for 32 inches, 296 fpm for 24 inches, and 640 fpm for 20 inches. These data clearly demonstrate the necessity for the spray gun operator to maintain the proper distance from the object being sprayed.

Describing the phenomena of "bounce-back," the authors note that "the paint particles of greater mass have sufficient kinetic energy to overcome air resistance and move toward and finally contact the object while smaller particles remain airborne and are deflected with the 'bounce'back' air stream. This results in paint fog of finely divided particles, to which the worker is exposed. Atomized droplets smaller than approximately 12 microns are generally considered of insufficient mass and momentum to overcome the drag forces of the air . . ." They attribute the reduced exposures from airless operations to the fact that "approximately 2 percent of the droplets in airless spray painting are smaller than 12 microns while approximately 20 percent of the droplets in conventional paint spraying are smaller . . ."

The current ventilation standard in 29 CFR 1910.94[121] is derived almost exclusively from the ANSI standard. For electrostatic and airless operations with a negligible crossdraft velocity, it gives design airflow velocities of 50 and 100 fpm for large and small booths respectively. With crossdrafts up to 50 fpm and air-operated guns, design velocities of 100 and 150 fpm are listed for large and small booths respectively. Crossdraft velocities up to 100 fpm where air-operated guns are used are noted as requiring design velocities of 150 fpm for large booths and 200 fpm for small booths for effective control.

The Michigan Department of Health[159] recommends booth face velocities of no significant difference from those suggested by the ACGIH and ANSI for air spray operations. For electrostatic paint coating, in states "there have been indications that the electrostatic process is so effective that there is little paint loss or overspray and no need for ventilation control. The latter supposition is decidely incorrect. Despite the effectiveness of the process, there are still overspray and solvent vapors to be captured and experience has indicated that a minimum control velocity of 100 feet per minute is not only necessary but will have no adverse effects on the operation."

Hemeon[113] recommends higher face velocities for spray painting booths for other sources, for some applications. Specifically, he recommends a face velocity of 125 fpm for booths with a face area greater than 50 square feet when the spray gun air pressure is less than 65 psi and one of 150 fpm when the pressure is 70 to 95 psi. With smaller booth face areas, he lists even higher velocities until at an area of 4 to 6 square feet he gives 200-225 fpm for spray guns with the lower air pressure and 275-300 fpm for the higher pressure guns. He notes that the values "are subject to considerable uncertainty where the objects being sprayed are of such a shape as to result in violent, direct rebound of the over-spray" and states that under certain circumstances "a spray gun with an extension arm which positions the worker backward . . . should be given consideration".

The Massachusetts Department of Labor and Industries[160] gives specific minimum velocity requirements for air spray operations as a function of the booth face area and the spray gun pressure. These performance criteria are almost exactly those recommended by Hemeon.

Baturin[103] reports that "the results of all investigations which have been carried out on the paint-spraying of small and medium objects in ventilated booths indicated that, provided the correct methods of working are observed and the extraction outlet is in the rear wall of the chamber, an air speed of 0.75 m/sec (147.65 fpm) in the working opening of the chamber is sufficient to protect the painter from harm due to the inhalation of paint or solvent vapours". He adds that "the same data show also that when spraying enamel primers and paints containing lead compounds, the cited rate is insufficient. The minimum suction rate for these cases should be taken as 1.1-1.3 m/sec (216.6-255.9 fpm)."

Discussion for Painting Operations

Air Spray-Painting Operations

The ACGIH manual and the ANSI standard are essentially similar in regard to suggested air velocities for air spray operations in which crossdrafts are not a significant factor. Where they can be significant, the ANSI standard suggests that the air velocity be increased by 50 fpm. From this we can conclude that ANSI recognizes the effect of crossdraft velocities in the selection of a design airflow as important whereas the ACGIH reasons that minimization of crossdrafts by baffling and other measures is a more logical approach than increasing ventilation system capacity.

A significant observation of Hemeon's recommendations is that he feels the spray gun air pressure is a significant factor. The air velocities he recommends for guns with 65 psi pressure are 20-40% lower than those for guns with 70 to 95 psi.

Of significance in Baturin's comments are the statements that 150 fpm is an adequate spray booth face velocity for most paint ingredients, but that 200-250 fpm is necessary for "enamel primers and paints containing lead compounds".

In summary, the above data allow us to make the following observations concerning air spray operations:

1. There is some difference in opinion in the literature as to the manner in which crossdraft effects can be negated. ANSI sees fit to provide users with design velocities which can overcome such effects. The ACGIH considers the elimination of such effects as being better practice.

2. There is some indication that the spray gun air pressure is a significant variable to be given consideration.

3. There are clear indications that the design criteria given in ANSI and the ACGIH manual for spray booths are inadequate for high toxicity substances.

4. It is evident that the distance of the spray gun from the object being treated is a critical factor to be given consideration.

5. The particle size distribution from the spray nozzle would appear to be a significant factor in determining the amount of paint which "bounces-back".

6. Ventilation is not considered solely adequate in many situations to maintain contaminant concentrations to levels at or below exposure limits.

7. The use of extension arms on spray guns may constitute an effective measure to move the operator backward from breathing zones of high contaminant concentration.

Airless Spray-Painting Operations

Recommendations and comments in the literature demonstrate that use of this method instead of air spray-finishing methods can in itself greatly reduce exposures for most substances. Brandt's comments, taken at face value, suggest that the benefits may not be so great where high toxicity substances are sprayed. Certain other advantages and disadvantages of the method are also worthy of mention.

Hama and Bonkowski state that since "the fluid delivery rate is higher than with conventional spraying, . . . airless spray is excellent for large areas and fast application". As a further advantage they report "because of the small rebound characteristic of airless spray painting, paint may be sprayed effectively into cavities and corners that cannot be painted with conventional spraying". Other advantages discussed include less dependence on the operator's skill and more uniform finishes. The only disadvantage discussed concerns the fact that not all materials can be conveyed by this method since "heavily pigmented, fiber filled, abrasive, or cohesive materials" may plug the spray gun's fine tip.

Spray-Finishing with Other Materials

Perry and Chilton[8] report that "flat, curved, and irregular surfaces such as tanks, vessels, boilers, and breechings are . . . insulated with . . . sprayed asbestos and/or mineral fiber-inorganic binder insulations". Benning[161] states that "the spray-in-place technique makes it possible to apply urethane foam in many types of commercial applications such as exterior tanks, ducts, roofs, and pipes of simple or complex design", and notes that epoxy, silicone, and pyranyl foams can also be sprayed-in-place. Sheinbaum[162] reports that "the plastering industry has developed and introduced on an increasing scale a pneumatic method of applying plasters. This method consists of pumping either plaster slurries or fluffs to the point of application and spraying them on the receiving surface". Thus, it is evident that a wide variety of substances other than paints can be sprayed and possibly released into the breathing zones of spray gun operators and other workers in their vicinity.

In 1963, the Michigan Department of Health[163] reported that asbestos is commonly applied to interior surfaces by pneumatic means and stated that "the principal hazard in spraying asbestos is excessive contamination of the air with aebestos dust particles". Tests conducted at sampling locations from 15 to 30 feet from the spraying operation revealed dust concentrations from 14 to 59 million particles per cubic foot, considerably above the maximum allowable concentration level of 5 million particles per cubic foot in effect at that time. Another study is described where similar high concentrations were found indoors.

By 1971, the hazards of asbestos were becoming highly appreciated, and in July of that year, the Philadelphia Board of Health prohibited the use of asbestos spraying in building construction. This was the first regulation of its kind in the USA. Gorson and Lieberman[164] give the account of a two-year investigation which led to this regulation. However, the current Federal regulation, 29 CFR 1910.93a, does not specifically prohibit the use of asbestos in this manner but requires the use of engineering controls where they are feasible and respiratory protective equipment where they are not.

Benning[161] discussed the various aspects of the formulation and spray-application of foams. Pertinent comments he makes for polyurethane are that "it is not practical for a man to spray at a rate of more than 5-6 lb/min unless he has a large open area and the appearance requirement is not critical . . ., urethane foam should be spray applied within an ambient temperature range of 65-100°F (since) loss of volatile components increases at higher temperatures, (and) . . . under high wind conditions care must be taken to prevent overspray and fumes from contaminating adjacent work areas". One of the common ingredients of urethan formulations is noted as being toluene diisocyanate (TDI).

The work of Peterson, Copeland, and Hoyle[165] is cited in Reference 163. These researchers indicated that hazardous concentrations of TDI can develop at varying distances downwind of the spray area depending upon "distance from the gun, spray gun elevation, formulation used, temperature and humidity of the air, air pressure of the spray gun, rate of adduct (reaction product of an excess of TDI with a polyol) used, configuration and size of the object being sprayed, wind velocity, and the use of tarpaulins to reduce air turbulence". From a review of their data, the Michigan Department of Health concluded "there is no question but that the immediate spray crew at the spray point is in a vulnerable position because of the presence of excessive concentrations of TDI . . ." They add that "the evolution of TDI vapor from the foam surface, once applied, was an insiginificant problem".

Among the other ingredients used to generate polyurethane foam, Benning includes water-soluble organosilicone surfactants, small amounts of catalysts such as 2,2,2-diazabicyclooctane and/or dilaurate or diacetate di-n-butyltin, and the prepolymer formed from reaction of an isocyanate and a polyol (resin). For low-density epoxy resin systems which can be sprayed in place, he includes a class of bisphenol A-epichlorohydrin polymers which forms the basis for a series of proprietary resins. Pyranyl foams, which may or may not still be marketed because of their expense, are noted as being generated from acrolein and methacrolein tetramers, strong Lewis acids, a surfactant mixture, and possibly other substances. Silicone foams are reported to be "essentially silicon-oxygen compounds which have the nature of organic polymers in which the silicone atoms are substituted for carbon atoms".

Benning indicates that epoxy and pyranyl foams can both be applied with modified urethane spray equipment. In discussing such equipment, he describes three types of spray guns. In the first, two reactant streams are pumped through the gun under pressure and ejected in separate streams. Mixing and atomization by air are accomplished simultaneously just exterior to the spray nozzle. The reactant streams are blended internally by an air-driven agitator in the second type of gun. The components are then pumped through the gun under pressure and are atomized as a single stream by air. The third type of gun uses airless atomization. Here, the reactant streams are internally mixed by passing them together through a labyrinth under high pressure, and atomization is accomplished on discharge by the pressure drop across the spray nozzle.

Sheinbaum notes that "epoxy resins may be used in paints". In this case, he reports "the hazards involved are due principally to the solvents and may be controlled in a manner identical to that used in spray painting with other materials". He makes this conclusion because of the large proportion of the paint which is made up of solvents. Where epoxy resins alone are used, he says "during polymerization various amines are given off which may cause severe dermatosis on contact with the skin or lung irritation if inhaled".

Spray-application of gypsum plaster coats are reported by Sheinbaum to utilize a "mixture of about 20% gypsum plaster, 65% sand, and 15% water with trace amount of additives". He adds that compressed air between 16 and 22 pounds pressure is used to force the slurry out of the nozzle and that "the distance between the spray gun and the surface being sprayed may vary from a minimum of six inches to a maximum of four feet". The operation is considered to be "extremely dirty because of the considerable quantities of underspray, rebound, and drippings. Air movement within the area affects the operation and *steps must be taken to close all exterior openings prior to spraying.*" Studies conducted by Sheinbaum to determine breathing zone dust counts in such operations revealed dust counts of 26-75 mppcf with a free silica concentration of 61%. Mean particle size in the dry state was 1.5 microns; 6.4 microns in the wet state.

Overall Discussion and Conclusions

For spraying of paints containing low or moderate toxicity ingredients, it is concluded that sufficient information exists for someone to design a spray booth or room which maintains exposures at or below permissible limits. Though the air velocities recommended in the literature are not universally consistent, the experience from which they are derived indicates that the most appropriate rate for any particular operation can be found in overall ranges proposed. Additionally, the use of airless spray methods instead of those using air atomization, where practical, appears to in itself be a useful engineering control. These conclusions do not in themselves, however, indicate that further research in this area is not worthwhile.

For spraying of paints with high toxicity ingredients, there does not appear to be any consensus of opinion regarding proper design velocities for spray booths. Indeed, the subject is hardly even mentioned, and thus, leads to the suspicion that an inexperienced person, using the most common recommended practices, may find results to be disappointing.

Airless spray techniques for paints, coupled with general ventilation, appear to be highly suitable for use with all types of paints for indoor applications which cannot practically be conducted in booths. Though specific data cannot be cited to prove the point, air spray techniques under similar circumstances must be considered inadequately controlled.

Virtually no information could be found which specifically demonstrates the availability or adequacy of engineering controls for spray-finishing operations utilizing substances other than paint. Though it can be assumed that indoor operations which are repeatedly performed in one location may be controlled with booths similar to those used for paint, it cannot be assumed that the same airflows recommended are appropriate. As for paint, where "one-shot" indoor and outdoor operations are conducted, only general mechanical or natural ventilation appears to be available or used. Engineering control design criteria for such operations must, therefore, be considered only partially available and/or adequate.

Recommendations—Based upon the information presented above, and our own familiarity with spray-finishing operations, the following recommendations are made for research activities.

Spray Booth Design Improvements

All exhaust ventilated spray booths require make-up air for proper operation. In some booths the air is simply drawn through the booth face, while in others, it is introduced at some intermediate point. With the assumption made that make-up air available is free of contaminants, there is cause to consider that the precise location in the booth at which air is introduced can have a significant effect upon breathing zone concentrations. It is therefore considered desirable to investigate the potential beneficence of various booth modifications involving the entrance location of make-up air and to provide "baseline" data for spray-finishing operations in general.

The "bounce-back" of coatings being sprayed from an object being treated is a significant cause of excessive exposures. There are clear indications that spray guns using airless atomization result in considerably less "bounce-back" than guns using air atomization techniques when substances of low or moderate toxicity are sprayed. For high toxicity substances, there appears to be a lack of data as to the air flow rates necessary to achieve adequate control with either type of gun. It is therefore additionally desirable to confirm and/or determine the relative advantages and disadvantages of the two techniques and to determine appropriate minimum air velocities for effective control of highly toxic substances.

"Extended arms" for spray guns have been suggested as being useful to force the operator out of breathing zones of excessively contaminated air. It is desirable to investigate the benefits of such devices.

Specifically, it is recommended that a study be performed in which researchers perform the following tasks:

a. Provide two spray booths; one typical of those in which the operator works inside, and one of smaller dimensions which do not allow entrance. These booths shall be designed in accordance with the NIOSH recommended ventilation guidelines for spray-finishing operations.

b. Provide the booths with an exhaust system capable of maintaining exhaust volumes of 0-200 cfm/sq ft of booth cross-section for the larger booth and of 0-300 cfm/sq for the smaller booth.

c. Provide examples of typically used air spray and airless spray gun systems for paints, examples of the various types of spray guns used for urethane and other foam applications, and examples of other such units which may be used in industry for manual spray-finishing operations.

d. Prepare a "paint" formulation consisting of a limited number of ingredients. The ingredients selected shall be typical of those used in "paints" and shall be of widely differing and known volatility.

e. Conduct, under controlled conditions of temperature, discharge time and rate, etc., a sufficient number of experiments to develop curves of breathing zone concentrations vs. exhaust volume for each of the spray guns. The special paint formulation shall be used for the paint spray guns; typical foam or other forumlations for the others as appropriate.

f. Conduct experiments to determine the effects of providing make-up air in various positions and in manners which can be expected to reduce breathing zone contaminant concentrations.

g. Investigate the feasibility and benefit of using "extension arms" or spray guns. Of possible usefulness might be the utilization of make-up air to form an air curtain in front of the operator while he uses a spray gun with an extended arm which penetrates the curtain.

h. Utilize the data obtained to develop specific ventilation system design criteria for coatings of varying physical, chemical, and toxicological characteristics.

A level of effort of about two man-years over a time span of 12 months is estimated as sufficient for such a program if the special "paint" formulation consists of ingredients whose concentration in air can simply be determined.

Engineering Controls for Exterior and "One-Shot" Interior Operations

It is difficult to conceive of engineering controls, other than those of substitution or use of airless spray guns, which can practically prevent operator exposures to excessive contaminant concentrations during exterior and "one-shot" interior operations. Research into substitution for the types of substances being considered must be considered not to be a feasible activity for NIOSH, and the use of airless spray techniques does not truly require further research if the first recommendation above is accepted. Personal protective and administrative controls must therefore continue to be relied upon.

Consolidation and Dissemination of Information

This evaluation of spray-finishing operation control technology has enumerated the various factors influencing worker exposures. Additionally, it has recommended research which may lead to a firm and accurate data base from which ventilation system performance criteria can be generated which is known to be applicable and adequate for a diverse set of circumstances. If the recommendation above is accepted, and there is confidence that the data generated is useful, it might then be considered advisable for NIOSH to produce a manual which reviews the hazards associated with these operations, the experience noted in the literature, the results of its own test program, and those "good practices" which can reduce worker health problems. Since no new data would need be generated, a time span of six months and a level of effort of six man-months would be expected to be adequate for this endeavor.

4.12 Textile Coating and Dyeing

Introduction—The equipment and procedures described and discussed in the following are normally considered representative of activities conducted by the textile industry. Nevertheless, many of them are also utilized for the treatment of paper, wood, plastic, and rubber products. Thus, although the following is written in the context of their use specifically in the textile industry, it is somewhat applicable to operations conducted in a number of others, particularly in regards to coating operations.

Dyeing operations are conducted to impart colors to fabrics for decorative purposes. Coatings are applied to textiles (and other flexible materials) to render them resistant against penetration by water and other liquids and solutions, to apply an adhesive of some sort before further processing, etc. Werber[166] states that "the field of chemical finishing of fabrics, in particular, has been blessed with a plethora of new chemicals, dyes, and techniques of application which have given us permanent press shirts and outerwear, washable wool, soil-repellent, water-repellent, and soil-release finishes, and a spectrum of bright colors in both men's and women's wear."

The plethora of new chemicals and dyes to which Werber refers, and the multitude of those whose use has continued from the past to the present, represent potential occupational health and industrial hygiene problems for the textile industry and others which have adopted its methods. In the following, we describe some of the basic processes, health hazards, and engineering controls associated with these operations.

Dyeing Processes[12,31,76,167-170]—The classes of dyes are many. Included are "basic, or cationic (dyes); acid and premetalized; chrome and mordant; direct and developed direct; sulfur, azoic, vat, disperse, and reactive" types.[31] Each of these classes is in turn comprised of a variety of chemical substances; the purpose of all of which is to produce a particular color in the fibers of a material by chemical reaction or strong physical binding forces. To promote the dyeing process, various substances other than dyes themselves may be found in dye formulations.

Simply stated, dyeing involves the introduction of a material into a vessel containing a dye solution. This may be accomplished either in a batch or continuous mode depending upon the demand for a given product. Introduction is achieved by means of rollers, racks containing skeins, or in loose mesh bags as appropriate. Depending upon the particular process being used, the operation may be conducted at various elevated pressures and temperatures or at atmospheric pressure and more moderate temperatures. The equipment used may be open or totally enclosed.

A considerable amount of chemical pretreatment of fabrics is necessary before the actual dyeing operation can take place. Wool must be scoured with a soap and soda ash solution, bleached with hydrogen peroxide or sulfur dioxide, passed through a sodium carbonate bath, and then washed thoroughly. Cotton is first desized in a diastase solution. Depending on the process used, it may then be subjected to sodium hydroxide, sodium carbonate, turkey red oil, hypochlorite solution, sodium bisulphite solution, and dilute hydrochloric or sulfuric acid. Even a synthetic such as nylon can require scouring, some form of setting treatment and, in some cases, bleaching.

Until recently, most dyeing operations were conducted with aqueous solutions of dyes. To reduce water pollution problems, however, solvents are being used to replace water. Howrey[169] describes some of the equipment and systems available for solvent dyeing, scouring and finishing. Many of the units are clearly stated as being totally enclosed and capable of solvent recovery.

A recent development in the dyeing of synthetic carpet yarns is the non-aqueous process of dyeing with ammonia invented by Arthur D. Little, Inc.[170] The only ingredients necessary for this process are the dye stuff and liquid ammonia. Available equipment can easily be converted to this system, and the process is inexpensive. Another benefit of the system is that the exhaust hoods which are vital to the system serve the dual purpose of protecting workers' health and recovering ammonia.

Coating Operations—The materials used to coat fabrics include cellulose nitrate, cellulose acetate, natural rubber, synthetic rubbers, polyvinyl chloride, polyvinyl chloride-acetate copolymers, polyvinyl chloride-vinylidene chloride copolymers, polyvinyl butyrol, drying oils, various varnishes, polyester resins, silicon resins, and numerous other substances. These may or may not be dissolved in organic solvents for ease of application.

For the purposes of this report, we shall discuss the two basic coating methods and the equipment involved in each. The first general method is "spread coating" and the second is "roll coating."[1]

Spread Coating[1]

Floating Knife Coaters

Viscous solutions with a high content of solids (i.e. cellulose nitrate and vinyl solutions) are applied to fabrics with floating knife coaters, as are other solutions such as organosols and plastisols. From a take-off roll, the fabric is fed over two support rolls between which a spreader knife is mounted. The coating material is applied in front of the knife and smoothly coats the fabric passing underneath the knife. The specific thickness of the coating depends upon the type of knife used, the angle at which it is set, and the tension on the fabric.

Rubber Spreaders

The rubber spreader is used to coat fabric with viscous rubber solutions as well as vinyl and other viscous solutions. Its mechanical configuration is generally similar to that of a floating knife coater with the exception that the knife is mounted directly above a roll made of rubber.

Blanket Coaters

Low-strength materials may be coated with shellac, vinyl, rubber, lacquer, or varnish on a blanket coater. The configuration of this coater is also similar to that of a floating knife coater with the exception that the two support rolls over which the fabric passes have a rubber sheet over them in an endless loop configuration. This sheet forms the "blanket" and supports the fabric passing underneath the knife.

Roll Coating[1]

Roll coaters are more efficient than spread coaters and are thus better suited to rapid, quantity production. Their output is limited only by the capacity of the dryer and the speed of coating solution feed. Generally speaking, all of these operate by the transfer of coating material to a roller, and then from a roller to the fabric.

Calenders are used to apply vinyl compounds, synthetic rubbers, and natural rubber. Reverse roll coaters coat varnishes, organosols, and lacquers. Roll kiss coaters are suitable for applying "hot-melt coatings," and air knife coaters are advantageous for coating vinyl and rubber latexes. Solvent solutions are not applied with air knife coaters since explosive solvent-air mixtures can easily be formed. Finally, dip coaters are particularly well suited to the application of oleoresinous varnishes. These coaters function by causing the fabric to actually "dip" into a bath of coating material.

Health Hazards Associated with Textile Dyeing and Coating

Dyeing

According to the ILO,[12] dyes do not pose a serious health hazard to workers under conditions of ordinary use in industry. Apparently the main risk is from the primary and intermediate materials, reagents, and solvents. Certain dyes are known to cause bladder cancer in humans, however, and other dyes are known to cause cancer in experimental animals (i.e. rhodamine B, magenta, 2-naphthylamine, and dianisidine).

The "fast salts," azoic dyestuffs, present special health problems because they can cause respiratory sensitization and asthma. These highly reactive compounds are not actually dyes themselves, but they react with chemicals in the fiber to form a dye on the fiber.

Many chemicals used in dyeing can cause skin irritation and dermatitis. Organic solvents and corrosive acids and alkalies are also capable of such effects.[171]

There is a potential for exposure to chlorine in many factories which use gaseous chlorine, bleaching powder, or hypochlorite solutions for bleaching. Chlorine, a powerful irritant, can cause skin irritation, eye irritation, and delayed pulmonary edema.

There are numerous chemicals used in the application of dyes, and depending upon the particular conditions (i.e. open systems vs. closed systems, spray applications vs. dipping operations, etc.), there are varying degrees of potential for worker exposure.

That open dye vats may be a source of airborne contaminants is evidenced by statements of Stevens.[172] He states that "odorous emissions from open dye vats can . . . create air pollution which can be difficult and expensive to control." Of interest is his remark that the variety of chemicals used in complex dyeing processes "makes evaluation of stack emissions difficult and air pollution control equipment not easy to select."

Coating

The tremendous diversity of chemicals, equipment, and processes for textile coating operations demonstrates that there is also considerable potential for these to present health hazards. Recently, a coated fabric plant had to cease the use of methyl butyl ketone when several employees displayed nerve illnesses believed to be caused by this compound.[173,174]

Many rubberized fabrics are made using the compound resorcinol formaldehyde. This chemical is put on by a weaver in a water system. Industrial weavers, however, have misgivings about the use of this chemical solvent because of explosion and fire hazards, toxic fume exhaust problems and other reasons.[171]

Controls—Concerning worker exposure to chlorine during bleaching operations, the ILO recommends that bleaching vats be constructed as closed vessels, "with vents that allow a minimum escape of chlorine . . . The valves and other controls of the tank in which the liquid chlorine is supplied to the dyeworks should be controlled by a competent operator since the possibilities of an uncontrolled leak could well be disastrous."[12]

Textile Technology, Inc., has developed solvent processing systems which use perchloroethylene and "are claimed to dye polyester and all other common fibers, as well as scour, bleach and finish on converted conventional equipment." In beam-and-package dyeing operations superheated steam is used to vaporize the perchloroethylene from the dyed fabrics. Pariser points out that "conversion of equipment for solvent systems does not eliminate its use for aqueous processing . . . Enclosure to contain fumes, mounting of a condenser and in-and-out piping for the perchloroethylene to and from the still is all that is involved."[176]

In the ammonia dyeing system discussed previously, canopy hoods, mounted over converted space dyeing equipment, serve the dual purpose of protecting the worker and recovering ammonia.

Discussion and Conclusions—This brief study of coating and dyeing operations provides an overview of an industry which utilizes a multitude of chemicals, equipment types, operating procedures and processes. It provides us with sufficient information to develop the following observations.

The first is that very little published information exists describing the specific exposures which may occur, certainly not enough to allow determination of all the individual processes which may require controls. Thus, it is impossible to determine whether controls ·are available and adequate for all. That they are available and used for some is, however, evident.

Another observation is that many of those operations conducted in the open involve the use of toxic substances at elevated temperatures or the use of volatile solvents. Stevens suggests that dye vats may pose a problem. That evaporated solvents from various coating apparatus might cause a health hazard can be safely assumed.

These observations alone are sufficient to lead to the conclusion that further study of these operations is warranted to more fully describe them, the substances used, the contaminants which might be evolved, and the controls, if any, which appear necessary.

Recommendation—It is recommended that NIOSH consider a survey of selected plants conducting dyeing and/or coating operations with the purpose of determining the need for, availability, and adequacy of engineering controls. The program should have three objectives; the classification of the various individual operations conducted; the characterization of harmful exposures; and the development of any needed engineering control design criteria. An effort of 18 man-months over a time span of 9 months is estimated as adequate for such a program.

4.13 Welding Operations

Introduction—The term "welding," as used by the American Welding Society,[177] includes a large number of operations by which materials - usually metals - are joined, cut, or surfaced. From the standpoint of ventilation requirements, the following specific operations are of concerrn:

Gas welding
Oxygen cutting and gouging
Torch brazing and soldering
Shielded metal-arc welding
Arc cutting and gouging
Gas shielded-arc welding
Plasma arc welding
Submerged arc welding
Flux cored arc welding
Thermal spraying

These operations are of concern because the potential exists with each for the generation of air contaminant concentrations in excess of the exposure limits. However, excessive air contaminant concentrations have not actually been demonstrated or reported with all of these operations.

Each of the above welding operations generally involves melting of a metal in the presence of a flux or a shielding gas by means of a flame or an electric arc. The operation may produce gases or fumes from the metal, the flux, metal surface coatings, or surface contaminants. Certain toxic gases such as ozone or nitrogen dioxide may also be formed by the flame or arc. In addition, certain of these operations—such as cutting, gouging, and spraying—release air contaminants in high velocity streams. As a result of these factors, a welder may be exposed to a wide variety of air contaminants. Many studies have been performed of the welder's environment, and a review of existing data on this subject is presented in Reference 178.

Ventilation is the principal method employed to control exposures of welders to air contaminants. Ventilation methods used include three conventional methods: general mechanical ventilation, free-standing, open exhaust hoods, and ventilated enclosures, as well as three methods developed primarily for welding operations: crossdraft tables, downdraft tables, and gun-mounted open hoods.

Existing Ventilation Crtieria—The most comprehensive set of guidelines for ventilation of welding operations is found in Reference 179 and has been incorporated without substantial changes into the General Industry Safety and Health Regulations.[180] These guidelines, which are formulated by the ANSI Standards Committee Z49, specify minimum ventilation requirements for general mechanical ventilation, free-standing open hoods, enclosures, and downdraft tables. Guidelines also are included for the control of certain toxic materials encountered in welding operations.

Another set of guidelines for ventilating welding operations is contained in the Industrial Ventilation Manual.[11] These guidelines cover the use of free-standing open hoods, crossdraft tables, and enclosures with certain welding operations.

A third set of guidelines is contained in a report of a NIOSH-sponsored research program on the control of welding fumes.[181] This report contains detailed performance data for a free-standing open hood and a crossdraft table used with arc welding operations.

These are the only publications found in this program which provide quantitative guidelines to ventilating welding operations. Of these three publications, only the third contains ventilation system performance data. The first two publications offer no supporting data for the ventilation guidelines presented.

Identification of Problem Areas—Problem areas in the ventilation of welding operations can be identified by a table defining combinations of specific welding operations and ventilation methods, as shown in Table 7. Entries are contained in the table where guidelines exist for specific operation-method combinations. The spaces in the table without entries represent problem areas for which ventilation method performance data and ventilation guidelines are not available or do not exist. However, not all of the open spaces in the table are of equal significance. For example, the use of enclosures is not common with welding operations except for thermal spraying which presents a unique problem in contaminant control.

The most obvious problem areas revealed by Table 7 are with the use of crossdraft tables and gun-mounted open hoods. Another problem area, which is less obvious, is the need to verify the ventilation guidelines for which no supporting data are available. These problem areas are discussed in the following section.

Discussion of Problem Areas

Performance of Crossdraft Tables

A crossdraft table is a horizontal work surface with a slot exhaust hood mounted on the edge opposite the worker's position. Varying degrees of shielding or flanging are used with the hood and table to increase the effectiveness of the hood in controlling contaminants generated by the operation performed on the table.

A detailed study of crossdraft table performance with gas shielded-arc welding is described in Reference 181. This particular welding operation was selected for the study because baseline tests of several different operations indicated that gas shielded-arc welding has a high potential for causing excessive exposures to welding fumes.

In the absence of experimental data on the performance of crossdraft tables with other welding operations, it is reasonable to use the guidelines developed for controlling fumes from gas shielded-arc welding. However, it is likely that these guidelines will require ventilation rates which are either too high or too low for other operations. Consequently, it is recommended that a research program be undertaken to evaluate the performance of crossdraft tables with welding operations other than gas shielded-arc welding. This program could be completed most economically by considering it as an extension of the investigation described in Reference 6. Using the same experimental procedures, other welding operations could be investigated and ventilation guidelines developed analogous to those developed for gas shielded-arc welding.

Performance of Gun-Mounted Exhaust Hoods

Gun-mounted exhaust hoods are small hoods attached directly to the welding guns used in gas shielded-arc welding and flux cored arc welding. The advantages of gun-mounted hoods are that they require relatively low ventilation rates since they are located close to the contaminant sources, and they interfere minimally with the welding process because of their small sizes. Disadvantages of gun-mounted hoods are that they tend to disrupt the performance of the shield gas, thereby requiring higher shield gas flows, and they add complexity and weight to the welding gun.

Data on the performance of gun-mounted hoods are limited in availability and consist essentially of manufacturer's promotional literature and the results of a single test series in Reference 181. An investigation of gun-mounted hood performance is necessary to determine whether the approach actually is effective in contaminant

TABLE 7—PROBLEM AREAS IN VENTILATION OF WELDING OPERATIONS

Operation	Gen. Mech. Ventilation	Open Hood	Crossdraft Table	Downdraft Table	Gun-Mounted Hood	Enclosure
Gas Welding	G(180)	G(180)		G(180)	NA	
Torch Brazing and Soldering	G(180)	G(11, 180)		G(180)	NA	
Shielded Metal Arc Welding	G(180)	GD(181)	G(11)	G(180)	NA	
Plasma Arc Welding	G(180)	G(180)		G(180)	NA	
Submerged Arc Welding	G(180)	G(180)		G(180)	NA	
Gas-Shielded Arc Welding	G(180)	G(180)	GD(181)	G(180)		
Flux Cored Arc Welding	G(180)	G(180)		G(180)		
Oxygen Cutting and Gouging	G(180)	NA	NA	G(180)	NA	
Thermal Spraying	NA	G(11)	NA	NA	NA	G(11)

Symbols: G = Guidelines only
          GD = Guidelines and supporting data

NA = Not applicable
Numbers in parentheses indicate references.

control and to determine design and operational guidelines for the use of the hoods. It is recommended that an investigation be initiated to accomplish these two purposes.

To examine the contaminant control effectiveness of gun-mounted hoods, it is suggested that an experimental investigation be conducted similar to that described in Reference 181 but broader in scope. Commercially available gun-hood combinations should be tested with a variety of work materials. Quality of weld and effectiveness of contaminant control should be measured as functions of shield gas and exhaust flow rates. The tests should include actual industrial production operations so that the influence of the welder's procedures can be observed.

The data derived through the first task may be sufficient to establish design and operational guidelines for gun-mounted hoods. Otherwise, additional tests will be necessary to determine the effects of gun and hood design and operational parameters. The test results should then be transformed to a set of guidelines for the use of gun-mounted hoods.

Verification of Existing Guidelines

Most of the ventilation guidelines indicated in Table 7 are of uncertain origin and are unsupported by experimental data. Consequently, the effectiveness of these guidelines in controlling exposures to welding fumes is entirely unknown to NIOSH and OSHA. Consequently, it would be prudent for these guidelines to be verified by a program of exposure measurements under actual production welding operations.

It is recommended that a program be undertaken to evaluate the validity of existing guidelines for ventilation of welding operations. The program should consist of field measurements of welder exposures under as wide a variety of conditions of welding operations as possible. The measurement program should include careful measurements of ventilation system parameters so that the measured exposures can be correlated with both welding operation and ventilation system parameters. The results of this program can be used to modify existing ventilation criteria as necessary to validate the criteria.

Summary of Program Recommendations—Three programs have been recommended for the development or validation of ventilation criteria for welding operations. Suggested durations and levels of effort for these programs are listed below.

Program	Duration	Effort
Crossdraft Table Performance	1 Year	1-2 man-years
Gun-Mounted Hood Performance	1 Year	1-2 man-years
Existing Guidelines Validation	1 Year	1-2 man-years

SECTION TWO

RECOMMENDATIONS

5. Summary and Priortization of Research Recommendations

5.1 Introduction

All research recommendations developed in this report represent problem areas where further study can provide benefits to the working population, but some recommendations can be expected to be more worthy for implementation than others. Therefore, there is definite advantage to their prioritization.

5.2 Approach

In the conceptualization of this program, it was assumed that any methodology to rank recommendations would somehow take into account numbers of employees exposed, the severity of the effects of exposures, and the effectiveness of all control measures available for use, including those other than engineering controls. This was based on the supposition that most problem areas identified would be associated with particular industries and would involve limited numbers of identifiable toxic substances. The a prior feeling was that those operations which are common to many industries, and which therefore involve large numbers and types of toxic substances, would be generally found to be adequately controllable with current engineering control techniques.

As can be discerned from the results of this study, this was not always found to be the case. Close study of the technology upon which popular guides for engineering control design are based resulted in the identification of significant inadequacies. The available resources for investigating operations requiring engineering controls were therefore mostly expended upon these common operation categories, categories which defy characterization by parameters enumerated above. It became obvious that any prioritization methodology utilized would need to be based upon other, yet valid, considerations.

In reviewing and closely inspecting the various recommendations made, and the reasons why they were proposed, it became evident they could be characterized by six distinct classifications. Comparison of these classifications indicated that these could logically be ordered to demonstrate the relative importance of the various recommendations.

Table 8 lists these classifications in their order of importance. The highest priority classification is 1, the lowest is 6. In the following, the reasoning behind their formulation is presented and discussed.

Table 8—Priority Classifications for Research Recommendations

Classification

1. Guidelines do not exist. Subject not previously given sufficient attention, but perceived to be important.

2. Guidelines have been shown to be contradictory, often capable of resulting in inadequate control, and not based upon firm scientific principles.

3. Guidelines exist but origin uncertain and adequacy in some situations unconfirmed by data in the public domain. Effective control can be assumed in most circumstances.

4. Numerous guidelines or methods of control available. Some may result in adequate control in some or all situations; some may not. Consolidation and dissemination of information appears warranted after determination and verification of adequacy of best ones.

5. Guidelines do not exist, but adequate methods of control available. Research would result in savings of materials or energy or would result in more convenient control measures.

6. Necessity of research not addressed by literature. Recommendation presented as an idea for further consideration. Findings may or may not be significant.

Classification 1 represents those research recommendations which would result in control technology not presently available, but for which a definite need can be envisioned and/or justified; i.e., these controls are unavailable, but needed.

Classification 2 represents recommendations with the purpose of examining guidelines commonly found in the literature which are contradictory, are not based upon firm scientific principles, and/or have been reported to result in inadequate control in various applications. The guidelines referred to are those generally based upon some researcher's personal experience. They are sometimes noted as being "general averages", but more often than not, are presented without detailed qualifying remarks on their usage; i.e., these controls are available, but *sometimes* inadequate. In comparison to the first classification, this one must obviously be given a lower priority.

The third category pertains to very specific guidelines which exist in the literature. While they may be somewhat contradictory in some aspects for some problem areas, they are generally quite consistent in their recommendations. Typically, there are no indications that their use results in inadequate control. Their major failing is, however, that they are of uncertain origin and based on data which is unavailable in the public domain; i.e., these controls are available, can be *presumed* to be adequate until proven otherwise, but should be studied to ensure that they are indeed adequate. Since they are available, and since they have not been shown to often lead to inadequate control, it is evident that this Classification deserves a lower priority than those preceeding it.

Classification 4 represents recommendations which suggest the consolidation and dissemination of control method guidelines to industry. These recommendations generally were derived from realizations that numerous methods of control exist for some operations, that some methods are better than others under some circumstances, and that no one source of information exists which broadly discusses the subjects of interest. Since the development of a manual or set of guidelines which satisfy the objectives of these recommendations is predicated upon the prior confirmation that control methods presented are adequate, the priority of this classification can also be considered lower than those preceding; i.e., these recommendations are for distributing information *after* such information is sufficiently available.

Some of the recommendations made involve the development of new methods for controlling operations which can be adequately controlled with existing technology. The desirability of such studies stems from their perceived capabilities to result in equally effective control techniques at a saving in materials utilized, energy expended, etc.; i.e., controls are generally available and adequate but "better" control measures appear to be feasible. These recommendations are thus represented by Classification 5.

The research recommendations presented by Classification 6 are felt to have considerable merit. However, since their need cannot be fully justified by comments in the literature or other data, and since they are simply presented as good ideas for further consideration, they are assigned the lowest priority. There is no assurance that this research would provide any tangible benefits.

5.3 Results

Table 9 summarizes the specific engineering control research programs recommended in this report and presents their classifications as determined from the procedure described above. Though certain other types of programs were recommended in Section 3, their significantly different nature does not allow their inclusion. Hence, their merit must be judged on an individual basis.

Classification of the 37 research recommendations by the scheme presented does not result in an ordering in which one recommendation is shown to be most important, another as second most important, etc. Indeed, 12 recommendations are assigned a classification of 1, 9 are assigned a classification of 2, and so forth. Attempts made at internally ordering these groups so that they could be ultimately assembled in an absolute order were not successful. Quantitative methods were stymied by the same lack of data which necessitated the classification procedure utilized. All qualitative methods could not be justified as being formulated on any other basis than personal opinion. Thus, it was concluded that more definitive prioritization of these recommendations requires data not available to this study.

5.4 Conclusions

It is evident that some degree of engineering judgement was required to prioritize the numerous recommendations into the groups presented. While not everyone may agree with the particular placement of any given recommendation, we are confident that they are ranked as best as can be accomplished using the information available to this study.

TABLE 9—PRIORITIZED SUMMARY OF RESEARCH RECOMMENDATIONS

Subject Area	Classification	Programs Recommended	Estimated Effort (man-months)	Estimated Time Span (months)	Comments
Abrasive Machining Operations	1	Investigate use of open and shaped hoods for abrasive machining of non-metals; develop ventilation criteria	24-36	12	Criteria available for woodworking operations, lacking for all other non-metals (i.e. plastics, ceramics asbestos products, etc.)
	5	Investigate performance characteristics of LVHV hoods; develop ventilation criteria	24-36	12	Present guides are non-specific, design and application are now trial-and-error processes
	3	Collect performance data from exhausted enclosures	12-24	9	Additional data would provide firmer base for developing ventilation guidelines
Chemical Processes	1	Review existing literature concerning types of equipment train fittings used in industry, their leak-resistance, reliability, etc. Attempt development of design guidelines for reducing exposures	24	12	Most valves, pumps, pipe joints, etc. leak to some extent, some more than others. Proper selection of fittings can significantly reduce exposures
Cleaning and Maintenance	1	Describe cleaning methods used and define their appropriateness under various conditions	8	6	Some methods are better than others under certain circumstances; when and where they are requires definition
	6	Investigate possibility that respirable toxic dust can escape filters in typical vacuum cleaners	2	2	Simply an idea; no knowledge if this is realistic problem area; might be a problem when highly toxic, fine dusts are collected

TABLE 9 (CONTINUED)

Subject Area	Classifi-cation	Programs Recommended	Estimated Effort (man-months)	Estimated Time Span (months)	Comments
Cleaning and Maintenance (Cont'd)	1	Perform user need and engineering feasibility study for a portable local exhaust system with an internal air cleaning device for vapors and fumes	12	9	Adequate engineering controls for maintenance work not available for controlling vapors and fumes released; suggested equipment may be useful for both EPA and OSHA compliance purposes, emergency leak control, etc.
Dryer and Oven Use	4	Review state of the art of oven ventilation design methods; develop ventilation guideline or manual which stresses all important factors	6	5	Many important factors do not appear to be properly stressed in literature; specific control criteria only available for continuous tunnel type dryers
Grinding, Crushing, and Screening Operations	1	Compile information on control of equipment; identify and fill data gaps; develop control guidelines	24	12	Guidelines exist only for a few types of equipment used in specific industries
Laboratory Operations	1	Evaluate state of the art of biological cabinet design; formulate design and operation guidelines	6	9	No generally accepted design or operational criteria exist presently
	3	Verify effectiveness of existing ventilation guidelines; devise performance measurement procedures; update guidelines	24-36	18	Guidelines exist but data which verifies their effectiveness are unavailable

TABLE 9 (CONTINUED)

Subject Area	Classifi-cation	Program Recommended	Estimated Effort (man-months)	Estimated Time Span (months)	Comments
Materials Handling **Operations**		**Belt Conveyors**			
	6	Investigate feasibility for material test method to determine need for controls for belt conveyor transfer points, etc.	6	6	No convenient test method exists to determine if conveyed substance will present respirable dust problem during transfer operations
	1	Develop and validate test method described above if need is identified	24	12	
	2	Critically review methods to determine amount of air entrained with falling material	3	4	Various methods proposed in literature – all different
	2	Develop method for air temperature corrections necessitated by heated substances	3	4	Unconfirmed empirical methods exist; method based on rigorous analytical study or adequate experimental data base unavailable
	2	Using methods developed in above, develop comprehensive calculation procedure for belt conveyor transfer point ventilation; determine necessary control air volumes, validate method	8	6	This program incorporates data from previous programs recommended; popular guidelines for controlling this problem are too simplistic, may lead to inadequate control
	6	**Other Conveyor Types** Conduct survey to determine extent to which use of other conveyor types can result in exposures	5	8	Much attention given to belt conveyors; no indication in literature that other open types require controls

TABLE 9 (CONTINUED)

Subject Area	Classifi-cation	Program Recommended	Estimated Effort (man-months)	Estimated Time Span (months)	Comments
Materials Handling Operations (Cont'd)	1	If need exists, extrapolate belt conveyor controls to other conveyor types	1	2	
	2	Bin Filling <u>Bin Filling</u> Apply technology developed for falling materials	2	2	Existing guidelines are too simplistic
	2	<u>Bucket Elevators</u> Apply technology developed for falling materials	2	2	Existing guidelines are too simplistic
	2	<u>Chutes</u> Apply technology developed for falling materials	2	2	Existing guidelines are too simplistic
	2	<u>Bag Filling</u> Classify types of bags and filling machines; determine how combinations best controlled	8	6	Operation repeatedly noted as being hazardous in many industries; adequacy of controls unknown; typical guides contradictory
	2	<u>Drum Filling</u> Apply technology developed for falling materials	2	2	Existing guidelines are too simplistic
	3	<u>Bag Integrity</u> Review DOT container regulations as to how they affect work place exposure probability	24	12	Objectives of DOT regulations are conceivably different than those of NIOSH or OSHA

TABLE 9 (CONTINUED)

Subject Area	Classifi-cation	Programs Recommended	Estimated Effort (man-months)	Estimated Time Span (months)	Comments
Materials Handling Operations (Cont'd)	2	Bag and Drum Emptying Survey handling procedures for packaged solids; develop guide for work practices and engineering controls	6	5-7	Noted as hazardous operation in many industries; adequacy of controls unknown; control guides are contradictory
	4	Overall Produce manual consolidating state-of-the-art technology and methods for controlling materials handling operations	24	12	Numerous control methods exist for each type of operation; no single guide to the subject exists
Non-spray application	1	Develop test method for coatings	8	6	Method would allow comparison of coating hazards, determination of "safe entry" times, and determination of ventilation criteria
	1	Develop local exhaust ventilation criteria for coating operations	8	6	Guidelines only found in Hemeon; need exists for consolidation, better definition, and validation
Open-Surface Tanks	5	Develop analytical method for determining evaporation and gassing rates; validate; incorporate into ventilation guidelines	9	6	Quantitative estimation technique is not commonly recognized; existing guidelines result in frequent overdesign of exhaust systems.

TABLE 9 (CONTINUED)

Subject Area	Classi-fication	Programs Recommended	Estimated Effort (man-months)	Estimated Time Span (months)	Comments
Open-surface Tanks (Cont'd)	1	Investigate rate of immersion and vapor "drag-out" effects on contaminant concentrations; recommend appropriate work practices, work speeds	15-18	9	Improper practices can easily liberate vapors from control envelopes of local exhaust systems
	5	Investigate effects of ancillary control measures; develop guidelines for their use	12	9	Methods potentially very useful; proper guidelines do not exist for their use; other adequate controls exist however.
Spray-finishing Operations	3	Investigate spray booth design criteria; develop firm data base; improve existing guidelines	24	12	Present guidelines for paints not adequate for high toxicity substances; relative degree of exposures from various application techniques unknown; criteria unavailable for substances other than paint.
	4	Produce manual for engineering control of such operations	6	6	
Textile Coating and Dyeing Operations	1	Survey plants conducting these operations; determine need for, availability, and adequacy of engineering controls	18	9	Potential for exposures is considerable; virtually no engineering control criteria are available in the literature

TABLE 9 (CONTINUED)

Subject Area	Classi-fication	Programs Recommended	Estimated Effort (man-months)	Estimated Time Span (months)	Comments
Welding Operations	5	Investigate performance characteristics of crossdraft tables	12-24	12	Criteria presently exists only for gas shielded-arc welding
	5	Investigate gun-mounted hood performance	12-24	12	Present data very limited; design and operational guidelines are incomplete
	3	Verify existing ventilation guidelines	12-24	12	Most common guidelines are of uncertain origin and are unsupported by experimental data

6. General Observations and Recommendations

6.1 Introduction

In the performance of this study, a number of concepts, ideas, observations etc. were developed or made which were not suited for inclusion into the preceding sections of this report. In the following, the most worthwhile of these are briefly presented and discussed.

6.2 Data Collection

To satisfy the requirements of OSHA and the EPA, industry has itself conducted considerable research to develop effective control measures. Some of this work has resulted in novel techniques which have been reported upon in the general literature. Much of it, however, has simply involved the taking of air samples or other observations to define the problem, and the subsequent design of some sort of process modification or other engineering control to ameliorate it. Where this has involved only the application of basic principles, the successful control measure has not necessarily been reported upon to the industrial community. It can be perceived that this individualized approach by industry in developing effective control measures can lead to duplication of effort and unnecessary expenditures of resources on the part of companies or individuals addressing a problem already solved elsewhere.

NIOSH is in an ideal position to act as the focal point of government, labor and industry for promoting the transfer of technology between individual companies or industries. It is therefore suggested that NIOSH more fully develop the necessary procedures and mechanisms to facilitate such technology transfer and to encourage its occurrence.

The means by which NIOSH may act are many and commonly utilized by other agencies involved in research and development activities. They include (1) periodic sponsorship of conferences or seminars involving parties interested in a particular subject area, (2) complete publication of conference proceedings, (3) development and maintenance of a data base which can be accessed freely, and (4) the performance of industry-wide surveys to ascertain the state-of-the-art of engineering control utilization.

6.3 Experimentation Techniques

Popular sources of ventilation system design criteria usually present a particular hood design and a particular suggested exhaust volume, hood face velocity, or capture velocity for each type of operation treated. The air velocities or volumes are usually either a single value or a range of values. Rarely, separate values or ranges may be given for use depending on whether a substance is simply a nuisance material or is toxic.

It is commonly understood by personnel trained or otherwise experienced in the field of industrial hygiene engineering that these guidelines are not universally capable of resulting in effective control of contaminants under all circumstances. Indeed, it is realized that the given criteria are presented as "averages" which have through long-term experience been found to be generally applicable to many specific operations and which probably are *not* completely adequate where substances of high toxicity are handled.

While there is great benefit to be found in the use of these guidelines, there is, however, cause to consider that they are lacking in many respects. Specifically, the point can be made that they can be faulted for not defining the specific circumstances under which they are applicable and the specific degree of control which they can be expected to provide. Without such data, there would appear to be a considerable potential for criteria to be applied to situations for which they are not appropriate, thereby resulting in ineffective control. It is our feeling, therefore, that NIOSH, in all of its research work, should ensure that the circumstances for which a particular control has been developed be completely defined and included as a part of any engineering control design criteria presented to industry.

Carrying this concept a step further, one can also see cause for recommending that research not be conducted solely to determine what hood dimensions, capture velocities, etc. are necessary to achieve a *specific* and desired degree of control. Rather, it should be conducted to define the degree of control which can be achieved by varying control parameters through practical ranges. This would result in design criteria which are not presented as a single hood design with a single associated air velocity or volume, but which allow selection of a design which more precisely provides the degree of control desired. When controls are developed for some particular substance of interest, cotton dust for example, this approach would greatly simplify determination of whether feasible controls exist if permissible exposure limits are lowered.

6.4 Eliminating the Need for Engineering Controls

It is our observation that engineering controls in the form of ventilation systems and jerry-built enclosures can often be found about operations or processes which would not have required such "add-on" devices if they had been properly designed in the first place. An underlying concept in many of the recommendations made in this report has therefore involved the provision of that technology necessary for the design of processes which do not require such controls by virtue of the fact that they do not result in excessive exposures.

This concept can be found to particularly pervade the recommendations made in the "chemical processes" and "material handling" sections of this report. In both of these areas, there would appear to be significant opportunities for developing precisely the sort of technology envisioned to be needed. Hence, it was suggested that the chemical process engineer be given the tools he needs to design a system which does not result in subsequent exposure problems. Without such tools, his new plant would probably end up with a series of exhaust fans placed in holes punched in the walls. Similarly, the test method for determining whether a material to be transported in some conveying system will present a dust hazard would permit mechanical engineers to weigh the costs of ventilated enclosures for belt conveyors against the additional costs to be incurred by specifying sealed screw conveyors. With today's costs of air cleaning equipment, ducting systems, and energy, elimination of the problem would probably be cheaper than attempting to control it.

This concept is presented here simply to allow its full appreciation. It is our belief that this sort of basic approach to developing "engineering controls" is the most correct one and that NIOSH should be alert to any and all future opportunities to support such research.

References

1. Stauden, A. (ed.), Kirk-Othmer Encyclopedia of Chemical Technology, 2nd Edition, Interscience Publishers, New York, 1972.
2. Watrous, R. M., "Health Hazards of the Pharmaceutical Industry," Brit. Journ. Indus. Med., 4, 111, 1947.
3. Bresler, R. R., "Nitroglycerin Reactions Among Pharmaceutical Workers," Ind. Med. and Surg., pg. 519, December 1949.
4. Chilson, F., "G. M. P.", Drug and Chemical Industry, February 1975.
5. Chilson, F., "The Dread Disease", Drug and Cosmetic Industry, December 1974.
6. Anon., "Taking the Headache Out of Aspirin Production," Drug and Cosmetic Industry, December 1967.
7. Gerhart, R., Baxter, T. H., Beeler, E. C., and Gannon, T. F., "Design for Potent Drug Production Areas," Chemical Engineering Progress, Vol. 67, No. 5, May 1971.
8. Perry, R. H. and Chilton, C. H., "Chemical Engineers' Handbook," 5th Edition, McGraw-Hill Book Co., New York, 1973.
9. Various issues of Drug and Cosmetic Industry.
10. Anon., "Multi-Layer Tablet Production," Drug and Cosmetic Industry, January 1968.
11. "Industrial Ventilation: A Manual of Recommended Practice," 13th Edition, Committee on Industrial Ventilation, American Conference of Governmental Industrial Hygienists, Lansing, Michigan, 1974.
12. "Encyclopedia of Occupational Safety and Health," International Labour Office, McGraw-Hill Book Co., New York, 1971.
13. LeFaux, R., "Practical Toxicology of Plastics," CRC Press, Cleveland, Ohio, 1968.
14. Modern Plastics Encyclopedia, McGraw-Hill Publications Co., Vol. 51, No. 10A, October 1974.
15. Golding, B., "Polymers and Resins: Their Chemistry and Chemical Engineering," C. Van Nostrand Co., New York, 1959.
16. Benning, C. J., "Plastic Foams: the physics and chemistry of project performance and process technology," Vol. 1, Wiley-Interscience, New York, 1969.
17. Billmeyer, F. W., "Textbook of Polymer Science," Interscience Publishers, New York, 1962.
18. Key, M. M., "Occupational Dermatitis from Plastics," J. M. A. Georgia, Vol. 57, September 1968.
19. Harris, D. K., "Health Problems in the Manufacture and Use of Plastics," Brit. J. Industr. Med., 10, 255, 1953.
20. Eckardt, R. E., and Hindin, R., "The Health Hazards of Plastics," Journ. of Occ. Med., Vol. 15, No. 10, October 1973.
21. Kingsley, W. H., "Health Hazards and Control of an Epoxy Resin Operation," Industrial Hygiene Journal, June 1958.
22. Bourne, L. B., and Milner, F. J. M., "Polyester Resin Hazards," Brit., J. Industr. Med., 20, 100, 1963.
23. Harris, D. K., "Some Hazards in the Manufacture and Use of Plastics," Brit. J. Industr. Med., 16, 221, 1959.
24. "Polyurethanes-Hazards and Control," Michigan's Occupational Health, Mich. Dept. of Public Health, Vol. 16, No. 1, Fall 1970.
25. Mack, W. A., "VCM Reduction and Control," Chemical Engineering Progress, Vo. 71, No. 9, September 1975.
26. Spence, J. W., and Haynie, F. H., "Paint Technology and Air Pollution: A Survey and Economic Assessment," EPA Office of Air Programs Publication No. AP-103, Feb. 1972.

27. "Economic Analysis of Proposed Effluent Guidelines—Paint and Allied Products and Printing Ink Industries," U.S. Environmental Protection Agency, EPA-230/1-74-052, August 1974.

28. Piper, R., "The Hazards of Painting and Varnishing 1965," Brit. J. Industr. Med., 22, 247, 1965.

29. Bidlack, V. C., and Fasig, E. W., (editors), "Paint and Varnish Production Manual," John Wiley and Sons, New York, 1951.

30. Martin, R., "Lacquer and Synthetic Enamel Finishes," D. Van Nostrand Co., New York, 1940.

31. Considine, D. M., ed., "Chemical and Process Technology Encyclopedia," McGraw Hill Book Co., New York, 1974.

32. Mattiello, J., ed., "Protective and Decorative Coatings," John Wiley and Sons, New York, 1943.

33. Rahter, J. M., "Ball and Pebble Mills," Paint and Varnish Production, May 1968.

34. Daniel, F. K., "High Speed Dispersers: Operating and Design Principles," Paint and Varnish Production, May 1970.

35. Patton, T. C., "Theory of High Speed Disk Impeller Dispersion," Journal of Paint Technology, Vol. 42, No. 550, November 1970.

36. Goldstein, G. F., "Pigment Dispersions for Aqueous Coatings," Modern Paint and Coatings, August 1975.

37. "Product Information-Resonant Mill," Journal of Paint Technology, pg. 66A.

38. "New Materials and Equipment," Paint and Varnish Production, August 1972.

39. Freriks, R. D., "Paint Plant Safety," Journal of Paint Technology, Vol. 38, No. 492, January 1966.

40. Stern, A. C., and Horowitz, L. D., "Industrial Hygiene in the Paint Factory," N.Y. State Dept. of Labor Monthly Review, Vol. 32, No. 2, February 1953.

41. Patty, F. A., "Industrial Hygiene and Toxicology," Second Revised Edition, Interscience Publishers, New York, 1963.

42. Arnstein, L. R., "Pumping Liquids Effectively—in paint production," Paint and Varnish Production, pg. 35, September 1965.

43. Martens, C. R., "Review of Latex and Water-Soluble Coatings," Journal of Paint Technology, Vol. 38, No. 494, March 1966.

44. McCormick, W. E., "Environmental Health Control for the Rubber Industry," Rubber Chemistry and Technology, V. 44, n. 2, April 1971, pp. 512-533.

45. Morris, G. E., MD., "Synthetic Rubbers; Their Chemistry and Dermatological Aspects," A.M.A. Archives of Industrial Hygiene and Occupational Health.

46. The Vanderbilt Rubber Handbook, G. G. Windspear, ed. R. T. Vanderbilt Company, Inc., New York, 1968.

47. Stephenson, R. M., "Introduction to the Chemical Process Industries," Reinhold Publishing Corp., N.Y., 1966.

48. Mallette, F. S., "Industrial Hygiene—In Synthetic Rubber Manufacture," Industrial Medicine, July, 1943, V. 12, n.7, pp. 495-499.

49. McCormick, W. W., "Industrial Health Problems in the Rubber Industry," Industrial Hygiene Quarterly, March, 1952, V. 13, n. 1, pp. 37-41.

50. Bourne, H. G., H. T. Yee, and S. Seferian, "The Toxicity of Rubber Additives," Archives of Environmental Health, May 1968, V. 16 pp. 700-705.

51. Schwartz, L., "Skin Hazards in America—Industry, Part I," Public Health Bulletin No. 215, October 1934, pp. 1-5.

52. Greenburn, L., and S. Moskowitz, "The Safe Use of Solvents for Synthetic Rubbers," New York State Department of Labor Monthly Review, June 1, 1946, V. 25, n. 11.

53. Wechsler, A. E., and Harrison, J., "Evaluation of the Possible Impact of Pesticide Legislation on Research and Development Activities of Pesticide Manufacturers," Report to Office of Pesticide Programs, Environmental Protection Agency (Contract 68-01-2219) by Arthur D. Little, February 1975.

54. Danielson, J. A., "Air Pollution Engineering Manual," U.S. Department of Health, Education, and Welfare, National Center for Air Pollution Control, Cincinnati, Ohio 1967.

55. U.S. Environmental Protection Agency, "EPA Compendium of Registered Pesticides," U.S.G.P.O., Washington, D.C.

56. Popendorf, W. J., and Spear, R. C., "Preliminary Survey of Factors Affecting the Exposure of Harvesters to Pesticide Residues" American Industrial Hygiene Association Journal 35: 374-378, 1974.

57. "Economic Analysis of Pesticide Disposal Methods," Report to Strategic Studies Unit, Environmental Protection Agency (Contract 68-01-2614) by Arthur D. Little, Inc. March 1975.

58. Hamilton, A., and Hardy, H. L., "Industrial Toxicology," Publishing Sciences Group, Inc., 3rd ed., 1974.

59. Bray, T. J., "Health Hazard. Chemical Firm's Study Underscores Problems of Cleaning Up Plants," The Wall Street Journal, Tuesday, December 2, 1975.

60. Wolf, H. R., and Armstrong, J. F., "Exposure of Formulating Plant Workers to DDT," Archives of Environmental Health 23: 169-176, 1971.

61. Danielson, J. A., (ed.), "Air Pollution Engineering Manual," Environmental Protection Agency Pub. No. AP-40, May 1973.

62. Jegier, Z., "Health Hazards in Insecticide Spraying of Crops," Archives of Environmental Health 8: 670-674, 1964.

63. Kleinman, G. D., Wet, I., and Augstine, M. S., "Occupational Disease in California Attributed to Pesticides and Agricultural Chemicals," Archives of Environmental Health 1:118-124, 1960.

64. "Work Injuries in California Agriculture 1970," State of California, Department of Industrial Relations, Division of Labor Statistics and Research, 1972.

65. "Adhesives", Plastic Trends, Predicasts, Inc., Cleveland, Ohio, January 31, 1972.

66. Contractor Estimates.

67. Hawley, G. G., ed., "The Condensed Chemical Dictionary", Van Nostrand Reinhold Company, New York, 1971.

68. Skeist, I., "Handbook of Adhesives", Reinhold Publishing Corporation, London, 1962.

69. Shreve, R. N., "The Chemical Process Industries", McGraw Hill Book Company, New York, 2nd Edition, 1956.

70. Hurd, J., "Adhesives Guide" British Scientific Instrument Research Association, Research Report M. 39, 1959.

71. McGuire, E. P., ed., "Adhesive Raw Material Handbook", Padric Publishing, Mountainside, New Jersey, 1964.

72. "The Synthetic Adhesives Industry", Industry Comment Letter, Arthur D. Little, Inc., Cambridge, Massachusetts, July 24, 1972.

73. Gaynes, N. I., Danziger, G. N., and F. C. Kinsler, "Formulation of Organic Coatings", D. Van Nostrand Company, Inc., Princeton, New Jersey, 1967.

74. Smith, F. C., "The New Hot Melts-An Equipment Approach," Adhesives Age. Vol. 18, No. 8, August, 1975.

75. Houwink, R., and Salomen, G. (ed.'s), "Adhesion and Adhesives," Vol. 1 & 2, Elsevier Publishing Co., New York, 1965.

76. Stinson, S. C., "Textile Dye Industry," Chemical and Engineering News, October 26, 1970.

77. Renson, J. E., "An Analysis of the Census of Manufacturers," American Ink Maker, June 1974.

78. Strauss, V. "The Printing Industry," Printing Industries of America, 1967.

79. Wolfe, H. J., "Printing and Litho Inks," MacNair-Dorland Company, 1949.

80. Cralley, L. V., Cralley, L. J., Clayton, G. D., and Jurgiel, J. A. (ed.'s), "Industrial Environmental Health," Academic Press, New York, 1972.

81. Poucher, W., "Perfumes, Cosmetics and Soaps," D. Van Nostrand, Co., Inc., New York, 1941.

82. Sax, N. I., "Dangerous Properties of Industrial Materials," Third Edition, Van Nostrand Reinhold Company, New York, 1968.

83. International Printing and Graphic Communications Union, Education and Research Department. News and Views on the Respiratory Disease Project, October 1970.

84. Marsailes, T. P., "Ventilation, Filtration, and Exhaust Techniques Applied to Printing Plant Operations," ASHRAE Journal, Dec. 1970, p. 27.

85. Fairley, M. C., "Materials Handling in the Printing Industry," Pergamon Press, Oxford, 1971.

86. Fairley, M. C., Safety, Health, and Welfare in the Printing Industry Pergamon Press, Oxford, 1969.

87. International Printing and Graphic Communications Union, Education and Research Department. News and Views on the Respiratory Diseases Project, August 1970.

88. International Printing and Graphic Communications Union, Education and Research Department. News and Views on the Respiratory Diseases Project, April 1971.

89. International Printing and Graphic Communications Union, Education and Research Department. News and Views on the Respiratory Diseases Project, May 1972.

90. International Printing and Graphic Communications Union, Education and Research Department. News and Views on the Respiratory Diseases Project, January 1971.

91. Greiner, T. G., "Dust Free Cutting," Am. Paper Industry, Sept. 1972, p. 28.

92. Gadomski, R. R., Gimbrone, A. V., Green, W. J., Reitz, R. J., Eisamer, P. R., and Dale, J. T., "An Evaluation of Emissions and Control Technologies for the Metal Decorating Process:" J.A.P.C.A. 24:579, June 1974.

93. Gadomski, R. R., Gimbrone, A. V., Green, W. J., Reitz, R. J., Eisamen, P. R., and Dale, J. T., "An Evaluation of Emissions and Control Technologies for the Web Offset (Lithograph) Process:" J.A.P.C.A., 24:484, May 1974.

94. E. K. Bastress, et al., "Ventilation Requirements for Grinding, Buffing, and Polishing Operations," Report No. 0213, IKOR Incorporated, Burlington, Massachusetts, June 1973.

95. A. J. Breslin and W. B. Harris, "Health Protection in Beryllium Facilities; Summary of Ten Years of Experience," Health and Safety Laboratory, U.S. Atomic Energy Commission, May 1, 1958.

96. Ludwig, E. G., "Designing Process Plants to Meet OSHA Standards," Chemical Engineering, Sept. 3, 1973, pg. 88-100.

97. Wiley, S. K., "Controlling Vapor Losses," Chemical Engineering, Sept. 17, 1973, pg. 116-119.

98. Rothman, R. A., "Gaskets and Packings," Chemical Engineering, Deskbook Issue, Feb. 26, 1973, pg. 69-74.

99. Browning, R. L., "Estimating Loss Probabilities," Chemical Engineering, Dec. 15, 1969, pg. 135-140.

100. Templeton, H. C., "Valve Installation, Operation and Maintenance," Chemical Engineering, Deskbook Issue, Oct. 11, 1971, pg. 141-149.

101. Russell, W. W., "Safety in Flange Joints," Chemical Engineering Progress, Vol. 70, No. 4, April 1974, pg. 68-72.

102. Fitzpatrick, K., "Closures for High Pressure Piping Systems," Chemical Engineering Progress, Vol. 70, No. 9, Sept. 1974, pg. 60.

103. Baturin, V. V., "Fundamentals of Industrial Ventilation," Pergamon Press, Elmsford, N.Y., 3rd Edition, 1972.

104. "Chementator," Chemical Engineering, April 14, 1975, pg. 33-34.

105. Iammartino, N. R., "PVC Makers to Mop Up Monomer Emissions," Chemical Engineering, November 24, 1975.

106. "National Fire Codes," Vol. 6, National Fire Protection Association, Boston, 1975.

107. Conway, M., "Brown Lung Increases Among Cotton Workers," The Boston Globe, November 27, 1975, pg. 65.

108. Wheeler, R. N., and Sutherland, M. E., "Control of In-Transit VCM," Chemical Engineering Progress, Vol. 71, No. 9, September 1975, pg. 48-53.

109. National Fire Codes, Vol. 8 NFPA 86A, National Fire Protection Association, Boston 1975.

110. Michigan's Occupational Health, Michigan Dept. of Health Winter 1959-60, Vol. 5, No. 2.

111. Feiner, Genjamin and Kingsley, Irving, "Ventilation of Industrial Ovens," Air Conditioning, Heating and Ventilating, Dec. 1956, pp. 82-89.

112. Constance, John D., "Estimating Exhaust-Air Requirements for Processes," Chemical Engineering, August 10, 1970, pp. 116-118.

113. Hemeon, W. C. L., Plant and Process Ventilation, 2nd edition, Industrial Press Inc., New York 1963.

114. Boyle, John P., and Nicholas P. Novak, "Predicting Ventilation Requirements for Coating Materials," Industrial Hygiene Journal, Nov.-Dec. 1963, pp. 606-610.

115. Matthews, Chris, W., "Trends in Size Reduction of Solids . . . Screening" Chemical Engineering, V. 79, n. 15, July 10, 1972, p. 76 ff.

116. AIHA Committee on Industrial Hygiene Practices, "Industrial Hygiene Practices Guide; Laboratory Hood Ventilation," AIHA Journal, November-December 1968, pp. 611-617.

117. R. S. Brief, et al., "Design and Selection of Laboratory Hoods," Air Engineering, October 1963.

118. "Fundamentals Governing the Design and Operation of Local Exhaust Systems," American National Standard Z9.2-1971, American National Standards Institute, New York, 1971.

119. "Industrial Exhaust Systems," Chapter 22, 1973 Systems Handbook, American Society of Heating, Refrigeration, and Air-Conditioning Engineers.

120. "Standards for Protection Against Radiation," Code of Federal Regulations, Title 10, Chapter 20, U.S. Federal Register, January 1, 1973.

121. "General Industry Safety and Health Regulations, Part 1910," U.S. Department of Labor, OSHA 2206, June 1974.

122. Ronald L. Akers, et al., "Development of a Laminar Air-Flow Biological Cabinet," AIHA Journal, March-April 1969, pp. 177-185.

123. Buffington, M. A., "Mechanical Conveyors and Elevators," Chemical Engineering Deskbook Issue, October 13, 1969.

124. Baumeister, T., ed., "Standard Handbook for Mechanical Engineers", 7th Edition, McGraw-Hill Book Co., New York, 1967.

125. Morrison, J. N., "Combatting Dust at Conveyor Transfer Points," Rock Products, November 1970.

126. 1967 Book of ASTM Standards, American Society of Testing and Materials, 1967, pp. 108-111.

127. "Steel Mill Ventilation", Committee on Industrial Hygiene, American Iron and Steel Institute, New York, May 1965.

128. Pring, R. T., Knudson, J. F., and Dennis, R., "Design of Exhaust Ventilation for Solid Materials Handling", Ind. Eng. Chem., 1949.

129. Anderson, D. M. "Dust Control by Air Induction Technique", Ind. Med. Surg., 1964.

130. Dunlop, P. J., "Dust Control at the Helinger Milling Plant," Trans Can. Inst. Mining Met., 1939.

131. Pring, R. T., "Dust Control in Ore Concentration Operations", Amer. Inst. Mining Met. Engr. Tech. Print, 1940.

132. Hatch, T., "Some Physical Principles in Analysis and Design of Dust Exhaust Systems", Pact Huitieme Annee, 1954.

133. Larson, S., "Air Induction by Falling Materials as a Basis for Exhaust Hood Design", M. Sc. Thesis, University of Pittsburgh, 1952.

134. Dennis, R., Lecture notes (unpublished) Harvard School of Public Health, as cited in Reference 6 (Steel Mill Ventilation).

135. Kruse, C. W., and Bianconi, W. O., "Air Flow Induced in Enclosed Inclined Chutes of Materials Handling Systems," AIHA Journ. May-June, 1966.

136. Goldfield, J., and Brandt, F. E., "Dust Control Techniques in the Asbestos Industry," AIHA Journal, December 1974.
137. Kraus, M. N., "Pneumatic Conveyors", Chemical Engineering Deskbook Issue, October 13, 1969.
138. "Fills Powders, Prevents Dusts," Food Engineering, 43:83, August 1971.
139. Hama, G. M., "Ventilation for Control of Dust from Bagging Operations," Heating and Ventilating, April 1948.
140. Kane, J. M., "Design of Exhaust Systems," Heating and Ventilating, November 1945.
141. Burke, A. J., "Weighing Bulk Materials in the Process Industries," Chemical Engineering, March 5, 1973.
142. Uncles, R. F., "Containers and Packaging," Chemical Engineering, Deskbook Issue, October 13, 1969.
143. Owen, L. T., "Selecting In-Plant Dust-Control Systems," Chemical Engineering, Oct. 14, 1974.
144. Goss, A. E., and Ross Jr., A. M., "Effective Control of Lead Dust in the Manufacture of Vinyl Plastics," AIHA Quarterly, March 1953.
145. "Design Digest-Get Rid of Phenolic Dust," Iron Age, November 11, 1971.
146. Hills, D. W., "Economics of Dust Control" Annals of the New York Academy of Sciences, Vol. 132, December 31, 1965.
147. "Foundry Environmental Control," Volume 1, American Foundrymen's Society, Des Plaines, Illinois, 1972.
148. Riley, E. C., "Estimation of Atmospheric Concentrations of Volatile Compounds from Surface Coatings by Means of a Laboratory Model," AIHA Journal, Sept.-Oct. 1968.
149. Flanigan, L. J.: Kim, B. C., Semones, D. E.; Talbert, S. G., Development of Design Criteria for Exhaust Systems for Open-Surface Tanks. Research Report Battelle Laboratories, February 1974, pg. 7-8.
150. "American National Standard Practices for Ventilation and Operation of Open-Surface Tanks:" American National Standards Institute, Inc., N.Y. ANSI Z9.1-1971.
151. Clearwater, Robert M.; Esmen, Nurton A.; Correlation between Breathing Zone Solvent Vapor Concentrations and Solvent Zones from Vapor Degreasers; University of Delaware, April 1974, pg. 22.
152. Hama, G. M., Frederick, W.; Millage, D., Brown, J., "Absolute Control of Chromic Acid Mist—Investigation of a New Surface-Active Agent," American Industrial Hygienic Association Quarterly, 15:3, Sept., 1954.
153. Skinner, J. B., "Control of Health Hazards in the Operation of Metal Degreasers", American Industrial Hygiene Association Quarterly, March, 1952, Vol. 13, No. 1, pp. 11-16.
154. Doolittle, A. K., "Lacquer Solvents in Commercial Use," Industrial and Engineering Chemistry, Vol. 27, pp. 1169-1179.
155. "American Standard Safety Code for the Design, Construction, and Ventilation of Spray Finishing Operations," ANSI A9.3-1964 (Reaffirmed 1971).
156. Reichenbach, Jr., G. S., "Ventilation-For Ship Construction and Repair," AIHA Quarterly, pg. 307, December 1953.
157. Hama, G. M., and Bonkowski, K. J., "Ventilation Requirements for Airless Spray Painting," Heating, Piping, & Air Conditioning, October 1970.
158. Brandt, A. D., "Controlling Air Contaminants in New Industrial Processes," Industrial Hygiene Review, N.Y. Dept. of Labor, Vol. 5, No. 1, May 1962.
159. "Spray Painting Ventilation," Michigan's Occupational Health, Michigan Department of Health, Vol. 2, No. 1, Fall, 1956.
160. "Spray Booths," Mass. Dept. of Labor and Industries, Div. of Occupational Hygiene, Ventilation Data Sheet No. 3, December 1971.
161. Benning, C. J., "Plastic Foams: the physics and chemistry of project performance and process technology," Vol. 1, Wiley-Interscience, N.Y., 1969.
162. Sheinbaum, M., "Some Health Hazards Associated with the Building Trades," AIHA Journal, 23:353, Sept.-Oct. 1962.
163. "Construction Health Hazards," Michigan's Occupational Health, Michigan Department of Health, Vol. 8, No. 3, Spring 1963.
164. Gorson, R. O., and Lieberman, J. L., "The Prohibition of the Use of Asbestos Spray in Building Construction," Journ. of Occup. Med., Vol. 15, No. 3, p. 260-261, March 1973.
165. Peterson, J. E., Copeland, R. A., Hoyle, H. R., "Health Hazards of Spraying Polyurethane Foam Out-of-Doors," AIHA Journal, Vol. 23, Sept.-Oct. 1962.
166. Werber, F. X., "Chemical Engineering In the Textile Industry," Chemical Engineering Progress, Vol. 66, No. 4, April, 1970, p. 57.
167. Weigold, S., "High-temperature Dyeing of Polyester," Textile Industries, Vol. 133, No. 8, August, 1969, p. 150.
168. Draft of Development Document for Effluent Limitations Guidelines and Standards of Performance for Environmental Protection Agency, June 1973 (unpublished).
169. Howry, K. A., "Solvent Dyeing Efforts Widespread," Modern Textiles, Vol. L11, No. 4, April 1972, p. 12.

170. "A Yarn First, Space Dyeing with Ammonia," Textile World, Sept. 1973, p. 108.

171. Schwartz, L., "Skin Hazards in American Industry," Public Health Bulletin No. 229, September 1936.

172. Stevens, C. H., "Economical Control of Air Pollution from Textile Dryers," Modern Textiles, February 1973.

173. "Chementator," Chemical Engineering, Aug. 5, 1974, p. 44.

174. Billmaier, D., et al., "Peripheral Neuropathy in a Coated Fabrics Plant," Journal of Occupational Med., Vol. 16, No. 10, October 1974, pp. 665-671.

175. Ewald, G. W., "Finishing Industrial Fabrics," Textile Industries, Vol. 134, No. 11, Nov. 1970, p. 79.

176. Pariser, A. A., "A New Era in Solvent Processing," Modern Textiles, Vol. L11, No. 4, April 1972, p. 11.

177. Welding Handbook, Sixth Edition, L. Griffing, Editor, American Welding Society, New York, 1971.

178. The Welding Environment, American Welding Society, Miami, FL, 1973.

179. "Safety in Welding and Cutting," American National Standard Z49.1-1973, American Welding Society, Miami, FL 1973.

180. 29 CFR 1910.252(f), Health Protection and Ventilation for Welding, Cutting, and Brazing.

181. W. J. Astleford, and J. W. Register, "Engineering Control of Welding Fumes," Southwest Research Institute, San Antonio, Texas, July 1973.

TABLE A
SUMMARY OF USE/EXPOSURE DATA
FOR CHEMICALS IN SCP SETS

TABLE A—1
SUMMARY OF USE/EXPOSURE DATA FOR CHEMICALS IN SCP SET A

Operation	Acetone	Antimony	2-Butanone	Cyclohexanone	Hexone	Hydrogen Sulfide	Manganese	MDI	p-Nitroaniline	2-Pentanone
Rubber production	X	X				X			X	
Spray-finishing	X		X	X	X					X
Non-spray application	X		X	X	X			X		X
Paint manufacture		X	X				X	X		
Cleaning and maintenance	X	X			X					X
Wood pulping										
Plastic & foam manufacture					X			X		X
Pharmaceutical manufacture					X					X
Cosmetic production										X
Textile coating	X	X	X					X		
Pesticide application										
Pesticide manufacture					X					
Ink production										
Ink application				X						
Textile dyeing	X	X								
Adhesive production and use	X		X		X					X
Welding		X					X			
Abrasive machining		X								
Dry cleaning				X	X					
Aerosol production and use										
Electric equipment manufacture										
Refrigerant use										
Extractant or solvent use				X	X					X
Chemical processes	X			X		X	X	X	X	
Mineral oil refining and dewaxing										
Photographic film and chemical production										
Artificial leather production										
Safety glass production										

Operation	Acetone	Antimony	2-Butanone	Cyclohexanone	Hexone	Hydrogen Sulfide	Manganese	MDI	p-Nitroaniline	2-Pentanone
Anesthetic use										
Heat transfer fluid use										
Mining operations		X				X	X			
Ceramics and glass production		X				X	X			
Petroleum refining			X			X				
Cleaning and degreasing	X		X	X						
Synthetic fiber manufacture	X		X			X				
Leather tanning and treatment						X				
Paint and varnish remover use										
Drying of paints, varnishes, etc.			X		X					X
Detergent production							X			
Polishing compound production										
Dye manufacture			X			X			X	
Fuel production									X	
Lubricant manufacture and use										
Synthetic pine oil production										
Perfume production										
Food additive use										
Smokeless powder production										
Straw hat production										
Fire extinguisher production and use										
Water purification										
Pyrotechnics and explosive making and use		X					X			
Foundry operations							X	X		
Paper impregnation										
Corrosion inhibitor making and use										
Food processing										
Ore refining and metal processing										

TABLE A—2
SUMMARY OF USE/EXPOSURE DATA FOR CHEMICALS IN SCP SET B

Operation \ Chemical	Camphor	Chloro-acetaldehyde	Alpha-Chloro-acetophenone	Ethyl Butyl Ketone	Mesityl Oxide	Methyl (n-amyl) Ketone	5-Methyl-3-Heptanone	Ozone	Pival
Rubber production									
Spray-finishing	X			X	X	X	X		
Non-spray application	X			X					
Paint manufacture	X				X	X			
Cleaning and maintenance				X					X
Wood pulping									
Plastic & foam manufacture									
Pharmaceutical manufacture	X								X
Cosmetic production									
Textile coating									
Pesticide application		X							
Pesticide manufacture									X
Ink production					X				
Ink application					X				
Textile dyeing									
Adhesive production and use				X					
Welding								X	
Abrasive machining									
Dry cleaning									
Aerosol production and use			X						
Electric equipment manufacture									
Refrigerant use									
Extractant or solvent use						X	X		
Chemical processes	X	X	X		X			X	X
Mineral oil refining and dewaxing								X	
Photographic film and chemical production	X								
Artificial leather production									
Safety glass production									

Operation \ Chemical	Camphor	Chloro-acetaldehyde	Alpha-Chloro-acetophenone	Ethyl Butyl Ketone	Mesityl Oxide	Methyl (n-amyl) Ketone	5-Methyl-3-Heptanone	Ozone	Pival
Anesthetic use									
Heat transfer fluid use									
Mining operations					X				
Ceramics and glass production									
Petroleum refining									
Cleaning and degreasing					X				
Synthetic fiber manufacture									
Leather tanning and treatment									
Paint and varnish remover use					X				
Drying of paints, varnishes, etc.				X					X
Detergent production	X								
Polishing compound production									
Dye manufacture									
Fuel production									
Lubricant manufacture and use									
Synthetic pine oil production								X	X
Perfume production	X								
Food additive use									
Smokeless powder production									
Straw hat production									
Fire extinguisher production and use									X
Water purification									
Pyrotechnics and explosive making and use	X								
Foundry operations									
Paper impregnation									
Corrosion inhibitor making and use									
Food processing									
Ore refining and metal processing									

TABLE A—3
SUMMARY OF USE/EXPOSURE DATA FOR CHEMICALS IN SCP SET C

Operation \ Chemical	Acrolein	p-tert-Butyltoluene	Cumene	Cyclohexane	Diphenyl	Ethyl Benzene	Furfural	Alpha-Methyl Styrene	Styrene	Terphenyls	Vinyl Toluene
Rubber production	X						X	X	X		X
Spray-finishing			X			X		X	X	X	X
Non-spray application			X	X		X	X	X	X	X	X
Paint manufacture		X				X	X	X	X	X	X
Cleaning and maintenance			X				X			X	
Wood pulping											
Plastic & foam manufacture					X		X	X	X	X	X
Pharmaceutical manufacture		X									
Cosmetic production											
Textile coating							X		X		
Pesticide application	X										
Pesticide manufacture											
Ink production											
Ink application											
Textile dyeing											
Adhesive production and use						X	X	X			
Welding	X										
Abrasive machining											
Dry cleaning											
Aerosol production and use											
Electric equipment manufacture									X		
Refrigerant use											
Extractant or solvent use			X	X							
Chemical processes	X	X	X	X	X	X	X	X	X	X	
Mineral oil refining and dewaxing							X				
Photographic film and chemical production											
Artificial leather production											
Safety glass production											

Operation \ Chemical	Acrolein	p-tert-Butyltoluene	Cumene	Cyclohexane	Diphenyl	Ethyl Benzene	Furfural	Alpha-Methyl Styrene	Styrene	Terphenyls	Vinyl Toluene
Anesthetic use											
Heat transfer fluid use					X	X				X	
Mining operations											
Ceramics and glass production											
Petroleum refining			X								
Cleaning and degreasing											
Synthetic fiber manufacture											
Leather tanning and treatment											
Paint and varnish remover use						X	X				
Drying of paints, varnishes, etc.											
Detergent production											
Polishing compound production											
Dye manufacture						X				X	
Fuel production			X								
Lubricant manufacture and use										X	
Synthetic pine oil production											
Perfume production				X							
Food additive use											
Smokeless powder production											
Straw hat production											
Fire extinguisher production and use											
Water purification											
Pyrotechnics and explosive making and use											
Foundry operations											
Paper impregnation					X						
Corrosion inhibitor making and use											
Food processing											
Ore refining and metal processing											

TABLE A—4
SUMMARY OF USE/EXPOSURE DATA FOR CHEMICALS IN SCP SET D

Operation	sec-Amyl Acetate	sec-Butyl Acetate	tert-Butyl Acetate	Butyl Acetate	Dibutyl phthalate	Dimethyl-phthalate	2-Ethoxy-ethyl acetate	Ethyl Acrylate	Ethyl Formate	sec-Hexyl Acetate	Isoamyl Acetate	Isobutyl Acetate	Methyl Acetate	Methyl Acrylate	Methyl Cellosolve Acetate	Methyl Methacrylate	di-sec-octyl phthalate	n-Propyl Acetate
Rubber production	X				X			X										
Spray-finishing	X		X		X	X			X	X								
Non-spray application	X				X	X			X	X								
Paint manufacture	X		X		X	X		X		X				X				
Cleaning and maintenance																		
Wood pulping																		
Plastic & foam manufacture					X	X		X										
Pharmaceutical manufacture																		
Cosmetic production																		
Textile coating	X																	
Pesticide application									X									
Pesticide manufacture						X												
Ink production						X												
Ink application						X												
Textile dyeing																		
Adhesive production and use					X	X		X						X				
Welding																		
Abrasive machining																		
Dry cleaning			X															
Aerosol production and use																		
Electric equipment manufacture																		
Refrigerant use																		
Extractant or solvent use									X									
Chemical processes			X		X	X		X	X	X				X				
Mineral oil refining and dewaxing					X			X										
Photographic film and chemical production																		
Artificial leather production									X									
Safety glass production									X									

Operation	n-Propyl Acetate	di-sec-octyl phthalate	Methyl Methacrylate	Methyl Cellosolve Acetate	Methyl Acrylate	Methyl Acetate	Isobutyl Acetate	Isoamyl Acetate	sec-Hexyl Acetate	Ethyl Formate	Ethyl Acrylate	2-Ethoxy-ethyl acetate	Dimethyl-phthalate	Dibutyl phthalate	Butyl Acetate	tert-Butyl Acetate	sec-Butyl Acetate	sec-Amyl Acetate
Anesthetic use																		
Heat transfer fluid use																		
Mining operations																		
Ceramics and glass production																		
Petroleum refining																		
Cleaning and degreasing																		
Synthetic fiber manufacture					X					X								
Leather tanning and treatment																		X
Paint and varnish remover use																		
Drying of paints, varnishes, etc.																		
Detergent production																		
Polishing compound production																		
Dye manufacture																		
Fuel production																X		
Lubricant manufacture and use																		
Synthetic pine oil production																		
Perfume production																		X
Food additive use																		
Smokeless powder production																		
Straw hat production																		
Fire extinguisher production and use																		
Water purification																		
Pyrotechnics and explosive making and use																		
Foundry operations																		
Paper impregnation																		
Corrosion inhibitor making and use																		
Food processing																		
Ore refining and metal processing																		

TABLE A–5
SUMMARY OF USE/EXPOSURE DATA FOR CHEMICALS IN SCP SET E

Operation \ Chemical	Allyl Alcohol	n-Amyl Acetate	sec-Butyl Alcohol	tert-Butyl Alcohol	Butyl Alcohol	Cyclohexanol	Diacetone Alcohol	Ethyl Acetate	Ethyl Alcohol	Hydroquinone	Isoamyl Alcohol	Isobutyl Alcohol	Isopropyl Acetate	Isopropyl Alcohol	Methyl Alcohol	Methyl Isobutyl Carbinol	Propyl Alcohol
Rubber production										X				X	X		
Spray-finishing		X	X		X				X		X	X	X	X		X	X
Non-spray application		X	X	X	X	X		X	X		X	X		X	X	X	X
Paint manufacture		X	X	X	X				X		X	X	X	X			X
Cleaning and maintenance		X							X		X		X	X			
Wood pulping																	
Plastic & foam manufacture		X					X	X				X	X		X		
Pharmaceutical manufacture		X		X				X	X								
Cosmetic production	X								X		X				X	X	X
Textile coating		X					X		X				X	X	X		
Pesticide application	X																
Pesticide manufacture																	
Ink production		X					X	X	X				X				
Ink application		X					X	X	X		X		X	X	X		X
Textile dyeing						X	X							X			X
Adhesive production and use		X	X		X				X	X				X			X
Welding														X			
Abrasive machining																	
Dry cleaning		X						X									
Aerosol production and use																	
Electric equipment manufacture																	
Refrigerant use																	
Extractant or solvent use		X		X			X			X	X		X	X		X	X
Chemical processes	X	X	X	X	X	X	X	X	X	X	X	X	X	X	X	X	X
Mineral oil refining and dewaxing	X											X			X	X	
Photographic film and chemical production		X			X		X	X	X	X						X	
Artificial leather production							X	X	X								
Safety glass production					X									X			

Chemical / Operation	Allyl Alcohol	n-Amyl Acetate	sec-Butyl Alcohol	tert-Butyl Alcohol	Butyl Alcohol	Cyclohexanol	Diacetone Alcohol	Ethyl Acetate	Ethyl Alcohol	Hydroquinone	Isoamyl Alcohol	Isobutyl Alcohol	Isopropyl Acetate	Isopropyl Alcohol	Methyl Alcohol	Methyl Isobutyl Carbinol	Propyl Alcohol
Anesthetic use																	
Heat transfer fluid use																	
Mining operations											X					X	
Ceramics and glass production																	
Petroleum refining															X		
Cleaning and degreasing			X	X		X						X		X			X
Synthetic fiber manufacture					X												
Leather tanning and treatment													X				X
Paint and varnish remover use			X					X				X			X		
Drying of paints, varnishes, etc.									X								
Detergent production						X								X			X
Polishing compound production		X													X		X
Dye manufacture		X	X					X						X	X		X
Fuel production									X					X			
Lubricant manufacture and use												X				X	
Synthetic pine oil production																	
Perfume production	X		X	X	X						X		X				
Food additive use		X															
Smokeless powder production								X	X								
Straw hat production		X															
Fire extinguisher production and use																	
Water purification																	
Pyrotechnics and explosive making and use																	
Foundry operations																	
Paper impregnation																	
Corrosion inhibitor making and use																	
Food processing																	
Ore refining and metal processing																	

TABLE A—6
SUMMARY OF USE/EXPOSURE DATA FOR CHEMICALS IN SCP SET F

Operation \ Chemical	2-Butoxy-Ethanol	n-Butyl Glycidyl Ether	Chlorinated Camphene	Cyclohexene	Cyclopenta-diene	Diglycidyl Ether	Dipropylene Glycol Methyl Ether	Ethyl Ether	Glycidol	Isopropyl Glycidyl Ether	Methyl Acetylene	Methyl Cellosolve	Methylal	Phenyl Ether Vapor	Phenyl Ether-Bi-phenyl Vapor Mixture	Phenyl Glycidyl Ether	Propylene Oxide	Tetrahydro-furan
Rubber production															X	X		
Spray-finishing	X											X						
Non-spray application	X						X					X						
Paint manufacture					X		X		X			X	X					X
Cleaning and maintenance		X		X			X		X				X	X	X	X	X	
Wood pulping																		
Plastic & foam manufacture	X				X			X	X	X							X	
Pharmaceutical manufacture								X			X						X	X
Cosmetic production							X											
Textile coating															X	X		
Pesticide application			X														X	
Pesticide manufacture	X		X															X
Ink production							X					X						X
Ink application							X					X						
Textile dyeing									X			X			X		X	
Adhesive production and use		X								X		X	X				X	X
Welding											X							
Abrasive machining																		
Dry cleaning																		
Aerosol production and use																		
Electric equipment manufacture																X		
Refrigerant use												X						
Extractant or solvent use	X						X	X					X					
Chemical processes	X	X	X	X	X	X	X	X	X	X	X	X	X	X	X	X	X	X
Mineral oil refining and dewaxing																		
Photographic film and chemical production																		
Artificial leather production																		
Safety glass production																		

Chemical \ Operation	Anesthetic use	Heat transfer fluid use	Mining operations	Ceramics and glass production	Petroleum refining	Cleaning and degreasing	Synthetic fiber manufacture	Leather tanning and treatment	Paint and varnish remover use	Drying of paints, varnishes, etc.	Detergent production	Polishing compound production	Dye manufacture	Fuel production	Lubricant manufacture and use	Synthetic pine oil production	Perfume production	Food additive use	Smokeless powder production	Straw hat production	Fire extinguisher production and use	Water purification	Pyrotechnics and explosive making and use	Foundry operations	Paper impregnation	Corrosion inhibitor making and use	Food processing	Ore refining and metal processing
Tetrahydro-furan														X			X											
Propylene Oxide											X						X											
Phenyl Glycidyl Ether																												
Phenyl Ether-Biphenyl Vapor Mixture		X																										
Phenyl Ether Vapor		X																X										
Methylal														X				X										
Methyl Cellosolve													X															
Methyl Acetylene																												
Isopropyl Glycidyl Ether																												
Glycidol																			X									
Ethyl Ether	X												X					X	X									
Dipropylene Glycol Methyl Ether		X						X				X	X				X											
Diglycidyl Ether																												
Cyclopentadiene										X			?		X													
Cyclohexene																												
Chlorinated Camphene																												
n-Butyl Glycidyl Ether																												
2-Butoxy-Ethanol											X																	

TABLE A—7
SUMMARY OF USE/EXPOSURE DATA FOR CHEMICALS IN SCP SET G

Operation \ Chemical	Butadiene	Dioxane	Heptane	Hexane	Ketene	LPG	MAPP	Methyl Cyclohexane	Coal Tar Naphtha	Octachloro-naphthalene	Octane	Pentachloro-naphthalene	Pentane	Petroleum Naphtha	Propane	Propylene Dichloride	Stoddard Solvent	Turpentine
Rubber production	X					X		X								X	X	
Spray-finishing		X	X					X						X			X	X
Non-spray application		X	X	X				X						X			X	X
Paint manufacture		X	X	X	X	X			X					X			X	X
Cleaning and maintenance		X	X	X				X			X	X	X		X		X	
Wood pulping		X																
Plastic & foam manufacture			X	X	X						X							
Pharmaceutical manufacture		X	X	X	X													
Cosmetic production		X																
Textile coating	X							X	X	X	X	X		X			X	
Pesticide application																	X	
Pesticide manufacture									X		X			X			X	X
Ink production			X	X				X			X						X	
Ink application			X	X				X									X	
Textile dyeing		X																
Adhesive production and use		X	X	X					X									
Welding							X											
Abrasive machining						X												
Dry cleaning														X			X	
Aerosol production and use						X									X		X	
Electric equipment manufacture										X		X						
Refrigerant use															X			
Extractant or solvent use			X	X	X			X					X	X	X	X		
Chemical processes	X	X	X	X	X	X		X	X	X	X	X		X	X	X	X	X
Mineral oil refining and dewaxing																		
Photographic film and chemical production																		
Artificial leather production																		
Safety glass production																		

Operation \ Chemical	Butadiene	Dioxane	Heptane	Hexane	Ketene	LPG	MAPP	Methyl Cyclohexane	Coal Tar Naphtha	Octachloro-naphthalene	Octane	Pentachloro-naphthalene	Pentane	Petroleum Naphtha	Propane	Propylene Dichloride	Stoddard Solvent	Turpentine
Anesthetic use																		
Heat transfer fluid use													X					
Mining operations																		
Ceramics and glass production																		
Petroleum refining			X															
Cleaning and degreasing		X	X					X						X		X	X	
Synthetic fiber manufacture																		
Leather tanning and treatment	X																X	
Paint and varnish remover use		X														X		
Drying of paints, varnishes, etc.		X																
Detergent production		X																
Polishing compound production		X									X							X
Dye manufacture					X													
Fuel production			X			X					X		X		X			
Lubricant manufacture and use										X		X				X		
Synthetic pine oil production																		X
Perfume production																		X
Food additive use																		
Smokeless powder production																		
Straw hat production																		
Fire extinguisher production and use																		
Water purification																		
Pyrotechnics and explosive making and use																		
Foundry operations																		
Paper impregnation																		
Corrosion inhibitor making and use																		
Food processing																		
Ore refining and metal processing																		

TABLE A—8
SUMMARY OF USE/EXPOSURE DATA FOR CHEMICALS IN SCP SET H

Operation \ Chemical	Benzyl Chloride	Bromoform	Chlorobromomethane	Chloroprene	Dichlorodifluoromethane	1,2-Dichloroethylene	Dichloromonofluoromethane	Dichlorotetrafluoroethane	Difluorodibromomethane	Ethyl Bromide	Ethyl Chloride	Ethylene Chlorohydrin	Ethylene Dibromide	Fluorotrichloromethane	Hexachloroethane	Hexachloronaphthalene	Methyl Chloride	Methyl Iodide
Rubber production	X	X		X		X									X		X	
Spray-finishing																		
Non-spray application																		
Paint manufacture				X		X												
Cleaning and maintenance		X	X	X	X	X	X	X	X						X			
Wood pulping										X						X		X
Plastic & foam manufacture					X	X	X	X						X				
Pharmaceutical manufacture		X								X	X	X					X	X
Cosmetic production																	X	
Textile coating		X		X									X					
Pesticide application	X		X							X						X		
Pesticide manufacture	X		X	X		X						X		X			X	X
Ink production																		
Ink application																		
Textile dyeing	X					X												
Adhesive production and use						X												
Welding																		
Abrasive machining															X	X		
Dry cleaning						X												
Aerosol production and use					X		X	X			X			X			X	
Electric equipment manufacture															X	X		
Refrigerant use					X		X	X	X	X	X	X	X	X				
Extractant or solvent use		X			X	X	X	X		X	X	X	X	X			X	
Chemical processes	X	X	X	X	X	X	X	X	X	X	X	X	X	X	X	X	X	X
Mineral oil refining and dewaxing																	X	
Photographic film and chemical production																		
Artificial leather production																		
Safety glass production																		

Operation \ Chemical	Benzyl Chloride	Bromoform	Chlorobromomethane	Chloroprene	Dichlorodifluoro-methane	1,2-Dichloro-ethylene	Dichloromono-fluoromethane	Dichlorotetra-fluoroethane	Difluoro-dibromomethane	Ethyl Bromide	Ethyl Chloride	Ethylene Chlorohydrin	Ethylene Dibromide	Fluorotri-chloromethane	Hexachloro-ethane	Hexachloro-naphthalene	Methyl Chloride	Methyl Iodide
Anesthetic use										X	X							
Heat transfer fluid use					X			X										
Mining operations		X	X															
Ceramics and glass production																		
Petroleum refining																	X	
Cleaning and degreasing												X		X	X			
Synthetic fiber manufacture																		
Leather tanning and treatment																		
Paint and varnish remover use																		
Drying of paints, varnishes, etc.																		
Detergent production	X																	
Polishing compound production																		
Dye manufacture					X					X	X	X	X				X	
Fuel production													X					
Lubricant manufacture and use	X														X	X		
Synthetic pine oil production															X	X		
Perfume production							X			X	X						X	
Food additive use																		
Smokeless powder production																		
Straw hat production																		
Fire extinguisher production and use			X		X			X	X				X		X			
Water purification																		
Pyrotechnics and explosive making and use															X			
Foundry operations																		
Paper impregnation																		
Corrosion inhibitor making and use																		
Food processing																		
Ore refining and metal processing																		

TABLE A—9
SUMMARY OF USE/EXPOSURE DATA FOR CHEMICALS IN SCP SET I

Operation \ Chemical	Acetylene Tetrabromide	Allyl Chloride	Chlorinated Diphenyl Oxide	Chlorobenzene	Chlorodiphenyl, 42% chlorine	Chlorodiphenyl, 54% chlorine	1,3-Dichloro-5,5-Dimethylhydantain	1,1-Dichloroethane	Epichlorhydrin	1,1,1,2-Tetrachloro-1,2-Difluoroethane	1,1,1,2-Tetrachloro-2,2-Difluoroethane	1,1,2,2-Tetra-Chloroethane	Tetrachloro-naphthalene	1,1,2-Trichloro-1,2,2 Trifluoroethane	Trichloro-naphthalene	1,2,3-Trichloro-propane	Trifluoromono-bromomethane
Rubber production				X	X	X	X	X	X								
Spray-finishing				X	X												
Non-spray application				X		X						X					
Paint manufacture				X	X	X	X		X								
Cleaning and maintenance	X			X	X	X	X			X	X	X	X	X	X		X
Wood pulping																	
Plastic & foam manufacture	X				X	X	X		X	X	X		X	X		X	X
Pharmaceutical manufacture		X		X					X	X	X			X			
Cosmetic production					X	X			X								
Textile coating				X	X	X			X				X		X		
Pesticide application				X	X	X	X	X	X			X					
Pesticide manufacture			X	X	X	X						X					
Ink production			X	X	X	X											
Ink application				X													
Textile dyeing				X	X	X	X		X								
Adhesive production and use				X	X	X				X	X	X		X			
Welding									X								
Abrasive machining					X								X		X		
Dry cleaning										X	X			X			
Aerosol production and use																	
Electric equipment manufacture			X	X	X	X				X	X		X		X		
Refrigerant use														X		X	X
Extractant or solvent use				X				X		X	X	X		X		X	
Chemical processes	X	X	X	X	X	X	X	X	X	X	X	X	X	X	X	X	X
Mineral oil refining and dewaxing								X									
Photographic film and chemical production												X					
Artificial leather production																	
Safety glass production																	

Operation \ Chemical	Acetylene Tetrabromide	Allyl Chloride	Chlorinated Diphenyl Oxide	Chlorobenzene	Chlorodiphenyl, 42% chlorine	Chlorodiphenyl, 54% chlorine	1,3-Dichloro-5,5 dimethylhydantain	1,1-Dichloroethane	Epichlorohydrin	1,1,2,2-Tetrachloro-1,2-Difluoroethane	1,1,2,2-Tetrachloro-2,2-Difluoroethane	1,1,2,2-Tetra-Chloroethane	Tetrachloro-naphthalene	1,1,2-Trichloro-1,2,2-Trifluoroethane	Trichloro-naphthalene	1,2,3-Trichloro-propane	Trifluoromono-bromomethane
Anesthetic use																	
Heat transfer fluid use					X	X											
Mining operations	X																
Ceramics and glass production																	
Petroleum refining																	
Cleaning and degreasing				X								X				X	
Synthetic fiber manufacture										X	X						
Leather tanning and treatment																	
Paint and varnish remover use												X				X	
Drying of paints, varnishes, etc.																	
Detergent production																	
Polishing compound production					X	X											
Dye manufacture				X													
Fuel production																	
Lubricant manufacture and use					X	X							X		X		
Synthetic pine oil production																	
Perfume production				X													
Food additive use																	
Smokeless powder production																	
Straw hat production																	
Fire extinguisher production and use																	X
Water purification																	
Pyrotechnics and explosive making and use																	
Foundry operations																	
Paper impregnation																	
Corrosion inhibitor making and use																	
Food processing																	
Ore refining and metal processing																	

TABLE A—10
SUMMARY OF USE/EXPOSURE DATA FOR CHEMICALS IN SCP SET J

Operation / Chemical	Butylamine	Carbon Tetrachloride	Chloroform	Diazomethane	o-Dichloro-benzene	Diethylamine	Diethylamino Ethanol	Diisopropyl-amine	Dimethylamine	Ethanolamine	Ethylene Dichloride	Methyl Bromide	Methyl Chloroform	Methylene Chloride	Phosgene	Tetrachloro-ethylene	1,1,2-Trichloro-ethane	Trichloro-ethylene
Rubber production	X								X							X		X
Spray-finishing					X													X
Non-spray application					X						X		X	X				X
Paint manufacture			X		X		X											X
Cleaning and maintenance	X	X	X	X	X	X		X	X	X	X		X			X		X
Wood pulping																		
Plastic & foam manufacture	X		X			X	X		X					X				
Pharmaceutical manufacture			X		X	X	X		X					X	X	X		X
Cosmetic production								X										
Textile coating						X	X									X		
Pesticide application	X	X	X		X						X							X
Pesticide manufacture	X				X							X	X			X		
Ink production		X				X	X	X	X									
Ink application													X	X				X
Textile dyeing					X		X											X
Adhesive production and use		X								X	X		X	X		X		X
Welding															X			
Abrasive machining																		
Dry cleaning			X		X											X		X
Aerosol production and use													X	X		X		
Electric equipment manufacture		X											X	X				
Refrigerant use														X				
Extractant or solvent use	X	X	X			X	X	X	X		X	X	X	X		X		X
Chemical processes	X	X	X	X	X	X	X	X	X		X	X	X	X		X	X	X
Mineral oil refining and dewaxing																		
Photographic film and chemical production							X		X				X	X				
Artificial leather production																		
Safety glass production																		

Operation \ Chemical	Trichloro-ethylene	1,1,2-Trichloro-ethane	Tetrachloro-ethylene	Phosgene	Methylene Chloride	Methyl Chloroform	Methyl Bromide	Ethylene Dichloride	Ethanolamine	Dimethylamine	Diisopropylamine	Diethylamino Ethanol	Diethylamine	0-Dichloro-benzene	Diazomethane	Chloroform	Carbon Tetrachloride	Butylamine
Anesthetic use	X															X		
Heat transfer fluid use			X			X								X		X		
Mining operations										X								
Ceramics and glass production				X														
Petroleum refining																		
Cleaning and degreasing	X		X	X	X	X								X				
Synthetic fiber manufacture					X					X								
Leather tanning and treatment										X				X				
Paint and varnish remover use	X		X		X									X				
Drying of paints, varnishes, etc.																		
Detergent production							X	X	X	X			X	X				X
Polishing compound production						X			X			X	X	X				
Dye manufacture										X			X	X		X		X
Fuel production				X						X		X						
Lubricant manufacture and use	X					X						X	X					
Synthetic pine oil production						X												
Perfume production																		
Food additive use					X													
Smokeless powder production																		
Straw hat production																		
Fire extinguisher production and use																		
Water purification				X														
Pyrotechnics and explosive making and use																		
Foundry operations																		
Paper impregnation																		
Corrosion inhibitor making and use																		
Food processing																		
Ore refining and metal processing																		

TABLE A—11
SUMMARY OF USE/EXPOSURE DATA FOR CHEMICALS IN SCP SET K

Operation \ Chemical	Acrylamide	Acrylonitrile	2-Aminopyridine	Aniline	OCBM	1,1-Dimethyl Hydrazine	Ethylamine	Ethylenediamine	n-Ethyl Morpholine	Isopropylamine	Methylamine	Monomethylaniline	Monomethyl Hydrazine	Morpholine	p-phenylene diamine	Propylene Imine	Tetramethyl Succinonitrile	Triethylamine
Rubber production				X		X	X	X		X	X			X	X			
Spray-finishing																		
Non-spray application		X																
Paint manufacture						X												X
Cleaning and maintenance	X	X	X		X	X	X		X	X	X		X	X	X			
Wood pulping	X							X								X		
Plastic & foam manufacture		X		X			X	X	X	X			X				X	X
Pharmaceutical manufacture		X	X	X		X	X	X	X	X	X		X	X		X		X
Cosmetic production														X				
Textile coating	X	X				X	X	X						X				X
Pesticide application		X	X				X											
Pesticide manufacture		X		X														
Ink production				X			X	X		X	X		X	X				X
Ink application				X														X
Textile dyeing	X														X			
Adhesive production and use	X			X														
Welding							X	X										
Abrasive machining																		
Dry cleaning																		
Aerosol production and use																		
Electric equipment manufacture						X												
Refrigerant use												X						
Extractant or solvent use						X	X	X		X	X		X	X	X			X
Chemical processes	X	X	X	X	X	X	X	X		X	X		X	X	X		X	X
Mineral oil refining and dewaxing										X	X	X				X		
Photographic film and chemical production	X			X									X		X			X
Artificial leather production																		
Safety glass production																		

Operation \ Chemical	Acrylamide	Acrylonitrile	2-Aminopyridine	Aniline	OCBM	1,1-Dimethyl Hydrazine	Ethylamine	Ethylenediamine	n-Ethyl Morpholine	Isopropylamine	Methylamine	Monomethylaniline	Monomethyl Hydrazine	Morpholine	p-Phenylene diamine	Propylene Imine	Tetramethyl Succinonitrile	Triethylamine
Anesthetic use																		
Heat transfer fluid use																		
Mining operations	X																	
Ceramics and glass production							X											
Petroleum refining							X											
Cleaning and degreasing								X					X					X
Synthetic fiber manufacture	X	X		X					X						X			
Leather tanning and treatment										X								
Paint and varnish remover use																		
Drying of paints, varnishes, etc.																		
Detergent production							X	X	X	X				X				
Polishing compound production														X				
Dye manufacture		X	X	X		X	X	X	X	X	X	X			X			X
Fuel production	X					X					X		X					X
Lubricant manufacture and use	X		X				X											
Synthetic pine oil production																		
Perfume production																		
Food additive use				X														X
Smokeless powder production																		
Straw hat production																		
Fire extinguisher production and use																		
Water purification	X							X										
Pyrotechnics and explosive making and use																		
Foundry operations											X							
Paper impregnation																		
Corrosion inhibitor making and use				X			X	X			X			X				X
Food processing																		
Ore refining and metal processing																		

TABLE A—12
SUMMARY OF USE/EXPOSURE DATA FOR CHEMICALS IN SCP SET L

Operation \ Chemical	Acetic Acid	Acetic Anhydride	Acetonitrile	Allyl propyl disulfide	Anisidine	Cresol	Dimethylaniline	Dinitro-o-cresol	Formic Acid	Hydrogen Fluoride	Oxalic Acid	Phenol	Phenylhydrazine	Pyridine	Sulfuric Acid	o-Toluidine	Xylidine
Rubber production	X					X			X				X		X	X	X
Spray-finishing																	
Non-spray application																	
Paint manufacture	X	X															
Cleaning and maintenance	X	X	X		X	X		X	X								
Wood pulping											X				X	X	X
Plastic & foam manufacture	X	X	X			X					X	X	X		X	X	X
Pharmaceutical manufacture	X	X	X			X	X		X	X	X	X	X	X	X		
Cosmetic production														X			X
Textile coating															X		
Pesticide application								X									
Pesticide manufacture		X	X			X			X		X		X		X	X	
Ink production						X						X					
Ink application						X											
Textile dyeing		X			X	X											
Adhesive production and use	X						X		X			X	X				
Welding																	
Abrasive machining																	
Dry cleaning	X											X					
Aerosol production and use										X	X						
Electric equipment manufacture																	
Refrigerant use									X	X	X						
Extractant or solvent use	X		X			X			X	X	X	X		X	X		
Chemical processes	X	X	X		X	X	X	X	X	X	X	X	X		X	X	X
Mineral oil refining and dewaxing													X	X			
Photographic film and chemical production	X	X				X				X		X	X				
Artificial leather production																	
Safety glass production																	

Operation \ Chemical	Acetic Acid	Acetic Anhydride	Acetonitrile	Allyl propyl disulfide	Anisidine	Cresol	Dimethylaniline	Dinitro-o-cresol	Formic Acid	Hydrogen Fluoride	Oxalic Acid	Phenol	Phenylhydrazine	Pyridine	Sulfuric Acid	o-Toluidine	Xylidine
Anesthetic use																	
Heat transfer fluid use											X					X	
Mining operations						X					X						
Ceramics and glass production											X		X			X	
Petroleum refining											X	X	X			X	
Cleaning and degreasing			X			X			X				X			X	
Synthetic fiber manufacture	X	X	X													X	
Leather tanning and treatment	X								X		X	X					
Paint and varnish remover use																	
Drying of paints, varnishes, etc.													X				
Detergent production						X											
Polishing compound production	X										X		X	X	X		X
Dye manufacture			X								X			X	X		X
Fuel production					X	X	X						X				
Lubricant manufacture and use																	
Synthetic pine oil production																	
Perfume production		X	X			X	X					X	X				
Food additive use	X					X			X								
Smokeless powder production																	
Straw hat production																	
Fire extinguisher production and use										X							
Water purification																	
Pyrotechnics and explosive making and use		X				X							X	X	X	X	
Foundry operations					X								X				
Paper impregnation											X						
Corrosion inhibitor making and use																	
Food processing				X												X	
Ore refining and metal processing																	

TABLE A—13
SUMMARY OF USE/EXPOSURE DATA FOR CHEMICALS IN SCP SET M

Operation \ Chemical	Copper Dusts & Mists	Crotonaldehyde	Hafnium	2-Hexanone	Maleic Anhydride	Molybdenum Soluble Compounds	Osmium Tetraoxide	Phosphoric Acid	Phosphorus, yellow	Phthalic Anhydride	Platinum Soluble Salt	Quinone	Rhodium	Selenium	Tellurium Hexafluoride	Tin (organic compounds)	Zirconium
Rubber production		X			X			X						X			
Spray-finishing																	
Non-spray application																	
Paint manufacture	X			X	X	X											
Cleaning and maintenance	X	X		X	X				X	X	X	X				X	
Wood pulping																	
Plastic & foam manufacture					X	X				X		X				X	
Pharmaceutical manufacture					X			X		X		X		X		X	X
Cosmetic production	X																
Textile coating	X				X												
Pesticide application																X	
Pesticide manufacture		X			X	X			X			X		X			
Ink production					X	X											
Ink application										X							
Textile dyeing						X											
Adhesive production and use					X			X									
Welding	X		X														
Abrasive machining X																	
Dry cleaning																	
Aerosol production and use																	
Electric equipment manufacture													X	X			X
Refrigerant use																	
Extractant or solvent use				X										X			X
Chemical processes	X	X		X	X	X	X	X	X	X	X	X		X	X	X	X
Mineral oil refining and dewaxing	X																
Photographic film and chemical production		X			X	X	X	X			X	X		X			
Artificial leather production																	
Safety glass production																	

Operation \ Chemical	Zirconium	Tin (organic compounds)	Tellurium Hexafluoride	Selenium	Rhodium	Quinone	Platinum Soluble Salt	Phthalic Anhydride	Phosphorus, yellow	Phosphoric Acid	Osmium Tetraoxide	Molybdenum Soluble Compounds	Maleic Anhydride	2-Hexanone	Hafnium	Crotonaldehyde	Copper Dusts & Mists
Anesthetic use																	
Heat transfer fluid use																	
Mining operations																	
Ceramics and glass production		X		X						X		X					
Petroleum refining												X					
Cleaning and degreasing										X							
Synthetic fiber manufacture													X				
Leather tanning and treatment		X				X						X	X			X	
Paint and varnish remover use																	
Drying of paints, varnishes, etc.																	
Detergent production				X						X		X	X				
Polishing compound production		X										X					X
Dye manufacture		X		X				X					X				
Fuel production			X														
Lubricant manufacture and use			X	X									X				X
Synthetic pine oil production																	
Perfume production								X									
Food additive use				X						X							
Smokeless powder production																	
Straw hat production																	
Fire extinguisher production and use																	
Water purification																	
Pyrotechnics and explosive making and use	X			X					X								
Foundry operations	X			X													
Paper impregnation													X				
Corrosion inhibitor making and use												X				X	
Food processing																	
Ore refining and metal processing	X			X	X		X		X	X	X	X			X		X

CHAPTER FIVE

TOXIC SUBSTANCES AND WORKER HEALTH

TOXIC SUBSTANCES POSING
CANCER RISK TO WORKERS[*]

Overview

It is proposed to deal with the identification, classification and regulation of toxic materials, for which there is evidence of carcinogenic potential to man by adding a new part to Title 29 of the Code of Federal Regulations ("CFR"). This Part would apply to all employments in all industries covered by the Act, including general industry, construction, maritime and agriculture. The Occupational Safety and Health Administration ("OSHA") requests the submission of written comments, data and arguments from interested persons on the variety of issues addressed or implicit in the proposal. In addition, an informal hearing has been scheduled, to provide an additional opportunity for discussion of the issues and to facilitate this rulemaking.

This proposed set of regulations, attempts to deal with one of the most important issues OSHA faces, namely the exposure of workers to toxic materials which may be potential or confirmed occupational carcinogenic substances in the workplace. At the outset, OSHA recognizes that some 1,500 agents have been identified by the National Institute for Occupational Safety and Health ("NIOSH") as being "suspect carcinogens". This means that NIOSH has found some scientific evidence, admittedly of varying quality, identifying those substances as having potential carcinogenic activity, based on observations in human populations or on results from experimentation with laboratory test animals. Yet, OSHA has completed regulatory activity for only 17 of those substances. Likewise, OSHA recognizes that in regulating occupational exposure to carcinogens, many gaps remain in our knowledge of cancer, its causes and prevention. However, to wait for years to scientifically resolve these issues without some consistent and workable system today for the regulation of toxic materials with evidence of a carcinogenic potential to man would be, we believe, inconsistent with OSHA's statutory obligations and unacceptable to all.

Thus, OSHA today proposes an orderly and comprehensive set of regulations to identify, classify and regulate potential carcinogens in American workplaces. In brief, this action is based upon three propositions, namely:

1. That the term "carcinogen", although perhaps difficult to define, as a matter of science, must be defined for purposes of overall regulatory activity.

2. That a toxic material that is confirmed as a carcinogen to a mammalian test animal species, as defined, is to be treated *as a policy matter* as posing a carcinogenic risk to man.

3. That when OSHA is dealing with a toxic material, identified as a carcinogen, as defined herein, the permissible exposure will be as low as feasible. In cases where there are suitable substitutes that are found to be less hazardous to the worker, no occupational exposure to the toxic material will be permitted. In other words, unless there is evidence sufficient to convince the Secretary of Labor that *the general policy* asserted herein is incorrect namely, that there is presently no means to determine a safe exposure level to a known carcinogen, the permissible exposure will be set as low as feasible or not permitted in certain cases.

In short, utilizing best available and generally accepted scientific knowledge, to establish the carcinogenicity of a substance, OSHA intends to rely on evidence from human epidemiological studies, adequately designed and conducted animal studies, or both. The degree of conclusiveness of such data will permit classification of the substance as a "confirmed" (classification I agent) carcinogen or a "suspect" (classification II agent) carcinogen, or neither (classification III agent) for which further analysis of the data may be needed. Such classifications will trigger appropriate regulatory action based on the quality of the scientific data as defined, presumptively, to limit or eliminate exposure to the toxic material whether by way of engineering controls or by way of work practices such as simple housekeeping procedures.

OSHA recognizes that some of the issues are still subject to controversy. Most, however, are not new but have, in fact, been the basis for policy decisions made by OSHA and other governmental agencies in the past. For example, see the preambles to OSHA's carcinogen standard, applicable to 14 selected substances, 29 CFR 1910.1003-1910.1016 (39 FR 3758), *aff'd Synthetic Organic Chemical Manufacturers v. Brennan* 503 F2d 1151 (3rd Cir., 1974); the vinyl chloride standard, 29 CFR 1910.1017 (39 FR 35892) *aff'd Society of the*

---

[*] Proposed Rulemaking: Identification, Classification and Regulation of Toxic Materials Posing a Potential Occupational Cancer Risk to Workers, Occupational Safety and Health Administration, January 1977.

Plastic Industry v. Departrtment of Labor (2nd Cir, 1975) *cert den. (95 S. C. 1998, 44 L ED. 2nd 482;* the inorganic arsenic proposal (40 FR 3392, 1975); the coke ovens emissions proposal (40 FR 322268, 1975) and final (41 FR 46742, 1976); the asbestos proposal (40 FR 47652, 1975); and the beryllium proposal (40 FR 48814, 1975). In addition to previous OSHA regulatory decisions, the Environmental Protection Agency ("EPA"), the Food and Drug Administration ("FDA"), and other governmental agencies and bodies have made regulatory decisions that substances pose carcinogenic risks to man, based primarily or solely upon evidence derived from studies conducted in laboratory animal populations. In particular, decisions made by EPA regarding the cancellation or suspension of certain chlorinated hydrocarbon pesticides involved extensive consideration of carcinogenicity based on laboratory animal data. See: *In Re Stevens,* (37 FR 13369, 1972) *aff'd Environmental Defense Fund* v. *EPA* 465 F2d 528 (DC Cir. 1973); *In Re Shell,* 39 FR 37265 (Oct. 18, 1974), *aff'd EDF* v. *EPA* 510 F2d 129 (DC Cir. 1975); and *In Re Velsicol,* (41 FR 7552, 1976) *aff'd EDF* v. *EPA* F2d (DC Cir. 1976). Further, FDA, in its decisions to order the elimination of chloroform from human drug and cosmetic products, relied explicitly on data supplied by the National Cancer Institute ("NCI")'' experiments showing chloroform to be carcinogenic in mice (41 FR 15026-29, April 9, 1976 and 41 FR 26842-46, June 29, 1976).

We know that some may question the policy decisions inherent in this proposal as going too far, while others may contend that they do not go far enough. We ask, however, that all concerned approach these issues in a spirit of co-operation and candor—and with the recognition that policy determinations cannot be based solely upon scientific fact where often factual data are incomplete. But insofar as the central issue of whether exposure to carcinogens should be regulated, we believe that the time has come to say that at least this issue has been resolved.

In proposing these regulations, OSHA is relying upon leading scientific evidence and opinions believed to reflect the research conclusions of individual cancer specialists and expert national and international cancer committees and agencies. In such a task we recognize that we are operating on the frontiers of knowledge. We rely however upon what we believe to be the best available evidence and interpretations and are prepared to modify our views in these regulations, if new evidence or future scientific advances show we are in error. But if we are, we hope that the error was in being too careful and overly protective of the human being for the alternative is unacceptable.

The Regulatory Dilemma

Any decision to regulate carcinogens is obviously complex. As the cancer rate increases, the causes elude us.

And with the increasing number of environmental chemicals, the number of carcinogens also increase, together with the size and complexity of OSHA's rulemakings. In its five year history, OSHA has concluded only 4 rulemaking proceedings concerning carcinogens, namely the asbestos standard in 1972, the carcinogen standard in January 1974 (regulating 14 substances), the vinyl chloride standard in October 1974 and the coke oven emissions standard in October 1976.

And, chemical or physical agents known or suspected to pose a risk of carcinogencity in man pose certain unique problems to the regulation of toxic materials in the workplace which are discussed below.

Thus, we propose today a system for regulatory action which will assure, we believe, an internal consistency of approach in regulating carcinogens, a speedy approach and an approach which will limit the size of OSHA's rulemakings which have grown far beyond the ability of OSHA's staff to handle by the present case-by-case approach.

Nature of the disease

Cancer is a particularly dreaded and costly disease. As pointed out below, the natural course of its development is generally irreversible and autonomous. It is a disease generally characterized in its advanced stages by the aggressive growth of abnormal (immature) populations of cells. The early cellular events of the disease and the causative factors leading to initiation of the disease are poorly understood although much progress has been made as a result of intensive research for several decades. There also is evidence—though not definitive—that at least some cancers may originate from a single transformed cell.

In any case, once the initial carcinogenic events have been triggered, the resulting aberrant cells are capable of progressing to stages where normal tissue is invaded, and cancerous cells are capable of spreading throughout the body, even though the agent or agents responsible for inducing the disease may be no longer present. It also appears that the effects of continuous or occasional exposure to different carcinogenic agents may be cumulative or synergistic. The development of cancer can take as long as twenty to thirty years before the disease progresses to the point where detection is possible. Thus, cancer seen today may be due to exposure occurring as little as five or as many as forty or more years ago.

The Increase In Cancer and Its Massive Economic Impact

Only a few generations ago, human life expectancy was 30 years. Today, however, human life expectancy and general health conditions prevailing in developed countries such as the United States reflect a striking—and comparatively recent—improvement over those that characterized most of man's long history.

Throughout the 19th century, the prime life limiting factor in this country was infectious disease. Since 1900 however, public health measures and medicine have made remarkable contributions to human longevity in the United States. A man born in Massachusetts in 1850 had a life expectancy of 38.3 years; a woman, of 40.5 years. As the decades passed, life expectancy at birth inched upward—by 1900 a man could expect to live to 46 years and a woman to nearly 50. Today, such expectancy is much greater.

A comparison of leading causes of death in 1900 with those in 1960 reveals a noteworthy shift in disease patterns. At the turn of the century, infectious diseases—pneumonia, influenza, tuberculosis, and gastroenteritis—were the leading killers, accounting for 31 percent of all deaths. Owing to such medical advances as the development of immunization techniques and antibiotics, tuberculosis and gastroenteritis are no longer significant causes of death in the United States, diphtheria and typhoid fever have been nearly eliminated, and even influenza and pneumonia have declined in significance to one-seventh of their previous level.

On the other hand, heart disease and cancer—the fourth and eighth leading causes of death, respectively, in 1900 now lead the nation's list of killers. In 1900, these two diseases were responsible for 12 percent of deaths; today, they account for more than one-half of U.S. mortality. Annual rates of death from heart disease rose from 137.4 per 100,000 population in 1900 to 362 in 1970, while deaths from cancer rose from 64 per 100,000 population to 168.

Some increase in the cancer mortality rate may have been predicted on the basis of demographic factors, such as population growth and general aging of the population, and hence on the success of medicine in reducing the significance of infectious diseases. Another unquantifiable portion is undoubtedly traceable to improved reporting methods, i.e., some deaths from cancer in earlier times were mistakenly ascribed to other causes. Even after these factors have been taken into account, however, the death rate for cancer sharply exceeds predictions.

Cancer killed a reported 358,400 U.S. citizens in 1974—almost 1,000 persons per day. Over one million are under treatment for the disease, and each year 900,000 new cases are diagnosed. Of these, about one-third are skin cancers: troublesome, sometimes painful, usually treatable, and with little significant impact on life expectancy; the other 600,000, however, are serious and are potentially fatal. The American Cancer Society now estimates that 25 percent of the United States' population will ultimately develop some form of cancer.

The economic and social impacts of cancer in the United States are massive and hard to estimate. Certainly the human anguish is. An estimated $1.8 billion per year is spent solely for hospital care of cancer patients. Additional costs, doctor bills, outpatient therapy, and other treatment-related fees, raise the direct expenditures for cancer well into the tens of billions of dollars. To these direct expenditures must be added indirect costs, such as the estimated 1.8 million work years lost to the national economy and to family income by unemployed or underemployed cancer victims. One estimate cited by the General Accounting Office of the United States Congress is that the annual cost of cancer is $15 billion, of which $3-5 billion is attributable to direct care and treatment and the remainder attributed to the loss of earning power and productivity (GAO, 1976, p. 1). This estimate is supported by a study prepared by the National Cancer Institute ("NCI") on the Economic Costs of Cancer to the United States, and to individual patients' families in particular. See "The National Cancer Program, the Strategic Plan", January 1973 ed.

Cancer rates vary significantly through the United States. In general, however, states with high rates are the industrial states. It has long been known that densely populated and industrialized areas have higher death rates from many causes than nearby rural areas. Although urbanization and mortality have been associated for heart disease, cancer of the respiratory and digestive systems, and many other diseases, the reasons for this association are not fully understood. This excess health risk may be related to lifestyle (urban dwellers use more tobacco, for example), to occupation (working with industrial pollutants), to the environment of cities (air pollution and water pollution), or other factors as yet unidentified, as well as combinations thereof.

The Recognition that the Development of Cancer is Influenced by Environmental and Occupational Factors

Most of the details of how and why cancers develop still elude scientists and physicians, so that the exact causative mechanisms of observed malignancies cannot be well defined. But a number of factors appear to be implicated, both separately and in combination, for many observed cancers.

Among the causes of cancer, those most prominent are believed to be traceable to environmental factors acting in conjunction with genetic susceptibilities. It should be remembered that the extensive exposure to cancer-causing environmental factors is relatively new in the history of man. Today, there is growing recognition that 60 to 90 percent of all cancer may be related to environmental factors. The basis for the estimates largely derives from large community studies over extended periods of time, which have revealed wide geographical variations in the incidence of cancer of various organs. Specific examples include, among others, the large component of lung cancer attributable to cigarette smoking, exposures to chemicals in the workplace, and cancers from physical agents such as solar and ionizing radiation, asbestos, and aflatoxins (a class of naturally occurring chemicals secreted by some molds on certain foodstuffs, such as peanuts).

The extent to which the observed incidence and rise in incidence of cancers are atrributable to manmade chemicals cannot be estimated with any precision however, but the tragic effects are evident. Recognition by cancer specialists that many of human cancers are influenced by environmental factors is of extraordinary significance to OSHA—it means that most occupational cancers may be preventable if the causative agents can be identified and human exposure to them eliminated or minimized.

Only some of these manmade agents have been identified. Unfortunately, our nation's capacity to develop new chemical substances far exceeds its present ability to determine their carcinogenic potential and, as pointed out below, this agency's inability to promulgate standards in any speedy fashion by way of the present case-by-case system. In the past 10 years, the production of synthetic organic chemicals has expanded by 255 percent; relatively few of the new compounds have been studied for their cancer-causing potential. Because of the typical latency period of 15-40 years for cancer, we must assume that any increase in cancer which may have been initiated by recent industrial development is not yet observable.

Most Agents Appear Not to be Carcinogens Even at High Doses

As additional studies are reported of the carcinogenicity of substances representing a wide range of chemical structures and used for a variety of purposes, there has been tendency for some to suspect that all chemicals are capable of causing cancer if administered to experimental animals at sufficiently high doses. This suspicion appears to be based on the belief that carcinogenesis is a nonspecific kind of biological process associated with any substance under the right conditions, and that animals bred for their sensitivity to carcinogens when fed maximum tolerated doses of a substance for their lifetime are very ill, they may be more susceptible to carcinogenic effects. However, there is evidence that renders this suspicion very unlikely to be true.

A publication of the U.S. Department of Health, Education, and Welfare compiles and abstracts all substances tested for carcinogenic effects. ("Survey of Compounds Which Have Been Tested For Carcinogenic Activity," Volumes I-VII (1947-1976)). Although these volumes are not intended to provide a critical evaluation of the data on each of the substances, only about 17% of

over 6000 were reported as showing tumorigenic effects and this in spite of the fact that most of the compounds were selected because of their suspicious nature.

In a study sponsored by the National Cancer Institute from 1965 to 1968 and performed by the Bionetics Research Laboratories under contracts PH 43-64-57 and PH 43-67-735, approximately 130 pesticide and industrial chemicals were tested for carcinogenic activity. Pesticide chemicals, of course, are known to possess high biological activity by virtue of their use as pesticides. In addition, the substances tested in the Bionetics study were selected on the basis of a number of criteria, among which were: (1) published evidence of toxicity that suggested potential hazards to man and (2) the similarity between possible test substances and substances of known carcinogenic activity. Each substance was fed to both sexes of two strains of mice at a maximum tolerated dose determined at a young age by continuous oral administration with both positive and negative controls for 18 months. Despite weighting in the choice of test substances, less than 10% of the substances tested produced tumorigenic effects. (R. Innes et al., "Bioassay of Pesticides and Industrial Chemicals for Tumorigenicity in Mice: A Preliminary Note." J. Natl. Cancer Inst 42, 1101-14, 1969.)

Furthermore, it has been recognized for many years that the negative findings of the NCI carcinogenic bioassays often are generally not published and are even rejected for publication (Report of the Panel on Carcinogencity, Int. Union Against Cancer, "Carcinogenicity Testing," edited by I. Berenblum, IUAC, Geneva, 1969).

During the hearings to consider the intent of the Administrator of the U.S. Environmental Protection Agency to cancel the registration of pesticides aldrin and dieldrin, testimony was presented by two government officials who have long experience in the testing of substances for carcinogenic activity. The Associate Director for Carcinogenesis of the Division of Cancer Cause and Prevention at NCI, Dr. Umberto Saffiotti, stated:

"Not all chemicals can cause cancers, contrary to some ill-informed belief; in fact, only a relatively small proportion of the chemical species that have been studied are capable of such activity. The activity is dependent on a number of specific chemical and physical properties of the molecules, which are gradually brought to light. It can be expected that, when thousands of new environmental chemicals will be tested, only a few hundred, or may be just a dozen, will be found to be carcinogenic, and it will be possible to concentrate our preventive measures on those." (Exhibit 40 of EPA, pp. 7-8, Public Hearings concerning the cancellation of the pesticide registration of Aldrin and Dieldrin, 1973).

Dr. Adrian Gross, Assistant Director for Scientific Coordination in the Office of Pharmacological Research and Testing, Bureau of Drugs, of FDA stated:

"... literally tens of thousands of agents have been tested for carcinogenic activity, many tens of thousands of agents. Only a miniscule fraction of these have been found to be carcinogenic. It is certainly much less than 1 percent." (p. 9222 of the Transcript of the Suspension Proceeding, Public Hearings concerning the suspension of the pesticide registrations of Aldrin and Dieldrin, 1974).

Thus, the available evidence provides strong indication indeed that all chemicals are not capable of causing cancer and that only a small number of the total is so capable. This apparent fact has also been recognized by two Administrators of the Environmental Protection Agency. In the DDT case, *In Re Stevens*, EPA IFR Docket No 63 *et al*, 37 F.R. 13369 (1972), *aff'd EDF* v *EPA* 465 F2d 528 (D.C. Cir. 1973) Administrator William D. Ruckelshaus stated:

"The 'everything is cancerous' argument fails because it ignores the fact that not all chemicals fed to animals in equally concentrated doses have produced the same tumorigenic results."

Likewise, in the Aldrin/Dieldrin case, *In Re Shell Chemical Co.* EPA FIFRA Docket No. 145, 39 F.R. 37265 (1974) *aff'd EDF* v *EPA* 510 F2d 129 (D.C. Cir. 1975), Administrator Russell E. Train stated:

"Carcinogenicity is a relatively rare phenomenon exhibited by only a few of the many hundreds of thousands of chemicals."

And in ordering the suspension of the pesticides Heptachlor and Chlordane in *In Re Velsicol Chemical Co. et. al.*, EPA FIFRA Docket No. 384, 41 F.R. 7572 (1976), *aff'd EDF v. EPA _____ F2d _____* (D.C. Cir 1976) Administrator Train also stated:

"Finally, I have noted some tendency, not entirely absent from the record, to assert that any chemical, if fed in sufficiently large amounts, will cause cancer in test animals. This is not true. A study sponsored by the National Cancer Institute tested 140 pesticides and industrial chemicals in two strains of mice, and less than ten percent of these were found to be carcinogenic."

Latency and Irreversibility of Effect

Another very troubling problem to the regulation of chemical carcinogens is posed by two biological characteristics of cancer that *are* well established and that distinguish it from other processes of chronic toxicity. These are the general irreversibility of effect and the generally long latent period between exposure to the carcinogen and manifestation of the effect, namely the tumor.

Irreversibility—Irreversibility of effect is fundamental to man's hoped for understanding of the mechanism of chemical carcinogenesis. The effect under consideration is the induction of a critical change in the target cells which determines their subsequent growth as tumor cells. The specific molecular target of a carcinogen within a cell is not yet fully identified, although it is known that carcinogens interact directly with the genetic material of the cells (DNA), as well as with other molecules that control cellular functions (RNA and proteins). (Cite: For instance, see I.B. Weinstein, et al. "Use of Epithelial Cell Cultures for Studies on the Mechanism of Transformation," *In Vitro*, 11, 130-141 (1975).)

Once a cell has been switched ("transformed") from its normal status to a neoplastic one, this cell can replicate and then will produce neoplastic daughter cells. Only a few initially transformed neoplastic cells may be sufficient to give rise to a growing tumor. There is experimental evidence that even a single cell can be transformed by chemicals to produce a malignant tumor. It is consistent with our knowledge of carcinogenesis that a relatively small number of molecules may be sufficient to trigger a neoplastic change in a target cell. Following such an initial event, exposure to the carcinogen is no longer required to maintain the new cancer cell, which is then "on its own." Thus a single biological event produced by a very small number of molecules of the carcinogen may be sufficient to initiate the irreversible development of a tumor.

Long Latent Period—The other important characteristic of the carcinogenic effect is that it is a *delayed effect*, one where the manifestation (the appearance of a tumor) follows the causative exposure by a long period of time, called the *latent period*, during which the action of the carcinogen on the host, although already effected, cannot be detected. In man as well as in experimental animals, the latent period of chemical carcinogens is often as long as a major portion of a lifetime. It has been found to be as long as 15, 20, 30 or more years in man, while it is usually from 1 to 2 years or even more in rodents, which live up to two or three years. It is generally uncommon for human cancers to appear within 5 to 10 years of exposure to a chemical carcinogen. In most cases the latent period for human carcinogenesis is 20 to 30 years. In case histories of human exposure to chemical carcinogens with such long latent periods, such as cigarette smoke or asbestos, conclusive epidemiological findings were not available until 20 or more years after the start of exposure, by which time large numbers of persons had been exposed and a substantial cancer epidemic had been caused. Thus, as pointed out below, the significance of the long latent period for chemical carcinogenesis is that *it is impractical and imprudent to wait for results of epidemiological studies in man;* prevention of chemical carcinogenesis in man requires identification of the carcinogens by experimentation in laboratory animals with a short lifetime and a correspondingly short latent period for tumor induction. As the Administrator of EPA pointed out in *In Re Velsicol, supra:*

"... the latency periods for cancer inducing chemicals tend to be long (in the tens of years), so that what results there are from the study of human exposure are difficult to gather and may well appear too late to save people from an already widespread chemical to which they are unavoidably exposed."

Known Human Carcinogens are also Carcinogens in Animals

At least 28 chemical substances or mixtures are known to cause cancer in man; in most cases this knowledge is derived from epidemiological studies. With one possible exception, all of these are known to cause cancer in experimental animals. In the text of a review published by Dr. U. Saffiotti:

"Over one thousand chemical substances have been shown to be carcinogenic by tests in animals (although some of the 'positive' tests reported in the literature—particularly the older ones—are not quite satisfactory by present standards). Several individual chemicals or mixtures of chemicals have also been shown conclusively to be carcinogenic by direct observation in man (Table 1). With the exception of arsenic, still under experimental study, all the main products that were found to be carcinogenic by direct evidence in man have also been proven carcinogenic in animals. On the other hand, proof that a substance, which had been recognized as carcinogenic in animals, actually causes cancer in man would require in most cases extremely complex and lengthy epidemiologic studies. In many cases, it may be impossible to obtain such proof because of the complexity of controls that would be needed for a satisfactory demonstration. Therefore, the only prudent course of action at the present state of our knowledge is to assume that chemicals which are carcinogenic in animals could also be so in man, although the direct demonstration in man is lacking."

In a report by a Committee of the National Research Council (Contemporary Pest Control Practices and Prospects, Vol. 1, 1975: "NRC Pest Control"), published data on 13 compounds known to cause cancer in man are tabulated: 12 of them are known to cause cancer in mice, rats, and/or hamsters (NRC Pest Control, pp. 67-73).

"Notwithstanding the difficulty of identifying specific carcinogens affecting man, about a dozen agents are now recognized as human carcinogens. All of these but one are definitely carcinogenic in tests on laboratory animals, as indicated in Table 4. The only exception is arsenic, which is still under test on animals. It may be noted that the organs affected in man are not always the same as those that are found to develop tumors in laboratory animals, nor is the organ specificity the same in different species of rodents. Nevertheless, the evidence, based on a limited number of carcinogens, suggests that most agents that pose a carcinogenic threat to man will be carcinogenic in laboratory tests on animals." (NRC Pest Control, p. 66).

The NRC Committee also tabulated numerical data on the doses of these chemicals associated with increased cancer incidence (*ibid.*, pp. 75-81, Table 5). They found an approximate correspondence between the sensitivities of humans and rodents when they were compared on the basis of cumulative lifetime exposure:

"Thus, as a working hypothesis, in the absence of countervailing evidence for the specific agent in question, it appears reasonable to assume that the lifetime cancer incidence induced by chronic exposure in man can be approximated by the lifetime incidence induced by similar exposure in laboratory animals at the same total dose per body weight." (*Ibid.*, p. 82).

Regulatory Approach

This proposal marks a departure from OSHA's usual pattern of health standard's substance-by-substance approach. One obvious result of OSHA's current case-by-case approach to the regulation of carcinogens has been the relitigation of certain issues in each and every rulemaking. This taxes witnesses who have in the past been willing to testify on prior occasions. In addition, the approach hardly guarantees a continuity of approach in every case. And finally, the manpower resources of this Agency are strapped by the present approach. It is OSHA's belief that if this proposal or something similar is not promulgated, with present resources the output of standards to protect American workers from carcinogens will never be adequate and may collapse by means of the futility of the effort. Indeed, to follow the present system and procedure for every individual substance and hazard would be, we contend, beyond the abilities of any agency, no matter how large a staff, it may have.

The vinyl chloride standard (29 CFR 1910.1017) provides a striking example of the effort which has been required for standard promulgation, under optimum circumstances pursuant to the *de novo* case-by-case approach. That standard, developed on a top priority basis with resources borrowed from other on-going projects, took the full statutory six months from the issuance of the emergency temporary standard to hold required hearings, adequately review the record and publish a final rule. The vinyl chloride rulemaking proceedings concerned a substance which undisputedly was a human carcinogen. Yet, controversy raged during the hearing concerning the precise level of exposure which posed a hazard and whether a "safe" exposure level existed. The comments and testimony concerning these issues were discussed in a significant part of the more than 600 written comments, 200 written and oral hearing submissions, the hearings, and the 4,000 page record.

Notwithstanding the bulk and comprehensiveness of the record, OSHA finally determined that the record did not provide definitive answers to these questions. The Agency declared, "we cannot wait until indisputable answers to these questions are available, because lives of employees are at stake," (29 FR 35892).

Certain issues predictably and inevitably reoccurred in the vinyl chloride proceeding and other proceedings concerning highly toxic, and carcinogenic substances, such as arsenic and coke oven emissions. Yet, the determinations on these issues have been made by the Agency as a matter of "policy", not necessarily with factual certainty. These determinations have been upheld as policy matters. For example, the Agency's determination in the standard regulating exposure to one of the 14 carcinogens (ethyleneimine "EI"), to the effect that if carcinogencity for kinds of animals studies is established, then the substance should be treated as a human carcinogen, was characterized by the Third Circuit as "not really a factual matter." Rather, the court stated, the determination "is in the nature of a recommendation for prudent legislative action," i.e., a policy decision. *SOCMA v. Brennan,* 503 F. 2d 1151 (1974).

To dedicate substantial resources to the rehearing, in each rulemaking, of these kinds of policy issues is truly non-productive.

OSHA, therefore, like other Administrative agencies in the past has determined that it is necessary, to the fulfilment of its statutory objectives to reshape the size and content of its rulemaking proceedings, at least insofar as potential carcinogens are concerned. This proposal incorporates policy determinations concerning how and when chemical or physical agents should be classified and consequently regulated as human carcinogens. It is OSHA's intention, once this proposal is duly promulgated, to foreclose in subsequent 6(b) rulemakings the rehearing of the validity of this classification scheme and most other policy determinations made in this proposal, including the procedural structure intended to be followed.

For example, one such foreclosed policy determination is that no employee shall be exposed to a "potential occupational carcinogen" above the lowest level feasible. Accordingly, during subsequent Section 6(b) proceedings, interested parties would be foreclosed from the present endless debate concerning whether permissible exposure limits to humans can be related or extrapolated to those levels of exposure in test animals that indicated evidence of carcinogenesis.

OSHA intends and expects that these and the other issues raised by this proposal will be fully discussed, debated and aired in this rulemaking proceeding. At the conclusion of these proceedings, these policy decisions, including the regulatory framework embodied in this new Part 1990, will have been subjected to public proceedings, their scientific bases debated, and their consequences delineated. Consequently, the Agency believes that this "generic" form of standard setting will not in any way short cut the statutory procedural requirements set forth in section 6(b) of the Act or in 5 U.S.C. 553, for every opportunity for notice, comment, and public participation in extensive hearings pursuant to those provisions will have been afforded.

It has long been recognized by the courts that regulatory agencies, such as OSHA, must and can use innovative rulemaking procedures to set policy, in order to comply with their statutory mandates, even at the expense of nominally depriving interested parties of a hearing on an individual claim which is contrary to the generally announced policy. Thus in 1956, the Supreme Court held that the Federal Communications Commission was not required to give a license applicant a statutorily required hearing, when the applicant did not qualify under the Commission's policy, adopted through rulemaking, that no person could own more than 5 stations. (*United States v. Storer Broadcasting Co.,* 351 U.S. 152, 202. *See also, FPC v. Texaco,* 377 U.S. 33 (1964) *Airlines Pilot Asso. v. Quesada,* 276 F. 2d 892 (2nd Cir. 1960)).

As stated above, the procedure involved in this proposal affords parties every statutory procedural right mentioned. It deprives parties only of the opportunity to have more than one chance to reargue contentions fully heard, evaluated and rejected by the Agency in its policy determinations, when no new reason nor scientific advance is shown to warrant a rehearing.

The overriding concern in these cases was whether parties had unreasonably been deprived of statutory procedural rights which were intended to assure that the agency would not act arbitraily in denying claims.

We believe, the combination of the broadly based rulemaking hearing and proceeding concerning this proposal and the subsequent section 6(b) proceedings on individual standards provide all parties with ample opportunity to present their views and comments at the most relevant procedural stages. Only those "legislative" policy determinations made herein, will be foreclosed from rehearing at the subsequent section 6(b) hearings. Those policy decisions based more on factual considerations, such as the content of specific protective provisions such as monitoring, respirator use and protective clothing and the lowest feasible exposure limit will always be open to discussion in the section 6(b) hearing.

In particular, the following questions will always be at issue in the subsequent proceedings relating to individual substances: (1) whether the Secretary correctly classified the toxic material according to the appropriate criteria, (2) whether the Secretary correctly decided that the classification should not be rebutted (3) the determination of the lowest feasible occupational exposure, or whether there are suitable substitutes found that are less hazardous to humans than the toxic materials, (4) the appropriateness and feasibility of the specific protective measures of the proposed standard and (5) the environmental impact arising from regulation of the toxic material.

Therefore, OSHA believes the flexibility and fairness of the individual substance by substance approach will be preserved by the procedures and policies incorporated in this proposal just as now, the feasibility and appropriateness of the protective provisions as applied to a given substance will be fully discussed at the rulemakings concerning a specific substances. Yet OSHA's policy judgments whose factual bases mainly are "on the frontiers of knowledge", will be established after opportunity for full public participation in this rulemaking, and any future amendments thereto, and will not be allowed to be relitigated" in the standard by standard process.

Subpart A
General

1990.1. Scope

This part provides for the regulation of certain toxic materials, as required by the Occupational Safety and Health Act of 1970 ("the Act").

Subpart B
Identification, Classification and Regulation of Toxic Materials Posing a Potential Occupational Carcinogenic Risk to Humans

1990.2. Scope

This subpart provides for the identification, classification and regulation of toxic materials which pose an occupational carcinogenic risk to humans.

1990.3. Definitions

Terms used in this subpart shall have the meanings set forth in the Act. In addition, as used in this subpart, the following terms shall have the meanings set forth below:

The term "Act" means the Occupational Safety and Health Act of 1970 (Pub.L. 91-596, 84 Stat. 1590).

The term "Administrator of EPA" means the Administrator of the United States Environmental Protection Agency.

The term "Director of NIOSH" means the Director of the National Institute for Occupational Safety and Health, United States Department of Health, Education and Welfare or his designee.

The term "Director of NCI" means the Director of the National Cancer Institute, United States Department of Health, Education, and Welfare or his designee.

The term "multi-test evidence" for mutagenicity includes, but is not limited to: (a) positive results in more than one of the following assays for: (1) the induction of DNA damage and repair; (2) mutagenesis in bacteria, yeast, *Drosophila melanogaster,* or in mammalian somatic cell cultures; (3) the dominant lethal test and (4) evidence of chromosomal damage; or (b) positive results in tests for neoplastic transformation of mammalian cells in culture.

The term "mutagenic" means the property of a toxic material to induce changes in the genetic material of either somatic or germinal cells or tissues in subsequent generations.

The term "potential occupational carcinogen" means any toxic material in the workplace, which (a) produces or induces, at any level of exposure or dose, as the result of any oral, respiratory or dermal exposure, or any other exposure which results in the systemic distribution in the organism of the substance under consideration, an increase of benign or malignant neoplasms or tumors, or a combination thereof, in (1) humans or (2) in one or more experimental mammalian species, or (b) in a statistically significant manner decreases the latency period between exposure and onset of neoplasm or tumor formation in either humans or in one or more experimental mammalian species.

The term "Secretary" means the Secretary of Labor or his designee.

The term "suggestive" means the degree of confidence in test animal or human data which is less than persuasive or not statistically significant but which nonetheless raises biological concerns that the toxic material may pose a carcinogenic risk to the exposed animals or exposed humans.

The term "toxic material" for the purposes of this subpart means any material in the workplace, either a single chemical substance or combination of substances, or any other agent which has been reported by scientific study to (a) produce or induce neoplasm or tumor formation in (1) humans or (2) in one or more experimental mammalian species, or (b) in a statistically significant manner decrease the latency period between exposure and onset of neoplasm or tumor formation in either humans or in one or more experimental mammalian species.

1990.4. Classification of materials

(a) General—Whenever the Secretary receives information, submitted to him in writing by any interested person, concerning any toxic material, he shall, within thirty (30) days publish a notice of receipt in the Federal Register, together with a short statement identifying the contents and sources of such information, and shall provide a period of at least thirty (30) days for public comment.

(b) Classification—Within thirty (30) days from the close of the comment period, the Secretary shall classify the toxic material and publish in the *Federal Register* a notice of his decision as to the classification of that material, any other relevant facts and any other consequences and actions required by this subpart.

Category I Toxic Materials

1990.10. Category I Toxic Materials

A presumption exists that a toxic material shall be classified by the Secretary as a "Category I Toxic Material" if: (a) the toxic material meets the definition of "potential occupational carcinogen" in (1) humans, or (2) two mammalian test species, or (3) a single mammalian species, if those results have been replicated in the same species in another experiment or (4) a single mammalian species if those results are supported by multi-test evidence for mutagenicity, or (b) if the Secretary finds that any other evidence is sufficient to convince him that the toxic material should be classified as a Category I toxic material.

1990.11. Rebuttal of a Category I Classification to a Category II or III Classification

(a) Within thirty (30) days from the close of the comment period referred to in section 1990.4(b), or at any

other time subsequent to a classification of a toxic material as a Category I toxic material the Secretary may rebut the presumtion of the Category I classification if he determines, after consultation with the Director of NIOSH, that scientifically:

(1) the carcinogenic evidence based on animal data clearly resulted from physical, rather than chemical induction, or

(2) the route of exposure in animals is grossly inappropriate relative to the potential occupational routes of human exposure, or

(3) the animal or human studies relied upon are only suggestive, or

(4) the animal or human studies relied upon are inadequate to establish any conclusion with respect to the carcinogencity or non-carcinogenicity of the toxic material, or

(5) for some other reason, the positive results in the experimental animal species are not scientifically relevant to man insofar as the toxic material in question is concerned.

(b) Such determinations and reasons shall be in writing and published in the *Federal Register*, pursuant to section 1990.4, and shall classify the material as a Category II toxic material if the determination was based on Section 1990.11(a)(3) in that the Secretary has determined that the animal or human studies relied upon are only suggestive. All other toxic materials for which the Secretary has determined that a Category I classification is inappropriate shall be classified as Category III.

1990.12. Consequences of a Category I Classification

In the event the Secretary determines that a toxic material shall be classified as a Category I Toxic Material, he shall require the initiation of the following actions:

(a) He shall, at the time he classifies the toxic material as Category I, issue an Emergency Temporary Standard, pursuant to 6(c) of the Act, which shall follow at the minimum the format and content of the model emergency temporary standard set forth in section 1990.40 hereof. Any significant deviation, addition or change from that format or content shall be explained, together with the reasons therefor.

(b) He shall, within sixty (60) days from such classification, issue a notice of proposed rulemaking, pursuant to 6(b) of the Act, which shall follow at the minimum the format and content of the model standard set forth in 1990.50 hereof. Any deviation, addition or change from that format shall be explained, together with the reasons therefor. The permissible exposure level, insofar as the 6(b) proposal is concerned, shall be as low as feasible. When it is determined by the Secretary that there are suitable substitutes for certain uses or classes of uses that are less hazardous to humans, on the basis of the best available evidence, the proposal shall permit no occupational exposure.

1990.13. Limitation of Issues at a Public Hearing or Comment

At any time after a request for a public hearing is made or upon his initiative, pursuant to the issuance of the documents referred to in Section 1990.12 and 29 CFR Part 1911, the Secretary shall provide for such a hearing in accordance with Section 6(b) of the Act and the regulations thereunder. The issues in the hearing or in comment shall be limited to the following: (1) whether the Secretary correctly classified the toxic material according to the Category I criteria (Section 1990.10), (2) whether the Secretary was correct in his determination that the Category I classification should not be rebutted (Section 1990.11), (3) the determination of the lowest feasible occupational exposure, or whether there are suitable substitutes that are found to be less hazardous to humans than the toxic material, (4) the appropriateness and feasibility of the specific protective measures of the proposed standard and (5) the environmental impact arising from regulation of the toxic material.

1990.14. Final Issuance of a Category I Standard

(a) Affirmance—At the time of the issuance of the final standard, if the Secretary determines based on the record that his classification of the toxic material as Category I was correct and that his decision not to rebut the presumption was also correct, he shall issue a standard which shall follow at the minimum the format and content of the model standard contained in section 1990.50 hereof. Any significant deviation, addition or change from that format and content shall be explained, together with the reasons therefor. However, the permissible exposure level shall be as low as feasible or, when it is determined by the Secretary that there are suitable substitutes that are less hazardous to humans, on the best available evidence, no occupational exposure shall be permitted.

(b) Change in Classification—At the time of the issuance of the final standard, if the Secretary determines based on the record that his classification of the toxic material is Category I or was incorrect or that his decision not to rebut the presumption was incorrect, he shall classify the toxic material as Category II or III. At the same time, if he classifies the toxic material as Category II, he shall issue a final standard which shall follow at the minimum, the format and content of the model standard contained in section 1990.50 hereof. Any significant

deviation, addition or change from that format and content shall be explained, together with the reasons therefor.

Category II Toxic Materials

Section 1990.20 Category II Toxic Materials

A toxic material shall be classified by the Secretary as a "Category II toxic material" if it is found to meet the definition of a "potential occupational carcinogen: (a) in test animals or humans, the results of which, however, were found by the Secretary, pursuant to Section 1990.11(a)(3), to be only suggestive or (b) in a single experiment in a single mammalian test speices.

Section 1990.21. Rebuttal of a Category II Classification to a Category III Classification

(a) Within thirty (30) days from the close of the comment period referred to in section 1990.4(b), or at any other time subsequent to a classification of a toxic material as a Category II toxic material, the Secretary may rebut the presumption of the Category II Classification if he determines, after consultation with the Director of NIOSH, that scientifically

(1) the carcinogenic evidence based on the animal data clearly resulted from physical, rather than chemical induction, or

(2) the route exposure in animals is grossly inappropriate relative to the potential occupational routes of human exposure, or

(3) the animal study relied upon is inadequate to establish any conclusion with respect to the carcinogenicity of noncarcinogenicity of the toxic material, or

(4) for some other reason, the positive results in the experimental animal species are not scientifically relevant to man insofar is the toxic material in question is concerned.

(b) Such determinations and reasons shall be in writing and published in the *Federal Register,* pursuant to section 1990.4, and shall classify the material as a Category II toxic material.

Section 1990.22. Consequences of a Category II Classification

In the event the Secretary determines that a toxic material shall be classified as a Category II Toxic Material suspect he shall require the initiation of the following actions:

(a) He shall, at the same time of the issuance of the determination also issue a notice of proposed rulemaking, pursuant to section 6(b) of the Act, which shall follow at the minimum the format, and content of the model standard set forth in section 1990.60 hereof. He shall also give notice that if, pursuant to this rule-making the toxic material should be classified as Category I, that 1990.12 herein would thereafter apply. Any significant deviation, additions or change from that format or content shall be explained, together with the reasons therefor. The permissible exposure level, insofar as the 6(b) proposal is concerned, shall be (i) the present OSHA standard or (ii) where none exists, an appropriate level based upon other acute or chronic effects of exposure to the toxic material or (iii) where other acute or chronic effects indicate that the present OSHA standard is inadequate, the exposure level shall be lowered to the level found appropriate by the Secretary.

(b) Immediately notify the applicable federal agencies, including the Administrator of EPA, the Director of NCI, and the Director of NIOSH, of his determination that the evidence is only "suggestive" or that the evidence consists of only positive results in one mammalian species and request that the applicable agencies engage in or stimulate further research to develop new

Section 1990.23. Limitation of Issues at a Public Hearing or in Comment

At any time after a request for a public hearing is made or upon his initiative, pursuant to the issuances of the document referred to in Section 1990.22, the Secretary shall provide for such a hearing in accordance with section 6(b) of the Act and the regulations thereunder. The issues in the hearing or in comment shall be limited to: (1) whether the Secretary correctly classified the toxic material according to the Category II criteria, (Section 1990.20) (2) whether the Secretary was correct in his determination that the Category II classification should not be rebutted (Section 1990.21), (3) if OSHA presently has no standard, the appropriate level of exposure, (4) the appropriateness and feasibility of the specified protective measures of the proposed standard and (5) environmental impact arising from regulation of the toxic taterial.

Section 1990.24. Final Issuance of a Category II Standard

(a) Affirmance—At the time of issuance of the final standard, if the Secretary determines based on the record that his classification of the toxic material as Category II is correct, he shall issue a standard which shall follow the format and content contained in section 1990.60 hereof. Any significant deviation, addition or change from that

format and content shall be explained, together with the reasons therefor.

(b) Change in Classification—At the time of the issuance of the final standard or at any other time subsequent to a classification to a classification of a toxic material as a Category II toxic material, if the Secretary determines based on the record that his classification of the toxic material as Category II was incorrect, he shall classify the toxic material as Category I or Category III.

(i) If the Secretary determines that the toxic material should be as classified as a Category I toxic material, he shall follow the procedure as provided by s1990.12 *et seq,*

(ii) If the Secretary determines that the toxic material should be classified as a Category III toxic material, he shall follow the procedure as provided by s1990.31.

Category III Toxic Materials

Section 1990.30 Category III. Toxic Materials

Any toxic material not classified by the Secretary as a Category I or a Category II toxic material, pursuant to this subpart, shall be classified as a Category III toxic material.

Section 1990.31. Immediate Consequences of Category III Classification

At the time a toxic material is classified as Category III, the Secretary shall transmit his findings to the Director of NCI, the Director of NIOSH and the Administrator of EPA with a request that each determine whether there is additional information previously unavailable to the Secretary which could have a bearing on reconsideration of the Category III classification of the toxic material.

Section 1990.32. Reconsideration of Category III Classification

If, at any time, any of the agencies, referred to in Section 1990.31 provide additional information having a bearing on the Secretary shall, within thirty (30) days after the receipt of such a response, make public the reeponse and reconsider his decision classification of the toxic material as Category III, then call a public meetint to discuss the issues. If at the conclusion of such a meeting the Secretary determines his original findings were in error, he shall classify the toxic material as a Category I or Category II toxic material.

Model Standards

1990.40. Section 6(c) Standard for Category I Toxic Material

1910.0000 xxxxx; Emergency Temporary Standard for Exposure to xxxx

(a) Scope and Application—This section applies to the production, release, packaging, repackaging, storage, transportation, handling, or use of xxxxx except that this section will not apply to working conditions with respect to which any other Federal Agency has exercised statutory authority to prescribe or enforce standards or regulations affecting occupational safety or health hazards covered by this section.

(b) Definitions—"Director" means the Director, National Institute for Occupational Safety and Health, U.S. Department of Health, Education and Welfare or his designee.

"Secretary" means the Secretary of Labor or his designee.

"xxxxx" means (*definition of substance to be regulated.*)

(c) Exposure Limits—(1) *Permissible exposure limits.* (i) The employer shall assure that no employee is exposed to an airborne concentration of xxxxx in excess of (insert appropriate exposure limit representing the lowest feasible level that can be complied with immediately) parts per million or billion parts of air (ppm or ppb), or (other appropriate designation of concentration such as micrograms per cubic meters of air) as determined on an eight-hour average, and

(ii) The employer shall assure that no employee is exposed to an airborne concentration of xxxxx in excess of (insert appropriate exposure limit representing the lowest feasible level that can te complied with immediately) parts per million parts of air (ppm) (or other designation of concentration such as micrograms per cubic meter of air as averaged over any 15-minute period during the working day.

(2) *Dermal and eye exposure limit.* No employee may be exposed to eye contact or repeated skin contact with xxxxx.

(d) Notification of Use— Within 30 days of the effective date of this section, every employer who has a place of employment in which xxxxx is present shall report the following information to the nearest OSHA Area Director for each such establishment.

(1) The address and location of each establishment in which employee exposure to xxxxx occurs.

(2) A brief description of each process or operation which may result in employee exposure to xxxxx.

(3) The number of employees engaged in each process or operation described in paragraph (d)(2) of this section and an estimate of the frequency and degree of exposure that results.

(4) A brief description of the employee safety and health program in effect in each establishment listed in paragraph (d)(1) of this section and a description of any specific measures or controls designed to limit employee exposure to xxxxx.

(e) Exposure monitoring and measurement—(1) *Initial determination.* Each employer who has a place of employment in which xxxxx is present shall monitor each such workplace and work operation to determine if any employee may be exposed to xxxxx above the permissible exposure limits, and to determine if any employee is exposed by skin contact with xxxxx. Such a determination shall be made by monitoring and measurements which are representative of each employee's exposure to xxxxx over an 8-hour period.

(2) *Frequency of measurements.* The measurements required under paragraph (e)(1) of this section shall be repeated at least quarterly.

(3) *Measurement above the Permissible Limits.* (i) Where the measurements reveal employee exposures to be in excess of the permissible exposure limits, the measurements required under paragraph (e)(1) of this section shall be repeated for each employee at least monthly. The employer shall continue measurements under this paragraph (e)(3) until at least two consecutive measurements at least seven (7) days apart are below the permissible exposure limits and thereafter the employer shall comply with paragraph (e)(2) of this section.

(ii) If the measurements reveal employee exposures to be above the permissible exposure limit, the employer shall, in addition to the requirement in paragraph (e)(3)(i):

(A) Inform each affected employee of the exposure as required by paragraph (e)(5) of the section;

(B) Institute control measures as required by paragraph (f) of this section; and

(C) provide personal protective equipment and clothing as required by paragraphs (g) and (i) of this section;

(4) *Additional monitoring.* Whenever there has been a production, process, or control change which may result in new or additional exposures, or whenever the employer has any other reason to suspect a change which may result in new or additional exposures, additional measurements in accordance with this paragraph shall be made.

(5) *Employee notification.* (i) The employer shall notify each employee in writing, of the exposure measurements which represent that employee's exposure within 5 working days after the receipt of measurement results, required by paragraphs (e)(1) and (e)(2) of this section.

(ii) Whenever such results indicate that the representative employee exposure exceeds the permissible exposure limit, the employer shall, in such notification, inform each employer of that fact and of the corrective action being taken to reduce exposure to or below the permissible exposure limit.

(6) *Accuracy of measurement.* The method of measurement shall have an accuracy (with a confidence limit of 95%) of not less than plus or minus (insert appropriate value) for concentrations of xxxxx greater than or equal to (insert the permissible exposure limit).

(7) *Employee exposure.* For the purposes of this section, employee exposure is that exposure which would occur if the employee were not using a respirator.

(f) *Methods of Compliance*—During the effective period of this standard, employee exposures shall be controlled to or below the permissible exposure limit by any practicable combination oengineering controls, work practices and personal protective devices as follows:

(1) Engineering controls to reduce the airborne concentration of xxxxx shall be instituted where feasible. Such controls include, among other things, substitution of a less hazardous material, enclosure of the process, and local exhaust ventilation.

(2) The employer shall examine each work area in which xxxxx is present to determine whether there are work practices shall that are appropriate to reduce employee exposures to xxxxx. Such work practices include, among other things, the following:

(i) Limiting access to work areas where xxxxx is present to authorized personnel only.

(ii) Written procedures and work practices for each operation which may result in employee exposure to xxxxx.

(iii) Prohibiting smoking and the consumption of food and beverages in work areas where xxxxx is present.

(iv) Maintenance of good housekeeping including the prompt cleanup of spills, repair of leaks, etc.

(v) Requiring the use of appropriate protective clothing to minimize skin contact with xxxxx.

(vi) Use of signs, labels, or other means to clearly designate all work areas where xxxxx may be preeent.

(3) Where engineering and work practice controls described in paragraphs f(1) and f(2) of this section are not adequate to reduce employee exposures to or below the permissible exposure limit, they shall be supplemented by the use of respirators in accordance with paragraph (g) of this section.

(g) *Respirators*—(1) *Required use.* Respirators shall be used where required under this section to reduce employee exposure to or below the permissible exposure limit.

(2) *Respirator selection.* (i) Where respirators are required under this section, the employer shall select and provide the appropriate respirator from Table 1 below and shall assure that the employee uses the respirator provided.

(ii) Respirators shall be selected from those approved by the Mining Enforcement and Safety Administration or by the National Institute for Occupational Safety and Health under the provision of 30 CFR Part 11.

TABLE 1—REQUIRED RESPIRATORY PROTECTION

(Table 1 will indicate which types of respirators are approved for various levels of exposure to xxxxx.)

(3) *Respirator program.* The employer shall institute a respiratory protection program in accordance with s1910.134 (b), (d), (e) and (f).

(h) *Medical Surveillance*—(1) *General requirements.* (i) Each employer shall institute a medical surveillance program for all employees, or above the permissible exposure level.

(ii) The program shall provide each affected employee with an opportunity for medical examinations in accordance with this paragraph.

(iii) All medical examinations and procedures shall be performed by or under the supervision of a licensed physician, and shall be provided without cost to the employee.

(iv) If an employee refuses any required medical examination, the employer shall inform the employee of the possible health consequences of such refusal and shall obtain a signed statement from the employee indicating that the employee understands the risk involved in the refusal to be examined.

(2) *Content.* Within thirty days of the effective date of this section, each affected employee shall be provided medical surveillance consisting of at least the following where applicable:

(i) History and physical examinations shall direct emphasis towards the pulmonary, renal, and hepatic systems, and shall include the personal history of the employee, family, and occupational background, including genetic and environmental factors. Additionally, such factors as the current systems review, pregnancy, current treatment with steroids or cytotoxic agents, and smoking habits should be considered.

(ii) The following tests shall be included in the examination:

(A) 14'' x 17'' chest X-Ray

(B) Laboratory examinations to include:
Complete Blood count; and
Blood Chemistry tests to include glutamic oxalacetic transaminase (SGOT), glutamic pyruvic transaminase (SGPT), alkaline phosphatase, total bilirubin, and serum glutamyl transpeptidase (GGTP); and
Complete urinalysis to include microscopic examination and cytologic examination for neoplastic cells.

(C) Additional tests may be considered by the examining physician including sputum cytology where medically indicated.

(3) *Information provided to the physician.* The employer shall provide the following information to the examining physician:

(i) A copy of this regulation and its appendixes;

(ii) A description of the affected employee's duties as they relate to the employee's exposure;

(iii) The employee's exposure level or anticipated exposure level;

(iv) A description of any personal protective equipment used or to be used.

(4) Physician's written opinion. (i) The employer shall obtain, a written opinion from the examining physician which shall include:

(A) The results of the medical examinations;

(B) The physician's opinion as to whether the employee has any detected medical condition which would place the employee at increased risk of material impairment of the employee's health from exposure to xxxxx;

(C) Any recommended limitations upon the employee's exposure to xxxxx or upon the use of protective clothing or equipment such as respirators; and

(D) A statement that the employee has been informed by the physician of any medical conditions which require further examination or treatment.

(ii) The employer shall instruct the physician not to reveal in the written opinion specific findings or diagnoses unrelated to occupational exposure.

(i) Employee Information and Training—(1) *Training Program.* The employer shall provide a training program for employees assigned to workplace areas where xxxxx is present and shall assure that each affected employee is informed of the following:

(i) The information contained in the substance data sheets for xxxxx which are contained in Appendixes A and B of this section; (omitted).

(ii) The quantity, location, manner of use, release or storage of xxxxx and the specific nature of operations which could result in exposure at or above the permissible exposure limit as well as necessary protective steps;

(iii) The purpose, proper use, and limitations of respiratory devices as specified in s1910.134;

(iv) The purpose and a description of the medical surveillance program required by paragraph (h) of this section and the information contained in Appendix C of this section; (omitted) and,

(v) A review of this standard.

(2) *Access to training materials.* (i) the employer shall make a copy of this standard and its appendixes readily available to all affected employees.

(ii) The employer shall provide upon request all materials relating to the employee information and training program to the Secretary and the Director.

(j) Recordkeeping—(1) *Measurements.* The employer shall establish and maintain an accurate record of all measurements taken to monitor employee exposure to xxxxx required in paragraph (e) of this section.

(i) This record shall include:

(A) The dates, number, duration and results of each of the samples taken, including a description of the sampling procedure used to determine representative employee exposure when applicable;

(B) A description of the sampling and analytical methods used and evidence of their accuracy;

(C) Type of respiratory protective devices worn, if any; and

(D) Name, social security number, and job classification of the employees monitored.

(ii) This record shall be maintained for at least 3 years.

(2) *Medical surveillance.* The employer shall establish and maintain an accurate record for each employee subject to medical surveillance as required by paragraph (h) of this section.

(i) This record shall include:

(a) A copy of the physician's written opinion;

(b) Any employee medical complaints related to exposure to xxxxx;

(c) A copy of the information provided to the physician as required by paragraph (h)(6) of this section; and

(d) A signed statement of any refusal to be examined.

(ii) This record shall be maintained for at least 3 years.

(3) *Availability.* (i) All records required to be maintained by this section shall be made available upon request to the Secretary and the Director for examination and copying.

(ii) Employee exposure measurement records as required by this section shall be made available for examination and copying to employees, former employees, and their designated representatives.

(iii) Employee medical records required to be maintained by this section shall be made available upon request for examination and copying to a physician designated by the employee or former employee.

(k) Observation of Monitoring—(1) *Employee observation.* The employer shall provide affected employees or their representatives, an opportunity to observe any measuring or monitoring of employee exposure to xxxxx conducted pursuant to paragraph (e) of this section.

(2) *Observation procedures.* (i) When observation of the measuring or monitoring of employee exposure to xxxxx requires entry into an area where the use of protective clothing or equipment is required, the employer shall provide the observer with and assure the use of such equipment and shall require the observer to comply with all other applicable safety and health procedures.

(ii) Without interfering with the measurement, observers shall be entitled to:

(A) An explanation of the measurement procedures;

(B) Observe all steps related to the measurement of airborne concentration of xxxxx performed at the place of exposure; and

(C) Record the results obtained.

(l) Effective Date—This standard shall become effective following publication of the standard in the Federal Register.

1990.50 Section 6(b) Model Standard for a Category I Toxic Material Exposure to xxxxx

1910.000xxxx

(a) Scope and Application—This section applies to the transportation, production, release, packaging, repackaging, storage, handling, or use of xxxxx except that this section will not apply to working conditions with respect to which any other Federal agency has excised statutory authority to prescribe or enforce standards or regulations affecting occupational safety or health hazards covered by this section.

(b) Definitions—"Authorized person" means any person specifically authorized by the employer whose duties require the person to enter a regulated area, or any person entering such an areas as a designated representative of employees for the purpose of exercising the opportunity to observe monitoring and measuring procedures under paragraph (d) of this section.

"xxxxx" means (definition of substance to be regulated).

"Director" means the Director, National Institute for Occupational Safety and Health, U.S. Department of Health, Education and Welfare, or his or her designee.

"Emergency" means any occurence such as, but not limited to equipment failure, rupture of containers, or failure of control equipment which does or is likely to, result in any massive release of xxxxx.

"Secretary" means the Secretary of Labor, U.S. Department of Labor, or his or her designee.

(c) Exposure Limits—(1) *Permissible exposure limits.* (i) The employer shall assure that no employee is exposed to an 8-hour average airborne concentration of xxxxx in excess of (insert appropriate exposure limit representing the lowest feasible level) or when it is determined by the Secretary that there are available substitutes for certain uses or classes of uses that are less hazardous to humans, the proposal shall permit no occupational exposure.

(ii) The employer shall assure that no employee is exposed to an airborne concentration of xxxxx (ceiling limit if appropriate) as averaged over any 15 minute period.

(2) *Dermal and eye exposure.* No employee may be exposed by skin or eye contact to xxxxx.

(d) Exposure Monitoring and Measurement—(1) *Initial monitoring.* Each employer who has a place of employment in which xxxxx is present shall monitor each such workplace and work operation to accurately measure if any employee may be exposed to xxxxx above the permissible exposure limits, and determine if any employee is exposed by skin contact with xxxxx. Such a determination shall be made by monitoring which is representative of each employee's exposure to xxxxx over an 8-hour period.

(2) *Measurements below the permissible limit.* If the measurements under paragraph (d)(1) of this section reveal employee exposure, to be below the permissible exposure limits the employer shall repeat the measurements for each employee at least quarterly.

(3) *Measurements above the permissible limits.* (i) Where the measurements reveal employee exposures to be in excess of the permissible exposure limits, the measurements required under paragraph (d)(1) of this section shall be repeated for each employee at least monthly. The employer shall continue measurements under this paragraph (d)(3) until at least two consecutive measurements are below the permissible exposure limits and thereafter the employer shall comply with paragraph (d)(2) of this section.

(ii) If exposure measurements reveal employee exposure to be above the permissible exposure limit or the ceiling limit, the employer shall, in addition to the requirement in paragraph (d)(3)(i):

(A) Inform the employee of the exposure as required by paragraph (d)(5) of this section;

(B) Institute control measures as required by paragraph (f) of this section; and

(C) Provide personal protective equipment and clothing as required by paragraphs (g) and (i) of this section.

(4) *Additional monitoring.* Whenever there has been a production, process, or control change which may result in new or additional exposures, or whenever the employer has any other reason to suspect a change which may result in new or additional exposures, additional measurements in accordance with this paragraph shall be made.

(5) *Employee notification.* (i) The employer shall notify each employee in writing of the exposure measurements which represent that employees exposure within 5 working days after the receipt of results, of measurements required by paragraphs (d)(1) and (d)(2) of this section.

(ii) Whenever such results indicate that the representative employee exposure exceeds the permissible exposure limit, the employer shall, in such notification, inform each employee of that fact and of the corrective action being taken to reduce exposure to or below the permissible exposure limit.

(6) *Accuracy of measurement.* The method of measurement shall have an accuracy (with a confidence limit of 95%) of not less than plus or minus (insert appropriate value) for concentrations of xxxxx greater than or equal to the permissible exposure limit.

(7) *Employee exposure.* For the purposes of this section, employee exposure is that exposure which would occur if the employee were not using a respirator.

(e) Regulated Areas—(1) The employer shall establish regulated areas where there is exposure to xxxxx

(2) Access to regulated areas shall be limited to authorized persons.

(f) Methods of Compliance—(1) *Engineering controls.* (i) The employer shall institute immediately engineering controls to reduce exposures to or below the permissible exposure limits, except to the extent that such controls are not feasible.

(ii) Wherever the engineering and work practice controls which can be instituted are not sufficient to reduce employee exposures to or below the permissible exposure limit, the employer shall nonetheless use them to reduce exposures to the lowest level achievable by such controls and shall supplement them by the use of respiratory pro-

tection which complies with the requirements of paragraph (g) of this section.

(2) *Work practice controls.* Wherever feasible engineering controls which can be instituted immediately are not sufficient to reduce exposures to or below the permissible exposure limits, they shall nonetheless be used to reduce exposures to the lowest practicable level, and shall be supplemented by work practice controls.

(3) *Respirators.* Where feasible engineering controls and supplemental work practice controls are insufficient to reduce exposures to or below permissible exposure limits they shall nonetheless be used to reduce exposures to the lowest practicable level and shall be supplemented by the use of respirators in accordance with paragraph (g) of this section as required.

(4) *Compliance program.* (i) Each employer shall establish and implement a written program to reduce exposure solely by means of engineering controls, as specified in paragraph (f)(1) of this section.

(ii) The written program shall include at least the following:

(a) A description of each operation or process resulting in employee exposure to xxxxx;

(b) Engineering plans and other studies used to determine the controls for each process;

(c) A report of the technology considered in meeting the permissible exposure limit;

(d) Monitoring data obtained in accordance with paragraph (d) of this section;

(e) A detailed schedule for the implementation of engineering controls,

(f) Other relevant information.

(iii) Written plans for such program shall be submitted, upon request, to the Secretary and the Director, and shall be available at the worksite for examination and copying by the Secretary, the Director, any affected employee or their representative. The plans required under paragraph (e)(4) of this section shall be revised and updated at least every six months to reflect the current status of the program.

(5) *Prohibited exposures.* (Where the Secretary determines), based on the record of public rulemaking, that there are materials less hazardous to employees and that those materials are suitable substitutes in certain applications where xxxxx might otherwise be used, the Secretary shall specify that no occupational exposure to xxxxx shall be permitted in those circumstances.)

(g) Respiratory Protection—(1) *Permitted use.* Where respirators are required under this section, compliance with the permissible exposure limit may not be achieved by the use of respirators except:

(i) During the time period necessary to install engineering controls; or

(ii) In work operations such as maintenance and repair activities in which engineering and work practice controls are technologically not feasible.

(iii) In work situations in which engineering controls and supplemental work practice controls are insufficient to reduce exposure to or below the permissible exposure limits; or

(iv) In emergencies.

(2) *Respirator selection.* (i) Where respirators are required under this section the employer shall select and provide the appropriate respirator from Table 1 and shall ensure the employee uses the respirator provided.

TABLE 1—RESPIRATORY PROTECTION FOR XXXXX

(The table will contain a listing of the appropriate type of respirator for various conditions of exposure).

(ii) Respirators shall be selected from those approved by the Mining Enforcement and Safety Administration or by the National Institute for Occupational Safety and Health under the provisions of 30 CFR Part 11.

(iii) Respirators prescribed for higher concentrations may be used for any lower concentration.

(3) *Respirator program.* (i) The employer shall institute a respiratory program in accordance with 1910.134 (b), (d), (e), and (f).

(ii) Employees who wear respirators shall be allowed to wash the face and respirator facepiece ot prevent potential skin irritation associated with respirator use.

(h) Emergency Situations—(1) *Written plans.* (i) A written plan for emergency situations shall be developed for each facility involved in a xxxxx operation in which there is a possibility of an emergency. Appropriate portions of the plan shall be implemented in the event of an emergency.

(ii) The plan shall specifically provide that employees engaged in correcting emergency conditions shall be equipped as required in paragraphs (g) and (i) of this section until the emergency is abated.

(iii) Employees not engaged in correcting the emergency shall be restricted from the area and normal operations in the affected area(s) shall not be resumed until the emergency is abated.

(2) *Alerting employees.* Where there is the possibility of employee exposure to xxxxx in excess of the ceiling limit due to the occurrence of an emergency, a general alarm shall be installed and maintained to promptly alert employees of such occurrences.

(i) Skin Protection and Work Clothing—(1) *Work clothing.* Where employees are exposed to airborne concentrations of xxxxx in excess of the permissible exposure limits, or are subject to skin contact with xxxxx the employer shall provide and assure that employees wear work clothing and other appropriate protective equipment in accordance with this paragraph.

(i) The employer shall provide each employee with coveralls or similar fullbody work clothing, headcoverings, and work shoes or shoe coverings. Resin-impregnated paper or similar disposable work clothing may be substituted for fabric-type clothing.

(ii) New or laundered work clothing shall be provided at least (insert appropriate interval) to each affected employee.

(2) *Skin and eye protection.* (1) Wherever employees are subject to skin contact with xxxxx the employer shall provide and assure that employees wear protective gloves,

(ii) Additional protection such as face shields, goggles, and gauntlets, which provide protection for eyes, face, neck, arms and other exposed skin areas, shall be provided if the operation results in such areas having contact with xxxxx.

(iii) Protective clothing and equipment required by this paragraph shall be supplied to each employee (insert appropriate interval) and shall be maintained in accordance with paragraph (j) of this section.

(j) Equipment and Clothing Laundering and Maintenance—(1) *Laundering.* (i) The employer shall launder, maintain, or dispose of skin protective devices and work clothing required by paragraph (i) of this section.

(ii) The employer shall inform any person who launders or cleans xxxxx contaminated protective devices or work clothing of the potentially harmful effects of exposure to xxxxx.

(2) *Removal and storage.* (i) the employer shall assure that employees remove contaminated work clothing only in change rooms as required by paragraph (m)(1) of this section.

(ii) The employer shall assure that no employee removes contaminated protective devices and work clothing from the change room except for those employees authorized to do so for the purpose of laundering, maintenance, or disposal.

(iii) xxxxx-contaminated protective devices and work clothing shall be placed and stored in closed containers which prevent dispersion of the xxxxx outside the container.

(iv) Containers of contaminated protective devices or work clothing which are to be removed from change rooms or from the work place for laundering or disposal, or for any other reason, shall bear labels in accordance with paragraph (p)(2) of this section.

(v) Dust removal by blowing or shaking of work clothing is prohibited.

(k) Housekeeping—(1) *Work surfaces.* (i) All external work surfaces shall be maintained free of accumulations of xxxxx.

(ii) Dry sweeping and the use of compressed air for the cleaning of floors and other surfaces where xxxxx dust is found is prohibited.

(iii) Where vacuuming methods are selected, either portable units or permanent systems may be used.

(A) If a portable unit is selected, the exhaust shall be attached to the general workplace exhaust ventilation system or collected within the vacuum unit, equipped with high efficiency filters or other appropriate means of contaminant removal, so that xxxxx is not reintroduced into the work place air; and

(B) Portable vacuum units used to collect xxxxx, may not be used for other cleaning purposes and shall be labeled as prescribed by paragraph (p)(2) of this section.

(iv) Cleaning of floors and other contaminated surfaces may not be performed by washing down with a hose, unless a fine spray has first been laid down.

(2) *Dust collection systems.* Periodic cleaning of dust collection systems, i.e. ducts, filters. etc. shall be performed to reduce xxxxx dust buildups.

(l) Waste disposal—xxxxx waste, scrap, debris, bags, containers or equipment, shall be disposed of in sealed bags or other closed containers which prevent dispersion of xxxxx outside the container.

(m) Hygiene Facilities and Practices—(1) *Change rooms.* The employer shall provide clean change rooms equipped with storage facilities for street clothes and separate storage facilities for protective clothing and equipment whenever employees are required to wear protective clothing and equipment in accordance with paragraph (i) of this section.

(2) *Showers.* (i) The employer shall assure that employees working in the regulated area shower at the end of the work shift.

(ii) The employer shall provide shower facilities in accordance with 29 CFR 1910.141(d)(3).

(3) *Lunchrooms.* The employer shall provide lunchroom facilities which have a temperature controlled, positive pressure, filtered air supply, and which are readily accessible to employees working in a regulated area.

(4) *Lavatories.* (i) The employer shall assure that employees working in the regulated area wash their hands and face prior to eating.

(ii) The employer shall provide lavatory facilities in accordance with 29 CFR Part 1910.141(d)(1) and (2).

(5) *Prohibition of activities in the regulated area.* The employer shall assure that, in the regulated area, food or beverages are not presented or consumed, smoking products are not present or used, and cosmetics are not applied, (except that these activities may be conducted in the lunchroom, change rooms and showers required under paragraphs (m)(1) - (m)(3) of this section.)

(n) Medical Surveillance—(1) *General requirements.* (i) Each employer who has a place of employment in which employees are exposed to xxxxx shall institute a medical surveillance program.

(ii) The program shall provide each affected employee with an opportunity for medical examinations in accordance with this paragraph (n).

(iii) If any employee refuses any required medical examination, the employer shall inform the employee of the possible health consequences of such refusal and obtain a signed statement from the employee indicating that the employee understands the risk involved by the refusal to be examined.

(iv) The employer shall assure that all medical examinations and procedures are performed by or under the supervision of a licensed physician, and are provided without cost to the employee.

(2) *Initial examinations.* At the time of initial assignment, or upon institution of the medical surveillance program, the employer shall provide each affected employee an opportunity for a medical examination including at least the following elements: (i) History and physical examinations shall direct emphasis towards the pulmonary, renal and hepatic systems, and shall include the personal history of the employee, family, and occupational background, including genetic and environmental factors. Additionally, such factors as the current systems review, pregnancy, current treatment with steriods or cytotoxic agents, and smoking habits should be considered.

(ii) The following tests shall be included in the examination:

(A) 14'' x 17'' chest X-Ray

(B) Laboratory examinations to include:

Complete Blood count; and blood chemistry tests to include glutamic oxalaretic
transaminase (SOGT), glutamic pyruvic transaminase (SGPT), alkaline phosphatase, total bilirubin, and serum
glutamyl transpeptidase (GGTP); and
Complete urinalysis to include microscopic examination and cytologic examination for neoplastic cells.

(C) Additional tests may be considered by the examining physician including sputum cytolgy when medically indicated.

(3) *Periodic examinations.* (i) The employer shall provide examinations specified in this paragraph at least annually for all employees specified in paragraph (n)(1) of this section.

(ii) If an employee has not had the examinations prescribed in paragraph (n)(2) of this section within 6 months of his termination of employment, the employer shall make such examination available to the employee.

(4) *Information provided to the physician.* The employer shall provide the following information to the examining physician:

(i) A copy of this regulation and its Appendixes;

(ii) A description of the affected employee's duties as they relate to the employee's exposure;

(iii) The employee's exposure level or anticipated exposure level.

(iv) A description of any personal protective equipment used or to be used; and

(v) Information from previous medical examinations of the affected employee which is not readily available to the examining physician.

(6) *Physician's written opinion.* (i) The employer shall obtain a written opinion from the examining physician which shall include:

(a) The results of the medical examination;

(b) The physician's opinion as to whether the employee has any detected medical condition which would place the employee at increased risk of material impairment of the employee's health from exposure to xxxxx;

(c) Any recommended limitations upon the employee's exposure to xxxxx or upon the use of protective clothing and equipment such as respirators; and

(d) A statement that the employee has been informed by the physician of the results of the medical examination and any medical conditions which required further examination or treatment.

(ii) The employer shall instruct the physician not to reveal in the written opinion specific findings or diagnoses unrelated to exposure to xxxxx.

(iii) The employer shall provide a copy of the written opinion to the affected employee.

(*o*) *Employee Information and Training*—(1) *Training program.* (i) The employer shall institute a training program for all employees assigned to workplace areas where any xxxxx is produced, released, packaged,

repackaged, stored, handled, or used and shall ensure their participation.

(ii) The training program shall be provided at the time of initial assignment and at least annually thereafter, and shall include informing each employee of:

(A) The information contained in the substance data sheets for xxxxx, which are contained in Appendixes A and B (omitted);

(B) The quantity, location, manner of use, release or storage of xxxxx and the specific nature of operations which could result in exposure to xxxxx (at or above the permissible limits) as well as any necessary protective steps;

(C) The purpose, proper use, and limitations of respiratory devices as specified in 1910.134 (b), (d), (e), and (f);

(D) The purpose and a description of the medical surveillance program ae required by paragraph (n) of this section and the information contained in Appendix C (omitted);

(E) Emergency procedures as required by paragraph (h) of this section; and

(F) A review of this standard.

(2) *Access to training materials.* (i) A copy of this standard and its appendices shall be readily available to all employees exposed to xxxxx.

(ii) All materials relating to the employee information and training program shall be provided upon request to the Assistant Secretary and the Director.

(*p*) *Precautionary Signs and Labels*—(1) *General.* (i) The employer may use labels or signs required by other statutes regulations, or ordinances in addition to, or in combination with, signs and labels required by this paragraph.

(ii) The employer shall assure that no statement appears on or near any sign required by this paragraph which contradicts or detracts from the effects of the required sign.

(iii) The employer shall assure that signs required by this paragraph are illuminated and cleaned as necessary so that the legend is readily visible.

(2) *Signs.* The employer shall post signs at entrance to the regulated area bearing the legends:

DANGER
CANCER HAZARD
AUTHORIZED PERSONNEL ONLY
RESPIRATOR REQUIRED
NO SMOKING OR EATING

(3) *Labels.* The employer shall apply precautionary labels to all containers or xxxxx. The labels shall bear the following legend:

DANGER
CANCER HAZARD
CONTAINS

(g) Recordkeeping—(1) *Exposure measurements.* The employer shall establish and maintain an accurate record of all measurements taken to monitor employee exposure to xxxxx as prescribed in paragraph (d) of this section.

(i) This record shall include:

(A) The dates number, duration, and results of each of the samples taken, including a description of the sampling procedure used to determine representative employee exposure where applicable.

(B) A description of the sampling and analytical methods used and evidence of their accuracy;

(C) Type of respiratory protective devices worn, if any; and

(D) Name and social security number and job classification of the employees monitored.

(ii) This record shall be maintained for at least 40 years.

(2) *Medical surveillance.* The employer shall establish and maintain an accurate record for each employee subject to medical surveillance required by paragraph (n) of this section.

(i) This record shall include:

(A) A copy of the physician's written opinion;

(B) Any employee medical complaints related to exposure to xxxxx;

(C) A copy of the information provided to the physician as required by paragraph (n)(6) of this section; and

(D) A signed statement of any refusal to be examined.

(ii) This record shall be maintained for at least 40 years, or for the duration of employment plus 20 years, whichever is longer.

(3) *Availability.* (i) The employer shall make available upon request all records required to be maintained by this section to the Secretary and the Director for examination and copying.

(ii) The employer shall make available upon request records of employee exposure measurements required by this section for examination and copying to employees, former employees, and their designated representatives.

(iii) The employer shall make available upon request employee medical records required to be maintained by this section for examination and copying to a physician designated by the employee or former employee.

(4) *Transfer of records.* (i) Whenever the employer ceases to do business, the successor employer shall receive and retain all records required to be maintained by this section.

(ii) Whenever the employer ceases to do business and there is no successor employer to receive and retain the records for the prescribed period, these records shall be transmitted by registered mail to the Director.

(r) Observation of Monitoring—(1) *Employee observation.* The employer shall provide affected employees or employee representatives, an opportunity to observe any measuring or monitoring of employee exposure to xxxxx conducted pursuant to this section.

(2) *Observation procedures.* (i) Whenever observation of the measuring or monitoring of employee exposure to xxxxx requires entry into an area where the use of protective clothing or equipment is required, the employer shall provide the observer with and assure the use of such equipment and shall require the observer to comply with all other applicable safety and health procedures.

(ii) Without interfering with the measurement, observers shall be entitled to;

(A) An explanation of the measurement procedures;

(B) Observe all steps to the measurement of xxxxx performed at the place of exposure; and

(C) Record the results obtained.

(s) Effective Date—This standard shall become effective no later than 90 days following publication of the final standard in the Federal Register.

OCCUPATIONS AND DISEASES*

SECTION ONE

A GUIDE TO TOXIC EXPOSURES
BY OCCUPATION

---

* A Guide to Work-Relatedness of Disease, National Institute of Occupational Safety and Health, 1976.

OCCUPATION	AGENT(S)	OCCUPATION	AGENT(S)
Abrasive blasters	Silica	Aniline makers	Benzene Nitrogen dioxide
Abrasives makers	Silica	Aniline color makers	Arsenic
Abrasion resistant rubber makers	Tol. diisocyanate	Aniline workers	Arsenic
Acetic acid makers	Carbon monoxide	Antimony Ore Smelters	Antimony
Acetylene workers	Arsenic	Antimony workers	Antimony
Acetylene purifiers	Chromic acid	Arc welders	Carbon monoxide
Acid finishers	Lead	Arsenic workers	Arsenic
Acid dippers	Arsenic nitrogen dioxide	Arsine workers	Arsenic
Acoustical product makers	Asbestos	Art Glass workers	Benzene
Acoustical Product installers	Asbestos	Artificial flower makers	Arsenic
Actors	Lead	Artificial leather makers	Benzene
Adhesive workers	Tol. diisocyanate	Artificial abrasive makers	Carbon monoxide
Adhesive makers	Benzene	Artificial gas workers	Carbon monoxide
Air filter makers	Asbestos	Asbestos-cement products makers	Asbestos
Aircraft burners	Tol. diisocyanate	Asbestos-cement products makers	Asbestos
Airplane Dope makers	Benzene	Asbestos-cement products users	Asbestos
Airplane pilots	Carbon monoxide	Asbestos-coating makers	Asbestos
Alcohol workers	Benzene	Asbestos-coating users	Asbestos
Alkali-salt makers	Sulfur dioxide	Asbestos-Grout makers	Asbestos
Alloy makers	Arsenic	Asbestos-Grout users	Asbestos
Aluminum anodizers	Chromic acid		
Aluminum hard coaters	Chromic acid		
Ammonia makers	Carbon monoxide		

OCCUPATION	AGENT(S)	OCCUPATION	AGENT(S)
Asbestos-millboard makers	Asbestos	Babbit metal workers	Antimony arsenic
Asbestos-millboard users	Asbestos	Bakers	Carbon monoxide
Asbestos-mortar makers	Asbestos	Battery makers	Lead
Asbestos-mortar users	Asbestos	Battery workers (storage)	Antimony
Asbestos millers	Asbestos	Battery (dry) makers	Benzene
Asbestos miners	Asbestos	Beaming operators (cotton mill)	Cotton dust
Asbestos-paper makers	Asbestos	Beet sugar bleachers	Sulfur dioxide
Asbestos-paper users	Asbestos	Belt scourers	Benzene
Asbestos-plaster makers	Asbestos	Benzene Hexachloride makers	Benzene
Asbestos-plaster users	Asbestos	Benzene workers	Benzene
Asbestos sprayers	Asbestos	Beryllium alloy machiners	Beryllium
Asbestos workers	Asbestos	Beryllium alloy makers	Beryllium
Asbestos product impregnators	Asbestos	Beryllium compound makers	Beryllium
Asphalt mixers	Benzene	Beryllium-copper founders	Beryllium
Auto garage workers	Asbestos Silica	Beryllium-copper grinders	Beryllium
Auto painters	Nitrogen dioxide	Beryllium-copper polishers	Beryllium
Automobile repair garage workers	Asbestos	Beryllium extractors	Beryllium
Automobile Users	Carbon monoxide	Beryllium metal machiners	Beryllium
Babbiters	Lead	Beryllium mineral miners	Beryllium

OCCUPATION	AGENT(S)	OCCUPATION	AGENT(S)
Beryllium phosphor makers	Beryllium	Brass makers	Arsenic
Beryllium workers	Beryllium	Braziers	Lead Nitrogen dioxide
Bisque-kiln workers	Carbon monoxide silica	Brewery workers	Sulfur dioxide
Blacksmiths	Carbon monoxide lead	Brewers	Carbon monoxide
		Brick burners	Carbon monoxide
Blast furnace workers	Carbon monoxide sulphur dioxide	Brick makers	Lead
Blast furnace gas users	Carbon monoxide	Brick layers	Lead Silica
		Brickmakers	Sulfur dioxide
Bleaching powder makers	Arsenic	Bright dip workers	Nitrogen dioxide
Blockers (felt hat)	Carbon monoxide	Britannia metal workers	Antimony
Blueprints	Nitrogen dioxide	Bronzers	Antimony Arsenic Benzene Lead
Boiler rooms	Noise		
Boiler water treaters	Sulfur dioxide	Bronze cleaners	Nitrogen dioxide
Boiler operators	Arsenic	Bronze makers	Arsenic
Boiler room workers	Carbon monoxide	Broommakers	Sulfur dioxide
Bone extractors	Sulfur dioxide	Brush makers	Lead
Book binders	Arsenic	Buffers	Silica
Bookbinders	Lead	Buhrstone workers	Silica
Bottle cap makers	Lead	Building demolition workers	Asbestos
Brake lining makers	Asbestos	Burnishers	Antimony Benzene
Brakelining makers	Benzene	Busdrivers	Carbon monoxide
Brass founders	Antimony carbon monoxide	Cable makers	Lead
Brass polishers	Lead	Cable splicers	Antimony Carbon monoxide Lead
Brass cleaners	Nitrogen dioxide		

OCCUPATION	AGENT(S)	OCCUPATION	AGENT(S)
Cadium workers	Arsenic	Cellulose makers	Sulfur dioxide
Can makers	Benzene	Cement makers	Carbon monoxide
Candle (colored) makers	Arsenic		Silica
Canners	Arsenic	Cement mixers	Silica
	Lead	Ceramic makers	Antimony
Carbide makers	Carbon monoxide		Arsenic
Carbolic acid makers	Benzene		Beryllium
	Sulfur dioxide		Lead
Carbon monoxide workers	Carbon monoxide	Ceramic workers	Silica
Carders (cotton mill)	Cotton dust	Ceramic enamel workers	Arsenic
Carders (asbestos)	Asbestos	Charcoal burners	Carbon monoxide
Carding machine operatirs (cotton mill)	Cotton dust	Chauffers	Carbon monoxide
Carper makers	Arsenic	Chemical equipment makers	Lead
Carborundum makers	Silica	Chemical products manufacture	Noise
Carroters (felt hat)	Arsenic	Chemical glass makers	Silica
Cartridge makers	Lead	Chimney masons	Carbon monoxide
Cast scrubbers (electroplating)	Benzene	Chimney sweepers	Carbon monoxide
Casting Cleaners (Foundry)	Silica	Chippers	Lead
			Silica
Cathode ray tube makers	Beryllium	Chlorinated paraffin makers	Lead
Cattle dip workers	Arsenic	Chlorobenzene makers	Benzene
Caulking compound makers	Asbestos	Chlorodiphenyl makers	Benzene
Caulking compound users	Asbestos	Chrome platers	Beryllium
			Chromic acid
Celluloid makers	Nitrogen dioxide	Chromic acid makers	Chromic acid
		Cigar makers	Lead

OCCUPATION	AGENT(S)	OCCUPATION	AGENT(S)
Cleaner operators (Cotton mill)	Cotton dust	Coke oven workers	Benzene / Carbon monoxide / Sulfur dioxide
Cleaners (Cotton mill)	Cotton dust	Colored glass maker	Chromic acid
Clutch facing makers	Asbestos	Combing machine operators (cotton mill)	Cotton dust
Clutch Disc impregnators	Benzene	Compositors	Antimony
Coal miners	Silica	Compressed air workers	Carbon monoxide
Coal tar refiners	Benzene	Computer parts makers	Beryllium
Coal tar workers	Benzene	Construction	Noise
Coal distillers	Carbon monoxide	Construct workers	Asbestos / Silica
Cobblers (Asbestos)	Asbestos	Copper cleaners	Nitrogen dioxide
Cobblers	Benzene	Copper Smelters	Arsenic / Sulfur dioxide
Coke oven door cleaners-luterman	Coke oven emissions	Copper refiners	Antimony
Coke oven door machine operators	Coke oven emissions	Copper strippers	Chromic acid
Coke oven heater	Coke oven emissions	Corrosion inhibitor workers	Chromic acid
Coke oven larry car operators	Coke oven emissions	Corrugated paper manufacture	Noise
Coke oven lidmen-larrymen	Coke oven emissions	Cosmetics makers	Silica
Coke oven maintenance men	Coke oven emissions	Cotton bleachers	Nitrogen dioxide
Coke oven patcher	Coke oven emissions	Crop dusters	Arsenic / Lead
Coke oven pusher operators	Coke oven emissions	Crushers (asbestos)	Asbestos
Coke oven quench car operators	Coke oven emissions	Cupola workers	Carbon monoxide
Coke oven tar chaser	Coke oven emissions		

OCCUPATION	AGENT(S)	OCCUPATION	AGENT(S)
Cutlery makers	Lead Silica	Dimethysulfate makers	Arsenic
Cyclohexane makers	Benzene	Diphenyl makers	Benzene
DDT najers	Benzene	Dippers (acid)	Chromic acid
Decorators (Pottery)	Lead	Disinfectant makers	Arsenic Benzene Sulfur dioxide
Defoliant applicators	Arsenic	Disinfectors	Sulfur dioxide
Defoliant makers	Arsenic	Divers	Carbon monoxide
Degreasers	Benzene	Dock workers	Carbon monoxide
Demolition workers	Lead	Drawing frame operators (cotton mill)	Cotton dust
Demolition	Noise	Drier workers	Carbon monoxide
Dental Technicians	Lead	Drug makers	Arsenic Benzene
Dental workers	Nitrogen dioxide	Dry cleaners	Benzene
Detergent makers	Benzene	Dryer operators (cotton mill)	Cotton dust
Diamond Polishers	Lead	Dye makers	Antimony Arsenic Benzene Lead Nitrogen dioxide Sulfur dioxide
Diatomaceous earth calciners	Silica	Dyers	Lead
Dichlorobenzene makers	Benzene	Earth moving equipment operators	Noise
Diesel Equipment Operators	Nitrogen dioxide	Electronic device makers	Lead
Diesel Engine Operators	Carbon monoxide Sulfur dioxide	Electroplaters	Antimony Chromic acid Arsenic Lead Benzene Nitrogen dioxide
Diesel Engine Repairmen	Sulfur dioxide		

OCCUPATION	AGENT(S)	OCCUPATION	AGENT(S)
Electrotypers	Lead	Explosive users	Nitrogen dioxide
Electric arc welders	Nitrogen dioxide	Explosive makers	Benzene
Electrical equipment manufacture	Noise	Exterminators	Arsenic Sulfur dioxide
Electronic equipment makers	Silica	Fabricated metal product manufacture	Noise
Electrolytic copper workers	Arsenic	Farm equipment operators	Noise
Electric equipment makers	Beryllium	Farmers	Arsenic Lead
Embroidery workers	Lead	Feather workers	Arsenic Benzene Sulfur dioxide
Emery Wheel makers	Lead		
Enamel burners	Lead	Ferrosilicon workers	Arsenic
Enamelers	Arsenic Benzene Carbon monoxide Lead	Fertilizer makers	Arsenic Nitrogen dioxide Silica Sulfur dioxide
Enamel makers	Arsenic Lead	Fettlers	Silica
		Fiberizers (Asbestos)	Asbestos
Enamellers	Silica	File Cutters	Lead
Engine rooms	Noise	Firemen	Asbestos Lead
Engravers	Benzene	Fireman	Carbon monoxide
Etchers	Arsenic Nitrogen dioxide	Fireproofers	Asbestos
Ethylbenzene makers	Benzene	Fireworks makers	Antimony Arsenic
Explosives makers	Antimony Lead Nitrogen dioxide	Fischer-Tropsch Process workers	Carbon monoxide
		Flameproofers	Antimony
		Flint workers	Silica

OCCUPATION	AGENT(S)	OCCUPATION	AGENT(S)
Flour bleachers	Nitrogen dioxide Sulfur dioxide	Fumigators	Sulfur dioxide
		Fungicide makers	Benzene
Flower makers (artificial)	Lead	Furnace liners	Silica
		Furnace operators	Sulfur dioxide
Flue cleaners	Sulfur dioxide	Furnace filter makers	Asbestos
Fluorescent screen makers	Beryllium	Furnace starters	Carbon monoxide
Flypaper	Arsenic	Furnace workers	Carbon monoxide
Food processing	Noise	Furniture manufacture	Noise
Food bleachers	Sulfur dioxide	Furniture finishers	Benzene
Formaldehyde makers	Carbon monoxide	Fused quartz workers	Silica
Foundry workers	Antimony Carbon monoxide Lead Silica Sulfur dioxide	Galvanizers	Arsenic Lead Sulfur dioxide
		Garage mechanics	Carbon monoxide Lead
Foundries	Noise	Gas mantle makers	Beryllium
Foundry workers (core making)	Tol. Diisocyanate	Gas shrinking operators	Nitrogen dioxide
Fruit bleachers	Sulfur dioxide	Gas Station attendants	Carbon monoxide
Fruit preserves	Sulfur dioxide	Gas workers (Illumination)	Benzene Carbon monoxide
Fumigant makers	Benzene Sulfur dioxide	Gasket makers	Asbestos
		Gasoline engine testers	Carbon monoxide
Fumigators	Sulfur dioxide	Gelatin bleachers	Sulfur dioxide
Fungicide makers	Benzene	Gin stand operators (cotton mill	Cotton dust

OCCUPATION	AGENT(S)	OCCUPATION	AGENT(S)
Ginners	Cotton dust	Hair remover makers	Arsenic
Glass makers	Antimony Arsenic Beryllium Lead Silica Sulfur dioxide	Hairdressers	Benzene
		Handpickers (cotton)	Cotton dust
		Hard rock miners	Silica
Glass polishers	Lead	Heat treaters	Nitrogen dioxide
Glass manufacture	Noise	Heat treaters (magnesium)	Sulfur dioxide
Glaze Mixers (Pottery)	Silica	Heat resistant clothing makers	Asbestos
Glaze dippers (Pottery)	Antimony	Heat treaters	Carbon monoxide
Glost-kiln workers	Lead	Herbicide makers	Arsenic Benzene
Glue bleachers	Sulfur dioxide		
Glue makers	Benzene	Hide preservers	Arsenic
Gold refiners	Antimony Lead	Histology technicians	Benzene
Gold extractors	Arsenic	Hydrochloric acid workers	Benzene
Gold refiners	Arsenic	Ice makers	Arsenic
Grain bleachers	Sulfur dioxide	Illuminating gas workers	Arsenic
Granite cutters	Silica	Incandescent lamp makers	Lead
Granite workers	Silica	Inert filter media workers	Asbestos
Grinders (Cotton mill)	Cotton dust	Ink makers	Arsenic Benzene Chromic Acid Lead
Grinding wheel makers	Silica		
Grindstone workers	Silica	Insecticide makers	Antimony Arsenic Benzene Lead Silica Sulfur dioxide
Gun barrel browners	Lead		
Gyroscope makers	Beryllium		

OCCUPATION	AGENT(S)	OCCUPATION	AGENT(S)
Insecticide users	Lead	Lacquer makers	Benzene Lead Nitrogen dioxide
Insulators	Silica		
Insulation workers	Asbestos Tol. Diisocyanate	Lacquer workers	Tol. diisocyanate
		Laggers	Asbestos
Insulators (Wire)	Antimony	Lake color makers	Antimony
Internal guidance system makers	Beryllium	Laundry workers	Carbon monoxide
Iron workers	Carbon monoxide	Lead burners	Lead
Ironing board cover makers	Asbestos	Lead counterweight makers	Lead
		Lead flooring makers	Lead
Isocyanate resin workers	Tol. diisocyanate	Lead foil makers	Lead
Japan makers	Lead	Lead mill workers	Lead
Japanners	Lead	Lead miners	Lead
Japan makers	Arsenic	Lead pipe makers	Lead
Japanners	Arsenic	Lead salt makers	Lead
Jet fuel makers	Nitrogen dioxide	Lead shield makers	Lead
Jewelers	Arsenic Lead Silica	Lead smelters	Lead
		Lead stearate makers	Lead
Junk metal refiners	Lead	Lead workers	Lead
Jute workers	Silica	Lead smelters	Sulfur Dioxide
Kiln liners	Silica	Lead burners	Antimony
Kraft recovery furnace workers	Carbon monoxide	Lead hardeners	Antimony
		Lead shot workers	Antimony
Labelers (paint can)	Lead	Lead burners	Arsenic
Laboratory hood installers	Asbestos	Lead shot makers	Arsenic

OCCUPATION	AGENT(S)	OCCUPATION	AGENT(S)
Lead smelters	Arsenic	Mercury smelters	Carbon monoxide Sulfur dioxide
Leather mordants	Antimony	Metal burners	Lead
Leather workers	Arsenic	Metal cutters	Lead
Leather makers	Benzene	Metal grinders	Lead
Lift truck operators (propane and gasoline fueled)	Carbon monoxide	Metal miners	Lead
		Metal polishers	Lead
Lime burners	Arsenic	Metal refiners	Lead
Lime kiln workers	Carbon monoxide	Metallizers	Lead
Linoleum makers	Benzene Lead	Metal forming	Noise
Linotypers	Antimony Lead	Metal machining	Noise
		Metal working	Noise
Linseed oil boilers	Lead	Metal buffers	Silica
Lint cleaner operators (cotton mill)	Cotton dust	Metal burnishers	Silica
		Metal polishers	Silica
Lithographers	Lead	Metal refiners	Arsenic Sulfur dioxide
Lithotransfer workers	Lead		
Lithographers	Benzene Silica	Metal bronzers	Antimony
		Metal cleaners	Arsenic
Lumbering	Noise	Metallurgists	Beryllium
Magnesium foundry workers	Sulfur dioxide	Metal oxide reducers	Carbon monoxide
Maleic acid makers	Benzene	Metal refiners	Carbon monoxide
Masons	Silica	Methanol makers	Carbon monoxide
Match makers	Antimony Lead	Microscopical preparation workers	Chromic acid
Meat preservers	Sulfur dioxide	Millinery workers	Benzene
Medical technicians	Nitrogen dioxide	Mine workers	Nitrogen dioxide

OCCUPATION	AGENT(S)	OCCUPATION	AGENT(S)
Mine-tunnel coaters	Tol. diisocyanate	Nitric acid workers	Lead Nitrogen dioxide
Miners	Antimony Arsenic Carbon monoxide Silica	Nitroglycerin makers	Lead
Mining open pit	Noise	Nitrogen dioxide workers	Nitrogen dioxide
Mining underground	Noise	Nitrocellulose makers	Arsenic
Mirror silverers	Benzene Lead	Nitrobenzene makers	Benzene
Missile technicians	Beryllium	Nitrocellulose makers	Arsenic
Mold makers (plastic)	Beryllium	Nitrobenzene makers	Benzene
Mond process workers	Carbon monoxide	Nitrocellulose workers	Benzene
Monotypers	Antimony Carbon monoxide	Nonsparking tool makers	Beryllium
Mordanters	Antimony Arsenic Benzene	Nuclear physicists	Beryllium
		Nuclear reactor workers	Beryllium
Mortar makers	Silica	Nurses	Nitrogen dioxide
Motorman	Silica	Nylon & makers	Tol. diisocyanate
Musical instrument makers	Lead	Oil bleachers	Sulfur dioxide
		Oil purifiers	Silica
Neon sign workers	Beryllium	Oil processors	Benzene Sulfur dioxide
Neon tube makers	Beryllium	Oil purifiers	Chromic acid
Nickel refiners	carbon monoxide	Oilcloth makers	Benzene
Nickel smelters	Carbon monoxide	Oilstone workers	Silica
Nitrate workers	Nitrogen dioxide	Openers (cotton mill)	Cotton dust

OCCUPATION	AGENT(S)	OCCUPATION	AGENT(S)
Optical equipment makers	Silica	Paper manufacture	Noise
Ordnance manufacturing	Noise	Paper products manufacture	Noise
Ore smelting workers	Sulfur dioxide	Paper makers	Arsenic Sulfur dioxide
Ore smelters	Arsenic	Paraffin processors	Benzene
Organic chemical synthesizers	Antimony Arsenic Benzene Carbon monoxide Chromic acid Nitrogen dioxide Tol. diisocyanate	Patent leather makers	Carbon monoxide Lead
Organic sulfonate makers	Sulfur dioxide	Pearl makers (imitation)	Lead
		Pencil makers	Benzene
Oxalic acid makers	Carbon monoxide	Perfume makers	Antimony Benzene
Oxidized cellulose compound makers	Nitrogen dioxide	Petroleum refinery workers	Benzene
Painters	Lead	Petroleum refining	Noise
Paint makers	Antimony Lead	Petroleum refinery workers	Arsenic Sulfur dioxide
Paint pigment makers	Lead	Petrochemical workers	Benzene
Paint mixers	Silica	Pewter workers	Antimony
Painters	Antimony Arsenic	Pharmaceutical makers	Lead
Paint makers	Arsenic Asbestos	Pharmaceutical workers	Antimony Arsenic Benzene
Painters	Benzene	Phenol makers	Benzene
Paint makers	Benzene	Phosphor makers	Antimony
Paper hangers	Arsenic Lead	Phosphate coating strippers	Chromic acid

OCCUPATION	AGENT(S)	OCCUPATION	AGENT(S)
Photography workers	Lead	Plastic makers	Abestos
Photoengravers	Nitrogen dioxide	Plastics workers	Beryllium
Photographic Chemical makers	Benzene	Plumbers	Arsenic Lead
Photography workers	Chromic acid	Policemen	Lead
Physicians	Nitrogen dioxide	Police	Carbon monoxide
Pickers (cotton mill)	Cotton dust	Polishing soap makers	Silica
Picklers	Chromic acid Nitrogen dioxide	Polish makers	Benzene
Picric acid makers	Benzene	Polyurethane foam makers	Tol. diisocyanate
Pigment makers	Antimony Arsenic	Polyurethane foam users	Tol. diisocyanate
Pipe fitters	Lead Nitrogen dioxide	Polyurethane sprayers	Tol. diisocyanate
Pipe insulators	Asbestos	Porcelain workers	Antimony Silica
Plasma torch operators	Nitrogen dioxide	Pottery glaze mixers	Lead
Plastic workers	Lead	Pottery glaze dippers	Lead
Plastics manufacture	Noise	Pottery workers	Antimony Lead Silica Sulfur dioxide
Plastic products manufacture	Noise	Pottery decorators	Benzene
Plastic foam makers	Tol. diisocyanate	Pottery kiln workers	Carbon monoxide
Plasticizer workers	Tol. diisocyanate	Pouncers (felt hat)	Silica
Plaster cast bronzers	Antimony	Power plant operators	Noise
Plastic workers	Arsenic		

OCCUPATION	AGENT(S)	OCCUPATION	AGENT(S)
Precision instrument makers	Beryllium	Quarrying	Noise
Preservative makers	Sulfur dioxide	Quarry workers	Silica
Press box operators (cotton mill)	Cotton dust	Quartz workers	Silica
		Raw silk bleachers	Nitrogen dioxide
Primary metal processing	Noise	Rayon makers	Arsenic
Printers	Lead	Reclaimers (rubber)	Benzene
Printing	Noise	Refractory Makers	Chromic acid Silica
Printers	Antimony		
Printing ink workers	Arsenic	Refractory material Makers	Beryllium
Printers	Benzene	Refrigeration workers	Sulfur Dioxide
Producer gas workers	Carbon monoxide	Resin Makers	Benzene
Protein makers (industrial)	Sulfur dioxide	Riveters	Lead
Protein makers (food)	Sulfur dioxide	Road Constructors	Silica
Pulpstone workers	Silica	Rock Crushers	Silica
Pump packing makers	Asbestos	Rock Cutters	Silica
		Rock Drillers	Silica
Putty makers	Benzene Lead	Rock Grinders	Silica
Pyrites burners	Sulfur dioxide	Rock Screeners	Silica
Pyrotechnics workers	Antimony Arsenic	Rocket Fuel makers	Nitrogen dioxide
		Rodenticide makers	Arsenic
		Roofers	Asbestos Lead
Pyroxylin-plastics workers	Lead	Roofing materials Makers	Asbestos
		Rotogravure printers	Benzene

OCCUPATION	AGENT(S)	OCCUPATION	AGENT(S)
Roving Frame Operators (cotton mill)	Cotton dust	Semiconductor workers	Antimony Lead
Rubber Buffers	Lead	Semiconductor compound makers	Arsenic
Rubber makers	Antimony Benzene Lead	Service station attendants	Lead
Rubber Reclaimers	Lead	Sewer workers	Carbon monoxide
Rubber manufacture	Noise	Sheep dip workers	Arsenic
		Sheet metal workers	Lead
Rubber products manufacture	Noise	Shellac makers	Benzene Lead
Rubber compound mixers	Silica	Shingle makers	Asbestos
Rubber compounders	Asbestos	Ship dismantlers	Lead
Rubber cementers	Benzene	Ship burners	Tol. diisocyanate
Rubber gasket makers	Benzene	Ship welders	Tol. diisocyanate
Sand cutters	Silica	Ship builders	Asbestos
Sand pulverizers	Silica	Ship demolition workers	Asbestos
Sandblasters	Silica	Shipbuilding	Noise
Sandpaper makers	Silica	Shoe stainers	Lead
Sandstone grinders	Silica	Shoe factory workers	Benzene
Sanitation workers	Carbon monoxide	Shoe finishers	Benzene
Sawyers	Silica	Shot makers	Lead
Scouring soap workers	Silica	Silica brick workers	Silica
		Silicon alloy makers	Silica
Scrap metal workers	Lead	Silk weighters	Lead
		Silo Fillers	Nitrogen dioxide
Sealing wax makers	Arsenic	Silver polishers	Silica

OCCUPATION	AGENT(S)	OCCUPATION	AGENT(S)
Silver Refiners	Arsenic	Spray painters	Tol. diisocyanate
Silver Platers	Beryllium	Stainers	Benzene
Slashing Operators (cotton mill)	Cotton dust	Stain makers	Benzene
Slate workers	Silica	Steel engravers	Lead
Slushers (porcelain enameling)	Lead	Steel making	Noise
Smelters	Silica	Steel makers	Carbon monoxide
Soap makers	Benzene	Stereotypers	Antimony Lead
Soda makers	Arsenic	Stokers	Carbon monoxide
Sodium silicate makers	Silica	Stone products industries (cement mills)	Noise
Sodium sulfite makers	Sulfur dioxide	Stone workers	Noise
Soil Sterilizer makers	Arsenic	Stone bedrubbers	Silica
Solderers	Arsenic Carbon monoxide Lead	Stone cutters	Silica
		Stone planers	Silica
Solder makers	Antimony Lead	Storage battery chargers	Sulfur dioxide
Solid Rocket Fuel Makers	Beryllium	Storage batter workers	Antimony
Solvent makers	Benzene	Straw bleachers	Sulfur dioxide
Spacecraft workers	Silica	Street sweepers	Silica
Spindle pickers (cotton)	Cotton dust	Stripper operators (cotton)	Cotton dust
Spinners (cotton mill)	Cotton dust	Stripper operators (cotton mill)	Cotton dust
Spinners (asbestos)	Asbestos	Strippers	Chromic acid
		Styrene makers	Benzene
Spooling operators (cotton mill)	Cotton dust	Submarine workers	Arsenic

OCCUPATION	AGENT(S)	OCCUPATION	AGENT(S)
Subway construction workers	Silica	Textile dryers	Antimony
Sugar refiners	Sulfur dioxide	Textile flame-proofers	Antimony Asbestos
Sulfite makers	Sulfur dioxide		
Sulfuric acid makers	Nitrogen dioxide Sulfur dioxide	Textile printers	Antimony Arsenic
		Textile workers	Asbestos
Sulfur dioxide workers	Sulfur dioxide	Textile mordants	Chromic acid
Sulfurers (malt and hops)	Sulfur dioxide	Thermometer makers (vapor pressure)	Sulfur dioxide
Sulfuric acid workers	Arsenic	Thionyl chloride makers	Sulfur dioxide
Synthetic fiber makers	Benzene	Tile makers	Lead Silica
Talc miners	Asbestos	Tin foil makers	Lead
Talc workers	Asbestos	Tinners	Arsenic Lead
Tannery workers	Lead Sulfur dioxide	Tobacco seedling treaters	Benzene
Tanners	Arsenic Chromic acid	Toll collectors (highway)	Carbon monoxide
Tar workers	Arsenic	Tooth paste makers	Silica
Taxidermists	Arsenic	Traffic controllers	Carbon monoxide
TDI workers	Tol. diisocyanate	Transportation equip-ment operators	Noise
Television picture tube makers	Lead	Tree Sprayers	Arsenic
Temperers	Lead	Trinitrotoluol makers	Benzene
Textile makers	Lead	Trucking	Noise
Textile manufacture	Noise	Tube mill liners	Silica
Textile bleachers	Sulfur dioxide	Tumbling barrel workers	Silica
Textile processors	Tol. diisocyanate		

OCCUPATION	AGENT(S)	OCCUPATION	AGENT(S)
Tunnel workers	Carbon monoxide Nitrogen dioxide	War gas makers	Benzene
Tunneling	Noise	Warehouse workers	Carbon monoxide
Tunnel construction workers	Silica	Warfare gas makers	Arsenic
		Water weed controllers	Arsenic
Tunnel attendants	Carbon monoxide	Water gas workers	Carbon monoxide
Twisters (cotton mill)	Cotton dust	Wax makers	Benzene
Type founders	Lead	Weavers (cotton mill)	Cotton dust
Type setters	Lead	Weavers (asbestos)	Asbestos
Type metal workers	Antimony Arsenic	Weed sprayers	Arsenic
Type cleaners	Benzene	Welders	Benzene Carbon monoxide Lead Nitrogen dioxide
Typesetters	Antimony		
Undercoaters	Asbestos	Whetstone workers	Silica
Upholstery makers	Tol. diisocyanate	Wicker ware bleachers	Sulfur dioxide
Vanadium compound makers	Lead	Window shade makers	Benzene
		Wine makers	Sulfur dioxide
Varnish makers	Arsenic Benzene Lead	Wire coating workers	Tol. diisocyanate
		Wire drawers	Arsenic
Vegetable preservers	Sulfur dioxide	Wire insulators	Benzene
Vehicle tunnel attendants	Lead	Wood stainers	Lead
Vinyl-asbestos tile makers	Asbestos	Wood products manufacture	Noise
		Wood filler workers	Silica
Vinyl-asbestos tile installers	Asbestos	Wood bleachers	Sulfur dioxide
Vulcanizers	Antimony Benzene Sulfur dioxide	Wood pulp bleachers	Sulfur dioxide
		Wood preservative makers	Arsenic
Wallpaper-printers	Arsenic Lead		

OCCUPATION	AGENT(S)
Wood preservers	Arsenic
Wood distillers	Carbon monoxide
X-ray tube makers	Beryllium
Zinc mill workers.	Lead
Zinc smelter chargers	Lead
Zinc Smelters	Sulfur dioxide
Zinc refiners	Antimony
Zinc chloride makers	Arsenic
Zinc miners	Arsenic
Zinc refiners	Arsenic
Zinc white makers	Carbon monoxide

SECTION TWO

ASBESTOS AND OCCUPATIONAL DISEASE

Introduction

Asbestos is a mineral fiber, and is the name given to about thirty silicate compounds. Of these, only the following 5 are of significance in industry:

Chrysotile (white asbestos)
Amosite
Tremolite
Crocidolite (blue asbestos)
Anthophyllite

Chrysotile accounts for about 97 percent of all the asbestos used in this country.

Asbestos is widespread in the environment because of its extensive use in industry and the home. Over 3,000 products contain asbestos.

Because of this wide usage, it may be difficult at times to determine if a disease arising from asbestos is occupational in origin. For example, the air of some relatively new apartment buildings has been found to contain more asbestos fibers than the maximum recommended levels in industry. The source of the fibers in the apartment buildings is the insulating materials used in the ventilating system.

Exposure to asbestos can produce a lung fibrosis called *asbestosis*. The onset of asbestosis is usually gradual, developing over a period of 10 to 30 years of exposure to significant concentrations of asbestos. Occasionally, from very massive exposures, it may develop more quickly.

Asbestos is also a *cancer* producing agent (bronchogenic carcinoma, mesothelioma) and can cause certain specific skin diseases (asbestotic subcutaneous granulomatosis and asbestotic cutaneous verruca). Heavy exposure to dust containing asbestos can cause *skin irritation*. Epidemiologic studies (experience with groups of people) and animal studies have shown that increased exposure to any of the types of asbestos increases the risk of lung cancer (bronchial carcinoma). This carcinoma appears to be related to the degree of exposure to asbestos, the type of asbestos and cigarette smoking. It is also significant that cigarette smoking in men and women greatly increase the risk of lung cancer in those who are exposed to asbestos. Smoking is a factor that should be considered when determining whether lung cancer is caused, wholly or in part, by an occupational exposure to asbestos.

Mesothelioma, a rare malignant tumor of the membrane which lines the chest cavity and the abdominal cavity, is occurring with increasing frequency in workers with exposure to asbestos. The development of this tumor apparently is not related to the amount of asbestos inhaled and it is found in persons not having asbestosis. Levels of exposure which are within accepted standards for protection against asbestosis, may not protect against mesothelioma.

An increased incidence of malignancy of the stomach and colon has been reported among insulation workers using asbestos.

Occupations with Potential Exposure to Asbestos

Acoustical Product Makers	Crushers (Asbestos)
Acoustical Product Installers	Fiberizers (Asbestos)
Air filter makers	Fireproofers
Asbestos-cement products makers	
Asbestos-cement products users	Firemen
Asbestos-coatings makers	Furnace filter makers
Asbestos-coatings users	Gasket makers
	Heat resistant clothing makers
Asbestos-grout makers	Insulation workers
Asbestos-grout users	Inert filter media workers
Asbestos-millboard makers	Ironing board cover makers
Asbestos-millboard users	Laboratory hood installers
Asbestos-mortar makers	Laggers
Asbestos-mortar users	Paint makers
Asbestos millers	Pipe insulators

Asbestos miners

Asbestos-paper makers

Asbestos-paper users

Asbestos-plaster makers

Asbestos-plaster users

Asbestos sprayers

Asbestos workers

Asphalt mixers

Automobile repair garage workers

Brake lining makers

Building demolition workers

Carders (asbestos)

Caulking compound makers

Caulking compound users

Clutch facing makers

Cobbers (asbestos)

Construction workers

Plastics makers

Pump packing makers

Roofers

Roofing materials makers

Rubber compounders

Shingle makers

Ship builders

Ship demolition workers

Spinners (Asbestos)

Talc miners

Talc workers

Textile flameproofers

Textile workers

Undercoaters

Vinyl-asbestos tile makers

Vinyl-asbestos tile installers

Weavers (asbestos)

Medical Evaluation

In addition to the usual medical history, the following should be considered:

1. Any history of diseases of the heart or lung or abnormal tissue growth should be carefully evaluated to determine the relationship between the previous disease and the claimant's present condition.

2. A respiratory questionnaire can be useful in evaluating the extent and importance of respiratory symptoms such as:

   —breathlessness
   —phelgm (sputum) production
   —chest pain
   —cough
   —wheezing

Asbestosis—Shortness of breath upon exertion is usually the first symptom, frequently accompanied by a dry cough. This symptom develops after several years of progressive pulmonary fibrosis. As asbestosis progresses, the following signs and symptoms are observed:

—cough with production of sputum
—anorexia (loss of appetite)
—secondary respiratory infections that are difficult to control
—rapid breathing
—repetitive end-inspiratory crackles (crackling sounds heard in the lower part of the lungs through stethoscope when employee completes each of a series of inhaled breaths)
—orthopnea (breathing difficulty in a recumbent position)

—cyanosis (change in skin color to bluish, grayish, slatelike or dark purple)
—decrease of chest expansion
—digital clubbing (rounding of the ends, and swelling of the fingers and/or toes)
—sequelae (other resultant diseases) including cor pulmonale (right heart failure), **bronchogenic carcinoma** (lung cancer), stomach or intestinal cancer, or pleural carcinoma (cancer of the membrane lining the chest)

Fibrosis results in alveolo-capillary block (impaired ability of the lungs to transfer oxygen into the blood). This impairment is often more severe than is indicated by chest x-rays.

Mesothelioma—In cases of mesothelioma, the rare malignancy noted above, there may be a long latent period, as much as 40 years, between initial exposure to asbestos and the development of the tumor.

Mesothelioma of the *peritoneum* (membrane surrounding the abdominal organs) is usually accompanied by abdominal swelling and pain that is not concentrated in a particular area. Signs and symptoms of this type of tumor (which *may* be associated with asbestos exposure) include:

—weight loss
—obstruction of the bowel
—excessive accumulation of fluid in the abdominal cavity (ascites) is almost always present

This malignant tumor of the peritoneum may spread to the chest cavity.

With mesothelioma of the *pleura,* complaints include chest pain and breathlessness. Signs and symptoms of pleural mesothelioma include:

—pleural effusion (accumulation of fluid in the space around the lungs)
—the tumor may grow outward through the chest wall in the form of a lump beneath the skin (subcutaneous lump)
—the tumor may spread to involve bone, lymph glands (nodes) mediastinum (area between the right and left lungs), and pericardium (the sac enclosing the heart). As a result, the supraclavicular nodes may become enlarged, ribs may develop tumors, and obstruction of the superior vena cava (major vein draining the upper portion of the body) may occur.
—in addition, pericardial effusion (fluid in the heart cavity) may occur, causing tamponade.

Additional tests which will assist in arriving at a correct diagnosis are:

Chest X-rays

Findings should be classified according to the ILO/UC 1971 Classification of the Radiographs of the Pneumoconioses.

Findings for asbestosis vary, but the usual picture shows a density in both lungs, with the lower one-third of the lungs involved. In the affected area there is a "ground glass" appearance.

As asbestosis progresses, more and more of the lung is involved, except the apices (tips of the lungs). The X-rays will show gradual obscuring of the border between the lungs and the diaphragm. It may show shadows from the presence of nodules.

X-ray findings usually will show the following as the asbestosis progresses:

- reduced radiographic volume
- formation of cysts combined with increased size of the heart, dilation (enlargement) of the proximal pulmonary arteries (arteries which lead from the heart to the lungs)

Lung Function Tests

Reduced lung capacities and other lung changes do not differ from those resulting from other forms of lung fibrosis, both occupational and nonoccupational. Therefore, the results of lung function tests alone or chest X-ray findings alone do not lead to diagnosis of asbestosis. Asbestos bodies in lymph nodes indicate exposure, but not necessarily asbestosis.

- Asbestosis causes a reduction in the vital capacity (VC) of the lungs and a reduction in total lung capacity (TLC). These capacities are further reduced as the disease progresses.
- The residual volume (RV) of the lungs will be normal or slightly increased.
- The lungs' diffusing capacity for carbon monoxide (D_L) will be reduced.

Other lung function test results which are found in asbestosis include:

- Increased minute ventilation (amount of air breathed in one minute)
- Reduced oxygenation of the arterial blood (arterial hypoxemia)
- Increased static transpulmonary pressures
- Decreased lung compliance

An exercise test will result in an increased amount of air required during physical effort, decreased oxygen in the blood, leading to cyanosis.

Sputum Examination

Asbestos fibers or bodies may be found in the sputum. These indicate asbestos exposure, but not necessarily asbestosis. Where cancer cells are present in the sputum, and chest X-ray findings are normal, bronchoscopy may be necessary to confirm and locate the lung tumor.

Skin Tests

The following tests should be performed by the physician to exclude possible infectious diseases:

1. PPD (tuberculin test) 3. histoplasmin
2. blastomycin   4. coccidioidin

Epidemiological Data

Various epidemiologic studies have demonstrated the relationship between asbestos and lung disease, including mesothelioma, in such trades and occupations as mining, insulation installation, textiles, paint, electrical industries, and many other occupations as a result of the widespread use of this substance.

The available information indicates evidence of a dose-response relationship for asbestos exposure and the risk of asbestosis and/or bronchogenic carcinoma. However, much of this information is epidemiological in nature and there is little correlation between epidemiologic data and environmental exposure data. For this reason and others, including the long latent period for the development of carcinomas, it is difficult to develop a specific dose-response relationship. This should be taken into consideration when referring to the following material:

Enterline has reported an exposure-response relationship between asbestos exposure (evaluated as millions of particles per cubic foot years) and the risk of malignant and nonmalignant respiratory disease. Enterline's data indicates that the risk of respiratory cancer increased from 166.7 (standardized mortality ratio) at minimum exposure to 555.6 at cumulative exposures exceeding 750 million particles per cubic foot years. Enterline's data is summarized in a table by NIOSH.

Murphy reported that asbestosis was 11 times more common among pipe coverers in new ship construction than in a control group. The first asbestosis was found after 13 years of exposure to an estimated cumulative dose of about 60 million particles per cubic foot years. After 20 years, asbestosis prevalence was 38%. Murphy reported no asbestosis for men exposed to 60 mppcf years but 20% asbestosis in men exposed to 75-100 mppcf years. Murphy reports atmospheric dust concentrations ranged from 0.8-10.0 mppcf depending on the different operations evaluated. Asbestosis was considered present if the worker had at least three of the following: vascular rales in two or more sites, clubbing of the fingers, vital capacity of less than 80% predicted, roentgenography consistent with moderately advanced or advanced asbestosis, shortness of breath on climbing one flight of stairs.

The Pennsylvania Department of Health reported a study of asbestos dust concentrations in two plants (one studies from 1930-1967 and the other from 1948-1968). 64 cases of asbestosis were reported. In the two plants, the study indicates that the air concentrations of particulates were generally less than five mppcf and in many cases less than two mppcf.

Epidemiological evidence is also available relating the development of mesothelioma with exposure to asbestos. Selikoff reported 14 deaths from mesothelioma in 532 asbestos insulation workers from 1943-1968. No deaths from mesothelioma would be expected from the same number of individuals in the general population.

Evidence of Exposure

Historically, there have been two air sampling and analysis methods to determine the quantity of asbestos in the workplace environment. The earlier light field impinger count method allowed only a measure of the overall dust level in the air rather than focusing on the amount of asbestos fibers in the air. The current fiber count method satisfactorily determines the amount of asbestos fibers in the air. It is performed by collecting airborne materials on a membrane filter and then counting the fibers using a phase contrast microscope at a 400 to 450 times magnification ratio (400X-450X).

Asbestos fibers occur in varying lengths and diameters. As of the publication of the guide, the Occupational Safety and Health Act (OSHA) establishes maximum allowable limits for asbestos fibers greater than five micrometers (μm) in length. OSHA limits such asbestos fibers to no more than five fibers per cubic centimeter of air (based on an eight hour time-weighted average exposure).

OSHA further requires that no workers be exposed to more than 10 asbestos fibers (greater than five μm in length) during any one 15 minute period of time.

For samples collected by the field impinger count method, results may be compared to the pre-1970 limit (TLV) of five million particles per cubic foot of air.

Occupational exposure to asbestos fibers five μm in length or greater, at quantities averaging more than five fibers per cubic centimeter of air or frequent exposures to more than 10 such fibers during a 15-minute period of time is evidence of a possible causal relationship between disease and occupation.

Conclusion

The diagnosis of occupational asbestosis is based on meeting the following criteria:

1. Confirmed history of occupational exposure to asbestos.
2. X-ray findings compatible with those indicating asbestosis according to ILO/UC 1971 "Classification of Radiographs of the Pneumoconioses."
3. Pulmonary impairment, particularly a decrease in lung diffusing capacity and an increase in alveolar-arterial oxygen difference, as demonstrated by lung function tests.

The diagnosis of occupational mesothelioma is based on meeting the following criteria:

1. Confirmed history of occupational exposure to asbestos.
2. Pathological evidence of mesothelioma.

SECTION THREE

INORGANIC LEAD AND OCCUPATIONAL DISEASE

Introduction

Lead is a naturally occurring element found in some quantity in the human body. This Guide discusses only inorganic lead which, in industry, is usually absorbed into the body through inhalation of dust or fumes. Nonoccupational exposure may occur through ingestion, e.g. lead etched from the glaze of pottery used for food, paint from water pipes, or from food contaminated with lead.

Although lead may be absorbed into the body, absorption does not necessarily constitute lead poisoning. Body burden will affect individual tolerance. At a given body burden, some persons may have signs and symptoms of lead poisoning while others do not.

Lead may be stored in the body and, following an illness or some stress factor, be released into the system and produce symptoms of lead poisoning.

Lead-Chemical and Common Names

Chemical name	Common Names
lead	plumbum
lead acetate	normal lead acetate, sugar of lead, salt of Saturn
lead antimonate	Naple's yellow, antimony yellow
lead borate	
lead bromide	
lead butyrate	butyric acid lead salt
lead carbonate	basic lead carbonate, lead subcarbonate, white lead, flake lead, ceruse, cerussa, cerussite
lead chlorate	
lead chloride	cotunnite, matlockite
lead chromate	chrome yellow, Cologne yellow, King's yellow, Leipzig yellow, Paris yellow, crocoite
lead chromate oxide	basic lead chromate, red lead chromate, chrome red, Persian Red, Austrian cinnabar
lead citrate	
lead cyanide	
lead dioxide	lead oxide brown, lead superoxide, lead peroxide, plumbic acid anhydride, plattnerite
lead fluoride	
lead hexafluorosilicate	lead fluorosilicate, lead silicofluoride
lead hydroxide	basic lead hydroxide, lead hydrate, hydrated lead oxide
lead iodide	
lead metaborate	
lead metasilicate	alamosite
lead molybdate	wulfenite
lead monoxide	lead oxide yellow, plumbus oxide, litharge, massicot, lead protoxide
lead nitrate	
lead nitrite	
lead phosphate	pyromorphite
lead phosphite	
lead sesquioxide	lead trioxide, plumbus plumbate
lead sulfate	anglisite
lead sulfide	galena, plumbous sulfide
lead tartrate	
lead telluride	altaite

lead tetrafluoride	plumbing fluoride
lead tetraoxide	lead oxide red, red lead, minimum, lead orthoplumbate, mineral orange, mineral red, Paris red, Saturn red
lead thiocyanate	lead sulfocyanate
lead thiosulfate	lead hyposulfate
lead tungstate	raspite, scheelite, stolzite, lead wulframate
lead vanadate	lead metavanadate, vanadinite

Some Occupations with Potential Lead Exposures

Acid finishers
Actors
Babbiters
Battery makers
Blacksmiths
Bookbinders
Bottle cap makers
Brass founders
Brass polishers
Braziers
Brick burners
Brick makers
Bronzers
Brushmakers
Cable makers
Cable splicers
Canners
Cartridge makers
Chemical equipment makers
Chlorinated Paraffin makers
Chippers
Cigar makers
Crop dusters
Cutlery makers
Decorators (pottery)
Demolition workers
Dental technicians
Diamond polishers
Dye makers
Dyers
Electronic device makers
Electroplaters
Electrotypers
Embroidery workers
Emery wheel makers
Enamel burners

Lead workers
Linoleum makers
Linotypers
Linseed oil boilers
Lithographers
Lithotransfer workers
Match makers
Metal burners
Metal cutters
Metal grinders
Metal polishers
Metal refiners
Metal refinishers
Metallizers
Mirror silverers
Musical instrument makers
Nitric acid workers
Nitroglycerin makers
Painters
Paint makers
Paint pigment makers
Paper hangers
Patent leather makers
Pearl makers (imitation)
Pharmaceutical makers
Photography workers
Pipe fitters
Plastic workers
Plumbers
Printers
Policemen
Pottery glaze mixers
Pottery glaze dippers
Pottery workers
Putty makers
Pyroxylin-plastics workers

Enamelers
Enamel makers
Explosives makers
Farmers
File cutters
Firemen
Flower makers (artificial)
Foundry workers
Galvanizers
Garage mechanics
Glass makers
Glass polishers
Glost kiln workers
Gold refiners
Gun barrel browners

Incandescent lamp makers
Ink makers
Insecticide makers
Insecticide users
Japan makers
Japanners

Jewellers
Junk metal refiners
Labelers (paint can)
Lacquer makers
Lead burners
Lead counterweight makers
Lead flooring makers
Lead foil makers

Lead Mill Workers
Lead miners
Lead pipe makers
Lead salt makers
Lead shield makers
Lead smelters
Lead stearate makers

Riveters
Roofers
Rubber buffers
Rubber makers
Rubber reclaimers
Scrap metal workers
Semiconductor workers
Service station attendants
Sheet metal workers
Shellac makers
Ship dismantlers
Shoe stainers
Shot makers
Silk weighters
Slushers (porcelain enameling)
Solderers
Solder makers
Steel engravers
Stereotypers
Tannery workers
Television picture tube makers
Temperers
Textile makers
Tile makers
Tin foil makers
Tinners
Type founders
Type setters
Vanadium compound makers
Varnish makers
Vehicle tunnel attendants
Wallpaper printers
Welders
Wood stainers
Zinc mill workers
Zinc smelter chargers

Medical Evaluation

In the personal history, consider the following:

—Lead is so widely used that a careful inquiry into hobbies and recreation is especially important. Chronic exposure to inorganic lead in hobbies can produce the same signs and symptoms as occupational lead poisoning, but it is not occupational. Common non-occupational lead exposures include:

—ceramics, pottery and related hobbies
—electronics and related hobbies involving extensive soldering

—firing ranges

—hunting (especially those who cast their own bullets)

—eating or drinking from improperly fired lead-glazed ceramic tableware

—eating lead-bearing paint (especially children)

—burning battery castings

—consuming illicitly distilled whiskey

—extensive auto driving (especially in cities)

—extensive work with motor fuels

—painting with lead-containing paints

—home plumbing repairs (lead pipe systems)

—exterminating

Signs and Symptoms

The early signs and symptoms of lead poisoning are nonspecific and may resemble many diseases including influenza. Early signs and symptoms are the following:

—malaise

—fatigue

—sleep disturbance

—headache

—nausea and vomiting

—loss of appetite

—irritability

—aching muscles and bones

—constipation

—abdominal cramps

In more advanced cases of lead poisoning, the above signs and symptoms progress and frequently involve the gastro-intestinal and neuro-muscular systems (both nerves and muscles), and the kidneys (Fanconi Syndrome).

Gastro-intestinal signs and symptoms are:

—severe abdominal pain (lead colic)

—constipation (never diarrhea)

—marked loss of appetite (anorexia) leading to weight loss

—characteristic lead line of the gums may be present, usually with pyorrhea

Neuro-muscular and neuro-behavioral symptoms are:

—generalized tenderness or pain in the muscles (myalgia)

—muscular weakness, especially of the most frequently used muscles

—tremors or palsy may be present

—decreased hand grip strength

—characteristic "wrist drop" (wrist flexed and cannot be extended because of nerve involvement)

—the peripheral nerves of the upper extremities are involved; rarely those of the lower extremities

Reproductive:

—decreased fertility in men and spontaneous abortion in women have been reported

A most severe form of lead poisoning is lead encephalopathy, impairment of the brain due to lead poisoning. Lead or Saturnine encephalopathy is rarely seen today because of improved techniques of handling lead. It is a vague term and includes coma, delirium, psychosis, convulsions; muscles affecting speech, eyes, and face are often involved. It can result in blindness and death.

Laboratory

Signs pertaining to lead's effect on the blood forming organs (hematopoietic system) are determined by laboratory analysis. These signs occur early with excess lead absorption—usually before the outward symptoms of poisoning appear. These tests are useful in the routine biological monitoring of persons exposed to lead.

Abnormal laboratory values that may be found in lead poisoning:

—decreased red blood count

—decreased hemoglobin and hematocrit

—decreased motor nerve conduction velocity

—increased urinary delta-aminolevulinic acid

—increased free erythrocyte protoporphyrin (FEP) and zinc protoporphyrin (ZP or ZPP)

—increased lead in blood

—increased lead in urine

NOTE: Results of blood and urine laboratory analyses for lead are subject to a 10-15 percent error factor. The normal values for the laboratory performing the tests should be ascertained. Blood lead determinations must be corrected for the mass of circulating red cells (hematocrit); and urinary lead determinations, for the specific gravity of the urine.

Epidemiologic Data

There is vast and detailed scientific literature providing evidence of lead poisoning in workers with significant exposure. Neuropathies, nephropathy and blood changes are well documented. Lead absorption, however, does not necessarily indicate poisoning.

Lane reported a study of nine lead workers in a storage battery industry who had been exposed to lead concentrations in air around 0.5 milligrams per cubic meter of air for over 20 years. All died from hypertension and renal failure between the ages of 42 and 52.

Williams, King and Walford report the following data taken in table form from the Criteria Document on lead, published by the National Institute for Occupational Safety and Health.

The figures in the top lines indicate mean values.

(It is recommended that blood lead levels greater than 0.060 milligrams lead per 100 grams whole blood is in-

AIR LEAD CONC. IN MILLIGRAMS PER CUBIC METER	BLOOD LEAD MILLIGRAMS PER 100 MILLILITERS	URINE LEAD MILLIGRAMS PER LITER	URINE COPRO-PORPHYRIN (DONATH)	URINE ALA* MILLIGRAMS PER 100 MILLILITERS
0.20	0.070 (0.048-0.092)	0.143 (0.056-0.230)	4.2 (2.4-6.0)	1.8 (0.3-3.3)
0.15	0.060 (0.038-0.082)	0.118 (0.031-0.205)	3.6 (1.8-5.4)	1.4 (0.1-2.9)

* ALA values were determined by a method which probably gives higher values than do other methods, thus a high "normal" value.

dicative of unacceptable lead absorption and that urine lead levels of 0.20 milligrams lead per liter of urine or greater is indicative of unacceptable lead absorption.)

Elkins assembled data available on lead in air and lead in urine and reported that a urinary lead level of 0.2 milligrams lead per liter of urine would, averaging, correspond to an air concentration of 0.2 milligrams lead per cubic meter of air.

Hartogenesis and Zielhuis report blood changes in workers exposed to lead chromate dust in concentrations greater than 0.2 milligrams per cubic meter of air as lead. They further report doubtful changes in blood at exposures to atmospheric concentrations between 0.1 and 0.2 milligrams per cubic meter as lead.

The above data relating average blood lead content with exposure and duration of employment has been adapted from Dreessen et al. Committee on Biologic Effects of Atmospheric Pollutants, and the National Institute of Occupational Safety and Health.

Air Sampling and Analysis

There are three commonly accepted methods of lead air sampling:

1. impingement
2. electrostatic precipitation
3. mechanical filtration

There are three commonly accepted methods to analyze the samples for presence of lead:

1. atomic absorption spectrophotometry
2. colorimetrically using the dithizone method
3. polarographically

DURATION OF LEAD EXPOSURE	AIR LEAD CONTENT MILLIGRAMS PER CUBIC METER			
	0-0.074	0.075-0.14	0.15-0.29	0.3 OR MORE
YEARS 0-4				
Number	17	16	32	20
Average	0.0187	0.0316	0.0378	0.0463
Median	0.021	0.030	0.038	0.050
YEARS 5-9				
Number	10	13	40	20
Average	0.0278	0.0405	0.0501	0.0505
Median	0.033	0.040	0.043	0.050
YEARS 10-14				
Number	23	24	30	32
Average	0.0198	0.0375	0.0502	0.0481
Median	0.018	0.038	0.046	0.048
YEARS 15+				
Number	44	30	59	45
Average	0.0293	0.0407	0.0457	0.0493
Median	0.023	0.036	0.045	0.045

These methods are not intended to be exclusive. but other methods should be justified.

The Occupational Safety and Health Adm. (OSHA) limits exposure to lead and its inorganic compounds (except lead arsenate) to 0.2 milligrams per cubic meter of air based on an eight hour time-weighted average exposure.

The American Conference of Governmental Industrial Hygienists threshold limit value for lead and its inorganic compounds (except lead arsenate) is 0.15 milligrams per cubic meter of air based on an eight hour time-weighted average exposure.

Conclusion

Diagnostic criteria for occupational lead poisoning are based on meeting the following:

1. confirmed history of occupational exposure to lead
2. findings compatible with lead poisoning
3. increased lead in blood and/or urine

NOTE: A diagnosis of lead poisoning does not necessarily mean that it is occupational in origin. Further, lead intoxication with symptoms can exist with normal laboratory test findings.

SECTION FOUR

CASE HISTORY OF CHRONIC ARSENIC INTOXICATION

Occupational Disease Case History

Complaint: Malaise, increasing fatigue and "pins and needles" sensation in the feet.

Medical Evaluation

Evaluation of complaint: Past few days noticed a "pins and needles" sensation in his feet and some weakness of the lower legs. For several weeks or longer he has generally felt weak and tired and not himself. In general he has not been feeling well for quite some time. He has had some weight loss but has not been eating well because of lack of appetite. For a time he has had intermittent periods of nausea and vomiting, but they "come and go." Insomnia and rather frequent headaches have been occurring. Remaining systemic review is negative.

Medical history.—General health has always been good. Tonsils and adenoids removed as a child; usual childhood diseases; occasional colds but nothing serious.

Personal history.—Age 36, white male, married with children, boy 13 and girl 11. Drinks 8 to 10 ounces of alcohol a day and smokes one pack of cigarettes a day. Lived all his life in Brooklyn, New York. Graduated from high school at age 18. Mother and father and two siblings living and well—mother has diabetes. As a hobby he gardens and has many house plants, but does not use insecticides.

Occupational history.—Present occupation: Handyman—works with five other people in a small shop where arts and crafts are made. The work entails mixing pigments and dyes used in printing textiles and for coloring enamels and glazes; generally keeps the shop clean and in order.

Previous occupations: Took two courses of arts and crafts, pottery-making and glazing in high school. Worked part-time as a grocery clerk while in school. After gradua-

tion worked for five years as a ship cutter; exposed to lead, asbestos and iron oxide.

Building superintendent, two years. No known exposures to agents but perhaps some polishes, detergents and disinfectants.

Painter, four years. Exposed to pigments found in paint such as lead, chromium and arsenic.

Gardener, three years. Exposed to insecticides and weed killers. Knows that some had pyrethrums, arsenic and parathion-like substances in them.

Present job, four years. Some of the pigments he mixes contain nickel, lead, arsenic, iron and other chemicals. He cleans with a vacuum cleaner, wears no protection and there is some dust. He has no secondary job.

Clinical Evaluation

The examination revealed a well developed male who appeared tired. His face was pale and the skin over the trunk appeared somewhat pigmented. Examination of the head, eyes, ears and throat showed them to be normal. The nasal septum was inflamed. No adenopathy. The thyroid was normal. Chest expanded symmetrically and percussion and ausculzation were normal.

The pulse was 78 and regular, the blood pressure was 128/82. Heart sounds were normal and no evidence of enlargement. There was slight tenderness on palpation of the right upper quadrant but the liver edge was not palpable.

External genitalia was normal. Peripheral circulation was normal. On examination of the extremities a hyperkeratosis of the palms of the hands and soles of the feet were found. There was decreased sensation to touch and vibration in the feet. Patella and ankle reflexes were decreased; those of the wrist and elbow were normal.

Laboratory Evaluation

CBC and Differential:	RBC 4.0 million/cubic mm
	Hb. 12 g/100 ml
	Hct. 40 percent
	WBC 4,000 per cubic mm
Chest X-Ray: 14" x 17"	Normal
Electrocardiogram:	Normal
SMA-12:	Normal
Urinalysis:	Normal
Thyroid function tests:	Normal

Blood Lead: 0.03 mg/100 gms

Urinary Arsenic: 0.9 mg/liter

Epidemiological Findings

The workplace was surveyed (see Table 1). It was found that the atmosphere contained levels of arsenic in excess of the Occupational Safety and Health Act (OSHA) standards. At breathing level, where the patient worked at mixing the pigments, arsenic levels often were much too high. Dust on the floor and walls contained arsenic and when cleaning, larger than recommended amounts of airborne arsenic were found. Even though pigment containing arsenic was not mixed daily, there was cumulative exposure.

The literature contains ample evidence to indicate that such exposure to arsenic dust could produce arsenic intoxication.

It can be clearly seen from Table I that the employee's exposure to arsenic was the only exposure evaluated which exceeded the allowable limit (in this case nearly twice the permitted exposure). Exposures to nickel, lead, and chromium were well within the eight hour time-weighted average limits and continued exposure at the levels evaluated should not result in any health hazards.

Contaminants in the Work Environment

Hyperpigmentation has been reported among employees exposed to arsenic concentrations ranging from 0.110-4.038 milligrams per cubic meter of air (0.562 milligrams per cubic meter was the mean exposure). (Dinman, B.D. 1960. *J. Occ. Med.* 2:137.) This would conform with the clinical evaluation in this specific case where the average exposure to arsenic was 0.94 milligrams per cubic meter of air and hyperpigmentation was observed.

Laboratory findings indicated absorption of arsenic by urinary arsenic levels of 0.9 milligrams per liter. Tox-

TABLE 1—ATMOSPHERIC METAL DUST AND FUME CONCENTRATIONS

October 1, 1975
ABC ARTS & CRAFTS
ANYTOWN, U.S.A.

SAMPLE NUMBER LOCATION	TIME START/ STOP	RESULTS IN MILLIGRAMS PER CUBIC METER OF AIR			
		ARSENIC	NICKEL	LEAD	CHROMIUM
OSHA ALLOWABLE LIMITS		0.5	1	0.2	0.5
Operator's Breathing Zone:					
1 John Doe—General Work in stockroom weighing pigments.	0700/ 1900	0.47	<0.001	<0.001	<0.001
2 John Doe—Weighing and mixing pigments.	0900/ 1100	1.33	<0.001	0.021	0.007
3 John Doe—Mixing and packaging pigments; 30 minute lunch.	1100/ 1300	1.21	<0.001	0.050	0.042
4 John Doe—Plant cleanup	1300/ 1500	0.75	<0.001	0.027	0.003
Time-Weighted Average Exposures:		0.94	<0.001	0.025	0.013

< Denotes less than.

icological data would also imply increased urinary arsenic levels at the atmospheric concentrations evaluated as indicated by the report of an average urinary arsenic level of 0.23 milligrams per liter in workers exposed to mean air concentrations of 0.562 milligrams arsenic per cubic meter.

Conclusion

The differential diagnosis would include lead poisoning, hypothyroidism, anemia and chronic arsenic poisoning; the laboratory findings rule out lead poisoning and hypothyroidism and indicate an absorption of arsenic. Anemia would not account for all of the symptoms and could be part of the pathology of arsenic intoxication.

This history of the complaint, the symptoms and signs along with the laboratory information ad the abnormal exposure to arsenic in the workplace, and no evidence of nonoccupational exposure make the diagnosis of chronic arsenic intoxication, occupational in origin.

CHAPTER SIX

SIGNIFICANT CASES INVOLVING TOXIC SUBSTANCES

SECRETARY OF LABOR v. THE PROKO COMPANY OF TEXAS, INC.

(OSAHRC Docket No. 12748, Review
Commission Decision, April, 1977)

Airborne Asbestos

Summary of Decision

An employer, whose employees' exposure to airborne asbestos fibers is within the limits permitted by 29 C.F.R. §1910.1001(b)(2), nevertheless violates 29 C.F.R. §1910.1001(j)(3) by failing to furnish free annual medical examinations to affected employees. This ruling reverses an earlier decision where an administrative law judge dismissed the charge. In the initial decision the judge held that because the paint and drywall manufacturer had provided controls that were adequate in keeping concentrations of airborne asbestos fibers at legally permissible levels, the requirement to provide annual medical examinations for affected employees did not apply. It was the opinion of OSAHRC that the requirement to provide free annual medical examinations applies to all employers (where concentrations of airborne asbestos exist) regardless of the fact that such concentrations are within the levels prescribed by law.

Full Text of Decision

CLEARY, Commissioner:

The decision of Administrative Law Judge J. Paul Brenton, dated November 4, 1975, is before the full Commission for review by separate directions for review issued by Chairman Barnako and me pursuant to section 12(j) of the Occupational Safety and Health Act of 1970, 29 U.S.C. §651 *et seq.*

The issue is whether the ALJ erred in vacating one item of a citation against respondent employer, a paint and drywall manufacturer using asbestos in its manufacturing process, by concluding that 29 CFR §1910.1001(j)(3) does not require an employer to furnish annual medical examinations to employees engaged in occupations exposed to airborne concentrations of asbestos fibers when the level of exposure is within that permitted by 29 CFR §1910.1001(b)(2).

We hold that the Judge did err, and reverse his action.

The Judge's decision preceded by ten days the Commission's decision in *GAF Corp. and United Engineers & Constructors, Inc.,* BNA 3 OSHC 1686, 1975-76 CCH OSHD para. 20,163 (Nos. 3203, 4008 and 7355, 1975).

There, we held that §1910.1001(j)(3) requires an employer to provide, or make available at its cost, medical examinations to employees engaged in occupations that require exposure to concentrations of airborne asbestos in any degree. *GAF Corp.* is controlling here.

Concentrations of asbestos were released into the air during respondent employer's manufacturing process, which includes the opening of bags of asbestos and the mixing of the asbestos with other materials. Respondent's president, Mr. H. M. Self, admitted that asbestos fibers were released in the air each time a bag was opened, and the asbestos poured into a mixing machine. Thus, the standard requires that comprehensive annual medical examinations be furnished to respondent's employees.

Accordingly, it is ORDERED, that the Judge's decision is reversed, and the item of the citation alleging noncompliance with 29 CFR §1910.1001(j)(3) is affirmed. The Secretary proposed no penalty for this item, and we assess none.

Dissenting Opinion

MORAN, Commissioner, dissenting:

Judge Brenton correctly decided this case. His well-reasoned decision, should therefore be affirmed.

The record establishes that respondent, through engineering and monitoring controls, had dutifully protected its employees from exposure to any harmful concentrations of asbestos fibers. I fully share Judge Brenton's view that it is sheer folly to interpret 29 C.F.R. §1910.93a(j) so as to require that medical examinations be given under these circumstances. *Secretary v. GAF Corporation,* OSAHRC Docket No. 3203 [3 OSHC 1686], November 14, 1975 (dissenting opinion).

Moreover, vacation of the charge is warranted because the cited standard was invalidly promulgated. As I indicated in *GAF,* substantive changes made by the Secretary of Labor in the text of §1910.93a(j) as initially proposed—without adherence to the proper rulemaking procedures required by 29 U.S.C. §655(b)—render that regulation unenforceable.

GAF CORPORATION v. OCCUPATIONAL SAFETY AND HEALTH REVIEW COMMISSION AND DUNLOP, SECRETARY OF LABOR

(OSAHRC Docket No. 76-1028 Court of
Appeals Denial of Petition to
Review, June, 1976)

Asbestos Fibers

Summary of Decision

In this case, the U.S. Court of Appeals denied a petition to review a decision of the Occupational Safety and Health Review Commission (3 OSHC 1686). By denying the petition, the Commission's decision stood, upholding three major points. First, the requirement that employers must provide medical examinations for all workers exposed to airborne asbestos, regardless of whether the amounts of exposure fall within acceptable limits, was upheld. Second, the asbestos standard of 29 C.F.R. 1910.1001 (j) does not mitigate OSHA provisions which allow the Secretary of HEW to pay for medical examinations conducted as medical research, and not solely for the purposes of employee health protection. Third, changes made in asbestos standards after their publication in the Federal Register do not automatically void the existing standards. GAF's main argument was based upon the contention that the word "concentration," as used in the regulation, implies that a specific quantitative amount of airborne asbestos must exist before medical examinations must be supplied by employers. The omission of a specific quantity in the final standard was deliberate, however, and the regulation's validity was upheld by the Commission's ruling, thus setting an important precedent for future cases (see Secretary of Labor v. Proko Company of Texas).

Full Text of Decision

ROBB, Circuit Judge:

GAF Corporation (GAF) petitions for review of an order of the Occupational Safety and Health Review Commission (the Commission). The challenged order found GAF in violation of a regulation requiring employers to provide medical examinations to workers exposed to airborne concentrations of asbestos. GAF employees are exposed to asbestos dust during the manufacture of floor coverings and fiber building products. GAF contends that the Commission misinterpreted the regulation and that in any event the regulation is void. We conclude that neither of these contentions is correct and accordingly we affirm.

I. Background—This dispute began in 1973 when the Secretary of Labor (the Secretary) issued citations charging GAF with violations of 29 C.F.R. §1910.93a(j) (subsequently renumbered to §1910.1001(j)). The cited regulation requires employers to provide certain pre-employment, annual, and separation medical examinations to all workers "in an occupation exposed to airborne concentrations of asbestos fibers . . . ."

GAF does not provide the specified examinations, but argues that the regulation does not require them, given the low levels of airborne asbestos at GAF's plants. At the time of the citation, which involved two of GAF's plants, the level at both plants was below the maximum permissible level. (The maximum permissible level at that time was 5 fibers, of more than 5 micrometers in length, per cubic centimeter of air. *See* 29 C.F.R. §1910.93a(b).) GAF contends that the regulations require medical examinations only if the concentration of airborne asbestos exceeds this maximum level or some other level higher than that found in GAF's plants. The Secretary argues that the presence of any airborne asbestos at all triggers the requirement for medical examinations.

GAF challenged the citations throughout the levels of administrative review provided by the Occupational Safety and Health Act of 1970 (the Act), 29 U.S.C. §659 (1970). In due course the Commission upheld the citations. *Secretary of Labor v. GAF Corp.,* [3 OSHC 1686] 75 OSAHRC Rep. 3/A2 (microfiche). GAF then petitioned this court seeking review of the Commission's order as provided by the Act. *See* 29 U.S.C. §660 (1970).

GAF contends that the Commission erred in interpreting the regulation to require medical examinations for employees exposed to the relatively low concentrations of airborne asbestos found in GAF's plants. GAF also contends that the regulation is void because it was improperly promulgated, is inconsistent with the Act, and is arbitrary, irrational, and unsupported by the evidence. We shall consider each of these contentions in turn.

A. Interpretation of the Regulation—We turn first to GAF's contention that the Commission erred in upholding the Secretary's interpretation of the disputed regulation. GAF is supported by neither the language nor the history of the regulation.

We note at the outset that the Secretary is charged by law with administering the Act and the regulations supporting it. 29 U.S.C. §§655 *et seq.* (1970). Hence, his interpretation of the regulation "becomes of controlling weight unless it is plainly erroneous or inconsistent with the regulation." *Udall v. Tallman,* 380 U.S. 1, 16-47 (1965). The Secretary's interpretation remains controlling "even though the chosen exegesis may not appear quite as reasonable as some other construction." [citation omitted] *Budd Co. v. Occupational Safety and Health Review Commission,* 513 F.2d 201, 205 [2 OSHC 1698] (3rd Cir. 1975); *accord, Clarkson Const'n Co. v. Occupational Safety and Health Review Commission,* 531 F.2d 451, 457 [3 OSHC 1880] (10th Cir. 1976). Furthermore, the standards must be construed, as they were construed by the Secretary, to protect the employees. *See Brennan v. Occupational Safety and Health Review Commission,* 491 F.2d 1340, 1344 [1 OSHC 1523] (2d Cir. 1974).

The Secretary's interpretation of the regulations not only satisfied the standards discussed above; it is far more reasonable than the interpretation urged by GAF.

GAF bases its argument upon the language of the regulation, which requires medical examinations for those "in an occupation exposed to airborne concentrations of asbestos fibers . . . ." 29 C.F.R. §1910.93a(j). The Secretary interprets this language as requiring examinations for those in occupations exposed to any level of asbestos. GAF, in contrast, argues that the word "concentration" implies a quantitative amount. Because the disputed section does not specify any quantity of asbestos, GAF reasons that the section should be read to include the 5-fiber limit of 29 C.F.R. §1910.93a(b), or at least some other quantitative limit greater than the level found in GAF's plants.

GAF's argument flies in the face of the language of the regulation. It would be a unique regulation indeed which managed to impose a quantitative limitation by not mentioning any quantity at all. Furthermore, a detailed examination of the regulation reveals that the omission of the quantitative limit was deliberate.

As originally proposed, the requirement of examinations was triggered by the presence of a specific quantity of asbestos:

Medical Examinations. The employer shall provide, or make available at his cost, appropriate medical examinations on a periodic basis to any employee who is exposed to asbestos dust in excess of the limits specified in paragraph (a) of this section [5 fibers per cubic centimeter.].

37 Fed. Reg. 468 (1972).

The final standard, however, deletes the reference to a specific concentration. The preamble to the final standard notes that several changes have been made to the proposed regulation, and then states that medical examinations are now required for "*every* employee exposed to airborne concentrations of asbestos." [emphasis supplied] 37 Fed. Reg. 11319 (1972). The preamble makes no mention of any specific quantity of asbestos. Nor does the final regulation itself. Instead, the final regulation requires examinations for those "in occupations exposed to airborne concentrations of asbestos fibers . . . ." The regulation states this requirement not once, but three times, in almost identical language: once each in the paragraphs requiring preplacement, annual, and termination examinations. *See* 29 C.F.R. §§1910.93a(j)(2)-(4). In contrast, 29 C.F.R. §1910.93a(g)(1)(i) requires posting of caution signs "*where airborne concentrations of asbestos fibers may be in excess of the exposure limits prescribed* in paragraph (b) of this section." [emphasis added]

[1] We think these considerations leave room for only one conclusion: the disputed regulations require medical examinations for all those in occupations exposed to airborne asbestos in any measurable concentration.

B. Validity of the Regulation—GAF contends that, even if the Secretary correctly interpreted the regulation, it is void on a number of grounds. We find this contention to be without merit.

Although we have carefully considered all GAF's arguments concerning the validity of the regulation, only four merit discussion here. These are GAF's contentions that the regulation is arbitrary and unsupported by evidence; that the regulation is inconsistent with the Act; that the Secretary's interpretation is inconsistent with his approval of a California occupational safety plan: and that the federal regulation was improperly promulgated.

GAF argues first that the regulation is arbitrary and unsupported by evidence. GAF bases this argument partly upon the recommendation of the National Institute for Occupational Safety and Health (NIOSH). NIOSH was created by the Act for the purpose of developing health and safety standards and recommending them to the Secretary. 29 U.S.C. §671 (1970). NIOSH recommended to the Secretary that medical examinations be required where airborne asbestos concentrations exceed 1 fiber per cubic centimeter of air. *National Institute for Occupational Safety and Health, criteria for a recommended standard . . . Occupational Exposure to Asbestos* at I-3 (1972) [cited hereafter as "NIOSH criteria"]. The Secretary rejected this recommendation and instead required medical examinations for those exposed to any concentration of airborne asbestos.

GAF contends that this rejection by the Secretary was arbitrary and unsupported by the evidence, and especially arbitrary because NIOSH stated that its recommended standard included a "Safety factor". NIOSH criteria at II-2.

[**2**] We have previously held that the Secretary is not bound by NIOSH recommendations. *Industrial Union Department v. Hodgson,* 162 U.S. App. D.C. 331, 340-41, 499 F.2d 467, 476-77 [1 OSHC 1631] (1974). And in this case the Secretary's departure from the NIOSH recommendation is reasonable. NIOSH noted that concentrations as low as 1.2 fibers per cubic centimeter have been known to cause serious diseases. Furthermore, NIOSH noted, the state of knowledge concerning asbestos-related diseases is such that no exposure standard other than zero would assure freedom from such diseases. NIOSH criteria at III-23, III-9. Hence the Secretary acted reasonably and on substantial evidence in requiring medical examinations for those exposed to any concentration of airborne asbestos rather than limiting examinations to those exposed to more than 1 fiber per cubic centimeter.

GAF argues next that even if the challenged regulation is supported by evidence, the regulation is invalid because it is inconsistent with the Act. GAF contends that the examinations are in effect being used as a form of medical research, because some useful knowledge will probably be gained from them. Under the Act, GAF continues, the Secretary of Health, Education, and Welfare (HEW) must pay for medical examinations used as research. Hence the challenged regulation, by requiring GAF to supply medical examinations at its own expense, contravenes the statute.

GAF misconstrues the Act and its effect upon the challenged regulation. The Act does not require the Secretary of HEW to pay for medical examinations used as research; it merely *permits* him to do so. The Act provides:

> In the event such medical examinations are in the nature of research, *as determined by the Secretary of Health, Education, and Welfare,* such examinations may be furnished at the expense of the Secretary of Health, Education, and Welfare. [emphasis supplied]

29 U.S.C. §655(b)(7) (1970). Thus, the Act clearly vests the Secretary of HEW with discretion to determine whether the examinations are in the nature of medical research and if so, whether to pay for them. His determination in this regard may be reversed only for an abuse of discretion.

We note initially that no decision of the Secretary of HEW in this matter is before the court. In any event, a review of the challenged regulation convinces us that the examinations in issue are not "in the nature of medical research" and are thus properly chargeable to GAF. GAF cannot seriously contend that the examinations are primarily designed to further medical research rather than to protect employees exposed to asbestos dust. As we have noted above, NIOSH found that non-zero exposure limits alone could not guarantee the prevention of asbestos-induced disease. NIOSH criteria at III-9. Consequently, NIOSH recommended periodic medical examinations for the protection of the workers, noting:

> The major objective of such surveillance will be to ensure proper medical management of individuals who show evidence of reaction to past dust exposures, either due to excessive exposures or unusual susceptibility. Medical management may range from recommendations as to job placement, improved work practices, cessation of smoking, to specific therapy for asbestos-related disease or its complications.

NIOSH criteria at I-3. And in promulgating the final regulation, the Secretary of Labor noted the hazards connected with long-term exposure to asbestos dust and concluded, "the conflict in the medical evidence is resolved in favor of the health of employees." 37 Fed. Reg. 11318 (1972). Finally, the Secretary specified in the regulation that records of the medical examinations in question would be kept by the employer, whereas the Act requires that records of examinations used as medical research be forwarded to the Secretary of HEW. 29 C.F.R. §1910.93a(j)(6); 29 U.S.C. §655(b)(7) (1970).

[**3**] These factors considered together clearly indicate that the examinations at issue are to be provided primarily for the protection of GAF's employees rather than for the purposes of medical research. Consequently, the challenged regulation does not contravene the section of the Act permitting the Secretary of HEW to pay for medical examinations conducted as medical research.

GAF argues that even if the challenged regulation is supported by evidence and consistent with the Act, the Secretary of Labor's interpretation of the regulation is inconsistent with his approval of a California occupational health and safety plan.

The alleged inconsistency arises because the California plan, like the NIOSH recommendation for a federal standard, requires medical examinations only if the employees may be exposed to concentrations of asbestos greater than 1 fiber per cubic centimeter. General Industry Safety Orders 8 CAL. ADMIN. CODE Chp. 4, Subch. 7 §5208(j). The Secretary may approve of a state plan and permit the state to take over some enforcement functions of the federal government, but only if the state plan "provides for the development and enforcement of safety and health standards ... which standards (and the enforcement of which standards) ... *will be at least as effective* in providing safe and healthful employment ... *as the standards promulgated under section 655* of this title . . . ." [emphasis supplied] 29 U.S.C. §667(c)(1970).

The Secretary has approved the California plan. 38 Fed. Reg. 10717-20 (1973); 41 Fed. Reg. 1904-06 (1976). This approval, GAF argues, indicates that the California plan is "at least as effective" as the federal scheme. Consequently, GAF contends, the Secretary's approval of the California plan is a tacit admission that the federal regulation does not require medical examinations when asbestos concentrations are less than 1 fiber per cubic centimeter.

[4] We cannot agree with GAF's reasoning. In approving the California plan, the Secretary clearly evaluated the plan as a whole, discussing not only limits for various toxins, but enforcement, notice, recordkeeping, and sanction requirements as well. 38 Fed. Reg. 10717-20 (1973); 41 Fed. Reg. 1904-06 (1976). The Secretary noted that California would provide more enforcement personnel than the federal program. 38 Fed. Reg. 10718 (1973). Hence, it was reasonable for the Secretary to conclude that the California standards *as enforced* would be "at least as effective" as the federal standards despite differences in the triggering concentrations of asbestos.

GAF argues finally that, even if the Secretary correctly interpreted the federal regulation, and even if the regulation is consistent with the California plan, the federal regulation is void because it was improperly promulgated. GAF's argument is as follows: the Act requires that each proposed regulation be published in the Federal Register to permit public comment. 29 U.S.C. §655(b) (1970). In this case the Secretary published a proposed regulation but then made changes in it before publishing it in final form. GAF argues that the amended rule should first have been published as a new proposed rule, with opportunity for public comment before being promulgated as a final regulation.

The short answer to this contention is that GAF never raised this objection before the Commission and is therefore precluded from raising it here. The judicial review provision of the Act provides that "no objection that has not been urged before the Commission shall be considered by the court . . . ." 29 U.S.C. §660(a) (1970). Furthermore, as this court has previously noted:

> The requirement of submission of a proposed rule for comment does not automatically generate a new opportunity for comment merely because the rule promulgated by the agency differs from the rule it proposed, partly at least in response to submissions.

International Harvester Co. v. Ruckelshaus, 155 U.S. App. D.C. 411, 428, 478 F.2d 615, 632, n. 51 and accompanying text (1973); *accord, South Terminal Corp. v. E.P.A.,* 504 F.2d 646, 659 (1st Cir. 1974). We think the same principle governs GAF's contention in this case and accordingly we reject GAF's argument.

In conclusion, we hold that the Secretary's interpretation of the disputed regulation is reasonable, and that the regulation is supported by the evidence, consistent with the California plan, and was properly promulgated. GAF's petition to review and reverse the decision of the Occupational Safety and Health Review Commission is denied.

Concurring Opinion

MacKINNON, Circuit Judge, concurring specially:

Because the "concentrations" of asbestos fibers found in the GAF plants in this case were not insignificant, and because the record shows that petitioner had adequate notice that the Commission would construe its regulation to require physical examinations at these "concentrations," I reluctantly concur in the result reached by the panel. I believe, however, that the Occupational Safety and Health Review Commission should clarify the regulation governing medical examinations so as to give other persons subject to the regulation reasonable notice of some measurable quantity of airborne asbestos the Commission intends to trigger the medical examination requirement. Merely to state that all employers must provide medical examinations whenever their employees are exposed to "concentrations" of asbestos fibers does not provide an ascertainable standard for those who wish to comply with the law.

The regulation provides that employers must provide medical examinations to each employee "in an occupation exposed to airborne *concentrations* of asbestos fibers," 29 C.F.R. §1910.1001(j)(2) and (3) (1976) (emphasis added). No person can discern from this language what degree of "concentration" must exist before medical examinations are required. Other asbestos exposure regulations by the Commission which do set ascertainable exposure limits provide:

(b) *Permissible exposure to airborne concentrations of asbestos fibers.* (1) *Standard effective July 7, 1972.* The 8-hour time-weighted average airborne concentrations of asbestos fibers to which any employee may be exposed shall not exceed five fibers, longer than 5 micrometers, per cubic centimeter of air, as determined by the method prescribed in paragraph (e) of this section.

(2) *Standard effective July 1, 1976.* The 8-hour time-weighted average airborne concentrations of asbestos fibers to which any employee may be exposed shall not exceed two fibers, longer than 5 micrometers, per cubic centimeter of air, as determined by the method prescribed in paragraph (e) of this section.

29 C.F.R. §1910.1001(b)(1) and (2) (1976). The prescribed standards in these regulations were: "five fibers, longer than 5 micrometers, per cubic centimeter of

air," and after July 7, 1976, "two fibers." These standards inform the nation's employers *precisely* what degree of exposure is excessive, but the regulation requiring medical examinations is almost completely deficient in this respect.

What number of fibers, of what length, per what volume of air will constitute a "concentration" is nowhere stated or even hinted at—and like the regulation, the court's opinion leaves the matter completely to conjecture. This is particularly unsettling to law abiding citizens when the agency is dealing with tremendously minute quantities of infinitesimally small particles. It makes compulsory law enforcement difficult and it lessens the likelihood of voluntary compliance with the law, a result devoutly to be wished. The first requirement for uniform and voluntary compliance with the law is a clear understandable statement of what conduct is required and this regulation falls woefully short of that minimal requirement.

Regulations that have the great importance that this regulation has to human life should be written in more precise terms—so people of ordinary understanding can determine what course of conduct is being required of them. What one person might consider to constitute a "concentration" will differ greatly from the interpretation that another would give to that term; and what one person today might regard *not* to constitute a concentration" might be considered tomorrow, on the basis of hindsight as medical knowledge increases, to come within that term.

The statute requires the Commission to "set [a] standard":

> (5) The Secretary, in promulgating standards dealing with toxic materials or harmful physical agents under this subsection, *shall set the standard* which most adequately assures, to the extent feasible, on the basis of the best available evidence, that no employee will suffer material impairment of health or functional capacity even if such employee has regular exposure to the hazard dealt with by such standard for the period of his working life. Development of standards under this subsection shall be based upon research, demonstrations, experiments, and such other information as may be appropriate. In addition to the attainment of the highest degree of health and safety protection for the employee, other considerations shall be the latest available scientific data in the field, the feasibility of the standards, and experience gained under this and other health and safety laws. Whenever practicable, the standard promulgated *shall* be expressed in terms of *objective criteria and of the performance desired.*

29 U.S.C. §655(b)(5) (1970) (emphasis added). The agency has promulgated standards that state "objective criteria and . . . the performance desired" for employee exposure on the job but it has not done as much for medi-

cal examinations. It is practical to state "objective criteria" for medical examinations and I believe the agency should do so. Merely requiring medical examinations whenever there is some indefinite, ambiguous, nonspecific degree of "concentration" does not meet the requirement for a feasible standard with *objective* criteria that Congress imposed on the agency. This is a serious dereliction because, if in the future medical knowledge progresses to the point where more minute quantities of asbestos fibers are found to be harmful than are presently so considered, then employers in the distant future may be held liable to the extent of millions of dollars in tort suits based on their alleged failure to provide the required medical examinations that would have prevented, or discovered at a preventable stage, what eventually turned out to be a fatal disease.

In such lawsuits for damages the present ambiguous regulation would have the status of a statute and failure to conform to some more strict *post hoc* interpretation of its provisions might constitute negligence per se. As Professor Prosser states the law:

> Once the statute is determined to be applicable—which is to say, once it is interpreted as designed to protect the class of persons in which the plaintiff is included, against the risk of the type of harm which has in fact occurred as a result of its violation—the great majority of the courts held that an unexcused violation is conclusive on the issue of negligence, and that the court must so direct the jury. The standard of conduct is taken over by the court from that fixed by the legislature, and "jurors have no dispensing power by which to relax it," except in so far as the court may recognize the possibility of a valid excuse for disobedience of the law. This usually is expressed by saying that the unexcused violation is negligence "per se," or in itself.

W. PROSSER TORTS 200 (4th ed. 1971). When it can be avoided, a statute or regulation should not by its mere indefiniteness expose employers to such great hazards. The agency should state in clear, understandable objective terms specifically what degree of "concentration" triggers the requirement for medical examinations—be it one fiber per cubic foot of air or one fiber per 100 cubic feet of air, or some other unambiguous, quantitative standard. Imposing a non-specific ambiguous standard is completely unsatisfactory. "Concentration" is a noun, the meaning of which varies greatly depending upon the context in which it is used. Concentration can be high, low, medium, dangerous, hazardous, insignificant, "negligible," etc.—standing alone its meaning is highly indefinite. It would not be difficult to state a standard with precision.

A recent case before the Connecticut Occupation Safety and Health Review Commission, *Tuttle & Bailey Div. of Allied Thermal Corp.,* No. 76-315 (Conn. OSAHRC, Dec. 3, 1976), involved a state medical examination requirement in language practically identical to that present here, and a similar factual situation. In that case the State Commission decided that the word "concentration" in the standard must have been intended to have some meaning, because if no threshold of exposure were intended, the word "concentration" could be omitted without changing the meaning of the section. It would then address itself to "airborne asbestos fibers." The State Commission reasoned:

> If this Commission were to adopt the interpretation as proposed by the Labor Commissioner, the effect would be to establish a requirement " . . . that whenever employees are exposed to any trace of asbestos—no matter how temporary or insignificant—their employer must furnish (and employees must undergo)—annual physical examination and tests as prescribed in [the regulation]." . . .

> This Review Commission cannot conclude that the intent of this regulation is to require such examination for the occasional "messenger" or "visitor" entering the exposure area, nor the employee far removed from the source whose exposure is in fact to *de minimis* concentration of airborne asbestos fibers. The Review Commission concludes that the word "concentration" has meaning as used within the regulation.

Id. at 5-6.

The Commission also noted:

> One can conceive of very different circumstances wherein employees would be exposed to airborne concentration of asbestos fiber, to varying degrees and for varying periods of time. For an employer to try to determine whether or not medical examination[s] as required by [the regulation] should be provided or not provided under a "reasonable man" theory would be most difficult. To attempt to apply a "reasonable person" standard in recognizing a hazard of microscopic concentration of asbestos fiber in the air, is not in itself reasonable. *Reasonable laymen cannot conceive a rule of conduct in an area where reasonable experts cannot agree.*

Id. at 6-7 (emphasis added). The Commission therefore concluded that it was

> unable to determine what [the] threshold [of concentration] is or should be, nor does this Commission accept that a reasonable employer could so determine.

Id. at 7. Finally, it held that until the regulation was redrafted to state its intended meaning in clear understandable terms, the Commission would assume that the 5 fiber standard applied as well to the physical examination provision. The State Commission appeared to recognize that this interpretation was not directly supported by the language of the regulation, but based its holding on the necessity for some understandable meaning. If the instant case were disposed of on a similar rationale, the two-fiber limit presently in force in 29 C.F.R. §1910.1001(b)(2) would also govern the medical examination requirement.

Recently, the Federal Commissioners attempted to distinguish their decision presently under review on the ground that "*GAF* did not involve the situation . . . where the concentration of the air contaminant is *negligible* compared to the *limit* established by the *standard.*" *Western Electric, Inc.,* No. 8902, at 7 [4 OSHC 2021] (OSAHRC, Jan. 24, 1977) (emphasis added). What "standard" the Commission had in contemplation was not disclosed, nor was there any reference to what "limit" they were applying. But the reference of the Commission to a *"negligible"* amount indicates *some quantitative standard* was intended. It would be helpful if the agency would state what that "standard" is. That is not a difficult task if they have one in mind. If they do not have one they should have. Maybe the Commission does not have substantial evidence to prove that the "standard" of "concentration" they are applyint constitutes a "hazard," and that may be the reason they chose to leave it in an ambiguous state. As it is now they are home free leaving the public in the dark as to what degree of "concentration" is more than "negligible" according to some unstated "standard" of "concentration."

Such imprecision should not be countenanced in such an important matter. Nothing would be lost by requiring the agency to state a specific objective standard—and everybody would benefit. It could be done quickly and there would not be any delay in enforcement.

It is also my opinion that permitting the entire state of California to have a specific standard that is different, and more liberal, from that applied to appellants in this case, and throughout the rest of the nation, is highly questionable. The mere availability of more investigators in California is *not* a permissible basis for allowing employees to be exposed to more asbestos fibers before a medical examination is required. *The number of investigators is not germane to the hazard produced by the added contamination of more asbestos fibers.*

The evidentiary record in this case does not in my opinion present a case of sufficient strength to justify setting aside the regulation but there is every indication that a proper case can be made to achieve such result. I accordingly strongly suggest that the Commission should amend the regulation to establish the definite workable and understandable "standard" with "objective criteria" that the statute requires.

SECRETARY OF LABOR v. GOODYEAR TIRE & RUBBER COMPANY

(OSAHRC Docket No. 13442,
Review Commission Decision,
May, 1977)

Asbestos Fibers

Summary of Decision

Before an employer can be found in violation of 29 C.F.R. § 1910.1001, a regulation that requires a company to use equipment to monitor concentrations of airborne asbestos fibers in a work area, there must be an actual release of such fibers, and not merely a "genuine possibility" of that condition occurring. In this case, the company's Houston, Texas plant was inspected by an OSHA compliance officer, who, during the inspection, found a few pieces of asbestos insulation in a storage loft. The compliance officer did not take any air samples, but based his citing the employer for a violation of 29 C.F.R. § 1910.1001 on the grounds that according to his experiences at other plants working with asbestos, there would have to have been a release of some fibers in the air. The compliance officer further testified that his Area Director and an OSHA industrial hygienist agreed with his opinion. The judge, however, did not find the compliance officer's testimony to be binding. He ruled that the monitoring requirement only applies where asbestos fibers are released, not where such fibers are merely present. Because evidence pointed to the fact that the company routinely exercises every reasonable precaution when working with asbestos insulation, and that no evidence presented could substantiate the allegation that asbestos fibers were released into the air, it was ruled that the company was not in violation by failing to provide monitoring equipment.

Full Text of Decision

BARNAKO, Chairman:

The issue in this case is whether Respondent violated the monitoring and recordkeeping requirements of the standard for asbestos. Respondent was cited for allegedly failing to monitor an operation wherein asbestos insulation on pipes was removed and replaced with non-asbestos insulation. Judge J. Paul Brenton dismissed the charges on the grounds that Complainant failed to establish a *prima facie* case that asbestos fibers were released during the operation. For the reasons that follow we affirm Judge Brenton's decision.

Respondent is engaged in the manufacture of petrochemicals and latex rubber at its Houston, Texas plant. The plant contains approximately 100 miles of pipe, most of it out-of-doors. One quarter of the piping is insulated with various materials, some of which is asbestos insulation. Most of the asbestos insulation is covered and contained with metal jacketing.

Respondent ceased installing asbestos insulation in 1972. Thereafter, whenever leaks in the piping occur and repairs are necessary, any asbestos insulation in the vicinity of the leaks is removed and replaced by non-asbestos insulation. Since most of the insulated pipe carries caustic liquids, a line is thoroughly flushed as part of the repair procedure, and this results in complete saturation of the insulation before its removal.

The job of removing insulation is performed by one of four different employees who are pipefitters. The operation is performed on the average of once monthly and takes between five and ten minutes. When changing insulation employees wear protective clothing and respirators in accordance with Respondent's instructions.

Infrequently, the pipefitters replace dry insulation in buildings where "dryers" are located. The amount of insulation removed is less than ten feet and the job is performed when no one else is in the building. Any dry asbestos insulation removed is placed in plastic bags and put in the trash.

During the inspection a few pieces of asbestos insulation were discovered in a storage loft contrary to Respondent's instructions. However, only the pipefitter working on replacement entered the loft while this insulation was stored there.

Complainant's compliance officer testified that, generally, whether fibers are released is determined by sampling the air in the vicinity of the asbestos. He stated he did not know if asbestos fibers were being released at Respondent's plant since he did not take any air samples. He opined that, from his past experience working with asbestos fibers at other plants, there was some release of asbestos fibers in the atmosphere of Respondent's plant. He also testified that his area director and one of Complainant's industrial hygienists were of the opinion that due to the type of operation, there was a release of fibers.

Respondent's pipe department foreman maintained that dust is not created when old insulation is removed as it is already saturated. He testified that all asbestos insulation is covered, the pipefitters performing this work always wear respiratory equipment, 90 percent of the work is out-of-doors. A former employee testified that he had always worn a respirator when changing insulation.

The Judge ruled that since the standard requires monitoring only "where asbestos fibers are released", it does not operate to require monitoring where asbestos is merely present and/or in use. He stated that one essential element of proof was "some fact or circumstance from which it may be reasonably rationalized that at least a trace of this kind of fiber was being released".

Regarding the asbestos insulation on the pipes, the Judge pointed out that most of it was completely saturated upon removal. He observed that any dry insulation that was replaced because it had been knocked loose from the pipes was contained with metal jacketing, and there was no evidence that this jacketing had been punctured so as to allow release of asbestos fibers. The Judge found that the mere presence of insulation, dry or wet, does not prove release. The evidence as a whole, he concluded, failed to sustain the inference that fibers were released into the atmosphere in the presence of employees. Accordingly, he vacated the alleged violations of the asbestos standard.

On review Complainant contends that a *prima facie* case is established by showing a genuine possibility of the release of asbestos fibers and employee access thereto. Complainant argues that the evidence of record establishes the possibility of release and employee access. Complainant also argues that the standard does not require proof of exposure inasmuch as the standard speaks only of release. Even if exposure is necessary to sustain a violation, Complainant contends that evidence of Respondent's precautionary measures is irrelevant because the standard is directed toward obtaining a reading of asbestos in the atmosphere and not the amount inhaled by employees. Finally, Complainant argues that Respondent, by alleging that asbestos release did not occur, has the burden of proof on this matter.

As noted above, Complainant would have us read the standard to require monitoring by a showing of "the genuine possibility of asbestos release in Respondent's plant and employee access thereto." The standard does not rely, however, on the mere possibility of release before triggering the monitoring requirement. It requires monitoring only "where asbestos fibers *are released*" (emphasis added). Complainant has the burden of proving that the standard was violated. 29 C.F.R. 2200.73. He must carry that burden by a preponderance of the evidence. *Armor Elevator Co.,* 5 OSAHRC 260, BNA 1 OSHC 1409, CCH OSHD para. 16,958 (1973). Thus, to prove a violation, Complainant must establish that it is more likely than not that fibers were released. McCor-

mick, *Law of Evidence,* Sec. 339, at 794 (2d ed. 1972). We therefore reject the argument that he needed only show a "genuine possibility" of release.

Complainant relies on *Amoco Oil Co.,* 76 OSAHRC 39/A2. BNA 3 OSHC 1745, CCH OSHD para. 20,183 (1975). In that case, the administrative law judge found a violation based on evidence that certain work practices at an Amoco refinery caused the airborne release of asbestos fibers. That is not the case here, and this case is distinguishable because release cannot be inferred from the nature of Respondent's work practices.

Complainant also relies on the compliance officer's testimony to establish that an inference of release can be drawn. We note, however, that the compliance officer's testimony was equivocal. In weighing the evidence, the Judge found his testimony insufficient. The Judge's evaluation is supported by the record and we accept it. *Okland Construction Co.,* 76 OSAHRC 30/F4, BNA 3 OSHC 2023, CCH OSHD para. 20,411 (1976).

Complainant also urges that we find a violation based on our decision in *GAF Corp.,* 75 OSAHRC 3 A2, BNA 3 OSHC 1686, CCH OSHD para. 20,163 (1975), pet. for review filed. No. 76-1028 (D.C. Cir., Jan. 13, 1976). In *GAF* a divided Commission affirmed citations for violations of the medical examination requirements of the asbestos standard upon a showing that employees were exposed to some concentration of airborne asbestos. In this case, there has been no similar showing. Furthermore, in placing the burden of proving release upon Complainant we do not go so far as to require that he prove the concentration of asbestos fibers has reached any specified level in order to establish a *prima facie* case. In this respect our decision in this case is consistent with *GAF Corp.* See *Western Electric, Inc.,* No. 8902, BNA 4 OSHC 2021. CCH OSHD para. 21,538 (January 24, 1977).

Since the evidence fails to establish that Respondent's process of replacing asbestos insulation resulted in the release of asbestos fibers, a violation of Section 1910.1001 (i)(1) has not been established. Consequently there is no violation of the recordkeeping requirements of Section 1910.1001 (i)(1).

Accordingly, the decision of Judge Brenton to vacate the alleged violations of the asbestos standard is affirmed.

Dissenting Opinion

CLEARY, Commissioner, dissenting:

Regardless of whether the asbestos standard requires monitoring when there is a "genuine possibility of asbestos release," as argued by the Secretary, or only when "it is more likely than not that fibers [are] released," as stated by the majority, the evidence adduced in this case shows a failure to comply with the standard published at 29 CFR § 1910.1001 (f)(1). The majority, therefore, errs when it accepts the Judge's evaluation of the evidence and affirms his disposition.

There was no dispute that respondent's pipefitters repaired leaks and replaced worn or damaged insulation on pipes, some of which were insulated with asbestos. It was complainant's position at the hearing that the removal of asbestos insulation would release asbestos fibers into the air. Although not urged by complainant during the course of these proceedings, there is a strong indication within the provisions of § 1910.1001 that the removal of asbestos insulation causes the release of asbestos fibers. Paragraph (c)(2)(iii) of § 1910.1001 requires the wearing of respirators and special clothing when employees are engaged " . . . in the removal or demolition of asbestos insulation or coverings. . . . " The presence of such a requirement strongly suggests a finding by the Secretary of Labor during rulemaking that the removal of asbestos insulation releases asbestos fibers.

Moreover, proof of complainant's position was adduced through testimony of the compliance officer and two of respondent's employees. Their testimony establishes a probability of asbestos release warranting monitoring of the operations involving the handling of asbestos insulation.

Before working for the Department of Labor, the compliance officer had worked for 21 years in the petrochemical industry. During this time he had handled asbestos insulation on numerous occasions. He testified that, when replacing worn and oftentimes crumbling asbestos insulation, fibers were released into the air. The majority notes, but fails to specify, equivocation in the compliance officer's testimony. My review of his testimony reveals nothing that could be considered equivocal.

In any event, the most persuasive testimony on asbestos release during the replacement process came from the foreman of respondent's pipe department. Part of his testimony was that most but not all the asbestos insulation was saturated with water before removal. He opined that saturated insulation would not release asbestos fibers during removal and replacement. Both the majority and the Judge attach significance to the fact that most of the insulation was saturated before replacement. Apparently, they regard the saturation process as obviating the requirement for monitoring. I would not.

Although saturation is a work practice which contributes to compliance under § 1910.1001 (c)(2)(i), it does not necessarily follow that employers who use the method are relieved from the duty to monitor under § 1910.1001 (f)(1). I submit that, by its terms, § 1910.1001 (f)(1) requires an employer to monitor exposure levels *before* instituting specific compliance methods. This is evidenced by the last sentence of § 1910.1001 (f)(1), which reads as follows:

> If the [permissible exposure] limits are exceeded, the employer shall immediately undertake a compliance program in accordance with paragraph (c) of this section.

Respondent did not monitor the vicinity of damaged asbestos insulation before saturating it. Moreover, even if I were to disregard the standards' mandate for preabatement monitoring, I submit that an employer who implements a compliance program must conduct personal monitoring under § 1910.1001 (f)(2) to determine the efficacy of the methods in use. This was not done by respondent.

As noted, respondent's foreman testified that some of the asbestos insulation was replaced without first being saturated. He also testified that dry, damaged insulation would release asbestos fibers into the air during replacement. Nevertheless, the Judge rejected this evidence of asbestos release because "the quantity [of asbestos dust released] would be so small and infrequent that exposure to asbestos fibers would be nil. . . ." In addition, he stressed the fact that the insulation was enclosed in a jacketing that was "pretty hard."

Both bases for ignoring the foreman's explicit statement that asbestos dust is released are incorrect. It is plain error to attach any significance to estimates of low concentrations in determining whether the duty to monitor attaches. Monitoring standards are intended to supplant such estimates, no matter how reliable they may be. *See Western Electric, Inc.,* 4 BNA OSHC 2021, 1976-77 CCH OSHD para. 21,538 (No. 8902, 1977) (Cleary, Commissioner, dissenting), *petition for review withdrawn,* No. 77-1252, 8th Cir., April 8, 1977. Under the facts of this case it is also error to regard the fact that the insulation was jacketed as negating a duty to monitor. The reason for replacing the insulation was that it was damaged. Such damage, according to both respondent's pipefitter and foreman, sometimes involved breaks or cracks in the jacketing. Thus, the Judge erred in ruling that respondent had no duty to monitor asbestos release during the handling of dry insulation.

Similarly, the majority errs when, in addition to accepting the Judge's erroneous analysis, it regards as probative the frequency and duration of exposure and the precautions taken when dry asbestos insulation is replaced. The monitoring standard does not permit these qualifications. Rather, § 1910.1001 (f)(1) requires that " . . . every employer shall cause every place of employment where asbestos fibers are released to be monitored. . . . "

Accordingly, I dissent from the majority's disposition of the citations at issue in this case.

SECRETARY OF LABOR v.
RESEARCH-COTTRELL, INC.

(OSAHRC Docket No. 76-73,
Review Commission, May, 1977)

Asbestos Fibers

Summary of Decision

The employer in this case was cited for violation of three safety standards involving employee exposure to asbestos fiber. The alleged violations included failure to monitor and to establish degree of hazard present, failure to post warning signs in the vicinity of the hazard, and failure to provide medical examinations for exposed employees. The Secretary of Labor failed, however, to prove that the asbestos fibers were even present in lengths of five micrometers (the size constituting a potential hazard, as specified by the regulations) or that the construction procedures used caused employee exposure to the fibers. The Review Commission held for the company since no hazard was proven to be present; therefore, the monitoring warning signs and medical examination requirements need not have been met.

Full Text of Administrative Law Judge's Decision

BRADY, Judge: This proceeding is brought pursuant to section 19 of the Occupational Safety and Health Act of 1970, 29 U.S.C. 651, et. seq., (hereinafter referred to as the Act) to contest a citation issued by the Secretary of Labor (hereinafter referred to as the Secretary) pursuant to section 9(a) of the Act. The citation which was issued December 12, 1975 alleges that as a result of an inspection of respondent's construction site on Geddie Road, Tallahassee, Florida, respondent violated section 5(a)(2) of the Act by failing to comply with specific Occupational Safety and Health Standards promulgated by the Secretary pursuant to section 6 thereof.

The Secretary alleges that on December 2, 1975 the respondent violated the standards codified in 29 C.F.R. 1910.1001(f)(1), 29 C.F.R. 1910.1001(g)(1)(i) and 29 C.F.R. 1910.1001(j)(2).

The facts are not in dispute that at the time the alleged violations occurred respondent was in the process of constructing a mechanical draft cooling tower for the purpose of cooling water to be used in the generation of electrical power. In the process of constructing the cooling tower

"fill sheets" were installed to provide surfaces over which water was to run for the purpose of cooling. The sheets measured approximately 10 feet long, 16 inches high and .165 inches in width. They were composed of concrete, sand and asbestos bonded together. The quantities of the mixture were unknown but the sheets were extremely hard with a compression strength of 10,000 lbs. per square inch. Each sheet is placed separately on support beams and at times it is necessary to modify the size which requires cutting or drilling holes (Exhibit R-1, Tr. 14, 18). It is estimated that less than 10 percent of the fill sheets require cutting or the drilling of holes which accounts for about 2 percent of an installer's time per day (Tr. 13).

The testimony is not disputed that installation of the fill sheets commenced approximately October 14, 1975 and on October 30, 1975 respondent requested that its insurance carrier monitor the worksite for airborne asbestos fibers. Pursuant to this request the worksite was monitored approximately six weeks subsequent to the date of the inspection.

The inspecting officer testified that he had conducted approximately 6 or 8 inspections of construction sites involving asbestos. He stated that he could not distinguish asbestos from sand and concrete in the air, and indicated that in his previous experience the loose asbestos was identified from the packaging containing it.

During construction of the tower and installation of the sheets the structure was not enclosed in any way.

Alleged Violation of 29 C.F.R. 1910.1001(f)(1)— The standard reads in pertinent part as follows:

"(f) Monitoring—(1) Initial determinations. Within 6 months of the publication of this section, every employer shall cause every place of employment where asbestos fibers are released to be monitored in such a way as to determine whether every employee's exposure to asbestos fibers is below the limits prescribed in paragraph (b) of this section . . . .

The alleged violation is described in the citation as follows:

"Failed to monitor and establish if the degree of hazard from airborne concentrations of asbestos fibers to employees is within the prescribed limits set forth in 29 C.F.R. 1910.1001(b)(1), therefore not establishing the need for a compliance program and exposing workers to unknown quantities to asbestos fibers, such as sheets of material that contain asbestos being installed in the cooling tower."

The Secretary's case is based solely on the testimony of respondent's site manager who testified employees installed fill sheets which contained a mixture of sand, concrete and asbestos, the quantities of which were unknown. For the reasons set forth below, it must be held that the Secretary has failed to prove essential elements necessary to establish that the standard had been violated.

The standard requires monitoring by every employer at the place of employment where "asbestos fibers" are released, to determine if an employee's exposure to asbestos fibers is below certain prescribed limits. "Asbestos fibers" are specifically defined under section 1910.1001 as only those fibers exceeding the length of 5 micrometers. The Secretary necessarily assumes in alleging the violation that there were asbestos fibers present as defined without attempting to offer any proof thereof. Further, the Secretary assumed, without offering proof, that the asbestos fibers were "released" as set forth in the standard. There is no evidence that the process of installing the sheets as reflected in the record caused asbestos fibers to be "released". It is also contended that the standard has been violated when an employer allows his employees to come in contact with asbestos, the sheets in this case, without monitoring. This is clearly an unreasonable interpretation of the standard because of the vague nature of its application. Under this view an employer could only speculate as to what activity comes within the purview of the standard. Would it apply to those employees placing the sheets on pallets and moving them into position for installation, or only those employees actually installing the sheets, or only those employees engaged in cutting or drilling the sheets?

The basic purpose of this standard is to determine if employees have been exposed to asbestos fibers under certain conditions and certainly the Secretary's argument that the standard is of a mandatory nature is unquestioned. However, it is only mandatory within the confines of its specific language and applicable definitions.

Based upon the facts of record, it is held that the Secretary has failed to prove that the procedure shown to be used by the employer constituted a violation of the standard.

In post hearing argument respondent asserts that a literal reading of the standard renders it impossible to comply with. The standard provides that within six months of the date of the publication of the regulation,

monitoring is to be effective in accordance with such standards. The publication occurred October 18, 1972, and respondent argues that its workplace did not exist until April 1975, with products containing asbestos being present only after mid October 1975. It is reasonable to construe the standard as allowing six months from the date asbestos had been introduced to the job site within which to provide the required monitoring. However, the matter of publication is not an issue for determination since complainant has failed to provide competent evidence of the essential elements necessary to show a violation occurred.

*Alleged Violation of 29 C.F.R. 1910.1001 (g)(1)(i)—*The standard states as follows:

"(g) Caution signs and labels. (1) Caution signs. (i) Posting. Caution signs shall be provided and displayed at each location where airborne concentrations of asbestos fibers may be in excess of the exposure limits prescribed in paragraph (b) of this section. Signs shall be posted at such a distance from such a location so that an employee may read the signs and take necessary protective steps before entering the area marked by the signs. Signs shall be posted at all approaches to areas containing excessive concentrations of airborne asbestos fibers."

The alleged violation is described in the citation as follows:

"Failed to provide and post signs at such a distance and location an employee may read and take necessary precautions to prevent exposure to asbestos fibers, exposing employees to hazard of asbestos inhalation in the area where the asbestos impregnated sheets are being installed."

As previously set forth, there is no evidence in the record to indicate there were "airborne concentrations of asbestos fibers" at the work site nor has it been shown that such "asbestos fibers" were present within the terms of the definition provided by the Secretary. It seems true, as respondent argues that the complainant has merely surmised from the presence of fill sheets containing asbestos at the work site, that airborne concentrations of asbestos fibers existed.

In addition to a clear showing of the presence of "excessive concentrations of airborne asbestos fibers", it is incumbent upon the Secretary to specify the location or areas of such concentrations in order to prove a violation of this standard.

Therefore, it is held that the Secretary has failed to carry his burden in proving the violation as alleged.

*Alleged Violation of 29 C.F.R. 1910.1001(j)(2)—*The standard provides in pertinent part as follows:

"(j) Medical examinations—(2) Preplacement. The employer shall provide or make available to each of his employees, within 30 calendar days following his first employment in an occupation exposed to airborne concentrations of asbestos fibers, a comprehensive medical examination . . . . ''

The alleged violation is described in the citation as follows:

"Failed to provide or make available to cost, medical examinations relative to exposure to asbestos exposing employees to hazard of insufficient medical data of asbestos exposure."

In addition to the reasons previously set forth, the evidence fails to establish a violation of this standard as alleged. The evidence in no way establishes the duties or occupation respondent's employees were engaged in which caused exposure to concentrations of asbestos. Any attempt to apply the standard in this case must necessarily be based upon speculation as to when any such employees would be first exposed to airborne concentrations of asbestos. Therefore, it is held that the Secretary has failed to carry its burden in proving that this standard has been violated as alleged.

This holding is contrary to the argument of counsel for the Secretary that there are no conditions preceding the mandatory requirement other than the fact that the respondent's employees use asbestos and there is a possibility of air borne concentrations of asbestos fibers. The standard can not be reasonably applied in this to the facts of this case as the standard specifically refers to employee exposure to air borne concentrations of asbestos fibers, which has not been established. The standard does not speak of employee exposure to materials containing asbestos as the criteria for requiring medical examinations.

Findings of Fact

1. Research-Cottrell, Inc., is a corporation having a place of business and doing business among other places at Route 4, Box 449, Tallahassee, Florida, where it is engaged in the business of construction contracting.

2. On December 2, 1975 respondent was engaged in the construction of a cooling tower which required the installation of sheets containing cement, sand and asbestos of unknown quantities.

3. Asbestos fibers, for the purposes of this case, means only those fibers exceeding five micrometers and such fibers were not shown to have been present at the worksite.

4. Respondent's employees were not shown to have been exposed to a hazard involving asbestos fibers in the course of their employment at the construction site.

Conclusions of Law

1. Research-Cottrell, Inc., at all times pertinent hereto, was an employer engaged in a business affecting commerce within the meaning of section 3(5) of the Occupational Safety and Health Act of 1970, and the Commission has jurisdiction of the parties and subject matter herein, pursuant to section 10(c) of the Act.

2. Respondent was not in violation of the standard at 29 C.F.R. 1910.1001(f)(1) as alleged in the citation.

3. Respondent was not in violation of the standard at 29 C.F.R. 1910.1001(g)(1)(i) as alleged in the citation.

4. Respondent was not in violation of the standard at 29 C.F.R. 1910.1001(j)(2) as alleged in the citation.

On the basis of the foregoing findings of fact and conclusions of law, and the entire record, it is

Ordered:—The citation which alleges violation of the standards at 29 C.F.R. 1910.1001(f)(1), 29 C.F.R. 1910.1001(g)(1)(i) and 29 C.F.R. 1910.1001(j)(2) is hereby vacated.

SECRETARY OF LABOR v. AMERACE CORPORATION

(OSAHRC Docket No. 7490,
Review Commission Decision,
April, 1977)

Chromic Acid

Summary of Decision

Employers are required to provide periodic medical examinations for all employees who are exposed to chromic acid mist or liquid as stipulated in 29 C.F.R. 1910.94(d)(9)(viii). This ruling is a Review Commission reversal of an earlier decision by an administrative law judge who had held the standard to be "unenforceably vague". The violating company manufactures chrome-plated plastic parts. As a part of the process, approximately twenty employees come into contact with chromic acid mists; these mists are known to cause ulcerations upon contact with the skin or upper respiratory tract. The company had failed to provide, at any time, even one medical examination for each affected employee, and was therefore found in violation of section 5(a)(2) of the Occupational Safety and Health Act of 1970.

Full Text of Decision

CLEARY, Commissioner:

On February 4, 1975, Administrative Law Judge Henry C. Winters issued a decision vacating a citation alleging that respondent failed to comply with 29 CFR §1910.94(d)(9)(viii) and thereby violated section 5(a)(2) of the Occupational Safety and Health Act of 1970, 29 U.S.C. §651 *et seq.* (hereinafter "the Act"). On February 14, 1975, the Secretary of Labor filed a petition for discretionary review of the decision, and the petition was granted. The primary issues raised by the petition are:

(1) Whether the ALJ erred in finding the standard at 29 CFR §1910.94(d)(9)(viii) unenforceably vague with respect to medical examinations of workers exposed to chromic acids.

(2) Whether, on the facts of this case, respondent complied with the cited standard.

Both parties have filed briefs before us.

The standard cited by the Secretary reads as follows:

§1910.94 *Ventilation.*

★ ★ ★

(d) Open surface tanks.

★ ★ ★

(9) Personal protection.

★ ★ ★

(viii) Operators with sores, burns, or other skin lesions requiring medical treatment shall not be allowed to work at their regular operations until so authorized by a physician. Any small skin abrasions, cuts, rash, or open sores which are found or reported shall be treated by a properly designated person so that chances of exposures to the chemicals are removed. *Workers exposed to chromic acids shall have a periodic examination made of the nostrils and other parts of the body, to detect incipient ulceration.*

The underscored portion of the standard is the portion which the Secretary has alleged that respondent has not met.

The important facts are essentially undisputed. Respondent manufactures chrome-plated plastic parts. As part of a computer-automated chrome-plating process, employees load plastic parts onto racks that are dipped into a series of acid washes and rinses. Employees then unload the parts. Chromic acid in solution is used in the acid washes. Chromic acid is a corrosive substance that, in liquid or mist form, produces ulcerations upon contact with the skin or upper respiratory tract. Approximately 20 employees work in close proximity to the plating area and are exposed to chromic acid mists. Respondent neither provides nor requires physical examinations of its employees and, despite repeated requests by the employees' union local president, has refused to do so.

The Administrative Law Judge held that the cited portion of the standard is unenforceably vague, and that in any event the evidence of a violation was insufficient.

Noting that the standard does not define the terms "exposed to chromic acids" and "periodic examination," the Judge held that the standard incorporates no external or objective tests that would afford reasonable warning of its prohibitions. We disagree with these holdings.

[1] First we point out that the cited portion of the standard is plainly preventive in nature. Its purpose is not to detect ulcerations that have become readily apparent, but "to detect incipient ulceration." We agree with the Secretary that in order to preserve for employees the protection which the standard expressly contemplates, the phrase "workers exposed to chromic acids" must be interpreted to include employees working in a workplace containing detectable levels of chromic acid in either mist or liquid form. *Cf. GAF Corporation.* BNA 3 OSHC 1686, CCH 1975-76 OSHD para. 20,163 (No. 3203, etc., 1975). As so construed, this term of the standard is not considered vague. *See Rose v. Locke,* 96 S.Ct. 243, 244 (1975).

With respect to the term "periodic examinations," it is unnecessary to give a comprehensive interpretation at this time. In the record as presently composed. Amerace has not given its employees even one examination even though it has known that its employees were exposed to a detectable level of chromic acid. The standard under any reading contemplates that employees will be given at least one examination, if for no other reason than to help ascertain whether incipient ulcerations have already occurred, and with what frequency any subsequent examinations must be given. Accordingly, the vagueness claim on the present record lacks merit. *See United States Steel Corp. v. O.S.H.R.C.* 537 F.2d 708 [4 OSHC 1424] (3d Cir. 1976).

[2] On the present record, we conclude that Amerace employees were exposed to a detectable level of chromic acid and did not receive any physical examinations to detect incipient ulceration. The Judge, however, granted the respondent's motion for involuntary dismissal at the close of the Secretary's case-in-chief. The case must therefore be remanded for further proceedings.

Accordingly, the case is remanded for further proceedings not inconsistent with this opinion. The Judge shall expedite the proceedings.

So ORDERED.

Concurring Opinion

BARNAKO, Chairman, concurring:

I concur in remanding the case for Respondent's evidence. I do not join in my colleague's discussion of the interpretation of the term "periodic examination" inasmuch as an interpretation at this time would be premature. I decline to attempt to define the term on the incomplete record which is before us, since such a definition would be based on an abstraction. See *Brennan v. OSHRC* (Santa Fe Trail Transport Co.), 505 F.2d 869 [2 OSHC 1274] (10th Cir. 1947). Moreover, evidence adduced by Respondent on remand may render a determination on the meaning of "periodic examination" unnecessary.

Dissenting Opinion

MORAN, Commissioner, dissenting:

Judge Winters correctly found in his decision that the occupational safety and health standard codified at 29 C.F.R. §1910.94(d)(9)(viii) is unenforceably vague. I agree with his reasoning and would affirm his decision.

The vagueness of the standard is well-illustrated by the inability of my colleagues to agree on a definition of the term "periodic examination." Although he declines to give "a comprehensive interpretation at this time," Commissioner Cleary states that it means "at least one examination." Chairman Barnako does not give a definition and states that he does not join in Commissioner Cleary's interpretation of the term because an interpretation is premature at this time. The Commission, however, should not have to speculate as to what was intended by the term, and I decline to participate in speculation. It is a matter which the Secretary of Labor can easily correct by modifying his regulation, and he should be required to do so.

If my colleagues cannot tell the employers of the country what is meant by "a periodic examination," how can they reverse the Judge's decision? The answer is obvious, and their holding is patently wrong.

At the hearing, complainant was unable to cure the ambiguity of what is meant by "exposed to chromic acid." He did not present any evidence of what the phrase would mean to a reasonable man in respondent's industry. Instead, complainant's witnesses expressed contradictory opinions on what level of chromic acid in the air would be harmful and would constitute exposure. They also disagreed on what employees in which parts of the plant would be "exposed to chromic acids." Complainant's inspectors did not take any air samples to determine whether chromic acid was actually in the air breathed by any of respondent's employees, nor did complainant present any expert testimony on whether the chromic acid would waft from the acid baths to the racking stations where employees were working. Instead of submitting persuasive evidence, complainant relied on showing the locations of the baths and the work stations and evidence that a few employees suffered various medical problems. He did not, however, present any expert evidence to show that the employees' problems resulted from exposure to chromic acids.

The major infirmity in §1910.94(d)(9)(viii) is that it "creates no specific standard" as to when a period examination must be given. *Hoffman Construction Company v. OSAHRC*, No. 75-1741 [4 OSHC 1813] (9th Cir., November 1, 1976). It is therefore subject to arbitrary application and multiple interpretations by govern-ment enforcement officials, as the testimony in this case so well illustrates. Thus, it is unenforceably vague on its face as well as "in the light of the conduct to which it is applied." *United States v. National Dairy Products Corp.*, 372 U.S. 29, 36 (1963). *See McLean Trucking Company v. OSAHRC*, 503 F.2d 8 [2 OSHC 1165] (4th Cir. 1974); *Ryder Truck Lines, Inc. v. Brennan*, 497 F.2d 230 [2 OSHC 1075] (5th Cir. 1974).

BULK TERMINALS ET AL. v. THE ENVIRONMENTAL PROTECTION AGENCY ET AL.

(Supreme Court of Illinois,
357 N.E. 2d 430)*

Hydrochloric Acid and Silicon Dioxide

Summary of Decision

The company in this case had been charged by the city and convicted of causing emissions of hydrochloric acid vapor and silicon dioxide. The EPA subsequently brought action against the company before the Pollution Control Board based on the same incident. The circuit court stated that the company's efforts to prevent the State of Illinois from prosecuting and fining them twice for the same offense should be allowed to take place in the courts and not be restricted to administrative remedies. In contrast, the appellate court held that the proceedings before the Pollution Control Board were prohibited by previous prosecution by the city, and reversed the lower court ruling. The State Supreme Court affirmed the circuit court ruling and reversed the appellate decision. The important precedent set here is that a company prosecuted and fined once for toxic emissions in a local action is not necessarily immune from prosecution by the EPA. The protection against double jeopardy does not guard against being punished (fined) twice, but rather against being tried twice. One violation, therefore, can, under certain circumstances, result in a double fine and costly appeals.

Company appealed from judgment of the Circuit Court, Cook County, which dismissed action for injunction or prohibition seeking to terminate proceedings commenced by Environmental Protection Agency. The Appellate Court, 29 Ill.App.3d 978, 331 N.E.2d 260, affirmed and petition for leave to appeal was allowed. The Supreme Court, Goldenhersh, J., held that complaint which alleged that company had been charged by city with causing emissions of hydrochloric acid vapor and silicon dioxide and had been convicted and that Evnironmental Protection Agency thereafter commenced action before

the Pollution Control Board based on the same incident and that the Board had overruled company's motion to dismiss based on res judicata and double jeopardy did not state any facts which would warrant an exception to exhaustion of administrative remedies doctrine.

Appellate Court reversed; Circuit Court affirmed.

Health and Environment

Complaint which alleged that company had been prosecuted by city for emission of hydrochloric acid vapor and silicon dioxide and that Environmental Protection Agency had nonetheless initiated proceedings against the company before the Pollution Control Board for violations based on the same incident and that the Board had overruled the company's motion to dismiss based on res judicata and double jeopardy did not state any facts which would warrant an exception to exhaustion of administrative remedies doctrine so that the company was required to exhaust its remedies with the Board before seeking relief in court.

Text of Decision
Goldenhersh, Justice.

Plaintiffs, Bulk Terminals Company, hereafter Bulk, and Gerald L. Spaeth, its president, appealed from the judgment of the circuit court of Cook County dismissing their action for injunction, or alternatively prohibition, seeking to terminate proceedings commenced by defendants Environmental Protection Agency and Citizens for a Better Environment before the defendant Pollution Control Board. The appellate court reversed (29 Ill.App.3d 978.331 N.E.2d 260) and we allowed petitions for leave to appeal filed by defendant Citizens for a Better Environment (No. 47746) and jointly by the other defendants (No. 47754).

In their complaint for injunction or prohibition plaintiffs alleged that a leak in a storage tank situated at Bulk's premises caused the emission of hydrochloric acid vapor and silicon dioxide; that the city of Chicago filed complaints charging Bulk with violations of section 17—2.6 of the Chicago Municipal Code; that Bulk was tried and found guilty of violations of the ordinance and that fines were assessed and paid; that defendant Citizens for a Better Environment filed a complaint before the defendant Pollution Control Board charging plaintiffs with violations of the Illinois Environmental Protection Act and of certain air pollution regulations; that defendant Environmental Protection Agency also filed a complaint before the Pollution Control Board charging similar violations of the Act and the regulations; that the violations charged in the proceedings before the Pollution Control Board involved the same emissions on the same dates as those for which Bulk was prosecuted under the Chicago ordinance; that plaintiffs filed answers to the complaints before the Pollution Control Board affirmatively setting forth "the facts in support of their constitutional and common law defenses" and a motion and amended motion to dismiss the proceedings; that the defendant Pollution Control Board has denied their motions to dismiss and to stay discovery; and that unless the proceedings before the Pollution Control Board are enjoined they will suffer irreparable loss and damage. In a second count they repeated the allegations and sought as alternative relief the issuance of a writ of prohibition. It is also alleged in the complaint that:

"10. On or before July 31, 1974, all silicon tetrachloride previously stored on the premises of Bulk had been removed by the owner thereof and transported away from Cook County, Illinois. Bulk has no present intention to store in the future silicon tetrachloride on its premises in Chicago, Illinois."

The circuit court dismissed the suit on the ground that "the complaint fails to establish that plaintiffs have exhausted their remedies under the Environmental Protection Act and the Administrative Review Act * * *." The appellate court, although recognizing that under the Administrative Review Act only final decisions of administrative agencies are reviewable, stated that in this action in which * * * plaintiffs seek in effect to prevent the State of Illinois from twice prosecuting and fining them for the same offense * * * to allow a remedy in a judicial forum only after the fact of double prosecution would be improper and could not be mandated by the Administrative Review Act." (29 Ill.App.3d 978, 982, 331 N.E.2d 260, 263.) The appellate court held that the proceedings before the Pollution Control Board were barred by the prior prosecution under the Chicago ordinance and reversed the judgment.

Although the parties and *amicus curiae,* the Illinois Manufacturer's Association, have briefed and argued a number of questions we need consider only whether plaintiffs, prior to seeking judicial relief, were required to exhaust the administrative remedies provided in section 41 of the Environmental Protection Act (Ill.Rev.Stat.1975, ch. 111½, par. 1041) and the Administrative Review Act (Ill.Rev.Stat.1975, ch. 110, par. 264 *et seq.*). It is defendants' position that the circuit court "is without jurisdiction to review interlocutory orders of the Pollution Control Board." Plaintiffs contend that "this is a proper case for the exercise of the circuit court's power to issue a writ of prohibition or to order injunctive relief. Plaintiffs have no other remedy for the wrongs being done to them." They argue that "judicial review of a final order of the Pollution Control Board is inadequate relief because the guarantee against double jeopardy precludes a second prosecution as well as a second punishment," that *"res judicata* not only precludes multiple liability, but subsequent actions to impose that liability as well" and that "the Administrative Review Act does not bar the relief plaintiffs seek; if it did, it would be unconstitutional."

In discussing the doctrine of exhaustion of remedies, in *Illinois Bell Telephone Co. v. Allphin,* 60 Ill.2d 350, 326 N.E.2d 737, we said:

"'* * * the doctrine of exhaustion has long been a basic principle of administrative law—a party aggrieved by administrative action ordinarily cannot seek review in the courts without first pursuing all administrative remedies available to him. (*Myers v. Bethlehem Shipbuilding Corp.* (1938), 303 U.S. 41, 58 S.Ct. 459, 82 L.Ed. 638.) The rule is the counterpart of the procedural rule which, with certain exceptions, precludes appellate review prior to a final judgment in the trial court, and the reasons for its existence are numerous: (1) it allows full development of the facts before the agency; (2) it allows the agency an opportunity to utilize its expertise; and (3) the aggrieved party may succeed before the agency, rendering judicial review unnecessary. 2 F. Cooper, State Administrative Law 572-574 (1965); L. Jaffe, Judicial Control of Administrative Action 424-426 (1965); 3 K. Davis, Administrative Law Treatise secs. 20.01-20.10 (1958), and 1970 Supplement at 642-669.

All jurisdictions have recognized that the exhaustion doctrine, if strictly applied, could sometimes produce very harsh and inequitable results. While our courts have required comparatively strict compliance with the exhaustion rule, exceptions have been recognized where an ordinance or statute is attacked as unconstitutional in its entirety (*Bright v. City of Evanston* (1956), 10 Ill.2d 178, 139 N.E.2d 270), or where multiple remedies exist before the same zoning board and at least one has been exhausted (*Herman v. Village of Hillside (1958), 15 Ill.2d 396, 155 N.E.2d 47),* or where irreparable harm will result from further pursuit of administrative remedies. (*Peoples Gas Light and Coke Co. v. Slattery* (1939), 373 Ill. 31, 25 N.E.2d 482.) It is not our intention by this opinion to affect these existing exceptions.

These exceptions to the exhaustion doctrine have been fashioned in recognition of the time-honored rule that equitable relief will be available if the remedy at law is inadequate. In those situations covered by these exceptions, further recourse to the administrative process would not, or cannot, for a variety of reasons, provide adequate relief." 60 Ill.2d 350, 357-59, 326 N.E.2d 737, 741.

The circuit court entered judgment upon allowance of defendants' motion to dismiss and all facts properly pleaded in the complaint are taken as true. (*Acorn Auto Driving School, Inc. v. Board of Education,* 27 Ill.2d 93, 96, 187 N.E.2d 722.) The question whether, upon those facts, the proceedings before the Pollution Control Board are barred by either the constitutional proscription of double jeopardy or the doctrine of *res judicata* is one of law which we need not and do not decide. Other than a conclusional allegation in the complaint that unless relief is granted they will suffer irreparable harm, plaintiffs do not contend that under the facts alleged this case falls within one of the exceptions recognized in *Allphin,* and we are presented the narrow question whether the allegation of double jeopardy, or alternatively, *res judicata,* should invoke an additional exception to the requirement that administrative remedies be exhausted prior to seeking judicial relief.

It is clear that the motions to dismiss filed before defendant Pollution Control Board were the appropriate method by which to present the issue and that the Board had jurisdiction to decide the question. (*Sugden v. Department of Public Welfare,* 20 Ill.2d 119, 121, 169 N.E.2d 248; *People ex rel. United Motor Coach Co. v. Carpentier,* 17 Ill.2d 303, 305, 161 N.E.2d 97.) Plaintiffs, conceding that the order denying the motions to dismiss was interlocutory in nature, argue that the denial of the plea of double jeopardy served to "finally determine rights separate from and collateral to the main action, determine collateral rights that are too important to be denied review and determine rights that will have been lost, probably irreparably, after judgment has been entered." In support of their contention that an interlocutory order denying a plea of double jeopardy is appealable they cite *United States v. Beckerman,* 516 F.2d 905 (2d Cir. 1975) wherein, at page 906, the court said:

"The issue of double jeopardy is collateral to the determination of whether the accused is innocent or guilty of the offense for which he has been indicted. The

constitutional protection against being twice put in jeopardy for the same offense is a 'valued right,' *Wade v. Hunter,* 336 U.S. 684, 689, 69 S.Ct. 834, 93 L.Ed. 974 (1949), that is too important to be denied review. The protection against double jeopardy guards 'not against being twice punished, but against being twice put in jeopardy,' *United States v. Ball,* 163 U.S. 662, 669, 16 S.Ct. 1192, 1194, 41 L.Ed. 300 (1896). The right will be invaded if an accused, who has properly invoked the Fifth Amendment protection against being twice put in jeopardy, is called upon to suffer the pain of two trials. *Green v. United States,* 355 U.S. 184, 187, 78 S.Ct. 221, 2 L.Ed.2d 199 (1957).

If an accused is to be afforded 'the full protection of the double jeopardy clause, a final determination of whether jeopardy has attached to the previous trial must, where possible, be determined prior to any retrial.' *United States v. Lansdown,* 460 F.2d 164, 171 (4th Cir. 1972). Contra, *Gilmore v. United States,* 264 F.2d 44 (5th Cir. 1959), cert. denied, 359 U.S. 994, 79 S.Ct. 1126, 3 L.Ed.2d 982 (1959). The reasoning of the opinion in *Lansdown* is persuasive here. We conclude that we have jurisdiction to review the denial of the defendant's motion to dismiss the indictment on the predicate of the double jeopardy clause."

The situation in *Beckerman,* however, is clearly distinguishable. Had plaintiffs, despite the statutory provision limiting review to "final" orders, sought review of the interlocutory order of the Pollution Control Board denying their motions, this case would present a question similar to that involved in *Beckerman.* This, however, they did not do. We note parenthetically that the Court of Appeals for the Fifth Circuit has decided the question contrary to *Beckerman* (see *United States v. Fanley* (1975), 512 F.2d 833; *Gilmore v. United States* (1959), 264 F.2d 44) and that in *People v. Miller,* 35 Ill.2d 62, 219 N.E.2d 475 it was held "that no appeal lies from an interlocutory order in the absence of a statute or rule specifically authorizing such review" (35 Ill.2d 62, 67, 219 N.E.2d 475, 478). We note, too, that a panel of the Second Circuit Court of Appeals in *United States v. Alessi,* 544 F.2d 1139 (1976), although stating that it would prefer to follow *Gilmore,* considered itself compelled to follow *Beckerman.*

We conclude, for the reasons stated, that this record presents no question of the appealability of an interlocutory order and that the complaint failed to state facts which would warrant an exception to the exhaustion doctrine. The judgment of the appellate court is therefore reversed and the judgment of the circuit court of Cook County, dismissing the action, is affirmed.

Appellate court reversed; circuit court affirmed.

NATURAL RESOURCES DEFENSE
COUNCIL, INC. v. TRAIN

(U.S. Court of Appeals,
545 F.2d 320)*

Lead

Summary of Decision

This case resolves an important issue involving the EPA and potentially harmful substances which federal regulations control. The EPA appealed an order from the district court which required EPA to place lead on its list of air pollutants under Section 108 (a)(1) of the Clean Air Act. EPA agreed that lead met the conditions for listing under the Clean Air Act, Section 108 (a)(1) (A) and (B), which basically require that a substance has: (1) an adverse effect on public health and welfare and (2) the presence of the substance results from diverse sources. However, EPA maintained that it is up to the Administrator's discretion whether to list a pollutant, and, subsequently, promulgate standards for it, once the criteria have been met. EPA argued that it could choose instead to control such emissions at the sources, regardless of ambient air concentration. The court of appeals, however, ruled against EPA, stating that once conditions of Section 108 (a) (1) (A) and (B) have been met, the listing of the substance and issuance of standards become mandatory. This affirmation of the lower court ruling has set an important precedent in that an appeal by industry to EPA would be futile in preventing the imposition of standards once the potential harm of a substance to the public has been established; such a substance will appear in revised pollutant listings and standards will be issued.

The United States District Court for the Southern District of New York, Charles E. Stewart, Jr., J., 411 F.Supp. 864, issued order granting relief, and Administrator of Environmental Protection Agency appealed. The Court of Appeals, J. Joseph Smith, Circuit Judge, held that phrase "but for which he plans to issue air quality criteria," within provision of Clean Air Act Amendments of 1970 requiring Administrator to publish a list of each air pollutant for which air quality criteria have not been issued, but for which he plans to issue air quality criteria, operates to require inclusion of those pollutants for which air quality criteria have not been issued but which Administrator has already found in his judgment to have an adverse effect on public health or welfare and to come from numerous or diverse mobile or stationary sources.

1. Health and Environment

Fuel controls set by Administrator of Environmental Protection Agency under Clean Air Act Amendments of 1970 were intended by Congress as a means for obtaining primary air quality standards rather than as an alternative to promulgation of such standards. Clean Air Act, § 108(a)(1), (a)(1)(A−C) as amended 42 U.S.C.A. § 1857c−3(a)(1), (a)(1)(A−C).

2. Health and Environment

Provision of Clean Air Amendments of 1970 governing promulgation of national primary and secondary ambient air quality standards by Administrator of Environmental Protection Agency require inclusion on initial list to be issued of those pollutants for which air quality criteria have not been issued but which Administrator has already found in his judgment to have an adverse effect on public health or welfare and to come from numerous or diverse mobile or stationary sources. Clean Air Act, §§ 108(a)(1), 304(a) as amended 42 U.S.C.A. §§ 1857c−3(a)(1), 1857h−2(a).

3. Health and Environment

Phrase "but for which he plans to issue air quality criteria," within provision of Clean Air Act Amendments of 1970 requiring Administrator of Environmental Protection Agency to publish a list of each air pollutant for which air quality criteria have not been issued, but for which he plans to issue air quality criteria, operates to require inclusion of those pollutants for which air quality criteria have not been issued but which Administrator has already found in his judgment to have an adverse effect on public health or welfare and to come from numerous or diverse mobile or stationary sources. Clean Air Act, §§

108(a)(1), 304(a) as amended 42 U.S.C.A. §§ 1857c—3(a)(1), 1857h—2(a).

4. Statutes

When a specific provision of a total statutory scheme may be construed to be in conflict with the congressional purpose expressed in an act, it becomes necessary to examine the act's legislative history to determine whether the specific provision is reconcilable with the intent of Congress.

5. Health and Environment

Once Administrator of Environmental Protection Agency determines that a particular pollutant has an adverse effect on public health or welfare and originates from one or more numerous or diverse mobile or stationary sources, provisions of Clean Air Act Amendments of 1970 are to be automatically invoked, and promulgation of national air quality standards and implementation thereof by states within a limited, fixed time schedule becomes mandatory. Clean Air Act, §§ 108(a)(1), (a)(1)(C), 109, 110 as amended 42 U.S.C.A. §§ 1857c—3(a)(1), (a)(1)(C), 1857c—4, 1857c—5.

6. Health and Environment

Language of Clean Air Act Amendments of 1970 requiring Administrator of Environmental Protection Agency to publish a list of air pollutants for which air quality criteria have not been issued "but for which he plans to issue air quality criteria" does not constitute a separate and third criterion to be met before Amendments require listing of lead as a pollutant and issuance of air quality standards and, hence, does not leave decision to list lead within discretion of Administrator. Clean Air Act, §§ 108(a)(1), 304(a) as amended 42 U.S.C.A. §§ 1857c—3(a)(1), 1857h—2(a).

7. Health and Environment

Language of Senate report relating to issuance of new source performance standards makes it clear that Senate intended such standards to be supplementary to, not in lieu of, ambient air quality standards contained in Clean Air Act Amendments of 1970. Clean Air Act, §§ 108(a)(1), 304 (a) as amended 42 U.S.C.A. §§1857c—3(a)(1), 1857h—2(a).

8. Health and Environment

Deliberate inclusion of a specific timetable for attainment of air quality standards incorporated by Congress in Clean Air Act Amendments of 1970 would become an exercise in futility if Administrator of Environmental Protection Agency could avoid listing pollutants simply by choosing not to issue air quality criteria. Clean Air Act, §§ 108—110, 211 as amended 42 U.S.C.A. §§ 1857c—3 to 1857c—5, 1857f—6c.

9. Health and Environment

Discretion given to Administrator of Environmental Protection Agency under Clean Air Act Amendments of 1970 pertains to review of state implementation plans and to regulation of fuel or fuel additives, but does not extend to issuance of air quality standards for substances derived from specified sources that Administrator has already adjudged injurious to health. Clean Air Act, §§ 108—110, 211 as amended 42 U.S.C.A. §§ 1857c—3 to 1857c—5, 1857f—6c.

10. Health and Environment

Provisions of Clean Air Act Amendments of 1970 did not authorize Administrator of Environmental Protection agency to order emission source controls instead of promulgating ambient air quality standards for substances such as lead. Clean Air Act, §§ 108(a)(1)(A, B), 111, 202, 211, 231 as amended 42 U.S.C.A. §§ 1857c—3(a)(1)(A, B), 1857c—6, 1857f—1, 1857f—6c, 1857f—9.

11. Health and Environment

Under scheme of Clean Air Act Amendments of 1970, emission source control is a supplement to air quality standards, not an alternative to them. Clean Air Act, §§ 108(a)(1)(A, B), 111, 202, 211, 231 as amended 42 U.S.C.A. §§ 1857c—3(a)(1)(A, B), and 1857c—6, 1857f—1, 1857f—6c, 1857f—9.

12. Health and Environment

Structure of Clean Air Amendments of 1970, their legislative history, and judicial gloss placed upon them leave no room for an interpretation that makes issuance of air quality standards for lead discretionary with Administrator of Environmental Protection Agency. Clean Air Act, §§ 108(a)(1)(A, B), 111, 202, 211, 231 as amended 42 U.S.C.A. §§ 1857c—3(a)(1)(A, B), 1857c—6, 1857f—1, 1857f—6c, 1857f—9.

Before SMITH, OAKES and MESKILL, Circuit Judges.

J. JOSEPH SMITH, Circuit Judge:

The Environmental Protection Agency, ("EPA"), and its Administrator, Russell Train, appeal from an order of the United States District Court for the Southern District of New York, Charles E. Stewart, Jr., Judge in an action under § 304 of the Clean Air Act, as amended, 42 U.S.C. § 1857h−2(a), requiring the Administrator of the EPA, within thirty days, to place lead on a list of air pollutants under § 108(a)(1) of the Clean Air Act, as amended, 42 U.S.C. § 1857c−3(a)(1), ("the Act"). We affirm the order of the district court.

The 1970 Clean Air Act Amendments provide two different approaches for controlling pollutants in the air. One approach, incorporated in §§ 108−110, 42 U.S.C. §§ 1857c−3 to c−5, provides for the publication of a list of pollutants adverse to public health or welfare, derived from "numerous or diverse" sources, the promulgation of national ambient air quality standards for listed pollutants, and subsequent implementation of these standards by the states. The second approach of the Act provides for control of certain pollutants at the source, pursuant to §§ 111, 112, 202, 211 and 231 (42 U.S.C. §§1857c−6, c−7, f−1, f−6c, f−9).

The relevant part of § 108 reads as follows:

(a)(1) For the purpose of establishing national primary and secondary ambient air quality standards, the Administrator shall within 30 days after December 31, 1970, publish, and shall from time to time thereafter revise, a list which includes each air pollutant—

(A) which in his judgment has an adverse effect on public health or welfare;

(B) the presence of which in the ambient air results from numerous or diverse mobile or stationary sources; and

(C) for which air quality criteria had not been issued before December 31, 1970, but for which he plans to issue air quality criteria under this section.

Once a pollutant has been listed under § 108(a)(1), §§ 109 and 110 of the Act are automatically invoked. These sections require that for any pollutant for which air quality criteria are issued under § 108(a)(1)(C) after the date of enactment of the Clean Air Amendments of 1970, the Administrator must simultaneously issue air quality standards. Within nine months of the promulgation of such standards, states are required to submit implementation plans to the Administrator. § 110(a)(1). The Administrator must approve or disapprove a state plan within four months. § 110(a)(2). If a state fails to submit an acceptable plan, the Administrator is required to prepare and publish such a plan himself. § 110(c). State imple-

mentation plans must provide for the attainment of primary ambient air quality standards no later than three years from the date of approval of a plan. § 110(a)(2)(A)(i). Extension of the three-year period for attaining the primary standard may be granted by the Administrator only in very limited circumstances, and in no case for more than two years. § 110(e).

The EPA concedes that lead meets the conditions of §§ 108(a)(1)(A) and (B)—that it has an adverse effect on public health and welfare, and that the presence of lead in the ambient air results from numerous or diverse mobile or stationary sources. The EPA maintains, however, that under § 108(a)(1)(C) of the Act, the Administrator retains discretion whether to list a pollutant, even though the pollutant meets the criteria of §§ 108(a)(1)(A) and (B). The EPA regards the listing of lead under § 108(a)(1) and the issuance of ambient air quality standards as one of numerous alternative control strategies for lead available to it. Listing of substances is mandatory, the EPA argues, only for those pollutants for which the Administrator "plans to issue air quality criteria." He may, it is contended, choose not to issue, i.e., not "plan to issue" such criteria, and decide to control lead solely by regulating emission at the source, regardless of the total concentration of lead in the ambient air. The Administrator argues that if he chooses to control lead (or other pollutants) under § 211, he is not required to list the pollutant under § 108(a)(1) or to set air quality standards.

The EPA advances three reasons for the position that the Administrator has discretion whether to list a pollutant even when the conditions of § 108(a)(1)(A) and (B) have been met: the plain meaning of § 108(a)(1)(C); the structure of the Clean Air Act as a whole; and the legislative history of the Act.

The issue is one of statutory construction. We agree with the district court and with appellees, National Resources Defense Council, Inc., et al., that the interpretation of the Clean Air Act advanced by the EPA is contrary to the structure of the Act as a whole, and that if accepted, it would vitiate the public policy underlying the enactment of the 1970 Amendments as set forth in the Act and in its legislative history. Recent court decisions are in accord, and have construed § 108(a)(1) to be mandatory if the criteria of subsections A and B are met.

Section 108(a)(1) and the Structure of the Clean Air Act

[1,2] Section 108(a)(1) contains mandatory language. It provides that "the Administrator *shall* . . . publish . . . a list . . . ." (Emphasis added.) If the EPA interpretation were accepted and listing were mandatory only for substances "for which [the Administrator] plans to issue air quality criteria . . . ", then the mandatory language of § 108(a)(1)(A) would become mere surplusage. The deter-

mination to list a pollutant and to issue air quality criteria would remain discretionary with the Administrator, and the rigid deadlines of § 108(a)(2), § 109, and § 110 for attaining air quality standards could be by-passed by him at will. If Congress had enacted § 211 as an alternative to, rather than as a supplement to, §§ 108—110, then one would expect a similar fixed timetable for implementation of the fuel control section. The absence of such a timetable for the enforcement of § 211 lends support to the view that fuel controls were intended by Congress as a means for attaining primary air quality standards rather than as an alternative to the promulgation of such standards.

The EPA Administrator himself initially interpreted § 108(a)(1) as requiring inclusion on the initial list to be issued of those pollutants for which air quality criteria had not been issued but which he had already found in his judgment to have an adverse effect on public health or welfare and to come from sources specified in § 108(a)(1)(B). 36 Fed.Reg. 1515 (1971).

[3] We agree with Judge Stewart that it is to the initial list alone that the phrase "but for which he plans to issue air quality criteria" is directed, and that the Administrator must list those pollutants which he has determined meet the two requisites set forth in section 108.

Legislative History

[4,5] When a specific provision of a total statutory scheme may be construed to be in conflict with the congressional purpose expressed in an act, it becomes necessary to examine the act's legislative history to determine whether the specific provision is reconcilable with the intent of Congress. Because state planning and implementation under the Air Quality Act of 1967 had made little progress by 1970, Congress reacted by "taking a stick to the States in the form of the Clean Air Amendments of 1970 . . . ." *Train v. Natural Resources Defense Council,* 421 U.S. 60, 64, 95 S.Ct. 1470, 1474, 43 L.Ed.2d 731 (1975). It enacted § 108(a)(1) which provides that the Administrator of the Environmental Protection Agency "shall" publish a list which includes each air pollutant which is harmful to health and originates from specified sources. Once a pollutant is listed under § 108(a)(1), §§ 109 and 110 are to be automatically invoked, and promulgation of national air quality standards and implementation thereof by the states within a limited, fixed time schedule becomes mandatory.

[6] The EPA contention that the language of § 108(a)(1)(C) "for which [the Administrator] plans to issue air quality criteria" is a separate and third criterion to be met before § 108 requires listing lead and issuing air quality standards, thereby leaving the decision to list lead within the discretion of the Administrator, finds no support in the legislative history of the 1970 Amendments to

the Act. The summary of the provisions of the conference agreement furnished the Senate by Senator Muskie contain the following language:

> The agreement requires issuance of remaining air quality criteria for major pollutants within 13 months of date of enactment.

and

> Within the 13-month deadline, the Congress expects criteria to be issued for nitrogen oxides, fluorides, lead, polynuclear organic matter, and odors, though others may be necessary.

The section-by-section analysis of the National Air Standards Act of 1970 in the Senate Report on S. 4358, 91st Cong., 2d Sess., contains this language describing § 108 (formerly designated § 109):

> This new section directs the Secretary to publish (initially 30 days after enactment) a list of air pollution agents or combination thereof for which air quality criteria will be issued. He can add to the list periodically. The agents on the initial list must include all those pollution agents or combinations of agents which have, or can be expected to have, an adverse effect on health and welfare and which are emitted from widely distributed mobile and stationary sources, and all those for which air quality criteria are planned.

> Twelve months after such initial list is published, the Secretary must issue air quality criteria for those listed agents.

> This section continues in effect those air quality criteria and information on pollution control techniques published prior to this section.

> This section provides that such criteria and information shall be published in the Federal Register and be available to the public.

The same Senate Report contains the following explicit language regarding §§ 108 and 109 (formerly § 109 and § 110):

> Air quality criteria for five pollution agents have already been issued (sulfur oxides, particulates, carbon monoxide, hydrocarbons, and photochemical oxidants). *Other contaminants of broad national impact include fluorides, nitrogen oxides, polynuclear organic matter, lead, and odors. Others may be added to this group as knowledge increases. The bill would require that air quality criteria for these and other pollutants be issued within 13 months from enactment.* If the [Secretary] subsequently should find that there are other pollution agents for which the ambient air quality standards procedure is appropriate, he could list those agents in the Federal Register, and repeat the criteria process.

.   .   .   .   .

Within 30 days after enactment the [Secretary] would be required to publish proposed national air quality standards for those pollutants covered by existing air quality criteria (sulfur oxides, particulate matter, carbon monoixide, hydrocarbons, and photochemical oxidants). Since these criteria have been available for some time, it is realistic to expect that proposed national standards for these five pollution agents would be published within the 30-day period. Proposed national air quality standards for pollutants for which criteria would be issued subsequent to enactment would be published simultaneously with the issuance of such criteria. *These pollutants would include nitrogen oxides, lead, polynuclear organics, odors, and fluorides.*

(Emphases added.)

[**7**] Language relating to the issuance of new source performance standards makes it clear that the Senate intended these standards to be supplementary to, not in lieu of, ambient air quality standards:

The committee recognizes that the construction of major new industrial facilities in some regions may conflict with implementation plans for national air quality standards and goals—even where such new facilities are designed, equipped, and operated so as to comply with applicable Federal standards of performance. This is most likely to occur in places where existing levels of air pollution are excessive. Accordingly, the bill would provide that new-source certification procedures must include pre-construction review of the location as well as the design of affected new facilities so that certified new sources would not hinder the implementation of air quality standards and goals.

[**8,9**] While the literal language of § 108(a)(1)(C) is somewhat ambiguous, this ambiguity is resolved when this section is placed in the context of the Act as a whole and in its legislative history. The deliberate inclusion of a specific timetable for the attainment of ambient air quality standards incorporated by Congress in §§ 108−110 would become an exercise in futility if the Administrator could avoid listing pollutants simply by choosing not to issue air quality criteria. The discretion given to the Administrator under the Act pertains to the review of state implementation plans under § 110, and to § 221 which authorizes but does not mandate the regulation of fuel or fuel additives. It does not extend to the issuance of air quality standards for substances derived from specified sources which the Administrator had already adjudged injurious to health.

Judicial Interpretations

The Supreme Court in *Union Electric Co. v. Environmental Protection Agency,* 427 U.S. 246, 256, 96 S.Ct.

2518, 2525, 49 L.Ed.2d 474 (1976), referred to the 1970 Amendments of the Clean Air Act as "a drastic remedy to what was perceived as a serious and otherwise uncheckable problem of air pollution." In the same opinion the Court described the three-year deadline for achieving primary air quality standards as "central to the Amendments' regulatory scheme." *Id.* at 258, 96 S.Ct. at 2526. Previously the Court had referred to the attainment of the national air quality standards within three years from the date of approval of state implementation plans as "the heart of the 1970 Amendments." *Train v. Natural Resources Defense Council,* 421 U.S. 60, 66, 95 S.Ct. 1470, 1475, 43 L Ed.2d 731 (1975). The EPA, the Court stated, "is plainly charged by the Act with the responsibility for setting the national ambient air standards. Just as plainly, however, it is relegated by the Act to a secondary role in the process of determining and enforcing specific, source-by-source emission limitations which are necessary if the national standards it has set are to be met." *Id.* at 79, 95 S.Ct. at 1481. Reference by the Court to EPA authority to regulate emissions at the source, under §§ 111, 202, 211, and 231, is restricted to a footnote. *Id.* at 79, n. 16, 95 S.Ct. 1470.

[**10—12**] The Court's language in *Train and Union Electric* lends no support to appellants' position that the EPA Administrator may order emission source controls instead of promulgating ambient air quality standards for substances, such as lead, which meet the criteria of §§ 108(a)(1)(A) and (B). Under the scheme of the Act, emission source control is a supplement to air quality standards, not an alternative to them.

When "interpreting a statute, the court will not look merely to a particular clause in which general words may be used, but will take in connection with it the whole statute . . . and the objects and policy of the law, as indicated by its various provisions, and give to it such a construction as will carry into execution the will of the Legislature. . . ."

Chief Justice Burger, in *Kokoszka v. Belford,* 417 U.S. 642, 650, 94 S.Ct. 2431, 41 L.Ed.2d 374 (1974), *reh. denied,* 419 U.S. 886, 95 S.Ct. 160, 42 L.Ed.2d 131 (1974) (citation omitted).

The structure of the Clean Air Act as amended in 1970, its legislative history, and the judicial gloss placed upon the Act leave no room for an interpretation which makes the issuance of air quality standards for lead under § 108 discretionary. The Congress sought to eliminate, not perpetuate, opportunity for adminstrative foot-dragging. Once the conditions of §§ 108(a)(1)(A) and (B) have been met, the listing of lead and the issuance of air quality standards for lead become mandatory.

The order of the district court is affirmed.

SECRETARY OF LABOR v.
REPUBLIC STEEL CORPORATION

(OSAHRC Docket No. 7801,
Review Commission Decision,
April, 1977)

Sodium Hydroxide

Summary of Decision

This case involves the necessity of employer providing proper ventilation in the workplace to control potential employee exposure to toxic hazards. The corporation processed steel tubing; during the treatment, the tubing was cleaned in open-surface tanks containing sodium hydroxide, thus exposing the tank operator to five percent of the threshold limit value (TLV) designated by 29 C.F.R. 1910.1000 for that substance. The ventilation provided by the company through the use of a lateral exhaust system, yielded a ventilation rate below the 10,800 cubic feet per minute as required by the regulations. The company was therefore cited for a non-serious violation of 29 C.F.R. 1910.94(d)(4) for providing a velocity rate of ventilation below the required level, and fined $55.00. The corporation appealed, and the administrative law judge agreed with their argument, that the ventilation provision of 1910.94(d)(4) is not an absolute rule, but rather applies only in cases where the TLV levels specified by 1910.1000 are exceeded. The Review Commission decision (Commissioner Clearly dissenting) concurred with the judge's ruling, stating again that if the levels specified by 1910.1000 are not exceeded, than it cannot be proven that a fire or explosion hazard exists; therefore ventilation velocity rates required by 1910.94(d)(4) are inapplicable.

Full Text of Decision

BARNAKO, Chairman:

This case presents the issue of whether Judge Robert P. Weil properly vacated a citation issued under the Occupational Safety and Health Act of 1970 (29 U.S.C. 651 *et. seq.,* hereinafter "the Act") which alleged noncompliance with 29 C.F.R. 1910.94(d)(4). Judge Weil determined, among other things, that the cited standard's ventilation provision applies only where the airborne concentration of a hazardous substance exceeds the threshold limit value (hereinafter "TLV") for that substance desig-

nated by 29 C.F.R. 1910.1000. For the reasons below, we affirm the Judge's vacation of the citation.

The facts surrounding the alleged violation are largely undisputed. Respondent was engaged in the processing of steel tubing during which the tubing was treated in a series of open-surface tanks. The open-surface tank in question contained 2500 gallons of a sodium hydroxide solution in which the tubing was cleaned. The tank operator stood approximately two to three inches from the tank. The parties agreed that the tank operator was exposed to only five percent of the TLV for airborne concentrations of sodium hydroxide as provided in 1910.1000.

The tank was equipped with a lateral exhaust system and the flow of air around the tank was not disturbed by local environmental conditions such as open windows or fans. The tank operation was classified under 1910.94(d)(2) as a C-2 operation. The parties stipulated that, assuming the applicability of 1910.94 where the concentration of airborne sodium hydroxide does not exceed the TLV, the minimum ventilation rate for the sodium hydroxide tank set by Tables G-14 and G-15 of 1910.94 is 10,800 cubic feet per minute. It was agreed that the ventilation rate at the tank was below that level.

The industrial hygienist who testified for the Secretary stated that the concentration of sodium hydroxide in the air, although below the TLV, presented potential hazards to employees in the area. He identified the potential hazards as corrosive action of employees' respiratory systems, dermatitis due to contact or condensation of sodium hydroxide on the skin, and slippery conditions due to condensation of vapors on walking surfaces. He did not testify that the concentration of sodium hydroxide created a fire or explosion hazard, or a hazard from decreased visibility or increased humidity.

For these reasons, Respondent was issued a citation alleging a nonserious violation of 29 C.F.R. 1910.94 (d)(4) in that the control velocity at the cited open-surface tank did not conform to the velocity rate set out in Table G-14. A $55 penalty was proposed.

Judge Weil granted Respondent's motion to dismiss the case at the end of the Secretary's case-in-chief, thereby vacating the citation. He adopted Respondent's argument that the ventilation provision in 1910.94(d)(4) is not an absolute mandate but rather applies only where the TLV levels specified by 1910.1000 are exceeded. In this regard, he read 1910.94(d)(4) as being limited by 1910.94(d)(3), which states that the purpose of the ventilation provision is to control potential exposures to workers as defined in 1910.94(d)(2)(iii). The Judge interpreted 1910.94 (d)(2)(iii) as defining potential exposures to mean the TLV levels set by 1910.1000. He was further persuaded to his interpretation because section 1910.94(d)(4) speaks in terms of the "flow of air past the breathing or working zone of the operator." Accordingly, since the concentration of sodium hydroxide in the working zone of the tank operator was only five percent of the TLV for that substance, the Judge found that additional ventilation was not needed.

On review, the Secretary objects to the Judge's conclusion that TLV levels must be exceeded before ventilation pursuant to 1910.94(d)(4) becomes necessary. He argues that such an interpretation renders 1910.94(d)(4) redundant to 1910.1000 and that, if 1910.94(d) had been intended to apply only where TLV levels were exceeded, this requirement would have been specifically set out in the standard. The Secretary urges that the ventilation provision in 1910.94(d)(4) applies independently of the TLV levels. He contends that the ventilation provision was meant to protect employees against hazards in addition to the breathing of toxic materials. In this regard, he notes that the ventilation rates in Table G-14 are calculated from the classifications of toxic materials as determined by 1910.94(d)(2) and that these classifications take into account not only the TLV, but also the rate of gas, vapor, or mist evolution. The Secretary also points to the source standard for 1910.94(d), ANSI Z9.1-1971, which lists as its purposes the prevention of emissions of a substance in quantities tending to injure health, condense upon workroom surfaces, or accumulate in the air so as to reduce visibility or increase humidity.

Respondent argues in favor of the Judge's interpretation of the standard. In support of this interpretation, it maintains that the source standard, ANSI Z9.1-1971, made the ventilation provision expressly contingent on a showing that the ventilation was insufficient to insure that employees were not exposed to excessive levels of a toxic substance, and stated that the escape of small amounts of a toxic substance was acceptable. It further contends that it is clear, when 1910.94(d)(4) is read in context with other provisions of the same section, that the standard's applicability is conditioned on the existence of concentrations of toxic substances in excess of the TLV levels.

We have carefully considered the Judge's decision and the parties' arguments and have reviewed the cited standard in context with other provisions in the same section. We conclude that the ventilation provision of 1910.94(d)(4) applies only where the Secretary proves that the concentration of a toxic substance in the air exceeds the TLV level or that there is a fire or explosion hazard from the concentration of the toxic substance in the air.

The cited standard is part of section 1910.94(d) which deals with ventilation and other means of reducing to nonhazardous levels concentrations of air contaminants around open-surface tanks. It sets out control velocities which are based upon the classification of the toxic substance involved. However, it is appropriate to read the cited section in conjunction with 1910.94(d)(3), which sets out the circumstances under which ventilation is to be used. Section 1910.94(d)(3) is entitled "Ventilation" and provides that:

> Where ventilation is used to control potential exposures to workers as defined in sub-paragraph (2)(iii) of this paragraph, it shall be adequate to reduce the concentration of the air contaminant to the degree that a hazard to the worker does not exist.

Section 1910.94(d)(2)(iii) defines hazard potential as the "hazard associated with the substance contained in the tank because of the toxic, flammable, or explosive nature of the vapor, gas, or mist produced therefrom." Section 1910.94(d)(2)(iii) provides that the toxic hazard shall be determined from the TLV concentrations in 1910.1000, while 1910.94(d)(2)(iv) states that the flammable and explosive hazard is to be measured in terms of the closed-cup flash point of the substance in the tank. Thus, we conclude that it is evident by reading 1910.94(d)(3) and (4) together that the control velocities set out in 1910.94(d)(4) were meant to be applicable only where there exists a toxic hazard due to airborne levels of a substance in excess of the TLV or a fire or explosion hazard as determined by the flashpoint of the substance. Cf. *Bethlehem Fabricators, Inc.,* 76 OSAHRC 62/C2, BNA 4 OSHC 1289, CCH OSHD para. 20,782 (1976).

We are not persuaded to a contrary interpretation of the ventilation provision by the Secretary's argument that the purpose for the underlying ANSI standard is very broad. The cited standard was originally published as a national consensus standard, ANSI Z9.1-1971, entitled "Practices for Ventilation and Operation of Open-Surface Tanks." The purpose of the entire ANSI standard as set out in section 1.2 is:

> ... to prevent the emission into the workroom atmosphere of gas, vapor, or mist from open-surface tank operations in quantities tending to:
>
> (1) Injure the health
>
> (2) Condense upon floors, ceilings, or walls of any workroom to produce an unsanitary or unsafe condition

(3) Accumulate in the air in a manner which will significantly reduce visibility or objectionably increase humidity.

These rules also are intended to prevent the accumulation of explosive concentrations of gases or vapors in any duct, hood, booth, or enclosure; and to protect the operator from splash or other contact with liquids injurious to his health.

The Secretary contends that this statement of purpose indicates that the ventilation requirement contained in the ANSI standard was designed to prevent not only toxic hazards, but also condensation and slipping hazards. However, the Secretary's argument disregards the fact that sections of the ANSI standard other than the section dealing with ventilation are specifically directed at such hazards. For example, section 9 deals generally with personal protective equipment designed to prevent liquid and mist from irritating an employee's skin, and section 11.1 requires that floors and platforms around tanks be prevented from becoming slippery both by their type of construction and by frequent flushing. The Secretary's argument also ignores the fact that the ANSI standard contains a specific purpose clause for the ventilation requirement. Section 4 of the ANSI standard states that the purpose for the control velocity requirement is "to reduce the concentration of the air contaminant in the vicinity of the worker below the limits set in accordance with 2.1.1." Section 2.1.1, like 1910.94(d)(2)(iii), sets out the limits as being determined by the toxic, flammable, or explosive nature of the substance. Thus, the ANSI ventilation requirement is of the same scope as the ventilation requirement contained in 1910.94(d)(4).

The Secretary also argues that the cited standard would have specified that it applies only when the limits of 1910.1000 have been exceeded had this been the intent of the drafters. In support of this assertion, he points to two other standards in the same section, 1910.94(a)(2)(ii) and 1910.94(b)(2), which expressly provide that ventilation is necessary only where TLV levels are exceeded. However, the two standards cited by the Secretary derive from ANSI standards other than the ANSI standard involved here. Moreover, 1910.94(b)(2) has been substantively amended by the Secretary. Since the standards were drafted by different persons at different times, the language of 1910.94(a)(2)(ii) and 1910.94(b)(2) cannot be used as a guide in interpreting 1910.94(d)(4). See 2A Sutherland Statutory Construction §§51.01-.03 (4th ed. 1973).

The Secretary further argues that the Judge's interpretation of 1910.94(d)(4) renders that section redundant to 1910.1000. Howeveer, as discussed above, the cited standard applies where the Secretary has shown either that the TLV level in 1910.1000 has been exceeded or that there is a fire or explosion hazard. The standard at 1910.1000 does not deal with hazards from fire or explo-

sion. Therefore, our interpretation of 1910.94(d)(4) does not render it redundant.

Applying the provisions of 1910.94(d)(4) to the facts of this case, we conclude that the Secretary has not proved that a hazard within the meaning of 1910.94(d)(3) exists. The TLV level prescribed for sodium hydroxide in 1910.1000 was not exceeded and the Secretary failed to show a fire or explosion hazard. Therefore, additional ventilation in conformance with Tables G-14 and G-15 of 1910.94 was unnecessary.

Accordingly, we vacate the citation which alleged a violation of 1910.94(d)(4) and the penalty proposed therefor. It is so ORDERED.

Dissenting Opinion

CLEARY, Commissioner, dissenting:

I dissent from the disposition ordered by my colleagues. Before deciding this important case involving the exposure of employees to sodium hydroxide I would invite briefs or oral argument on the application of section 1910.94 from qualified amici such as ACGIH, ANSI, and NIOSH (National Institute for Occupational Safety and Health). See 3A C.J.S. §3 at 424, Amicus Curiae. Cf. 1 CFR §305.71.6. Recommendation No. 71-6 of the Administrative Conference of the United States. I would also be receptive to a motion for reconsideration filed by the Secretary indicating that qualified persons or organizations are willing to express their views as amici upon rehearing.

To illustrate the need for briefing of this kind I note that Appendix H to the source standard, ANSI Z9.1—1971, gives sample calculations for minimum ventilation rates that are not triggered by the presence of amounts of airborne contaminants in excess of the threshold limit values. See particularly example H.4, which deals specifically with a sodium hydroxide solution. I would also want to see or hear argument on the Secretary's contention that the ventilation requirement cited here is in part directed against sudden, temporary high levels of airborne contaminants in excess of the TLV. No discussion or refutation of the argument is given in the lead opinion.

If no standard is found applicable, I would not, in view of the testimony that concentrations of sodium hydroxide not in excess of the TLV may present hazards to employees, dispose of this case until consideration is given to remanding for further proceedings to investigate the possible application of section 5(a)(1), *Dunlop v. Uriel G. Ashworth,* 538 F.2d 562, 564 [3 OSHC 2065] (4th Cir. 1976). There is testimony by an expert industrial hygienist pointing out that simply because the TLV was not exceeded does not mean that employees are not subjected to health hazards. The inhalation of even small quantities of sodium hydroxide, a caustic substance, could injure employees, and skin contact could cause skin ulcerations, defattening and emulsifying of skin tissue, burns, holes in the nasal septum and dermatitis.

SECRETARY OF LABOR v.
SPENCER LEATHERS

(OSAHRC Docket No. 13720,
Review Commission ORDER,
April, 1977)

Toxic Gases

Summary of Decision

Employers are required to provide instructions and training of employees for the purpose of preventing improper mixing of chemicals that can yield toxic gases; employers are also required to properly label pipes receiving chemicals. Failure to comply will result in violation of Section 5(a)(1) of the Occupational Safety and Health Act. In this case, the employer was cited for violation of these two provisions following an inspection of their tannery, conducted the day after an incident which proved fatal to two employees and hospitalized twenty-five. Although the hazard of mixing sulfuric acid and sodium sulphydrate, which causes the formation of toxic hydrogen sulfide gas, was known to the employer, the company had provided no instructions to employees as to how the pipes receiving the chemicals were to be identified and isolated to prevent hazardous mixing. In addition, no physical barriers, such as locks, had been supplied to assure that improper mixing would not occur. The administrative law judge held the company at fault, and imposed a $1,000.00 fine; the Review Commission concurred with the judge's decision.

Full Text of Administrative Law Judge's Decision

George W. Otto, Judge

This is a proceeding pursuant to Section 10 of the Occupational Safety and Health Act of 1970 (29 U.S.C. 651 et seq., herein referred to as the Act). On May 2, 1975 the Secretary of Labor (Secretary) made an inspection and investigation of a May 1, 1975 accident resulting in two fatalities and hospitalization of twenty-five employees of Spencer Leathers, Division of Spencer Foods, Inc. (respondent) in a tannery located at 1830 South Third Street, Milwaukee, Wisconsin. On May 23 the Secretary issued a citation charging respondent with serious violation of 29 U.S.C. 654(a)(1). A penalty of $800 was proposed. By letter of June 13 respondent contested the citation; complaint and answer were duly served and hearing was held in Milwaukee, Wisconsin on September 24, 1975.

Respondent denies violation, questions the abatement dates and disputes the proposed penalty.

The tragic event of May 1 was caused by exposure of employees to hydrogen sulfide gas, produced by the mixing of sulfuric acid with sodium sulphydrate.

The sodium sulphydrate intake and distribution system was installed between January 20, 1975 and February 4, 1975 (Transcript pages 153, 154). The sodium sulphydrate intake pipe was located on the inner side of an outside wall, about one foot from a loading dock doorway (T.18, 20, 33, 175, 181, Secretary's Exhibit C-1: photo 1-B, 1-D). There was no label or other identification either on the pipe or adjacent wall (T. 181).

The sulfuric acid intake pipe was not over 18 to 20 feet away from the sodium sulphydrate pipe, on the outer wall of the building and west of the doorway (T. 18, 112, 120). The words "DANGER ACID" were printed on the adjacent wall (T. 18, 34, 119, 181, C-1: 1-C).

Another outside pipe was east of the doorway about 15 to 20 feet, with the word "CHROME" printed on the adjacent wall (T. 18, 32, 181, C-1: 1-A).

Sodium sulphydrate deliveries were made on March 6, 1975 and April 29, 1975 (T. 141, 142, C-5). Sulfuric acid deliveries were made in 1975 on January 8 and 12, February 19, March 5, April 3 and 18, May 1 (T. 134, 135, 138-141, C-4, C-5).

Stanley Maxwell, employee of Schneider Tank Lines, arrived at respondent's tannery the afternoon of May 1, 1975 to deliver a tank truck load of sulfuric acid. He hooked up his hose to the sodium sulphydrate pipe after it had been identified as the sulfuric acid pipe by Charles Jordan, respondent employee, and by an unidentified man on the loading dock. Mr. Jordan identified O. B.

Jamison, respondent employee, as the man to see and sign for the load. Mr. Jamison was in charge because the foreman was absent that afternoon (T. 108-112, 143).

Referring to Mr. Jamison, Stanley Maxwell testified "We were standing right by this pipe where I was hooked, and I asked him if that was the unloading for the sulfuric acid and he said, yes, and I handed him the bill of lading for the load, and he walked out to the fender of the tank which was approximately fifteen foot away from the building, and he signed the bill of lading. And at that time, I questioned him, the capacity of the tank and where the tank was located, and he told me that the tank was on the third floor and that there was plenty of room and there was no problem. So again, I asked him if it was okay to let the product go and he said, yes, and he left. And then I proceeded to build my air pressure up and let the product go. So then approximately fifteen, twenty minutes later, a gentleman by the name of Mr. Hagen came out yelling to shut the product—to shut it off, and I shut the product off at that time." (T. 108-112, 113, 114).

29 U.S.C. 654(a)(1) provides as follows:

Each employer shall furnish to each of his employees employment and a place of employment which are free from recognized hazards that are causing or are likely to cause death or serious physical harm to his employees.

The alleged violation is described in the citation and complaint as follows:

Employer failed to furnish to each of his employees employment free from recognized hazards that are causing or likely to cause death or serious physical harm to his employees in the following separate manners: a) employer failed to provide proper training and supervision of employees in the handling and receiving of dangerous chemicals; e.g., shipping employees allowed mixing of Sulfuric Acid with Sodium Sulphydrate, b) employer failed to properly supervise receipt of Sulfuric Acid; e.g., supervisory employee did not oversee the receipt of Sulfuric Acid so as to prevent its mixing with Sodium Sulphydrate, c) employer failed to provide proper storage facility; e.g., overflow pipe on Sodium Sulphydrate tank vented into the workplace; and d) employer failed to label receiving pipe for dangerous chemicals; e.g., no label on receiving pipe for Sodium Sulphydrate so as to prevent accidental mixing of incompatible chemicals.

In order to establish a violation of section 5(a)(1), the Secretary must prove: (1) that the employer failed to render its workplace 'free' of a hazard which was (2) 'recognized' and (3) 'causing or likely to cause death or serious physical harm.' National Realty and Constr. Co., Inc. v. O.S.H.R.C., 489 F.2d 1257, 1265 (D.C. Cir. 1973).

Both sulfuric acid and sodium sulphydrate are commonly used in the tanning industry (T. 97, C-3). Any hazard related to the handling or use of either chemical, alone and unmixed, is not the hazard in issue herein. The mixing of these two chemicals produced the hazard.

A mixing of sodium sulphydrate and sulfuric acid reacts to form hydrogen sulfide gas and sodium sulfate. Hydrogen sulfide gas is dangerous. Its lethal level has been fixed at about .07 percent, or 700 cubic feet of hydrogen sulfide mixed with a million cubic feet of air. This level makes the atmosphere unsafe for human life (T. 57, 62, 106, 107).

This hazard was recognized by respondent. In the course of the investigation and inspection, Ernest Hagen, plant manager, stated the two chemicals will have to be kept separated, because if they come in contact, hydrogen sulfide would come out as a result (T. 15, 16). His orders to evacuate the plant and to disconnect the sulfuric acid hookup to the sodium sulphydrate line confirm his recognition of the hazard.

This hazard is recognized by the tanning industry. At least half the tanners in America belong to the Tanners Council of America; about a third are members of the leather industry division of the National Safety Council (T. 68, 69, 70). The two Councils jointly prepared the Tanners Safety Manual, published in 1952 (T. 72, 73, 74, C-3). Respondent is not a member of the National Safety Council; the plant manager did not know if it was a member of the Tanners Council of America (T. 183, 184). Respondent is a member of the tanning industry, engaged in tanning leather and related types of activities (complaint para. II, answer para. II). The Manual states, pages 48 and 54, "Potential hazards to life are hydrogen sulfide gas and oxygen deficiency. Hydrogen sulfide is an exceedingly toxic, as well as highly inflammable gas . . ." "Care must be used . . . in using sodium sulfide where it may come in contact with an acid because, on contact with acid, it forms hydrogen sulfide (sometimes called "rotten eggs" gas) which is fatally toxic. Any place where either sodium sulfide or sulphydrate is used should be well ventilated, and acids should be kept away."

Originally the sodium sulphydrate overflow line was to be installed running from the third floor tank to the roof area but was rerouted by respondent and installed to run down the inside of the building (T. 152, 153). Whether ventilation of this line to the outside would have prevented employee exposure to the hazard is not established. It would not have prevented the hazard. The location of the outlet pipe about ten feet from the sewer appears proper (T. 94, 184, C-1: 1-F, 1-G).

At least two respondent employees, including one supervisory, incorrectly identified the sodium sulphydrate pipe as the sulfuric acid line; acting upon such misinformation the truck driver began delivery and thereby the hazard was created.

It was well within the capability of respondent to furnish employment and a place of employment free from this recognized hazard. There were no instructions, written or otherwise, to identify and isolate the proper pipe for the chemical delivered. There could have been specific identification of the line to declare its sole function, intake of sodium sulphydrate. There was none. The words "DANGER ACID" at the sulfuric acid pipe did not serve to identify the sodium sulphydrate line, nor did the word "CHROME" at the third pipe. There was no training of supervisory personnel, no detailed procedure to follow to assure correct delivery. There were no instructions to employees generally, designed to emphasize the hazards inherent in the chemicals and the necessity of precise, accurate delivery. There was no physical barrier to pipe use, such as a lock. In short, there was no enforced and practiced safety program on May 1, 1975 sufficient to preclude the mixing of the two chemicals.

Respondent is not charged with failure to issue and enforce instructions, identify pipes, provide locks or to train employees. The section of the Act charged does not so specify. No occupational safety and health standard has been cited by the Secretary. The employer is not an absolute insurer. He may not be responsible for an isolated occurrence beyond his contemplation or control. However, the method of receiving chemical deliveries was, at the best, casual. Mr. Jamison's name and/or signature appears on each bill of lading covering the two deliveries of sodium sulphydrate and seven deliveries of sulfuric acid between January 8, 1975 and May 1, 1975 (C-4, C-5). It was not his practice to check the loads. He did not make a practice of asking what line deliveries were hooked up to; at times he signed bills of lading while in the plant away from the receiving dock area (T. 137, 138). An employer is entitled to assume the item ordered is the item delivered. However, bills of lading were signed without first checking the delivery point to assure the identification and use of the proper intake pipe. Mr. Jamison represented the respondent in authorizing the use of the sodium sulphydrate pipe for delivery of sulfuric acid.

Respondent is charged with the responsibility of knowing the properties of chemicals used in his production process, at least to an extent necessary to protect his employees from hazards associated with proper or improper use. Any of several safety methods were available. The hazard resulting from the improper delivery in this case could have been eliminated by precautionary measures assuring proper delivery.

The violation was serious. Section 29 U.S.C. 666(j) provides that a serious violation shall be deemed to exist in a place of employment if there is a substantial probability that death or serious physical harm could result from a condition which exists, or from one or more practices, means, methods, operations, or processes which

have been adopted or are in use, in such place of employment unless the employer did not, and could not with the exercise of reasonable diligence, know of the presence of the violation. Respondent had the requisite knowledge. The violation caused death to his employees.

The abatement dates as set forth in the citation and complaint were proper and reasonable. Respondent was allowed four weeks to provide proper training and supervision of employees in the handling and receiving of the chemicals. Such period was sufficient to permit training of all involved employees. Also the violation was subject to correction by proper supervision of receipt of chemicals, by labeling receiving pipes and with associated procedures and safeguards designed to eliminate the improper mixing of chemicals; these abative steps could have been invoked immediately.

Section 29 U.S.C. 666(b) provides that a penalty shall be assessed of up to $1,000 for a serious violation. The Secretary proposes a penalty of $800, based upon a 20% credit for no history of previous violations. Section 29 U.S.C. 666(i) provides that the Commission shall have authority to assess all civil penalties provided in this section, giving due consideration to the appropriateness of the penalty with respect to the size of the business of the employer being charged, the gravity of the violation, the good faith of the employer, and the history of previous violations. Each of the four statutory factors is evaluated separately.

In Secretary v. Nacirema Operating Company, Inc., 1 OSAHRC 33, 37, the Commission stated "We believe that the four criteria to be considered in assessing penalties cannot always be given equal weight. Obviously, for example, a particular violation may be so grave as to warrant the assessment of the maximum penalty, even though employer may rate perfect marks on the other three criteria." It does not necessarily follow that every violation established under the Act resulting in a fatality shall be considered of sufficient gravity to require the assessment of the maximum penalty. Nor does it follow that in every case where there has been no prior investigation or inspection or where there has been no history of previous violation that the maximum penalty is thereby automatically reduced to $800 instead of $1,000. Considering the inevitable result of death or serious physical harm to the employee exposed to the hazard created by this violation, the factor of gravity herein requires assessment of the maximum penalty of $1,000 for this serious violation.

Findings of Fact

1. On May 1, 1975 respondent was engaged in a business affecting commerce while operating a tannery in Milwaukee, Wisconsin.

2. On said date an order of sulfuric acid was delivered into a sodium sulphydrate line, with mixture of the two chemicals producing hydrogen sulfide gas.

3. Respondent employees were exposed to the hydrogen sulfide gas with resulting fatality.

4. The hydrogen sulfide created a hazard recognized by the respondent and by respondent's industry.

5. Respondent failed to provide controls sufficient to assure the delivery of sulfuric acid through the sulfuric acid intake pipe.

6. Respondent failed to furnish to each of his employees employment and place of employment free from said recognized hazard.

7. The violation was serious.

8. The abatement dates set forth in the citation and complaint herein are reasonable.

Conclusions of Law

1. The parties are subject to the Act and the Review Commission has jurisdiction.

2. Respondent failed in its duty under Section 29 U.S.C. 654(a)(1) to furnish to each of his employees employment and a place of employment free from a recognized hazard causing or likely to cause death or serious physical harm.

Now therefore it is ORDERED that the citation issued May 23, 1975 be and the same is hereby affirmed. Penalty is assessed in the sum of $1,000.

APPENDIX

APPENDIX A

CURRENT FEDERAL RESEARCH ON TEN MAJOR TOXIC SUBSTANCES

(Identification of Selected Federal Activities
Directed toward Chemicals of Near-Term
Concern, Office of Toxic Substances, U.S.
Environmental Protection Agency July, 1976)

Abbreviations of Organizations

Environmental Protection Agency

OAQPS	—Office of Air Quality Planning and Standards
OE	—Office of Enforcement
OIA	—Office of International Activities
OPP	—Office of Pesticide Programs
ORD	—Office of Research and Development
OSWMP	—Office of Solid Waste Management Program
OTS	—Office of Toxic Substances
OWPS	—Office of Water Planning and Standards
OWS	—Office of Water Supply
Region II	—Regional Office, New York, New York
Region V	—Regional Office, Chicago, Illinois
Region VI	—Regional Office, Dallas, Texas
Region X	—Regional Office, Seattle, Washington

Other Organizations

CDC	—Center for Disease Control
DOA	—Department of Agriculture
DOT	—Department of Transportation
FDA	—Food and Drug Administration
MESA	—Mining Enforcement and Safety Administration
NCI	—National Cancer Institute
NIEHS	—National Institute for Environmental Health Sciences
NIOSH	—National Institute for Occupational Safety and Health
ORNL	—Oak Ridge National Laboratory
OSHA	—Occupational Safety and Health Administration

Asbestos

General Studies

Review of Environmental Effects—A draft report of the effects of asbestos on the environment has been prepared and reviewed by the EPA Science Advisory Board. The report will be completed by the end of 1976. Dr. Gerald Stara, ORD, (513) 684-7407.

Health and Ecological Effects and Environmental Behavior

Carcinogenic Potential via Ingestion—Two studies of the oral carcinogenic potential of asbestiform fibers of various types and configurations will begin in September 1976. Each study will use one animal type (rats or hamsters) for the anticipated three-year study period. Tremolite will be the first fiber type administered to test animals. Robert Tardiff, ORD, (513) 684-7213 (EPA participant with NIEHS).

Cellular Transformation—A three-year study to determine if cells in the large intestine of hamsters are transformed as a result of asbestiform fiber ingestion began in the fall of 1975. Robert Tardiff, ORD, (513) 684-7213.

Mutagenesis and Co-Carcinogenesis—A variety of fiber types will be tested *in vitro* to document mutagenesis and co-carcinogenesis potential. Results are expected by 1979. Robert Tardiff, ORD, (513) 684-7213.

Health Effects of Mine Samples—Asbestos samples from a Minnesota mine will be administered to rats by various exposure routes. Carcinogenicity will be documented, as will effects on organs, especially the lungs and pleura. The three to four-year study has just begun. David Coffin, ORD, (919) 549-8411.

Impact of A/C Pipe—A study of several communities in Connecticut may help determine if the use of asbestos/cement pipe in transporting drinking water has any impact on increased cancer mortality. This study was prompted by the identification of 24 cases of peritoneal mesothelioma. Results are expected by the end of the year. Dr. Gunther Graun, ORD, (513) 684-7217.

Regional Epidemiological Survey—The San Francisco Bay area will be surveyed to determine if patterns of cancer occurrence can be correlated with asbestos in drinking water sources. This two-year project should be reported in early 1978. Lee McCabe, ORD, (513) 684-7211.

Talc Refining—A detailed study has been undertaken in upstate New York to document environmental levels of asbestiform fibers near a talc facility and mortality and morbidity data on workers. John Dement, NIOSH, (513) 684-3191 and H.P. Richardson, MESA, (202) 235-8132.

Study of Workers in Underground Mines—An epidemiology study of workers in underground mines is being conducted. Environmental, mortality, and morbidity data will be collected and correlations sought with mined substances, including asbestos. Data should be reported in mid-1977. John Dement, NIOSH, (513) 684-3191 and H.P. Richardson, MESA, (202) 235-8132.

Impact on Aquatic Species—Bioassay experiments to determine if the asbestos fiber content of Lake Superior water may affect the results of toxicity tests will be completed in 1976. Philip Cook, ORD, (218) 727-6692.

Current and Projected Sources, Environmental Levels, and Exposed Populations

Production and Use Trends—The 1976 report on asbestos mining and use is expected in September. R.A. Clifton, Bureau of Mines, (202) 634-1206.

Materials Balance—A study of the commercial movement of asbestos from mining to disposal should be completed in mid-1977. Robert Carton, OTS, (202) 755-0300.

Leaching from A/C Pipe—The potential for asbestos to leach from asbestos/cement pipe in a variety of circumstances is being studied. The results of this series of six-month tests will be reported in late 1977. Earl McFarren, ORD, (513) 684-7236.

Runoff from Roads—Six sites are being selected for the identification and enumeration of asbestos fibers by type in road dust. Data are expected to be reported in early 1978. Byron Lord, DOT, (202) 426-4980.

National Monitoring Program—Water supplies of selected U.S. cities, effluents from selected mining and manufacturing sites, runoff from selected natural sites, and emissions from selected taconite milling plants will be sampled and analyzed for asbestos fibers. Phase I (water) of the study will be completed by mid-1976. Phase II (air) will be completed by December 1976. Robert Carton, OTS, (202) 755-0300.

Levels in Lake Superior—Levels of asbestos in Lake Superior, rivers emptying into Lake Superior, and water intakes from Lake Superior are being established. Samples were collected in 1974, and a report is due later this year. William Fairless, Region V, (312) 353-8370.

Fugitive Dust Study—A pilot study is being conducted to analyze fugitive dust from asbestos/cement waste piles and milling waste piles. The first phase of this study has been completed, and an expanded study on asbestos/cement wastes is slated for completion later this year. David Oestreich, ORD, (919) 549-8411.

Interim Method for Analysis of Water Samples—An interim method for measurement of asbestos fibers in water is being prepared for promulgation in late 1976. Charles Anderson, ORD, (404) 546-3525.

Rapid Method for Analysis of Fibers in Water—A two-year contract was initiated in 1975 to develop a rapid method for determining asbestos fiber levels in water. The report is due in 1977. Charles Anderson, ORD, (404) 546-3525.

Sample Storage Conditions and Sample Preparation—A study of the changes in asbestos samples resulting from storage conditions and sample preparation is underway. Charles Anderson, ORD, (404) 546-3525.

Evaluation of Electron Microscope Methods for Analysis—An evaluation of electron microscope methods for measurement of airborne asbestos concentrations and the development of an optimal measurement procedure are under study. This activity will be completed in 1977. Jack Wagman, ORD, (919) 549-8411.

Improved Method of Measurement in Air—A method is being refined for the improvement in accuracy of measurements of asbestos fibers in air. Philip Cook, ORD, (218) 727-6692.

Improved Method of Sampling in Stack Gases—the development of an improved sampling method for asbestos fibers in stack gases and a refined analytical method are being developed. This effort should be completed in 1976. Louis Paley, OE, (202) 755-8137.

Substitutes, Control Technology, and Related Costs and Economic Factors

Water Supply Treatment Demonstration Facility—A demonstration facility for full-scale removal of asbestiform fibers is due to be completed in Duluth in early 1977. It will be operated as an experiment for three years. Earl McFarren, ORD, (513) 684-7236.

Pilot Plant Studies—A pilot plant (15-20 gpm) for removal of chrysotile asbestos from water is scheduled to be completed later this year. Testing at this facility will be based on the experiences gained in Duluth. Gary Logsdon, ORD, (513) 684-7228.

Control of Discharges from Solid Wastes—Studies to develop techniques for control of emissions from asbestos-containing waste piles at asbestos/cement pipe manufacturing sites are due to be reported by the end of the year. Mary Stinson, ORD, (201) 548-3414.

Tracer Fibers—A pilot study of incorporating radioactive tritium into asbestiform fibers has been successfully completed. Tritiated chrysotile fibers will be available for research activities by the end of the year. Richard Bull, ORD, (513) 684-7217.

Economic Impact of Controls—The potential economic impact of several possible approaches for controlling asbestos will be studied, beginning later this year. Robert Carton, OTS, (202) 755-0300.

Control Options, Regulatory Actions, and Attendant Impacts

Drinking Water Standard—Asbestos is one of the contaminants being considered in a study by the National Academy of Sciences on the health effects of contaminants in drinking water as a requirement of the Safe Drinking Water Act. The report is due December 15, 1976. Edgar Jeffrey, WSD, (202) 426-8877.

Hazardous Air Pollutant Standard—Iron ore beneficiation plants are being studied to determine the feasibility and desirability of extending coverage of current Hazardous Air Pollutant Standards to this possible source of asbestos. Gilbert Wood, OAQPS, (919) 688-8146 X-295.

Workplace Standard—A downward revision of the workplace exposure limit has been proposed. After economic impact studies are completed and public hearings have been held, the revised standard may be promulgated. William Tarren, OSHA, (202) 523-7177.

Workplace Studies—The brake lining and clutch rebuilding industries are being studied to determine the best means for protecting workers. This classification of workers is not presently covered by workplace standards, and recommendations may be sent to OSHA before the end of the year. John Dement NIOSH, (513) 684-3191.

Mine Safety Standard—The mine safety standards for metal and nonmetal industries, including asbestos, will probably be revised later in 1976. H.P. Richardson, MESA, (202) 235-8307.

Arsenic

General Studies

Overview of Environmental Considerations—A report has been prepared describing four major areas: industrial sources of arsenic emissions, commercial flow of arsenic trioxide and its derivatives, hazards presented to man and the environment, and preliminary assessments of possible controls. Robert Carton, OTS, (202) 755-0300.

Review of Literature on Environmental Hazards—The National Academy of Sciences is preparing a critical review of the existing literature on the environmental hazards of arsenic. The report is expected in August 1976. Dr. Orin Stopinski, ORD, (919) 549-8611, X-266.

Environmental Hazard Assessment—A Scientific and Technical Assessment Review of arsenic is being developed. Dr. Orin Stopinski, ORD, (919) 549-8611, X-266.

Review of Arsenical Pesticides—A broad, multidisciplinary in-house review report on arsenical pesticides will be published later in 1976. Dr. Robert Potrepka, ORD (202) 557-7480.

Review of Arsenical Pesticides—A second internal review of arsenical pesticides will also be released for publication in 1976. Dr. Homer Fairchild, OPP, (202) 557-7725.

Review of Environmental Effects—A literature search is documenting available information on chemical and physical properties, health effects on humans and other organisms, analytical methods, and media distribution of arsenic and its compounds. A draft report has been prepared, and will soon be published. Dr. Gerald Stara, ORD, 513 684-7407.

Health and Ecological Effects and Environmental Behavior

Acute Oral, Dermal, and Intratracheal Toxicity—Studies documenting acute toxicity by several routes of exposure are scheduled to be reported in June 1977. Cacodylic acid, monosodium metharsonate (MSMA), and disodium metharsonate (DSMA) are being used. Weanling and adult rodents will be tested, and the results will be reported for any impact of age or sex on toxicity. Lawrence Hall, ORD, (919) 549-8411.

Subacute Toxicity—The subacute toxicity of cacodylic acid, MSMA, and DSMA will be determined, using 90-day LD_{50} and chronicity factors. Rodents will first be fed pesticide formulations and then pure compounds. The cacodylic acid report is expected by January 1977, and the MSMA/DSMA reports in January 1978. Lawrence Hall, ORD (919) 549-8411, X-606.

Absorption, Excretion, Distribution, and Metabolism—The absorption, excretion, distribution, and metabolism of cacodylic acid, MSMA, and DSMA will be studied. These results should be available in 1978. Lawrence Hall, ORD, (919) 549-8411, X-606.

Fetotoxicity—An assessment and quantification of fetotoxic effects of cacodylic acid in rats and mice is being made. The report is due in September. Neil Chernoff, ORD, (919) 549-8411, X-327.

Acute Inhalation Toxicity—Inhalation exposures of rats and mice to cacodylic acid, MSMA, and DSMA have been completed, and the acute broncho-pulmonary effects are being evaluated. A full report is expected shortly. James Stevens, ORD, (919) 549-8411, X-233.

Subacute and Chronic Inhalation Toxicity—Evaluations of the subacute and chronic broncho-pulmonary effects of exposure to fractions of acute LC_{50}'s of cacodylic acid, MSMA, and DSMA are scheduled in the future. James Stevens, ORD, (919) 549-8411, X-233.

Arsenic in Drinking Water—The drinking water of a community with a high rate of skin cancer is being analyzed to determine the presence of chemical species of arsenic. A control drinking water supply is also being analyzed. The species of arsenic identified will then be used in animal studies directed to carcinogenic potential. The final report is scheduled for January 1979. Robert Tardiff, ORD, (513) 684-7213.

Epidemiological Studies in Baltimore and Tacoma—Epidemiological studies of individuals living near industrial sources of arsenic in Baltimore, Maryland, and Tacoma, Washington, are being conducted. These are expected to be reported by the end of the year. Robert Carton, OTS, (202) 755-0300.

Study of Children Near Smelters—An epidemiological study of children around twenty-two smelters which discharge arsenic was completed in June 1976. Blood, urine, hair, and dust samples were collected and analyzed. The report is in draft, and will soon be finalized. Carl Hays, ORD, (919) 549-8411, X-674. (In cooperation with CDC)

In-Depth Epidemiological Studies Near Six Smelters—Six smelters will be selected for in-depth epidemiological and monitoring studies. Routes of exposure will be determined for individuals in all age groups. The activity will include 30 to 45 days of environmental monitoring. Warren Gelke, ORD (919) 549-8411, X-861.

Bioaccumulation Potential—A series of tests is being conducted to determine the bioaccumulation potential of four arsenical compounds (trioxide, pentoxide, sodium methyl arsenate, and dimethyl arsenate) in four aquatic species (snails, Gammarus, rainbow trout, and Daphnia). The final report is due in June 1977. Robert Spehar, ORD, (218) 727-6692.

Transport and Fate—The behavior of arsenic in a terrestrial (grassland and forest floor) and an aquatic (littoral zone of a lake) ecosystem will be documented in a three-year study scheduled for completion in June 1978. Site-specific protocols are due to be developed by May 1977. Robert Van Hook, ORNL, (615) 483-8611, X-36488.

Transport and Fate in River Systems—Studies on the fate and transport of arsenic in river systems will be initiated in October 1976. George Baughman, ORD, (404) 546-3145.

Impact on Agriculture—Ongoing studies address (1) arsenite, HSMA, and cacodylic acid in farm ponds, (2) MSMA and cacodylic acid in vegetation control of forests, and (3) MSMA in soils and crops. Edward Woolsen, USDA, (301) 344-3076.

Ecological Impact of Smelter Emissions—The environmental and ecological impacts of arsenic emissions from smelters are being evaluated. At the same time, the impacts of heavy metal emissions will be analyzed. The report is due in January 1977. Robert Carton, OTS, (202) 755-0300.

Current and Projected Sources, Environmental Levels, and Exposed Populations

Meteorological Modelling—A meteorological modelling study is underway to help determine the ambient air concentration of arsenic released from the Tacoma smelter. The ambient air concentration will be determined from stack emissions and fugitive low level emissions. Kenneth Lepic, Region X, (206) 442-1125.

Measurement Capability—A state-of-the-art assessment of measurement capability for arsenic and its compounds will address techniques for measuring ambient levels, effluents, sediments, and biota. A report is expected later this year. Charles Plost, ORD, (202) 426-2026.

Substitutes, Control Technology, and Related Costs and Economic Factors

Smelter Control Technology and Costs—A review of technology available to give a higher degree of control of arsenic from smelters is underway. Also, the economic impact of implementing this technology will be studied. A report on the control technology is expected by September 1976. A report on the economic impact is due by December 1976. Kenneth Kepic, Region X, (202) 422-1125.

Japanese Emission Control Technology—A study to evaluate emission sources, emissions, and control technology of Japanese smelters with regard to sulfur oxides, particulates, and trace elements will start soon. A final report is expected by January 1977. Conrad Kleveno, OIA, (202) 755-0533.

Electrostatic Precipitator Applications—A design manual will be developed for ESP applications in non-ferrous industries. It will include evaluations of ESP performance in at least three copper smelters, two zinc smelters, and one lead smelter. Sampling and analysis for trace metal constituents will be conducted at the same time. Margaret Stasikowski, ORD, (513) 684-4491.

Leaching and Fixation Techniques—A grant has been awarded to Montana Tech Foundation Mineral Research Center to develop a number of leaching and fixation techniques for arsenic-bearing solid wastes from smelters. Margaret Stasikowski, ORD, (513) 684-4491.

Disposal of Arsenic-Bearing Wastes—A laboratory study will determine the effectiveness of fixation processes on arsenic-bearing wastes. Donald Sanning, ORD, (513) 684-7871.

Waste Disposal Technology— A review of waste disposal technology has been conducted. Guidelines may be issued in the future. Fred Lindsey, OSWMP, (202) 755-9206.

Control Options, Regulatory Actions, and Attendant Impacts

Review of Arsenical Pesticides— Arsenic is a candidate for rebuttable presumption proceeding under Section 3 of FIFRA. A determination under this proceeding will be made by May 1977. Ronald Dreer, OPP, (202) 755-5687.

Interim Drinking Water Standards— A maximum permissible concentration of 0.05 mg/1 for arsenic in drinking water has been promulgated. This concentration is currently being reviewed in connection with the development of additional standards in 1977. Joseph Cotruvo, OWS, (202) 755-5643.

Water Quality Criteria— A concentration of 50 μg/1 has been proposed for total arsenic as a water quality criterion. David Critchfield, OWPS, (202) 245-3042.

Effluent Guidelines— The revision of best available technology limitations will include considerations of arsenic. A broad examination is being directed to the best approach for controlling arsenic. Guidelines for some industrial categories can be expected within the next three years. Ernst Hall, OWPS, (202) 426-2576.

Hazardous Material Spills— Arsenic is included in the preliminary listing of hazardous chemicals under Section 311 of FWPCA. Mandatory reporting of any spill and clean-up and civil penalties are contemplated. Promulgation of the final regulation is being considered for late 1976. Allen Jennings, OWPS, (202) 245-0607.

Air Pollution Assessment— An assessment of arsenic as an an air contaminant will include a summary of the analysis of the National Air Sampling Network samples and other air samples around nine smelters. A final report is due in mid-1976. Josephine Cooper, OAQPS, (919) 688-8146, X-501.

New Source Performance Standards— Arsenic data are being collected from process sources at primary copper, zinc, and lead smelters. Whether standards are set under Section 111 of the Clean Air Act is contingent on these data and the air pollution assessment. The overall study will take two years. Allen Vervaert, OAOPS, (919) 549-8411, X-301.

Workplace Standards— Revised arsenic workplace standards were proposed in January 1975. The final review of the inflationary impact statement is being com-

pleted. After this review and hearings, the final standard may be promulgated. Gerald Weinstein, OSHA, (202) 523-7186.

Benzidine

Health and Ecological Effects and Environmental Behavior

Carcinogenic Potential at Low Dose Levels— Animal feeding studies, together with metabolism studies, are underway at the National Center for Toxicological Research. These studies are intended to demonstrate improved testing approaches to characterizing carcinogens at low dose levels. William Marcus, OTS, (202) 755-0300.

Current and Projected Sources, Environmental Levels, and Exposed Populations

Monitoring Method Development— Appropriate analytical methods for benzidine will be selected by August 1976. Three EPA laboratories are considering available methods for reliability, detection limits, and feasibility in all media. Initial evaluations were completed in June, and the recommendations will soon follow. John Moran, ORD, (202) 426-2026.

Field Monitoring— Monitoring activities will be considered when an appropriate method is available. Vincent DeCarlo, OTS, (202) 755-6956.

Control Options, Regulatory Actions, and Attendant Impacts

Toxic Effluent Standard— A toxic effluent standard under Section 307(a) of FWPCA was proposed in June 1976, and hearings have begun. Kent Ballentine, OWPS, (202) 245-3030.

Effluent Guidelines— Industrial categories for which effluent guidelines were established under Section 304 of FWPCA will be reviewed to determine the potential for benzidine release in their effluents. Walter Hunt, OWPS, (202) 426-2724.

Hazardous Spills— Benzidine is being studied for possible inclusion under the hazardous spill provisions of the Federal Water Pollution Control Act (Sec. 311). Michael Flaherty, OWPS, (202) 245-3047.

Ethylene Dibromide

Health and Ecological Effects and Environmental Behavior

Health Effects on Workers—Background information is being gathered on residue levels, epidemiology, carcinogenicity, and inhalation toxicity for a report to be completed in 1976. Dr. Roscoe Moore, NIOSH, (301) 443-3843.

Teratogenicity—A laboratory experiment addressed the teratogenic potential to mice via inhalation. Positive results were obtained at 32 ppm. The final report has just become available. William Coniglio, OTS, (202) 755-0300.

Carcinogenicity—Animal inhalation studies began in April 1976. Dr. Cipriano Cueto, NCI, (202) 496-4875.

Monograph on Carcinogenic Risk—The International Agency for Research on Cancer is preparing a monograph on carcinogenic potential with publication estimated for February 1978. Dr. H. Kraybill, NCI, (301) 496-1625.

Current and Projected Sources, Environmental Levels, and Exposed Populations

Vegetation Exposure—Land use patterns and non-point source emissions are being studied in an assessment of the impact of production facilities in Magnolia and El Dorado, Arkansas. A report will be available in mid-1976. In addition, infrared color aerial photos are being taken over El Dorado to obtain indications of vegetation stress that may be resulting from fallout attributable to the plants. The results should be available in early July. Allen Waters, ORD, (703) 347-6224.

Monitoring Near Production Plants—This study will determine ambient air concentrations (if any) in the vicinity of two production plants near Magnolia and El Dorado, Arkansas. Frank Hall, Region VI, (214) 749-3971.

Environmental Levels Associated with Pesticide Usage—This program is developing data on environmental levels near sites where EDB is used as a pesticide. G. Rohwer, USDA, (202) 436-8261 and William Coniglio, OTS, (202) 755-0300.

Environmental Levels Resulting from Use in Gasoline—Environmental levels in air and water are being determined at urban and rural sites to help delineate the zone of impact from suspected sources, e.g., gas stations, vehicular traffic, and storage facilities. Sampling sites are located in Arizona, California, Kansas, New Jersey, and Oklahoma. William Coniglio, OTS, (202) 755-0300.

Control Options, Regulatory Actions, and Attendant Impacts

Review of Registered Pesticides—Registered pesticide uses are being reviewed in connection with possible limitations under the rebuttable presumption procedure. A decision under this procedure is expected in late 1976. H. Hall, OPP, (202) 755-8053.

Air Pollution Assessment—A review of the general literature and modelling of anticipated ambient air levels have been completed. This report will be used as a basis for developing recommendations concerning possible regulatory action. Richard Johnson, OAQPS, (919) 688-8146, X-501.

Criteria Document for Workplace Exposure—A criteria document is scheduled for initiation in early 1977 with a nine-month completion time. Dr. Robert Mason, NIOSH, (513) 684-8209.

Hexachlorobenzene

General Studies

Toxicity and Environmental Exposure—An overview of the hazards, types and levels of exposure, and potential sources is being developed. William Coniglio, OTS, (202) 755-0300.

Health and Ecological Effects and Environmental Behavior

Reproduction Study in Rats—A three-generation rat reproduction study to determine the impact of low levels of HCB exposure has been completed. The report will be available in a few weeks. Dr. August Curley, ORD, (919) 549-8411, X-655.

Chronic Effects—A two-year feeding study to establish the types of chronic effects has been completed. Histopathological findings should be available by December 1976. Dr. Harold Smalley, USDA, (713) 846-1371.

Chronic Toxicity in Dogs—A twelve-month dog feeding study will be completed in September 1976. Analyses of blood sera for immunological changes are in process. William Coniglio, OTS, (202) 755-0300.

Carcinogenicity Testing—Two carcinogenicity studies have been undertaken, including determination of the carcinogenicity in animals receiving a vitamin-deficient diet. Dr. S. Charbonneau, Canadian Ministry of Health, (613) 996-3117.

Toxicity Studies on Swine, Dogs, and Poultry— The oral toxicity using multiple exposure levels is being studied to provide information on the impact of residue levels in feed. Gross effects have been noted, as well as the details of histopathology investigations and blood examinations. Dr. Richard Teske, FDA, (202) 344-2556.

Effects on Aquatic Organisms—Acute and chronic toxicity studies of the toxicity of HCB to crayfish and fish as well as accumulation and depuration rates in aquatic organisms have been conducted. The final report has just become available. William Coniglio, OTS, (202) 755-0300.

Accumulation and Excretion in Cattle—The rate of accumulation and excretion in beef cattle is being studied at Louisiana State University. Histopathological evaluations of tissues collected during feeding trials are being completed. Dr. Edwin Goode, USDA, (202) 344-2714.

Toxicity in Fish—Thirty-day bioconcentration tests, as well as toxicity bioassays, have been carried out in selected species of fish. The results should be available in October 1976. Gilman Veith, ORD, (281) 727-6692.

Transport and Fate—The behavior in a terrestrial (grassland and forest floor) and an aquatic (littoral zone of a lake) ecosystem will be documented in a three-year study due to be completed in July 1978. Site-specific protocols are scheduled for development by May 1977. Robert Van Hook, ORNL, (615) 483-8611, X-36488.

Current and Projected Sources, Environmental Levels, and Exposed Populations

Environmental Concentrations Associated with Production Activities—The level of environmental contamination immediately adjacent to production facilities has been determined. Levels in air, water, and soil will be presented in a report that should be completed by August 1976. William Coniglio, OTS, (202) 755-0300.

Levels in Sediments and Fish—Routine monitoring downstream of industrial facilities has identified levels of up to 1.9 ppm in fish and up to 30 ppm in sediments in Michigan. John Hesse, State of Michigan, (517) 343-0927.

Levels in Drinking Water—The National Organics Reconnaissance Survey has found HCB in the ppb range in five of the supplies being monitored. Joseph Cotruvo, OWS, (202) 755-5643.

Levels in Human Tissue—Seventeen hundred human adipose tissue samples are being analyzed for HCB as well as other contaminants. Several new methods (Derivitization Technique, Coulson Technique) are being used to confirm ppb levels measured by an electron capture mode. Dr. Frederick Kutz, OPP, (202) 755-8060.

Levels in Mother's Milk—HCB has been found in more than 70 percent of the mother's milk recently sampled in the nation-wide survey. The mean level was 87 ppb and the maximum 260 ppb. A formal report has not been prepared. Dr. Jack Griffith, OOP, (202) 755-2778.

Occurrence in the Domestic Meat Supply— Measurements are being performed on the domestic and imported meat supply. Periodic reports are issued. Dr. John Spalding, USDA, (202) 447-2807.

Food Contamination—Periodic reports are issued on measurements in routinely collected food samples. Paul Corneliussen, FDA, (202) 245-1152.

Control Options, Regulatory Actions, and Attendant Impacts

Water Quality Criteria—HCB is one of the chlorinated benzene compounds which will be studied to develop a basis for water quality criteria. The report and recommendations are expected in early 1978. John Carroll, OWPS, (202) 245-3042.

Development of Solid Waste Disposal Guidelines—Guidelines for the disposal of HCB are being developed. Studies of volatilization and leaching characteristics are providing part of the basis for the recommendations. John Lehman, OSWMP, (202) 755-9185.

Hexachlorobutadiene

General Studies

General Literature Review—An overview of low molecular weight haloalkenes, to be completed in November 1976, will include consideration of chemical and physical properties, methods of analysis, sources and background levels, and control technology. Emily Copenhaver, ORNL, (615) 483-8611, X-36823.

Health and Ecological Effects and Environmental Behavior

Review of Russian Publications—A literature review has resulted in a compilation of mostly Russian-language articles. Translations have been obtained, and analysis of the information is in process. William Coniglio, OTS, (202) 755-0300.

Environmental Effects—Aquatic ecosystems studies have recently been completed. Mortality, histopathology, uptake, and bioaccumulation were investigated. William Coniglio, OTS, (202) 755-0300.

Current and Projected Sources, Environmental Levels, and Exposed Populations

Industrial Release Potential—Several synthetic organic chemical industries were evaluated for potential release of chemicals into the environment. The report indicates formation of HCBD as a waste from several industries, including perchloroethylene, trichloroethylene, and carbon tetrachloride manufacturing. William Coniglio, OTS, (202) 755-0300.

Monitoring Near Industrial Sites—Monitoring of water, air, sediment, and biota near industrial areas in Michigan has recently been completed. HCBD was detected in several media. John Hesse, Michigan Department of Water Resources, (517) 373-0927.

Levels in Food—Food samples monitored since 1974, indicate the presence of HCBD in a few fish samples taken in the Mississippi River delta. Paul Corneliussen, FDA, (202) 245-1152.

Polybrominated Biphenyls

Health and Ecological Effects and Environmental Behavior

Health Effects Summary—A summary of current health effects data, including populations exposed and levels of exposure, has just been released. This report is designed to assist regulatory agencies. Dr. Albert Kolby, FDA, (202) 245-1301.

Teratology and Fertility Concerns—The effects on litter size and off-spring of rats are being studied. A report should be available in late 1976. Dr. Elaine Cecil, USDA, (301) 344-2099.

Rat Feeding Study—A rat feeding study, scheduled for completion in August 1976, is examining residue concentrations in various organs and characterizing pathological findings. Dr. Elaine Cecil, USDA, (301) 344-2099.

Single-Dose Investigations in Rats—Adverse effects of single oral doses on rats will be investigated. A report is due in early 1977. Dr. Renate Kimbrough, CDC, (404) 633-3311, X-5235.

Epidemiological Studies—A total of 4,000 persons will be monitored over the next several years to document the effects of human ingestion of these chemicals. Phil Landrigan, CDC, (404) 633-3311, X-3166.

Cattle Studies—A two-year study to document toxicity, distribution, and excretion in cattle was initiated in mid-1975. Dr. Richard Teske, FDA, (301) 344-2556.

Chicken Reproduction—Two studies document the effects of ingestion. The first study focuses on the impacts on feeding, egg production, chicken embryos, chick growth, and chick viability, as well as bioaccumulation in eggs. The second entails pathological examination of the organs of young chickens and hematological tests. Dr. Robert Ringer, Michigan State University, (517) 353-8414.

Egg Production—The effects of exposure on laying hens, including residues in eggs, residue depletion, and hatch-ability, should be reported within the month. Dr. Elaine Cecil, USDA, (301) 344-3099.

Behavior in Soils—Detailed analyses of leaching, absorption, microbial degradation, and plant uptake in soils will be completed by mid-1977. Dr. Lee Jacobs, Michigan State University, (517) 353-7273.

Degradation Products—The occurrence of furans as a contaminant resulting from PBB degradation is being investigated. A report is due in late 1976. Dr. George Fries, USDA, (301) 344-3076.

Current and Projected Sources, Environmental Levels, and Exposed Populations

Estimation of Exposed Population—Ten thousand individuals have been identified in demographic studies as having been impacted by the Michigan feed contamination incident. Dr. Harold Humphrey, Michigan Department of Public Health, (517) 373-2037.

Exposure Through Milk Consumption—The distribution of milk products from four affected herds was studied to determine the extent of contamination. The results are in preparation. Dr. Mary Zabik, Michigan State University, (517) 353-5251.

Environmental Levels—The Pine River area is being monitored to determine trends in contamination levels. John Hesse, State of Michigan, (517) 343-0927.

Monitoring Methods—Assistance is being provided on methods for identifying sources, techniques for eliminating them, and protocols for analyzing samples. Karl Bremmer, Region V, (312) 353-1459.

Levels in Human Adipose Tissues—Data on levels in adipose tissues are being obtained. Dr. Frederick Kutz, OPP, (202) 755-8060.

Substitutes, Control Technology, and Related Costs and Economic Factors

Effects of Cooking—A study to determine the possibility that cooking will drive PBB's out of chicken meat is in progress. The results are expected within the next month. Dr. Mary Zabik, Michigan State University, (517) 353-5251.

Polychlorinated Biphenyls

General Studies

Activities of Government Agencies Regarding Inventories, Substitutes, and Housekeeping—Meetings have been held with senior officials of several Government agencies which (a) own or operate electrical equipment containing PCB's, or (b) are responsible for Governmental procurement of PCB-containing products. These agencies have been urged to provide an inventory of the current Governmental usage of such products and an evaluation of the steps they are taking to (a) introduce substitutes, and (b) reduce PCB discharges into the environment. George Wirth, OTS, (202) 755-6179.

Technical Information Exchange with States—The Governors of the States have been asked to work with EPA in reducing environmental discharges of PCB's. To assist, EPA is developing a technical information exchange program. George Wirth, OTS, (202) 755-6179.

Consultations with OECD on Uses, Labeling, and Reporting—Discussions are continuing with the Organization for Economic Cooperation and Development concerning the efforts of member countries to reduce the need for PCB's. Also of concern is the consistency of labeling and reporting activities directed to PCB's and products containing PCB's. At the most recent meeting in June 1976, recommendations were forwarded to the OECD Environmental Council for consideration at its meeting in July. Jack Thompson, OIA, (202) 755-0430.

Industrial Task Forces on Substitutes and Housekeeping Procedures—Meetings have been held with senior representatives of firms which manufacture PCB's and transformers and capacitors containing PCB's, as well as with railway, transit, and electrical utility organizations. Discussions have considered replacements for PCB's, effluent controls, housekeeping, and other immediate steps to reduce environmental discharges of PCB's. Three industry task forces have completed their tasks and prepared their reports which will be published in August 1976. George Wirth, OTS, (202) 755-6179.

Health and Ecological Effects and Environmental Behavior

Health Effects Summary—A summary of current health effects data, including populations exposed and levels of exposure, has just been released. Dr. Albert Kolby, FDA, (202) 245-1301.

Drinking Water Contamination—The National Academy of Sciences has been requested to review the health effects of PCB's in drinking water. The Academy will provide information on the dose-response characteristics of PCB's by December 1976, as a basis for determining whether a standard is appropriate. Edgar Jeffrey, OWS, (202) 426-8877.

Behavior in Soils—A review of existing literature to determine the interactions of PCB's in soils is underway. Also, as an adjunct to previously scheduled investigations of leachates from selected landfills, investigations of possible PCB leakage are being concluded. A preliminary report was published in May 1976. Alan Corson, OSWMP, (202) 755-9187.

Current and Projected Sources, Environmental Levels, and Exposed Populations

Hydraulic and Heat Transfer Fluids—Preliminary assessments indicate that PCB's have been widely used as hydraulic fluids and heat transfer fluids and in other similar applications. An assessment of these uses, including current recycling and reclamation practices, will assist in determining actions needed to curtail significant environmental discharges. These studies are slated for completion in December 1976. Thomas Kopp, OTS, (202) 755-0300.

Investment Casting—An analysis of the use of PCB's in the investment casting industry will help clarify the extent of PCB usage, discharges from users, the industry's dependency on PCB's, and the feasibility of substitutes. Recommendations for follow-up actions will be made. These are expected to be completed in December 1976. Thomas Kopp, OTS, (202) 755-0300.

Pulp and Paper Industry—Based on a preliminary survey of the paper recycling industry, an attempt will be made to assess the scope of environmental problems associated with the occurrence of paper impregnated with PCB's. This task is to be completed in December 1976. Thomas Kopp, OTS, (202) 755-0300.

Inventory of PCB-Containing Carbonless Copy Paper—An inventory of GSA and DSA supply warehouses to determine the quantities of used and unused PCB-contaminated carbonless copy paper is being conducted. Thomas Kopp, OTS, (202) 755-0300.

*Contamination of Imported Products—*Once an analytical procedure has been identified and standardized, selected imported products will be tested for contamination by PCB's and related materials. John Moran, ORD, (202) 426-2026.

*Ambient Levels and Trends—*Limited monitoring of air, water, and soil at selected sites is being carried out to document current and ambient levels of PCB's. Monitoring is also conducted near specialized users of PCB's. Preliminary results were correlated and published in May. Vincent DeCarlo, OTS, (202) 755-6956.

*Drinking Water Surveys—*Drinking water supplies in 122 cities are being surveyed for various organic constituents, including PCB's. The results will be available by February 1977. Edgar Jeffrey, OWS, (202) 426-8877.

*Point-Source Information—*Additional information concerning the uses and environmental discharges of PCB's from point-sources are being sought from industrial organizations pursuant to Secs. 308 of the Federal Water Pollution Control Act, and 114 of the Clean Air Act. Carl Schafer, OE, (202) 755-0750.

*Human Adipose Tissues—*PCB's are among the environmental contaminants being measured in the 1700 adipose tissue samples taken from selected cities each year. Annual reports are prepared in June. Dr. Frederick Kutz, EPA/OPP, (202) 755-8060.

*Levels in Air—*Limited preliminary air monitoring data are being obtained near several facilities. These data will be considered in determining whether a more intensive monitoring effort is in order as a step to determining the desirability, feasibility, and impact of an air standard. Implementation of this effort awaits approval of an interim monitoring method. Mike Jones, OAQPS, (919) 688-8501.

*Consultations with Canada on Levels in Great Lakes—*Consultations with Canada on the contamination of the fishing resources of the Great Lakes are continuing. The most recent session was held in March. Conrad Kleveno, EPA/OIA, (202) 755-8712.

*Analytical Method—*Although monitoring has been undertaken, the analytical procedures used have not been standardized. Available methods are being screened for reliability and feasibility, and recommendations will be made for their use in various media. John Moran, ORD, (202) 426-2026.

*Predictive Models—*Mathematical models to predict environmental distribution, levels, and transport based on release data are being developed. Thomas Kopp, OTS, (202) 755-0300.

Substitutes, Control Technology, and Related Costs and Economic Factors

Solid Waste Management— A continuing program of technical assistance provides guidance to public and private organizations faced with practical disposal problems involving hazardous substances, including PCB's. Advice is available on recycling, burial, incineration, and other possible disposal techniques. Recommended procedures concerning environmentally acceptable waste disposal practices for PCB's and PCB-containing materials were published in the *Federal Register* on April 1. William Wallace, OSWMP, (202) 755-9190.

*Guidelines of the American National Standards Institute—*The ANSI Standards on PCB's are designed to improve the handling, maintenance, disposal, and related environmental aspects of PCB's and equipment containing PCB's. EPA is actively participating in the current revision of this standard which should be widely accepted throughout the industrial sector. An initial draft of the revision is in review. Thomas Kopp, OTS, (202) 755-0300.

*Incineration of Capacitors—*Test burns of PCB-containing capacitors in a high-temperature incinerator were made this spring. By-product formation at various temperatures will be among the factors evaluated in the final report due in August. John Schaum, OSWMP, (202) 755-9200.

*Destruction Dechlorination—*Several control technologies for removal of PCB's from effluent streams were evaluated. Destructive dechlorination was determined to have the greatest potential utility. A demonstration grant is expected to be made within the next few months to provide working scale application. Paul DesRosiers, ORD, (202) 755-9014.

*Replacement Fluids—*A continuing assessment of substitutes for PCB's, including development of environmentally acceptable replacement fluids, is necessary as PCB uses are phased-out. A variety of substitutes currently under development are being considered as well as substitutes which might be suggested at a later date. Two requests for an evaluation of possible substitute fluids have resulted in responses published in the June 9 *Federal Register.* Thomas Kopp, OTS, (202) 755-0300.

*Harbor Studies—*Studies under Section 115 of the Federal Water Pollution Control Act concerning the removal of in-place pollutants in harbors will give special consideration to PCB's. Kenneth Mackenthun, OWPS, (202) 755-0100.

*Lake Contamination—*The Clean Lakes program provides a mechanism to encourage States to consider contamination from PCB's and their removal. A letter on this topic was sent to all States in February 1976. Kenneth Mackenthun, OWPS, (202) 755-0100.

Dredging of Sediments—The Corps of Engineers is consulting with EPA on controlling possible transference of PCB's from sediments to the water during dredging operations. Vance Hughes, OWPS, (202) 755-0100.

Control Options, Regulatory Actions, and Attendant Impacts

Water Quality Criteria—A maximum level of 1 ppt in navigable waters has been proposed as a Water Quality Criterion under Sec. 304 of the Federal Water Pollution Control Act. Leonard Guarraia, OWPS, (202) 245-3042.

Toxic Effluent Standard—A national water effluent standard for PCB's under Section 307(a) of the Federal Water Pollution Control Act has been proposed. Pretreatment standards pursuant to Section 307(b) are under study. Kent Ballentine, OWPS, (202) 245-3030.

Effluent Guidelines—The need for and feasibility of PCB effluent guidelines based on Best Practical Technology for those industries involved in the manufacturing of machinery and mechanical products are being studied. If such guidelines are warranted, steps will be initiated in 1976 to develop and incorporate discharge limitations into regulations as appropriate. DeVereaux Barnes, OWPS, (202) 426-2727.

Spill Prevention—Regulations designed to prevent spills of PCB's into navigable waterways will be developed pursuant to Section 311 of the Federal Water Pollution Control Act. The proposed regulations should be ready for publication in late 1976. Henry VanCleave, OWPS, (202) 245-3045.

Drinking Water Standard—If determined appropriate, a drinking water standard will be proposed in early 1977. Charles Hendricks, OWS, (202) 426-8877.

Tolerances in Food—Recent health effects reports are being reviewed to determine if the temporary food tolerance for PCB's should be revised. Dr. Albert Kolby, FDA, (202) 245-1301.

Workplace Criteria Document—A criteria document which will recommend to OSHA an appropriate workplace exposure standard is in preparation. Richard Rhoden, NIOSH, (301) 443-3680.

State Regulatory Actions—Five states have adopted PCB control acts. Others are examining the possibility of establishing pretreatment standards. Thomas Kopp, OTS, (202) 755-0300.

Vinyl Chloride

Health and Ecological Effects and Environmental Behavior

Carcinogenic Mechanisms—Two studies are being undertaken in an effort to document the mechanism of cancer induction. Joseph McLaughlin, CPSC, (202) 245-1445, and Robert Dixon, NIEHS, (919) 688-8146, X-333.

Epidemiological Investigations in New England—An epidemiological study of the population living near industries using vinyl chloride should be reported in late 1976. George Wirth, OTS, (202) 755-6179.

Epidemiological Investigations in Canada—An epidemiological study is being conducted in Shawinigan, Quebec, which has a high rate of liver cancer and a number of plastics manufacturers and users. Robert McGaughy, ORD, (202) 426-4637.

Current and Projected Sources, Environmental Levels, and Exposed Populations

Residual Unreacted Monomer in Consumer Products—Air samples taken from automobiles, homes, and other sites where PVC materials are used have been analyzed for unreacted monomer. William Coniglio, OTS, (202) 755-0300.

Emissions From Landfills—Landfills where PVC and related wastes are disposed are being monitored for possible emissions. Emery Lazar, OSWMP, (202) 755-9206.

Monitoring of Chemical Landfills—Monitoring of selected chemical landfill sites in New Jersey is in progress. Henry Gluckstern, Region II, (212) 264-4430.

Levels in Food—Food samples are being analyzed with the results expected shortly. Joseph Conrey, USDA, (202) 447-2807.

Migration from PVC Pipe—Studies are being conducted to determine the potential for migration from PVC pipe into drinking water. Preliminary data were reported in January 1976. Joseph Cotruvo, OWS, (202) 755-5643.

Analytical Method Selection—The evaluation of analytical methods for measurements in various media in progress. Recommendations are expected in August 1976. John Moran, ORD, (202) 426-2026.

Substitutes, Control Technology, and Related Costs and Economic Factors

Solid Waste Disposal Guidelines—Guidelines for

disposal of aerosol containers have been prepared and are in the final stages of review. Walter Kovalek, OSWMP, (202) 755-9187.

Control Options, Regulatory Actions, and Attendant Impacts

Air Standard—A National Emission Standard for Hazardous Air Pollutants under Sec. 112 of the Clean Air Act was proposed in December 1975. The final standard has been drafted, and after necessary reviews, should be promulgated in the fall. Susan Wyatt, OAQPS, (919) 688-8146, X-305.

Control of Leaching from Drinking Water Pipes—FDA and EPA are reviewing their authorities which might be used to control leaching from pipes and other equipment used in transporting drinking water. Joseph Cotruvo, OWS, (202) 755-5643.

Vinylidene Chloride

Health and Ecological Effects and Environmental Behavior

Inhalation Toxicology in Mice—Inhalation toxicology studies using mice are in progress. Preliminary results have been reviewed, and the final results should be published in late 1976. Dr. Joseph Seifter, OTS, (202) 755-4803.

Inhalation Toxicology in Rats—Industry-sponsored inhalation toxicology studies using rats are expected to be completed in 1976. The studies simulate workplace exposures. Jessie Norris, Dow Chemical Company, (517) 636-1527.

Worker Exposure Incident in New Jersey—A report on medical investigations of workers exposed to VDC will be issued in the fall. Dr. Bobby Craft, NIOSH, (513) 684-2427.

Current and Projected Sources, Environmental Levels, and Exposed Populations

Analytical Methods—The most feasible methods for analysis in various media are under investigation. Three field laboratories are evaluating the methods and are expected to make recommendations in August 1976. John Moran, ORD, (202) 426-2026.

Monitoring Near Industrial Sites—Monitoring near selected industrial sites will begin as soon as appropriate sampling and analysis techniques have been identified. Perry Brunner, OTS, (202) 755-6956.

Materials Balance Studies—Studies of the flow of VDC through the economy and into the environment are scheduled for the near future. Perry Brunner, OTS, (202) 755-6956.

Control Options, Regulatory Actions, and Attendant Impacts

Regulation Under the Clean Air Act—A preliminary air pollution assessment of VDC indicates that regulation under Sections 111 or 112 of the Clean Air Act is not warranted at this time. Robert Kellum, OAQPS, (919) 688-8146, X-501.

Worker Exposure Criteria Document—Development of a criteria document is scheduled to begin in March 1977. Vernon Rose, NIOSH, (301) 443-3680.

APPENDIX B

TOXIC SUBSTANCES CONTROL ACT, 1976: PUBLIC LAW 94-469

SECTION 1. SHORT TITLE AND TABLE OF CONTENTS.
This Act may be cited as the "Toxic Substances Control Act".

Toxic Substances
Control Act.
15 USC 2601
note.

TABLE OF CONTENTS

Sec. 1. Short title and table of contents.
Sec. 2. Findings, policy, and intent.
Sec. 3. Definitions.
Sec. 4. Testing of chemical substances and mixtures.
Sec. 5. Manufacturing and processing notices.
Sec. 6. Regulation of hazardous chemical substances and mixtures.
Sec. 7. Imminent hazards.
Sec. 8. Reporting and retention of information.
Sec. 9. Relationship to other Federal laws.
Sec. 10. Research, development, collection, dissemination, and utilization of data.
Sec. 11. Inspections and subpoenas.
Sec. 12. Exports.
Sec. 13. Entry into customs territory of the United States.
Sec. 14. Disclosure of data.
Sec. 15. Prohibited acts.
Sec. 16. Penalties.
Sec. 17. Specific enforcement and seizure.
Sec. 18. Preemption.
Sec. 19. Judicial review.
Sec. 20. Citizens' civil actions.
Sec. 21. Citizens' petitions.
Sec. 22. National defense waiver.
Sec. 23. Employee protection.
Sec. 24. Employment effects.
Sec. 25. Studies.
Sec. 26. Administration of the Act.
Sec. 27. Development and evaluation of test methods.
Sec. 28. State programs.
Sec. 29. Authorization for appropriations.
Sec. 30. Annual report.
Sec. 31. Effective date.

SEC. 2. FINDINGS, POLICY, AND INTENT.

(a) Findings.—The Congress finds that—   15 USC 2601.

(1) human beings and the environment are being exposed each year to a large number of chemical substances and mixtures;

(2) among the many chemical substances and mixtures which are constantly being developed and produced, there are some whose manufacture, processing, distribution in commerce, use, or disposal may present an unreasonable risk of injury to health or the environment; and

(3) the effective regulation of interstate commerce in such chemical substances and mixtures also necessitates the regulation of intrastate commerce in such chemical substances and mixtures.

(b) Policy.—It is the policy of the United States that—

(1) adequate data should be developed with respect to the effect of chemical substances and mixtures on health and the environ-

ment and that the development of such data should be the responsibility of those who manufacture and those who process such chemical substances and mixtures;

(2) adequate authority should exist to regulate chemical substances and mixtures which present an unreasonable risk of injury to health or the environment, and to take action with respect to chemical substances and mixtures which are imminent hazards; and

(3) authority over chemical substances and mixtures should be exercised in such a manner as not to impede unduly or create unnecessary economic barriers to technological innovation while fulfilling the primary purpose of this Act to assure that such innovation and commerce in such chemical substances and mixtures do not present an unreasonable risk of injury to health or the environment.

(c) INTENT OF CONGRESS.—It is the intent of Congress that the Administrator shall carry out this Act in a reasonable and prudent manner, and that the Administrator shall consider the environmental, economic, and social impact of any action the Administrator takes or proposes to take under this Act.

SEC. 3. DEFINITIONS.

15 USC 2602.

As used in this Act:

(1) the term "Administrator" means the Administrator of the Environmental Protection Agency.

(2)(A) Except as provided in subparagraph (B), the term "chemical substance" means any organic or inorganic substance of a particular molecular identity, including—

(i) any combination of such substances occurring in whole or in part as a result of a chemical reaction or occurring in nature, and

(ii) any element or uncombined radical.

(B) Such term does not include—

(i) any mixture,

7 USC 136 note.

(ii) any pesticide (as defined in the Federal Insecticide, Fungicide, and Rodenticide Act) when manufactured, processed, or distributed in commerce for use as a pesticide,

(iii) tobacco or any tobacco product,

(iv) any source material, special nuclear material, or byproduct material (as such terms are defined in the Atomic Energy Act of 1954 and regulations issued under such Act),

42 USC 2011 note.

26 USC 4181.

(v) any article the sale of which is subject to the tax imposed by section 4181 of the Internal Revenue Code of 1954 (determined without regard to any exemptions from such tax provided by section 4182 or 4221 or any other provision of such Code), and

26 USC 4182, 4221.

21 USC 321.

(vi) any food, food additive, drug, cosmetic, or device (as such terms are defined in section 201 of the Federal Food, Drug, and Cosmetic Act) when manufactured, processed, or distributed in commerce for use as a food, food additive, drug, cosmetic, or device.

The term "food" as used in clause (vi) of this subparagraph includes poultry and poultry products (as defined in sections 4(e) and 4(f)

21 USC 453.

21 USC 601.

of the Poultry Products Inspection Act), meat and meat food products (as defined in section 1(j) of the Federal Meat Inspection Act), and eggs and egg products (as defined in section 4 of the Egg Products

21 USC 1033.

Inspection Act).

(3) The term "commerce" means trade, traffic, transportation, or other commerce (A) between a place in a State and any place outside

of such State, or (B) which affects trade, traffic, transportation, or commerce described in clause (A).

(4) The terms "distribute in commerce" and "distribution in commerce" when used to describe an action taken with respect to a chemical substance or mixture or article containing a substance or mixture mean to sell, or the sale of, the substance, mixture, or article in commerce; to introduce or deliver for introduction into commerce, or the introduction or delivery for introduction into commerce of, the substance, mixture, or article; or to hold, or the holding of, the substance, mixture, or article after its introduction into commerce.

(5) The term "environment" includes water, air, and land and the interrelationship which exists among and between water, air, and land and all living things.

(6) The term "health and safety study" means any study of any effect of a chemical substance or mixture on health or the environment or on both, including underlying data and epidemiological studies, studies of occupational exposure to a chemical substance or mixture, toxicological, clinical, and ecological studies of a chemical substance or mixture, and any test performed pursuant to this Act.

(7) The term "manufacture" means to import into the customs territory of the United States (as defined in general headnote 2 of the Tariff Schedules of the United States), produce, or manufacture. **19 USC 1202.**

(8) The term "mixture" means any combination of two or more chemical substances if the combination does not occur in nature and is not, in whole or in part, the result of a chemical reaction; except that such term does include any combination which occurs, in whole or in part, as a result of a chemical reaction if none of the chemical substances comprising the combination is a new chemical substance and if the combination could have been manufactured for commercial purposes without a chemical reaction at the time the chemical substances comprising the combination were combined.

(9) The term "new chemical substance" means any chemical substance which is not included in the chemical substance list compiled and published under section 8(b). *Post,* p. 2027.

(10) The term "process" means the preparation of a chemical substance or mixture, after its manufacture, for distribution in commerce—

　(A) in the same form or physical state as, or in a different form or physical state from, that in which it was received by the person so preparing such substance or mixture, or

　(B) as part of an article containing the chemical substance or mixture.

(11) The term "processor" means any person who processes a chemical substance or mixture.

(12) The term "standards for the development of test data" means a prescription of—

　(A) the—

　　(i) health and environmental effects, and

　　(ii) information relating to toxicity, persistence, and other characteristics which affect health and the environment,

for which test data for a chemical substance or mixture are to be developed and any analysis that is to be performed on such data, and

　(B) to the extent necessary to assure that data respecting such effects and characteristics are reliable and adequate—

　　(i) the manner in which such data are to be developed,

　　(ii) the specification of any test protocol or methodology to be employed in the development of such data, and

(iii) such other requirements as are necessary to provide such assurance.

(13) The term "State" means any State of the United States, the District of Columbia, the Commonwealth of Puerto Rico, the Virgin Islands, Guam, the Canal Zone, American Samoa, the Northern Mariana Islands, or any other territory or possession of the United States.

(14) The term "United States", when used in the geographic sense, means all of the States.

SEC. 4. TESTING OF CHEMICAL SUBSTANCES AND MIXTURES.

15 USC 2603.

(a) TESTING REQUIREMENTS.—If the Administrator finds that—

(1) (A) (i) the manufacture, distribution in commerce, processing, use, or disposal of a chemical substance or mixture, or that any combination of such activities, may present an unreasonable risk of injury to health or the environment,

(ii) there are insufficient data and experience upon which the effects of such manufacture, distribution in commerce, processing, use, or disposal of such substance or mixture or of any combination of such activities on health or the environment can reasonably be determined or predicted, and

(iii) testing of such substance or mixture with respect to such effects is necessary to develop such data; or

(B) (i) a chemical substance or mixture is or will be produced in substantial quantities, and (I) it enters or may reasonably be anticipated to enter the environment in substantial quantities or (II) there is or may be significant or substantial human exposure to such substance or mixture,

(ii) there are insufficient data and experience upon which the effects of the manufacture, distribution in commerce, processing, use, or disposal of such substance or mixture or of any combination of such activities on health or the environment can reasonably be determined or predicted, and

(iii) testing of such substance or mixture with respect to such effects is necessary to develop such data; and

(2) in the case of a mixture, the effects which the mixture's manufacture, distribution in commerce, processing, use, or disposal or any combination of such activities may have on health or the environment may not be reasonably and more efficiently determined or predicted by testing the chemical substances which comprise the mixture;

Rules.

the Administrator shall by rule require that testing be conducted on such substance or mixture to develop data with respect to the health and environmental effects for which there is an insufficiency of data and experience and which are relevant to a determination that the manufacture, distribution in commerce, processing, use, or disposal of such substance or mixture, or that any combination of such activities, does or does not present an unreasonable risk of injury to health or the environment.

(b) (1) TESTING REQUIREMENT RULE.—A rule under subsection (a) shall include—

(A) identification of the chemical substance or mixture for which testing is required under the rule,

Standards for development of test data. Data, submittal to Administrator.

(B) standards for the development of test data for such substance or mixture, and

(C) with respect to chemical substances which are not new chemical substances and to mixtures, a specification of the period (which period may not be of unreasonable duration) within which the persons required to conduct the testing shall submit to the Administrator data developed in accordance with the standards referred to in subparagraph (B).

In determining the standards and period to be included, pursuant to subparagraphs (B) and (C), in a rule under subsection (a), the Administrator's considerations shall include the relative costs of the various test protocols and methodologies which may be required under the rule and the reasonably foreseeable availability of the facilities and personnel needed to perform the testing required under the rule. Any such rule may require the submission to the Administrator of preliminary data during the period prescribed under subparagraph (C).

(2)(A) The health and environmental effects for which standards for the development of test data may be prescribed include carcinogenesis, mutagenesis, teratogenesis, behavioral disorders, cumulative or synergistic effects, and any other effect which may present an unreasonable risk of injury to health or the environment. The characteristics of chemical substances and mixtures for which such standards may be prescribed include persistence, acute toxicity, subacute toxicity, chronic toxicity, and any other characteristic which may present such a risk. The methodologies that may be prescribed in such standards include epidemiologic studies, serial or hierarchical tests, in vitro tests, and whole animal tests, except that before prescribing epidemiologic studies of employees, the Administrator shall consult with the Director of the National Institute for Occupational Safety and Health.

(B) From time to time, but not less than once each 12 months, the Administrator shall review the adequacy of the standards for development of data prescribed in rules under subsection (a) and shall, if necessary, institute proceedings to make appropriate revisions of such standards.

Review of standards.

(3)(A) A rule under subsection (a) respecting a chemical substance or mixture shall require the persons described in subparagraph (B) to conduct tests and submit data to the Administrator on such substance or mixture, except that the Administrator may permit two or more of such persons to designate one such person or a qualified third party to conduct such tests and submit such data on behalf of the persons making the designation.

(B) The following persons shall be required to conduct tests and submit data on a chemical substance or mixture subject to a rule under subsection (a):

(i) Each person who manufactures or intends to manufacture such substance or mixture if the Administrator makes a finding described in subsection (a)(1)(A)(ii) or (a)(1)(B)(ii) with respect to the manufacture of such substance or mixture.

(ii) Each person who processes or intends to process such substance or mixture if the Administrator makes a finding described in subsection (a)(1)(A)(ii) or (a)(1)(B)(ii) with respect to the processing of such substance or mixture.

(iii) Each person who manufactures or processes or intends to manufacture or process such substance or mixture if the Administrator makes a finding described in subsection (a)(1)(A)(ii) or (a)(1)(B)(ii) with respect to the distribution in commerce, use, or disposal of such substance or mixture.

(4) Any rule under subsection (a) requiring the testing of and submission of data for a particular chemical substance or mixture shall expire at the end of the reimbursement period (as defined in subsection (c)(3)(B)) which is applicable to test data for such substance or mixture unless the Administrator repeals the rule before such date;

and a rule under subsection (a) requiring the testing of and submission of data for a category of chemical substances or mixtures shall expire with respect to a chemical substance or mixture included in the category at the end of the reimbursement period (as so defined) which is applicable to test data for such substance or mixture unless the Administrator before such date repeals the application of the rule to such substance or mixture or repeals the rule.

(5) Rules issued under subsection (a) (and any substantive amendment thereto or repeal thereof) shall be promulgated pursuant to section 553 of title 5, United States Code, except that (A) the Administrator shall give interested persons an opportunity for the oral presentation of data, views, or arguments, in addition to an opportunity to make written submissions; (B) a transcript shall be made of any oral presentation; and (C) the Administrator shall make and publish with the rule the findings described in paragraph (1)(A) or (1)(B) of subsection (a) and, in the case of a rule respecting a mixture, the finding described in paragraph (2) of such subsection.

(c) EXEMPTION.—(1) Any person required by a rule under subsection (a) to conduct tests and submit data on a chemical substance or mixture may apply to the Administrator (in such form and manner as the Administrator shall prescribe) for an exemption from such requirement.

(2) If, upon receipt of an application under paragraph (1), the Administrator determines that—

(A) the chemical substance or mixture with respect to which such application was submitted is equivalent to a chemical substance or mixture for which data has been submitted to the Administrator in accordance with a rule under subsection (a) or for which data is being developed pursuant to such a rule, and

(B) submission of data by the applicant on such substance or mixture would be duplicative of data which has been submitted to the Administrator in accordance with such rule or which is being developed pursuant to such rule,

the Administrator shall exempt, in accordance with paragraph (3) or (4), the applicant from conducting tests and submitting data on such substance or mixture under the rule with respect to which such application was submitted.

(3)(A) If the exemption under paragraph (2) of any person from the requirement to conduct tests and submit test data on a chemical substance or mixture is granted on the basis of the existence of previously submitted test data and if such exemption is granted during the reimbursement period for such test data (as prescribed by subparagraph (B)), then (unless such person and the persons referred to in clauses (i) and (ii) agree on the amount and method of reimbursement) the Administrator shall order the person granted the exemption to provide fair and equitable reimbursement (in an amount determined under rules of the Administrator)—

(i) to the person who previously submitted such test data, for a portion of the costs incurred by such person in complying with the requirement to submit such data, and

(ii) to any other person who has been required under this subparagraph to contribute with respect to such costs, for a portion of the amount such person was required to contribute.

In promulgating rules for the determination of fair and equitable reimbursement to the persons described in clauses (i) and (ii) for costs incurred with respect to a chemical substance or mixture, the Administrator shall, after consultation with the Attorney General

---

Margin notes:

Oral presentation and written submissions.

Transcript.

Publication.

Application.

Fair and equitable reimbursement.

Rules.

and the Federal Trade Commission, consider all relevant factors, including the effect on the competitive position of the person required to provide reimbursement in relation to the person to be reimbursed and the share of the market for such substance or mixture of the person required to provide reimbursement in relation to the share of such market of the persons to be reimbursed. An order under this subparagraph shall, for purposes of judicial review, be considered final agency action.

(B) For purposes of subparagraph (A), the reimbursement period for any test data for a chemical substance or mixture is a period—

Reimbursement period.

    (i) beginning on the date such data is submitted in accordance with a rule promulgated under subsection (a), and

    (ii) ending—

        (I) five years after the date referred to in clause (i), or

        (II) at the expiration of a period which begins on the date referred to in clause (i) and which is equal to the period which the Administrator determines was necessary to develop such data,

whichever is later.

(4)(A) If the exemption under paragraph (2) of any person from the requirement to conduct tests and submit test data on a chemical substance or mixture is granted on the basis of the fact that test data is being developed by one or more persons pursuant to a rule promulgated under subsection (a), then (unless such person and the persons referred to in clauses (i) and (ii) agree on the amount and method of reimbursement) the Administrator shall order the person granted the exemption to provide fair and equitable reimbursement (in an amount determined under rules of the Administrator)—

    (i) to each such person who is developing such test data, for a portion of the costs incurred by each such person in complying with such rule, and

    (ii) to any other person who has been required under this subparagraph to contribute with respect to the costs of complying with such rule, for a portion of the amount such person was required to contribute.

In promulgating rules for the determination of fair and equitable reimbursement to the persons described in clauses (i) and (ii) for costs incurred with respect to a chemical substance or mixture, the Administrator shall, after consultation with the Attorney General and the Federal Trade Commission, consider the factors described in the second sentence of paragraph (3)(A). An order under this subparagraph shall, for purposes of judicial review, be considered final agency action.

(B) If any exemption is granted under paragraph (2) on the basis of the fact that one or more persons are developing test data pursuant to a rule promulgated under subsection (a) and if after such exemption is granted the Administrator determines that no such person has complied with such rule, the Administrator shall (i) after providing written notice to the person who holds such exemption and an opportunity for a hearing, by order terminate such exemption, and (ii) notify in writing such person of the requirements of the rule with respect to which such exemption was granted.

(d) Notice.—Upon the receipt of any test data pursuant to a rule under subsection (a), the Administrator shall publish a notice of the receipt of such data in the Federal Register within 15 days of its receipt. Subject to section 14, each such notice shall (1) identify the chemical substance or mixture for which data have been received; (2) list the uses or intended uses of such substance or mixture and the

Publication in Federal Register.

information required by the applicable standards for the development of test data; and (3) describe the nature of the test data developed. Except as otherwise provided in section 14, such data shall be made available by the Administrator for examination by any person.

Post, p. 2034.

Committee to make recommendations to Administrator.

(e) PRIORITY LIST.— (1) (A) There is established a committee to make recommendations to the Administrator respecting the chemical substances and mixtures to which the Administrator should give priority consideration for the promulgation of a rule under subsection (a). In making such a recommendation with respect to any chemical substance or mixture, the committee shall consider all relevant factors, including—

(i) the quantities in which the substance or mixture is or will be manufactured,

(ii) the quantities in which the substance or mixture enters or will enter the environment,

(iii) the number of individuals who are or will be exposed to the substance or mixture in their places of employment and the duration of such exposure,

(iv) the extent to which human beings are or will be exposed to the substance or mixture,

(v) the extent to which the substance or mixture is closely related to a chemical substance or mixture which is known to present an unreasonable risk of injury to health or the environment,

(vi) the existence of data concerning the effects of the substance or mixture on health or the environment,

(vii) the extent to which testing of the substance or mixture may result in the development of data upon which the effects of the substance or mixture on health or the environment can reasonably be determined or predicted, and

(viii) the reasonably foreseeable availability of facilities and personnel for performing testing on the substance or mixture.

Recommendations, list of chemical substances and mixtures.

The recommendations of the committee shall be in the form of a list of chemical substances and mixtures which shall be set forth, either by individual substance or mixture or by groups of substances or mixtures, in the order in which the committee determines the Administrator should take action under subsection (a) with respect to the substances and mixtures. In establishing such list, the committee shall give priority attention to those chemical substances and mixtures which are known to cause or contribute to or which are suspected of causing or contributing to cancer, gene mutations, or birth defects. The committee shall designate chemical substances and mixtures on the list with respect to which the committee determines the Administrator should, within 12 months of the date on which such substances and mixtures are first designated, initiate a proceeding under subsection (a). The total number of chemical substances and mixtures on the list which are designated under the preceding sentence may not, at any time, exceed 50.

Publication in Federal Register; transmittal to Administrator.

(B) As soon as practicable but not later than nine months after the effective date of this Act, the committee shall publish in the Federal Register and transmit to the Administrator the list and designations required by subparagraph (A) together with the reasons for the committee's inclusion of each chemical substance or mixture on the list. At least every six months after the date of the transmission to the Administrator of the list pursuant to the preceding sentence, the committee shall make such revisions in the list as it determines to be necessary and shall transmit them to the Administrator together with the committee's reasons for the revisions. Upon receipt of any such revision,

List revision, publication in Federal Register.

the Administrator shall publish in the Federal Register the list with such revision, the reasons for such revision, and the designations made under subparagraph (A). The Administrator shall provide reasonable opportunity to any interested person to file with the Administrator written comments on the committee's list, any revision of such list by the committee, and designations made by the committee, and shall make such comments available to the public. Within the 12-month period beginning on the date of the first inclusion on the list of a chemical substance or mixture designated by the committee under subparagraph (A) the Administrator shall with respect to such chemical substance or mixture either initiate a rulemaking proceeding under subsection (a) or if such a proceeding is not initiated within such period, publish in the Federal Register the Administrator's reason for not initiating such a proceeding.

Comments.

Publication in Federal Register.

(2) (A) The committee established by paragraph (1) (A) shall consist of eight members as follows:

Membership.

(i) One member appointed by the Administrator from the Environmental Protection Agency.

(ii) One member appointed by the Secretary of Labor from officers or employees of the Department of Labor engaged in the Secretary's activities under the Occupational Safety and Health Act of 1970.

(iii) One member appointed by the Chairman of the Council on Environmental Quality from the Council or its officers or employees.

(iv) One member appointed by the Director of the National Institute for Occupational Safety and Health from officers or employees of the Institute.

(v) One member appointed by the Director of the National Institute of Environmental Health Sciences from officers or employees of the Institute.

(vi) One member appointed by the Director of the National Cancer Institute from officers or employees of the Institute.

(vii) One member appointed by the Director of the National Science Foundation from officers or employees of the Foundation.

(viii) One member appointed by the Secretary of Commerce from officers or employees of the Department of Commerce.

(B) (i) An appointed member may designate an individual to serve on the committee on the member's behalf. Such a designation may be made only with the approval of the applicable appointing authority and only if the individual is from the entity from which the member was appointed.

(ii) No individual may serve as a member of the committee for more than four years in the aggregate. If any member of the committee leaves the entity from which the member was appointed, such member may not continue as a member of the committee, and the member's position shall be considered to be vacant. A vacancy in the committee shall be filled in the same manner in which the original appointment was made.

(iii) Initial appointments to the committee shall be made not later than the 60th day after the effective date of this Act. Not later than the 90th day after such date the members of the committee shall hold a meeting for the selection of a chairperson from among their number.

(C) (i) No member of the committee, or designee of such member, shall accept employment or compensation from any person subject to any requirement of this Act or of any rule promulgated or order issued thereunder, for a period of at least 12 months after termination of service on the committee.

(ii) No person, while serving as a member of the committee, or designee of such member, may own any stocks or bonds, or have any pecuniary interest, of substantial value in any person engaged in the manufacture, processing, or distribution in commerce of any chemical substance or mixture subject to any requirement of this Act or of any rule promulgated or order issued thereunder.

(iii) The Administrator, acting through attorneys of the Environmental Protection Agency, or the Attorney General may bring an action in the appropriate district court of the United States to restrain any violation of this subparagraph.

(D) The Administrator shall provide the committee such administrative support services as may be necessary to enable the committee to carry out its function under this subsection.

(f) REQUIRED ACTIONS.—Upon the receipt of—

(1) any test data required to be submitted under this Act, or

(2) any other information available to the Administrator,

which indicates to the Administrator that there may be a reasonable basis to conclude that a chemical substance or mixture presents or will present a significant risk of serious or widespread harm to human beings from cancer, gene mutations, or birth defects, the Administrator shall, within the 180-day period beginning on the date of the receipt of such data or information, initiate appropriate action under section 5, 6, or 7 to prevent or reduce to a sufficient extent such risk or publish in the Federal Register a finding that such risk is not unreasonable. For good cause shown the Administrator may extend such period for an additional period of not more than 90 days. The Administrator shall publish in the Federal Register notice of any such extension and the reasons therefor. A finding by the Administrator that a risk is not unreasonable shall be considered agency action for purposes of judicial review under chapter 7 of title 5, United States Code. This subsection shall not take effect until two years after the effective date of this Act.

Publication in Federal Register.

5 USC 701.

(g) PETITION FOR STANDARDS FOR THE DEVELOPMENT OF TEST DATA.— A person intending to manufacture or process a chemical substance for which notice is required under section 5(a) and who is not required under a rule under subsection (a) to conduct tests and submit data on such substance may petition the Administrator to prescribe standards for the development of test data for such substance. The Administrator shall by order either grant or deny any such petition within 60 days of its receipt. If the petition is granted, the Administrator shall prescribe such standards for such substance within 75 days of the date the petition is granted. If the petition is denied, the Administrator shall publish, subject to section 14, in the Federal Register the reasons for such denial.

Infra.

Publication in Federal Register.
***Post,* p. 2034.**

SEC. 5. MANUFACTURING AND PROCESSING NOTICES.

15 USC 2604.

(a) IN GENERAL.—(1) Except as provided in subsection (h), no person may—

(A) manufacture a new chemical substance on or after the 30th day after the date on which the Administrator first publishes the list required by section 8(b), or

(B) manufacture or process any chemical substance for a use which the Administrator has determined, in accordance with paragraph (2), is a significant new use,

unless such person submits to the Administrator, at least 90 days before such manufacture or processing, a notice, in accordance with subsection (d), of such person's intention to manufacture or process such substance and such person complies with any applicable requirement of subsection (b).

(2) A determination by the Administrator that a use of a chemical substance is a significant new use with respect to which notification is required under paragraph (1) shall be made by a rule promulgated after a consideration of all relevant factors, including—

(A) the projected volume of manufacturing and processing of a chemical substance,

(B) the extent to which a use changes the type or form of exposure of human beings or the environment to a chemical substance,

(C) the extent to which a use increases the magnitude and duration of exposure of human beings or the environment to a chemical substance, and

(D) the reasonably anticipated manner and methods of manufacturing, processing, distribution in commerce, and disposal of a chemical substance.

(b) SUBMISSION OF TEST DATA.—(1)(A) If (i) a person is required by subsection (a)(1) to submit a notice to the Administrator before beginning the manufacture or processing of a chemical substance, and (ii) such person is required to submit test data for such substance pursuant to a rule promulgated under section 4 before the submission of such notice, such person shall submit to the Administrator such data in accordance with such rule at the time notice is submitted in accordance with subsection (a)(1).

(B) If—

(i) a person is required by subsection (a)(1) to submit a notice to the Administrator, and

(ii) such person has been granted an exemption under section 4(c) from the requirements of a rule promulgated under section 4 before the submission of such notice,

such person may not, before the expiration of the 90 day period which begins on the date of the submission in accordance with such rule of the test data the submission or development of which was the basis for the exemption, manufacture such substance if such person is subject to subsection (a)(1)(A) or manufacture or process such substance for a significant new use if the person is subject to subsection (a)(1)(B).

(2)(A) If a person—

(i) is required by subsection (a)(1) to submit a notice to the Administrator before beginning the manufacture or processing of a chemical substance listed under paragraph (4), and

(ii) is not required by a rule promulgated under section 4 before the submission of such notice to submit test data for such substance,

such person shall submit to the Administrator data prescribed by subparagraph (B) at the time notice is submitted in accordance with subsection (a)(1).

(B) Data submitted pursuant to subparagraph (A) shall be data which the person submitting the data believes show that—

(i) in the case of a substance with respect to which notice is required under subsection (a)(1)(A), the manufacture, processing, distribution in commerce, use, and disposal of the chemical substance or any combination of such activities will not present an unreasonable risk of injury to health or the environment, or

(ii) in the case of a chemical substance with respect to which notice is required under subsection (a)(1)(B), the intended significant new use of the chemical substance will not present an unreasonable risk of injury to health or the environment.

Post, p. 2034.

(3) Data submitted under paragraph (1) or (2) shall be made available, subject to section 14, for examination by interested persons.

(4)(A)(i) The Administrator may, by rule, compile and keep current a list of chemical substances with respect to which the Administrator finds that the manufacture, processing, distribution in commerce, use, or disposal, or any combination of such activities, presents or may present an unreasonable risk of injury to health or the environment.

(ii) In making a finding under clause (i) that the manufacture, processing, distribution in commerce, use, or disposal of a chemical substance or any combination of such activities presents or may present an unreasonable risk of injury to health or the environment, the Administrator shall consider all relevant factors, including—

(I) the effects of the chemical substance on health and the magnitude of human exposure to such substance; and

(II) the effects of the chemical substance on the environment and the magnitude of environmental exposure to such substance.

(B) The Administrator shall, in prescribing a rule under subparagraph (A) which lists any chemical substance, identify those uses, if any, which the Administrator determines, by rule under subsection (a)(2), would constitute a significant new use of such substance.

(C) Any rule under subparagraph (A), and any substantive amendment or repeal of such a rule, shall be promulgated pursuant to the procedures specified in section 553 of title 5, United States Code, except that (i) the Administrator shall give interested persons an opportunity for the oral presentation of data, views, or arguments, in addition to an opportunity to make written submissions, (ii) a transcript shall be kept of any oral presentation, and (iii) the Administrator shall make and publish with the rule the finding described in subparagraph (A).

Oral presentation. Transcript. Publication.

(c) EXTENSION OF NOTICE PERIOD.—The Administrator may for good cause extend for additional periods (not to exceed in the aggregate 90 days) the period, prescribed by subsection (a) or (b) before which the manufacturing or processing of a chemical substance subject to such subsection may begin. Subject to section 14, such an extension and the reasons therefor shall be published in the Federal Register and shall constitute a final agency action subject to judicial review.

Publication in Federal Register.

(d) CONTENT OF NOTICE; PUBLICATIONS IN THE FEDERAL REGISTER.—
(1) The notice required by subsection (a) shall include—

(A) insofar as known to the person submitting the notice or insofar as reasonably ascertainable, the information described in subparagraphs (A), (B), (C), (D), (F), and (G) of section 8(a)(2), and

(B) in such form and manner as the Administrator may prescribe, any test data in the possession or control of the person giving such notice which are related to the effect of any manufacture, processing, distribution in commerce, use, or disposal of such substance or any article containing such substance, or of any combination of such activities, on health or the environment, and

(C) a description of any other data concerning the environmental and health effects of such substance, insofar as known to the person making the notice or insofar as reasonably ascertainable.

Such a notice shall be made available, subject to section 14, for examination by interested persons.

(2) Subject to section 14, not later than five days (excluding Saturdays, Sundays and legal holidays) after the date of the receipt of a

notice under subsection (a) or of data under subsection (b), the Administrator shall publish in the Federal Register a notice which—

(A) identifies the chemical substance for which notice or data has been received;

(B) lists the uses or intended uses of such substance; and

(C) in the case of the receipt of data under subsection (b), describes the nature of the tests performed on such substance and any data which was developed pursuant to subsection (b) or a rule under section 4.

A notice under this paragraph respecting a chemical substance shall identify the chemical substance by generic class unless the Administrator determines that more specific identification is required in the public interest.

(3) At the beginning of each month the Administrator shall publish a list in the Federal Register of (A) each chemical substance for which notice has been received under subsection (a) and for which the notification period prescribed by subsection (a), (b), or (c) has not expired, and (B) each chemical substance for which such notification period has expired since the last publication in the Federal Register of such list.

(e) Regulation Pending Development of Information.—(1)(A) If the Administrator determines that—

(i) the information available to the Administrator is insufficient to permit a reasoned evaluation of the health and environmental effects of a chemical substance with respect to which notice is required by subsection (a); and

(ii)(I) in the absence of sufficient information to permit the Administrator to make such an evaluation, the manufacture, processing, distribution in commerce, use, or disposal of such substance, or any combination of such activities, may present an unreasonable risk of injury to health or the environment, or

(II) such substance is or will be produced in substantial quantities, and such substance either enters or may reasonably be anticipated to enter the environment in substantial quantities or there is or may be significant or substantial human exposure to the substance,

the Administrator may issue a proposed order, to take effect on the expiration of the notification period applicable to the manufacturing or processing of such substance under subsection (a), (b), or (c), to prohibit or limit the manufacture, processing, distribution in commerce, use, or disposal of such substance or to prohibit or limit any combination of such activities. **Proposed order.**

(B) A proposed order may not be issued under subparagraph (A) respecting a chemical substance (i) later than 45 days before the expiration of the notification period applicable to the manufacture or processing of such substance under subsection (a), (b), or (c), and (ii) unless the Administrator has, on or before the issuance of the proposed order, notified, in writing, each manufacturer or processor, as the case may be, of such substance of the determination which underlies such order.

(C) If a manufacturer or processor of a chemical substance to be subject to a proposed order issued under subparagraph (A) files with the Administrator (within the 30-day period beginning on the date such manufacturer or processor received the notice required by subparagraph (B)(ii)) objections specifying with particularity the provisions of the order deemed objectionable and stating the grounds therefor, the proposed order shall not take effect.

Injunction, application.

(2) (A) (i) Except as provided in clause (ii), if with respect to a chemical substance with respect to which notice is required by subsection (a), the Administrator makes the determination described in paragraph (1) (A) and if—

(I) the Administrator does not issue a proposed order under paragraph (1) respecting such substance, or

(II) the Administrator issues such an order respecting such substance but such order does not take effect because objections were filed under paragraph (1) (C) with respect to it,

the Administrator, through attorneys of the Environmental Protection Agency, shall apply to the United States District Court for the District of Columbia or the United States district court for the judicial district in which the manufacturer or processor, as the case may be, of such substance is found, resides, or transacts business for an injunction to prohibit or limit the manufacture, processing, distribution in commerce, use, or disposal of such substance (or to prohibit or limit any combination of such activities).

(ii) If the Administrator issues a proposed order under paragraph (1) (A) respecting a chemical substance but such order does not take effect because objections have been filed under paragraph (1) (C) with respect to it, the Administrator is not required to apply for an injunction under clause (i) respecting such substance if the Administrator determines, on the basis of such objections, that the determinations under paragraph (1) (A) may not be made.

(B) A district court of the United States which receives an application under subparagraph (A) (i) for an injunction respecting a chemical substance shall issue such injunction if the court finds that—

(i) the information available to the Administrator is insufficient to permit a reasoned evaluation of the health and environmental effects of a chemical substance with respect to which notice is required by subsection (a) ; and

(ii) (I) in the absence of sufficient information to permit the Administrator to make such an evaluation, the manufacture, processing, distribution in commerce, use, or disposal of such substance, or any combination of such activities, may present an unreasonable risk of injury to health or the environment, or

(II) such substance is or will be produced in substantial quantities, and such substance either enters or may reasonably be anticipated to enter the environment in substantial quantities or there is or may be significant or substantial human exposure to the substance.

(C) Pending the completion of a proceeding for the issuance of an injunction under subparagraph (B) respecting a chemical substance, the court may, upon application of the Administrator made through attorneys of the Environmental Protection Agency, issue a temporary restraining order or a preliminary injunction to prohibit the manufacture, processing, distribution in commerce, use, or disposal of such a substance (or any combination of such activities) if the court finds that the notification period applicable under subsection (a), (b), or (c) to the manufacturing or processing of such substance may expire before such proceeding can be completed.

(D) After the submission to the Administrator of test data sufficient to evaluate the health and environmental effects of a chemical substance subject to an injunction issued under subparagraph (B) and the evaluation of such data by the Administrator, the district court of the United States which issued such injunction shall, upon petition, dissolve the injunction unless the Administrator has initiated a proceeding for the issuance of a rule under section 6(a) respecting the substance. If such a proceeding has been initiated, such court shall continue the injunction in effect until the effective date of the rule promulgated in such proceeding or, if such proceeding is terminated without the promulgation of a rule, upon the termination of the proceeding, whichever occurs first.

(f) Protection Against Unreasonable Risks.—(1) If the Administrator finds that there is a reasonable basis to conclude that the manufacture, processing, distribution in commerce, use, or disposal of a chemical substance with respect to which notice is required by subsection (a), or that any combination of such activities, presents or will present an unreasonable risk of injury to health or environment before a rule promulgated under section 6 can protect against such risk, the Administrator shall, before the expiration of the notification period applicable under subsection (a), (b), or (c) to the manufacturing or processing of such substance, take the action authorized by paragraph (2) or (3) to the extent necessary to protect against such risk.

(2) The Administrator may issue a proposed rule under section 6(a) to apply to a chemical substance with respect to which a finding was made under paragraph (1)— *Proposed rule.*

(A) a requirement limiting the amount of such substance which may be manufactured, processed, or distributed in commerce,

(B) a requirement described in paragraph (2), (3), (4), (5), (6), or (7) of section 6(a), or

(C) any combination of the requirements referred to in subparagraph (B).

Such a proposed rule shall be effective upon its publication in the Federal Register. Section 6(d)(2)(B) shall apply with respect to such rule. *Publication in Federal Register.*

(3)(A) The Administrator may—

(i) issue a proposed order to prohibit the manufacture, processing, or distribution in commerce of a substance with respect to which a finding was made under paragraph (1), or *Proposed order.*

(ii) apply, through attorneys of the Environmental Protection Agency, to the United States District Court for the District of Columbia or the United States district court for the judicial district in which the manufacturer, or processor, as the case may be, of such substance, is found, resides, or transacts business for an injunction to prohibit the manufacture, processing, or distribution in commerce of such substance. *Injunction application.*

A proposed order issued under clause (i) respecting a chemical substance shall take effect on the expiration of the notification period applicable under subsection (a), (b), or (c) to the manufacture or processing of such substance.

(B) If the district court of the United States to which an application has been made under subparagraph (A)(ii) finds that there is a reasonable basis to conclude that the manufacture, processing, distribution in commerce, use, or disposal of the chemical substance with respect to which such application was made, or that any combination of such activities, presents or will present an unreasonable risk of injury to health or the environment before a rule promulgated under section 6 can protect against such risk, the court shall issue an injunction to prohibit the manufacture, processing, or distribution in commerce of such substance or to prohibit any combination of such activities.

(C) The provisions of subparagraphs (B) and (C) of subsection (e)(1) shall apply with respect to an order issued under clause (i) of subparagraph (A); and the provisions of subparagraph (C) of subsection (e)(2) shall apply with respect to an injunction issued under subparagraph (B).

(D) If the Administrator issues an order pursuant to subparagraph (A)(i) respecting a chemical substance and objections are filed in accordance with subsection (e)(1)(C), the Administrator shall seek an injunction under subparagraph (A)(ii) respecting such substance unless the Administrator determines, on the basis of such objections, that such substance does not or will not present an unreasonable risk of injury to health or the environment.

(g) STATEMENT OF REASONS FOR NOT TAKING ACTION.—If the Administrator has not initiated any action under this section or section 6 or 7 to prohibit or limit the manufacture, processing, distribution in commerce, use, or disposal of a chemical substance, with respect to which notification or data is required by subsection (a)(1)(B) or (b), before the expiration of the notification period applicable to the manufacturing or processing of such substance, the Administrator shall publish a statement of the Administrator's reasons for not initiating such action. Such a statement shall be published in the Federal Register before the expiration of such period. Publication of such statement in accordance with the preceding sentence is not a prerequisite to the manufacturing or processing of the substance with respect to which the statement is to be published.

Publication in Federal Register.

(h) EXEMPTIONS.—(1) The Administrator may, upon application, exempt any person from any requirement of subsection (a) or (b) to permit such person to manufacture or process a chemical substance for test marketing purposes—

> (A) upon a showing by such person satisfactory to the Administrator that the manufacture, processing, distribution in commerce, use, and disposal of such substance, and that any combination of such activities, for such purposes will not present any unreasonable risk of injury to health or the environment, and

> (B) under such restrictions as the Administrator considers appropriate.

(2)(A) The Administrator may, upon application, exempt any person from the requirement of subsection (b)(2) to submit data for a chemical substance. If, upon receipt of an application under the preceding sentence, the Administrator determines that—

> (i) the chemical substance with respect to which such application was submitted is equivalent to a chemical substance for which data has been submitted to the Administrator as required by subsection (b)(2), and

> (ii) submission of data by the applicant on such substance would be duplicative of data which has been submitted to the Administrator in accordance with such subsection,

the Administrator shall exempt the applicant from the requirement to submit such data on such substance. No exemption which is granted under this subparagraph with respect to the submission of data for a chemical substance may take effect before the beginning of the reimbursement period applicable to such data.

Fair and equitable reimbursement.

(B) If the Administrator exempts any person, under subparagraph (A), from submitting data required under subsection (b)(2) for a chemical substance because of the existence of previously submitted data and if such exemption is granted during the reimbursement period for such data, then (unless such person and the persons referred to in clauses (i) and (ii) agree on the amount and method of reimbursement) the Administrator shall order the person granted the exemption to provide fair and equitable reimbursement (in an amount determined under rules of the Administrator)—

(i) to the person who previously submitted the data on which the exemption was based, for a portion of the costs incurred by such person in complying with the requirement under subsection (b)(2) to submit such data, and

(ii) to any other person who has been required under this subparagraph to contribute with respect to such costs, for a portion of the amount such person was required to contribute.

In promulgating rules for the determination of fair and equitable reimbursement to the persons described in clauses (i) and (ii) for costs incurred with respect to a chemical substance, the Administrator shall, after consultation with the Attorney General and the Federal Trade Commission, consider all relevant factors, including the effect on the competitive position of the person required to provide reimbursement in relation to the persons to be reimbursed and the share of the market for such substance of the person required to provide reimbursement in relation to the share of such market of the persons to be reimbursed. For purposes of judicial review, an order under this subparagraph shall be considered final agency action.

(C) For purposes of this paragraph, the reimbursement period for any previously submitted data for a chemical substance is a period—

Reimbursement period.

(i) beginning on the date of the termination of the prohibition, imposed under this section, on the manufacture or processing of such substance by the person who submitted such data to the Administrator, and

(ii) ending—

(I) five years after the date referred to in clause (i), or

(II) at the expiration of a period which begins on the date referred to in clause (i) and is equal to the period which the Administrator determines was necessary to develop such data,

whichever is later.

(3) The requirements of subsections (a) and (b) do not apply with respect to the manufacturing or processing of any chemical substance which is manufactured or processed, or proposed to be manufactured or processed, only in small quantities (as defined by the Administrator by rule) solely for purposes of—

(A) scientific experimentation or analysis, or

(B) chemical research on, or analysis of such substance or another substance, including such research or analysis for the development of a product,

if all persons engaged in such experimentation, research, or analysis for a manufacturer or processor are notified (in such form and manner as the Administrator may prescribe) of any risk to health which the manufacturer, processor, or the Administrator has reason to believe may be associated with such chemical substance.

(4) The Administrator may, upon application and by rule, exempt the manufacturer of any new chemical substance from all or part of the requirements of this section if the Administrator determines that the manufacture, processing, distribution in commerce, use, or disposal of such chemical substance, or that any combination of such activities, will not present an unreasonable risk of injury to health or the environment. A rule promulgated under this paragraph (and any substantive amendment to, or repeal of, such a rule) shall be promulgated in accordance with paragraphs (2) and (3) of section 6(c).

(5) The Administrator may, upon application, make the requirements of subsections (a) and (b) inapplicable with respect to the manufacturing or processing of any chemical substance (A) which exists temporarily as a result of a chemical reaction in the manufacturing or processing of a mixture or another chemical substance, and (B) to which there is no, and will not be, human or environmental exposure.

Publication in
Federal Register.
Comments.

(6) Immediately upon receipt of an application under paragraph (1) or (5) the Administrator shall publish in the Federal Register notice of the receipt of such application. The Administrator shall give interested persons an opportunity to comment upon any such application and shall, within 45 days of its receipt, either approve or deny the application. The Administrator shall publish in the Federal Register notice of the approval or denial of such an application.

Publication in
Federal Register.

(i) DEFINITION.—For purposes of this section, the terms "manufacture" and "process" mean manufacturing or processing for commercial purposes.

SEC. 6. REGULATION OF HAZARDOUS CHEMICAL SUBSTANCES AND MIXTURES.

15 USC 2605.

(a) SCOPE OF REGULATION.—If the Administrator finds that there is a reasonable basis to conclude that the manufacture, processing, distribution in commerce, use, or disposal of a chemical substance or mixture, or that any combination of such activities, presents or will present an unreasonable risk of injury to health or the environment, the Administrator shall by rule apply one or more of the following requirements to such substance or mixture to the extent necessary to protect adequately against such risk using the least burdensome requirements:

(1) A requirement (A) prohibiting the manufacturing, processing, or distribution in commerce of such substance or mixture, or (B) limiting the amount of such substance or mixture which may be manufactured, processed, or distributed in commerce.

(2) A requirement—

(A) prohibiting the manufacture, processing, or distribution in commerce of such substance or mixture for (i) a particular use or (ii) a particular use in a concentration in excess of a level specified by the Administrator in the rule imposing the requirement, or

(B) limiting the amount of such substance or mixture which may be manufactured, processed, or distributed in commerce for (i) a particular use or (ii) a particular use in a concentration in excess of a level specified by the Administrator in the rule imposing the requirement.

(3) A requirement that such substance or mixture or any article containing such substance or mixture be marked with or accompanied by clear and adequate warnings and instructions with respect to its use, distribution in commerce, or disposal or with respect to any combination of such activities. The form and content of such warnings and instructions shall be prescribed by the Administrator.

(4) A requirement that manufacturers and processors of such substance or mixture make and retain records of the processes used to manufacture or process such substance or mixture and monitor or conduct tests which are reasonable and necessary to assure compliance with the requirements of any rule applicable under this subsection.

(5) A requirement prohibiting or otherwise regulating any manner or method of commercial use of such substance or mixture.

(6) (A) A requirement prohibiting or otherwise regulating any manner or method of disposal of such substance or mixture, or of any article containing such substance or mixture, by its manufacturer or processor or by any other person who uses, or disposes of, it for commercial purposes.

(B) A requirement under subparagraph (A) may not require any person to take any action which would be in violation of any law or requirement of, or in effect for, a State or political subdivision, and shall require each person subject to it to notify each State and political subdivision in which a required disposal may occur of such disposal.

(7) A requirement directing manufacturers or processors of such substance or mixture (A) to give notice of such unreasonable risk of injury to distributors in commerce of such substance or mixture and, to the extent reasonably ascertainable, to other persons in possession of such substance or mixture or exposed to such substance or mixture, (B) to give public notice of such risk of injury, and (C) to replace or repurchase such substance or mixture as elected by the person to which the requirement is directed.

Any requirement (or combination of requirements) imposed under this subsection may be limited in application to specified geographic areas.

(b) Quality Control.—If the Administrator has a reasonable basis to conclude that a particular manufacturer or processor is manufacturing or processing a chemical substance or mixture in a manner which unintentionally causes the chemical substance or mixture to present or which will cause it to present an unreasonable risk of injury to health or the environment—

(1) the Administrator may by order require such manufacturer or processor to submit a description of the relevant quality control procedures followed in the manufacturing or processing of such chemical substance or mixture; and

(2) if the Administrator determines—

(A) that such quality control procedures are inadequate to prevent the chemical substance or mixture from presenting such risk of injury, the Administrator may order the manufacturer or processor to revise such quality control procedures to the extent necessary to remedy such inadequacy; or

(B) that the use of such quality control procedures has resulted in the distribution in commerce of chemical substances or mixtures which present an unreasonable risk of injury to health or the environment, the Administrator may order the manufacturer or processor to (i) give notice of such risk to processors or distributors in commerce of any such substance or mixture, or to both, and, to the extent reasonably ascertainable, to any other person in possession of or exposed to any such substance, (ii) to give public notice of such risk, and (iii) to provide such replacement or repurchase of any such substance or mixture as is necessary to adequately protect health or the environment.

A determination under subparagraph (A) or (B) of paragraph (2) shall be made on the record after opportunity for hearing in accordance with section 554 of title 5, United States Code. Any manufacturer

Hearing.

or processor subject to a requirement to replace or repurchase a chemical substance or mixture may elect either to replace or repurchase the substance or mixture and shall take either such action in the manner prescribed by the Administrator.

Statement, publication.

(c) PROMULGATION OF SUBSECTION (a) RULES.—(1) In promulgating any rule under subsection (a) with respect to a chemical substance or mixture, the Administrator shall consider and publish a statement with respect to—

(A) the effects of such substance or mixture on health and the magnitude of the exposure of human beings to such substance or mixture,

(B) the effects of such substance or mixture on the environment and the magnitude of the exposure of the environment to such substance or mixture,

(C) the benefits of such substance or mixture for various uses and the availability of substitutes for such uses, and

(D) the reasonably ascertainable economic consequences of the rule, after consideration of the effect on the national economy, small business, technological innovation, the environment, and public health.

If the Administrator determines that a risk of injury to health or the environment could be eliminated or reduced to a sufficient extent by actions taken under another Federal law (or laws) administered in whole or in part by the Administrator, the Administrator may not promulgate a rule under subsection (a) to protect against such risk of injury unless the Administrator finds, in the Administrator's discretion, that it is in the public interest to protect against such risk under this Act. In making such a finding the Administrator shall consider (i) all relevant aspects of the risk, as determined by the Administrator in the Administrator's discretion, (ii) a comparison of the estimated costs of complying with actions taken under this Act and under such law (or laws), and (iii) the relative efficiency of actions under this Act and under such law (or laws) to protect against such risk of injury.

5 USC 556, 557.
Notice, publication.
Written data, views, arguments, submittal.
Hearing.
Final rule.

(2) When prescribing a rule under subsection (a) the Administrator shall proceed in accordance with section 553 of title 5, United States Code (without regard to any reference in such section to sections 556 and 557 of such title), and shall also (A) publish a notice of proposed rulemaking stating with particularity the reason for the proposed rule; (B) allow interested persons to submit written data, views, and arguments, and make all such submissions publicly available; (C) provide an opportunity for an informal hearing in accordance with paragraph (3); (D) promulgate, if appropriate, a final rule based on the matter in the rulemaking record (as defined in section 19(a)), and (E) make and publish with the rule the finding described in subsection (a).

Informal hearings.

(3) Informal hearings required by paragraph (2)(C) shall be conducted by the Administrator in accordance with the following requirements:

(A) Subject to subparagraph (B), an interested person is entitled—

(i) to present such person's position orally or by documentary submissions (or both), and

(ii) if the Administrator determines that there are disputed issues of material fact it is necessary to resolve, to present such rebuttal submissions and to conduct (or have conducted under subparagraph (B)(ii)) such cross-examination of persons as the Administrator determines (I) to be appropriate, and (II) to be required for a full and true disclosure with respect to such issues.

(B) The Administrator may prescribe such rules and make such rulings concerning procedures in such hearings to avoid unnecessary costs or delay. Such rules or rulings may include (i) the imposition of reasonable time limits on each interested person's oral presentations, and (ii) requirements that any cross-examination to which a person may be entitled under subparagraph (A) be conducted by the Administrator on behalf of that person in such manner as the Administrator determines (I) to be appropriate, and (II) to be required for a full and true disclosure with respect to disputed issues of material fact. **Rules.**

(C) (i) Except as provided in clause (ii), if a group of persons each of whom under subparagraphs (A) and (B) would be entitled to conduct (or have conducted) cross-examination and who are determined by the Administrator to have the same or similar interests in the proceeding cannot agree upon a single representative of such interests for purposes of cross-examination, the Administrator may make rules and rulings (I) limiting the representation of such interest for such purposes, and (II) governing the manner in which such cross-examination shall be limited.

(ii) When any person who is a member of a group with respect to which the Administrator has made a determination under clause (i) is unable to agree upon group representation with the other members of the group, then such person shall not be denied under the authority of clause (i) the opportunity to conduct (or have conducted) cross-examination as to issues affecting the person's particular interests if (I) the person satisfies the Administrator that the person has made a reasonable and good faith effort to reach agreement upon group representation with the other members of the group and (II) the Administrator determines that there are substantial and relevant issues which are not adequately presented by the group representative.

(D) A verbatim transcript shall be taken of any oral presentation made, and cross-examination conducted in any informal hearing under this subsection. Such transcript shall be available to the public. **Verbatim transcript.**

(4) (A) The Administrator may, pursuant to rules prescribed by the Administrator, provide compensation for reasonable attorneys' fees, expert witness fees, and other costs of participating in a rulemaking proceeding for the promulgation of a rule under subsection (a) to any person— **Compensation.**

(i) who represents an interest which would substantially contribute to a fair determination of the issues to be resolved in the proceeding, and

(ii) if—

(I) the economic interest of such person is small in comparison to the costs of effective participation in the proceeding by such person, or

(II) such person demonstrates to the satisfaction of the Administrator that such person does not have sufficient resources adequately to participate in the proceeding without compensation under this subparagraph.

In determining for purposes of clause (i) if an interest will substantially contribute to a fair determination of the issues to be resolved in a proceeding, the Administrator shall take into account the number and complexity of such issues and the extent to which representation of such interest will contribute to widespread public participation in the proceeding and representation of a fair balance of interests for the resolution of such issues.

(B) In determining whether compensation should be provided to a person under subparagraph (A) and the amount of such compensation, the Administrator shall take into account the financial burden which will be incurred by such person in participating in the rule-making proceeding. The Administrator shall take such action as may be necessary to ensure that the aggregate amount of compensation paid under this paragraph in any fiscal year to all persons who, in rulemaking proceedings in which they receive compensation, are persons who either—

(i) would be regulated by the proposed rule, or

(ii) represent persons who would be so regulated,

may not exceed 25 per centum of the aggregate amount paid as compensation under this paragraph to all persons in such fiscal year.

(5) Paragraph (1), (2), (3), and (4) of this subsection apply to the promulgation of a rule repealing, or making a substantive amendment to, a rule promulgated under subsection (a).

(d) EFFECTIVE DATE.— (1) The Administrator shall specify in any rule under subsection (a) the date on which it shall take effect, which date shall be as soon as feasible.

Publication in Federal Register. (2) (A) The Administrator may declare a proposed rule under subsection (a) to be effective upon its publication in the Federal Register and until the effective date of final action taken, in accordance with subparagraph (B), respecting such rule if—

(i) the Administrator determines that—

(I) the manufacture, processing, distribution in commerce, use, or disposal of the chemical substance or mixture subject to such proposed rule or any combination of such activities is likely to result in an unreasonable risk of serious or widespread injury to health or the environment before such effective date; and

(II) making such proposed rule so effective is necessary to protect the public interest; and

(ii) in the case of a proposed rule to prohibit the manufacture, processing, or distribution of a chemical substance or mixture because of the risk determined under clause (i)(I), a court has in an action under section 7 granted relief with respect to such risk associated with such substance or mixture.

Such a proposed rule which is made so effective shall not, for purposes of judicial review, be considered final agency action.

Notice. (B) If the Administrator makes a proposed rule effective upon its publication in the Federal Register, the Administrator shall, as expeditiously as possible, give interested persons prompt notice of such action, provide reasonable opportunity, in accordance with paragraphs (2) and (3) of subsection (c), for a hearing on such rule, and either promulgate such rule (as proposed or with modifications) or revoke it; and if such a hearing is requested, the Administrator shall commence the hearing within five days from the date such request is made unless the Administrator and the person making the request agree upon a later date for the hearing to begin, and after the hearing is concluded the Administrator shall, within ten days of the conclusion of the hearing, either promulgate such rule (as proposed or with modifications) or revoke it.

(e) POLYCHLORINATED BIPHENYLS.—(1) Within six months after the effective date of this Act the Administrator shall promulgate rules to—

    (A) prescribe methods for the disposal of polychlorinated biphenyls, and

    (B) require polychlorinated biphenyls to be marked with clear and adequate warnings, and instructions with respect to their processing, distribution in commerce, use, or disposal or with respect to any combination of such activities.

Requirements prescribed by rules under this paragraph shall be consistent with the requirements of paragraphs (2) and (3).

    (2)(A) Except as provided under subparagraph (B), effective one year after the effective date of this Act no person may manufacture, process, or distribute in commerce or use any polychlorinated biphenyl in any manner other than in a totally enclosed manner.

    (B) The Administrator may by rule authorize the manufacture, processing, distribution in commerce or use (or any combination of such activities) of any polychlorinated biphenyl in a manner other than in a totally enclosed manner if the Administrator finds that such manufacture, processing, distribution in commerce, or use (or combination of such activities) will not present an unreasonable risk of injury to health or the environment.

    (C) For the purposes of this paragraph, the term "totally enclosed manner" means any manner which will ensure that any exposure of human beings or the environment to a polychlorinated biphenyl will be insignificant as determined by the Administrator by rule.

    (3)(A) Except as provided in subparagraphs (B) and (C)—

    (i) no person may manufacture any polychlorinated biphenyl after two years after the effective date of this Act, and

    (ii) no person may process or distribute in commerce any polychlorinated biphenyl after two and one-half years after such date.

    (B) Any person may petition the Administrator for an exemption from the requirements of subparagraph (A), and the Administrator may grant by rule such an exemption if the Administrator finds that—

    (i) an unreasonable risk of injury to health or environment would not result, and

    (ii) good faith efforts have been made to develop a chemical substance which does not present an unreasonable risk of injury to health or the environment and which may be substituted for such polychlorinated biphenyl.

An exemption granted under this subparagraph shall be subject to such terms and conditions as the Administrator may prescribe and shall be in effect for such period (but not more than one year from the date it is granted) as the Administrator may prescribe.

    (C) Subparagraph (A) shall not apply to the distribution in commerce of any polychlorinated biphenyl if such polychlorinated biphenyl was sold for purposes other than resale before two and one half years after the date of enactment of this Act.

    (4) Any rule under paragraph (1), (2)(B), or (3)(B) shall be promulgated in accordance with paragraphs (2), (3), and (4) of subsection (c).

    (5) This subsection does not limit the authority of the Administrator, under any other provision of this Act or any other Federal law, to take action respecting any polychlorinated biphenyl.

SEC. 7. IMMINENT HAZARDS.

Civil action.
15 USC 2606.

(a) ACTIONS AUTHORIZED AND REQUIRED.—(1) The Administrator may commence a civil action in an appropriate district court of the United States—

(A) for seizure of an imminently hazardous chemical substance or mixture or any article containing such a substance or mixture,

(B) for relief (as authorized by subsection (b)) against any person who manufactures, processes, distributes in commerce, or uses, or disposes of, an imminently hazardous chemical substance or mixture or any article containing such a substance or mixture, or

(C) for both such seizure and relief.

A civil action may be commenced under this paragraph notwithstanding the existence of a rule under section 4, 5, or 6 or an order under section 5, and notwithstanding the pendency of any administrative or judicial proceeding under any provision of this Act.

(2) If the Administrator has not made a rule under section 6(a) immediately effective (as authorized by subsection 6(d)(2)(A)(i)) with respect to an imminently hazardous chemical substance or mixture, the Administrator shall commence in a district court of the United States with respect to such substance or mixture or article containing such substance or mixture a civil action described in subparagraph (A), (B), or (C) of paragraph (1).

Jurisdiction.

(b) RELIEF AUTHORIZED.—(1) The district court of the United States in which an action under subsection (a) is brought shall have jurisdiction to grant such temporary or permanent relief as may be necessary to protect health or the environment from the unreasonable risk associated with the chemical substance, mixture, or article involved in such action.

(2) In the case of an action under subsection (a) brought against a person who manufactures, processes, or distributes in commerce a chemical substance or mixture or an article containing a chemical substance or mixture, the relief authorized by paragraph (1) may include the issuance of a mandatory order requiring (A) in the case of purchasers of such substance, mixture, or article known to the defendant, notification to such purchasers of the risk associated with it; (B) public notice of such risk; (C) recall; (D) the replacement or repurchase of such substance, mixture, or article; or (E) any combination of the actions described in the preceding clauses.

(3) In the case of an action under subsection (a) against a chemical substance, mixture, or article, such substance, mixture, or article may be proceeded against by process of libel for its seizure and condemnation. Proceedings in such an action shall conform as nearly as possible to proceedings in rem in admiralty.

(c) VENUE AND CONSOLIDATION.—(1)(A) An action under subsection (a) against a person who manufactures, processes, or distributes a chemical substance or mixture or an article containing a chemical substance or mixture may be brought in the United States District Court for the District of Columbia or for any judicial district in which any of the defendants is found, resides, or transacts business; and process in such an action may be served on a defendant in any other district in which such defendant resides or may be found. An action under subsection (a) against a chemical substance, mixture, or article may be brought in any United States district court within the jurisdiction of which the substance, mixture, or article is found.

(B) In determining the judicial district in which an action may be brought under subsection (a) in instances in which such action may be brought in more than one judicial district, the Administrator shall take into account the convenience of the parties.

(C) Subpeonas requiring attendance of witnesses in an action brought under subsection (a) may be served in any judicial district.

(2) Whenever proceedings under subsection (a) involving identical chemical substances, mixtures, or articles are pending in courts in two or more judicial districts, they shall be consolidated for trial by order of any such court upon application reasonably made by any party in interest, upon notice to all parties in interest.

(d) ACTION UNDER SECTION 6.—Where appropriate, concurrently with the filing of an action under subsection (a) or as soon thereafter as may be practicable, the Administrator shall initiate a proceeding for the promulgation of a rule under section 6(a).

(e) REPRESENTATION.—Notwithstanding any other provision of law, in any action under subsection (a), the Administrator may direct attorneys of the Environmental Protection Agency to appear and represent the Administrator in such an action.

(f) DEFINITION.—For the purposes of subsection (a), the term "imminently hazardous chemical substance or mixture" means a chemical substance or mixture which presents an imminent and unreasonable risk of serious or widespread injury to health or the environment. Such a risk to health or the environment shall be considered imminent if it is shown that the manufacture, processing, distribution in commerce, use, or disposal of the chemical substance or mixture, or that any combination of such activities, is likely to result in such injury to health or the environment before a final rule under section 6 can protect against such risk.

SEC. 8. REPORTING AND RETENTION OF INFORMATION.

(a) REPORTS.—(1) The Administrator shall promulgate rules under which—

> Rules.
> 15 USC 2607.

> (A) each person (other than a small manufacturer or processor) who manufactures or processes or proposes to manufacture or process a chemical substance (other than a chemical substance described in subparagraph (B)(ii)) shall maintain such records, and shall submit to the Administrator such reports, as the Administrator may reasonably require, and

> (B) each person (other than a small manufacturer or processor) who manufactures or processes or proposes to manufacture or process—

>> (i) a mixture, or

>> (ii) a chemical substance in small quantities (as defined by the Administrator by rule) solely for purposes of scientific experimentation or analysis or chemical research on, or analysis of, such substance or another substance, including any such research or analysis for the development of a product,

> shall maintain records and submit to the Administrator reports but only to the extent the Administrator determines the maintenance of records or submission of reports, or both, is necessary for the effective enforcement of this Act.

The Administrator may not require in a rule promulgated under this paragraph the maintenance of records or the submission of reports with respect to changes in the proportions of the components of a mixture unless the Administrator finds that the maintenance of such records or the submission of such reports, or both, is necessary for the effective enforcement of this Act. For purposes of the compilation of the list of chemical substances required under subsection (b), the Administrator shall promulgate rules pursuant to this subsection not later than 180 days after the effective date of this Act.

(2) The Administrator may require under paragraph (1) maintenance of records and reporting with respect to the following insofar as known to the person making the report or insofar as reasonably ascertainable:

(A) The common or trade name, the chemical identity, and the molecular structure of each chemical substance or mixture for which such a report is required.

(B) The categories or proposed categories of use of each such substance or mixture.

(C) The total amount of each such substance and mixture manufactured or processed, reasonable estimates of the total amount to be manufactured or processed, the amount manufactured or processed for each of its categories of use, and reasonable estimates of the amount to be manufactured or processed for each of its categories of use or proposed categories of use.

(D) A description of the byproducts resulting from the manufacture, processing, use, or disposal of each such substance or mixture.

(E) All existing data concerning the environmental and health effects of such substance or mixture.

(F) The number of individuals exposed, and reasonable estimates of the number who will be exposed, to such substance or mixture in their places of employment and the duration of such exposure.

(G) In the initial report under paragraph (1) on such substance or mixture, the manner or method of its disposal, and in any subsequent report on such substance or mixture, any change in such manner or method.

To the extent feasible, the Administrator shall not require under paragraph (1), any reporting which is unnecessary or duplicative.

(3)(A)(i) The Administrator may by rule require a small manufacturer or processor of a chemical substance to submit to the Administrator such information respecting the chemical substance as the Administrator may require for publication of the first list of chemical substances required by subsection (b).

(ii) The Administrator may by rule require a small manufacturer or processor of a chemical substance or mixture—

(I) subject to a rule proposed or promulgated under section 4, 5(b)(4), or 6, or an order in effect under section 5(e), or

(II) with respect to which relief has been granted pursuant to a civil action brought under section 5 or 7,

to maintain such records on such substance or mixture, and to submit to the Administrator such reports on such substance or mixture, as the Administrator may reasonably require. A rule under this clause requiring reporting may require reporting with respect to the matters referred to in paragraph (2).

Standards.   (B) The Administrator, after consultation with the Administrator of the Small Business Administration, shall by rule prescribe standards for determining the manufacturers and processors which qualify as small manufacturers and processors for purposes of this paragraph and paragraph (1).

(b) INVENTORY.—(1) The Administrator shall compile, keep current, and publish a list of each chemical substance which is manufactured or processed in the United States. Such list shall at least include each chemical substance which any person reports, under section 5 or subsection (a) of this section, is manufactured or processed in the United States. Such list may not include any chemical substance which was not manufactured or processed in the United States within three years before the effective date of the rules promulgated pursuant to the last sentence of subsection (a) (1). In the case of a chemical substance for which a notice is submitted in accordance with section 5, such chemical substance shall be included in such list as of the earliest date (as determined by the Administrator) on which such substance was manufactured or processed in the United States. The Administrator shall first publish such a list not later than 315 days after the effective date of this Act. The Administrator shall not include in such list any chemical substance which is manufactured or processed only in small quantities (as defined by the Administrator by rule) solely for purposes of scientific experimentation or analysis or chemical research on, or analysis of, such substance or another substance, including such research or analysis for the development of a product.

(2) To the extent consistent with the purposes of this Act, the Administrator may, in lieu of listing, pursuant to paragraph (1), a chemical substance individually, list a category of chemical substances in which such substance is included.

(c) RECORDS.—Any person who manufactures, processes, or distributes in commerce any chemical substance or mixture shall maintain records of significant adverse reactions to health or the environment, as determined by the Administrator by rule, alleged to have been caused by the substance or mixture. Records of such adverse reactions to the health of employees shall be retained for a period of 30 years from the date such reactions were first reported to or known by the person maintaining such records. Any other record of such adverse reactions shall be retained for a period of five years from the date the information contained in the record was first reported to or known by the person maintaining the record. Records required to be maintained under this subsection shall include records of consumer allegations of personal injury or harm to health, reports of occupational disease or injury, and reports or complaints of injury to the environment submitted to the manufacturer, processor, or distributor in commerce from any source. Upon request of any duly designated representative of the Administrator, each person who is required to maintain records under this subsection shall permit the inspection of such records and shall submit copies of such records.

(d) HEALTH AND SAFETY STUDIES.—The Administrator shall promulgate rules under which the Administrator shall require any person who manufactures, processes, or distributes in commerce or who proposes to manufacture, process, or distribute in commerce any chemical substance or mixture (or with respect to paragraph (2), any person who has possession of a study) to submit to the Administrator—   Rules.

(1) lists of health and safety studies (A) conducted or initiated by or for such person with respect to such substance or mixture at any time, (B) known to such person, or (C) reasonably ascertainable by such person, except that the Administrator may exclude certain types or categories of studies from the requirements of this subsection if the Administrator finds that submission of lists of such studies are unnecessary to carry out the purposes of this Act; and

(2) copies of any study contained on a list submitted pursuant to paragraph (1) or otherwise known by such person.

(e) NOTICE TO ADMINISTRATOR OF SUBSTANTIAL RISKS.—Any person who manufactures, processes, or distributes in commerce a chemical substance or mixture and who obtains information which reasonably supports the conclusion that such substance or mixture presents a substantial risk of injury to health or the environment shall immediately inform the Administrator of such information unless such person has actual knowledge that the Administrator has been adequately informed of such information.

(f) DEFINITIONS.—For purposes of this section, the terms "manufacture" and "process" mean manufacture or process for commercial purposes.

SEC. 9. RELATIONSHIP TO OTHER FEDERAL LAWS.

15 USC 2608.

(a) LAWS NOT ADMINISTERED BY THE ADMINISTRATOR.—(1) If the Administrator has reasonable basis to conclude that the manufacture, processing, distribution in commerce, use, or disposal of a chemical substance or mixture, or that any combination of such activities, presents or will present an unreasonable risk of injury to health or the environment and determines, in the Administrator's discretion, that such risk may be prevented or reduced to a sufficient extent by action taken under a Federal law not administered by the Administrator,

Report.

the Administrator shall submit to the agency which administers such law a report which describes such risk and includes in such description a specification of the activity or combination of activities which the Administrator has reason to believe so presents such risk. Such report shall also request such agency—

(A)(i) to determine if the risk described in such report may be prevented or reduced to a sufficient extent by action taken under such law, and

(ii) if the agency determines that such risk may be so prevented or reduced, to issue an order declaring whether or not the activity or combination of activities specified in the description of such risk presents such risk; and

(B) to respond to the Administrator with respect to the matters described in subparagraph (A).

Publication in Federal Register.

Any report of the Administrator shall include a detailed statement of the information on which it is based and shall be published in the Federal Register. The agency receiving a request under such a report shall make the requested determination, issue the requested order, and make the requested response within such time as the Administrator specifies in the request, but such time specified may not be less than 90 days from the date the request was made. The response of an agency shall be accompanied by a detailed statement of the findings and conclusions of the agency and shall be published in the Federal Register.

(2) If the Administrator makes a report under paragraph (1) with respect to a chemical substance or mixture and the agency to which such report was made either—

(A) issues an order declaring that the activity or combination of activities specified in the description of the risk described in the report does not present the risk described in the report, or

(B) initiates, within 90 days of the publication in the Federal Register of the response of the agency under paragraph (1), action under the law (or laws) administered by such agency to protect against such risk associated with such activity or combination of activities,

the Administrator may not take any action under section 6 or 7 with respect to such risk.

(3) If the Administrator has initiated action under section 6 or 7 with respect to a risk associated with a chemical substance or mixture which was the subject of a report made to an agency under paragraph (1), such agency shall before taking action under the law (or laws)

administered by it to protect against such risk consult with the Administrator for the purpose of avoiding duplication of Federal action against such risk.

(b) Laws Administered by the Administrator.—The Administrator shall coordinate actions taken under this Act with actions taken under other Federal laws administered in whole or in part by the Administrator. If the Administrator determines that a risk to health or the environment associated with a chemical substance or mixture could be eliminated or reduced to a sufficient extent by actions taken under the authorities contained in such other Federal laws, the Administrator shall use such authorities to protect against such risk unless the Administrator determines, in the Administrator's discretion, that it is in the public interest to protect against such risk by actions taken under this Act. This subsection shall not be construed to relieve the Administrator of any requirement imposed on the Administrator by such other Federal laws.

(c) Occupational Safety and Health.—In exercising any authority under this Act, the Administrator shall not, for purposes of section 4(b)(1) of the Occupational Safety and Health Act of 1970, be deemed to be exercising statutory authority to prescribe or enforce standards or regulations affecting occupational safety and health.   *29 USC 651 note.*

(d) Coordination.—In administering this Act, the Administrator shall consult and coordinate with the Secretary of Health, Education, and Welfare and the heads of any other appropriate Federal executive department or agency, any relevant independent regulatory agency, and any other appropriate instrumentality of the Federal Government for the purpose of achieving the maximum enforcement of this Act while imposing the least burdens of duplicative requirements on those subject to the Act and for other purposes. The Administrator shall, in the report required by section 30, report annually to the Congress on actions taken to coordinate with such other Federal departments, agencies, or instrumentalities, and on actions taken to coordinate the authority under this Act with the authority granted under other Acts referred to in subsection (b).

SEC. 10. RESEARCH, DEVELOPMENT, COLLECTION, DISSEMINATION, AND UTILIZATION OF DATA.

(a) Authority.—The Administrator shall, in consultation and cooperation with the Secretary of Health, Education, and Welfare and with other heads of appropriate departments and agencies, conduct such research, development, and monitoring as is necessary to carry out the purposes of this Act. The Administrator may enter into contracts and may make grants for research, development, and monitoring under this subsection. Contracts may be entered into under this subsection without regard to sections 3648 and 3709 of the Revised Statutes (31 U.S.C. 529, 14 U.S.C. 5).   *15 USC 2609.*

(b) Data Systems.—(1) The Administrator shall establish, administer, and be responsible for the continuing activities of an interagency committee which shall design, establish, and coordinate an efficient and effective system, within the Environmental Protection Agency, for the collection, dissemination to other Federal departments and agencies, and use of data submitted to the Administrator under this Act.

(2)(A) The Administrator shall, in consultation and cooperation with the Secretary of Health, Education, and Welfare and other heads of appropriate departments and agencies design, establish, and coordinate an efficient and effective system for the retrieval of toxicological and other scientific data which could be useful to the Administrator in carrying out the purposes of this Act. Systematized retrieval shall be developed for use by all Federal and other departments and agencies with responsibilities in the area of regulation or study of chemical substances and mixtures and their effect on health or the environment.

(B) The Administrator, in consultation and cooperation with the Secretary of Health, Education, and Welfare, may make grants and enter into contracts for the development of a data retrieval system described in subparagraph (A). Contracts may be entered into under this subparagraph without regard to sections 3648 and 3709 of the Revised Statutes (31 U.S.C. 529, 41 U.S.C. 5).

(c) SCREENING TECHNIQUES.—The Administrator shall coordinate, with the Assistant Secretary for Health of the Department of Health, Education, and Welfare, research undertaken by the Administrator and directed toward the development of rapid, reliable, and economical screening techniques for carcinogenic, mutagenic, teratogenic, and ecological effects of chemical substances and mixtures.

(d) MONITORING.—The Administrator shall, in consultation and cooperation with the Secretary of Health, Education, and Welfare, establish and be responsible for research aimed at the development, in cooperation with local, State, and Federal agencies, of monitoring techniques and instruments which may be used in the detection of toxic chemical substances and mixtures and which are reliable, economical, and capable of being implemented under a wide variety of conditions.

(e) BASIC RESEARCH.—The Administrator shall, in consultation and cooperation with the Secretary of Health, Education, and Welfare, establish research programs to develop the fundamental scientific basis of the screening and monitoring techniques described in subsections (c) and (d), the bounds of the reliability of such techniques, and the opportunities for their improvement.

(f) TRAINING.—The Administrator shall establish and promote programs and workshops to train or facilitate the training of Federal laboratory and technical personnel in existing or newly developed screening and monitoring techniques.

(g) EXCHANGE OF RESEARCH AND DEVELOPMENT RESULTS.—The Administrator shall, in consultation with the Secretary of Health, Education, and Welfare and other heads of appropriate departments and agencies, establish and coordinate a system for exchange among Federal, State, and local authorities of research and development results respecting toxic chemical substances and mixtures, including a system to facilitate and promote the development of standard data format and analysis and consistent testing procedures.

SEC. 11. INSPECTIONS AND SUBPOENAS.

15 USC 2610.

(a) IN GENERAL.—For purposes of administering this Act, the Administrator, and any duly designated representative of the Administrator, may inspect any establishment, facility, or other premises in which chemical substances or mixtures are manufactured, processed, stored, or held before or after their distribution in commerce and any conveyance being used to transport chemical substances, mixtures, or such articles in connection with distribution in commerce. Such an inspection may only be made upon the presentation of appropriate credentials and of a written notice to the owner, operator, or agent in charge of the premises or conveyance to be inspected. A separate notice shall be given for each such inspection, but a notice shall not be required for each entry made during the period covered by the inspection. Each such inspection shall be commenced and completed with reasonable promptness and shall be conducted at reasonable times, within reasonable limits, and in a reasonable manner.

(b) SCOPE.—(1) Except as provided in paragraph (2), an inspection conducted under subsection (a) shall extend to all things within the premises or conveyance inspected (including records, files, papers, processes, controls, and facilities) bearing on whether the requirements of this Act applicable to the chemical substances or mixtures within such premises or conveyance have been complied with.

(2) No inspection under subsection (a) shall extend to—

    (A) financial data,

    (B) sales data (other than shipment data),

    (C) pricing data,

    (D) personnel data, or

    (E) research data (other than data required by this Act or under a rule promulgated thereunder),

unless the nature and extent of such data are described with reasonable specificity in the written notice required by subsection (a) for such inspection.

(c) SUBPOENAS.—In carrying out this Act, the Administrator may by subpoena require the attendance and testimony of witnesses and the production of reports, papers, documents, answers to questions, and other information that the Administrator deems necessary. Witnesses shall be paid the same fees and mileage that are paid witnesses in the courts of the United States. In the event of contumacy, failure, or refusal of any person to obey any such subpoena, any district court of the United States in which venue is proper shall have jurisdiction to order any such person to comply with such subpoena. Any failure to obey such an order of the court is punishable by the court as a contempt thereof.

SEC. 12. EXPORTS.

(a) IN GENERAL.—(1) Except as provided in paragraph (2) and subsection (b), this Act (other than section 8) shall not apply to any chemical substance, mixture, or to an article containing a chemical substance or mixture, if—   15 USC 2611.

    (A) it can be shown that such substance, mixture, or article is being manufactured, processed, or distributed in commerce for export from the United States, unless such substance, mixture, or article was, in fact, manufactured, processed, or distributed in commerce, for use in the United States, and

    (B) such substance, mixture, or article (when distributed in commerce), or any container in which it is enclosed (when so distributed), bears a stamp or label stating that such substance, mixture, or article is intended for export.

(2) Paragraph (1) shall not apply to any chemical substance, mixture, or article if the Administrator finds that the substance, mixture, or article will present an unreasonable risk of injury to health within the United States or to the environment of the United States. The Administrator may require, under section 4, testing of any chemical substance or mixture exempted from this Act by paragraph (1) for the purpose of determining whether or not such substance or mixture presents an unreasonable risk of injury to health within the United States or to the environment of the United States.

(b) NOTICE.—(1) If any person exports or intends to export to a foreign country a chemical substance or mixture for which the submission of data is required under section 4 or 5(b), such person shall notify the Administrator of such exportation or intent to export and the Administrator shall furnish to the government of such country notice of the availability of the data submitted to the Administrator under such section for such substance or mixture.

(2) If any person exports or intends to export to a foreign country a chemical substance or mixture for which an order has been issued

under section 5 or a rule has been proposed or promulgated under section 5 or 6, or with respect to which an action is pending, or relief has been granted under section 5 or 7, such person shall notify the Administrator of such exportation or intent to export and the Administrator shall furnish to the government of such country notice of such rule, order, action, or relief.

SEC. 13. ENTRY INTO CUSTOMS TERRITORY OF THE UNITED STATES.

15 USC 2612.

19 USC 1202.

(a) IN GENERAL.—(1) The Secretary of the Treasury shall refuse entry into the customs territory of the United States (as defined in general headnote 2 to the Tariff Schedules of the United States) of any chemical substance, mixture, or article containing a chemical substance or mixture offered for such entry if—

(A) it fails to comply with any rule in effect under this Act, or

(B) it is offered for entry in violation of section 5 or 6, a rule or order under section 5 or 6, or an order issued in a civil action brought under section 5 or 7.

Notification.

(2) If a chemical substance, mixture, or article is refused entry under paragraph (1), the Secretary of the Treasury shall notify the consignee of such entry refusal, shall not release it to the consignee, and shall cause its disposal or storage (under such rules as the Secretary of the Treasury may prescribe) if it has not been exported by the consignee within 90 days from the date of receipt of notice of such refusal, except that the Secretary of the Treasury may, pending a review by the Administrator of the entry refusal, release to the consignee such substance, mixture, or article on execution of bond for the amount of the full invoice of such substance, mixture, or article (as such value is set forth in the customs entry), together with the duty thereon. On failure to return such substance, mixture, or article for any cause to the custody of the Secretary of the Treasury when demanded, such consignee shall be liable to the United States for liquidated damages equal to the full amount of such bond. All charges for storage, cartage, and labor on and for disposal of substances, mixtures, or articles which are refused entry or release under this section shall be paid by the owner or consignee, and in default of such payment shall constitute a lien against any future entry made by such owner or consignee.

(b) RULES.—The Secretary of the Treasury, after consultation with the Administrator, shall issue rules for the administration of subsection (a) of this section.

SEC. 14. DISCLOSURE OF DATA.

15 USC 2613.

(a) IN GENERAL.—Except as provided by subsection (b), any information reported to, or otherwise obtained by, the Administrator (or any representative of the Administrator) under this Act, which is exempt from disclosure pursuant to subsection (a) of section 552 of title 5, United States Code, by reason of subsection (b)(4) of such section, shall, notwithstanding the provisions of any other section of this Act, not be disclosed by the Administrator or by any officer or employee of the United States, except that such information—

(1) shall be disclosed to any officer or employee of the United States—

(A) in connection with the official duties of such officer or employee under any law for the protection of health or the environment, or

(B) for specific law enforcement purposes;

(2) shall be disclosed to contractors with the United States and employees of such contractors if in the opinion of the Administra-

tor such disclosure is necessary for the satisfactory performance by the contractor of a contract with the United States entered into on or after the date of enactment of this Act for the performance of work in connection with this Act and under such conditions as the Administrator may specify;

(3) shall be disclosed if the Administrator determines it necessary to protect health or the environment against an unreasonable risk of injury to health or the environment; or

(4) may be disclosed when relevant in any proceeding under this Act, except that disclosure in such a proceeding shall be made in such manner as to preserve confidentiality to the extent practicable without impairing the proceeding.

In any proceeding under section 552(a) of title 5, United States Code, to obtain information the disclosure of which has been denied because of the provisions of this subsection, the Administrator may not rely on section 552(b)(3) of such title to sustain the Administrator's action.

(b) DATA FROM HEALTH AND SAFETY STUDIES.—(1) Subsection (a) does not prohibit the disclosure of—

(A) any health and safety study which is submitted under this Act with respect to—

(i) any chemical substance or mixture which, on the date on which such study is to be disclosed has been offered for commercial distribution, or

(ii) any chemical substance or mixture for which testing is required under section 4 or for which notification is required under section 5, and

(B) any data reported to, or otherwise obtained by, the Administrator from a health and safety study which relates to a chemical substance or mixture described in clause (i) or (ii) of subparagraph (A).

This paragraph does not authorize the release of any data which discloses processes used in the manufacturing or processing of a chemical substance or mixture or, in the case of a mixture, the release of data disclosing the portion of the mixture comprised by any of the chemical substances in the mixture.

(2) If a request is made to the Administrator under subsection (a) of section 552 of title 5, United States Code, for information which is described in the first sentence of paragraph (1) and which is not information described in the second sentence of such paragraph, the Administrator may not deny such request on the basis of subsection (b)(4) of such section.

(c) DESIGNATION AND RELEASE OF CONFIDENTIAL DATA.—(1) In submitting data under this Act, a manufacturer, processor, or distributor in commerce may (A) designate the data which such person believes is entitled to confidential treatment under subsection (a), and (B) submit such designated data separately from other data submitted under this Act. A designation under this paragraph shall be made in writing and in such manner as the Administrator may prescribe.

(2)(A) Except as provided by subparagraph (B), if the Administrator proposes to release for inspection data which has been designated under paragraph (1)(A), the Administrator shall notify, in writing and by certified mail, the manufacturer, processor, or distributor in commerce who submitted such data of the intent to release such data. If the release of such data is to be made pursuant to a request made under section 552(a) of title 5, United States Code, such notice shall be given immediately upon approval of such request by the Administrator. The Administrator may not release such data until the expiration of 30 days after the manufacturer, processor, or distributor in commerce submitting such data has received the notice required by this subparagraph.

(B)(i) Subparagraph (A) shall not apply to the release of information under paragraph (1), (2), (3), or (4) of subsection (a), except that the Administrator may not release data under paragraph (3) of subsection (a) unless the Administrator has notified each manufacturer, processor, and distributor in commerce who submitted such data of such release. Such notice shall be made in writing by certified mail at least 15 days before the release of such data, except that if the Administrator determines that the release of such data is necessary to protect against an imminent, unreasonable risk of injury to health or the environment, such notice may be made by such means as the Administrator determines will provide notice at least 24 hours before such release is made.

(ii) Subparagraph (A) shall not apply to the release of information described in subsection (b)(1) other than information described in the second sentence of such subsection.

(d) CRIMINAL PENALTY FOR WRONGFUL DISCLOSURE.—(1) Any officer or employee of the United States or former officer or employee of the United States, who by virtue of such employment or official position has obtained possession of, or has access to, material the disclosure of which is prohibited by subsection (a), and who knowing that disclosure of such material is prohibited by such subsection, willfully discloses the material in any manner to any person not entitled to receive it, shall be guilty of a misdemeanor and fined not more than $5,000 or imprisoned for not more than one year, or both. Section 1905 of title 18, United States Code, does not apply with respect to the publishing, divulging, disclosure, or making known of, or making available, information reported or otherwise obtained under this Act.

(2) For the purposes of paragraph (1), any contractor with the United States who is furnished information as authorized by subsection (a)(2), and any employee of any such contractor, shall be considered to be an employee of the United States.

(e) ACCESS BY CONGRESS.—Notwithstanding any limitation contained in this section or any other provision of law, all information reported to or otherwise obtained by the Administrator (or any representative of the Administrator) under this Act shall be made available, upon written request of any duly authorized committee of the Congress, to such committee.

SEC. 15. PROHIBITED ACTS.

It shall be unlawful for any person to—

(1) fail or refuse to comply with (A) any rule promulgated or order issued under section 4, (B) any requirement prescribed by section 5 or 6, or (C) any rule promulgated or order issued under section 5 or 6;

(2) use for commercial purposes a chemical substance or mixture which such person knew or had reason to know was manufactured, processed, or distributed in commerce in violation of section 5 or 6, a rule or order under section 5 or 6, or an order issued in action brought under section 5 or 7;

(3) fail or refuse to (A) establish or maintain records, (B) submit reports, notices, or other information, or (C) permit access to or copying of records, as required by this Act or a rule thereunder; or

(4) fail or refuse to permit entry or inspection as required by section 11.

SEC. 16. PENALTIES.

(a) CIVIL.—(1) Any person who violates a provision of section 15 shall be liable to the United States for a civil penalty in an amount not to exceed $25,000 for each such violation. Each day such a violation continues shall, for purposes of this subsection, constitute a separate violation of section 15.

15 USC 2615.

(2) (A) A civil penalty for a violation of section 15 shall be assessed by the Administrator by an order made on the record after opportunity (provided in accordance with this subparagraph) for a hearing in accordance with section 554 of title 5, United States Code. Before issuing such an order, the Administrator shall give written notice to the person to be assessed a civil penalty under such order of the Administrator's proposal to issue such order and provide such person an opportunity to request, within 15 days of the date the notice is received by such person, such a hearing on the order.

Hearing.

(B) In determining the amount of a civil penalty, the Administrator shall take into account the nature, circumstances, extent, and gravity of the violation or violations and, with respect to the violator, ability to pay, effect on ability to continue to do business, any history of prior such violations, the degree of culpability, and such other matters as justice may require.

(C) The Administrator may compromise, modify, or remit, with or without conditions, any civil penalty which may be imposed under this subsection. The amount of such penalty, when finally determined, or the amount agreed upon in compromise, may be deducted from any sums owing by the United States to the person charged.

(3) Any person who requested in accordance with paragraph (2) (A) a hearing respecting the assessment of a civil penalty and who is aggrieved by an order assessing a civil penalty may file a petition for judicial review of such order with the United States Court of Appeals for the District of Columbia Circuit or for any other circuit in which such person resides or transacts business. Such a petition may only be filed within the 30-day period beginning on the date the order making such assessment was issued.

Petition for judicial review.

(4) If any person fails to pay an assessment of a civil penalty—

(A) after the order making the assessment has become a final order and if such person does not file a petition for judicial review of the order in accordance with paragraph (3), or

(B) after a court in an action brought under paragraph (3) has entered a final judgment in favor of the Administrator,

the Attorney General shall recover the amount assessed (plus interest at currently prevailing rates from the date of the expiration of the 30-day period referred to in paragraph (3) or the date of such final judgment, as the case may be) in an action brought in any appropriate district court of the United States. In such an action, the validity, amount, and appropriateness of such penalty shall not be subject to review.

(b) CRIMINAL.—Any person who knowingly or willfully violates any provision of section 15 shall, in addition to or in lieu of any civil penalty which may be imposed under subsection (a) of this section for such violation, be subject, upon conviction, to a fine of not more than $25,000 for each day of violation, or to imprisonment for not more than one year, or both.

SEC. 17. SPECIFIC ENFORCEMENT AND SEIZURE.

(a) SPECIFIC ENFORCEMENT.—(1) The district courts of the United States shall have jurisdiction over civil actions to—

15 USC 2616.

(A) restrain any violation of section 15,

(B) restrain any person from taking any action prohibited by section 5 or 6 or by a rule or order under section 5 or 6,

(C) compel the taking of any action required by or under this Act, or

(D) direct any manufacturer or processor of a chemical substance or mixture manufactured or processed in violation of section 5 or 6 or a rule or order under section 5 or 6 and distributed in commerce, (i) to give notice of such fact to distributors in commerce of such substance or mixture and, to the extent reasonably ascertainable, to other persons in possession of such substance or mixture or exposed to such substance or mixture, (ii) to give public notice of such risk of injury, and (iii) to either replace or repurchase such substance or mixture, whichever the person to which the requirement is directed elects.

(2) A civil action described in paragraph (1) may be brought—

(A) in the case of a civil action described in subparagraph (A) of such paragraph, in the United States district court for the judicial district wherein any act, omission, or transaction constituting a violation of section 15 occurred or wherein the defendant is found or transacts business, or

(B) in the case of any other civil action described in such paragraph, in the United States district court for the judicial district wherein the defendant is found or transacts business.

In any such civil action process may be served on a defendant in any judicial district in which a defendant resides or may be found. Subpoenas requiring attendance of witnesses in any such action may be served in any judicial district.

(b) SEIZURE.—Any chemical substance or mixture which was manufactured, processed, or distributed in commerce in violation of this Act or any rule promulgated or order issued under this Act or any article containing such a substance or mixture shall be liable to be proceeded against, by process of libel for the seizure and condemnation of such substance, mixture, or article, in any district court of the United States within the jurisdiction of which such substance, mixture, or article is found. Such proceedings shall conform as nearly as possible to proceedings in rem in admiralty.

SEC. 18. PREEMPTION.

15 USC 2617.

(a) EFFECT ON STATE LAW.—(1) Except as provided in paragraph (2), nothing in this Act shall affect the authority of any State or political subdivision of a State to establish or continue in effect regulation of any chemical substance, mixture, or article containing a chemical substance or mixture.

(2) Except as provided in subsection (b)—

(A) if the Administrator requires by a rule promulgated under section 4 the testing of a chemical substance or mixture, no State or political subdivision may, after the effective date of such rule, establish or continue in effect a requirement for the testing of such substance or mixture for purposes similar to those for which testing is required under such rule; and

(B) if the Administrator prescribes a rule or order under section 5 or 6 (other than a rule imposing a requirement described in subsection (a)(6) of section 6) which is applicable to a chemical substance or mixture, and which is designed to protect against a risk of injury to health or the environment associated with such substance or mixture, no State or political subdivision of a State may, after the effective date of such requirement, establish or continue in effect, any requirement which is applicable to such substance or mixture, or an article containing such substance or mixture, and which is designed to protect against such risk unless such requirement (i) is identical to the requirement prescribed by the Administrator, (ii) is adopted under the authority of the Clean Air Act or any other Federal law, or (iii) prohibits the use of such substance or mixture in such State or political subdivision (other than its use in the manufacture or processing of other substances or mixtures).

(b) EXEMPTION.—Upon application of a State or political subdivision of a State the Administrator may by rule exempt from subsection (a)(2), under such conditions as may be prescribed in such rule, a requirement of such State or political subdivision designed to protect against a risk of injury to health or the environment associated with a chemical substance, mixture, or article containing a chemical substance or mixture if— *[Application.]*

> (1) compliance with the requirement would not cause the manufacturing, processing, distribution in commerce, or use of the substance, mixture, or article to be in violation of the applicable requirement under this Act described in subsection (a)(2), and
>
> (2) the State or political subdivision requirement (A) provides a significantly higher degree of protection from such risk than the requirement under this Act described in subsection (a)(2) and (B) does not, through difficulties in marketing, distribution, or other factors, unduly burden interstate commerce.

SEC. 19. JUDICIAL REVIEW.

(a) IN GENERAL.—(1)(A) Not later than 60 days after the date of the promulgation of a rule under section 4(a), 5(a)(2), 5(b)(4), 6(a), 6(e), or 8, any person may file a petition for judicial review of such rule with the United States Court of Appeals for the District of Columbia Circuit or for the circuit in which such person resides or in which such person's principal place of business is located. Courts of appeals of the United States shall have exclusive jurisdiction of any action to obtain judicial review (other than in an enforcement proceeding) of such a rule if any district court of the United States would have had jurisdiction of such action but for this subparagraph. *[Petition. 15 USC 2618.]*

(B) Courts of appeals of the United States shall have exclusive jurisdiction of any action to obtain judicial review (other than in an enforcement proceeding) of an order issued under subparagraph (A) or (B) of section 6(b)(1) if any district court of the United States would have had jurisdiction of such action but for this subparagraph. *[Jurisdiction.]*

(2) Copies of any petition filed under paragraph (1)(A) shall be transmitted forthwith to the Administrator and to the Attorney General by the clerk of the court with which such petition was filed. The provisions of section 2112 of title 28, United States Code, shall apply to the filing of the rulemaking record of proceedings on which the Administrator based the rule being reviewed under this section and to the transfer of proceedings between United States courts of appeals. *[Petition copies, transmittal to Administrator and Attorney General.]*

(3) For purposes of this section, the term "rulemaking record" means— *["Rulemaking record."]*

> (A) the rule being reviewed under this section;
>
> (B) in the case of a rule under section 4(a), the finding required by such section, in the case of a rule under section 5(b)(4), the finding required by such section, in the case of a rule under section 6(a) the finding required by section 5(f) or 6(a), as the case may be, in the case of a rule under section 6(a), the statement required by section 6(c)(1), and in the case of a rule under section 6(e), the findings required by paragraph (2)(B) or (3)(B) of such section, as the case may be;

(C) any transcript required to be made of oral presentations made in proceedings for the promulgation of such rule;

(D) any written submission of interested parties respecting the promulgation of such rule; and

Notice, publication in Federal Register.

(E) any other information which the Administrator considers to be relevant to such rule and which the Administrator identified, on or before the date of the promulgation of such rule, in a notice published in the Federal Register.

(b) ADDITIONAL SUBMISSIONS AND PRESENTATIONS; MODIFICATIONS.—If in an action under this section to review a rule the petitioner or the Administrator applies to the court for leave to make additional oral submissions or written presentations respecting such rule and shows to the satisfaction of the court that such submissions and presentations would be material and that there were reasonable grounds for the submissions and failure to make such submissions and presentations in the proceeding before the Administrator, the court may order the Administrator to provide additional opportunity to make such submissions and presentations. The Administrator may modify or set aside the rule being reviewed or make a new rule by reason of the additional submissions and presentations and shall file such modified or new rule with the return of such submissions and presentations. The court shall thereafter review such new or modified rule.

Review.

(c) STANDARD OF REVIEW.—(1)(A) Upon the filing of a petition under subsection (a)(1) for judicial review of a rule, the court shall have jurisdiction (i) to grant appropriate relief, including interim relief, as provided in chapter 7 of title 5, United States Code, and (ii) except as otherwise provided in subparagraph (B), to review such rule in accordance with chapter 7 of title 5, United States Code.

(B) Section 706 of title 5, United States Code, shall apply to review of a rule under this section, except that—

(i) in the case of review of a rule under section 4(a), 5(b)(4), 6(a), or 6(e), the standard for review prescribed by paragraph (2)(E) of such section 706 shall not apply and the court shall hold unlawful and set aside such rule if the court finds that the rule is not supported by substantial evidence in the rulemaking record (as defined in subsection (a)(3)) taken as a whole;

(ii) in the case of review of a rule under section 6(a), the court shall hold unlawful and set aside such rule if it finds that—

(I) a determination by the Administrator under section 6(c)(3) that the petitioner seeking review of such rule is not entitled to conduct (or have conducted) cross-examination or to present rebuttal submissions, or

(II) a rule of, or ruling by, the Administrator under section 6(c)(3) limiting such petitioner's cross-examination or oral presentations,

has precluded disclosure of disputed material facts which was necessary to a fair determination by the Administrator of the rulemaking proceeding taken as a whole; and section 706(2)(D) shall not apply with respect to a determination, rule, or ruling referred to in subclause (I) or (II); and

(iii) the court may not review the contents and adequacy of—

(I) any statement required to be made pursuant to section 6(c)(1), or

(II) any statement of basis and purpose required by section 553(c) of title 5, United States Code, to be incorporated in the rule

except as part of a review of the rulemaking record taken as a whole.

The term "evidence" as used in clause (i) means any matter in the rulemaking record. **"Evidence."**

(C) A determination, rule, or ruling of the Administrator described in subparagraph (B)(ii) may be reviewed only in an action under this section and only in accordance with such subparagraph.

(2) The judgment of the court affirming or setting aside, in whole or in part, any rule reviewed in accordance with this section shall be final, subject to review by the Supreme Court of the United States upon certiorari or certification, as provided in section 1254 of title 28, United States Code.

(d) Fees and costs.—The decision of the court in an action commenced under subsection (a), or of the Supreme Court of the United States on review of such a decision, may include an award of costs of suit and reasonable fees for attorneys and expert witnesses if the court determines that such an award is appropriate.

(e) Other remedies.—The remedies as provided in this section shall be in addition to and not in lieu of any other remedies provided by law.

SEC. 20. CITIZENS' CIVIL ACTIONS.

(a) In General.—Except as provided in subsection (b), any person may commence a civil action— **15 USC 2619.**

   (1) against any person (including (A) the United States, and (B) any other governmental instrumentality or agency to the extent permitted by the eleventh amendment to the Constitution) who is alleged to be in violation of this Act or any rule promulgated under section 4, 5, or 6 or order issued under section 5 to restrain such violation, or

   (2) against the Administrator to compel the Administrator to perform any act or duty under this Act which is not discretionary.

Any civil action under paragraph (1) shall be brought in the United States district court for the district in which the alleged violation occurred or in which the defendant resides or in which the defendant's principal place of business is located. Any action brought under paragraph (2) shall be brought in the United States District Court for the District of Columbia, or the United States district court for the judicial district in which the plaintiff is domiciled. The district courts **Jurisdiction.** of the United States shall have jurisdiction over suits brought under this section, without regard to the amount in controversy or the citizenship of the parties. In any civil action under this subsection process may be served on a defendant in any judicial district in which the defendant resides or may be found and subpoenas for witnesses may be served in any judicial district.

(b) Limitation.—No civil action may be commenced—

   (1) under subsection (a)(1) to restrain a violation of this Act or rule or order under this Act—

      (A) before the expiration of 60 days after the plaintiff **Notice.** has given notice of such violation (i) to the Administrator, and (ii) to the person who is alleged to have committed such violation, or

      (B) if the Administrator has commenced and is diligently prosecuting a proceeding for the issuance of an order under section 16(a)(2) to require compliance with this Act or with such rule or order or if the Attorney General has commenced and is diligently prosecuting a civil action in a court of the United States to require compliance with this Act or with such rule or order, but if such proceeding or civil action is commenced after the giving of notice, any person giving such notice may intervene as a matter of right in such proceeding or action; or

Notice.

(2) under subsection (a) (2) before the expiration of 60 days after the plaintiff has given notice to the Administrator of the alleged failure of the Administrator to perform an act or duty which is the basis for such action or, in the case of an action under such subsection for the failure of the Administrator to file an action under section 7, before the expiration of ten days after such notification.

Rule.

Notice under this subsection shall be given in such manner as the Administrator shall prescribe by rule.

(c) GENERAL.—(1) In any action under this section, the Administrator, if not a party, may intervene as a matter of right.

(2) The court, in issuing any final order in any action brought pursuant to subsection (a), may award costs of suit and reasonable fees for attorneys and expert witnesses if the court determines that such an award is appropriate. Any court, in issuing its decision in an action brought to review such an order, may award costs of suit and reasonable fees for attorneys if the court determines that such an award is appropriate.

(3) Nothing in this section shall restrict any right which any person (or class of persons) may have under any statute or common law to seek enforcement of this Act or any rule or order under this Act or to seek any other relief.

(d) CONSOLIDATION.—When two or more civil actions brought under subsection (a) involving the same defendant and the same issues or violations are pending in two or more judicial districts, such pending actions, upon application of such defendants to such actions which is made to a court in which any such action is brought, may, if such court in its discretion so decides, be consolidated for trial by order (issued after giving all parties reasonable notice and opportunity to be heard) of such court and tried in—

(1) any district which is selected by such defendant and in which one of such actions is pending,

(2) a district which is agreed upon by stipulation between all the parties to such actions and in which one of such actions is pending, or

(3) a district which is selected by the court and in which one of such actions is pending.

The court issuing such an order shall give prompt notification of the order to the other courts in which the civil actions consolidated under the order are pending.

SEC. 21. CITIZENS' PETITIONS.

15 USC 2620.

(a) IN GENERAL.—Any person may petition the Administrator to initiate a proceeding for the issuance, amendment, or repeal of a rule under section 4, 6, or 8 or an order under section 5(e) or (6)(b)(2).

(b) PROCEDURES.—(1) Such petition shall be filed in the principal office of the Administrator and shall set forth the facts which it is claimed establish that it is necessary to issue, amend, or repeal a rule under section 4, 6, or 8 or an order under section 5(e), 6(b)(1)(A), or 6(b)(1)(B).

Public hearing.

(2) The Administrator may hold a public hearing or may conduct such investigation or proceeding as the Administrator deems appropriate in order to determine whether or not such petition should be granted.

(3) Within 90 days after filing of a petition described in paragraph (1), the Administrator shall either grant or deny the petition. If the Administrator grants such petition, the Administrator shall promptly

commence an appropriate proceeding in accordance with section 4, 5, 6, or 8. If the Administrator denies such petition, the Administrator shall publish in the Federal Register the Administrator's reasons for such denial.

(4)(A) If the Administrator denies a petition filed under this section (or if the Administrator fails to grant or deny such petition within the 90-day period) the petitioner may commence a civil action in a district court of the United States to compel the Administrator to initiate a rulemaking proceeding as requested in the petition. Any such action shall be filed within 60 days after the Administrator's denial of the petition or, if the Administrator fails to grant or deny the petition within 90 days after filing the petition, within 60 days after the expiration of the 90-day period.

(B) In an action under subparagraph (A) respecting a petition to initiate a proceeding to issue a rule under section 4, 6, or 8 or an order under section 5(e) or 6(b)(2), the petitioner shall be provided an opportunity to have such petition considered by the court in a de novo proceeding. If the petitioner demonstrates to the satisfaction of the court by a preponderance of the evidence that—

(i) in the case of a petition to initiate a proceeding for the issuance of a rule under section 4 or an order under section 5(e)—

(I) information available to the Administrator is insufficient to permit a reasoned evaluation of the health and environmental effects of the chemical substance to be subject to such rule or order; and

(II) in the absence of such information, the substance may present an unreasonable risk to health or the environment, or the substance is or will be produced in substantial quantities and it enters or may reasonably be anticipated to enter the environment in substantial quantities or there is or may be significant or substantial human exposure to it; or

(ii) in the case of a petition to initiate a proceeding for the issuance of a rule under section 6 or 8 or an order under section 6(b)(2), there is a reasonable basis to conclude that the issuance of such a rule or order is necessary to protect health or the environment against an unreasonable risk of injury to health or the environment.

the court shall order the Administrator to initiate the action requested by the petitioner. If the court finds that the extent of the risk to health or the environment alleged by the petitioner is less than the extent of risks to health or the environment with respect to which the Administrator is taking action under this Act and there are insufficient resources available to the Administrator to take the action requested by the petitioner, the court may permit the Administrator to defer initiating the action requested by the petitioner until such time as the court prescribes.

(C) The court in issuing any final order in any action brought pursuant to subparagraph (A) may award costs of suit and reasonable fees for attorneys and expert witnesses if the court determines that such an award is appropriate. Any court, in issuing its decision in an action brought to review such an order, may award costs of suit and reasonable fees for attorneys if the court determines that such an award is appropriate.

(5) The remedies under this section shall be in addition to, and not in lieu of, other remedies provided by law.

SEC. 22. NATIONAL DEFENSE WAIVER.

15 USC 2621.

The Administrator shall waive compliance with any provision of this Act upon a request and determination by the President that the requested waiver is necessary in the interest of national defense. The Administrator shall maintain a written record of the basis upon which such waiver was granted and make such record available for in camera examination when relevant in a judicial proceeding under this Act. Upon the issuance of such a waiver, the Administrator shall publish in the Federal Register a notice that the waiver was granted for national defense purposes, unless, upon the request of the President, the Administrator determines to omit such publication because the publication itself would be contrary to the interests of national defense, in which event the Administrator shall submit notice thereof to the Armed Services Committees of the Senate and the House of Representatives.

Publication in Federal Register. Notice to congressional committee.

SEC. 23. EMPLOYEE PROTECTION.

15 USC 2622.

(a) IN GENERAL.—No employer may discharge any employee or otherwise discriminate against any employee with respect to the employee's compensation, terms, conditions, or privileges of employment because the employee (or any person acting pursuant to a request of the employee) has—

(1) commenced, caused to be commenced, or is about to commence or cause to be commenced a proceeding under this Act;

(2) testified or is about to testify in any such proceeding; or

(3) assisted or participated or is about to assist or participate in any manner in such a proceeding or in any other action to carry out the purposes of this Act.

(b) REMEDY.—(1) Any employee who believes that the employee has been discharged or otherwise discriminated against by any person in violation of subsection (a) of this section may, within 30 days after such alleged violation occurs, file (or have any person file on the employee's behalf) a complaint with the Secretary of Labor (hereinafter in this section referred to as the "Secretary") alleging such discharge or discrimination. Upon receipt of such a complaint, the Secretary shall notify the person named in the complaint of the filing of the complaint.

Notification.

(2)(A) Upon receipt of a complaint filed under paragraph (1), the Secretary shall conduct an investigation of the violation alleged in the complaint. Within 30 days of the receipt of such complaint, the Secretary shall complete such investigation and shall notify in writing the complainant (and any person acting on behalf of the complainant) and the person alleged to have committed such violation of the results of the investigation conducted pursuant to this paragraph. Within ninety days of the receipt of such complaint the Secretary shall, unless the proceeding on the complaint is terminated by the Secretary on the basis of a settlement entered into by the Secretary and the person alleged to have committed such violation, issue an order either providing the relief prescribed by subparagraph (B) or denying the complaint. An order of the Secretary shall be made on the record after notice and opportunity for agency hearing. The Secretary may not enter into a settlement terminating a proceeding on a complaint without the participation and consent of the complainant.

Investigation.

Notification.

Notice, hearing.

(B) If in response to a complaint filed under paragraph (1) the Secretary determines that a violation of subsection (a) of this section has occurred, the Secretary shall order (i) the person who committed such violation to take affirmative action to abate the violation, (ii) such person to reinstate the complainant to the complainant's former position together with the compensation (including back pay), terms, conditions, and privileges of the complainant's employment, (iii) compensatory damages, and (iv) where appropriate, exemplary damages. If such an order issued, the Secretary, at the request of the complainant, shall assess against the person against whom the order is issued a sum equal to the aggregate amount of all costs and expenses (including attorney's fees) reasonably incurred, as determined by the Secretary, by the complainant for, or in connection with, the bringing of the complaint upon which the order was issued.

(c) REVIEW.—(1) Any employee or employer adversely affected or aggrieved by an order issued under subsection (b) may obtain review of the order in the United States Court of Appeals for the circuit in which the violation, with respect to which the order was issued, allegedly occurred. The petition for review must be filed within sixty days from the issuance of the Secretary's order. Review shall conform to chapter 7 of title 5 of the United States Code.

(2) An order of the Secretary, with respect to which review could have been obtained under paragraph (1), shall not be subject to judicial review in any criminal or other civil proceeding.

(d) ENFORCEMENT.—Whenever a person has failed to comply with an order issued under subsection (b)(2), the Secretary shall file a civil action in the United States district court for the district in which the violation was found to occur to enforce such order. In actions brought under this subsection, the district courts shall have jurisdiction to grant all appropriate relief, including injunctive relief and compensatory and exemplary damages. Civil actions brought under this subsection shall be heard and decided expeditiously. *Civil action.* *Jurisdiction.*

(e) EXCLUSION.—Subsection (a) of this section shall not apply with respect to any employee who, acting without direction from the employee's employer (or any agent of the employer), deliberately causes a violation of any requirement of this Act.

SEC. 24. EMPLOYMENT EFFECTS.

(a) IN GENERAL.—The Administrator shall evaluate on a continuing basis the potential effects on employment (including reductions in employment or loss of employment from threatened plant closures) of— *Evaluation.* *15 USC 2623.*

> (1) the issuance of a rule or order under section 4, 5, or 6, or
> (2) a requirement of section 5 or 6.

(b)(1) INVESTIGATIONS.—Any employee (or any representative of an employee) may request the Administrator to make an investigation of—

> (A) a discharge or layoff or threatened discharge or layoff of the employee, or
> (B) adverse or threatened adverse effects on the employee's employment,

allegedly resulting from a rule or order under section 4, 5, or 6 or a requirement of section 5 or 6. Any such request shall be made in writing, shall set forth with reasonable particularity the grounds for the request, and shall be signed by the employee, or representative of such employee, making the request.

(2)(A) Upon receipt of a request made in accordance with paragraph (1) the Administrator shall (i) conduct the investigation requested, and (ii) if requested by any interested person, hold public hearings on any matter involved in the investigation unless the Administrator, by order issued within 45 days of the date such hearings are *Public hearings.*

requested, denies the request for the hearings because the Administrator determines there are no reasonable grounds for holding such hearings. If the Administrator makes such a determination, the Administrator shall notify in writing the person requesting the hearing of the determination and the reasons therefor and shall publish the determination and the reasons therefor in the Federal Register.

Notification.

Publication in Federal Register.

(B) If public hearings are to be held on any matter involved in an investigation conducted under this subsection—

    (i) at least five days' notice shall be provided the person making the request for the investigation and any person identified in such request,

    (ii) such hearings shall be held in accordance with section 6(c)(3), and

    (iii) each employee who made or for whom was made a request for such hearings and the employer of such employee shall be required to present information respecting the applicable matter referred to in paragraph (1)(A) or (1)(B) together with the basis for such information.

Recommendations.

(3) Upon completion of an investigation under paragraph (2), the Administrator shall make findings of fact, shall make such recommendations as the Administrator deems appropriate, and shall make available to the public such findings and recommendations.

(4) This section shall not be construed to require the Administrator to amend or repeal any rule or order in effect under this Act.

SEC. 25. STUDIES.

15 USC 2624.

(a) INDEMNIFICATION STUDY.—The Administrator shall conduct a study of all Federal laws administered by the Administrator for the purpose of determining whether and under what conditions, if any, indemnification should be accorded any person as a result of any action taken by the Administrator under any such law. The study shall—

    (1) include an estimate of the probable cost of any indemnification programs which may be recommended;

    (2) include an examination of all viable means of financing the cost of any recommended indemnification; and

Submittal to Congress.
GAO review.

    (3) be completed and submitted to Congress within two years from the effective date of enactment of this Act.

The General Accounting Office shall review the adequacy of the study submitted to Congress pursuant to paragraph (3) and shall report the results of its review to the Congress within six months of the date such study is submitted to Congress.

Consultation.

(b) CLASSIFICATION, STORAGE, AND RETRIEVAL STUDY.—The Council on Environmental Quality, in consultation with the Administrator, the Secretary of Health, Education, and Welfare, the Secretary of Commerce, and the heads of other appropriate Federal departments or agencies, shall coordinate a study of the feasibility of establishing (1) a standard classification system for chemical substances and related substances, and (2) a standard means for storing and for obtaining rapid access to information respecting such substances. A report on such study shall be completed and submitted to Congress not later than 18 months after the effective date of enactment of this Act.

Report to Congress.

SEC. 26. ADMINISTRATION OF THE ACT.

15 USC 2625.

(a) COOPERATION OF FEDERAL AGENCIES.—Upon request by the Administrator, each Federal department and agency is authorized—

    (1) to make its services, personnel, and facilities available (with or without reimbursement) to the Administrator to assist the Administrator in the administration of this Act; and

(2) to furnish to the Administrator such information, data, estimates, and statistics, and to allow the Administrator access to all information in its possession as the Administrator may reasonably determine to be necessary for the administration of this Act.

(b) FEES.—(1) The Administrator may, by rule, require the payment of a reasonable fee from any person required to submit data under section 4 or 5 to defray the cost of administering this Act. Such rules shall not provide for any fee in excess of $2,500 or, in the case of a small business concern, any fee in excess of $100. In setting a fee under this paragraph, the Administrator shall take into account the ability to pay of the person required to submit the data and the cost to the Administrator of reviewing such data. Such rules may provide for sharing such a fee in any case in which the expenses of testing are shared under section 4 or 5.

(2) The Administrator, after consultation with the Administrator of the Small Business Administration, shall by rule prescribe standards for determining the persons which qualify as small business concerns for purposes of paragraph (1). **Consultation. Rule.**

(c) ACTION WITH RESPECT TO CATEGORIES.—(1) Any action authorized or required to be taken by the Administrator under any provision of this Act with respect to a chemical substance or mixture may be taken by the Administrator in accordance with that provision with respect to a category of chemical substances or mixtures. Whenever the Administrator takes action under a provision of this Act with respect to a category of chemical substances or mixtures, any reference in this Act to a chemical substance or mixture (insofar as it relates to such action) shall be deemed to be a reference to each chemical substance or mixture in such category.

(2) For purposes of paragraph (1): **Definitions.**

(A) The term "category of chemical substances" means a group of chemical substances the members of which are similar in molecular structure, in physical, chemical, or biological properties, in use, or in mode of entrance into the human body or into the environment, or the members of which are in some other way suitable for classification as such for purposes of this Act, except that such term does not mean a group of chemical substances which are grouped together solely on the basis of their being new chemical substances.

(B) The term "category of mixtures" means a group of mixtures the members of which are similar in molecular structure, in physical, chemical, or biological properties, in use, or in the mode of entrance into the human body or into the environment, or the members of which are in some other way suitable for classification as such for purposes of this Act.

(d) ASSISTANCE OFFICE.—The Administrator shall establish in the Environmental Protection Agency an identifiable office to provide technical and other nonfinancial assistance to manufacturers and processors of chemical substances and mixtures respecting the requirements of this Act applicable to such manufacturers and processors, the policy of the Agency respecting the application of such requirements to such manufacturers and processors, and the means and methods by which such manufacturers and processors may comply with such requirements. **Establishment.**

(e) FINANCIAL DISCLOSURES.—(1) Except as provided under paragraph (3), each officer or employee of the Environmental Protection Agency and the Department of Health, Education, and Welfare who—

(A) performs any function or duty under this Act, and

(B) has any known financial interest (i) in any person subject to this Act or any rule or order in effect under this Act, or (ii) in any person who applies for or receives any grant or contract under this Act,

shall, on February 1, 1978, and on February 1 of each year thereafter, file with the Administrator or the Secretary of Health, Education, and Welfare (hereinafter in this subsection referred to as the "Secretary"), as appropriate, a written statement concerning all such interests held by such officer or employee during the preceding calendar year. Such statement shall be made available to the public.

(2) The Administrator and the Secretary shall—

(A) act within 90 days of the effective date of this Act—

(i) to define the term "known financial interests" for purposes of paragraph (1), and

(ii) to establish the methods by which the requirement to file written statements specified in paragraph (1) will be monitored and enforced, including appropriate provisions for review by the Administrator and the Secretary of such statements; and

Report to Congress.

(B) report to the Congress on June 1, 1978, and on June 1 of each year thereafter with respect to such statements and the actions taken in regard thereto during the preceding calendar year.

(3) The Administrator may by rule identify specific positions with the Environmental Protection Agency, and the Secretary may by rule identify specific positions with the Department of Health, Education, and Welfare, which are of a nonregulatory or nonpolicymaking nature, and the Administrator and the Secretary may by rule provide that officers or employees occupying such positions shall be exempt from the requirements of paragraph (1).

(4) This subsection does not supersede any requirement of chapter 11 of title 18, United States Code.

Penalty.

(5) Any officer or employee who is subject to, and knowingly violates, this subsection or any rule issued thereunder, shall be fined not more than $2,500 or imprisoned not more than one year, or both.

(f) STATEMENT OF BASIS AND PURPOSE.—Any final order issued under this Act shall be accompanied by a statement of its basis and purpose. The contents and adequacy of any such statement shall not be subject to judicial review in any respect.

Appointment.

(g) ASSISTANT ADMINISTRATOR.—(1) The President, by and with the advice and consent of the Senate, shall appoint an Assistant Administrator for Toxic Substances of the Environmental Protection Agency. Such Assistant Administrator shall be a qualified individual who is, by reason of background and experience, especially qualified to direct a program concerning the effects of chemicals on human health and the environment. Such Assistant Administrator shall be responsible for (A) the collection of data, (B) the preparation of studies, (C) the making of recommendations to the Administrator for regulatory and other actions to carry out the purposes and to facilitate the administration of this Act, and (D) such other functions as the Administrator may assign or delegate.

5 USC app. II.

(2) The Assistant Administrator to be appointed under paragraph (1) shall (A) be in addition to the Assistant Administrators of the Environmental Protection Agency authorized by section 1(d) of Reorganization Plan No. 3 of 1970, and (B) be compensated at the rate of pay authorized for such Assistant Administrators.

SEC. 27. DEVELOPMENT AND EVALUATION OF TEST METHODS.

(a) In General.—The Secretary of Health, Education, and Welfare, in consultation with the Administrator and acting through the Assistant Secretary for Health, may conduct, and make grants to public and nonprofit private entities and enter into contracts with public and private entities for, projects for the development and evaluation of inexpensive and efficient methods (1) for determining and evaluating the health and environmental effects of chemical substances and mixtures, and their toxicity, persistence, and other characteristics which affect health and the environment, and (2) which may be used for the development of test data to meet the requirements of rules promulgated under section 4. The Administrator shall consider such methods in prescribing under section 4 standards for the development of test data.

Consultation.
15 USC 2626.

(b) Approval by Secretary.—No grant may be made or contract entered into under subsection (a) unless an application therefor has been submitted to and approved by the Secretary. Such an application shall be submitted in such form and manner and contain such information as the Secretary may require. The Secretary may apply such conditions to grants and contracts under subsection (a) as the Secretary determines are necessary to carry out the purposes of such subsection. Contracts may be entered into under such subsection without regard to sections 3648 and 3709 of the Revised Statutes (31 U.S.C. 529; 41 U.S.C. 5).

Grants or contracts, application.

(c) Annual Reports.—(1) The Secretary shall prepare and submit to the President and the Congress on or before January 1 of each year a report of the number of grants made and contracts entered into under this section and the results of such grants and contracts.

Report to President and Congress.

(2) The Secretary shall periodically publish in the Federal Register reports describing the progress and results of any contract entered into or grant made under this section.

Publication in Federal Register.

SEC. 28. STATE PROGRAMS.

(a) In General.—For the purpose of complementing (but not reducing) the authority of, or actions taken by, the Administrator under this Act, the Administrator may make grants to States for the establishment and operation of programs to prevent or eliminate unreasonable risks within the States to health or the environment which are associated with a chemical substance or mixture and with respect to which the Administrator is unable or is not likely to take action under this Act for their prevention or elimination. The amount of a grant under this subsection shall be determined by the Administrator, except that no grant for any State program may exceed 75 per centum of the establishment and operation costs (as determined by the Administrator) of such program during the period for which the grant is made.

15 USC 2627.

(b) Approval by Administrator.—(1) No grant may be made under subsection (a) unless an application therefor is submitted to and approved by the Administrator. Such an application shall be submitted in such form and manner as the Administrator may require and shall—

Grants, application.

(A) set forth the need of the applicant for a grant under subsection (a),

(B) identify the agency or agencies of the State which shall establish or operate, or both, the program for which the application is submitted,

(C) describe the actions proposed to be taken under such program,

(D) contain or be supported by assurances satisfactory to the Administrator that such program shall, to the extent feasible, be integrated with other programs of the applicant for environmental and public health protection,

(E) provide for the making of such reports and evaluations as the Administrator may require, and

(F) contain such other information as the Administrator may prescribe.

Application approval.

(2) The Administrator may approve an application submitted in accordance with paragraph (1) only if the applicant has established to the satisfaction of the Administrator a priority need, as determined under rules of the Administrator, for the grant for which the application has been submitted. Such rules shall take into consideration the seriousness of the health effects in a State which are associated with chemical substances or mixtures, including cancer, birth defects, and gene mutations, the extent of the exposure in a State of human beings and the environment to chemical substances and mixtures, and the extent to which chemical substances and mixtures are manufactured, processed, used, and disposed of in a State.

Report to Congress.

(c) ANNUAL REPORTS.—Not later than six months after the end of each of the fiscal years 1979, 1980, and 1981, the Administrator shall submit to the Congress a report respecting the programs assisted by grants under subsection (a) in the preceding fiscal year and the extent to which the Administrator has disseminated information respecting such programs.

(d) AUTHORIZATION.—For the purpose of making grants under subsection (a) there are authorized to be appropriated $1,500,000 for the fiscal year ending September 30, 1977, $1,500,000 for the fiscal year ending September 30, 1978, and $1,500,000 for the fiscal year ending September 30, 1979. Sums appropriated under this subsection shall remain available until expended.

SEC. 29. AUTHORIZATION FOR APPROPRIATIONS.

15 USC 2628.

There are authorized to be appropriated to the Administrator for purposes of carrying out this Act (other than sections 27 and 28 and subsections (a) and (c) through (g) of section 10 thereof) $10,100,000 for the fiscal year ending September 30, 1977, $12,625,000 for the fiscal year ending September 30, 1978, $16,200,000 for the fiscal year ending September 30, 1979. No part of the funds appropriated under this section may be used to construct any research laboratories.

SEC. 30. ANNUAL REPORT.

Report to President and Congress.
5 USC 2629.

The Administrator shall prepare and submit to the President and the Congress on or before January 1, 1978, and on or before January 1 of each succeeding year a comprehensive report on the administration of this Act during the preceding fiscal year. Such report shall include—

(1) a list of the testing required under section 4 during the year for which the report is made and an estimate of the costs incurred during such year by the persons required to perform such tests;

(2) the number of notices received during such year under section 5, the number of such notices received during such year under such section for chemical substances subject to a section 4 rule, and a summary of any action taken during such year under section 5(g);

(3) a list of rules issued during such year under section 6;

(4) a list, with a brief statement of the issues, of completed or pending judicial actions under this Act and administrative actions under section 16 during such year;

(5) a summary of major problems encountered in the administration of this Act; and

(6) such recommendations for additional legislation as the Administrator deems necessary to carry out the purposes of this Act.

Recommendations.

SEC. 31. EFFECTIVE DATE.

Except as provided in section 4(f), this Act shall take effect on January 1, 1977.

15 USC 2601 note.

Approved October 11, 1976.

APPENDIX C

**THE FEDERAL INSECTICIDE,
FUNGICIDE, AND RODENTICIDE ACT,
AS AMENDED, 1975: PUBLIC LAWS 92-516, 94-140**

"SECTION 1. SHORT TITLE AND TABLE OF CONTENTS.

"(a) Short Title.—This Act may be cited as the 'Federal Insecticide, Fungicide, and Rodenticide Act'.

"(b) Table of Contents.—

"Section 1. Short title and table of contents.
    "(a) Short title.
    "(b) Table of contents.
"Sec. 2. Definitions.
    "(a) Active ingredient.
    "(b) Administrator.
    "(c) Adulterated.
    "(d) Animal.
    "(e) Certified applicator, etc.
      "(1) Certified applicator.
      "(2) Private applicator.
      "(3) Commercial applicator.
      "(4) Under the direct supervision of a certified applicator.
    "(f) Defoliant.
    "(g) Desiccant.
    "(h) Device.
    "(i) District court.
    "(j) Environment.
    "(k) Fungus.
    "(l) Imminent hazard.
    "(m) Inert ingredient.
    "(n) Ingredient statement.
    "(o) Insect.
    "(p) Label and labeling.
      "(1) Label.
      "(2) Labeling.
    "(q) Misbranded.
    "(r) Nematode.
    "(s) Person.
    "(t) Pest.
    "(u) Pesticide.
    "(v) Plant regulator.
    "(w) Producer and produce.
    "(x) Protect health and the environment.
    "(y) Registrant.
    "(z) Registration.
    "(aa) State.
    "(bb) Unreasonable adverse effects on the environment.
    "(cc) Weed.
    "(dd) Establishment.
"Sec. 3. Registration of pesticides.
    "(a) Requirement.
    "(b) Exemptions.
    "(c) Procedure for registration.
      "(1) Statement required.
      "(2) Data in support of registration.
      "(3) Time for acting with respect to application.
      "(4) Notice of application.
      "(5) Approval of registration.
      "(6) Denial of registration.
    "(d) Classification of pesticides.
      "(1) Classification for general use, restricted use, or both.
      "(2) Change in classification.
    "(e) Products with same formulation and claims.
    "(f) Miscellaneous.
      "(1) Effect of change of labeling or formulation.
      "(2) Registration not a defense.
      "(3) Authority to consult other Federal agencies.
"Sec. 4. Use of restricted use pesticides; certified applicators.
    "(a) Certification procedure.
      "(1) Federal certification.
      "(2) State certification.
    "(b) State plans.
    "(c) Instruction in integrated pest management techniques.
"Sec. 5. Experimental use permits.
    "(a) Issuance.
    "(b) Temporary tolerance level.
    "(c) Use under permit.
    "(d) Studies.
    "(e) Revocation.

"(f) State issuance of permits.
"(g) Exemption for agricultural research agencies.
"Sec. 6. Administrative review; suspension.
    "(a) Cancellation after five years.
        "(1) Procedure.
        "(2) Information.
    "(b) Cancellation and change in classification.
    "(c) Suspension.
        "(1) Order.
        "(2) Expedite hearing.
        "(3) Emergency order.
        "(4) Judicial review.
    "(d) Public hearings and scientific review.
    "(e) Judicial review.
"Sec. 7. Registration of establishments.
    "(a) Requirement.
    "(b) Registration.
    "(c) Information required.
    "(d) Confidential records and information.
"Sec. 8. Books and records.
    "(a) Requirements.
    "(b) Inspection.
"Sec. 9. Inspection of establishments, etc.
    "(a) In general.
    "(b) Warrants.
    "(c) Enforcement.
        "(1) Certification of facts to Attorney General.
        "(2) Notice not required.
        "(3) Warning notices.
"Sec. 10. Protection of trade secrets and other information.
    "(a) In general.
    "(b) Disclosure.
    "(c) Disputes.
"Sec. 11. Standards applicable to pesticide applicators.
    "(a) In general.
    "(b) Separate standards.
"Sec. 12. Unlawful acts.
    "(a) In general.
    "(b) Exemptions.
"Sec. 13. Stop sale, use, removal, and seizure.
    "(a) Stop sale, etc., orders.
    "(b) Seizure.
    "(c) Disposition after condemnation.
    "(d) Court costs, etc.
"Sec. 14. Penalties.
    "(a) Civil penalties.
        "(1) In general.
        "(2) Private applicator.
        "(3) Hearing.
        "(4) References to Attorney General.
    "(b) Criminal penalties.
        "(1) In general.
        "(2) Private applicator.
        "(3) Disclosure of information.
        "(4) Acts of officers, agents, etc.
"Sec. 15. Indemnities.
    "(a) Requirement.
    "(b) Amount of payment.
        "(1) In general.
        "(2) Special rule.
"Sec. 16. Administrative procedure; judicial review.
    "(a) District court review.
    "(b) Review by Court of Appeals.
    "(c) Jurisdiction of district courts.
    "(d) Notice of judgments.
"Sec. 17. Imports and exports.
    "(a) Pesticides and devices intended for export.
    "(b) Cancellation notices furnished to foreign governments.
    "(c) Importation of pesticides and devices.
    "(d) Cooperation in international efforts.
    "(e) Regulations.
"Sec. 18. Exemption of Federal agencies.
"Sec. 19. Disposal and transportation.
    "(a) Procedures.
    "(b) Advice to Secretary of Transportation.
"Sec. 20. Research and monitoring.
    "(a) Research.
    "(b) National monitoring plan.
    "(c) Monitoring.
"Sec. 21. Solicitation of public comments; notice of public hearings.
"Sec. 22. Delegation and cooperation.
    "(a) Delegation.
    "(b) Cooperation.
"Sec. 23. State cooperation, aid, and training.
    "(a) Cooperative agreements.
    "(b) Contracts for training.
"Sec. 24. Authority of States.
"Sec. 25. Authority of Administrator.
    "(a)(1) Regulations.
      "(2) Procedure.
        "(A) Proposed regulations.
        "(B) Final regulations.
        "(C) Time requirements.
        "(D) Publication in the Federal Register.
      "(3) Congressional committees.
    "(b) Exemption of pesticides.
    "(c) Other authority.
"Sec. 26. Severability.
"Sec. 27. Authorization for appropriations.

"SEC. 2. DEFINITIONS.

"For purposes of this Act—

"(a) Active Ingredient.—The term 'active ingredient' means—

"(1) in the case of a pesticide other than a plant regulator, defoliant, or desiccant, an ingredient which will prevent, destroy, repel, or mitigate any pest;

"(2) in the case of a plant regulator, an ingredient which, through physiological action, will accelerate or retard the rate of growth or rate of maturation or otherwise alter the behavior of ornamental or crop plants or the product thereof;

"(3) in the case of a defoliant, an ingredient which will cause the leaves or foliage to drop from a plant; and

"(4) in the case of a desiccant, an ingredient which will artificially accelerate the drying of plant tissue.

"(b) Administrator.—The term 'Administrator' means the Administrator of the Environmental Protection Agency.

"(c) Adulterated.—The term 'adulterated' applies to any pesticide if:

"(1) its strength or purity falls below the professed standard of quality as expressed on its labeling under which it is sold;

"(2) any substance has been substituted wholly or in part for the pesticide; or

"(3) any valuable constituent of the pesticide has been wholly or in part abstracted.

"(d) Animal.—The term 'animal' means all vertebrate and invertebrate species, including but not limited to man and other mammals, birds, fish, and shellfish.

"(e) Certified Applicator, Etc.—

"(1) Certified applicator.—The term 'certified applicator' means any individual who is certified under section 4 as authorized to use or supervise the use of any pesticide which is classified for restricted use.

"(2) Private applicator.—The term 'private applicator' means a certified applicator who uses or supervises the use of any pesticide which is classified for restricted use for purposes of producing any agricultural commodity on property owned or rented by him or his employer or (if applied without compensation other than trading of personal services between producers of agricultural commodities) on the property of another person.

"(3) Commercial applicator.—The term 'commercial applicator' means a certified applicator (whether or not he is a private applicator with respect to some uses) who uses or supervises the use of any pesticide which is classified for restricted use for any purpose or on any property other than as provided by paragraph (2).

"(4) Under the direct supervision of a certified applicator.—Unless otherwise prescribed by its labeling, a pesticide shall be considered to be applied under the direct supervision of a certified applicator if it is applied by a competent person acting under the instructions and control of a certified applicator who is available if and when needed, even though such certified applicator is not physically present at the time and place the pesticide is applied.

"(f) Defoliant.—The term 'defoliant' means any substance or mixture of substances intended for causing the leaves or foliage to drop from a plant, with or without causing abscission.

"(g) Desiccant.—The term 'desiccant' means any substance or mixture of substances intended for artificially accelerating the drying of plant tissue.

"(h) Device.—The term 'device' means any instrument or contrivance (other than a firearm) which is intended for trapping, destroying, repelling, or mitigating any pest or any other form of plant or animal life (other than man and other than bacteria, virus, or other microorganism on or in living man or other living animals); but not including equipment used for the application of pesticides when sold separately therefrom.

"(i) District Court.—The term 'district court' means a United States district court, the District Court of Guam, the District Court of the Virgin Islands, and the highest court of American Samoa.

"(j) Environment.—The term 'environment' includes water, air, land, and all plants and man and other animals living therein, and the inter-relationships which exist among these.

"(k) Fungus.—The term 'fungus' means any non-chlorophyll-bearing thallophyte (that is, any non-chlorophyll-bearing plant of a lower order than mosses and liverworts), as for example, rust, smut, mildew, mold, yeast, and bacteria, except those on or in living man or other animals and those on or in processed food, beverages, or pharmaceuticals.

"(l) Imminent Hazard.—The term 'imminent hazard' means a situation which exists when the continued use of a pesticide during the time required for cancellation proceeding would be likely to result in unreasonable adverse effects on the environment or will involve unreasonable hazard to the survival of a species declared endangered by the Secretary of the Interior under Public Law 91–135.

83 Stat. 275.
16 USC 668cc-1.

"(m) INERT INGREDIENT.—The term 'inert ingredient' means an ingredient which is not active.

"(n) INGREDIENT STATEMENT.—The term 'ingredient statement' means a statement which contains—

"(1) the name and percentage of each active ingredient, and the total percentage of all inert ingredients, in the pesticide; and

"(2) if the pesticide contains arsenic in any form, a statement of the percentages of total and water soluble arsenic, calculated as elementary arsenic.

"(o) INSECT.—The term 'insect' means any of the numerous small invertebrate animals generally having the body more or less obviously segmented, for the most part belonging to the class insecta, comprising six-legged, usually winged forms, as for example, beetles, bugs, bees, flies, and to other allied classes of arthropods whose members are wingless and usually have more than six legs, as for example, spiders, mites, ticks, centipedes, and wood lice.

"(p) LABEL AND LABELING.—

"(1) LABEL.—The term 'label' means the written, printed, or graphic matter on, or attached to, the pesticide or device or any of its containers or wrappers.

"(2) LABELING.—The term 'labeling' means all labels and all other written, printed, or graphic matter—

"(A) accompanying the pesticide or device at any time; or

"(B) to which reference is made on the label or in literature accompanying the pesticide or device, except to current official publications of the Environmental Protection Agency, the United States Departments of Agriculture and Interior, the Department of Health, Education, and Welfare, State experiment stations, State agricultural colleges, and other similar Federal or State institutions or agencies authorized by law to conduct research in the field of pesticides.

"(q) MISBRANDED.—

"(1) A pesticide is misbranded if—

"(A) its labeling bears any statement, design, or graphic representation relative thereto or to its ingredients which is false or misleading in any particular;

"(B) it is contained in a package or other container or wrapping which does not conform to the standards established by the Administrator pursuant to section 25(c)(3);

"(C) it is an imitation of, or is offered for sale under the name of, another pesticide;

"(D) its label does not bear the registration number assigned under section 7 to each establishment in which it was produced;

"(E) any word, statement, or other information required by or under authority of this Act to appear on the label or labeling is not prominently placed thereon with such conspicuousness (as compared with other words, statements, designs, or graphic matter in the labeling) and in such terms as to render it likely to be read and understood by the ordinary individual under customary conditions of purchase and use;

"(F) the labeling accompanying it does not contain directions for use which are necessary for effecting the purpose for which the product is intended and if complied with, together with any requirements imposed under section 3(d) of this Act, are adequate to protect health and the environment;

"(G) the label does not contain a warning or caution statement which may be necessary and if complied with, together with any requirements imposed under section 3(d) of this Act, is adequate to protect health and the environment.

"(2) A pesticide is misbranded if—

"(A) the label does not bear an ingredient statement on that part of the immediate container (and on the outside container or wrapper of the retail package, if there be one, through which the ingredient statement on the immediate container cannot be clearly read) which is presented or displayed under customary conditions of purchase, except that a pesticide is not misbranded under this subparagraph if:

"(i) the size of form of the immediate container, or the outside container or wrapper of the retail package, makes it impracticable to place the ingredient statement on the part which is presented or displayed under customary conditions of purchase; and "(ii) the ingredient statement appears prominently on another part of the immediate container, or outside container or wrapper, permitted by the Administrator;

"(B) the labeling does not contain a statement of the use classification under which the product is registered;

"(C) there is not affixed to its container, and to the outside container or wrapper of the retail package, if there be one, through which the required information on the immediate container cannot be clearly read, a label bearing—

"(i) the name and address of the producer, registrant, or person for whom produced;

"(ii) the name, brand, or trademark under which the pesticide is sold;

"(iii) the net weight or measure of the content: *Provided,* That the Administrator may permit reasonable variations; and

"(iv) when required by regulation of the Administrator to effectuate the purposes of this Act, the registration number assigned to the pesticide under this Act, and the use classification; and

"(D) the pesticide contains any substance or substances in quantities highly toxic to man, unless the label shall bear, in addition to any other matter required by this Act—

"(i) the skull and crossbones;

"(ii) the word 'poison' prominently in red on a background of distinctly contrasting color; and

"(iii) a statement of a practical treatment (first aid or otherwise) in case of poisoning by the pesticide.

"(r) NEMATODE.—The term 'nematode' means invertebrate animals of the phylum nemathelminthes and class nematoda, that is, unsegmented round worms with elongated, fusiform, or saclike bodies covered with cuticle, and inhabiting soil, water, plants, or plant parts; may also be called nemas or eelworms.

"(s) PERSON.—The term 'person' means any individual, partnership, association, corporation, or any organized group of persons whether incorporated or not.

"(t) PEST.—The term 'pest' means (1) any insect, rodent, nematode, fungus, weed, or (2) any other form of terrestrial or aquatic plant or animal life or virus, bacteria, or other micro-organism (except viruses, bacteria, or other micro-organisms on or in living man or other living animals) which the Administrator declares to be a pest under section 25(c)(1).

"(u) PESTICIDE.—The term 'pesticide' means (1) any substance or mixture of substances intended for preventing, destroying, repelling, or mitigating any pest, and (2) any substance or mixture of substances intended for use as a plant regulator, defoliant, or desiccant: *Provided,* That the term 'pesticide' shall not include any article (1)(a) that is a 'new animal drug' within the meaning of section 201(w) of the Federal Food, Drug, and Cosmetic Act (21 U.S.C. 321(w)), or (b) that has been determined by the Secretary of Health, Education, and Welfare not to be a new animal drug by a regulation establishing conditions of use for the article, or (2) that is an animal feed within the meaning of section 201(x) of such Act (21 U.S.C. 321(x)) bearing or containing an article covered by clause (1) of this proviso." P.L. 94-140
89 Stat. 754

"(v) PLANT REGULATOR.—The term 'plant regulator' means any substance or mixture of substances intended, through physiological action, for accelerating or retarding the rate of growth or rate of maturation, or for otherwise altering the behavior of plants or the produce thereof, but shall not include substances to the extent that they are intended as plant nutrients, trace elements, nutritional chemicals, plant inoculants, and soil amendments. Also, the term 'plant regulator' shall not be required to include any of such of those nutrient mixtures or soil amendments as are commonly known as vitamin-hormone horticultural products, intended for improvement, maintenance, survival, health, and propagation of plants, and as are not for pest destruction and are nontoxic, nonpoisonous in the undiluted packaged concentration.

"(w) PRODUCER AND PRODUCE.—The term 'producer' means the person who manufactures, prepares, compounds, propagates, or processes any pesticide or device. The term 'produce' means to manufacture, prepare, compound, propagate, or process any pesticide or device.

"(x) PROTECT HEALTH AND THE ENVIRONMENT.—The terms 'protect health and the environment' and 'protection of health and the environment' mean protection against any unreasonable adverse effects on the environment.

"(y) REGISTRANT.—The term 'registrant' means a person who has registered any pesticide pursuant to the provisions of this Act.

"(z) REGISTRATION.—The term 'registration' includes reregistration.

"(aa) STATE.—The term 'State' means a State, the District of Columbia, the Commonwealth of Puerto Rico, the Virgin Islands, Guam, the Trust Territory of the Pacific Islands, and American Samoa.

"(bb) UNREASONABLE ADVERSE EFFECTS ON THE ENVIRONMENT.—The term 'unreasonable adverse effects on the environment' means any unreasonable risk to man or the environment, taking into account the economic, social, and environmental costs and benefits of the use of any pesticide.

"(cc) WEED.—The term 'weed' means any plant which grows where not wanted.

"(dd) ESTABLISHMENT.—The term 'establishment' means any place where a pesticide or device is produced, or held, for distribution or sale.

"SEC. 3. REGISTRATION OF PESTICIDES.

"(a) REQUIREMENT.—Except as otherwise provided by this Act, no person in any State may distribute, sell, offer for sale, hold for sale, ship, deliver for shipment, or receive and (having so received) deliver or offer to deliver, to any person any pesticide which is not registered with the Administrator.

"(b) EXEMPTIONS.—A pesticide which is not registered with the Administrator may be transferred if—

"(1) the transfer is from one registered establishment to another registered establishment operated by the same producer solely for packaging at the second establishment or for use as a constituent part of another pesticide produced at the second establishment; or

"(2) the transfer is pursuant to and in accordance with the requirements of an experimental use permit.

"(c) PROCEDURE FOR REGISTRATION.—

(1) STATEMENT REQUIRED.—Each applicant for registration of a pesticide shall file with the Administrator a statement which includes—

"(A) the name and address of the applicant and of any other person whose name will appear on the labeling;

"(B) the name of the pesticide;

"(C) a complete copy of the labeling of the pesticide, a statement of all claims to be made for it, and any directions for its use;

**P.L. 94-140
89 Stat. 755**

Test data.

(D) if requested by the Administrator, a full description of the tests made and the results thereof upon which the claims are based, except that data submitted on or after January 1, 1970, in support of an application shall not, without permission of the applicant, be considered by the Administrator in support of any other application for registration unless such other applicant shall have first offered to pay reasonable compensation for producing the test data to be relied upon and such data is not protected from disclosure by section 10(b). This provision with regard to compensation for producing the test data to be relied upon shall apply with respect to all applications for registration or reregistration submitted on or after October 21, 1972. If the parties cannot agree on the amount and method of payment, the Administrator shall make such determination and may fix such other terms and conditions as may be reasonable under the circumstances. The Administrator's determination shall be made on the record after notice and opportunity for hearing. If either party does not agree with said determination, he may, within thirty days, take an appeal to the Federal district court for the district in which he resides with respect to either the amount of the payment or the terms of payment, or both. Registration shall not be delayed pending the determination of reasonable compensation between the applicants, by the Administrator or by the court.

Appeal.

"(E) the complete formula of the pesticide; and

"(F) a request that the pesticide be classified for general use, for restricted use, or for both.

"(2) DATA IN SUPPORT OF REGISTRATION.—The Administrator shall publish guidelines specifying the kinds of information which will be required to support the registration of a pesticide and shall revise such guidelines from time to time. If thereafter he requires any additional kind of information he shall permit sufficient time for applicants to obtain such additional information. Except as provided by subsection (c)(1)(D) of this section and section 10, within 30 days after the Administrator registers a pesticide under this Act he shall make available to the public the data called for in the registration statement together with such other scientific information as he deems relevant to his decision.

"(3) TIME FOR ACTING WITH RESPECT TO APPLICATION.—The Administrator shall review the data after receipt of the application and shall, as expeditiously as possible, either register the pesticide in accordance with paragraph (5), or notify the applicant of his determination that it does not comply with the provisions of the Act in accordance with paragraph (6).

"(4) Notice of application.—The Administrator shall publish in the Federal Register, promptly after receipt of the statement and other data required pursuant to paragraphs (1) and (2), a notice of each application for registration of any pesticide if it contains any new active ingredient or if it would entail a changed use pattern. The notice shall provide for a period of 30 days in which any Federal agency or any other interested person may comment.

Publication in Federal Register.

"(5) Approval of registration.—The Administrator shall register a pesticide if he determines that, when considered with any restrictions imposed under subsection (d)—

"(A) its composition is such as to warrant the proposed claims for it;

"(B) its labeling and other material required to be submitted comply with the requirements of this Act;

"(C) it will perform its intended function without unreasonable adverse effects on the environment; and

"(D) when used in accordance with widespread and commonly recognized practice it will not generally cause unreasonable adverse effects on the environment.

The Administrator shall not make any lack of essentiality a criterion for denying registration of any pesticide. Where two pesticides meet the requirements of this paragraph, one should not be registered in preference to the other.

"(6) Denial of registration.—If the Administrator determines that the requirements of paragraph (5) for registration are not satisfied, he shall notify the applicant for registration of his determination and of his reasons (including the factual basis) therefor, and that, unless the applicant corrects the conditions and notifies the Administrator thereof during the 30-day period beginning with the day after the date on which the applicant receives the notice, the Administrator may refuse to register the pesticide. Whenever the Administrator refuses to register a pesticide, he shall notify the applicant of his decision and of his reasons (including the factual basis) therefor. The Administrator shall promptly publish in the Federal Register notice of such denial of registration and the reasons therefor. Upon such notification, the applicant for registration or other interested person with the concurrence of the applicant shall have the same remedies as provided for in section 6.

Publication in Federal Register.

"(d) Classification of Pesticides. —

"(1) Classification for general use, restricted use, or both.—

"(A) As a part of the registration of a pesticide the Administrator shall classify it as being for general use or for restricted use, provided that if the Administrator determines that some of the uses for which the pesticide is registered should be for general use and that other uses for which it is registered should be for restricted use, he shall classify it for both general use and restricted use. If some of the uses of the pesticide are classified for general use and other uses are classified for restricted use, the directions relating to its general uses shall be clearly separated and distinguished from those directions relating to its restricted uses: *Provided, however,* That the Administrator may require that its packaging and labeling for restricted uses shall be clearly distinguishable from its packaging and labeling for general uses.

"(B) If the Administrator determines that the pesticide, when applied in accordance with its directions for use, warnings and cautions and for the uses for which it is registered, or for one or more of such uses, or in accordance with a widespread and commonly recognized practice, will not generally cause unreasonable adverse effects on the environment, he will classify the pesticide, or the particular use or uses of the pesticide to which the determination applies, for general use.

"(C) If the Administrator determines that the pesticide, when applied in accordance with its directions for use, warnings and cautions and for the uses for which it is registered, or for one or more of such uses, or in accordance with a widespread and commonly recognized practice, may generally cause, without additional regulatory restrictions, unreasonable adverse effects on the environment, including injury to the applicator, he shall classify the pesticide, or the particular use or uses to which the determination applies, for restricted use:

"(i) If the Administrator classifies a pesticide, or one or

Appendix 495

more uses of such pesticide, for restricted use because of a determination that the acute dermal or inhalation toxicity of the pesticide presents a hazard to the applicator or other persons, the pesticide shall be applied for any use to which the restricted classification applies only by or under the direct supervision of a certified applicator.

"(ii) If the Administrator classifies a pesticide, or one or more uses of such pesticide, for restricted use because of a determination that its use without additional regulatory restriction may cause unreasonable adverse effects on the environment, the pesticide shall be applied for any use to which the determination applies only by or under the direct supervision of a certified applicator, or subject to such other restrictions as the Administrator may provide by regulation. Any such regulation shall be reviewable in the appropriate court of appeals upon petition of a person adversely affected filed within 60 days of the publication of the regulation in final form.

Publication in Federal Register.

"(2) CHANGE IN CLASSIFICATION.—If the Administrator determines that a change in the classification of any use of a pesticide from general use to restricted use is necessary to prevent unreasonable adverse effects on the environment, he shall notify the registrant of such pesticide of such determination at least 30 days before making the change and shall publish the proposed change in the Federal Register. The registrant, or other interested person with the concurrence of the registrant, may seek relief from such determination under section 6(b).

"(e) PRODUCTS WITH SAME FORMULATION AND CLAIMS.—Products which have the same formulation, are manufactured by the same person, the labeling of which contains the same claims, and the labels of which bear a designation identifying the product as the same pesticide may be registered as a single pesticide; and additional names and labels shall be added to the registration by supplemental statements.

"(f) MISCELLANEOUS.—

"(1) EFFECT OF CHANGE OF LABELING OR FORMULATION.—If the labeling or formulation for a pesticide is changed, the registration shall be amended to reflect such change if the Administrator determines that the change will not violate any provision of this Act.

"(2) REGISTRATION NOT A DEFENSE.—In no event shall registration of an article be construed as a defense for the commission of any offense under this Act: *Provided,* That as long as no cancellation proceedings are in effect registration of a pesticide shall be prima facie evidence that the pesticide, its labeling and packaging comply with the registration provisions of the Act.

"(3) AUTHORITY TO CONSULT OTHER FEDERAL AGENCIES.—In connection with consideration of any registration or application for registration under this section, the Administrator may consult with any other Federal agency.

"SEC. 4. USE OF RESTRICTED USE PESTICIDES; CERTIFIED APPLICATORS.

"(a) CERTIFICATION PROCEDURE.—

Standards.

"(1) FEDERAL CERTIFICATION.—Subject to paragraph (2), the Administrator shall prescribe standards for the certification of applicators of pesticides. Such standards shall provide that to be certified, an individual must be determined to be competent with respect to the use and handling of pesticides, or to the use and handling of the pesticide or class of pesticides covered by such individual's certification: *Provided, however,* That the certification standard for a private applicator shall, under a State plan submitted for approval, be deemed fulfilled by his completing a certification form. The Administrator shall further assure that such form contains adequate information and affirmations to carry out the intent of this Act, and may include in the form an affirmation that the private applicator has completed a training program approved by the Administrator so long as the program does not require the private applicator to take, pursuant to a requirement prescribed by the Administrator, any examination to establish competency in the use of the pesticide. The Administrator may require any pesticide dealer participating in a certification program to be licensed under a State licensing program approved by him.

P.L. 94–140 89 Stat. 753

"(2) STATE CERTIFICATION.—If any State, at any time, desires to certify applicators of pesticides, the Governor of such State shall submit a State plan for such purpose. The Administrator shall approve the plan submitted by any State, or any modification thereof, if such plan in his judgment—

"(A) designates a State agency as the agency responsible for administering the plan throughout the State;

"(B) contains satisfactory assurances that such agency has or will have the legal authority and qualified personnel necessary to carry out the plan;

"(C) gives satisfactory assurances that the State will devote adequate funds to the administration of the plan;

"(D) provides that the State agency will make such reports to the Administrator in such form and containing such information as the Administrator may from time to time require; and

"(E) contains satisfactory assurances that State standards for the certification of applicators of pesticides conform with those standards prescribed by the Administrator under paragraph (1).

Any State certification program under this section shall be maintained in accordance with the State plan approved under this section.

"(b) STATE PLANS.—If the Administrator rejects a plan submitted under this paragraph, he shall afford the State submitting the plan due notice and opportunity for hearing before so doing. If the Administrator approves a plan submitted under this paragraph, then such State shall certify applicators of pesticides with respect to such State. Whenever the Administrator determines that a State is not administering the certification program in accordance with the plan approved under this section, he shall so notify the State and provide for a hearing at the request of the State, and, if appropriate corrective action is not taken within a reasonable time, not to exceed ninety days, the Administrator shall withdraw approval of such plan.

Hearing.

"(c) INSTRUCTION IN INTEGRATED PEST MANAGEMENT TECHNIQUES.— Standards prescribed by the Administrator for the certification of applicators of pesticides under subsection (a), and State plans submitted to the Administrator under subsections (a) and (b), shall include provisions for making instructional materials concerning integrated pest management techniques available to individuals at their request in accordance with the provisions of section 23(c) of this Act, but such plans may not require that any individual receive instruction concerning such techniques or be shown to be competent with respect to the use of such techniques. The Administrator and States implementing such plans shall provide that all interested individuals are notified of the availability of such instructional materials."

P.L. 94–140
89 Stat. 754

7 USC 136u.

"SEC. 5. EXPERIMENTAL USE PERMITS.

"(a) ISSUANCE.—Any person may apply to the Administrator for an experimental use permit for a pesticide. The Administrator may issue an experimental use permit if he determines that the applicant needs such permit in order to accumulate information necessary to register a pesticide under section 3. An application for an experimental use permit may be filed at the time of or before or after an application for registration is filed.

"(b) TEMPORARY TOLERANCE LEVEL.—If the Administrator determines that the use of a pesticide may reasonably be expected to result in any residue on or in food or feed, he may establish a temporary tolerance level for the residue of the pesticide before issuing the experimental use permit.

"(c) USE UNDER PERMIT.—Use of a pesticide under an experimental use permit shall be under the supervision of the Administrator, and shall be subject to such terms and conditions and be for such period of time as the Administrator may prescribe in the permit.

"(d) STUDIES.—When any experimental use permit is issued for a pesticide containing any chemical or combination of chemicals which has not been included in any previously registered pesticide, the Administrator may specify that studies be conducted to detect whether the use of the pesticide under the permit may cause unreasonable adverse effects on the environment. All results of such studies shall be reported to the Administrator before such pesticide may be registered under section 3.

"(e) REVOCATION.—The Administrator may revoke any experimental use permit, at any time, if he finds that its terms or conditions are being violated, or that its terms and conditions are inadequate to avoid unreasonable adverse effects on the environment.

"(f) STATE ISSUANCE OF PERMITS.— Notwithstanding the foregoing provisions of this section, the Administrator may, under such terms and conditions as he may by regulations prescribe, authorize any State to issue an experimental use permit for a pesticide. All provisions of section 4 relating to State plans shall apply with equal force to a State plan for the issuance of experimental use permits under this section.

"(g) EXEMPTION FOR AGRICULTURAL RESEARCH AGENCIES.—Notwithstanding the foregoing provisions of this section, the Administrator may issue an experimental use permit for a pesticide to any public or private agricultural research agency or educational institution which applies for such permit. Each permit shall not exceed more than a one-year period or such other specific time as the Administrator may prescribe. Such permit shall be issued

P.L. 94–140
89 Stat. 754

under such terms and conditions restricting the use of the pesticide as the Administrator may require: *Provided,* That such pesticide may be used only by such research agency or educational institution for purposes of experimentation."

"SEC. 6. ADMINISTRATIVE REVIEW; SUSPENSION.

"(a) CANCELLATION AFTER FIVE YEARS—

"(1) PROCEDURE.—The Administrator shall cancel the registration of any pesticide at the end of the five-year period which begins on the date of its registration (or at the end of any five-year period thereafter) unless the registrant, or other interested person with the concurrence of the registrant, before the end of such period, requests in accordance with regulations prescribed by the Administrator that the registration be continued in effect: *Provided,* That the Administrator may permit the continued sale and use of existing stocks of a pesticide whose registration is canceled under this subsection or subsection (b) to such extent, under such conditions, and for such uses as he may specify if he determines that such sale or use is not inconsistent with the purposes of this Act and will not have unreasonable adverse effects on the environment. The Administrator shall publish in the Federal Register, at least 30 days prior to the expiration of such five-year period, notice that the registration will be canceled if the registrant or other interested person with the concurrence of the registrant does not request that the registration be continued in effect.

Publication in Federal Register.

"(2) INFORMATION.—If at any time after the registration of a pesticide the registrant has additional factual information regarding unreasonable adverse effects on the environment of the pesticide, he shall submit such information to the Administrator.

"(b) CANCELLATION AND CHANGE IN CLASSIFICATION.—If it appears to the Administrator that a pesticide or its labeling or other material required to be submitted does not comply with the provisions of this Act or, when used in accordance with widespread and commonly recognized practice, generally causes unreasonable adverse effects on the environment, the Administrator may issue a notice of his intent either—

"(1) to cancel its registration or to change its classification together with the reasons (including the factual basis) for his action, or

Hearing.

"(2) to hold a hearing to determine whether or not its registration should be canceled or its classification changed.

P.L. 94–140
89 Stat. 751

Such notice shall be sent to the registrant and made public. In determining whether to issue any such notice, the Administrator shall include among those factors to be taken into account the impact of the action proposed in such notice on production and prices of agricultural commodities, retail food prices, and otherwise on the agricultural economy. At least 60 days prior to sending such notice to the registrant or making public such notice, whichever occurs first, the Administrator shall provide the Secretary of Agriculture with a copy of such notice and an analysis of such impact on the agricultural economy. If the Secretary comments in writing to the Administrator regarding the notice and analysis within 30 days after receiving them, the Administrator shall publish in the Federal Register (with the notice) the comments of the Secretary and the response of the Administrator with regard to the Secretary's comments. If the Secretary does not comment in writing to the Administrator regarding the notice and analysis within 30 days after receiving them, the Administrator may notify the registrant and make public the notice at any time after such 30-day period notwithstanding the foregoing 60-day time requirement. The time requirements imposed by the preceding 3 sentences may be waived or modified to the extent agreed upon by the Administrator and the Secretary. Notwithstanding any other provision of this subsection (b) and section 25(d), in the event that the Administrator determines that suspension of a pesticide registration is necessary to prevent an imminent hazard to human health, then upon such a finding the Administrator may waive the requirement of notice to and consultation with the Secretary of Agriculture pursuant to subsection (b) and of submission to the Scientific Advisory Panel pursuant to section 25(d) and proceed in accordance with subsection (c). The proposed action shall become final and effective at the end of 30 days from receipt by the registrant, or publication, of a notice issued under paragraph (1), whichever occurs later, unless within that time either (i) the registrant makes the necessary corrections, if possible, or (ii) a request for a hearing is made by a person adversely affected by the notice. In the event a hearing is held pursuant to such a request or to the Administrator's determination under paragraph (2), a decision pertaining to registration or classification issued after completion of such hearing shall be final.

Publication in Federal Register.

Ante, p. 753.

P.L. 94–140
89 Stat. 751

In taking any final action under this subsection, the Administrator shall include among those factors to be taken into account the impact of such final action on production and prices of agricultural commodities, retail food prices, and otherwise on the agricultural economy, and he shall publish in the Federal Register an analysis of such impact.

"(c) SUSPENSION.—

"(1) ORDER.—If the Administrator determines that action is necessary to prevent an imminent hazard during the time required for cancellation or change in classification proceedings he may, by order, suspend the registration of the pesticide immediately. No order of suspension may be issued unless the Administrator has issued or at the same time issues notice of his intention to cancel the registration or change the classification of the pesticide.

"Except as provided in paragraph (3), the Administrator shall notify the registrant prior to issuing any suspension order. Such notice shall include findings pertaining to the question of 'imminent hazard'. The registrant shall then have an opportunity, in accordance with the provisions of paragraph (2), for an expedited hearing before the Agency on the question of whether an imminent hazard exists.

"(2) EXPEDITE HEARING.—If no request for a hearing is submitted to the Agency within five days of the registrant's receipt of the notification provided for by paragraph (1), the suspension order may be issued and shall take effect and shall not be reviewable by a court. If a hearing is requested, it shall commence within five days of the receipt of the request for such hearing unless the registrant and the Agency agree that it shall commence at a later time. The hearing shall be held in accordance with the provisions of subchapter II of title 5 of the United States Code, except that the presiding officer need not be a certified hearing examiner. The presiding officer shall have ten days from the conclusion of the presentation of evidence to submit recommended findings and conclusions to the Administrator, who shall then have seven days to render a final order on the issue of suspension. 80 Stat. 381; 81 Stat. 54. 5 USC 551.

"(3) EMERGENCY ORDER.—Whenever the Administrator determines that an emergency exists that does not permit him to hold a hearing before suspending, he may issue a suspension order in advance of notification to the registrant. In that case, paragraph (2) shall apply except that (i) the order of suspension shall be in effect pending the expeditious completion of the remedies provided by that paragraph and the issuance of a final order on suspension, and (ii) no party other than the registrant and the Agency shall participate except that any person adversely affected may file briefs within the time allotted by the Agency's rules. Any person so filing briefs shall be considered a party to such proceeding for the purpose of section 16(b).

"(4) JUDICIAL REVIEW.—A final order on the question of suspension following a hearing shall be reviewable in accordance with Section 16 of this Act, notwithstanding the fact that any related cancellation proceedings have not been completed. Petitions to review orders on the issue of suspension shall be advanced on the docket of the courts of appeals. Any order of suspension entered prior to a hearing before the Administrator shall be subject to immediate review in an action by the registrant or other interested person with the concurrence of the registrant in an appropriate district court, solely to determine whether the order of suspension was arbitrary, capricious or an abuse of discretion, or whether the order was issued in accordance with the procedures established by law. The effect of any order of the court will be only to stay the effectiveness of the suspension order, pending the Administrator's final decision with respect to cancellation or change in classification. This action may be maintained simultaneously with any administrative review proceeding under this section. The commencement of proceedings under this paragraph shall not operate as a stay of order, unless ordered by the court.

"(d) PUBLIC HEARINGS AND SCIENTIFIC REVIEW.—In the event a hearing is requested pursuant to subsection (b) or determined upon by the Administrator pursuant to subsection (b), such hearing shall be held after due notice for the purpose of receiving evidence relevant and material to the issues raised by the objections filed by the applicant or other interested parties, or to the issues stated by the Administrator, if the hearing is called by the Administrator rather than by the filing of objections. Upon a showing of relevance and reasonable scope of evidence sought by any party to a public hearing, the Hearing Examiner shall issue a subpena to compel testimony or production of documents from any person. The Hearing Examiner shall be guided by the principles of the Federal Rules of Civil Procedure in making any order for the protection of the witness or the content of documents produced and shall order the payment of reasonable fees and expenses as a condition to requiring testimony of the witness. On contest, the subpena may be enforced by an appropriate United States district court Subpena.

28 USC app.

in accordance with the principles stated herein. Upon the request of any party to a public hearing and when in the Hearing Examiner's judgment it is necessary or desirable, the Hearing Examiner shall at any time before the hearing record is closed refer to a Committee of the National Academy of Sciences the relevant questions of scientific fact involved in the public hearing. No member of any committee of the National Academy of Sciences established to carry out the functions of this section shall have a financial or other conflict of interest with respect to any matter considered by such **Report.** committee. The Committee of the National Academy of Sciences shall report in writing to the Hearing Examiner within 60 days after such referral on these questions of scientific fact. The report shall be made public and shall be considered as part of the hearing record. The Administrator shall enter into appropriate arrangements with the National Academy of Sciences to assure an objective and competent scientific review of the questions presented to Committees of the Academy and to provide such other scientific advisory services as may be required by the Administrator for carrying out the purposes of this Act. As soon as practicable after completion of the hearing (including the report of the Academy) but not later than 90 days thereafter, the Administrator shall evaluate the data and reports before him and issue an order either revoking his notice of intention issued pursuant to this section, or shall issue an order either canceling the registration, changing the classification, denying the registration, or requiring modification of the labeling or packaging of the article. Such order shall be based only on substantial evidence of record of such hearing and shall set forth detailed findings of fact upon which the order is based.

"(e) JUDICIAL REVIEW.—Final orders of the Administrator under this section shall be subject to judicial review pursuant to section 16.

"SEC. 7. REGISTRATION OF ESTABLISHMENTS.

"(a) REQUIREMENT.—No person shall produce any pesticide subject to this Act in any State unless the establishment in which it is produced is registered with the Administrator. The application for registration of any establishment shall include the name and address of the establishment and of the producer who operates such establishment.

"(b) REGISTRATION.—Whenever the Administrator receives an application under subsection (a), he shall register the establishment and assign it an establishment number.

"(c) INFORMATION REQUIRED.—

"(1) Any producer operating an establishment registered under this section shall inform the Administrator within 30 days after it is registered of the types and amounts of pesticides—

"(A) which he is currently producing;

"(B) which he has produced during the past year; and

"(C) which he has sold or distributed during the past year.

The information required by this paragraph shall be kept current and submitted to the Administrator annually as required under such regulations as the Administrator may prescribe.

"(2) Any such producer shall, upon the request of the Administrator for the purpose of issuing a stop sale order pursuant to section 13, inform him of the name and address of any recipient of any pesticide produced in any registered establishment which he operates.

"(d) CONFIDENTIAL RECORDS AND INFORMATION.—Any information submitted to the Administrator pursuant to subsection (c) shall be considered confidential and shall be subject to the provisions of section 10.

"SEC. 8. BOOKS AND RECORDS.

Regulations. "(a) REQUIREMENTS.—The Administrator may prescribe regulations requiring producers to maintain such records with respect to their operations and the pesticides and devices produced as he determines are necessary for the effective enforcement of this Act. No records required under this subsection shall extend to financial data, sales data other than shipment data, pricing data, personnel data, and research data (other than data relating to registered pesticides or to a pesticide for which an application for registration has been filed).

"(b) INSPECTION.—For the purposes of enforcing the provisions of this Act, any producer, distributor, carrier, dealer, or any other person who sells or offers for sale, delivers or offers for delivery any pesticide or device subject to this Act, shall, upon request of any officer or employee of the Environmental Protection Agency or of any State or political subdivision, duly designated by the Administrator, furnish or permit such person at all reasonable times to have access to, and to copy: (1) all records showing the delivery, movement, or holding of such pesticide or device, including the quantity, the date of shipment and receipt, and the name of the consignor and consignee; or (2) in the event of the inability of any person to produce records containing such information, all other records and information relating to such delivery, movement, or holding of the pesticide or device. Any inspection with respect to any records and information referred to in this subsection shall not extend to financial data, sales data other than shipment data, pricing data, personnel data, and research data (other than data relating to registered pesticides or to a pesticide for which an application for registration has been filed).

"SEC. 9. INSPECTION OF ESTABLISHMENTS, ETC.

"(a) IN GENERAL.—For purposes of enforcing the provisions of this Act, officers or employees duly designated by the Administrator are authorized to enter at reasonable times, any establishment or other place where pesticides or devices are held for distribution or sale for the purpose of inspecting and obtaining samples of any pesticides or devices, packaged, labeled, and released for shipment, and samples of any containers or labeling for such pesticides or devices.

Before undertaking such inspection, the officers or employees must present to the owner, operator, or agent in charge of the establishment or other place where pesticides or devices are held for distribution or sale, appropriate credentials and a written statement as to the reason for the inspection, including a statement as to whether a violation of the law is suspected. If no violation is suspected, an alternate and sufficient reason shall be given in writing. Each such inspection shall be commenced and completed with reasonable promptness. If the officer or employee obtains any samples, prior to leaving the premises, he shall give to the owner, operator, or agent in charge a receipt describing the samples obtained and, if requested, a portion of each such sample equal in volume or weight to the portion retained. If an analysis is made of such samples, a copy of the results of such analysis shall be furnished promptly to the owner, operator, or agent in charge.

"(b) WARRANTS.— For purposes of enforcing the provisions of this Act and upon a showing to an officer or court of competent jurisdiction that there is reason to believe that the provisions of this Act have been violated, officers or employees duly designated by the Administrator are empowered to obtain and to execute warrants authorizing—

"(1) entry for the purpose of this section;

"(2) inspection and reproduction of all records showing the quantity, date of shipment, and the name of consignor and consignee of any pesticide or device found in the establishment which is adulterated, misbranded, not registered (in the case of a pesticide) or otherwise in violation of this Act and in the event of the inability of any person to produce records containing such information, all other records and information relating to such delivery, movement, or holding of the pesticide or device; and

"(3) the seizure of any pesticide or device which is in violation of this Act.

"(c) ENFORCEMENT.—

"(1) CERTIFICATION OF FACTS TO ATTORNEY GENERAL.—The examination of pesticides or devices shall be made in the Environmental Protection Agency or elsewhere as the Administrator may designate for the purpose of determining from such examinations whether they comply with the requirements of this Act. If it shall appear from any such examination that they fail to comply with the requirements of this Act, the Administrator shall cause notice to be given to the person against whom criminal or civil proceedings are contemplated. Any person so notified shall be given an opportunity to present his views, either orally or in writing, with regard to such contemplated proceedings, and if in the opinion of the Administrator it appears that the provisions of this Act have been violated by such person, then the Administrator shall certify the facts to the Attorney General, with a copy of the results of the analysis or the examination of such pesticide for the institution of a criminal proceeding pursuant to section 14(b) or a civil proceeding under section 14(a), when the Administrator determines that such action will be sufficient to effectuate the purposes of this Act.

"(2) NOTICE NOT REQUIRED.—The notice of contemplated proceedings and opportunity to present views set forth in this subsection are not prerequisites to the institution of any proceeding by the Attorney General.

"(3) WARNING NOTICES.— Nothing in this Act shall be construed as requiring the Administrator to institute proceedings for prosecution of minor violations of this Act whenever he believes that the public interest will be adequately served by a suitable written notice of warning.

"SEC. 10. PROTECTION OF TRADE SECRETS AND OTHER INFORMATION.

"(a) IN GENERAL.—In submitting data required by this Act, the applicant may (1) clearly mark any portions thereof which in his opinion are trade secrets or commercial or financial information and (2) submit such marked material separately from other material required to be submitted under this Act.

"(b) DISCLOSURE.—Notwithstanding any other provision of this Act, the Administrator shall not make public information which in his judgment contains or relates to trade secrets or commercial or financial information obtained from a person and privileged or confidential, except that, when necessary to carry out the provisions of this Act, information relating to formulas of products acquired by authorization of this Act may be revealed to any Federal agency consulted and may be revealed at a public hearing or in findings of fact issued by the Administrator.

"(c) DISPUTES.—If the Administrator proposes to release for inspection information which the applicant or registrant believes to be protected from disclosure under subsection (b), he shall notify the applicant or registrant, in writing, by certified mail. The Administrator shall not thereafter make available for inspection such data until thirty days after receipt of the notice by the applicant or registrant. During this period, the applicant or registrant may institute an action in an appropriate district court for a declaratory judgment as to whether such information is subject to protection under subsection (b).

"SEC. 11. STANDARDS APPLICABLE TO PESTICIDE APPLICATORS.

"(a) IN GENERAL.—No regulations prescribed by the Administrator for carrying out the provisions of this Act shall require any private applicator to maintain any records or file any reports or other documents.

"(b) SEPARATE STANDARDS.—When establishing or approving standards for licensing or certification, the Administrator shall establish separate standards for commercial and private applicators.

"SEC. 12. UNLAWFUL ACTS.

"(a) IN GENERAL.—

"(1) Except as provided by subsection (b), it shall be unlawful for any person in any State to distribute, sell, offer for sale, hold for sale, ship, deliver for shipment, or receive and (having so received) deliver or offer to deliver, to any person—

"(A) any pesticide which is not registered under section 3, except as provided by section 6(a)(1);

"(B) any registered pesticide if any claims made for it as a part of its distribution or sale substantially differ from any claims made for it as a part of the statement required in connection with its registration under section 3;

"(C) any registered pesticide the composition of which differs at the time of its distribution or sale from its composition as described in the statement required in connection with its registration under section 3;

"(D) any pesticide which has not been colored or discolored pursuant to the provisions of section 25(c)(5);

"(E) any pesticide which is adulterated or misbranded; or

"(F) any device which is misbranded.

"(2) It shall be unlawful for any person—

"(A) to detach, alter, deface, or destroy, in whole or in part, any labeling required under this Act;

"(B) to refuse to keep any records required pursuant to section 8, or to refuse to allow the inspection of any records or establishment pursuant to section 8 or 9, or to refuse to allow an officer or employee of the Environmental Protection Agency to take a sample of any pesticide pursuant to section 9;

"(C) to give a guaranty or undertaking provided for in subsection (b) which is false in any particular, except that a person who receives and relies upon a guaranty authorized under subsection (b) may give a guaranty to the same effect, which guaranty shall contain, in addition to his own name and address, the name and address of the person residing in the United States from whom he received the guaranty or undertaking;

"(D) to use for his own advantage or to reveal, other than to the Administrator, or officials or employees of the Environmental Protection Agency or other Federal executive agencies, or to the courts, or to physicians, pharmacists, and other qualified persons, needing such information for the performance of their duties, in accordance with such directions as the Administrator may prescribe; any information acquired by authority of this Act which is confidential under this Act;

"(E) who is a registrant, wholesaler, dealer, retailer, or other distributor to advertise a produce registered under this Act for restricted use without giving the classification of the product assigned to it under section 3;

"(F) to make available for use, or to use, any registered pesticide classified for restricted use for some or all purposes other than in accordance with section 3(d) and any regulations thereunder;

"(G) to use any registered pesticide in a manner inconsistent with its labeling;

"(H) to use any pesticide which is under an experimental use permit contrary to the provisions of such permit;

"(I) to violate any order issued under section 13;

"(J) to violate any suspension order issued under section 6;

"(K) to violate any cancellation of registration of a pesticide under section 6, except as provided by section 6(a)(1);

"(L) who is a producer to violate any of the provisions of section 7;

"(M) to knowingly falsify all or part of any application for registration, application for experimental use permit, any information submitted to the Administrator pursuant to section 7, any records required to be maintained pursuant to section 8, any report filed under this Act, or any information marked as confidential and submitted to the Administrator under any provision of this act;

"(N) who is a registrant, wholesaler, dealer, retailer, or other distributor to fail to file reports required by this Act;

"(O) to add any substance to, or take any substance from, any pesticide in a manner that may defeat the purpose of this Act; or

"(P) to use any pesticide in tests on human beings unless such human beings (i) are fully informed of the nature and purposes of the test and of any physical and mental health consequences which are reasonably foreseeable therefrom, and (ii) freely volunteer to participate in the test.

"(b) EXEMPTIONS.—The penalties provided for a violation of paragraph (1) of subsection (a) shall not apply to—

"(1) any person who establishes a guaranty signed by, and containing the name and address of, the registrant or person residing in the United States from whom he purchased or received in good faith the pesticide in the same unbroken package, to the effect that the pesticide was lawfully registered at the time of sale and delivery to him, and that it complies with the other requirements of this Act, and in such case the guarantor shall be subject to the penalties which would otherwise attach to the person holding the guaranty under the provisions of this Act;

"(2) any carrier while lawfully shipping, transporting, or delivering for shipment any pesticide or device, if such carrier upon request of any officer or employee duly designated by the Administrator shall permit such officer or employee to copy all of its records concerning such pesticide or device;

"(3) any public official while engaged in the performance of his official duties;

"(4) any person using or possessing any pesticide as provided by an experimental use permit in effect with respect to such pesticide and such use or possession; or

"(5) any person who ships a substance or mixture of substances being put through tests in which the purpose is only to determine its value for pesticide purposes or to determine its toxicity or other properties and from which the user does not expect to receive any benefit in pest control from its use.

"SEC. 13. STOP SALE, USE, REMOVAL, AND SEIZURE.

"(a) STOP SALE, ETC., ORDERS.—Whenever any pesticide or device is found by the Administrator in any State and there is reason to believe on the basis of inspection or tests that such pesticide or device is in violation of any of the provisions of this Act, or that such pesticide or device has been or is intended to be distributed or sold in violation of any such provisions, or when the registration of the pesticide has been canceled by a final order or has been suspended, the Administrator may issue a written or printed 'stop sale, use, or removal' order to any person who owns, controls, or has custody of such pesticide or device, and after receipt of such order no person shall sell, use, or remove the pesticide or device described in the order except in accordance with the provisions of the order.

"(b) SEIZURE.—Any pesticide or device that is being transported or, having been transported, remains unsold or in original unbroken packages, or that is sold or offered for sale in any State, or that is imported from a foreign country, shall be liable to be proceeded against in any district court in the district where it is found and seized for confiscation by a process in rem for condemnation if—

"(1) in the case of a pesticide—

"(A) it is adulterated or misbranded;

"(B) it is not registered pursuant to the provisions of section 3;

"(C) its labeling fails to bear the information required by this Act;

"(D) it is not colored or discolored and such coloring or discoloring is required under this Act; or

"(E) any of the claims made for it or any of the directions for its use differ in substance from the representations made in connection with its registration;

"(2) in the case of a device, it is misbranded; or

"(3) in the case of a pesticide or device, when used in accordance with the requirements imposed under this Act and as directed by the labeling, it nevertheless causes unreasonable adverse effects on the environment. In the case of a plant regulator, defoliant, or desiccant, used in accordance with the label claims and recommendations, physical or physiological effects on plants or parts thereof shall not be deemed to be injury, when such effects are the purpose for which the plant regulator, defoliant, or desiccant was applied.

"(c) DISPOSITION AFTER CONDEMNATION.--If the pesticide or device is condemned it shall, after entry of the decree, be disposed of by destruction or sale as the court may direct and the proceeds, if sold, less the court costs, shall be paid into the Treasury of the United States, but the pesticide or device shall not be sold contrary to the provisions of this Act or the laws of the jurisdiction in which it is sold: *Provided,* That upon payment of the costs of the condemnation proceedings and the execution and delivery of a good and sufficient bond conditioned that the pesticide or device shall not be sold or otherwise disposed of contrary to the provisions of the Act or the laws of any jurisdiction in which sold, the court may direct that such pesticide or device be delivered to the owner thereof. The proceedings of such condemnation cases shall conform, as near as may be to the proceedings in admiralty, except that either party may demand trial by jury of any issue of fact joined in any case, and all such proceedings shall be at the suit of and in the name of the United States.

"(d) COURT COSTS, ETC.—When a decree of condemnation is entered against the pesticide or device, court costs and fees, storage, and other proper expenses shall be awarded against the person, if any, intervening as claimant of the pesticide or device.

"SEC. 14. PENALTIES.

"(a) CIVIL PENALTIES.—

"(1) IN GENERAL.—Any registrant, commercial applicator, wholesaler, dealer, retailer, or other distributor who violates any provision of this Act may be assessed a civil penalty by the Administrator of not more than $5,000 for each offense.

"(2) PRIVATE APPLICATOR.—Any private applicator or other person not included in paragraph (1) who violates any provision of this Act subsequent to receiving a written warning from the Administrator or following a citation for a prior violation, may be assessed a civil penalty by the Administrator of not more than $1,000 for each offense.

"(3) HEARING.—No civil penalty shall be assessed unless the person charged shall have been given notice and opportunity for a hearing on such charge in the county, parish, or incorporated city of the residence of the person charged. In determining the amount of the penalty the Administrator shall consider the appropriateness of such penalty to the size of the business of the person charged, the effect on the person's ability to continue in business, and the gravity of the violation.

"(4) REFERENCES TO ATTORNEY GENERAL.—In case of inability to collect such civil penalty or failure of any person to pay all, or such portion of such civil penalty as the Administrator may determine, the Administrator shall refer the matter to the Attorney General, who shall recover such amount by action in the appropriate United States district court.

"(b) CRIMINAL PENALTIES.—

"(1) IN GENERAL.—Any registrant, commercial applicator, wholesaler, dealer, retailer, or other distributor who knowingly violates any provision of this Act shall be guilty of a misdemeanor and shall on conviction be fined not more than $25,000, or imprisoned for not more than one year, or both.

"(2) PRIVATE APPLICATOR.—Any private applicator or other person not included in paragraph (1) who knowingly violates any provision of this Act shall be guilty of a misdemeanor and shall on conviction be fined not more than $1,000, or imprisoned for not more than 30 days, or both.

"(3) DISCLOSURE OF INFORMATION.—Any person,, who, with intent to defraud, uses or reveals information relative to formulas of products acquired under the authority of section 3, shall be fined not more than $10,000, or imprisoned for not more than three years, or both.

"(4) ACTS OF OFFICERS, AGENTS, ETC.—When construing and enforcing the provisions of this Act, the act, omission, or failure of any officer, agent, or other person acting for or employed by any person shall in every case be also deemed to be the act, omission, or failure of such person as well as that of the person employed.

"SEC. 15. INDEMNITIES.

"(a) REQUIREMENT.—If—

"(1) the Administrator notifies a registrant that he has suspended the registration of a pesticide because such action is necessary to prevent an imminent hazard;

"(2) the registration of the pesticide is canceled as a result of a final determination that the use of such pesticide will create an imminent hazard; and

"(3) any person who owned any quantity of such pesticide immediately before the notice to the registrant under paragraph (1) suffered losses by reason of suspension or cancellation of the registration,

the Administrator shall make an indemnity payment to such person, unless the Administrator finds that such person (i) had knowledge of facts which, in themselves, would have shown that such pesticide did not meet the requirements of section 3(c)(5) for registration, and (ii) continued thereafter to produce such pesticide without giving timely notice of such facts to the Administrator.

"(b) AMOUNT OF PAYMENT.—

"(1) IN GENERAL.—The amount of the indemnity payment under subsection (a) to any person shall be determined on the basis of the cost of the pesticide owned by such person immediately before the notice to the registrant referred to in subsection (a)(1); except that in no event shall an indemnity payment to any person exceed the fair market value of the pesticide owned by such person immediately before the notice referred to in subsection (a)(1).

"(2) SPECIAL RULE.—Notwithstanding any other provision of this Act, the Administrator may provide a reasonable time for use or other disposal of such pesticide. In determining the quantity of any pesticide for which indemnity shall be paid under this subsection, proper adjustment shall be made for any pesticide used or otherwise disposed of by such owner.

"SEC. 16. ADMINISTRATIVE PROCEDURE; JUDICIAL REVIEW.

"(a) DISTRICT COURT REVIEW.—Except as is otherwise provided in this Act, Agency refusals to cancel or suspend registrations or change classifications not following a hearing and other final Agency actions not committed to Agency discretion by law are judicially reviewable in the district courts.

"(b) REVIEW BY COURT OF APPEALS.—In the case of actual controversy as to the validity of any order issued by the Administrator following a public hearing, any person who will be adversely affected by such order and who had been a party to the proceedings may obtain judicial review by filing in the United States court of appeals for the circuit wherein such person resides or has a place of business, within 60 days after the entry of such order, a petition praying that the order be set aside in whole or in part. A copy of the petition shall be forthwith transmitted by the clerk of the court to the Administrator or any officer designated by him for that purpose, and thereupon the Administrator shall file in the court the record of the proceedings on which he based his order, as provided in section 2112 of title 28, United States Code. Upon the filing of such petition the court shall have exclusive jurisdiction to affirm or set aside the order complained of in whole or in part. The court shall consider all evidence of record. The order of the Administrator shall be sustained if it is supported by substantial evidence when considered on the record as a whole. The judgment of the court affirming or setting aside, in whole or in part, any order under this section shall be final, subject to review by the Supreme Court of the United States upon certiorari or certification as provided in section 1254 of title 28 of the United States Code. The commencement of proceedings under this section shall not, unless specifically ordered by the court to the contrary, operate as a stay of an order. The court shall advance on the docket and expedite the disposition of all cases filed therein pursuant to this section.

72 Stat. 941;
80 Stat. 1323.

62 Stat. 928.

"(c) JURISDICTION OF DISTRICT COURTS.—The district courts of the United States are vested with jurisdiction specifically to enforce, and to prevent and restrain violations of, this Act.

"(d) NOTICE OF JUDGMENTS.—The Administrator shall, by publication in such manner as he may prescribe, give notice of all judgments entered in actions instituted under the authority of this Act.

"SEC. 17. IMPORTS AND EXPORTS.

"(a) PESTICIDES AND DEVICES INTENDED FOR EXPORT.—Notwithstanding any other provision of this Act, no pesticide or device shall be deemed in violation of this Act when intended solely for export to any foreign country and prepared or packed according to the specifications or directions of the foreign purchaser, except that producers of such pesticides and devices shall be subject to section 8 of this Act.

(b) CANCELLATION NOTICES FURNISHED TO FOREIGN GOVERNMENTS.— Whenever a registration, or a cancellation or suspension of the registration of a pesticide becomes effective, or ceases to be effective, the Administrator shall transmit through the State Department notification thereof to the governments of other countries and to appropriate international agencies.

"(c) IMPORTATION OF PESTICIDES AND DEVICES.—The Secretary of the Treasury shall notify the Administrator of the arrival of pesticides and devices and shall deliver to the Administrator, upon his request, samples of pesticides or devices which are being imported into the United States, giving notice to the owner or consignee, who may appear before the Administrator and have the right to introduce testimony. If it appears from the examination of a sample that it is adulterated, or misbranded or otherwise violates the provisions set forth in this Act, or is otherwise injurious to health or the environment, the pesticide or device may be refused admission, and the Secretary of the Treasury shall refuse delivery to the consignee and shall cause the destruction of any pesticide or device refused delivery which shall not be exported by the consignee within 90 days from the date of notice of such refusal under such regulations as the Secretary of the Treasury may prescribe: *Provided,* That the Secretary of the Treasury may deliver to the consignee such pesticide or device pending examination and decision in the matter on execution of bond for the amount of the full invoice value of such pesticide or device, together with the duty thereon, and on refusal to return such pesticide or device for any cause to the custody of the Secretary of the Treasury, when demanded, for the purpose of excluding them from the country, or for any other purpose, said consignee shall forfeit the full amount of said bond: *And provided further,* That all charges for storage, cartage, and labor on pesticides or devices which are refused admission or delivery shall be paid by the owner or consignee, and in default of such payment shall constitute a lien against any future importation made by such owner or consignee.

"(d) COOPERATION IN INTERNATIONAL EFFORTS.—The Administrator shall, in cooperation with the Department of State and any other appropriate Federal agency, participate and cooperate in any international efforts to develop improved pesticide research and regulations.

"(e) REGULATIONS.—The Secretary of the Treasury, in consultation with the Administrator, shall prescribe regulations for the enforcement of subsection (c) of this section.

"SEC. 18. EXEMPTION OF FEDERAL AGENCIES.

"The Administrator may, at his discretion, exempt any Federal or State agency from any provision of this Act if he determines that emergency conditions exist which require such exemption.

"The Administrator, in determining whether or not such emergency conditions exist, shall consult with the Secretary of Agriculture and the Governor of any State concerned if they request such determination."

"SEC. 19. DISPOSAL AND TRANSPORTATION.

"(a) PROCEDURES.—The Administrator shall, after consultation with other interested Federal agencies, establish procedures and regulations for the disposal or storage of packages and containers of pesticides and for disposal or storage of excess amounts of such pesticides, and accept at convenient locations for safe disposal a pesticide the registration of which is canceled under section 6(c) if requested by the owner of the pesticide.

"(b) ADVICE TO SECRETARY OF TRANSPORTATION.—The Administrator shall provide advice and assistance to the Secretary of Transportation with respect to his functions relating to the transportation of hazardous materials under the Department of Transportation Act (49 U.S.C. 1657), the Transportation of Explosives Act (18 U.S.C. 831–835), the Federal Aviation Act of 1958 (49 U.S.C. 1421–1430, 1472 II), and the Hazardous Cargo Act (46 U.S.C. 170, 375, 416).

P.L. 94–140
89 Stat. 754

Regulations.

80 Stat. 944.
74 Stat. 808;
79 Stat. 286.
72 Stat. 775;
85 Stat. 481.
Contract
authority.

"SEC. 20. RESEARCH AND MONITORING.

"(a) RESEARCH.—The Administrator shall undertake research, including research by grant or contract with other Federal agencies, universities, or others as may be necessary to carry out the purposes of this Act, and he shall give priority to research to develop biologically integrated alternatives for pest control. The Administrator shall also take care to insure that such research does not duplicate research being undertaken by any other Federal agency.

"(b) NATIONAL MONITORING PLAN.—The Administrator shall formulate and periodically revise, in cooperation with other Federal, State, or local agencies, a national plan for monitoring pesticides.

"(c) MONITORING.—The Administrator shall undertake such monitoring activities, including but not limited to monitoring in air, soil, water, man, plants, and animals, as may be necessary for the implementation of this Act and of the national pesticide monitoring plan. Such activities shall be carried out in cooperation with other Federal, State, and local agencies.

"SEC. 21. SOLICITATION OF COMMENTS; NOTICE OF PUBLIC HEARINGS.

"(a) The Administrator, before publishing regulations under this Act, shall solicit the views of the Secretary of Agriculture in accordance with the procedure described in section 25(a).

P.L. 94–140
89 Stat. 752.

"(b) In addition to any other authority relating to public hearings and solicitation of views, in connection with the suspension or cancellation of a pesticide registration or any other actions authorized under this Act, the Administrator may, at his discretion, solicit the views of all interested persons, either orally or in writing, and seek such advice from scientists, farmers, farm organizations, and other qualified persons as he deems proper.

"(c) In connection with all public hearings under this Act the Administrator shall publish timely notice of such hearings in the Federal Register.

Publication
in Federal
Register.

"SEC. 22. DELEGATION AND COOPERATION.

"(a) DELEGATION.—All authority vested in the Administrator by virtue of the provisions of this Act may with like force and effect be executed by such employees of the Environmental Protection Agency as the Administrator may designate for the purpose.

"(b) COOPERATION.—The Administrator shall cooperate with the Department of Agriculture, any other Federal agency, and any appropriate agency of any State or any political subdivision thereof, in carrying out the provisions of this Act, and in securing uniformity of regulations.

"SEC. 23. STATE COOPERATION, AID, AND TRAINING.

"(a) COOPERATIVE AGREEMENTS.—The Administrator is authorized to enter into cooperative agreements with States—

"(1) to delegate to any State the authority to cooperate in the enforcement of the Act through the use of its personnel or facilities, to train personnel of the State to cooperate in the enforcement of this Act, and to assist States in implementing cooperative enforcement programs through grants-in-aid; and

"(2) to assist State agencies in developing and administering State programs for training and certification of applicators consistent with the standards which he prescribes.

"(b) CONTRACTS FOR TRAINING.—In addition, the Administrator is authorized to enter into contracts with Federal or State agencies for the purpose of encouraging the training of certified applicators.

"(c) The Administrator may, in cooperation with the Secretary of Agriculture, utilize the services of the Cooperative State Extension Services for informing farmers of accepted uses and other regulations made pursuant to this Act.

"SEC. 24. AUTHORITY OF STATES.

"(a) A State may regulate the sale or use of any pesticide or device in the State, but only if and to the extent the regulation does not permit any sale or use prohibited by this Act;

"(b) such State shall not impose or continue in effect any requirements for labeling and packaging in addition to or different from those required pursuant to this Act; and

"(c) a State may provide registration for pesticides formulated for distribution and use within that State to meet special local needs if that State is certified by the Administrator as capable of exercising adequate controls to assure that such registration will be in accord with the purposes of this Act and if registration for such use has not previously been denied, disapproved, or canceled by the Administrator. Such registration shall be deemed registration under section 3 for all purposes of this Act, but shall authorize distribution and use only within such State and shall not be effective for more than 90 days if disapproved by the Administrator within that period.

"SEC. 25. AUTHORITY OF ADMINISTRATOR.

P.L. 94–140
89 Stat. 751.

"(a) (1) REGULATIONS.—The Administrator is authorized in accordance with the procedure described in paragraph (2), to prescribe regulations to carry out the provisions of this Act. Such regulations shall take into account the difference in concept and usage between various classes of pesticides.

P.L. 94–140
89 Stat. 752

(2) PROCEDURE.——

Publications in
Federal Register.

"(A) PROPOSED REGULATIONS.—At least 60 days prior to signing any proposed regulation for publication in the Federal Register, the Administrator shall provide the Secretary of Agriculture with a copy of such regulation. If the Secretary comments in writing to the Administrator regarding any such regulation within 30 days after receiving it, the Administrator shall publish in the Federal Register (with the proposed regulation) the comments of the Secretary and the response of the Administrator with regard to the Secretary's comments. If the Secretary does not comment in writing to the Administrator regarding the regulation within 30 days after receiving it, the Administrator may sign such regulation for publication in the Federal Register any time after such 30-day period notwithstanding the foregoing 60-day time requirement.

Publications in
Federal Register.

"(B) FINAL REGULATIONS.—At least 30 days prior to signing any regulation in final form for publication in the Federal Register, the Administrator shall provide the Secretary of Agriculture with a copy of such regulation. If the Secretary comments in writing to the Administrator regarding any such final regulation within 15 days after receiving it, the Administrator shall publish in the Federal Register (with the final regulation) the comments of the Secretary, if requested by the Secretary, and the response of the Administrator concerning the Secretary's comments. If the Secretary does not comment in writing to the Administrator regarding the regulation within 15 days after receiving it, the Administrator may sign such regulation for publication in the Federal Register at any time after such 15-day period notwithstanding the foregoing 30-day time requirement.

"(C) TIME REQUIREMENTS.—The time requirements imposed by subparagraphs (A) and (B) may be waived or modified to the extent agreed upon by the Administrator and the Secretary.

"(D) PUBLICATION IN THE FEDERAL REGISTER.—The Administrator shall, simultaneously with any notification to the Secretary of Agriculture under this paragraph prior to the issuance of any proposed or final regulation, publish such notification in the Federal Register.".

P.L. 94–140
89 Stat. 753

"(3) CONGRESSIONAL COMMITTEES.—At such time as the Administrator is required under paragraph (2) of this subsection to provide the Secretary of Agriculture with a copy of proposed regulations and a copy of the final form of regulations, he shall also furnish a copy of such regulations to the Committee on Agriculture of the House of Representatives and the Committee on Agriculture and Forestry of the Senate."

"(b) EXEMPTION OF PESTICIDES.—The Administrator may exempt from the requirements of this Act by regulation any pesticide which he determines either (1) to be adequately regulated by another Federal agency, or (2) to be of a character which is unnecessary to be subject to this Act in order to carry out the purposes of this Act.

"(c) OTHER AUTHORITY.—The Administrator, after notice and opportunity for hearing, is authorized—

"(1) to declare a pest any form of plant or animal life (other than man and other than bacteria, virus, and other micro-organisms on or in living man or other living animals) which is injurious to health or the environment:

"(2) to determine any pesticide which contains any substance or substances in quantities highly toxic to man;

84 Stat. 1670.
15 USC 1471
note.

"(3) to establish standards (which shall be consistent with those established under the authority of the Poison Prevention Packaging Act (Public Law 91–601)) with respect to the package, container, or wrapping in which a pesticide or device is enclosed for use or consumption, in order to protect children and adults from serious injury or illness resulting from accidental ingestion or contact with pesticides or devices regulated by this Act as well as to accomplish the other purposes of this Act;

"(4) to specify those classes of devices which shall be subject to any provision of paragraph 2(q)(1) or section 7 of this Act upon his determination that application of such provision is necessary to effectuate the purposes of this Act;

"(5) to prescribe regulations requiring any pesticide to be colored or discolored if he determines that such requirement is feasible and is necessary for the protection of health and the environment; and

"(6) to determine and establish suitable names to be used in the ingredient statement.

"(d) SCIENTIFIC ADVISORY PANEL.—The Administrator shall submit to an advisory panel for comment as to the impact on health and the environment of the action proposed in notices of intent issued under section 6(b) and of the proposed and final form of regulations issued under section 25(a) within the same time periods as provided for the comments of the Secretary of Agriculture under such sections. The time requirements for notices of intent and proposed and final forms of regulation may not be modified or waived unless in addition to meeting the requirements of section 6(b) or 25(a), as applicable, the advisory panel has failed to comment on the proposed action within the prescribed time period or has agreed to the modification or waiver. The comments of the advisory panel and the response of the Administrator shall be published in the Federal Register in the same manner as provided for publication of the comments of the Secretary of Agriculture under such sections. The panel referred to in this subsection shall consist of seven members appointed by the Administrator from a list of 12 nominees, six nominated by the National Institutes of Health, and six by the National Science Foundation. The Administrator may require such information from the nominees to the advisory panel as he deems necessary, and he shall publish in the Federal Register the name, address, and professional affiliations of each nominee. Each member of the panel shall receive per diem compensation at a rate not in excess of that fixed for GS–18 of the General Schedule as may be determined by the Administrator, except that any such member who holds another office or position under the Federal Government the compensation for which exceeds such rate may elect to receive compensation at the rate provided for such other office or position in lieu of the compensation provided by this subsection. In order to assure the objectivity of the advisory panel, the Administrator shall promulgate regulations regarding conflicts of interest with respect to the members of the panel.".

*P.L. 94–140
89 Stat. 753*

*7 USC 136d.
7 USC 136w.*

*Publication in
Federal Register.*

Members.

*Publication in
Federal Register.*

*Compensation.
5 USC 5332
note.*

*P.L. 94-140
89 Stat. 754*

Regulations.

"SEC. 26. SEVERABILITY.

"If any provision of this Act or the application thereof to any person or circumstance is held invalid, the invalidity shall not affect other provisions or applications of this Act which can be given effect without regard to the invalid provision or application, and to this end the provisions of this Act are severable.

"SEC. 27. AUTHORIZATION FOR APPROPRIATIONS.

"There is authorized to be appropriated such sums as may be necessary to carry out the provisions of this Act for each of the fiscal years ending June 30, 1973, June 30, 1974, and June 30, 1975. The amounts authorized to be appropriated for any fiscal year ending after June 30, 1975, shall be the sums hereafter provided by law."

"There are hereby authorized to be appropriated to carry out the provisions of this Act for the period beginning October 1, 1975, and ending September 30, 1976, the sum of $47,868,000, and for the period beginning October 1, 1976, and ending March 31, 1977, the sum of $23,600,000."

*P.L. 94–140
89 Stat. 752*

AMENDMENTS TO OTHER ACTS

SEC. 3. The following Acts are amended by striking out the terms "economic poisons" and "an economic poison" wherever they appear and inserting in lieu thereof "pesticides" and "a pesticide" respectively:

(1) The Federal Hazardous Substances Act, as amended (15 U.S.C. 1261 et seq.);

(2) The Poison Prevention Packaging Act, as amended (15 U.S.C. 1471 et seq.); and

(3) The Federal Food, Drug, and Cosmetic Act, as amended (21 U.S.C. 301 et seq.).

74 Stat. 1305.

84 Stat. 1670.

52 Stat. 1040.

EFFECTIVE DATES OF PROVISIONS OF ACT

SEC. 4. (a) Except as otherwise provided in the Federal Insecticide, Fungicide, and Rodenticide Act, as amended by this Act, and as otherwise provided by this section, the amendments made by this Act shall take effect at the close of the date of the enactment of this Act, provided if regulations are necessary for the implementation of any provision that becomes effective on the date of enactment, such regulations shall be promulgated and shall become effective within 90 days from the date of enactment of this Act.

Savings provision. 61 Stat. 163. 7 USC 135 note. P.L. 94–140 89 Stat. 752

(b) The provisions of the Federal Insecticide, Fungicide, and Rodenticide Act and the regulations thereunder as such existed prior to the enactment of this Act shall remain in effect until superseded by the amendments made by this Act and regulations thereunder: *Provided,* That all provisions made by these amendments and all regulations thereunder shall be effective within five years after the enactment of this Act.

(c) (1) Two years after the enactment of this Act the Administrator shall have promulgated regulations providing for the registration and classification of pesticides under the provisions of this Act and thereafter shall register all new applications under such provisions.

P.L. 94-140 89 Stat. 752

61 Stat. 163. 7 USC 135 note.

(2) After two years but within five years after the enactment of this Act the Administrator shall register and reclassify pesticides registered under the provisions of the Federal Insecticide, Fungicide, and Rodenticide Act prior to the effective date of the regulations promulgated under subsection (c)(1).

P.L. 94-140 89 Stat. 752

(3) Any requirements that a pesticide be registered for use only by a certified applicator shall not be effective until five years from the date of enactment of this Act.

P.L. 94-140 89 Stat. 753

(4) A period of five years from date of enactment shall be provided for certification of applicators.

(A) One year after the enactment of this Act the Administrator shall have prescribed the standards for the certification of applicators.

P.L. 94-140 89 Stat. 753

(B) Within four years after the enactment of this Act each State desiring to certify applicators shall submit a State plan to the Administrator for the purpose provided by section 4(b).

(C) As promptly as possible but in no event more than one year after submission of a State plan, the Administrator shall approve the State plan or disapprove it and indicate the reasons for disapproval. Consideration of plans resubmitted by States shall be expedited.

(5) One year after the enactment of this Act the Administrator shall have promulgated and shall make effective regulations relating to the registration of establishments, permits for experimental use, and the keeping of books and records under the provisions of this Act.

(d) No person shall be subject to any criminal or civil penalty imposed by the Federal Insecticide, Fungicide, and Rodenticide Act, as amended by this Act, for any act (or failure to act) occurring before the expiration of 60 days after the Administrator has published effective regulations in the Federal Register and taken such other action as may be necessary to permit compliance with the provisions under which the penalty is to be imposed.

(e) For purposes of determining any criminal or civil penalty or liability to any third person in respect of any act or omission occurring before the expiration of the periods referred to in this section, the Federal Insecticide, Fungicide, and Rodenticide Act shall be treated as continuing in effect as if this Act had not been enacted.

APPENDIX D

**MARINE PROTECTION, RESEARCH,
AND SANCTUARIES ACT OF 1972:
PUBLIC LAW 92-532 (EXCERPTS)**

TITLE I—OCEAN DUMPING

PROHIBITED ACTS

SEC. 101. (a) No person shall transport from the United States any radiological, chemical, or biological warfare agent or any high-level radioactive waste, or except as may be authorized in a permit issued under this title, and subject to regulations issued under section 108 hereof by the Secretary of the Department in which the Coast Guard is operating, any other material for the purpose of dumping it into ocean waters.

(b) No person shall dump any radiological, chemical, or biological warfare agent or any high-level radioactive waste, or, except as may be authorized in a permit issued under this title, any other material, transported from any location outside the United States, (1) into the territorial sea of the United States, or (2) into a zone contiguous to the territorial sea of the United States, extending to a line twelve nautical miles seaward from the base line from which the breadth of the territorial sea is measured, to the extent that it may affect the territorial sea or the territory of the United States.

(c) No officer, employee, agent, department, agency, or instrumentality of the United States shall transport from any location outside the United States any radiological, chemical, or biological warfare agent or any high-level radioactive waste, or, except as may be authorized in a permit issued under this title, any other material for the purpose of dumping it into ocean waters.

ENVIRONMENTAL PROTECTION AGENCY PERMITS

SEC. 102. (a) Except in relation to dredged material, as provided for in section 103 of this title, and in relation to radiological, chemical, and biological warfare agents and high-level radioactive waste, as provided for in section 101 of this title, the Administrator may issue permits, after notice and opportunity for public hearings, for the transportation from the United States or, in the case of an agency or instrumentality of the United States, for the transportation from a location outside the United States, of material for the purpose of dumping it into ocean waters, or for the dumping of material into the waters described in section 101(b), where the Administrator determines that such dumping will not unreasonably degrade or endanger human health, welfare, or amenities, or the marine environment, ecological systems, or economic potentialities. The Administrator shall establish and apply criteria for reviewing and evaluating such permit applications, and, in establishing or revising such criteria, shall consider, but not be limited in his consideration to, the following:

(A) The need for the proposed dumping.

(B) The effect of such dumping on human health and welfare, including economic, esthetic, and recreational values.

(C) The effect of such dumping on fisheries resources, plankton, fish, shellfish, wildlife, shore lines and beaches.

(D) The effect of such dumping on marine ecosystems, particularly with respect to—

(i) the transfer, concentration, and dispersion of such material and its byproducts through biological, physical, and chemical processes.

(ii) potential changes in marine ecosystem diversity, productivity, and stability, and

(iii) species and community population dynamics.

(E) The persistence and permanence of the effects of the dumping.

(F) The effect of dumping particular volumes and concentrations of such materials.

(G) Appropriate locations and methods of disposal or recycling, including land-based alternatives and the probable impact of requiring use of such alternate locations or methods upon considerations affecting the public interest.

(H) The effect on alternate uses of oceans, such as scientific study, fishing, and other living resource exploitation, and non-living resource exploitation.

(I) In designating recommended sites. the Administrator shall utilize wherever feasible locations beyond the edge of the Continental Shelf.

In establishing or revising such criteria. the Administrator shall consult with Federal. State. and local officials. and interested members of the general public. as may appear appropriate to the Administrator. With respect to such criteria as may affect the civil works program of the Department of the Army. the Administrator shall also consult with the Secretary. In reviewing applications for permits. the Administrator shall make such provision for consultation with interested Federal and State agencies as he deems useful or necessary. No permit shall be issued for a dumping of material which will violate applicable water quality standards.

(b) The Administrator may establish and issue various categories of permits, including the general permits described in section 104(c).

(c) The Administrator may, considering the criteria established pursuant to subsection (a) of this section, designate recommended sites or times for dumping and, when he finds it necessary to protect critical areas, shall, after consultation with the Secretary, also designate sites or times within which certain materials may not be dumped.

(d) No permit is required under this title for the transportation for dumping or the dumping of fish wastes, except when deposited in harbors or other protected or enclosed coastal waters, or where the Administrator finds that such deposits could endanger health, the environment, or ecological systems in a specific location. Where the Administrator makes such a finding, such material may be deposited only as authorized by a permit issued by the Administrator under this section.

CORPS OF ENGINEERS PERMITS

SEC. 103. (a) Subject to the provisions of subsections (b), (c), and (d) of this section, the Secretary may issue permits, after notice and opportunity for public hearings, for the transportation of dredged material for the purpose of dumping it into ocean waters, where the Secretary determines that the dumping will not unreasonably degrade or endanger human health, welfare, or amenities, or the marine environment, ecological systems, or economic potentialities.

(b) In making the determination required by subsection (a), the Secretary shall apply those criteria, established pursuant to section 102(a), relating to the effects of the dumping. Based upon an evaluation of the potential effect of a permit denial on navigation, economic and industrial development, and foreign and domestic commerce of the United States, the Secretary shall make an independent determination as to the need for the dumping. The Secretary shall also make an independent determination as to other possible methods of disposal and as to appropriate locations for the dumping. In considering appropriate locations, he shall, to the extent feasible, utilize the recommended sites designated by the Administrator pursuant to section 102(c).

(c) Prior to issuing any permit under this section, the Secretary shall first notify the Administrator of his intention to do so. In any case in which the Administrator disagrees with the determination of the Secretary as to compliance with the criteria established pursuant to section 102(a) relating to the effects of the dumping or with the restrictions established pursuant to section 102(c) relating to critical areas, the determination of the Administrator shall prevail. Unless the Administrator grants a waiver pursuant to subsection (d), the Secretary shall not issue a permit which does not comply with such criteria and with such restrictions.

Waiver.

(d) If, in any case, the Secretary finds that, in the disposition of dredged material, there is no economically feasible method or site available other than a dumping site the utilization of which would result in non-compliance with the criteria established pursuant to section 102(a) relating to the effects of dumping or with the restrictions established pursuant to section 102(c) relating to critical areas, he shall so certify and request a waiver from the Administrator of the specific requirements involved. Within thirty days of the receipt of the waiver request, unless the Administrator finds that the dumping of the material will result in an unacceptably adverse impact on municipal water supplies, shell-fish beds, wildlife, fisheries (including spawning and breeding areas), or recreational areas, he shall grant the waiver.

(e) In connection with Federal projects involving dredged material, the Secretary may, in lieu of the permit procedure, issue regulations which will require the application to such projects of the same criteria, other factors to be evaluated, the same procedures, and the same requirements which apply to the issuance of permits under subsections (a), (b), (c), and (d) of this section.

PERMIT CONDITIONS

SEC. 104. (a) Permits issued under this title shall designate and include (1) the type of material authorized to be transported for dumping or to be dumped; (2) the amount of material authorized to be transported for dumping or to be dumped; (3) the location where such transport for dumping will be terminated or where such dumping will occur; (4) the length of time for which the permits are valid and their expiration date; (5) any special provisions deemed necessary by the Administrator or the Secretary, as the case may be, after consultation with the Secretary of the Department in which the Coast Guard is operating, for the monitoring and surveillance of the transportation or dumping; and (6) such other matters as the Administrator or the Secretary, as the case may be, deems appropriate.

(b) The Administrator or the Secretary, as the case may be, may prescribe such processing fees for permits and such reporting requirements for actions taken pursuant to permits issued by him under this title as he deems appropriate.

(c) Consistent with the requirements of sections 102 and 103, but in lieu of a requirement for specific permits in such case, the Administrator or the Secretary, as the case may be, may issue general permits for the transportation for dumping, or dumping, or both, of specified materials or classes of materials for which he may issue permits, which he determines will have a minimal adverse environmental impact.

(d) Any permit issued under this title shall be reviewed periodically and, if appropriate, revised. The Administrator or the Secretary, as the case may be, may limit or deny the issuance of permits, or he may alter or revoke partially or entirely the terms of permits issued by him under this title, for the transportation for dumping, or for the dumping, or both, of specified materials or classes of materials, where he finds that such materials cannot be dumped consistently with the criteria and other factors required to be applied in evaluating the permit application. No action shall be taken under this subsection unless the affected person or permittee shall have been given notice and opportunity for a hearing on such action as proposed.

Review.

(e) The Administrator or the Secretary, as the case may be, shall require an applicant for a permit under this title to provide such information as he may consider necessary to review and evaluate such application.

(f) Information received by the Administrator or the Secretary, as the case may be, as a part of any application or in connection with any permit granted under this title shall be available to the public as a matter of public record, at every stage of the proceeding. The final determination of the Administrator or the Secretary, as the case may be, shall be likewise available.

Public information.

(g) A copy of any permit issued under this title shall be placed in a conspicuous place in the vessel which will be used for the transportation or dumping authorized by such permit, and an additional copy shall be furnished by the issuing official to the Secretary of the department in which the Coast Guard is operating, or its designee.

PENALTIES

Sec. 105. (a) Any person who violates any provision of this title, or of the regulations promulgated under this title, or a permit issued under this title shall be liable to a civil penalty of not more than $50,000 for each violation to be assessed by the Administrator. No penalty shall be assessed until the person charged shall have been given notice and an opportunity for a hearing of such violation. In determining the amount of the penalty, the gravity of the violation, prior violations, and the demonstrated good faith of the person charged in attempting to achieve rapid compliance after notification of a violation shall be considered by said Administrator. For good cause shown, the Administrator may remit or mitigate such penalty. Upon failure of the offending party to pay the penalty, the Administrator may request the Attorney General to commence an action in the appropriate district court of the United States for such relief as may be appropriate.

(b) In addition to any action which may be brought under subsection (a) of this section, a person who knowingly violates this title, regulations promulgated under this title, or a permit issued under this title shall be fined not more than $50,000, or imprisoned for not more than one year, or both.

(c) For the purpose of imposing civil penalties and criminal fines under this section, each day of a continuing violation shall constitute a separate offense as shall the dumping from each of several vessels, or other sources.

(d) The Attorney General or his delegate may bring actions for equitable relief to enjoin an imminent or continuing violation of this title, of regulations promulgated under this title, or of permits issued under this title, and the district courts of the United States shall have jurisdiction to grant such relief as the equities of the case may require.

Liability.

Ante, p..816.

(e) A vessel, except a public vessel within the meaning of section 13 of the Federal Water Pollution Control Act, as amended (33 U.S.C. 1163), used in a violation, shall be liable in rem for any civil penalty assessed or criminal fine imposed and may be proceeded against in any district court of the United States having jurisdiction thereof; but no vessel shall be liable unless it shall appear that one or more of the owners, or bareboat charterers, was at the time of the violation a consenting party or privy to such violation.

Ante, pp. 1054, 1055.

(f) If the provisions of any permit issued under section 102 or 103 are violated, the Administrator or the Secretary, as the case may be, may revoke the permit or may suspend the permit for a specified period of time. No permit shall be revoked or suspended unless the permittee shall have been given notice and opportunity for a hearing on such violation and proposed suspension or revocation.

(g)(1) Except as provided in paragraph (2) of this subsection any person may commence a civil suit on his own behalf to enjoin any person, including the United States and any other governmental instrumentality or agency (to the extent permitted by the eleventh amendment to the Constitution), who is alleged to be in violation of any prohibition, limitation, criterion, or permit established or issued by or under this title. The district courts shall have jurisdiction, without regard to the amount in controversy or the citizenship of the parties, to enforce such prohibition, limitation, criterion, or permit, as the case may be.

(2) No action may be commenced—

(A) prior to sixty days after notice of the violation has been given to the Administrator or to the Secretary, and to any alleged violator of the prohibition, limitation, criterion, or permit; or

(B) if the Attorney General has commenced and is diligently prosecuting a civil action in a court of the United States to require compliance with the prohibition, limitation, criterion, or permit; or

(C) if the Administrator has commenced action to impose a penalty pursuant to subsection (a) of this section, or if the Administrator, or the Secretary, has initiated permit revocation or suspension proceedings under subsection (f) of this section; or

(D) if the United States has commenced and is diligently prosecuting a criminal action in a court of the United States or a State to redress a violation of this title.

(3)(A) Any suit under this subsection may be brought in the judicial district in which the violation occurs.

(B) In any such suit under this subsection in which the United States is not a party, the Attorney General, at the request of the Administrator or Secretary, may intervene on behalf of the United States as a matter of right.

(4) The court, in issuing any final order in any suit brought pursuant to paragraph (1) of this subsection may award costs of litigation (including reasonable attorney and expert witness fees) to any party, whenever the court determines such award is appropriate.

(5) The injunctive relief provided by this subsection shall not restrict any right which any person (or class of persons) may have under any statute or common law to seek enforcement of any standard or limitation or to seek any other relief (including relief against the Administrator, the Secretary, or a State agency).

(h) No person shall be subject to a civil penalty or to a criminal Exception. fine or imprisonment for dumping materials from a vessel if such materials are dumped in an emergency to safeguard life at sea. Any such emergency dumping shall be reported to the Administrator under such conditions as he may prescribe.

RELATIONSHIP TO OTHER LAWS

Sec. 106. (a) After the effective date of this title, all licenses, permits, and authorizations other than those issued pursuant to this title shall be void and of no legal effect, to the extent that they purport to authorize any activity regulated by this title, and whether issued before or after the effective date of this title.

(b) The provisions of subsection (a) shall not apply to actions taken before the effective date of this title under the authority of the Rivers and Harbors Act of 1899 (30 Stat. 1151), as amended (33 U.S.C. 401 et. seq.).

(c) Prior to issuing any permit under this title, if it appears to the Administrator that the disposition of material, other than dredged material, may adversely affect navigation in the territorial sea of the United States, or in the approaches to any harbor of the United States, or may create an artificial island on the Outer Continental Shelf, the Administrator shall consult with the Secretary and no permit shall be issued if the Secretary determines that navigation will be unreasonably impaired.

(d) After the effective date of this title, no State shall adopt or enforce any rule or regulation relating to any activity regulated by this title. Any State may, however, propose to the Administrator criteria relating to the dumping of materials into ocean waters within its jurisdiction, or into other ocean waters to the extent that such dumping may affect waters within the jurisdiction of such State, and if the Administrator determines, after notice and opportunity for hearing, that the proposed criteria are not inconsistent with the purposes of this title, may adopt those criteria and may issue regulations to implement such criteria. Such determination shall be made by the Administrator within one hundred and twenty days of receipt of the proposed criteria. For the purposes of this subsection, the term "State" means any State, interstate or regional authority, Federal territory or Commonwealth or the District of Columbia.

"State."

(e) Nothing in this title shall be deemed to affect in any manner or to any extent any provision of the Fish and Wildlife Coordination Act as amended (16 U.S.C. 661–666c).

60 Stat. 1080;
72 Stat. 563.

ENFORCEMENT

SEC. 107. (a) The Administrator or the Secretary, as the case may be, may, whenever appropriate, utilize by agreement, the personnel, services and facilities of other Federal departments, agencies, and instrumentalities, or State agencies or instrumentalities, whether on a reimbursable or a nonreimbursable basis, in carrying out his responsibilities under this title.

(b) The Administrator or the Secretary may delegate responsibility and authority for reviewing and evaluating permit applications, including the decision as to whether a permit will be issued, to an officer of his agency, or he may delegate, by agreement, such responsibility and authority to the heads of other Federal departments or agencies, whether on a reimbursable or nonreimbursable basis.

(c) The Secretary of the department in which the Coast Guard is operating shall conduct surveillance and other appropriate enforcement activity to prevent unlawful transportation of material for dumping, or unlawful dumping. Such enforcement activity shall include, but not be limited to, enforcement of regulations issued by him pursuant to section 108, relating to safe transportation, handling, carriage, storage, and stowage. The Secretary of the Department in which the Coast Guard is operating shall supply to the Administrator and to the Attorney General, as appropriate, such information of enforcement activities and such evidentiary material assembled as they may require in carrying out their duties relative to penalty assessments, criminal prosecutions, or other actions involving litigation pursuant to the provisions of this title.

Infra.

APPENDIX E

SAFE DRINKING WATER ACT, 1974: PUBLIC LAW 93-523 (EXCERPTS)

"National Drinking Water Regulations

"Sec. 1412. (a)(1) The Administrator shall publish proposed national interim primary drinking water regulations within 90 days after the date of enactment of this title. Within 180 days after such date of enactment, he shall promulgate such regulations with such modifications as he deems appropriate. Regulations under this paragraph may be amended from time to time. **42 USC 300g-1.**

"(2) National interim primary drinking water regulations promulgated under paragraph (1) shall protect health to the extent feasible, using technology, treatment techniques, and other means, which the Administrator determines are generally available (taking costs into consideration) on the date of enactment of this title. **88 STAT. 1662**

"(3) The interim primary regulations first promulgated under paragraph (1) shall take effect eighteen months after the date of their promulgation. **88 STAT. 1663**

"(b)(1)(A) Within 10 days of the date the report on the study conducted pursuant to subsection (e) is submitted to Congress, the Administrator shall publish in the Federal Register, and provide opportunity for comment on, the— **Publication in Federal Register.**

"(i) proposals in the report for recommended maximum contaminant levels for national primary drinking water regulations, and

"(ii) list in the report of contaminants the levels of which in drinking water cannot be determined but which may have an adverse effect on the health of persons.

"(B) Within 90 days after the date the Administrator makes the publication required by subparagraph (A), he shall by rule establish recommended maximum contaminant levels for each contaminant which, in his judgment based on the report on the study conducted pursuant to subsection (e), may have any adverse effect on the health of persons. Each such recommended maximum contaminant level shall be set at a level at which, in the Administrator's judgment based on such report, no known or anticipated adverse effects on the health of persons occur and which allows an adequate margin of safety. In addition, he shall, on the basis of the report on the study conducted pursuant to subsection (e), list in the rules under this subparagraph any contaminant the level of which cannot be accurately enough measured in drinking water to establish a recommended maximum contaminant level and which may have any adverse effect on the health of persons. Based on information available to him, the Administrator may by rule change recommended levels established under this subparagraph or change such list. **Recommended maximum contaminant levels.**

"(2) On the date the Administrator establishes pursuant to paragraph (1)(B) recommended maximum contaminant levels he shall publish in the Federal Register proposed revised national primary drinking water regulations (meeting the requirements of paragraph (3)). Within 180 days after the date of such proposed regulations, he shall promulgate such revised drinking water regulations with such modifications as he deems appropriate. **Publication in Federal Register.**

"(3) Revised national primary drinking water regulations promulgated under paragraph (2) of this subsection shall be primary drinking water regulations which specify a maximum contaminant level or require the use of treatment techniques for each contaminant for which a recommended maximum contaminant level is established or which is listed in a rule under paragraph (1)(B). The maximum contaminant level specified in a revised national primary drinking water regulation for a contaminant shall be as close to the recommended maximum contaminant level established under paragraph (1)(B) for such contaminant as is feasible. A required treatment technique for a contaminant for which a recommended maximum contaminant level has been established under paragraph (1)(B) shall reduce such contaminant to a level which is as close to the recommended maximum contaminant level for such contaminant as is feasible. A required treatment technique for a contaminant which is listed under paragraph (1)(B) shall require treatment necessary in the Administrator's judgment to prevent known or anticipated adverse effects on the health of persons to the extent feasible. For purposes of this paragraph, the term 'feasible' means feasible with the use of the best technology, treatment techniques, and other means, which the Administrator finds are generally available (taking cost into consideration).

"(4) Revised national primary drinking water regulations shall be amended whenever changes in technology, treatment techniques, and other means permit greater protection of the health of persons, but in any event such regulations shall be reviewed at least once every 3 years.

"(5) Revised national primary drinking water regulations promulgated under this subsection (and amendments thereto) shall take effect eighteen months after the date of their promulgation. Regulations under subsection (a) shall be superseded by regulations under this subsection to the extent provided by the regulations under this subsection.

"(6) No national primary drinking water regulation may require the addition of any substance for preventive health care purposes unrelated to contamination of drinking water.

"(c) The Administrator shall publish proposed national secondary drinking water regulations within 270 days after the date of enactment of this title. Within 90 days after publication of any such regulation, he shall promulgate such regulation with such modifications as he deems appropriate. Regulations under this subsection may be amended from time to time.

"(d) Regulations under this section shall be prescribed in accordance with section 553 of title 5, United States Code (relating to rulemaking), except that the Administrator shall provide opportunity for public hearing prior to promulgation of such regulations. In proposing and promulgating regulations under this section, the Administrator shall consult with the Secretary and the National Drinking Water Advisory Council.

"(e)(1) The Administrator shall enter into appropriate arrangements with the National Academy of Sciences (or with another independent scientific organization if appropriate arrangements cannot be made with such Academy) to conduct a study to determine (A) the maximum contaminant levels which should be recommended under subsection (b)(2) in order to protect the health of persons from any known or anticipated adverse effects, and (B) the existence of any contaminants the levels of which in drinking water cannot be determined but which may have an adverse effect on the health of persons.

"(2) The result of the study shall be reported to Congress no later than 2 years after the date of enactment of this title. The report shall contain (A) a summary and evaluation of relevant publications and unpublished studies; (B) a statement of methodologies and assumptions for estimating the levels at which adverse health effects may occur; (C) a statement of methodologies and assumptions for estimating the margin of safety which should be incorporated in the national primary drinking water regulations; (D) proposals for recommended maximum contaminant levels for national primary drinking water regulations, based on the methodologies, assumptions, and studies referred to in clauses (A), (B), and (C) and in paragraph (4); (E) a list of contaminants the level of which in drinking water cannot be determined but which may have an adverse effect on the health of persons; and (F) recommended studies and test protocols for future research on the health effects of drinking water contaminants, including a list of the major research priorities and estimated costs necessary to conduct such priority research.

"(3) In developing its proposals for recommended maximum contaminant levels under paragraph (2)(D) the National Academy of Sciences (or other organization preparing the report) shall evaluate and explain (separately and in composite) the impact of the following considerations:

"(A) The existence of groups or individuals in the population which are more susceptible to adverse effects than the normal healthy adult.

"(B) The exposure to contaminants in other media than drinking water (including exposures in food, in the ambient air, and in occupational settings) and the resulting body burden of contaminants.

"(C) Synergistic effects resulting from exposure to or interaction by two or more contaminants.

"(D) The contaminant exposure and body burden levels which alter physiological function or structure in a manner reasonably suspected of increasing the risk of illness.

"(4) In making the study under this subsection, the National Academy of Sciences (or other organization) shall collect and correlate (A) morbidity and mortality data and (B) monitored data on the quality of drinking water. Any conclusions based on such correlation shall be included in the report of the study.

"(5) Neither the report of the study under this subsection nor any draft of such report shall be submitted to the Office of Management and Budget or to any other Federal agency (other than the Environmental Protection Agency) prior to its submission to Congress.   Report, submittal to OMB.

"(6) Of the funds authorized to be appropriated to the Administrator by this title, such amounts as may be required shall be available to carry out the study and to make the report directed by paragraph (2) of this subsection.   Funds.

APPENDIX F

THE FEDERAL WATER POLLUTION CONTROL ACT, 1972: PUBLIC LAW 92-500 (EXCERPTS)

"RESEARCH, INVESTIGATIONS, TRAINING, AND INFORMATION

"SEC. 104. (a) The Administrator shall establish national programs for the prevention, reduction, and elimination of pollution and as part of such programs shall—

"(1) in cooperation with other Federal, State, and local agencies, conduct and promote the coordination and acceleration of, research, investigations, experiments, training, demonstrations, surveys, and studies relating to the causes, effects, extent, prevention, reduction, and elimination of pollution;

"(2) encourage, cooperate with, and render technical services to pollution control agencies and other appropriate public or private agencies, institutions, and organizations, and individuals, including the general public, in the conduct of activities referred to in paragraph (1) of this subsection;

"(3) conduct, in cooperation with State water pollution control agencies and other interested agencies, organizations and persons, public investigations concerning the pollution of any navigable waters, and report on the results of such investigations;

"(4) establish advisory committees composed of recognized experts in various aspects of pollution and representatives of the public to assist in the examination and evaluation of research progress and proposals and to avoid duplication of research;

Water quality surveillance system, report.

"(5) in cooperation with the States, and their political subdivisions, and other Federal agencies establish, equip, and maintain a water quality surveillance system for the purpose of monitoring the quality of the navigable waters and ground waters and the contiguous zone and the oceans and the Administrator shall, to the extent practicable, conduct such surveillance by utilizing the resources of the National Aeronautics and Space Administration, the National Oceanic and Atmospheric Administration, the Geological Survey, and the Coast Guard, and shall report on such quality in the report required under subsection (a) of section 516; and

Report to Congress.

"(6) initiate and promote the coordination and acceleration of research designed to develop the most effective practicable tools and techniques for measuring the social and economic costs and benefits of activities which are subject to regulation under this Act; and shall transmit a report on the results of such research to the Congress not later than January 1, 1974.

"(b) In carrying out the provisions of subsection (a) of this section the Administrator is authorized to—

"(1) collect and make available, through publications and other appropriate means, the results of and other information, including appropriate recommendations by him in connection therewith, pertaining to such research and other activities referred to in paragraph (1) of subsection (a);

"(2) cooperate with other Federal departments and agencies, State water pollution control agencies, interstate agencies, other public and private agencies, institutions, organizations, industries involved, and individuals, in the preparation and conduct of such research and other activities referred to in paragraph (1) of subsection (a);

"(3) make grants to State water pollution control agencies, interstate agencies, other public or nonprofit private agencies. institutions, organizations, and individuals, for purposes stated in paragraph (1) of subsection (a) of this section;

"(4) contract with public or private agencies, institutions, organizations, and individuals, without regard to sections 3648 and 3709 of the Revised Statutes (31 U.S.C. 529; 41 U.S.C. 5), referred to in paragraph (1) of subsection (a);

"(5) establish and maintain research fellowships at public or nonprofit private educational institutions or research organizations;

"(6) collect and disseminate, in cooperation with other Federal departments and agencies, and with other public or private agencies, institutions, and organizations having related responsibilities, basic data on chemical, physical, and biological effects of varying water quality and other information pertaining to pollution and the prevention, reduction, and elimination thereof; and

"(7) develop effective and practical processes, methods, and prototype devices for the prevention, reduction, and elimination of pollution.

"(c) In carrying out the provisions of subsection (a) of this section the Administrator shall conduct research on, and survey the results of other scientific studies on, the harmful effects on the health or welfare of persons caused by pollutants. In order to avoid duplication of effort, the Administrator shall, to the extent practicable, conduct such research in cooperation with and through the facilities of the Secretary of Health, Education, and Welfare.

Pollutant effects, study.

HEW, cooperation.

"(d) In carrying out the provisions of this section the Administrator shall develop and demonstrate under varied conditions (including conducting such basic and applied research, studies, and experiments as may be necessary):

"(1) Practicable means of treating municipal sewage, and other waterborne wastes to implement the requirements of section 201 of this Act;

"(2) Improved methods and procedures to identify and measure the effects of pollutants, including those pollutants created by new technological developments; and

"(3) Methods and procedures for evaluating the effects on water quality of augmented streamflows to control pollution not susceptible to other means of prevention, reduction, or elimination.

"(e) The Administrator shall establish, equip, and maintain field laboratory and research facilities, including, but not limited to, one to be located in the northeastern area of the United States, one in the Middle Atlantic area, one in the southeastern area, one in the midwestern area, one in the southwestern area, one in the Pacific Northwest, and one in the State of Alaska, for the conduct of research. investigations, experiments, field demonstrations and studies, and training relating to the prevention, reduction and elimination of pollution. Insofar as practicable, each such facility shall be located near institutions of higher learning in which graduate training in such research might be carried out. In conjunction with the development of criteria under section 403 of this Act, the Administrator shall construct the facilities authorized for the National Marine Water Quality Laboratory established under this subsection.

Field research laboratories.

"(f) The Administrator shall conduct research and technical development work, and make studies, with respect to the quality of the waters of the Great Lakes, including an analysis of the present and

Great Lakes, water quality research.

projected future water quality of the Great Lakes under varying conditions of waste treatment and disposal, an evaluation of the water quality needs of those to be served by such waters, an evaluation of municipal, industrial, and vessel waste treatment and disposal practices with respect to such waters, and a study of alternate means of solving pollution problems (including additional waste treatment measures) with respect to such waters.

Treatment works, pilot training programs.

"(g)(1) For the purpose of providing an adequate supply of trained personnel to operate and maintain existing and future treatment works and related activities, and for the purpose of enhancing substantially the proficiency of those engaged in such activities, the Administrator shall finance pilot programs, in cooperation with State and interstate agencies, municipalities, educational institutions, and other organizations and individuals, of manpower development and training and retraining of persons in, on entering into, the field of operation and maintenance of treatment works and related activities. Such program and any funds expended for such a program shall supplement, not supplant, other manpower and training programs and funds available for the purposes of this paragraph. The Administrator is authorized, under such terms and conditions as he deems appropriate, to enter into agreements with one or more States, acting jointly or severally, or with other public or private agencies or institutions for the development and implementation of such a program.

Employment needs, forecasting.

"(2) The Administrator is authorized to enter into agreements with public and private agencies and institutions, and individuals to develop and maintain an effective system for forecasting the supply of, and demand for, various professional and other occupational categories needed for the prevention, reduction, and elimination of pollution in each region, State, or area of the United States and, from time to time, to publish the results of such forecasts.

"(3) In furtherance of the purposes of this Act, the Administrator is authorized to—

"(A) make grants to public or private agencies and institutions and to individuals for training projects, and provide for the conduct of training by contract with public or private agencies and institutions and with individuals without regard to sections 3648

**31 USC 529.
41 USC 5.**

and 3709 of the Revised Statutes;

"(B) establish and maintain research fellowships in the Environmental Protection Agency with such stipends and allowances, including traveling and subsistence expenses, as he may deem necessary to procure the assistance of the most promising research fellows; and

"(C) provide, in addition to the program established under paragraph (1) of this subsection, training in technical matters relating to the causes, prevention, reduction, and elimination of pollution for personnel of public agencies and other persons with suitable qualifications.

Report to President, transmittal to Congress.

"(4) The Administrator shall submit, through the President, a report to the Congress not later than December 31, 1973, summarizing the actions taken under this subsection and the effectiveness of such actions, and setting forth the number of persons trained, the occupational categories for which training was provided, the effectiveness of other Federal, State, and local training programs in this field, together with estimates of future needs, recommendations on improving training programs, and such other information and recommendations, including legislative recommendations, as he deems appropriate.

Lake pollution.

"(h) The Administrator is authorized to enter into contracts with, or make grants to, public or private agencies and organizations and individuals for (A) the purpose of developing and demonstrating new or improved methods for the prevention, removal, reduction, and elimination of pollution in lakes, including the undesirable effects of nutrients and vegetation, and (B) the construction of publicly owned research facilities for such purpose.

"(i) The Administrator, in cooperation with the Secretary of the department in which the Coast Guard is operating, shall— Oil pollution control, studies.

"(1) engage in such research, studies, experiments, and demonstrations as he deems appropriate, relative to the removal of oil from any waters and to the prevention, control, and elimination of oil and hazardous substances pollution;

"(2) publish from time to time the results of such activities; and

"(3) from time to time, develop and publish in the Federal Register specifications and other technical information on the various chemical compounds used in the control of oil and hazardous substances spills. Publication in Federal Register.

In carrying out this subsection, the Administrator may enter into contracts with, or make grants to, public or private agencies and organizations and individuals.

"(j) The Secretary of the department in which the Coast Guard is operating shall engage in such research, studies, experiments, and demonstrations as he deems appropriate relative to equipment which is to be installed on board a vessel and is designed to receive, retain, treat, or discharge human body wastes and the wastes from toilets and other receptacles intended to receive or retain body wastes with particular emphasis on equipment to be installed on small recreational vessels. The Secretary of the department in which the Coast Guard is operating shall report to Congress the results of such research, studies, experiments, and demonstrations prior to the effective date of any regulations established under section 312 of this Act. In carrying out this subsection the Secretary of the department in which the Coast Guard is operating may enter into contracts with, or make grants to, public or private organizations and individuals. Vessels, solid waste disposal equipment.

Report to Congress.

"(k) In carrying out the provisions of this section relating to the conduct by the Administrator of demonstration projects and the development of field laboratories and research facilities, the Administrator may acquire land and interests therein by purchase, with appropriated or donated funds, by donation, or by exchange for acquired or public lands under his jurisdiction which he classifies as suitable for disposition. The values of the properties so exchanged either shall be approximately equal, or if they are not approximately equal, the values shall be equalized by the payment of cash to the grantor or to the Administrator as the circumstances require. Land acquisition.

"(l)(1) The Administrator shall, after consultation with appropriate local, State, and Federal agencies, public and private organizations, and interested individuals, as soon as practicable but not later than January 1, 1973, develop and issue to the States for the purpose of carrying out this Act the latest scientific knowledge available in indicating the kind and extent of effects on health and welfare which may be expected from the presence of pesticides in the water in varying quantities. He shall revise and add to such information whenever necessary to reflect developing scientific knowledge. Pesticides, effects and control.

"(2) The President shall, in consultation with appropriate local, State, and Federal agencies, public and private organizations, and interested individuals, conduct studies and investigations of methods to control the release of pesticides into the environment which study shall include examination of the persistency of pesticides in the water

Reports to Congress.

environment and alternatives thereto. The President shall submit reports, from time to time, on such investigations to Congress together with his recommendations for any necessary legislation.

Waste oil disposal.

"(m)(1) The Administrator shall, in an effort to prevent degradation of the environment from the disposal of waste oil, conduct a study of (A) the generation of used engine, machine, cooling, and similar waste oil, including quantities generated, the nature and quality of such oil, present collecting methods and disposal practices, and alternate uses of such oil; (B) the long-term, chronic biological effects of the disposal of such waste oil; and (C) the potential market for such oils, including the economic and legal factors relating to the sale of products made from such oils, the level of subsidy, if any, needed to encourage the purchase by public and private nonprofit agencies of products from such oil, and the practicability of Federal procurement, on a priority basis, of products made from such oil. In conducting such study, the Administrator shall consult with affected industries and other persons.

Reports to Congress.

"(2) The Administrator shall report the preliminary results of such study to Congress within six months after the date of enactment of the Federal Water Pollution Control Act Amendments of 1972, and shall submit a final report to Congress within 18 months after such date of enactment.

Estuaries, pollution effects.

"(n)(1) The Administrator shall, in cooperation with the Secretary of the Army, the Secretary of Agriculture, the Water Resources Council, and with other appropriate Federal, State, interstate, or local public bodies and private organizations, institutions, and individuals, conduct and promote, and encourage contributions to, continuing comprehensive studies of the effects of pollution, including sedimentation, in the estuaries and estuarine zones of the United States on fish and wildlife, on sport and commercial fishing, on recreation, on water supply and water power, and on other beneficial purposes. Such studies shall also consider the effect of demographic trends, the exploitation of mineral resources and fossil fuels, land and industrial development, navigation, flood and erosion control, and other uses of estuaries and estuarine zones upon the pollution of the waters therein.

"(2) In conducting such studies, the Administrator shall assemble, coordinate, and organize all existing pertinent information on the Nation's estuaries and estuarine zones; carry out a program of investigations and surveys to supplement existing information in representative estuaries and estuarine zones; and identify the problems and areas where further research and study are required.

Reports to Congress.

"(3) The Administrator shall submit to Congress, from time to time, reports of the studies authorized by this subsection but at least one such report during any three year period. Copies of each such report shall be made available to all interested parties, public and private.

"Estuarine zones."

"(4) For the purpose of this subsection, the term 'estuarine zones' means an environmental system consisting of an estuary and those transitional areas which are consistently influenced or affected by water from an estuary such as, but not limited to, salt marshes, coastal and intertidal areas, bays, harbors, lagoons, inshore waters, and channels,

"Estuary."

and the term 'estuary' means all or part of the mouth of a river or stream or other body of water having unimpaired natural connection with open sea and within which the sea water is measurably diluted with fresh water derived from land drainage.

Water, unnecessary consumption.

"(o)(1) The Administrator shall conduct research and investigations on devices, systems, incentives, pricing policy, and other methods of reducing the total flow of sewage, including, but not limited to, unnecessary water consumption in order to reduce the requirements for, and the costs of, sewage and waste treatment services. Such research and investigations shall be directed to develop devices, systems, policies, and methods capable of achieving the maximum reduction of unnecessary water consumption.

"(2) The Administrator shall report the preliminary results of such studies and investigations to the Congress within one year after the date of enactment of the Federal Water Pollution Control Act Amendments of 1972, and annually thereafter in the report required under subsection (a) of section 516. Such report shall include recommendations for any legislation that may be required to provide for the adoption and use of devices, systems, policies, or other methods of reducing water consumption and reducing the total flow of sewage. Such report shall include an estimate of the benefits to be derived from adoption and use of such devices, systems, policies, or other methods and also shall reflect estimates of any increase in private, public, or other cost that would be occasioned thereby. *(Reports to Congress.)*

"(p) In carrying out the provisions of subsection (a) of this section the Administrator shall, in cooperation with the Secretary of Agriculture, other Federal agencies, and the States, carry out a comprehensive study and research program to determine new and improved methods and the better application of existing methods of preventing, reducing, and eliminating pollution from agriculture, including the legal, economic, and other implications of the use of such methods. *(Agricultural pollution.)*

"(q)(1) The Administrator shall conduct a comprehensive program of research and investigation and pilot project implementation into new and improved methods of preventing, reducing, storing, collecting, treating, or otherwise eliminating pollution from sewage in rural and other areas where collection of sewage in conventional, community-wide sewage collection systems is impractical, uneconomical, or otherwise infeasible, or where soil conditions or other factors preclude the use of septic tank and drainage field systems. *(Rural sewage.)*

"(2) The Administrator shall conduct a comprehensive program of research and investigation and pilot project implementation into new and improved methods for the collection and treatment of sewage and other liquid wastes combined with the treatment and disposal of solid wastes.

"(r) The Administrator is authorized to make grants to colleges and universities to conduct basic research into the structure and function of fresh water aquatic ecosystems, and to improve understanding of the ecological characteristics necessary to the maintenance of the chemical, physical, and biological integrity of freshwater aquatic ecosystems. *(Colleges, research grants.)*

"(s) The Administrator is authorized to make grants to one or more institutions of higher education (regionally located and to be designated as 'River Study Centers') for the purpose of conducting and reporting on interdisciplinary studies on the nature of river systems, including hydrology, biology, ecology, economics, the relationship between river uses and land uses, and the effects of development within river basins on river systems and on the value of water resources and water related activities. No such grant in any fiscal year shall exceed $1,000,000. *("River Study Centers.")*

"(t) The Administrator shall, in cooperation with State and Federal agencies and public and private organizations, conduct continuing comprehensive studies of the effects and methods of control of thermal discharges. In evaluating alternative methods of control the studies shall consider (1) such data as are available on the latest available technology, economic feasibility including cost-effec- *(Thermal discharges.)*

Public
information.

tiveness analysis, and (2) the total impact on the environment, considering not only water quality but also air quality, land use, and effective utilization and conservation of fresh water and other natural resources. Such studies shall consider methods of minimizing adverse effects and maximizing beneficial effects of thermal discharges. The results of these studies shall be reported by the Administrator as soon as practicable, but not later than 270 days after enactment of this subsection, and shall be made available to the public and the States, and considered as they become available by the Administrator in carrying out section 316 of this Act and by the States in proposing thermal water quality standards.

Appropriations.

"(u) There is authorized to be appropriated (1) $100,000,000 per fiscal year for the fiscal year ending June 30, 1973, and the fiscal year ending June 30, 1974, for carrying out the provisions of this section other than subsections (g) (1) and (2), (p), (r), and (t); (2) not to exceed $7,500,000 for fiscal year 1973 for carrying out the provisions of subsection (g) (1); (3) not to exceed $2.500,000 for fiscal year 1973 for carrying out the provisions of subsection (g)(2); (4) not to exceed $10,000,000 for each of the fiscal years ending June 30, 1973, and June 30, 1974, for carrying out the provisions of subsection (p); (5) not to exceed $15,000,000 per fiscal year for the fiscal years ending June 30, 1973, and June 30, 1974, for carrying out the provisions of subsection (r); and (6) not to exceed $10,000,000 per fiscal year for the fiscal years ending June 30, 1973, and June 30, 1974, for carrying out the provisions of subsection (t).

"IN-PLACE TOXIC POLLUTANTS

Appropriation.

"SEC. 115. The Administrator is directed to identify the location of in-place pollutants with emphasis on toxic pollutants in harbors and navigable waterways and is authorized, acting through the Secretary of the Army, to make contracts for the removal and appropriate disposal of such materials from critical port and harbor areas. There is authorized to be appropriated $15,000,000 to carry out the provisions of this section, which sum shall be available until expended.

"TOXIC AND PRETREATMENT EFFLUENT STANDARDS

"SEC. 307. (a)(1) The Administrator shall, within ninety days after the date of enactment of this title, publish (and from time to time thereafter revise) a list which includes any toxic pollutant or combination of such pollutants for which an effluent standard (which may include a prohibition of the discharge of such pollutants or combination of such pollutants) will be established under this section. The Administrator in publishing such list shall take into account the toxicity of the pollutant, its persistence, degradability, the usual or potential presence of the affected organisms in any waters, the importance of the affected organisms and the nature and extent of the effect of the toxic pollutant on such organisms.

Proposed
effluent
standard,
publication.
80 Stat. 383.

"(2) Within one hundred and eighty days after the date of publication of any list, or revision thereof, containing toxic pollutants or combination of pollutants under paragraph (1) of this subsection, the Administrator, in accordance with section 553 of title 5 of the United States Code, shall publish a proposed effluent standard (or a prohibition) for such pollutant or combination of pollutants which shall take into account the toxicity of the pollutant, its persistence, degradability,

Hearing.

Revised
effluent
standard.

Effective
date.

Pretreatment
standards,
proposed
regulations,
publication.

the usual or potential presence of the affected organisms in any waters, the importance of the affected organisms and the nature and extent of the effect of the toxic pollutant on such organisms, and he shall publish a notice for a public hearing on such proposed standard to be held within thirty days. As soon as possible after such hearing, but not later than six months after publication of the proposed effluent standard (or prohibition), unless the Administrator finds, on the record, that a modification of such proposed standard (or prohibition) is justified based upon a preponderance of evidence adduced at such hearings, such standard (or prohibition) shall be promulgated.

"(3) If after a public hearing the Administrator finds that a modification of such proposed standard (or prohibition) is justified, a revised effluent standard (or prohibition) for such pollutant or combination of pollutants shall be promulgated immediately. Such standard (or prohibition) shall be reviewed and, if appropriate, revised at least every three years.

"(4) Any effluent standard promulgated under this section shall be at that level which the Administrator determines provides an ample margin of safety.

"(5) When proposing or promulgating any effluent standard (or prohibition) under this section, the Administrator shall designate the category or categories of sources to which the effluent standard (or prohibition) shall apply. Any disposal of dredged material may be included in such a category of sources after consultation with the Secretary of the Army.

"(6) Any effluent standard (or prohibition) established pursuant to this section shall take effect on such date or dates as specified in the order promulgating such standard, but in no case more than one year from the date of such promulgation.

"(7) Prior to publishing any regulations pursuant to this section the Administrator shall, to the maximum extent practicable within the time provided, consult with appropriate advisory committees, States, independent experts, and Federal departments and agencies.

"(b)(1) The Administrator shall, within one hundred and eighty days after the date of enactment of this title and from time to time thereafter, publish proposed regulations establishing pretreatment standards for introduction of pollutants into treatment works (as defined in section 212 of this Act) which are publicly owned for those pollutants which are determined not to be susceptible to treatment by such treatment works or which would interfere with the operation of such treatment works. Not later than ninety days after such publication, and after opportunity for public hearing, the Administrator shall promulgate such pretreatment standards. Pretreatment standards under this subsection shall specify a time for compliance not to exceed three years from the date of promulgation and shall be established to prevent the discharge of any pollutant through treatment works (as defined in section 212 of this Act) which are publicly owned, which pollutant interferes with, passes through, or otherwise is incompatible with such works.

"(2) The Administrator shall, from time to time, as control technology, processes, operating methods, or other alternatives change, revise such standards following the procedure established by this subsection for promulgation of such standards.

"(3) When proposing or promulgating any pretreatment standard under this section, the Administrator shall designate the category or categories of sources to which such standard shall apply.

"(4) Nothing in this subsection shall affect any pretreatment requirement established by any State or local law not in conflict with any pretreatment standard established under this subsection.

"(c) In order to insure that any source introducing pollutants into a publicly owned treatment works, which source would be a new source subject to section 306 if it were to discharge pollutants, will not cause a violation of the effluent limitations established for any such treatment works, the Administrator shall promulgate pretreatment standards for the category of such sources simultaneously with the promulgation of standards of performance under section 306 for the equivalent category of new sources. Such pretreatment standards shall prevent the discharge of any pollutant into such treatment works, which pollutant may interfere with, pass through, or otherwise be incompatible with such works.

"(d) After the effective date of any effluent standard or prohibition or pretreatment standard promulgated under this section, it shall be unlawful for any owner or operator of any source to operate any source in violation of any such effluent standard or prohibition or pretreatment standard.

"INSPECTIONS, MONITORING AND ENTRY

"SEC. 308. (a) Whenever required to carry out the objective of this Act, including but not limited to (1) developing or assisting in the development of any effluent limitation, or other limitation, prohibition, or effluent standard, pretreatment standard, or standard of performance under this Act; (2) determining whether any person is in violation of any such effluent limitation, or other limitation, prohibition or effluent standard, pretreatment standard, or standard of performance; (3) any requirement established under this section; or (4) carrying out sections 305, 311, 402, and 504 of this Act—

"(A) the Administrator shall require the owner or operator of any point source to (i) establish and maintain such records, (ii) make such reports, (iii) install, use, and maintain such monitoring equipment or methods (including where appropriate, biological monitoring methods), (iv) sample such effluents (in accordance with such methods, at such locations, at such intervals, and in such manner as the Administrator shall prescribe), and (v) provide such other information as he may reasonably require; and

Recordkeeping; reports.

"(B) the Administrator or his authorized representative, upon presentation of his credentials—

"(i) shall have a right of entry to, upon, or through any premises in which an effluent source is located or in which any records required to be maintained under clause (A) of this subsection are located, and

"(ii) may at reasonable times have access to and copy any records, inspect any monitoring equipment or method required under clause (A), and sample any effluents which the owner or operator of such source is required to sample under such clause.

"(b) Any records, reports, or information obtained under this section (1) shall, in the case of effluent data, be related to any applicable effluent limitations, toxic, pretreatment, or new source performance standards, and (2) shall be available to the public, except that upon a showing satisfactory to the Administrator by any person that records, reports, or information, or particular part thereof (other than effluent data),·to which the Administrator has access under this section, if made public would divulge methods or processes entitled to protection as trade secrets of such person, the Administrator shall consider such record, report, or information, or particular portion thereof confidential in accordance with the purposes of section 1905

62 Stat. 791. of title 18 of the United States Code, except that such record, report, or information may be disclosed to other officers, employees, or authorized representatives of the United States concerned with carrying out this Act or when relevant in any proceeding under this Act.

"(c) Each State may develop and submit to the Administrator procedures under State law for inspection, monitoring, and entry with respect to point sources located in such State. If the Administrator finds that the procedures and the law of any State relating to inspection, monitoring, and entry are applicable to at least the same extent as those required by this section, such State is authorized to apply and enforce its procedures for inspection, monitoring, and entry with respect to point sources located in such State (except with respect to point sources owned or operated by the United States).

"FEDERAL ENFORCEMENT

"SEC. 309. (a) (1) Whenever, on the basis of any information available to him, the Administrator finds that any person is in violation of any condition or limitation which implements section 301, 302, 306, 307, or 308 of this Act in a permit issued by a State under an approved permit program under section 402 of this Act, he shall proceed under his authority in paragraph (3) of this subsection or he shall notify the person in alleged violation and such State of such finding. If beyond the thirtieth day after the Administrator's notification the State has not commenced appropriate enforcement action, the Administrator shall issue an order requiring such person to comply with such condition or limitation or shall bring a civil action in accordance with subsection (b) of this section.

"(2) Whenever, on the basis of information available to him, the Administrator finds that violations of permit conditions or limitations as set forth in paragraph (1) of this subsection are so widespread that such violations appear to result from a failure of the State to enforce such permit conditions or limitations effectively, he shall so notify the State. If the Administrator finds such failure extends beyond the thirtieth day after such notice, he shall give public notice of such finding. During the period beginning with such public notice and ending when such State satisfies the Administrator that it will enforce such conditions and limitations (hereafter referred to in this section as the period of 'federally assumed enforcement'), the Administrator shall enforce any permit condition or limitation with respect to any person—

"(A) by issuing an order to comply with such condition or limitation, or

"(B) by bringing a civil action under subsection (b) of this section.

"(3) Whenever on the basis of any information available to him the Administrator finds that any person is in violation of section 301, 302, 306, 307, or 308 of this Act, or is in violation of any permit condition or limitation implementing any of such sections in a permit issued under section 402 of this Act by him or by a State, he shall issue an order requiring such person to comply with such section or requirement, or he shall bring a civil action in accordance with subsection (b) of this section.

"(4) A copy of any order issued under this subsection shall be sent immediately by the Administrator to the State in which the violation occurs and other affected States. Any order issued under this subsection shall be by personal service and shall state with reasonable specificity the nature of the violation, specify a time for compliance, not to exceed thirty days, which the Administrator determines is reasonable, taking into account the seriousness of the violation and any good faith efforts to comply with applicable requirements. In any case in which an order under this subsection (or notice to a violator under paragraph (1) of this subsection) is issued to a corporation, a copy of such order (or notice) shall be served on any appropriate corporate officers. An order issued under this subsection relating to a violation of section 308 of this Act shall not take effect until the person to whom it is issued has had an opportunity to confer with the Administrator concerning the alleged violation.

"(b) The Administrator is authorized to commence a civil action for appropriate relief, including a permanent or temporary injunction, for any violation for which he is authorized to issue a compliance order under subsection (a) of this section. Any action under this subsection may be brought in the district court of the United States for the district in which the defendant is located or resides or is doing business, and such court shall have jurisdiction to restrain such violation and to require compliance. Notice of the commencement of such action shall be given immediately to the appropriate State.

"(c) (1) Any person who willfully or negligently violates section 301, 302, 306, 307, or 308 of this Act, or any permit condition or limitation implementing any of such sections in a permit issued under section 402 of this Act by the Administrator or by a State, shall be punished by a fine of not less than $2,500 nor more than $25,000 per day of violation, or by imprisonment for not more than one year, or by both. If the conviction is for a violation committed after a first conviction of such person under this paragraph, punishment shall be by a fine of not more than $50,000 per day of violation, or by imprisonment for not more than two years, or by both.

Penalties.

"(2) Any person who knowingly makes any false statement, representation, or certification in any application, record, report, plan, or other document filed or required to be maintained under this Act or who falsifies, tampers with, or knowingly renders inaccurate any monitoring device or method required to be maintained under this Act, shall upon conviction, be punished by a fine of not more than $10,000, or by imprisonment for not more than six months, or by both.

"(3) For the purposes of this subsection, the term 'person' shall mean, in addition to the definition contained in section 502(5) of this Act, any responsible corporate officer.

"Person."

"(d) Any person who violates section 301, 302, 306, 307, or 308 of this Act, or any permit condition or limitation implementing any of such sections in a permit issued under section 402 of this Act by the Administrator, or by a State, and any person who violates any order issued by the Administrator under subsection (a) of this section, shall be subject to a civil penalty not to exceed $10,000 per day of such violation.

"(e) Whenever a municipality is a party to a civil action brought by the United States under this section, the State in which such municipality is located shall be joined as a party. Such State shall be liable for payment of any judgment, or any expenses incurred as a result of complying with any judgment, entered against the municipality in such action to the extent that the laws of that State prevent the municipality from raising revenues needed to comply with such judgment.

"INTERNATIONAL POLLUTION ABATEMENT

"SEC. 310. (a) Whenever the Administrator, upon receipts of reports, surveys, or studies from any duly constituted international agency, has reason to believe that pollution is occurring which endangers the health or welfare of persons in a foreign country, and the Secretary of State requests him to abate such pollution, he shall give formal notification thereof to the State water pollution control agency of the State or States in which such discharge or discharges originate and to the appropriate interstate agency, if any. He shall also promptly call such a hearing, if he believes that such pollution is occurring in sufficient quantity to warrant such action, and if such foreign country has given the United States essentially the same rights with respect to the prevention and control of pollution occurring in that country as is given that country by this subsection. The Administrator, through the Secretary of State, shall invite the foreign country which may be adversely affected by the pollution to attend and participate in the hearing, and the representative of such country shall, for the purpose of the hearing and any further proceeding resulting from such hearing, have all the rights of a State water pollution control agency. Nothing in this subsection shall be construed to modify, amend, repeal, or otherwise affect the provisions of the 1909 Boundary Waters Treaty between Canada and the United States or the Water Utilization Treaty of 1944 between Mexico and the United States (59 Stat. 1219), relative to the control and abatement of pollution in waters covered by those treaties.

Hearing.

36 Stat. 2448.

"(b) The calling of a hearing under this section shall not be construed by the courts, the Administrator, or any person as limiting, modifying, or otherwise affecting the functions and responsibilities of the Administrator under this section to establish and enforce water quality requirements under this Act.

Notice, publication in Federal Register.

"(c) The Administrator shall publish in the Federal Register a notice of a public hearing before a hearing board of five or more persons appointed by the Administrator. A majority of the members of the board and the chairman who shall be designated by the Administrator shall not be officers or employees of Federal, State, or local governments. On the basis of the evidence presented at such hearing, the board shall within sixty days after completion of the hearing make findings of fact as to whether or not such pollution is occurring and shall thereupon by decision, incorporating its findings therein, make such recommendations to abate the pollution as may be appropriate and shall transmit such decision and the record of the hearings to the Administrator. All such decisions shall be public. Upon receipt of such decision, the Administrator shall promptly implement the board's decision in accordance with the provisions of this Act.

Report.

"(d) In connection with any hearing called under this subsection, the board is authorized to require any person whose alleged activities result in discharges causing or contributing to pollution to file with it in such forms as it may prescribe, a report based on existing data, furnishing such information as may reasonably be required as to the character, kind, and quantity of such discharges and the use of facilities or other means to prevent or reduce such discharges by the person filing such a report. Such report shall be made under oath or otherwise, as the board may prescribe, and shall be filed with the board within such reasonable period as it may prescribe, unless additional time is granted by it. Upon a showing satisfactory to the board by the person filing such report that such report or portion thereof (other than effluent data), to which the Administrator has access under this section, if made public would divulge trade secrets or secret processes of such person, the board shall consider such report or portion thereof confidential for the purposes of section 1905 of title 18 of the United States Code. If any person required to file any report under this paragraph shall fail to do so within the time fixed by the board for filing the same, and such failure shall continue for thirty days after notice of such default, such person shall forfeit to the United States the sum

62 Stat. 791.
Penalty.

of $1,000 for each and every day of the continuance of such failure, which forfeiture shall be payable into the Treasury of the United States, and shall be recoverable in a civil suit in the name of the United States in the district court of the United States where such person has his principal office or in any district in which he does business. The Administrator may upon application therefor remit or mitigate any forfeiture provided for under this subsection.

"(e) Board members, other than officers or employees of Federal, State, or local governments, shall be for each day (including travel-time) during which they are performing board business, entitled to receive compensation at a rate fixed by the Administrator but not in excess of the maximum rate of pay for grade GS–18, as provided in the General Schedule under section 5332 of title 5 of the United States Code, and shall, notwithstanding the limitations of sections 5703 and 5704 of title 5 of the United States Code, be fully reimbursed for travel, subsistence, and related expenses.

80 Stat. 499;
83 Stat. 190.

"(f) When any such recommendation adopted by the Administrator involves the institution of enforcement proceedings against any person to obtain the abatement of pollution subject to such recommendation, the Administrator shall institute such proceedings if he believes that the evidence warrants such proceedings. The district court of the United States shall consider and determine de novo all relevant issues, but shall receive in evidence the record of the proceedings before the conference or hearing board. The court shall have jurisdiction to enter such judgment and orders enforcing such judgment as it deems appropriate or to remand such proceedings to the Administrator for such further action as it may direct.

"OIL AND HAZARDOUS SUBSTANCE LIABILITY

"SEC. 311. (a) For the purpose of this section, the term—   Definitions.

"(1) 'oil' means oil of any kind or in any form, including, but not limited to, petroleum, fuel oil, sludge, oil refuse, and oil mixed with wastes other than dredged spoil;

"(2) 'discharge' includes, but is not limited to, any spilling, leaking, pumping, pouring, emitting, emptying or dumping;

"(3) 'vessel' means every description of watercraft or other artificial contrivance used, or capable of being used, as a means of transportation on water other than a public vessel;

"(4) 'public vessel' means a vessel owned or bareboat-chartered and operated by the United States, or by a State or political subdivision thereof, or by a foreign nation, except when such vessel is engaged in commerce;

"(5) 'United States' means the States, the District of Columbia, the Commonwealth of Puerto Rico, the Canal Zone, Guam, American Samoa, the Virgin Islands, and the Trust Territory of the Pacific Islands;

"(6) 'owner or operator' means (A) in the case of a vessel, any person owning, operating, or chartering by demise, such vessel, and (B) in the case of an onshore facility, and an offshore facility, any person owning or operating such onshore facility or offshore facility, and (C) in the case of any abandoned offshore facility, the person who owned or operated such facility immediately prior to such abandonment;

"(7) 'person' includes an individual, firm, corporation, association, and a partnership;

"(8) 'remove' or 'removal' refers to removal of the oil or hazardous substances from the water and shorelines or the taking of such other actions as may be necessary to minimize or mitigate damage to the public health or welfare, including, but not limited to, fish, shellfish, wildlife, and public and private property, shorelines, and beaches;

"(9) 'contiguous zone' means the entire zone established or to be established by the United States under article 24 of the Convention on the Territorial Sea and the Contiguous Zone;

"(10) 'onshore facility' means any facility (including, but not limited to, motor vehicles and rolling stock) of any kind located in, on, or under, any land within the United States other than submerged land;

"(11) 'offshore facility' means any facility of any kind located in, on, or under, any of the navigable waters of the United States other than a vessel or a public vessel;

"(12) 'act of God' means an act occasioned by an unanticipated grave natural disaster;

"(13) 'barrel' means 42 United States gallons at 60 degrees Fahrenheit;

"(14) 'hazardous substance' means any substance designated pursuant to subsection (b) (2) of this section.

Prohibition.

"(b)(1) The Congress hereby declares that it is the policy of the United States that there should be no discharges of oil or hazardous substances into or upon the navigable waters of the United States, adjoining shorelines, or into or upon the waters of the contiguous zone.

Regulations.

"(2) (A) The Administrator shall develop, promulgate, and revise as may be appropriate, regulations designating as hazardous substances, other than oil as defined in this section, such elements and compounds which, when discharged in any quantity into or upon the navigable waters of the United States or adjoining shorelines or the waters of the contiguous zone, present an imminent and substantial danger to the public health or welfare, including, but not limited to, fish, shellfish, wildlife, shorelines, and beaches.

Hazardous substances, removability determination.

"(B) (i) The Administrator shall include in any designation under subparagraph (A) of this subsection a determination whether any such designated hazardous substance can actually be removed.

"(ii) The owner or operator of any vessel, onshore facility, or offshore facility from which there is discharged during the two-year period beginning on the date of enactment of the Federal Water Pollution Control Act Amendments of 1972, any hazardous substance determined not removable under clause (i) of this subparagraph shall be liable, subject to the defenses to liability provided under subsection (f) of this section, as appropriate, to the United States for a civil penalty per discharge established by the Administrator based on toxicity, degradability, and dispersal characteristics of such substance, in an amount not to exceed $50,000, except that where the United States can show that such discharge was a result of willful negligence or willful misconduct within the privity and knowledge of the owner, such owner or operator shall be liable to the United States for a civil penalty in such amount as the Administrator shall establish, based upon the toxicity, degradability, and dispersal characteristics of such substance.

"(iii) After the expiration of the two-year period referred to in clause (ii) of this subparagraph, the owner or operator of any vessel, onshore facility, or offshore facility, from which there is discharged any hazardous substance determined not removable under clause (i) of this subparagraph shall be liable, subject to the defenses to liability provided in subsection (f) of this section, to the United States for either one or the other of the following penalties, the determination of which shall be in the discretion of the Administrator:

"(aa) a penalty in such amount as the Administrator shall establish, based on the toxicity, degradability, and dispersal characteristics of the substance, but not less than $500 nor more than $5,000; or

"(bb) a penalty determined by the number of units discharged multiplied by the amount established for such unit under clause (iv) of this subparagraph, but such penalty shall not be more than $5,000,000 in the case of a discharge from a vessel and $500,000 in the case of a discharge from an onshore or offshore facility.

"(iv) The Administrator shall establish by regulation, for each hazardous substance designated under subparagraph (A) of this paragraph, and within 180 days of the date of such designation, a unit of measurement based upon the usual trade practice and, for the purpose of determining the penalty under clause (iii)(bb) of this subparagraph, shall establish for each such unit a fixed monetary amount which shall be not less than $100 nor more than $1,000 per unit. He shall establish such fixed amount based on the toxicity, degradability, and dispersal characteristics of the substance.

"(3) The discharge of oil or hazardous substances into or upon the navigable waters of the United States, adjoining shorelines, or into or upon the waters of the contiguous zone in harmful quantities as determined by the President under paragraph (4) of this subsection, is prohibited, except (A) in the case of such discharges of oil into the waters of the contiguous zone, where permitted under article IV of the International Convention for the Prevention of Pollution of the Sea by Oil, 1954, as amended, and (B) where permitted in quantities and at times and locations or under such circumstances or conditions as the President may, by regulation, determine not to be harmful. Any regulations issued under this subsection shall be consistent with maritime safety and with marine and navigation laws and regulations and applicable water quality standards.

12 UST 2989.

"(4) The President shall by regulation, to be issued as soon as possible after the date of enactment of this paragraph, determine for the purposes of this section, those quantities of oil and any hazardous substance the discharge of which, at such times, locations, circumstances, and conditions, will be harmful to the public health or welfare of the United States, including, but not limited to, fish, shellfish, wildlife, and public and private property, shorelines, and beaches except that in the case of the discharge of oil into or upon the waters of the contiguous zone, only those discharges which threaten the fishery resources of the contiguous zone or threaten to pollute or contribute to the pollution of the territory or the territorial sea of the United States may be determined to be harmful.

Discharges, harmful quantities, determination.

"(5) Any person in charge of a vessel or of an onshore facility or an offshore facility shall, as soon as he has knowledge of any discharge of oil or a hazardous substance from such vessel or facility in violation of paragraph (3) of this subsection, immediately notify the appropriate agency of the United States Government of such discharge. Any such person who fails to notify immediately such agency of such discharge shall, upon conviction, be fined not more than $10,000, or imprisoned for not more than one year, or both. Notification received pursuant to this paragraph or information obtained by the exploitation of such notification shall not be used against any such person in any criminal case, except a prosecution for perjury or for giving a false statement.

Violation notification.

Penalty.

"(6) Any owner or operator of any vessel, onshore facility, or offshore facility from which oil or a hazardous substance is discharged in violation of paragraph (3) of this subsection shall be assessed a civil penalty by the Secretary of the department in which the Coast Guard is operating of not more than $5,000 for each offense. No penalty shall be assessed unless the owner or operator charged shall have been given notice and opportunity for a hearing on such charge. Each violation is a separate offense. Any such civil penalty may be compromised by such Secretary. In determining the amount of the penalty, or the amount agreed upon in compromise, the appropriateness of such penalty to the size of the business of the owner or operator charged, the effect on the owner or operator's ability to continue in business, and the gravity of the violation, shall be considered by such Secretary. The Secretary of the Treasury shall withhold at the request of such Secretary the clearance required by section 4197 of the Revised Statutes of the United States, as amended (46 U.S.C. 91), of any vessel the owner or operator of which is subject to the foregoing penalty. Clearance may be granted in such cases upon the filing of a bond or other surety satisfactory to such Secretary.

Discharge into U.S. navigable waters, removal. "(c)(1) Whenever any oil or a hazardous substance is discharged, into or upon the navigable waters of the United States, adjoining shorelines, or into or upon the waters of the contiguous zone, the President is authorized to act to remove or arrange for the removal of such oil or substance at any time, unless he determines such removal will be done properly by the owner or operator of the vessel, onshore facility, or offshore facility from which the discharge occurs.

National Contingency Plan. "(2) Within sixty days after the effective date of this section, the President shall prepare and publish a National Contingency Plan for removal of oil and hazardous substances, pursuant to this subsection. Such National Contingency Plan shall provide for efficient, coordinated, and effective action to minimize damage from oil and hazardous substance discharges, including containment, dispersal, and removal of oil and hazardous substances, and shall include, but not be limited to—

"(A) assignment of duties and responsibilities among Federal departments and agencies in coordination with State and local agencies, including, but not limited to, water pollution control, conservation, and port authorities;

"(B) identification, procurement, maintenance, and storage of equipment and supplies;

"(C) establishment or designation of a strike force consisting of personnel who shall be trained, prepared, and available to provide necessary services to carry out the Plan, including the establishment at major ports, to be determined by the President, of emergency task forces of trained personnel, adequate oil and hazardous substance pollution control equipment and material, and a detailed oil and hazardous substance pollution prevention and removal plan;

"(D) a system of surveillance and notice designed to insure earliest possible notice of discharges of oil and hazardous substances to the appropriate Federal agency;

"(E) establishment of a national center to provide coordination and direction for operations in carrying out the Plan;

"(F) procedures and techniques to be employed in identifying, containing, dispersing, and removing oil and hazardous substances;

"(G) a schedule, prepared in cooperation with the States, identifying (i) dispersants and other chemicals, if any, that may be used in carrying out the Plan, (ii) the waters in which such dispersants and chemicals may be used, and (iii) the quantities of such dispersant or chemical which can be used safely in such waters, which schedule shall provide in the case of any dispersant, chemical, or waters not specifically identified in such schedule that the President, or his delegate, may, on a case-by-case basis, identify the dispersants and other chemicals which may be used, the waters in which they may be used, and the quantities which can be used safely in such waters; and

"(H) a system whereby the State or States affected by a discharge of oil or hazardous substance may act where necessary to remove such discharge and such State or States may be reimbursed from the fund established under subsection (k) of this section for the reasonable costs incurred in such removal.

The President may, from time to time, as he deems advisable revise or otherwise amend the National Contingency Plan. After publication of the National Contingency Plan, the removal of oil and hazardous substances and actions to minimize damage from oil and hazardous substance discharges shall, to the greatest extent possible, be in accordance with the National Contingency Plan.

"(d) Whenever a marine disaster in or upon the navigable waters of the United States has created a substantial threat of a pollution hazard to the public health or welfare of the United States, including, but not limited to, fish, shellfish, and wildlife and the public and private shorelines and beaches of the United States, because of a discharge, or an imminent discharge, of large quantities of oil, or of a hazardous substance from a vessel the United States may (A) coordinate and direct all public and private efforts directed at the removal or elimination of such threat; and (B) summarily remove, and, if necessary, destroy such vessel by whatever means are available without regard to any provisions of law governing the employment of personnel or the expenditure of appropriated funds. Any expense incurred under this subsection shall be a cost incurred by the United States Government for the purposes of subsection (f) in the removal of oil or hazardous substance. *[marginal note: Maritime disaster discharge.]*

"(e) In addition to any other action taken by a State or local government, when the President determines there is an imminent and substantial threat to the public health or welfare of the United States, including, but not limited to, fish, shellfish, and wildlife and public and private property, shorelines, and beaches within the United States, because of an actual or threatened discharge of oil or hazardous substance into or upon the navigable waters of the United States from an onshore or offshore facility, the President may require the United States attorney of the district in which the threat occurs to secure such relief as may be necessary to abate such threat, and the district courts of the United States shall have jurisdiction to grant such relief as the public interest and the equities of the case may require. *[marginal notes: Relief. / Jurisdiction.]*

"(f)(1) Except where an owner or operator can prove that a discharge was caused solely by (A) an act of God, (B) an act of war, (C) negligence on the part of the United States Government, or (D) an act or omission of a third party without regard to whether any such act or omission was or was not negligent, or any combination of the foregoing clauses, such owner or operator of any vessel from which oil or a hazardous substance is discharged in violation of subsection (b)(2) of this section shall, notwithstanding any other provision of law, be liable to the United States Government for the actual costs incurred under subsection (c) for the removal of such oil or substance by the United States Government in an amount not to exceed $100 per gross ton of such vessel or $14,000,000, whichever is lesser, except that where the United States can show that such discharge was the result of willful negligence or willful misconduct within the privity and knowledge of the owner, such owner or operator shall be liable to the United States Government for the full amount of such costs. Such costs shall constitute a maritime lien on such vessel which may be recovered in an action in rem in the district court of the United *[marginal note: Liability.]*

States for any district within which any vessel may be found. The United States may also bring an action against the owner or operator of such vessel in any court of competent jurisdiction to recover such costs.

"(2) Except where an owner or operator of an onshore facility can prove that a discharge was caused solely by (A) an act of God, (B) an act of war, (C) negligence on the part of the United States Government, or (D) an act or omission of a third party without regard to whether any such act or omission was or was not negligent, or any combination of the foregoing clauses, such owner or operator of any such facility from which oil or a hazardous substance is discharged in violation of subsection (b)(2) of this section shall be liable to the United States Government for the actual costs incurred under subsection (c) for the removal of such oil or substance by the United States Government in an amount not to exceed $8,000,000, except that where the United States can show that such discharge was the result of willful negligence or willful misconduct within the privity and knowledge of the owner, such owner or operator shall be liable to the United States Government for the full amount of such costs. The United States may bring an action against the owner or operator of such facility in any court of competent jurisdiction to recover such costs. The Secretary is authorized, by regulation, after consultation with the Secretary of Commerce and the Small Business Administration, to establish reasonable and equitable classifications of those onshore facilities having a total fixed storage capacity of 1,000 barrels or less which he determines because of size, type, and location do not present a substantial risk of the discharge of oil or a hazardous substance in violation of subsection (b)(2) of this section, and apply with respect to such classifications differing limits of liability which may be less than the amount contained in this paragraph.

"(3) Except where an owner or operator of an offshore facility can prove that a discharge was caused solely by (A) an act of God, (B) an act of war, (C) negligence on the part of the United States Government, or (D) an act or omission of a third party without regard to whether any such act or omission was or was not negligent, or any combination of the foregoing clauses, such owner or operator of any such facility from which oil or a hazardous substance is discharged in violation of subsection (b)(2) of this section shall, notwithstanding any other provision of law, be liable to the United States Government for the actual costs incurred under subsection (c) for the removal of such oil or substance by the United States Government in an amount not to exceed $8,000,000, except that where the United States can show that such discharge was the result of willful negligence or willful misconduct within the privity and knowledge of the owner, such owner or operator shall be liable to the United States Government for the full amount of such costs. The United States may bring an action against the owner or operator of such a facility in any court of competent jurisdiction to recover such costs.

"(g) In any case where an owner or operator of a vessel, of an onshore facility, or of an offshore facility, from which oil or a hazardous substance is discharged in violation of subsection (b)(2) of this section, proves that such discharge of oil or hazardous substance was caused solely by an act or omission of a third party, or was caused solely by such an act or omission in combination with an act of God, an act of war, or negligence on the part of the United States Government, such third party shall, notwithstanding any other provision of law, be liable to the United States Government for the actual costs incurred under subsection (c) for removal of such oil or substance by the United States Government, except where such third party can prove that such discharge was caused solely by (A) an act of God, (B) an act of war, (C) negligence on the part of the United States Government, or (D) an act or omission of another party without regard to whether such act or omission was or was not negligent, or any combination of the foregoing clauses. If such third party was the owner or operator of a vessel which caused the discharge of oil or a hazardous substance in violation of subsection (b)(2) of this section, the liability of such third party under this subsection shall not exceed $100 per gross ton of such vessel or $14,000,000, whichever is the lesser. In any other case the liability of such third party shall not exceed the limitation which would have been applicable to the owner or operator of the vessel or the onshore or offshore facility from which the discharge actually occurred if such owner or operator were liable. If the United States can show that the discharge of oil or a hazardous substance in violation of subsection (b)(2) of this section was the result of willful negligence or willful misconduct within the privity and knowledge of such third party, such third party shall be liable to the United States Government for the full amount of such removal costs. The United States may bring an action against the third party in any court of competent jurisdiction to recover such removal costs.

"(h) The liabilities established by this section shall in no way affect any rights which (1) the owner or operator of a vessel or of an onshore facility or an offshore facility may have against any third party whose acts may in any way have caused or contributed to such discharge, or (2) The United States Government may have against any third party whose actions may in any way have caused or contributed to the discharge of oil or hazardous substance.

"(i)(1) In any case where an owner or operator of a vessel or an onshore facility or an offshore facility from which oil or a hazardous substance is discharged in violation of subsection (b)(2) of this section acts to remove such oil or substance in accordance with regulations promulgated pursuant to this section, such owner or operator shall be entitled to recover the reasonable costs incurred in such removal upon establishing, in a suit which may be brought against the United States Government in the United States Court of Claims, that such discharge was caused solely by (A) an act of God, (B) an act of war, (C) negligence on the part of the United States Government, or (D) an act or omission of a third party without regard to whether such act or omission was or was not negligent, or of any combination of the foregoing causes.

Removal costs, recovery.

"(2) The provisions of this subsection shall not apply in any case where liability is established pursuant to the Outer Continental Shelf Lands Act.

67 Stat. 462. 43 USC 1331 note.

"(3) Any amount paid in accordance with a judgment of the United States Court of Claims pursuant to this section shall be paid from the funds established pursuant to subsection (k).

"(j)(1) Consistent with the National Contingency Plan required by subsection (c)(2) of this section, as soon as practicable after the effective date of this section, and from time to time thereafter, the President shall issue regulations consistent with maritime safety and with marine and navigation laws (A) establishing methods and procedures for removal of discharged oil and hazardous substances, (B) establishing criteria for the development and implementation of local and regional oil and hazardous substance removal contingency plans, (C) establishing procedures, methods, and equipment and other requirements for equipment to prevent discharges of oil and hazardous substances from vessels and from onshore facilities and offshore facilities, and to contain such discharges, and (D) governing the inspection of vessels carrying cargoes of oil and hazardous substances

Regulations.

and the inspection of such cargoes in order to reduce the likelihood of discharges of oil from vessels in violation of this section.

Penalty.

"(2) Any owner or operator of a vessel or an onshore facility or an offshore facility and any other person subject to any regulation issued under paragraph (1) of this subsection who fails or refuses to comply with the provisions of any such regulations, shall be liable to a civil penalty of not more than $5,000 for each such violation. Each violation shall be a separate offense. The President may assess and compromise such penalty. No penalty shall be assessed until the owner, operator, or other person charged shall have been given notice and an opportunity for a hearing on such charge. In determining the amount of the penalty, or the amount agreed upon in compromise, the gravity of the violation, and the demonstrated good faith of the owner, operator, or other person charged in attemping to achieve rapid compliance, after notification of a violation, shall be considered by the President.

Appropriation.

"(k) There is hereby authorized to be appropriated to a revolving fund to be established in the Treasury not to exceed $35,000,000 to carry out the provisions of subsections (c), (d), (i), and (l) of this section. Any other funds received by the United States under this section shall also be deposited in said fund for such purposes. All sums appropriated to, or deposited in, said fund shall remain available until expended.

Administration.

"(l) The President is authorized to delegate the administration of this section to the heads of those Federal departments, agencies, and instrumentalities which he determines to be appropriate. Any moneys in the fund established by subsection (k) of this section shall be available to such Federal departments, agencies, and instrumentalities to carry out the provisions of subsections (c) and (i) of this section. Each such department, agency, and instrumentality, in order to avoid duplication of effort, shall, whenever appropriate, utilize the personnel, services, and facilities of other Federal departments, agencies, and instrumentalities.

"(m) Anyone authorized by the President to enforce the provisions of this section may, except as to public vessels, (A) board and inspect any vessel upon the navigable waters of the United States or the waters of the contiguous zone, (B) with or without a warrant arrest any person who violates the provisions of this section or any regulation issued thereunder in his presence or view, and (C) execute any warrant or other process issued by an officer or court of competent jurisdiction.

Jurisdiction.

"(n) The several district courts of the United States are invested with jurisdiction for any actions, other than actions pursuant to subsection (i)(1), arising under this section. In the case of Guam and the Trust Territory of the Pacific Islands, such actions may be brought in the district court of Guam, and in the case of the Virgin Islands such actions may be brought in the district court of the Virgin Islands. In the case of American Samoa and the Trust Territory of the Pacific Islands, such actions may be brought in the District Court of the United States for the District of Hawaii and such court shall have jurisdiction of such actions. In the case of the Canal Zone, such actions may be brought in the United States District Court for the District of the Canal Zone.

"(o)(1) Nothing in this section shall affect or modify in any way the obligations of any owner or operator of any vessel, or of any owner or operator of any onshore facility or offshore facility to any person or agency under any provision of law for damages to any publicly owned or privately owned property resulting from a discharge of any oil or hazardous substance or from the removal of any such oil or hazardous substance.

"(2) Nothing in this section shall be construed as preempting any State or political subdivision thereof from imposing any requirement or liability with respect to the discharge of oil or hazardous substance into any waters within such State.

"(3) Nothing in this section shall be construed as affecting or modifying any other existing authority of any Federal department, agency, or instrumentality, relative to onshore or offshore facilities under this Act or any other provision of law, or to affect any State or local law not in conflict with this section.

"(p)(1) Any vessel over three hundred gross tons, including any barge of equivalent size, but not including any barge that is not self-propelled and that does not carry oil or hazardous substances as cargo or fuel, using any port or place in the United States or the navigable waters of the United States for any purpose shall establish and maintain under regulations to be prescribed from time to time by the President, evidence of financial responsibility of $100 per gross ton, or $14,000,000 whichever is the lesser, to meet the liability to the United States which such vessel could be subjected under this section. In cases where an owner or operator owns, operates, or charters more than one such vessel, financial responsibility need only be established to meet the maximum liability to which the largest of such vessels could be subjected. Financial responsibility may be established by any one of, or a combination of, the following methods acceptable to the President: (A) evidence of insurance, (B) surety bonds, (C) qualification as a self-insurer, or (D) other evidence of financial responsibility. Any bond filed shall be issued by a bonding company authorized to do business in the United States. Certain vessels, financial res- ponsibility.

"(2) The provisions of paragraph (1) of this subsection shall be effective April 3, 1971, with respect to oil and one year after the date of enactment of this section with respect to hazardous substances. The President shall delegate the responsibility to carry out the provisions of this subsection to the appropriate agency head within sixty days after the date of enactment of this section. Regulations necessary to implement this subsection shall be issued within six months after the date of enactment of this section. Effective date.

"(3) Any claim for costs incurred by such vessel may be brought directly against the insurer or any other person providing evidence of financial responsibility as required under this subsection. In the case of any action pursuant to this subsection such insurer or other person shall be entitled to invoke all rights and defenses which would have been available to the owner or operator if an action had been brought against him by the claimant, and which would have been available to him if an action had been brought against him by the owner or operator.

"(4) Any owner or operator of a vessel subject to this subsection, who fails to comply with the provisions of this subsection or any regulation issued thereunder, shall be subject to a fine of not more than $10,000. Penalty.

"(5) The Secretary of the Treasury may refuse the clearance required by section 4197 of the Revised Statutes of the United States, as amended (4 U.S.C. 91), to any vessel subject to this subsection, which does not have evidence furnished by the President that the financial responsibility provisions of paragraph (1) of this subsection have been complied with. 46 USC 91.

"(6) The Secretary of the Department in which the Coast Guard is operated may (A) deny entry to any port or place in the United States or the navigable waters of the United States, to, and (B) detain at the port or place in the United States from which it is about to depart for any other port or place in the United States, any vessel subject to this subsection, which upon request, does not produce evidence furnished by the President that the financial responsibility provisions of paragraph (1) of this subsection have been complied with.

APPENDIX G

THE CLEAN AIR ACT,
AS AMENDED, 1974 (EXCERPTS)

NATIONAL EMISSION STANDARDS FOR HAZARDOUS AIR POLLUTANTS

"SEC. 112. (a) For purposes of this section—

"(1) The term 'hazardous air pollutant' means an air pollutant to which no ambient air quality standard is applicable and which in the judgment of the Administrator may cause, or contribute to, an increase in mortality or an increase in serious irreversible, or incapacitating reversible, illness.

"(2) The term 'new source' means a stationary source the construction or modification of which is commenced after the Administrator proposes regulations under this section establishing an emission standard which will be applicable to such source.

"(3) The terms 'stationary source,' 'modification,' 'owner or operator' and 'existing source' shall have the same meaning as such terms have under section 111(a).

"(b) (1) (A) The Administrator shall, within 90 days after the date of enactment of the Clean Air Amendments of 1970, publish (and shall from time to time thereafter revise) a list which includes each hazardous air pollutant for which he intends to establish an emission standard under this section.

"(B) Within 180 days after the inclusion of any air pollutant in such list, the Administrator shall publish proposed regulations establishing emission standards for such pollutant together with a notice of a public hearing within thirty days. Not later than 180 days after such publication, the Administrator shall prescribe an emission standard for such pollutant, unless he finds, on the basis of information presented at such hearings, that such pollutant clearly is not a hazardous air pollutant. The Administrator shall establish any such standard at the level which in his judgment provides an ample margin of safety to protect the public health from such hazardous air pollutants.

"(C) Any emission standard established pursuant to this section shall become effective upon promulgation.

"(2) The Administrator shall, from time to time, issue information on pollution control techniques for air pollutants subject to the provisions of this section.

"(c) (1) After the effective date of any emission standard under this section—

"(A) no person may construct any new source or modify any existing source which, in the Administrator's judgment, will emit an air pollutant to which such standard applies unless the Administrator finds that such source if properly operated will not cause emissions in violation of such standard, and

"(B) no air pollutant to which such standard applies may be emitted from any stationary source in violation of such standard, except that in the case of an existing source—

"(i) such standard shall not apply until 90 days after its effective date, and

"(ii) the Administrator may grant a waiver permitting such source a period of up to two years after the effective date of a standard to comply with the standard, if he finds that such period is necessary for the installation of controls and that steps will be taken during the period of the waiver to assure that the health of persons will be protected from imminent endangerment.

"(2) The President may exempt any stationary source from compliance with paragraph (1) for a period of not more than two years if he finds that the technology to implement such standards is not available and the operation of such source is required for reasons of national security. An exemption under this paragraph may be extended for one or more additional periods, each period not to exceed two years. The President shall make a report to Congress with respect to each exemption (or extension thereof) made under this paragraph.

"(d) (1) Each State may develop and submit to the Administrator a procedure for implementing and enforcing *emission standards for hazardous air pollutants* for stationary sources located in such State. If the Administrator finds the State procedure is adequate, he shall delegate to such State any authority he has under this Act to implement and enforce such standards (except with respect to stationary sources owned or operated by the United States).

"(2) Nothing in this subsection shall prohibit the Administrator from enforcing any applicable *emission* standard under this section.

FEDERAL ENFORCEMENT

"SEC. 113. (a)(1) Whenever, on the basis of any information available to him, the Administrator finds that any person is in violation of any requirement of an applicable implementation plan, the Administrator shall notify the person in violation of the plan and the State in which the plan applies of such finding. If such violation extends beyond the 30th day after the date of the Administrator's notification, the Administrator may issue an order requiring such person to comply with the requirements of such plan or he may bring a civil action in accordance with subsection (b).

"(2) Whenever, on the basis of information available to him, the Administrator finds that violations of an applicable implementation plan are so widespread that such violations appear to result from a failure of the State in which the plan applies to enforce the plan effectively, he shall so notify the State. If the Administrator finds such failure extends beyond the thirtieth day after such notice, he shall give public notice of such finding. During the period beginning with such public notice and ending when such State satisfies the Administrator that it will enforce such plan (hereafter referred to in this section as 'period of Federally assumed enforcement') the Administrator may enforce any requirement of such plan with respect to any person—

"(A) by issuing an order to comply with such requirement, or

"(B) by bringing a civil action under subsection (b).

"(3) Whenever, on the basis of any information available to him, the Administrator finds that any person is in violation of section 111(e) (relating to new source performance standards), 112(c) (relating to standards for hazardous emissions), or 119(g) (relating to energy-related authorities), or is in violation of any requirement of section 114 (relating to inspections, etc.), he may issue an order requiring such person to comply with such section or requirement, or he may bring a civil action in accordance with subsection (b).

"(4) An order issued under this subsection (other than an order relating to a violation of section 112) shall not take effect until the person to whom it is issued has had an opportunity to confer with the Administrator concerning the alleged violation. A copy of any order issued under this subsection shall be sent to the State air pollution control agency of any State in which the violation occurs. Any order issued under this subsection shall state with reasonable specificity the nature of the violation, specify a time for compliance which the Administrator determines is reasonable, taking into account the seriousness of the violation and any good faith efforts to comply with applicable requirements. In any case in which an order under this subsection (or notice to a violator under paragraph (1)) is issued to a corporation, a copy of such order (or notice) shall be issued to appropriate corporate officers.

"(b) The Administrator may commence a civil action for appropriate relief, including a permanent or temporary injunction, whenever any person—

"(1) violates or fails or refuses to comply with any order issued under subsection (a) ; or

"(2) violates any requirement of an applicable implementation plan (A) during any period of Federally assumed enforcement, or (B) more than 30 days after having been notified by the Administrator under subsection (a)(1) of a finding that such person is violating such requirement; or

"(3) violates section 111(e), 112(c), or 119(g) ; or

"(4) fails or refuses to comply with any requirement of section 114.

Any action under this subsection may be brought in the district court of the United States for the district in which the defendant is located or resides or is doing business, and such court shall have jurisdiction to restrain such violation and to require compliance. Notice of the commencement of such action shall be given to the appropriate State air pollution control agency.

"(c) (1) Any person who knowingly—

"(A) violates any requirement of an applicable implementation plan (i) during any period of Federally assumed enforcement, or (ii) more than 30 days after having been notified by the Administrator under subsection (a)(1) that such person is violating such requirement, or

"(B) violates or fails or refuses to comply with any order issued by the Administrator under subsection (a), or

"(C) violates section 111(e), section 112(c), or section 119(g) shall be punished by a fine of not more than $25,000 per day of violation, or by imprisonment for not more than one year, or by both. If the conviction is for a violation committed after the first conviction of such person under this paragraph, punishment shall be by a fine of not more than $50,000 per day of violation, or by imprisonment for not more than two years, or by both.

"(2) Any person who knowingly makes any false statement, representation, or certification in any application, record, report, plan, or other document filed or required to be maintained under this Act or who falsifies, tampers with, or knowingly renders inaccurate any monitoring device or method required to be maintained under this Act, shall upon conviction, be punished by a fine of not more than $10,000, or by imprisonment for not more than six months, or by both.

INSPECTIONS, MONITORING, AND ENTRY

"SEC. 114. (a) For the purpose (i) of developing or assisting in the development of any implementation plan under section 110 or 111(d), any standard of performance under section 111, or any emission standard under section 112,(ii) of determining whether any person is in violation of any such standard or any requirement of such a plan, or (iii) carrying out section 119 or 303—

"(1) the Administrator may require the owner or operator of any emission source to (A) establish and maintain such records, (B) make such reports, (C) install, use, and maintain such monitoring equipment or methods, (D) sample such emissions (in accordance with such methods, at such locations, at such intervals, and in such manner as the Administrator shall prescribe), and (E) provide such other information, as he may reasonably require; and

"(2) the Administrator or his authorized representative, upon presentation of his credentials—

"(A) shall have a right of entry to, upon, or through any premises in which an emission source is located or in which any records required to be maintained under paragraph (1) of this section are located, and

"(B) may at reasonable times have access to and copy any records, inspect any monitoring equipment or method required under paragraph (1), and sample any emissions which the owner or operator of such source is required to sample under paragraph (1).

"(b)(1) Each State may develop and submit to the Administrator a procedure for carrying out this section in such State. If the Administrator finds the State procedure is adequate, he may delegate to such State any authority he has to carry out this section (except with respect to new sources owned or operated by the United States).

"(2) Nothing in this subsection shall prohibit the Administrator from carrying out this section in a State.

"(c) Any records, reports or information obtained under subsection (a) shall be available to the public, except that upon a showing satisfactory to the Administrator by any person that records, reports, or information, or particular part thereof, (other than emission data) to which the Administrator has access under this section if made public, would divulge methods or processes entitled to protection as trade secrets of such person, the Administrator shall consider such record, report, or information or particular portion thereof confidential in accordance with the purposes of section 1905 of title 18 of the United States Code, except that such record, report, or information may be disclosed to other officers, employees, or authorized representatives of the United States concerned with carrying out this Act or when relevant in any proceeding under this Act.

APPENDIX H

RESOURCE CONSERVATION AND RECOVERY ACT OF 1976, PUBLIC LAW 94-580 (EXCERPTS)

"Subtitle C—Hazardous Waste Management

"IDENTIFICATION AND LISTING OF HAZARDOUS WASTE

42 USC 6921.

"SEC. 3001. (a) CRITERIA FOR IDENTIFICATION OR LISTING.—Not later than eighteen months after the date of the enactment of this Act, the Administrator shall, after notice and opportunity for public hearing, and after consultation with appropriate Federal and State agencies, develop and promulgate criteria for identifying the characteristics of hazardous waste, and for listing hazardous waste, which should be subject to the provisions of this subtitle, taking into account toxicity, persistence, and degradability in nature, potential for accumulation in tissue, and other related factors such as flammability, corrosiveness, and other hazardous characteristics. Such criteria shall be revised from time to time as may be appropriate.

Regulations.

"(b) IDENTIFICATION AND LISTING.—Not later than eighteen months after the date of enactment of this section, and after notice and opportunity for public hearing, the Administrator shall promulgate regulations identifying the characteristics of hazardous waste, and listing

Ante, p. 2799.

particular hazardous wastes (within the meaning of section 1004(5)), which shall be subject to the provisions of this subtitle. Such regulations shall be based on the criteria promulgated under subsection (a) and shall be revised from time to time thereafter as may be appropriate.

"(c) PETITION BY STATE GOVERNOR.—At any time after the date eighteen months after the enactment of this title, the Governor of any State may petition the Administrator to identify or list a material as a hazardous waste. The Administrator shall act upon such petition within ninety days following his receipt thereof and shall notify the Governor of such action. If the Administrator denies such petition because of financial considerations, in providing such notice to the Governor he shall include a statement concerning such considerations.

"STANDARDS APPLICABLE TO GENERATORS OF HAZARDOUS WASTE

Regulations.
42 USC 6922.

"SEC. 3002. Not later than eighteen months after the date of the enactment of this section, and after notice and opportunity for public hearings and after consultation with appropriate Federal and State agencies, the Administrator shall promulgate regulations establishing such standards, applicable to generators of hazardous waste identified or listed under this subtitle, as may be necessary to protect human health and the environment. Such standards shall establish requirements respecting—

"(1) recordkeeping practices that accurately identify the quantities of such hazardous waste generated, the constituents thereof which are significant in quantity or in potential harm to human health or the environment, and the disposition of such wastes;

"(2) labeling practices for any containers used for the storage, transport, or disposal of such hazardous waste such as will identify accurately such waste;

"(3) use of appropriate containers for such hazardous waste;

"(4) furnishing of information on the general chemical composition of such hazardous waste to persons transporting, treating, storing, or disposing of such wastes;

"(5) use of a manifest system to assure that all such hazardous waste generated is designated for treatment, storage, or disposal in treatment, storage, or disposal facilities (other than facilities on the premises where the waste is generated) for which a permit has been issued as provided in this subtitle; and

"(6) submission of reports to the Administrator (or the State agency in any case in which such agency carries out an authorized permit program pursuant to this subtitle at such times as the Administrator (or the State agency if appropriate) deems necessary, setting out—

Reports.

"(A) the quantities of hazardous waste identified or listed under this subtitle that he has generated during a particular time period; and

"(B) the disposition of all hazardous waste reported under subparagraph (A).

"STANDARDS APPLICABLE TO TRANSPORTERS OF HAZARDOUS WASTE

"SEC. 3003. (a) STANDARDS.—Not later than eighteen months after the date of enactment of this section, and after opportunity for public hearings, the Administrator, after consultation with the Secretary of Transportation and the States, shall promulgate regulations establishing such standards, applicable to transporters of hazardous waste identified or listed under this subtitle, as may be necessary to protect human health and the environment. Such standards shall include but need not be limited to requirements respecting—

Regulations.
42 USC 6923.

"(1) recordkeeping concerning such hazardous waste transported, and their source and delivery points;

"(2) transportation of such waste only if properly labeled;

"(3) compliance with the manifest system referred to in section 3002(5); and

"(4) transportation of all such hazardous waste only to the hazardous waste treatment, storage, or disposal facilities which the shipper designates on the manifest form to be a facility holding a permit issued under this subtitle.

"(b) COORDINATION WITH REGULATIONS OF SECRETARY OF TRANSPORTATION.—In case of any hazardous waste identified or listed under this subtitle which is subject to the Hazardous Materials Transportation Act (88 Stat. 2156; 49 U.S.C. 1801 and following), the regulations promulgated by the Administrator under this subtitle shall be consistent with the requirements of such Act and the regulations thereunder. The Administrator is authorized to make recommendations to the Secretary of Transportation respecting the regulations of such hazardous waste under the Hazardous Materials Transportation Act and for addition of materials to be covered by such Act.

Recommendations.

"STANDARDS APPLICABLE TO OWNERS AND OPERATORS OF HAZARDOUS WASTE TREATMENT, STORAGE, AND DISPOSAL FACILITIES

"SEC. 3004. Not later than eighteen months after the date of enactment of this section, and after opportunity for public hearings and after consultation with appropriate Federal and State agencies, the Administrator shall promulgate regulations establishing such performance standards, applicable to owners and operators of facilities for the treatment, storage, or disposal of hazardous waste identified or

Regulations.
42 USC 6924.

listed under this subtitle, as may be necessary to protect human health and the environment. Such standards shall include, but need not be limited to, requirements respecting—

"(1) maintaining records of all hazardous wastes identified or listed under this title which is treated, stored, or disposed of, as the case may be, and the manner in which such wastes were treated, stored, or disposed of;

"(2) satisfactory reporting, monitoring, and inspection and compliance with the manifest system referred to in section 3002(5);

"(3) treatment, storage, or disposal of all such waste received by the facility pursuant to such operating methods, techniques, and practices as may be satisfactory to the Administrator;

"(4) the location, design, and construction of such hazardous waste treatment, disposal, or storage facilities;

"(5) contingency plans for effective action to minimize unanticipated damage from any treatment, storage, or disposal of any such hazardous waste;

"(6) the maintenance of operation of such facilities and requiring such additional qualifications as to ownership, continuity of operation, training for personnel, and financial responsibility as may be necessary or desirable; and

"(7) compliance with the requirements of section 3005 respecting permits for treatment, storage, or disposal.

No private entity shall be precluded by reason of criteria established under paragraph (6) from the ownership or operation of facilities providing hazardous waste treatment, storage, or disposal services where such entity can provide assurances of financial responsibility and continuity of operation consistent with the degree and duration of risks associated with the treatment, storage, or disposal of specified hazardous waste.

"PERMITS FOR TREATMENT, STORAGE, OR DISPOSAL OF HAZARDOUS WASTE

42 USC 6925.   "SEC. 3005. (a) PERMIT REQUIREMENTS.—Not later than eighteen months after the date of the enactment of this section, the Administrator shall promulgate regulations requiring each person owning or operating a facility for the treatment, storage, or disposal of hazardous waste identified or listed under this subtitle to have a permit issued pursuant to this section. Such regulations shall take effect on the date provided in section 3010 and upon and after such date the disposal of any such hazardous waste is prohibited except in accordance with such a permit.

"(b) REQUIREMENTS OF PERMIT APPLICATION.—Each application for a permit under this section shall contain such information as may be required under regulations promulgated by the Administrator, including information respecting—

"(1) estimates with respect to the composition, quantities, and concentrations of any hazardous waste identified or listed under this subtitle, or combinations of any such hazardous waste and any other solid waste, proposed to be disposed of, treated, transported, or stored, and the time, frequency, or rate of which such waste is proposed to be disposed of, treated, transported, or stored; and

"(2) the site at which such hazardous waste or the products of treatment of such hazardous waste will be disposed of, treated, transported to, or stored.

"(c) PERMIT ISSUANCE.—Upon a determination by the Administrator (or a State, if applicable), of compliance by a facility for which a permit is applied for under this section with the requirements of this section and section 3004, the Administrator (or the State) shall issue a permit for such facilities. In the event permit applicants propose modification of their facilities, or in the event the Administrator (or the State) determines that modifications are necessary to conform to the requirements under this section and section 3004, the permit shall specify the time allowed to complete the modifications.

"(d) PERMIT REVOCATION.—Upon a determination by the Administrator (or by a State, in the case of a State having an authorized hazardous waste program under section 3006) of noncompliance by a facility having a permit under this title with the requirements of this section or section 3004, the Administrator (or State, in the case of a State having an authorized hazardous waste program under section 3006) shall revoke such permit.

"(e) INTERIM STATUS.—Any person who—

"(1) owns or operates a facility required to have a permit under this section which facility is in existence on the date of enactment of this Act,

"(2) has complied with the requirements of section 3010(a), and

"(3) has made an application for a permit under this section shall be treated as having been issued such permit until such time as final administrative disposition of such application is made, unless the Administrator or other plaintiff proves that final administrative disposition of such application has not been made because of the failure of the applicant to furnish information reasonably required or requested in order to process the application.

"AUTHORIZED STATE HAZARDOUS WASTE PROGRAMS

"SEC. 3006. (a) FEDERAL GUIDELINES.—Not later than eighteen months after the date of enactment of this Act, the Administrator, after consultation with State authorities, shall promulgate guidelines to assist States in the development of State hazardous waste programs.

42 USC 6926.

"(b) AUTHORIZATION OF STATE PROGRAM.—Any State which seeks to administer and enforce a hazardous waste program pursuant to this subtitle may develop and, after notice and opportunity for public hearing, submit to the Administrator an application, in such form as he shall require, for authorization of such program. Within ninety days following submission of an application under this subsection, the Administrator shall issue a notice as to whether or not he expects such program to be authorized, and within ninety days following such notice (and after opportunity for public hearing) he shall publish his findings as to whether or not the conditions listed in items (1), (2), and (3) below have been met. Such State is authorized to carry out such program in lieu of the Federal program under this subtitle in such State and to issue and enforce permits for the storage, treatment, or disposal of hazardous waste unless, within ninety days following submission of the application the Administrator notifies such State that such program may not be authorized and, within ninety days following such notice and after opportunity for public hearing, he finds that (1) such State program is not equivalent to the Federal program under this subtitle, (2) such program is not consistent with the Federal or State programs applicable in other States, or (3) such program does not provide adequate enforcement of compliance with the requirements of this subtitle.

"(c) Interim Authorization.—Any State which has in existence a hazardous waste program pursuant to State law before the date ninety days after the date required for promulgation of regulations under sections 3002, 3003, 3004, and 3005, submit to the Administrator evidence of such existing program and may request a temporary authorization to carry out such program under this subtitle. The Administrator shall, if the evidence submitted shows the existing State program to be substantially equivalent to the Federal program under this subtitle, grant an interim authorization to the State to carry out such program in lieu of the Federal program pursuant to this subtitle for a twenty-four month period beginning on the date six months after the date required for promulgation of regulations under sections 3002 through 3005.

"(d) Effect of State Permit.—Any action taken by a State under a hazardous waste program authorized under this section shall have the same force and effect as action taken by the Administrator under this subtitle.

"(e) Withdrawal of Authorization.—Whenever the Administrator determines after public hearing that a State is not administering and enforcing a program authorized under this section in accordance with requirements of this section, he shall so notify the State and, if appropriate corrective action is not taken within a reasonable time, not to exceed ninety days, the Administrator shall withdraw authorization of such program and establish a Federal program pursuant to this subtitle. The Administrator shall not withdraw authorization of any such program unless he shall first have notified the State, and made public, in writing, the reasons for such withdrawal.

"INSPECTIONS

42 USC 6927.   "Sec. 3007. (a) Access Entry.—For purposes of developing or assisting in the development of any regulation or enforcing the provisions of this subtitle, any person who generates, stores, treats, transports, disposes of, or otherwise handles hazardous wastes shall, upon request of any officer or employee of the Environmental Protection Agency, duly designated by the Administrator, or upon request of any duly designated officer employee of a State having an authorized hazardous waste program, furnish or permit such person at all reasonable times to have access to, and to copy all records relating to such wastes. For the purposes of developing or assisting in the development of any regulation or enforcing the provisions of this title, such officers or employees are authorized—

"(1) to enter at reasonable times any establishment or other place maintained by any person where hazardous wastes are generated, stored, treated, or disposed of;

"(2) to inspect and obtain samples from any person of any such wastes and samples of any containers or labeling for such wastes. Each such inspection shall be commenced and completed with reasonable promptness. If the officer or employee obtains any samples, prior to leaving the premises, he shall give to the owner, operator, or agent in charge a receipt describing the sample obtained and if requested a portion of each such sample equal in volume or weight to the portion retained. If any analysis is made of such samples, a copy of the results of such analysis shall be furnished promptly to the owner, operator, or agent in charge.

"(b) AVAILABILITY TO PUBLIC.—Any records, reports, or information obtained from any person under this section shall be available to the public, except that upon a showing satisfactory to the Administrator (or the State, as the case may be) by any person that records, reports, or information, or particular part thereof, to which the Administrator (or the State, as the case may be) has access under this section if made public, would divulge information entitled to protection under section 1905 of title 18 of the United States Code, the Administrator (or the State, as the case may be) shall consider such information or particular portion thereof confidential in accordance with the purposes of that section, except that such record, report, document, or information may be disclosed to other officers, employees, or authorized representatives of the United States concerned with carrying out this Act, or when relevant in any proceeding under this Act.

"FEDERAL ENFORCEMENT

"SEC. 3008. (a) COMPLIANCE ORDERS.—(1) Except as provided in paragraph (2), whenever on the basis of any information the Administrator determines that any person is in violation of any requirement of this subtitle, the Administrator shall give notice to the violator of his failure to comply with such requirement. If such violation extends beyond the thirtieth day after the Administrator's notification, the Administrator may issue an order requiring compliance within a specified time period or the Administrator may commence a civil action in the United States district court in the district in which the violation occurred for appropriate relief, including a temporary or permanent injunction.

42 USC 6928.

"(2) In the case of a violation of any requirement of this subtitle where such violation occurs in a State which is authorized to carry out a hazardous waste program under section 3006, the Administrator shall give notice to the State in which such violation has occurred thirty days prior to issuing an order or commencing a civil action under this section.

"(3) If such violator fails to take corrective action within the time specified in the order, he shall be liable for a civil penalty of not more than $25,000 for each day of continued noncompliance and the Administrator may suspend or revoke any permit issued to the violator (whether issued by the Administrator or the State).

Penalty.

"(b) PUBLIC HEARING.—Any order or any suspension or revocation of a permit shall become final unless, no later than thirty days after the order or notice of the suspension or revocation is served, the person or persons named therein request a public hearing. Upon such request the Administrator shall promptly conduct a public hearing. In connection with any proceeding under this section the Administrator may issue subpenas for the attendance and testimony of witnesses and the production of relevant papers, books, and documents, and may promulgate rules for discovery procedures.

Subpenas.

"(c) REQUIREMENTS OF COMPLIANCE ORDERS.—Any order issued under this section shall state with reasonable specificity the nature of the violation and specify a time for compliance and assess a penalty, if any, which the Administrator determines is reasonable taking into account the seriousness of the violation and any good faith efforts to comply with the applicable requirements.

Penalty.

"(d) CRIMINAL PENALTY.—Any person who knowingly—

"(1) transports any hazardous waste listed under this subtitle to a facility which does not have a permit under section 3005 (or 3006 in the case of a State program),

"(2) disposes of any hazardous waste listed under this subtitle without having obtained a permit therefor under this subtitle,

"(3) makes any false statement or representation in any application, label, manifest, record, report, permit or other document filed, maintained, or used for purposes of compliance with this subtitle.

shall, upon conviction, be subject to a fine of not more than $25,000 for each day of violation, or to imprisonment not to exceed one year, or both. If the conviction is for a violation committed after a first conviction of such person under this paragraph, punishment shall be by a fine of not more than $50,000 per day of violation, or by imprisonment for not more than two years, or by both.

APPENDIX I

REPORTING FOR THE CHEMICAL
SUBSTANCE INVENTORY*

The regulations discussed in this appendix are the first general reporting regulations issued by the U.S. Environmental Protection Agency under the Toxic Substances Control Act. (For a full text presentation of the regulations see page 151.) They govern the reporting of chemical substances to create a national inventory of chemical substances manufactured, imported, or processed in the United States for commercial purposes since January 1, 1975.

---

* Office of Toxic Substances, U.S. Environmental Protection Agency,
  December 1977.

SECTION I

THE CHEMICAL SUBSTANCE INVENTORY

The Toxic Substances Control Act (TSCA) requires the U.S. Environmental Protection Agency (EPA) to compile and publish an Inventory of chemical substances manufactured, imported, or processed in the United States for commercial purposes. To ensure a complete and reliable inventory, EPA has issued Inventory Reporting Regulations (40 CFR 710), as required by TSCA. These regulations govern reporting for the Inventory.

The Inventory will be compiled from reports which manufacturers, importers, processors, or users of chemical substances prepare and submit to EPA, in accordance with the regulations. Under a two-phase reporting schedule designed to minimize duplicative reporting, all such persons will have an opportunity to enter reportable chemical substances on the Inventory.

During an initial reporting period, ending May 1, 1978, some manufacturers and importers are required to report. Other manufacturers or importers, while not required to, may report or authorize another person to report on their behalf in order to be sure that the chemical substances they manufacture or import for a commercial purpose are included on the Inventory. For specific details, see Chapter III of this booklet describing who must report. Persons who have only processed or used a chemical substance, i.e., who have not manufactured or imported the substance, should NOT report the substance during the initial reporting period which ends May 1, 1978. Such persons will have an opportunity to report during a second reporting period. Some manufacturers and importers can also report after May 1. See Chapter IV of this booklet for a description of the conditions under which such reports may be submitted.

EPA will compile an Initial Inventory which will include those substances reported by manufacturers or importers, or by their duly-authorized agents during the initial reporting period. EPA expects that the Initial Inventory will be published some time near the end of 1978. Thirty days after its publication, the premanufacture notification provisons of TSCA, which requires notification to EPA at least 90 days in advance of manufacture or importation, will become effective for persons intending to manufacture or import (in bulk form) for a commercial purpose any chemical substance not identified on the Initial Inventory.

A second, 210-day reporting period will begin when EPA publishes the Initial Inventory. During this period, importers of chemical substances as part of mixtures or articles, and persons who have only processed or used, since January 1, 1975, a reportable chemical substance which did not appear on the Initial Inventory, may report such substance for inclusion in a Revised Inventory. This Revised Inventory will be published as soon as possible after the end of the second reporting period. Of course, EPA will add new chemical substances to the Inventory after they have satisfied the premanufacture notification provisions of Section 5 of TSCA.

Only nonconfidential chemical substance identities will appear on the Inventory. Generic names applied to chemical substance identities which are confidential will appear in an appendix to the Inventory. Neither the names of the manufacturers, importers, processors, or users who report chemical substances, nor production ranges or other reported information, will appear on the Inventory.

Chemical substances are often commercially distributed in products bearing trademarks or commercial names. In some cases, persons who process or use such products for commercial purposes do not know the product's chemical composition and will need assurance that the identity of all reportable chemical substances of which these products are comprised have been submitted to EPA for inclusion on the Inventory. During the initial reporting period, manufacturers and importers of such products will have the opportunity to report the trademarks or commercial names of their products on a separate form which permits listing of such names without accompanying information on product composition. In order to do so, however, they must certify that all reportable chemical substances comprising these products have been reported for the Inventory.

EPA will compile and publish a product trademark list from reports submitted by manufacturers and importers. This list will not be a part of the Inventory. During the second reporting period, this list will provide an easy means for processors and users to determine whether or not the chemical substances which comprise trademarked products have been reported for the Inventory.

SECTION II

CHEMICAL SUBSTANCES

The Inventory Reporting Regulations (40 CFR 710) govern reporting of certain substances for inclusion on the Inventory. Each substance reported must satisfy the following three criteria:

1. It must be a "chemical substance" as defined by section 710.2(h) of the regulations;

2. It must have been manufactured, imported, or processed for commercial purposes in the United States since January 1, 1975; and

3. It must not be excluded from the Inventory by any provision of section 710.4 of the regulations.

A "reportable chemical substance" is one which satisfies all of these criteria; a substance which fails to meet one or more of these criteria must not be reported for the Inventory.

This chapter discusses the three criteria. The first section of this chapter, Reportable Chemical Substances, presents and discusses the term "chemical substance" and the phrase "manufacture or import 'for commercial purposes.'" This section also describes how some particular classes of chemical substances, including polymers, should be identified for inclusion on the Inventory. The second section, Excluded Substances, presents and comments on some of the exclusions contained in section 710.4 of the regulations. These exclusions identify certain chemical substances which must not be reported for the Inventory.

IMPORTANT: Many terms used in the regulations and in this chapter (for example: "chemical substance," "mixture," "article," "intermediate," "manufacture," and "process") have very specific meanings and are defined in the regulations. To aid your understanding of this chapter, pertinent sections of the regulations to which you should refer for additional clarification are cited in square brackets, e.g. [710.2(h)].

"Appendix A: Significant Comments and Responses," which accompanies the regulations published in the December 23, 1977 FEDERAL REGISTER and begins on page 64580, discusses many aspects of the regulations in detail. Comments 29-82 are particularly relevant to the subjects discussed in this chapter. Pertinent Comments to which you should refer are cited in this chapter by their enclosure in braces, e.g., {73}.

Reportable Chemical Substances

Definition of "Chemical Substance"

The Toxic Substances Control Act (TSCA) identifies
three types of materials: (1) chemical substances, (2)
mixtures of chemical substances, and (3) articles comprised
of chemical substances and/or mixtures. The Inventory will
list only chemical substances. It will not list mixtures or
articles. It will list, however, chemical substances of
which mixtures and articles are comprised.

"Chemical substance" is defined in section 710.2(h) of
the regulations by chemical composition, by source or origin,
and by identification of certain categories of materials
which are not considered "chemical substances":

"Chemical substance" means any organic or inorganic
substance of a particular molecular identity, including
any combination of such substances occurring in whole
or in part as a result of a chemical reaction or occurring
in nature, and any chemical element or uncombined
radical; except that "chemical substance" does not
include:

1)   any mixture [710.2(q)]{31-36}.

2)   any pesticide when manufactured,
     processed, or distributed in commerce
     for use as a pesticide [710.2(b)]
     {37-39}.

3)   tobacco or any tobacco product, but not
     including any derivative products,

4)   any source material, special nuclear
     material, or byproduct material
     [710.2(c)],

5)   any pistol, firearm, revolver, shells and
     cartridges, and

6)   any food, food additive, drug, cosmetic
     or device, when manufactured, processed, or
     distributed in commerce for use as a food,
     food additive, drug, cosmetic, or device
     [710.2(a)] {37, 40-42}.

Composition: Except for its impurities [710.2(m)], a
chemical substance may be comprised of a single organic or
inorganic species, element, or free radical or a combination
of such entities. Substances whose composition can be
represented by definite chemical structure diagrams are
denoted Class 1 substances. Examples of Class 1 substances

are: acetone, iron, benzene, and sodium chloride. Substances which are combinations of different known or unknown species or whose composition cannot be represented by definite chemical structure diagrams are denoted Class 2 substances. Examples of Class 2 substances are: crude oil, superphosphate (fertilizer), tall oil, and coconut oil acids. Therefore, a chemical substance may be a "pure" compound, but does not necessarily have to be. A chemical substance may be a complex combination comprised of known or unknown chemical species. (See Appendix 5 of this booklet for a specification of what must be reported to identify a Class 1 or Class 2 chemical substance in a manner suitable for including its identity on the Inventory.)

Source or Origin: Whether or not a material, and in particular a complex combination of species, is a "chemical substance" or "mixture" depends upon its source, origin, or method of preparation. The definition of "chemical substance" clearly identifies "any combination...occurring...in nature" as a "chemical substance." Therefore, any material extracted or removed from nature is a "chemical substance" and is, by definition, not a "mixture" [710.2(q)]. In addition, if such a material is further separated into component parts, each component, as separated, is a "chemical substance." Separating a naturally occurring material into component parts does not cause such a material to lose its status as a "chemical substance."

A combination which is produced by a chemical reaction calls for a common sense determination as to its status as a "chemical substance" [710.2(h)] or "mixture" [710.2 (q)] based on the following consideration:

Could the combination have been prepared at this time for commercial purposes by combining commercially-available ingredients which do not chemically react when mixed?

o   If the answer is NO, the combination manufactured is a "chemical substance," and is subject to the Inventory Reporting Regulations.

o   If the answer is YES, the combination manufactured is a "mixture" of the chemical substances. Although the combination, in this case, must not be reported, the chemical substances which were in fact manufactured by the chemical reaction are subject to the regulations [See Note at 710.4 (c)(2)].

EXAMPLE: If commercially-available chemical substances A, B, and C are mixed, without chemical reaction, a combination of A, B, and C is produced which is a "mixture." Alternatively, if that combination was prepared by mixing chemical substances A, B, D, and E, and D and E chemically reacted to form C, the combination (A, B, and C) is a "mixture." However, chemical substance C has been manufactured.

Materials Not Considered "Chemical Substances": The six categories of materials listed by number in the definition of "chemical substance" [710.2(h)] are not considered chemical substances. Any material identified in that list is not a "chemical substance" and must not be reported for the Inventory.

Definition of "Manufacture or Import 'For Commercial Purposes'"

The phrase "manufacture or import for commercial purposes'" is important for determining whether or not a manufactured or imported chemical substance is a reportable chemical substance. Section 710.2(p) of the regulations defines the phrase:

"Manufacture or import 'for commercial purposes'" means to manufacture or import:

1) For distribution in commerce [710.2(j)] including for test marketing purposes [710.2(bb)] {64-66}, or

2) For use by the manufacturer, including for use as an intermediate [710.2(n)] {67-71}.

Thus, the Inventory will be comprised of not only chemical substances not otherwise excluded by section 710.4 of the regulations, which have been manufactured or imported since January 1, 1975, for "distribution in commerce" but also of those which persons have manufactured for their own use, including use as an "intermediate".

Chemical substances, not otherwise excluded by section 710.4 of the regulations, may also be reported if they have been processed for commercial purposes [710.2(u)] since January 1, 1975. A special reporting period for processors will be provided after publication of an Initial Inventory, as noted in Chapter IV of this booklet.

"Special Case" Chemical Substances

1. Reporting Polymers

Section 710.5(c) of the regulations specifies how to identify polymers for inclusion on the Inventory:

1)   To report a polymer, a person must list in the description of the polymer composition

at least those monomers used at greater than
two percent (by weight) in the manufacture of
the polymer.

2)   Those monomers used at two percent (by
weight) or less in the manufacture of the
polymer may be included as part of the
description of the polymer composition.

NOTE.--The "percent (by weight)" of a monomer is
the weight of the monomer charged expressed as a
percentage of the weight of the polymeric chemical
substance manufactured.

For example, if ten (10) pounds of one monomer is
charged into a reactor, along with other reactive
ingredients, and 100 pounds of "dry" weight copolymer
is manufactured, the monomer was used at ten (10)
percent (by weight) in the manufacture of the copolymer. The monomer, therefore, must be identified in the
description of the copolymer. (See Appendix 5 of this
booklet for additional information on how to identify
polymeric chemical substances.)

Although monomers used at two percent (by weight)
or less in the manufacture of a polymer are not required to be included as part of the description of
the polymer, such monomers, like other "intermediates,
are subject to the Inventory Reporting Regulations
regardless of their end use in the manufacture of
polymers.

The polymer description should identify only
monomers and other reactive ingredients such as chain-
transfer or crosslinking substances. Other additives,
such as emulsifiers and plasticizers, which are not
chemically a part of the polymeric composition should
not be identified in the description of the polymer,
and their weight should not be included in estimating
the "dry" weight of the polymer.

2.   Naturally Occurring Chemical Substances

Section 710.4(b) of the regulations defines a
category of chemical substances, "Naturally Occurring
Chemical Substances," which will appear on the Inventory. Persons who manufacture, import, or process
chemical substances which are included within that
category should not report such substances for inclusion on the Inventory because they are considered to
be automatically included. The category includes:

Any chemical substance which is naturally
occurring and

(1) which is (i) unprocessed or (ii) processed only by manual, mechanical, or gravitational means; by dissolution in water; by
flotation; or by heating solely to remove
water; or

(2) which is extracted from air by any means.

o   The category includes chemical substances which
are derived from nature (including the land, water,
atmosphere and life forms which naturally inhabit
the earth) by the means specified.

3.   Class 2 Chemical Substances Known Commercially
by Class 1 Names

Some reportable Class 2 chemical substances, which
are combinations of several different chemical species,
are known in commerce by specific chemical names that
identify a principal chemical species of the combination, for example, commercial "stearic acid."
Although the chemical name may incorrectly suggest that
such a chemical substance is a Class 1 substance, in
those cases where the name is actually used in commerce
to identify the chemical substance, it may also be used
to identify the chemical substance in reporting for the
Inventory.

Chapter V of this booklet specifies the report
forms to be used in reporting chemical substances for
the Inventory. For chemical substances of the type
described in this "special case", Report Form A may be
used to report the substance if the identity of the
principal species by which the chemical substance is
known commercially appears on the TSCA Candidate List
of Chemical Substances. If the identity of the principal species by which the chemical substance is known
commercially is not on the Candidate List but has a
known Chemical Abstracts Service (CAS) Registry Number,
the chemical substance may be reported using Form B.
Otherwise, the chemical substance must be reported
using Form C and be identified according to the procedures specified in Appendix 5 of this booklet for
reporting Class 2 chemical substances.

4.   Chemical Substances Which Are Fractionated Into
Component Chemical Substances

Some Class 2 chemical substances are complex
combinations of different chemical species and are
fractionated, in whole or in part, into component
chemical substances (fractions). In this "special
case," the unfractionated chemical substance need not

be reported for the Inventory if it is completely separated by its manufacturer into its fractions, and each fraction which is manufactured for commercial purposes is reported instead. On the other hand, the unfractionated chemical substance should be reported along with the relevant production range (if reported) for that amount which is not fractionated.

Excluded Substances

Some materials which are "chemical substances" [710.2(h)] and which have been manufactured, imported, or processed for commerical purposes since January 1, 1975, are excluded from the Inventory, and must not be reported. A chemical substance is excluded if it is, or has been:

1. Manufactured, imported, or processed solely in small quantities for research and development [710.4(c)(3), 710.2(y)] {29, 43-51}.

o   The NOTE appearing at section 710.2(y) of the regulations states that any chemical substance which is manufactured, imported, or processed in quantities of less than 1,000 pounds annually is presumed to be an R&D chemical substance. Such a chemical substance can be reported for the Inventory, however, if the manufacturer, importer, or processor can <u>certify</u> that the chemical substance was not manufactured, imported, or processed solely in "small quantities for research and development."

2. An impurity [710.2(m)], 710.4(d)(1)] {61}.

o   By this exclusion, impurities are not reportable, and, furthermore, no chemical substance which is reported for the Inventory should be identified in terms of its impurities, or by its commercial grades.

3. A byproduct [710.2(q)] which has no commercial purpose.

    NOTE.--A byproduct which has commercial value only to municipal or private organizations who (i) burn it as a fuel, (ii) dispose of it as a waste, including in a landfill or for enriching soil, or (iii) extract component chemical substances which have commercial value, may be reported for the Inventory, but will not be subject to premanufacturing notification under section 5 of TSCA if not included [710.4(d)(2)] {52-55 }.

o   Byproducts which have commercial value for reasons
     other than those specified in the NOTE are not
excluded from the Inventory [see 710.2(p)].

4.   A chemical substance which results from a chem-
     ical reaction that occurs incidental to exposure
     of another chemical substance, mixture, or article
     to environmental factors such as air, moisture,
     microbial organisms, or sunlight [710.4(d)(3)].

o   Chemical substances, such as rust on iron, or
     other corrosion or degradation products, which
form incidental to environmental exposure are excluded
from the Inventory.

5.   A chemical substance which results from a chemical
     reaction that occurs incidental to storage of
     another chemical substance, mixture, or article
     [710.4(d)(4)].

o   Degradation products which form incidental to the
     storage of a chemical substance, such as the
     partial polymerization of a drying oil, are excluded
     from the Inventory.

6.   A chemical substance which results from a chemical
     reaction that occurs upon end use of other chem-
     ical substances, mixtures, or articles such as ad-
     hesives, paints, miscellaneous cleansers or other
     housekeeping products, fuels and fuel additives,
     water softening and treatment agents, photographic
     films, batteries, matches, and safety flares, and
     which is not itself manufactured for distribution
     in commerce or for use as an intermediate [710.4(d)(5)].

o   Chemical substances which are the components of
     adhesives, paints, miscellaneous cleansers, etc.
are not excluded from the Inventory by this provision;
only the chemical substances which form upon their end
use are excluded.

7.   A chemical substance which results from a chemical
     reaction that occurs upon use of curable plastic
     or rubber molding compounds, inks, drying oils,
     metal finishing compounds, adhesives, or paints;
     or other chemical substances formed during manu-
     facture of an article destined for the marketplace
     without further chemical change of the chemical
     substances except for those chemical changes that
     may occur as described in section 710.4(d) of the
     regulations [710.4(d)(6)].

o   Chemical substances which are the components of
     curable plastic or rubber molding compounds, inks,
etc. are not excluded from the Inventory by this provision; only the chemical substances which are formed
upon the use of such materials are excluded.

8.   A chemical substance which results from a chem-
     ical reaction that occurs when (i) a stabilizer,
     colorant, odorant, antioxidant, filler, solvent,
     carrier, surfactant, plasticizer, corrosion
     inhibitor, antifoamer or defoamer, dispersant,
     precipitation inhibitor, binder, emulsifier, de-
     emulsifier, dewatering agent, agglomerating agent,
     adhesion promoter, flow modifier, pH neutralizer,
     sequesterant, coagulant, flocculant, fire retard-
     ant, lubricant, chelating agent, or quality
     control reagent functions as intended or (ii) a
     chemical substance, solely intended to impart a
     specific physico-chemical characteristic, functions
     as intended [710.4(d)(7)].

o   The substances which comprise the various mater-
     ials listed above are not excluded from the
Inventory; only the chemical substances which are
formed upon use of such materials are excluded.

9.   A chemical substance which is not intentionally
     removed from the equipment in which it was manu-
     factured.
     NOTE.--The "equipment in which it was manufac-
     tured" includes the reaction vessel in which the
     chemical substance was manufactured and other
     equipment which is strictly ancillary to the
     reaction vessel, and any other equipment through
     which the chemical substance may flow during a
     continuous flow process, but does not include
     tanks or other vessels in which the chemical
     substance is stored after its manufacture [710.4(d)(8
     710.2(n)] {67-71}.

SECTION III

REPORTING FOR THE INITIAL INVENTORY

How To Determine Who Must Report and
What Must Be Reported for the Initial Inventory

The Inventory Reporting Regulations require certain manufacturers and importers of chemical substances to report for the Initial Inventory, and permit optional reporting by others. This chapter can help you determine:

o   whether or not you are required to report
     for the Initial Inventory;

o   what information must be reported; and

o   what information may be reported voluntarily.

Section 710.3(a) of the regulations specify who is required to report for the Initial Inventory and what they must report. Specifically, a manufacturer whose plant site meets the following criteria must report all chemical substances manufactured for commercial purposes in 1977 at the plant site if:

1.   thirty percent or more by net weight of the products distributed from the plant site during calendar year 1977 were products within SIC groups 28 (Chemicals and Allied Products) or 2911 (Petroleum Refining Products), or

2.   the total pounds of reportable chemical substances manufactured at the plant site during calendar year 1977 equaled one million or more pounds.

In addition, manufacturers must report <u>any</u> chemical substance not reported under (1) or (2) that was manufactured for commercial purposes in quantities of 100,000 pounds or greater at a plant site during calendar year 1977.

The reporting requirements for importers of chemical substances in bulk form are parallel to these, except importers do not report by plant site.

Decision Flow-Charts

This chapter contains two decision flow-charts which should help you to determine your reporting requirements. One is for use by domestic manufacturers and the other by importers. Some terms have been defined specifically for use in these decision flow-charts. These terms are fully capitalized and are defined in the glossaries which appear with each chart.

For each chart there are four steps to follow. At each step you are asked a question, to which you would respond either YES or NO. These questions are based upon the same criteria for determining reporting requirements as those contained in the regulations.

As you progress from step to step, follow the arrow leading from one answer to the next question. Eventually, the arrow will lead to a numbered group of reporting requirements. On the pages following the chart, locate this specific group of reporting requirements to determine what information you must report.

Separate flow-charts are provided for domestic manufacturers and importers. Although the reporting requirements for domestic manufacturers and importers are similar, it is important to use the flow-chart appropriate to each activity. If you both manufactured and imported chemical substances, you should use the flow-chart appropriate to each activity separately to determine your particular reporting requirements. For example, suppose Company X is a manufacturing and importing company whose total annual sales exceeded $5 million in 1977. Company X owns only one CHEMICAL MANUFACTURING PLANT SITE, and does not qualify as either a CHEMICAL IMPORTER or VOLUME IMPORTER. However, Company X imported, in bulk form, ten chemical substances in 1977, three of which were imported in amounts greater than 100,000 pounds. By referring to the flow-chart for domestic manufacturers, Company X determines that it must report according to the Group 1 Reporting Requirements. Therefore, it reports concerning all chemical substances manufactured for commerial purposes in 1977 at its CHEMICAL MANUFACTURING PLANT SITE. Because the company did not qualify as a CHEMICAL IMPORTER or VOLUME IMPORTER, but did import three HIGH VOLUME CHEMICAL SUBSTANCES, it determines, by referring to the decision flow-chart for importers, that it is also subject to Group 9 Reporting Requirements and, therefore, reports as required concerning the three HIGH VOLUME CHEMICAL SUBSTANCES it imported in 1977. After referring to the "Optional Reporting Provisions," Company X determines it will exercise its option, and reports concerning the seven other chemical substances it imported in bulk form so as to ensure their inclusion on the Initial Inventory.

If you determine that you are not required to report
any information under these regulations (Group 5 for manufacturers and Group 10 for importers), you are encouraged to
read the section entitled "Optional Reporting Provisions,"
and report if necessary to ensure that the chemical substances you manufactured or imported are included on the
Inventory.

Reporting by Plant Site, Headquarters, or Business Address:

The group reporting requirements specify whether
required chemical substance reporting is to be done by
plant site, headquarters, or business address. Reporting
by plant site means that a chemical substance is reported
for the Inventory on a report form which identifies its
site of manufacture. By contrast, reporting by corporate
headquarters or by business address means that a chemical
substance is reported on a report form which identifies the
name and address of the business which is responsible for
the manufacture or importation of the substance. Reporting
of imported chemical substances will be done by the business
address of the importer. Although EPA encourages all domestic
manufacturers to report by plant site, manufacturers who
optionally report chemical substances they manufacture, or
who qualify as small manufacturers may report by corporate
headquarters. However, no person is a small manufacturer
with respect to a chemical substance which he or she manufactured in amounts equal to or greater than 100,000 pounds
at one plant site during 1977, and therefore must report that
substance by plant site.

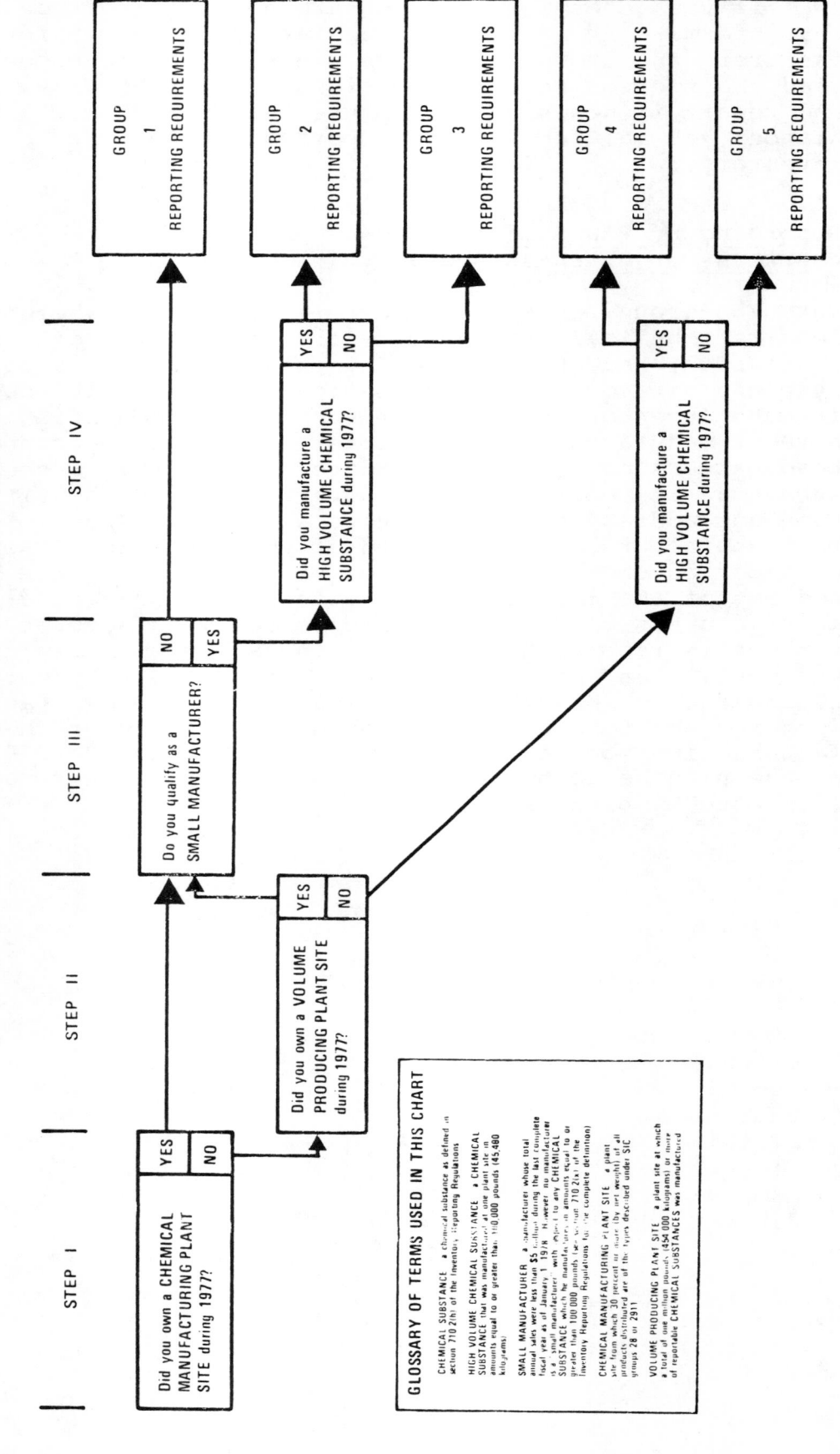

DECISION FLOW—CHART TO DETERMINE REPORTING REQUIREMENTS FOR DOMESTIC MANUFACTURERS OF CHEMICAL SUBSTANCES

STEP I   STEP II   STEP III   STEP IV

Did you own a CHEMICAL MANUFACTURING PLANT SITE during 1977?
YES / NO

Did you own a VOLUME PRODUCING PLANT SITE during 1977?
YES / NO

Do you qualify as a SMALL MANUFACTURER?
NO / YES

Did you manufacture a HIGH VOLUME CHEMICAL SUBSTANCE during 1977?
YES / NO

Did you manufacture a HIGH VOLUME CHEMICAL SUBSTANCE during 1977?
YES / NO

GROUP 1 — REPORTING REQUIREMENTS

GROUP 2 — REPORTING REQUIREMENTS

GROUP 3 — REPORTING REQUIREMENTS

GROUP 4 — REPORTING REQUIREMENTS

GROUP 5 — REPORTING REQUIREMENTS

GLOSSARY OF TERMS USED IN THIS CHART

CHEMICAL SUBSTANCE a chemical substance as defined in section 710.2(h) of the Inventory Reporting Regulations

HIGH VOLUME CHEMICAL SUBSTANCE a CHEMICAL SUBSTANCE that was manufactured at one plant site in amounts equal to or greater than 100,000 pounds (45,400 kilograms)

SMALL MANUFACTURER a manufacturer whose total annual sales were less than $5 million during the last complete fiscal year as of January 1, 1978. However, no manufacturer is a small manufacturer with respect to any CHEMICAL SUBSTANCE which he manufactures in amounts equal to or greater than 100,000 pounds (see section 710.2(x) of the Inventory Reporting Regulations for the complete definition)

CHEMICAL MANUFACTURING PLANT SITE a plant site from which 50 percent or more (by net weight) of all products distributed are of the types described under SIC groups 28 or 2911

VOLUME PRODUCING PLANT SITE a plant site at which a total of one million pounds (454,000 kilograms) or more of reportable CHEMICAL SUBSTANCES was manufactured

DECISION FLOW-CHART TO DETERMINE
REPORTING REQUIREMENTS FOR DOMESTIC
MANUFACTURERS OF CHEMICAL SUBSTANCES

Step I: Did you own a CHEMICAL MANUFACTURING PLANT SITE
during 1977?

CHEMICAL MANUFACTURING PLANT SITE means a plant site
from which 30 percent or more (by net weight) of all products distributed are of the types described under Standard
Industrial Classification (SIC) groups 28 or 2911 (see
Appendix 2).

Although these SIC groups include categories of products, such as pesticides and drugs, which are specifically
excluded from the Inventory, these products should be included in determining whether or not a plant site is a CHEM-
ICAL MANUFACTURING PLANT SITE. The regulations, however, do
not permit you to report excluded chemical substances, i.e.,
substances excluded from the Inventory by section 710.4 of
the regulations.

Step II: Did you own a VOLUME PRODUCING PLANT SITE during
1977?

VOLUME PRODUCING PLANT SITE means a plant site at which
a total of one (1) million pounds (454,000 kilograms) or
more of reportable chemical substances was manufactured.

This criteria should be used completely independently
of that considered in Step I. A substance may be a reportable chemical substance whether or not it is listed under
SIC groups 28 or 2911. Section 710.4 of the regulations
specifies and chapter II of this booklet discusses what is
a reportable chemical substance.

Step III: Do you qualify as a SMALL MANUFACTURER?

SMALL MANUFACTURER, as defined in Section 710.2(x) of
the regulations, means a manufacturer whose total annual
sales are less than $5 million based upon the manufacturer's
latest complete fiscal year as of January 1, 1978. However,
no manufacturer is a "small manufacturer" with respect to
any chemical substance which such person manufactured in 1977
at one site in amounts equal to or greater than 100,000 pounds
(45,400 kilograms).

Calculations for the $5 million criterion should be based
upon the total sales of all products, whether or not they
are chemical substances. In the case of a company which is
owned or controlled by another company, the $5 million cri-
terion applies to the total annual sales of the owned or con-

trolled company, the parent company, and all companies owned or controlled by the parent company taken together.

Step IV: Did you manufacture a HIGH VOLUME CHEMICAL SUB-STANCE during 1977?

HIGH VOLUME CHEMICAL SUBSTANCE means any chemical substance that was manufactured at one plant site in amounts equal to or greater than 100,000 pounds (45,400 kilograms).

GROUP 1 REPORTING REQUIREMENTS

Report by Plant Site:

a.   Identity: Report the identity of each reportable CHEMICAL SUBSTANCE you manufactured during 1977 at each CHEMICAL MANUFACTURING and/or VOLUME PRODUC-ING PLANT SITE. Also report the identity of each HIGH VOLUME CHEMICAL SUBSTANCE manufactured at any other plant site in 1977. (Separate reports must be submitted for each plant site.)

b.   Production Range: Report the 1977 production range of manufacture for each CHEMICAL SUBSTANCE reported under (a).

c.   Activity: Report that you manufactured each CHEMICAL SUBSTANCE reported under (a).

d.   Site-Limited: Report "site-limited" for each CHEMICAL SUBSTANCE reported under (a) which was manufactured and processed only within a plant site and was not distributed for commercial pur-poses as a substance or as part of a mixture or article outside the plant site.

See page 26 , "Optional Reporting Provisions," if you wish to report other chemical substances for the Inventory.

GROUP 2 REPORTING REQUIREMENTS

Report by Plant Site:

a.   Identity: Report the identity of each reportable HIGH VOLUME CHEMICAL SUBSTANCE you manufactured during 1977. (Separate reports must be submitted for each plant site.)

b.   Production Range: Report the 1977 production range of manufacture for each CHEMICAL SUBSTANCE reported under (a).

c.   Activity: Report that you manufactured each CHEMICAL SUBSTANCE reported under (a).

d.   Site-Limited: Report "site-limited" for each CHEMICAL SUBSTANCE reported under (a) which was manufactured and processed only within a plant site and was not distributed for commercial purposes as a substance or as part of a mixture or article outside the plant site.

Report by Headquarters:

e.   Identity: Report the identity of each reportable CHEMICAL SUBSTANCE, other than a HIGH VOLUME CHEMICAL SUBSTANCE, you manufactured during 1977 at each CHEMICAL MANUFACTURING and/or VOLUME PRODUCING PLANT SITE. (You may submit one report, or separate reports for each plant site. Although it is not mandatory that you report by plant site, EPA encourages you to do so.)

f.   Activity: Report that you manufactured each CHEMICAL SUBSTANCE reported under (e).

g.   Site-Limited: Report "site-limited" for each CHEMICAL SUBSTANCE reported under (e) which was manufactured and processed only within a plant site and was not distributed for commercial purposes as a substance or as part of a mixture or article outside the plant site.

Although it is not mandatory, EPA encourages you also to report:

h.   Production Range: Report the 1977 production range of manufacture for each CHEMICAL SUBSTANCE reported under (e).

See page 26 , "Optional Reporting Provisions," if you wish to report other chemical substances for the Inventory.

GROUP 3 REPORTING REQUIREMENTS

Report by Headquarters:

a.   Identity: Report the identity of each reportable CHEMICAL SUBSTANCE you manufactured during 1977 at

each CHEMICAL MANUFACTURING and/or VOLUME PRODUC-
ING PLANT SITE.

b.   Activity: Report that you manufactured each
      CHEMICAL SUBSTANCE reported under (a).

c.   Site-Limited: Report "site-limited" for each
      CHEMICAL SUBSTANCE reported under (a) which was
      manufactured and processed only within a plant
      site and was not distributed for commercial pur-
      poses as a substance or as part of a mixture or
      article outside the plant site.

Although it is not mandatory, EPA encourages you to
report by plant site, and to report in addition:

d.   Production Range: Report the 1977 production
      range of manufacture for each CHEMICAL SUBSTANCE
      reported under (a).

See page 26 , "Optional Reporting Provisions," if you
wish to report other chemical substances for the Inventory.

GROUP 4 REPORTING REQUIREMENTS

Report by Plant Site:

a.   Identity: Report the identity of each reportable
      HIGH VOLUME CHEMICAL SUBSTANCE you manufactured
      during 1977. (Separate reports must be submitted
      for each plant site.)

b.   Production Range: Report the 1977 production range
      of manufacture for each CHEMICAL SUBSTANCE reported
      under (a).

c.   Activity: Report that you manufactured each
      CHEMICAL SUBSTANCE reported under (a).

d.   Site-Limited: Report "site-limited" for each
      CHEMICAL SUBSTANCE reported under (a) which was
      manufactured and processed only within a plant
      site and was not distributed for commercial pur-
      poses as a chemical substance or as part of a
      mixture or article outside the plant site.

See page 26, "Optional Reporting Provisions," if you
wish to report other chemical substances for the Inventory.

GROUP 5 REPORTING REQUIREMENTS

You are not required to report for the Inventory under the regulations.

However, if you manufactured a chemical substance since January 1, 1975, and wish to ensure its inclusion on the Inventory, see page 26 , "Optional Reporting Provisions."

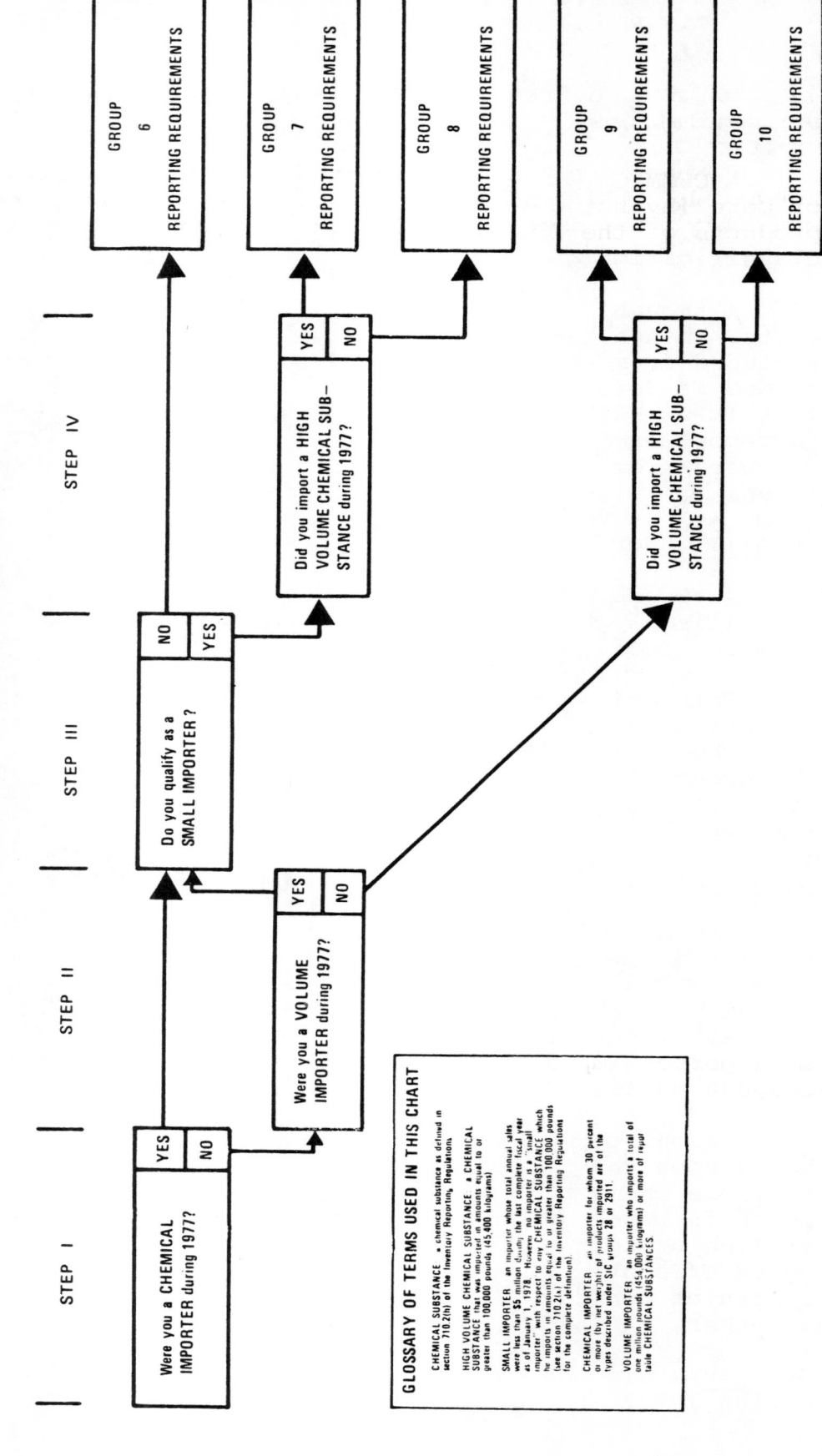

DECISION FLOW—CHART TO DETERMINE REPORTING REQUIREMENTS FOR IMPORTERS OF CHEMICAL SUBSTANCES

STEP I   STEP II   STEP III   STEP IV

Were you a CHEMICAL IMPORTER during 1977?   YES / NO

Were you a VOLUME IMPORTER during 1977?   YES / NO

Do you qualify as a SMALL IMPORTER?   NO / YES

Did you import a HIGH VOLUME CHEMICAL SUB— STANCE during 1977?   YES / NO

Did you import a HIGH VOLUME CHEMICAL SUB— STANCE during 1977?   YES / NO

GROUP 6 REPORTING REQUIREMENTS

GROUP 7 REPORTING REQUIREMENTS

GROUP 8 REPORTING REQUIREMENTS

GROUP 9 REPORTING REQUIREMENTS

GROUP 10 REPORTING REQUIREMENTS

GLOSSARY OF TERMS USED IN THIS CHART

CHEMICAL SUBSTANCE – a chemical substance as defined in section 710.2(h) of the Inventory Reporting Regulation.

HIGH VOLUME CHEMICAL SUBSTANCE – a CHEMICAL SUBSTANCE that was imported in amounts equal to or greater than 100,000 pounds (45,400 kilograms)

SMALL IMPORTER – an importer whose total annual sales were less than $5 million during the last complete fiscal year as of January 1, 1978. However, no importer is a "small importer" with respect to any CHEMICAL SUBSTANCE which he imports in amounts equal to or greater than 100,000 pounds (see section 710.2(x) of the Inventory Reporting Regulations for the complete definition)

CHEMICAL IMPORTER – an importer for whom 30 percent or more (by net weight) of products imported are of the types described under SIC groups 28 or 2911.

VOLUME IMPORTER – an importer who imports a total of one million pounds (454,000 kilograms) or more of total table CHEMICAL SUBSTANCES

DECISION FLOWCHART TO DETERMINE
REPORTING REQUIREMENTS FOR IMPORTERS OF
CHEMICAL SUBSTANCES (IN BULK FORM)

Step I: Were you a CHEMICAL IMPORTER during 1977?

A CHEMICAL IMPORTER is an importer for whom 30 percent or more (by net weight) of products imported consists of products of the types described under Standard Industrial Classification (SIC) groups 28 or 2911. (See Appendix 2.)

Although these SIC groups include categories of products, such as pesticides and drugs, which are specifically excluded from the inventory, these products should be included in determining whether or not an importer is a CHEMICAL IMPORTER. The regulations, however, do not permit you to report excluded chemical substances, i.e., substances excluded from the Inventory by section 710.4 of the regulations.

Step II: Were you a VOLUME IMPORTER during 1977?

A VOLUME IMPORTER is an importer who imports a total of one million pounds (454,000 kilograms) or more of reportable chemical substances.

This criterion should be used completely independently of that considered in Step I. A substance may be a reportable chemical substance whether or not it is listed under SIC groups 28 or 2911. Section 710.4 of the regulations specifies and chapter II of this booklet discusses what is a reportable chemical substance.

Step III: Do you qualify as a SMALL IMPORTER?

SMALL IMPORTER, as defined in Section 710.2(x) of the regulations, is an importer whose total annual sales are less than $5 million based on the importer's latest complete fiscal year as of January 1, 1978. However, no importer is a "small importer" with respect to chemical substances which such person imported in amounts equal to or greater than 100,000 pounds (45,400 kilograms).

Calculations for the $5 million criterion should be based upon the total sales of all products, whether or not they are chemical substances. In the case of a company which is owned or controlled by another company, the $5 million criterion applies to the total annual sales of the owned or controlled company, the parent company, and all companies owned or controlled by the parent company taken together.

Step IV: Did you import a HIGH VOLUME CHEMICAL SUBSTANCE during 1977?

HIGH VOLUME CHEMICAL SUBSTANCE means any chemical substance that was imported in bulk form by a company in amounts equal to or greater than 100,000 pounds (45,400 kilograms).

GROUP 6 REPORTING REQUIREMENTS

Report by Business Address:

a.   Identity: Report the identity of each reportable CHEMICAL SUBSTANCE, including HIGH VOLUME CHEMICAL SUBSTANCES, you imported in bulk form during 1977.

b.   Production Range: Report the 1977 production range of importation for each CHEMICAL SUBSTANCE reported under (a).

c.   Activity: Report that you imported each CHEMICAL SUBSTANCE reported under (a).

See page 26, "Optional Reporting Provisions," if you wish to report other chemical substances for the Inventory.

GROUP 7 REPORTING REQUIREMENTS

Report by Business Address:

a.   Identity: Report the identity of each reportable HIGH VOLUME CHEMICAL SUBSTANCE you imported in bulk form during 1977.

b.   Production Range: Report the 1977 production range of importation for each CHEMICAL SUBSTANCE reported under (a).

c.   Activity: Report that you imported each CHEMICAL SUBSTANCE reported under (a).

Also report by Business Address:

d.   Identity: Report the identity of each reportable CHEMICAL SUBSTANCE, other than a HIGH VOLUME CHEMICAL SUBSTANCE, you imported in bulk form during 1977.

e.   Activity: Report that you imported each CHEMICAL SUBSTANCE reported under (d).

Although it is not mandatory, EPA encourages you to report, in addition:

    f.   <u>Production Range</u>: Report the 1977 production range of importation for each CHEMICAL SUBSTANCE reported under (d).

See page 26 , "Optional Reporting Provisions," if you wish to report other chemical substances for the Inventory.

GROUP 8 REPORTING REQUIREMENTS

Report by Business Address:

    a.   <u>Identity</u>: Report the identity of each reportable CHEMICAL SUBSTANCE you imported in bulk form during 1977.

    b.   <u>Activity</u>: Report that you imported each CHEMICAL SUBSTANCE reported under (a).

Although it is not mandatory, EPA encourages you to report, in addition:

    c.   <u>Production Range</u>: Report the 1977 production range of importation for each CHEMICAL SUBSTANCE reported under (a).

See page 26 , "Optional Reporting Provisions," if you wish to report other chemical substances for the Inventory.

GROUP 9 REPORTING REQUIREMENTS

Report by Business Address:

    a.   <u>Identity</u>: Report the identity of each reportable HIGH VOLUME CHEMICAL SUBSTANCE you imported in bulk form during 1977.

    b.   <u>Production Range</u>: Report the 1977 production range of importation for each CHEMICAL SUBSTANCE reported under (a).

    c.   <u>Activity</u>: Report that you imported each CHEMICAL SUBSTANCE reported under (a).

See page 26 , "Optional Reporting Provisions," if you wish to report other chemical substances for the Inventory.

GROUP 10 REPORTING REQUIREMENT

You are not required to report for the Inventory under the regulations.

However, if you imported a reportable chemical substance (including a chemical substance as part of a mixture or article) since January 1, 1975, and wish to ensure its inclusion on the Inventory, see page 26, "Optional Reporting Provisions."

OPTIONAL REPORTING PROVISIONS

In order to ensure that a CHEMICAL SUBSTANCE(s) is included on the Inventory, any person who has manufactured or imported a reportable CHEMICAL SUBSTANCE(s) (including the importation of a CHEMICAL SUBSTANCE as part of a mixture or an article) for a commercial purpose since January 1, 1975, may report concerning that CHEMICAL SUBSTANCE for the Initial Inventory during the initial reporting period. This includes CHEMICAL SUBSTANCES manufactured or imported for the first time after December 31, 1977 (see chapter IV).

For each CHEMICAL SUBSTANCE that you report under these provisions, you must report:

a.   the identity of the CHEMICAL SUBSTANCE,

b.   your activity (manufacture and/or import) with respect to the CHEMICAL SUBSTANCE, and

c.   for domestic manufacturers, site-limited for each CHEMICAL SUBSTANCE you manufactured and processed at a plant site and did not distribute for commercial purposes as a chemical substance or as part of a mixture or article outside the plant site.

In addition, EPA encourages manufacturers to report by plant site, and encourages both manufacturers and importers to report:

d.   the 1977 Production Range for each CHEMICAL SUBSTANCE reported under these provisions.

Under these provisions you may either:

o   submit your own report; or

o   authorize a trade association or other agent to report on your behalf.

SECTION IV

WHEN TO REPORT
Section 710.6

Initial Reporting Period: Manufacturers and importers of chemical substances may report for the Initial Inventory until May 1, 1978. Chemical substances reported by persons who only process and use such substances for commercial purposes will not be included on the Initial Inventory.

Reporting of Chemical Substances Manufactured or Imported (in Bulk Form) for the First Time Between May 1, 1978 and the Effective Date of Premanufacture Notification Requirements: Premanufacture notification requirements for manufacturers of chemical substances and importers of chemical substances in bulk form will become effective 30 days after publication of the Initial Inventory. Any reportable chemical substance manufactured or imported for the first time prior to the effective date of premanufacture notification requirements is eligible for inclusion on the Inventory and will not be subject to premanufacture notification requirements if it is reported on Form A, B, or C as soon as manufacture or import begins.

Reporting Period for Revised Inventory: Persons who only process or use chemical substances for commercial purposes may report during a special 210-day reporting period which will begin on the date of publication of the Initial Inventory. Processors and users are not required to report. They are, however, permitted to report any chemical substance which they processed or used for commercial purposes.

> IMPORTANT: In order to avoid unnecessarily duplicative reporting, processors and users should not report any chemical substance which appears on the Initial Inventory. Processors and users should search the Initial Inventory and the TSCA Product Trademark List (which will be published in conjunction with the Initial Inventory) for the chemical substances (or products) they process or use, before reporting any chemical substance.

Special Reporting Rules for Importers of Chemical Substances as Part of Mixtures or Articles: Importers of chemical substances as part of mixtures or articles may report either during the initial reporting period, ending May 1, 1978, or during the 210-day reporting period for the Revised Inventory. Premanufacture notification requirements for importers of chemical substances as part of mixtures will begin 30 days after publication of the Revised Inventory. See Comment 21 in Appendix A to the Inventory Reporting Regulations, as published in the FEDERAL REGISTER, for discussion of premanufacture notification requirements which may apply to importers of chemical substances as part of articles [42 FR 64582].

IMPORTANT DATES

End of reporting period for for manufacturers, and importers of chemical substances in bulk form	May 1, 1978
Publication of the Initial Inventory	Near the end of 1978
Beginning of premanufacture notification requirements for manufacturers, and importers of chemical substances in bulk form	30 days after publication of the Initial Inventory
Reporting period for processors, users, and some importers	A 210-day period starting with the date of publication of the Initial Inventory
Beginning of premanufacture notification for some importers. Enforcement of TSCA as to processors and users of chemical substances not on the Inventory.	30 days after publication of the Revised Inventory

SECTION V

GENERAL INFORMATION ON
REPORTING FOR THE INITIAL INVENTORY

The Report Forms

There are four different kinds of Initial Inventory report forms, identified as Forms A, B, C, and D. It is important that you use the appropriate form to report chemical substances for the Initial Inventory. Different forms will be provided by EPA at a later date for use in submitting reports for the Revised Inventory.

Form A

Use Form A only to report chemical substances which appear in the "Toxic Substances Control Act Candidate List of Chemical Substances" or any addendum to that list for which a notice of availability is published in the FEDERAL REGISTER.

All chemical substances appearing on the Candidate List and addenda have Chemical Abstracts Service (CAS) Registry Numbers and valid EPA Code Designations. (See Appendix 3, "Guide to the Use of the TSCA Candidate List of Chemical Substances.")

As many as 26 chemical substances can be reported on each Form A.

Form B

Use Form B only to report chemical substances with known CAS Registry Numbers which do not appear in the TSCA Candidate List of Chemical Substances.

As many as ten chemical substances can be reported on each Form B.

Form C

Form C must be used to report chemical substances which have no known CAS Registry Numbers, and to report chemical substances whose identities for purposes of the Inventory are claimed confidential. Also, importers who are assisted in reporting by foreign suppliers must use Form C for each chemical substance they jointly report.

Only one chemical substance can be reported on each Form C.

Form D

Form D is a voluntary, supplemental form and cannot be used to report chemical substances for the Inventory. It does not replace Forms A, B, or C. Use Form D if you are a manufacturer or importer and wish to ensure persons who process or use your products for commercial purposes that all reportable chemical substances contained in these products have been reported for the Inventory. Product trademarks will not be included on the Inventory. EPA will publish a separate document, along with the Initial Inventory, that will list those product trademarks reported.

The sole purpose of Form D is to provide a means for you to assure processors and users of your products during the second 210-day reporting period that they may continue to process or use them without notifying EPA. No purpose is served by reporting trademarked products which are not processed or used for commercial purposes after distribution in commerce.

In order to report a product trademark on Form D, you must certify that, to the best of your knowledge and belief, all reportable chemical substances which are part of the trademarked product have been reported for the Inventory by you or someone else. Trademarked products which may be reported on Form D include chemical substances, mixtures, or articles.

Tips on Filling Out the Report Forms

o   Carefully read the Inventory Reporting Regulations and this instruction booklet, including the appendices, before attempting to fill out the forms.

o   Be sure to use the appropriate Inventory Report Form to report each chemical substance.

o   Type or print legibly using a black ball point pen -- press firmly to ensure that the carbon copies are legible.

o   If you make a mistake on a line, cross out the entry and start over on the next line.

o   Be sure that the appropriate person signs the certification statement(s) on each form.

o   Use only official TSCA Chemical Substances Inventory Report Forms. Chemical substances reported in letter form or on unofficial duplicates of the official report forms will not be processed by EPA in compiling the Inventory. However, chemical substances may be reported by computer tape (see How to Report by Computer Tape, appearing later in the chapter). EPA will provide additional copies of Forms A, B, C, or D upon request (see How to Get Additional Copies of Report Forms).

o   Retain the third copy of each form, marked "Submitting Company Copy" for your records.

o   Mail the remaining copies of each form and the attached postcard to:

> U.S. Environmental Protection Agency
> Office of Toxic Substances
> P.O. Box 02201
> Columbus, Ohio   43202

o   EPA will acknowledge receipt of each form by returning the postcard which is attached to each form to the addressee identified in Block III of each form. The first line of the address (the line directly under the plant site, headquarters, or business name) may be used to enter the name of the person or office to which the card should be sent.

How to Get Additional Copies of Report Forms

Before you order additional report forms, estimate how many copies of each form you will need.

1.   All EPA Regional offices have an ample supply of Forms A, B, C, and D. You should arrange to pick up these forms at a Regional Office (see page 34 for addresses) as Regional offices are not equipped to fill mail orders.

2.   You may order report forms by phone from the EPA's Office of Industry Assistance at (800) 424-9065. Allow two (2) weeks for delivery.

How to Get a Copy of the TSCA Candidate List of Chemical Substances

o   EPA will make available one free copy of the Candidate List to any interested organization or individual as long as supplies last. A request for either a printed or microfiche copy should be sent to:

> Candidate List, OTS (TS-799)
> U.S. Environmental Protection Agency
> 401 M Street, S.W.
> Washington, DC, 20460.

o   Both printed and microfiche copies may be picked up in person at all EPA regional offices. See page 34 for addresses.

o   The free copy of the Candidate List may be ordered by telephone by calling (800) 424-9065, EPA's Office of Industry Assistance. Allow two (2) weeks for delivery.

o   Additional printed copies may be obtained by written request from:

            Superintendent of Documents
            Government Printing Office (GPO)
            Washington, DC, 20402.

Requests should specify the document number (GPO No. 055-007-00001-2) and be accompanied by check or money order in the amount of $14.00 per copy.

o   Additional microfiche sets may be obtained from:

            National Technical Information Service (NTIS)
            5285 Port Royal Road
            Springfield, Virginia, 22161.

These requests should indicate No. PB 265-371 and be accompanied by check or money order in the amount of $9.00 per microfiche set.

o   A computer-readable version of the Candidate List may be obtained by written request to:

            Computer List, OTS (WH-557)
            Attention: Kenneth Olsen
            Environmental Protection Agency
            401 M Street, S.W.
            Washington, DC, 20460.

Persons requesting the computer-readable version of the Candidate List must comply with provisions set forth in the April 28, 1977 FEDERAL REGISTER, pages 21639-40 and the July 8, 1977 FEDERAL REGISTER, page 31583.

For copies of these FEDERAL REGISTER notices, contact the EPA Industry Assistance Office at (800)424-9065.

o   The Government Printing Office has arranged to place a copy of the Candidate List in each of its Regional Depository Libraries and in the more than 1,000 depository libraries throughout the country. A State librarian or local library can assist in identifying the location of the nearest depository library.

How to Report by Computer Tape

For special instructions on how to report by computer tape, contact Mr. Kenneth Olsen at (202) 755-2890 or write:

Instructions for Reporting by Computer Tape
Attention: Kenneth Olsen
Office of Toxic Substances (WH-557)
U.S. Environmental Protection Agency
401 M Street, S.W.
Washington, DC, 20460

Further Assistance

For further assistance in filling out the report forms or interpreting the regulations, contact:

1.   Your regional EPA office. Each region is staffed by persons who can respond to your questions concerning the Inventory Reporting Regulations. The person to contact in each regional office is identified on page 34.

2.   EPA's Office of Industry Assistance at (800) 424-9065. In addition, written inquiries may be addressed to:

Office of Industry Assistance (TS-788)
U.S. Environmental Protection Agency
401 M Street, S.W.
Washington, DC, 20460

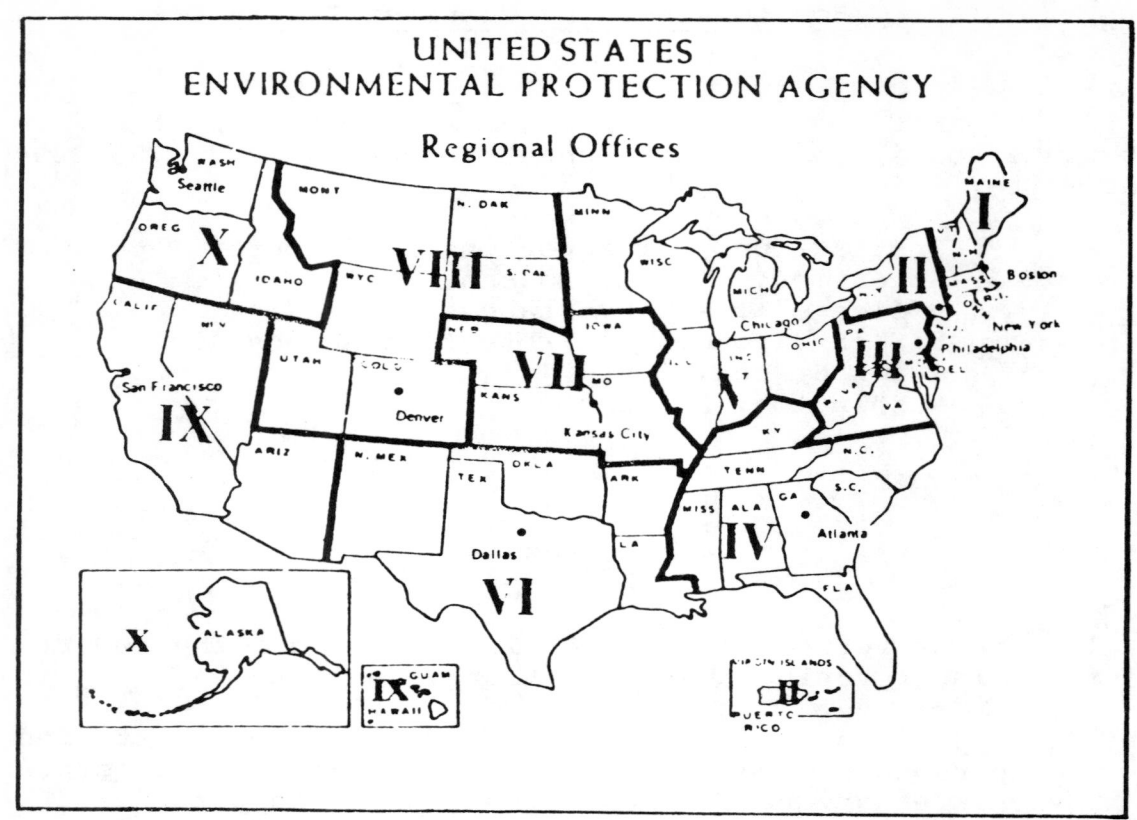

NATIONWIDE TOLL-FREE NUMBER: 800-424-9065

*If calling from Washington, D.C: 554-1404

Region I

Mr. Robert Dangel*
John F. Kennedy Federal Building
Boston, Massachusetts   02203
617-223-0585

Region II

Mr. William Librizzi*
Raritan Depot, Building 209
Edison, New Jersey   08817
201-321-6673

Region III

Mr. Edward Cohen*
Curtis Building
6th & Walnut Streets
Philadelphia, PA   19106
215-597-7668

*Toxic Substances Coordinator
 Air and Hazardous Materials Division

Region IV

Mr. Ralph W. Jennings, Rm. 345
Toxic Substances Section Chief
Air & Hazardous Materials Div.
345 Courtland Street, N.E.
Atlanta, Georgia   30308
404-881-3864

Region V

Mr. Karl E. Bremer*
230 South Dearborn Street
Chicago, Illinois   60604
312-353-2291

Region VI

Dr. Larry Thomas
Assistant TSCA Coordinator
First International Building
1201 Elm Street
Dallas, Texas   75270
214-767-2734

Region VII

Dr. Maxwell Wilcomb*
1735 Baltimore Street
Kansas City, Missouri   64108
816-374-3036

Region VIII

Mr. Ralph Larsen*
1860 Lincoln Street
Denver, Colorado   80203
303-837-3926

Region IX

Mr. Jerry Gavin*, Rm. 215
215 Freemont Street
San Francisco, California   94105

Region X

Dr. James Evert*
1200 6th Avenue
Seattle, Washington   98101
206-442-1090

INDEX

A

Acetone 296-97
Acetylene 233, 345
Acetic acid 58, 111, 318-319, 345
Acid finishers 371
Acids 111
Acnegenic tests 127
Acoustical products 345, 367
Acorn Auto Driving School, Inc. v. Board of Education 405
Acrolein 300-01
Acrylamide 316-17
Acrylic acid 61
Acute dermal toxicity 125
Adhesives 156, 225-29, 296, 298, 300, 302, 304, 306, 308, 310, 312, 314, 316, 318, 320, 345
Adipose tissues 87, 431-33
Aerosols, 122, 125-26, 296, 298, 300, 302, 304, 306, 308, 310, 312, 314, 316, 318, 320
Agitated pan dryers 248
Agriculture 224, 427, 429, 530
    See also Pesticides
Air quality 52, 53, 81-82, 98, 103, 237, 252, 407, 410, 425, 426, 428, 429, 435, 547-52
    See also Ventilation
Air spray painting operations 269-70
    See also Spray painting
Aircraft 226, 228, 345
Airless spray painting, 268
    See also Spray painting
Alcohols 217, 345
Aldrin 48, 51, 98, 328
Alfatoxins 328
Alkali 230, 232
Alkali bicarbonate 210
Alkoxy alkanols 59
Alkyds 217
Alkyl adipates 59
Alkyl amines 59
Alkyl phthalates 60
Alkyl sulfates and sulfonates 60
Allergenics and sensitizers 95, 123, 233, 234
Allergic dermatitis 123
Allied Chemical Company 81
Alloys 345
Allyl alcohol 304-05
Allyl chloride 312-13
Alpha-chloroacetophenone 298-99
Alpha-methyl styrene 300-01
Aluminum 111, 218, 345

Amerace Corporation 399
American Cancer Society 355
American Conference of Governmental Industrial Hygienists 100, 102, 375
American Industrial Hygiene Association 110
American Insurance Association 110
American National Standards Institute 47, 433
Ames bioassay 101
Ammonia 110, 273-74, 275, 345
Amyl acetate 302-03, 304-05
Anaerobic microbial action 93
Anemia 390
Anesthetics 297, 299, 301, 303, 305, 307, 309, 311, 313, 316, 317, 319
Aniline 230, 316-17, 345
Animal feed 96
Animal glues 226, 228
Anisidine 318-19
Anorexia 97, 373
Antioxidants 230, 232
Apparel 228
    See also Shoes, Textiles
Applicators 495-96
Appropriate lethal dose 124, 125
Approximate lethal concentration 126
Aquatic life and toxicity 41, 44, 48, 50, 131, 148, 424, 427, 430
Aqueous adhesives 226
Arc welders 345, 351
Aromatic coal tar 233
Arsenic 49, 50, 81-82, 233, 326, 345, 347, 348, 349, 350, 351, 352, 353, 354, 355, 356-58, 359, 360, 361, 362, 363, 364, 378, 379, 424-426
Arsine 62, 111
Arthur D. Little, Inc. 204, 273
Asbestos 3, 17, 47, 52, 83-84, 218, 270-71, 326, 328, 345, 346, 347, 348, 349, 350, 351, 352, 353, 354, 359, 362, 367-70, 383-385, 424-26
Asphalt 227
Aspirin 209
Atomic absorption spectrophotometry 375
Atomic Energy Act of 1954 151
Attritors 218

B

Babbiters 347, 371
Bacterial mutagenesis 130
Bag filling 259-60, 263, 284
Bag houses 82
Bakers 346

Balloon flues 82

Barium 49, 50

Barium sulphate 217

Batteries 88, 106, 372, 442

Beaming operators 346

Belt conveyors 254, 256-57, 283

Benzene 63, 84-85, 112, 217, 229, 258, 259, 345, 346, 347, 348, 349, 350, 351, 352, 353, 355, 356, 357, 358, 359, 360, 361, 362, 363

Benzidine 86-87, 428

Beryllium 50, 52, 64, 112, 239, 326, 346, 347, 348, 349, 353, 354, 355, 356, 358, 359, 361

Bin filling 258-59, 284

Bioaccumulation properties 45, 427

Bioconcentration 40

Biological cabinets 253-54

Bismuth compounds 64

Bitumens 217

Blacksmiths 347, 371

Bladder cancer 86

Blanket coaters 274

Blast furnaces 347

Bleachers, bleaching, and bleaching compounds 112, 274, 346, 349, 352, 356, 363, 383

Blenders 212

Blow molding 213

Blueprints 347

Booiler rooms 347

Booz-Allen Applied Research 39, 40, 41

Brake linings 83, 84, 347

Brass 347, 371, 372

Braziers 347, 372

Breweries 347

Bricks 347

Bright v. City of Evanston 404

Bright dip workers 347

Bromine 112

Bromoform 310, 311

Bronchogenic disorders. See Lung cancer and diseases

Bronze 237, 347, 372

Bucket elevators and chutes 259

Buffers 347

Building demolition 347

Bulk handling systems 254

Bulk Terminals et al. v. The Environmental Protection Agency 403-405

Bureau of Mines 425

Burnishers 347

Butadiene 221, 308-09

Butadiene-styrene copolymer 220

Butane 112

Butoxy ethanol 306-07

Butyl acetate 302-03

Butyl alcohol 304-05

Butylamine 314-15

C

Cables 347

Cacodylic acid 426

Cadmium 48, 49, 50, 87-89, 112, 230, 248

Cadmium red 218

Cadmium selenide 217

Calcium carbonate 217

Calender roll exhaust ventilation 216

Calendering 213, 220, 221

Camphor 298-99

Cancer 325-343
    See also Carcinogens

Candidate List of Chemical Substances 73, 165-68, 591-92

Candles 348

Canners 348

Capacitators 433

Capsules. See Pharmaceutical industry

Carbide makers 348

Carbon black 217, 220

Carbon monoxide 249, 346, 347, 348, 349, 350, 351, 352, 353, 354, 355, 356, 358, 359, 360, 361, 362, 363, 410, 411

Carbon tetrachloride 217, 314-15

Carbonates 217

Carbons 65, 99, 112

Carboxy methylcellulose 230

Carders 348

Cardiac failure 89

Cardiac sensitization 130

Casein glue 227

Casting 214

Catalytic converters 93, 96

Cathode ray tube 348

Caulking 348

Caustic potash 230

"Caution" 104, 396

Cellophane 214

Cellular transformation 424

Celluloid 348

Cellulose 227, 228, 348

Cements 227, 345, 348, 367

Center for Disease Control 427

Central nervous system 80, 99, 101, 224, 373

Ceramics 297, 299, 301, 303, 305, 307, 309, 311, 313, 315, 317, 319, 348, 350, 351, 353, 358

Ceruse 371

Charcoal 348

Chelating compounds 230

Chemical Abstracts Service Registry 78, 156, 165, 170-72

Chemical Industry Institute of Toxicology 4, 34

Chemical industry and processing 3, 5, 208-38, 241-44, 281, 296-320, 577-78

Chemicals. See Toxic substances

Chest X-rays 369
Chippers 348
Chlorofluorocarbons 9, 32, 33, 37
Chlorates 112
Chloradane 49, 51
Chlorinated hydrocarbon 49, 51
Chlorine 50, 112, 274, 275
Chloroacetaldehyde 298-99
Chlorobenzene 312-13, 348
Chlorobromomenthane 310
Chlorodiphenyl 348
Chloroform 100, 112, 314-15, 326
Chloronaphthalenes 127
Chlorophenoxys 49
Chloroprene 310, 311
Chrome 348
Chromic acid 65, 345, 355, 357, 358, 362, 379, 388
Chrome yellow 371
Chromium 49, 50, 65
Chromosomal aberrations 130
Chronic studies 128-29
Cigar makers 349, 372
Cismene 300-01
Citizen role 7, 19, 23, 32, 79-80
Class B Poison Tests 122, 125, 126
Classification systems for toxic substances 39-42, 43-45, 46-47, 48, 107-09, 494, 567-70
Clean Air Act 33, 40, 52, 81, 82, 87, 98, 243, 408, 409, 410, 411, 428, 435, 547-52
Cleaning and maintenance operations 156, 220, 223, 225, 245-47, 296-320
Closed-cup flash point 105
Clothing 226, 228
    See also Apparel and Textiles
Coal 81, 84, 349
Coal gasification 82
Coal tar 349
Coal tar naphtha 308-09
Coast Guard 39, 375, 512
Coatings 214, 217-19, 237, 265, 273-75, 285
Cobalt 65
Cobblers 349
    See also Shoes
Coke 93, 98
Coke ovens 326, 331, 349
Colcemid 130
Combustible materials. *See* Explosives and explosivity
Common carriers 42
Compositors 349
Compressing machines 210, 211
Compression molding 213, 215
Computer tapes (reporting by) 593
Condensation 109
Confidentiality 7, 13, 19, 156, 158-59, 162-63, 168, 171, 175

Conflict of interest 18
Conjunctivitis 93
Construction work 349, 359, 363, 368, 397
Consumer Product Safety Act 39-42, 103
Control and monitoring technology 211-12, 222-23, 237, 238, 425-428, 431, 433, 434, 455
Copper, 50, 81-82, 320-22, 346, 349, 351
Corrosion inhibitors 232, 300, 302, 304, 306, 308, 310, 312, 314, 316, 318, 320
Corrosive materials 40, 42, 43, 55, 106, 108
Cosmetics 39, 300, 302, 304, 306, 308, 312, 314, 316, 318, 320, 326, 349, 396, 398, 410
Cotton 81, 227, 273, 349, 350, 352, 355, 356, 357, 358, 359, 360, 361, 363, 375, 385
Coumarin 233
Cresol 113, 318-19
Crossdraft table 276-78
Crushing 212, 217, 220, 223, 229, 250-52, 282, 349
Curing 214, 215, 219, 221, 222, 227, 228, 247-50
Cyanide 49, 50
Cyclohexane 300-01, 350
Cyclohexanone 66, 296-97
Cyclohexene 66, 306-07
Cyclopentadiene 306-07
Cylinder dryers 248
Cylinder mills 250
Cytogenetics 130

D

DDT 48, 51, 223, 224, 350
"Danger" 100
Dichlorination 433
Decision flow charts 574-81
Decomposition 39, 109
Defense waivers 18
Defoliant 490
Demolition 350
Dinitrochlorobenzene 220
Department of the Army 40, 55
Department of Commerce 4
Department of Defense 55
Department of Health, Education, and Welfare 4, 8, 45-47, 101, 328
Department of Labor. *See* Occupational Safety and Health Administration and Secretary of Labor
Department of the Navy 40
Department of Transportation 33, 39, 42-45, 263, 425, 513
Department of Transportation Class B Poison and Skin Corrosion Test 123-24
Dermal toxicities 41, 42, 43, 44, 48, 107, 108, 121-25, 127, 147, 215, 220, 221, 224-25, 233, 336, 367, 369, 399, 401, 426

Dermatitis 215, 220, 221
　See also Dermal toxicity
Desiccant 490
Detergents 230-33, 297, 299, 301, 303, 305, 307, 309, 311, 313, 315, 317, 319, 350
Dextrine 226
Diacetone alcohol 304-05
Diamonds 350
Diazomethane 314-315
Dichlorobenzene 350
Dieldrin 48, 328
Dielectric heat dryers 248
Diesel-powered engines 96, 98, 350
Diethylamine 314-15
Dimethysulfate 350
Dioxane 308-09
Diphenyl 233, 300-01, 350
Detergents 230
Dispersion units 218
Disposal regulations 428-436
　See also Ocean dumping and Solid wastes
Dough mixers 218
Dow Chemical Company 81, 135-40
Drag conveyors 257
Dredged materials 513
Drinking water 17, 40, 48-49, 53, 82, 83, 84, 85, 86, 89, 424, 425-26, 427, 429, 433, 434, 519-22
　See also Safe Drinking Water Act
Dross 236
Drugs. See Pharmaceutical industry
Drum dryers 248
Drum filling 260, 284
Dry cleaning 296, 298, 300, 302, 304, 306, 308, 310, 312, 314, 316, 318, 320
Drywall 383
Drying and dryers 156, 213, 220, 223, 229, 247-50, 281, 283
Dumping 511-18
Dusts 106, 108, 209, 211, 212, 213, 219, 223, 224, 236, 237, 245, 249, 251, 254, 255, 259, 260, 261, 369-70, 425
Dyes 213, 229-30, 273, 286, 297, 299, 301, 303, 305, 307, 309, 311, 313, 315, 317, 319

E

Economic considerations and factors 2, 29, 37-38, 82, 83, 85, 87, 88, 90, 91, 93, 96, 97, 99, 100, 101, 102, 126-39, 134-139, 326-27, 425, 427, 433
Edema 122, 123
Effluent standards 531-32
Egg Products Inspection Act 425
Elastomers 226
Electrical equipment 296, 298, 300, 302, 304, 306, 308, 310, 312, 314, 316, 318, 320, 351

Electrical industry 369
Electroplating 88, 348, 350
Electrostatic precipitators 80, 374, 427
Electrotyping, 235, 351, 372
Emergency Temporary Standard 333
Emission standards 52, 548-52
Emphysema 88, 235
Employee protection 7
　See also Medical examinations, Protective clothing, Training, and Workplace
En masse conveyors 257
Enamel 272, 351
Endosulfan 51
Environmental considerations 5-6, 21, 24, 27, 29, 32, 81, 103, 407, 408, 424, 426, 427, 428, 429, 430-432, 435
Environmental Defense Fund 84, 101, 326
Environmental Protection Agency 3-38, 47-55, 134-39, 150-79, 218, 288, 326, 329, 403-05, 409, 424-435, 438-85, 513-14, 569-95
Enzymes 232, 233
Epichlorohydrin 312-13
Epidemiological studies 424, 427, 434
Epinephrine 130
Epoxies 217, 271
Epoxy resin systems 114, 214
Equipment trains 244
Erythema 106, 107
Estuaries 529
Esters 217, 229
Ethanol 67
Ethanolamine 314-15
Ethers 67, 229
Ethyl acetate 304-06
Ethyl alcohol 233, 304-05
Ethylamine 316-17
Ethylbenzene 300-01, 351
Ethyl bromide 310, 311
Ethyl butyl ketone 298-99
Ethyl chloride 310, 311
Ethylene dibromide 89-90
Etiologic agents 43
Exhaust systems 215, 216, 217, 219, 220, 221, 249, 252, 259, 260, 265, 276, 285
　See also Ventilation
Experimental use permits 496
Explosives and explosivity 40, 43, 105, 107, 108, 111, 120, 297, 299, 301, 303, 305, 307, 309, 311, 313, 315, 317, 319, 351, 372
Exports 18, 437-38
Exposure and exposure limits, 206-07, 220-22, 231, 234, 326
Extruders 216, 221-22
Eye irritants 42, 122-25, 148, 341

F

Fat extraction 233-34
Fatty acid 230
Feather workers 351
Federal Food, Drug and Cosmetics Act 40, 45, 151, 430, 435
Federal Hazardous Substances Act 39, 122-25
Federal Insecticide, Fungicide, Rodenticide Act 151, 193-200, 488-509
Federal Meat Inspection Act 151
Federal Register 385, 396, 446, 447, 449, 452, 454, 463, 464, 475, 477, 480, 494, 495, 497, 506, 507, 508, 520-21, 528, 564
Federal Trade Commission 18, 444
Federal Water Pollution Control Act 7, 33, 41, 49, 84, 424-45
Felt 83
Ferric ferrocyanide 217
Ferrosilicon 351
Fertilizer 88, 351
Fetotoxicity 426
Fiberglass 215
Fibrosis 368
Fireproof textiles 83, 351
   See also Flame retardants
Fischer-Tropsch Process 351
Fixation techniques 427
Fixatives 233
Flame retardants 67-68, 97-99, 362
Flammability 39, 40, 41, 42, 43, 48, 55, 105-07, 111-20, 251
   See also Volatile materials
Flash point 41, 42, 105, 108, 111-20
Flatbed cylinder presser 235
Floating knife coaters 274
Flourine 115
Fluorescent brightening agents 68
Fluoride 49, 115, 410
Fluorocarbons 68
Flux cored arc welding 277
Foam 213-17
Food 39, 431, 439
Food contamination 430
Food and Drug Administration 24, 40, 45, 326, 430, 435
Food, Drug, and Cosmetic Act. *See* Federal Food, Drug, and Cosmetic Act
Food industry 121-22
Formaldehydes 217, 227, 352
Forms 173, 177, 193-99, 589-93
Foster D. Snell, Inc. 141
Freedom of Information Act 18, 19
Fuel-oxider combinations 109
Fuels 39, 115
   See also Petroleum
Fumigants 352

Fungicides 71, 218, 223-30, 487-509
Fungus 490
Furfural 300-01
Furnaces 352
Furniture 226, 228, 352

G

Gaf Corporation 383, 384-390
Gas shielded-arc welding 276, 277
Gases 43, 44, 108, 112
   See also Vapor
Gasoline 17, 84, 89, 98, 111, 352
Gelatin 210
Generic names 158-59
Genetic toxicology 129-30
Geothermal energy 82
Ginners 353
Glass 81, 96, 296-320, 345, 348, 349, 353
Glass insulation 226
Glove box 253, 254
Glue 353
   See also Adhesive materials
Glycerinated gelatin 211
Glycerine 115, 230
Glycerol 210, 230
Glycol esters 217
Gold 353, 372
Gordon plasticators 220, 221
Governmental Industrial Hygienists 93
Granite 353
Gravure printing 236
Great Lakes 33, 526-27
Grinding 213, 217, 218, 220, 223, 225, 229, 250-252, 282, 353
Gundraft table 277
Gun-mounted exhaust hoods 276, 278, 287
Guthion 51
Gypsum 271

H

Hafnium 320-21
Hair removers 353
Halogenated aromatics 127
Hammer crushers 250
Haskell Laboratory for Toxicology and Industrial Medicine 121, 560
Hazardous Materials Transportation Act 55
Hazardous substances. *See* Toxic substances
Health considerations 3, 5-6, 7, 18, 21, 27, 38, 39, 43, 81, 83, 84, 86, 89, 91, 92, 94, 95, 96, 98, 99-100, 101, 107, 121-131, 209, 214-215, 219, 224-25, 230, 236, 238, 325-43, 373-75, 407-08, 424, 425-26, 427-28, 429-30, 431, 432, 434
Health signal 107

Heat transfer fluid 297, 299, 301, 303, 305, 307, 309, 311, 313, 315, 317, 319
Heptachlor 49, 51
Heptachlor epoxide 49
Heptane 308-09
Herbicides 223-25, 353
Herman v. Village of Hillside 404
Hexachloride 346
Hexachlorobenzene 89-90, 429-30
Hexachlorobutadiene 430-31
Hexone 296-97, 308-09
Hexyl acetate 302-03
Highway transport 42
Hoffman Construction Company v. OSAHRC 401
Human patch test 123
Human studies 123, 131
Hydrated lead oxide 371
Hydroblasting 245
Hydrocarbons 69, 93, 98, 217, 220, 221, 229, 410
Hydrochloric acid 353, 403-05
Hydrofluoric acid 115
Hydrogen sulfide 92, 296-97, 417
Hydroquinine 304-05
Hyperpigmentation 378
Hypersensitivity 42
Hypertension 88
Hypothyroidism 379

I

Illinois Bell Telephone v. Allphin 404
Imports 6, 18, 154, 583-86
Indirect dryers 248
"Industrial Data Sheets, Chemical Series" 110
Industrial disposal 53
Industries. *See* various types of industries, *i.e.* Pharmaceutical Industry
Inflation 145
Information collection and dissemination 9, 11, 13, 14, 16-26, 333, 432, 461-64, 468-70, 561-95
    acquisition 18-26
    chemical industry 243-45
    data availability 36
    data banks 36
    data evaluation 21
    data reliability 21
    data system operations 34-36
    laboratory operations 252-54
    needs 241
    ocean dumping 515
    pesticides 493, 499
    procedures 24-26
    quality control 19, 24
    research 281-321
    retrieval 20, 36
    standardization of data activities 36

water pollution 525-26, 528, 531
    See also Recording and recordkeeping and Reporting requirements
Ingestion. *See* Oral toxicity
Inhalation toxicity 41, 42, 43, 44, 48, 92, 108, 121-22, 125-26, 127-28, 129, 147, 205, 213, 224-25, 237, 399-401, 426, 435
Ink 156, 229-30, 296, 298, 300, 302, 304, 306, 308, 310, 312, 314, 316, 318, 320, 352, 372
Insecticides 51, 83, 223-25, 353, 372, 487-509
Insufflation 128
Insulation 226, 354, 369, 388-90
Insurance coverage 110
Interagency Advisory Committee 24
Interagency Testing Committee 17, 34
Intermediate chemicals 26-27
International Agency for Research on Cancer 98
International Union of Rubber Workers 221
Inventory of existing chemicals 26, 36
Investment casting 432
Iodine 116
Ionizing radiation 328
Iron 50, 354
Irritants 39, 40, 42, 43, 48
Isoamyl acetate 302-03
Isoamyl alcohol 304-05
Isobutyl acetate 205, 302-03
Isobutyl alcohol 304-05
Isocyanates 214, 216, 354
Isoprene 220
Isopropyl alcohol 304-05
Isopropyl glycidyl ether 306-07
Isopropylamine 316-17
Itai-Itai disease 88

J

Japan makers 354
Jaw crushers 250
Jet fuel 354
Jewelry 96
Jute oil 354

K

Kady mills 218
Kepone 224
Kerosene 116
Ketones 70, 217, 229
Kidneys and kidney disorders 88, 89, 353
Kilns 353

L

Labeling 39, 186-188, 342-43, 491-92
Laboratories and laboratory operations 213, 217, 220, 223, 229, 252-254, 282

Lacquers 217-223, 354, 372
Laggers 354, 368
Lamination 214
Lamp black 217
Land disposal 82, 86
Landfills 53, 424
Latent period 329, 330
Latex 227
Laundering 341
Leaching 425, 427, 435
Lead 17, 49, 50, 70, 85, 116, 221-22, 230, 236-37, 345, 346, 347, 348, 349, 350, 351, 352, 353, 354, 355, 356, 357, 358, 359, 360, 3t1, 362, 363, 364, 371-79, 407-11
Lead oxide 220
Leakage 243, 281
Leather 86, 297, 299, 301, 303, 305, 307, 309, 311, 313, 315, 317, 319, 353, 355, 362
Letterpress 234-35
Leukemia 84
Lime 116, 355
Lindane 49, 51
Linoleum 355, 371
Linotypers 355, 371
Linseed oil 217, 355
Liquid processing 210-211
Lithium hydride 70
Lithography 235-36
Liver diseases and cancers 81, 83, 89, 95, 97, 99, 102, 224
Low volume high velocity (LVHV) hoods 240
Lubricants 297, 299, 301, 303, 305, 307, 309, 311, 313, 315, 317, 319
Lung cancer and diseases 81, 98, 99, 122, 328, 367, 369
Lymphatic cancers 81

M

MCA study 110, 135-40
McLean Trucking Company v. OSAHRC 401
Mackey apparatus 43
Magnesium 116, 355
Magnesium silicate 217, 232
Magnetic properties 55
Maintenance operations 246-47
Malathion 51
Maleic anhydride 320-21
Mammary tumors 98
Manganese 50, 70, 296-97
Manufacturing
    adhesives 225-26
    chemical processes 241-45
    detergents 230-33
    perfume 233-34
    pesticides 223-25
    plastics 213-17, 226
    soap 230-33
    See also various industries, *i.e.* Rubber industry
Manufacturing Chemists' Association (MCA) 110, 135-40
Marine Protection, Research, and Sanctuaries Act of 1972 48, 511-518
Masons 355
Master-batching, 220, 221
Material handling operations 213, 220, 223, 225, 229, 254, 283-85
Maximum contaminant level 520, 522
Mechanical filtration 374
Mechanics 352
Median lethal dose 124-25
Medical examinations and surveillance 337-39, 341-42, 385-90
Medicines. *See* Pharmaceutical industry
Mercury 3, 48, 49, 50, 52, 94-95, 116, 355
Mercury aryl compounds 218
Mesityl oxide 298-99
Mesothelioma 83, 267-68
Metal primers 218
Metals and metal processing 156, 239, 240, 297, 299, 301, 303, 305, 307, 309, 311, 313, 315, 317, 319, 346, 351, 355, 359, 360, 372
Metallic oxides 217
Metaphase study 130
Methanes 70, 71, 116
Methyl acetate 302-03
Methyl alcohol 304-05
Methyl bromide 314-15
Methyl cellosolve 306-07
Methyl chloride 310-311
Methyl chloroform 100, 314-15
Methyl iodine 310, 311
Methylene chloride 314-15
Michigan Department of Health 249, 250
Micronucleus Test 130
Milk 32, 430, 431
Millboard 346
Mineral mining 346
Mining 257, 269, 297, 301, 303, 305, 307, 309, 311, 313, 315, 317, 319, 346, 353, 354, 355, 369, 424, 425, 426
Mining Enforcement and Safety Administration 36, 341, 424
Mirex 51
Miticides 85
Mold cure 221
Mond process 356
Monitoring 338-339, 344, 371-73, 428, 466, 534-535, 551-52
    See also Control and monitoring
Monosodium metharsonate 426

Mordants 356
Municipal disposal 53
Musical instruments 226, 356, 372
Mutagenesis 17, 24, 40, 47, 87, 129, 148, 424

N

Nader, Ralph 101
Naphtha 117, 308-09
Naphthalene 230, 231
National Academy of Sciences 33, 40, 41, 43-45, 425, 520-21
National Air Standards Act of 1970 410
National Ambient Air Quality Standard 85
National Cancer Institute 4, 40, 47, 89, 101, 326, 327, 328, 429
National Center for Toxicological Research 87, 428
National Contingency Plan 40, 41
National disposal sites 53
National Electrical Code 244
National Fire Protection Association 102, 107-09
National Institute for Environmental Health Sciences 4, 84, 424
National Institute for Occupational Safety and Health 33, 39, 40, 41, 45-47, 84, 85, 88, 101, 102-03, 204, 221, 239, 246, 252, 288, 325, 332, 390, 426, 429
National Library of Medicine 36
"N.S.C. Data Sheet" 109-10
National Science Foundation 4
Natural adhesives 226, 227
Natural gas 93
Natural Resources Defense Council, Inc. v. Train 407-11
Necrosis 122-23
Nematocide 223, 492
Neon 356
Neoplastigenic 47
New chemicals 9, 13, 16, 26, 27, 29
Nickel 356
Nicotine 117
Nitrate 50, 356
Nitric acid 116, 356, 372
Nitrile rubbers 220, 221
Nitroaniline 296-97
Nitrobenzene 356
Nitrocellulose 219, 356
Nitrogen 118
Nitrogen dioxide 345, 346, 347, 348, 349, 350, 351, 352, 353, 354, 355, 356, 357, 358, 359, 360, 362, 363
Nitroglycerin 207, 356, 372
Nitrous acid 87
Noise 347, 348, 349, 350, 353, 355, 357, 358, 359, 361, 363
Non spray applications 217, 220, 229, 296, 298, 300, 302, 304, 306, 308, 310, 312, 314, 316, 318, 320

Nuclear materials 3, 356
Nylon 356

O

OECD 432
Occupational Safety and Health Act of 1970 332-39, 395, 399
Occupational Safety and Health Administration 33, 36, 47, 81, 83, 88, 93, 100, 102, 204, 288, 325, 326, 328, 357-59, 370, 383-405, 426
Occupational hazards 345-63, 371-79, 325-43
    See also Workplace environment
Ocean dumping 40, 48, 512-18
Octane 71, 118, 308-09
Oil 48, 82, 93, 118, 356, 528, 529, 537
Ointments 211
Open-cup flash point 104
Open hoods 223, 239, 241, 252, 253, 254, 277, 281
Open-surface tank use 213, 217, 220, 223, 225, 229, 266-67, 285-86
Oral contraceptives 212
Oral toxicity 41, 43, 48, 55, 106, 107, 111, 121-28, 205, 424, 426
Ore 357
Organic chemical synthesizers 357
Organic lead 371-75
Organic sulfonate makers 357
Organohalogens 48
Organophosphate insecticides 51
Othmer, Kirk 212
Oxalic acid 318-19
Oxidizing materials 40, 42, 43
Oxygen 118
Ozone 118, 298-99

P

Packaging 211, 226, 260
Paints and painting 156, 217-19, 267-73, 296, 298, 300, 302, 304, 306, 308, 310, 312, 314, 316, 318, 320, 346, 357, 369, 372, 383
    See also Spray painting
Paint and varnish removers 297, 299, 301, 303, 305, 307, 309, 311, 313, 315, 317, 319
Pan conveyors 257
Paper 86, 92, 297, 299, 301, 303, 305, 307, 309, 311, 313, 315, 317, 319, 346, 349, 357, 368, 432
Paraffin 348, 357, 372
Parathion 51
Particle board 226
Particulates, 410, 411
Pebble mills 219
Penalties 71, 503-04, 535, 539, 543, 544, 559
Pentachlorophenol 218

Pentane 308-09

Pentanone 296-97

Perchlorethylene 100, 275

Perfumes 230, 232-33, 297, 299, 301, 303, 305, 307, 309, 311, 313, 315, 317, 319, 357

Pesticides 40, 41, 45, 47-48, 49, 81, 89, 131, 193-202, 223-25, 296, 298, 300, 302, 304, 306, 308, 310, 312, 314, 316, 318, 320, 328, 426, 428, 429, 439

Petroleum 72, 85, 297, 299, 301, 303, 305, 307, 309, 311, 313, 315, 317, 319, 357

Pharmaceutical industry 208-213, 272, 297, 299, 201, 303, 305, 307, 309, 311, 313, 315, 317, 319, 357

Phenol 118, 227, 318-19

Phenol resin 227

Phenyl ether vapor 306-07

Phosgene 314-315

Phosphate 357

Phosphorus 118

Phosphoric acid 320-21

Photographic film 272, 296, 298, 300, 302, 304, 306, 308, 310, 312, 314, 316, 318, 320

Picric acid 358

Pigments 217, 219, 220, 221, 229, 230, 358

"Pins and needles" sensation 377

Pival 298-99

Pivoted bucket carriers 258

Plasma arc welding 277

Plaster 271, 346, 358

Plastics 3, 88, 102, 213-15, 296, 298, 300, 302, 304, 306, 308, 310, 312, 314, 316, 318, 320, 358, 372

Platinosis 95

Plumbers 358

Plumbum 312

Plywood 226-28

Pneumatic conveyors 285

Pneumonia 93

Poisonous gases 43

Polargraphics 374

Polishing compounds 297, 299, 301, 303, 305, 307, 309, 311, 313, 315, 317, 319

Pollution. *See* Air quality, Water pollution, and Drinking water

Pollution Control Board 405

Polybrominated biphenyls, 3, 5, 9, 32, 33-34, 37, 50, 96, 431-32

Polychlorinated biphenyls 127

Polyesters 217

Polyester resins 215

Polytetrafluoroethylene 215

Polymers and polymerization 45, 109, 157, 213, 214, 215, 217, 226, 227, 567-68

Polynuclear aromatic hydrocarbons 98-99

Polyurethane foam 271

Polyurethanes 215, 217, 258

Polyvinyl acetate 227

Polyvinyl chloride 215-16

Porcelain 358, 361

Potassium hydroxide 230

Pottery 358

Poultry Products Inspection Act 161

Pouring stations 262

Premarket notification and review 9, 13, 15, 16, 19, 27, 28-30, 36, 161

Preservatives 232, 355, 359

Pressure flash tank 22, 220

Primary irritation 122-23

Printing industry 234-38, 359, 372

Processing. *See* Chemical processing

Production and use studies 425

Proko Company 383

Propanes 73, 308-09

Propyl acetate 302-03

Propyl alcohol 304-05

Propylene 119

Protective clothing and equipment 102, 223-25

Protein and protein products 226, 359

Pthalate ester 50

Pulmonary edema 93

Pumps 242-43

Putty 359

Pyorrhea 372

Pyridine 318-19

Pyroxylin 359

Pyrolysis products

Q

Quarries 359

Quartz 359

Quinone 320-21

R

Radioactive waste 512

Radioactive material and reactivity 40, 42, 55, 109

Rayon 93, 359

Reactive materials 109

Reactivity signal 107

Reactors 220-21

Record keeping and reporting 5-6, 19, 22, 24, 138, 338-39, 391-93, 454, 461-64, 499-500, 533-34, 554-55, 561-595

Red lead 371

Refrigerants 296, 298, 300, 302, 304, 306, 308, 310, 312, 314, 316, 318, 320

Regulation 370, 428

Reproductive system 89, 130, 224, 373, 429, 431

Research 145, 279-321

analytic methods 131

control and monitoring technology 279-321, 425-26, 428, 429, 430, 431, 433, 434-35
  detection methods 131
  drinking water 521-22
  dust control 254-55
  environmental considerations 425, 426, 429, 430-31, 432
  experimentation techniques 288
  federal 424-35
  grants 527
  health considerations 424, 426-47, 429, 430, 431
  information collection and dissemination 131, 288
  oil pollution 528
  pesticides 224-25, 506
  paints and painting 219-20
  pharmaceuticals 210
  plastics 214-16
  spray-paint 272
  volatile substances 264-66
  water pollution 528-31
Reserve Mining Company 83
Reserve Mining Task Force 84
Resins 213, 217, 218, 219, 220, 226, 228, 271
Resorcinol formaldehyde 275
Respirators 212
Respiratory diseases and malfunctions 92, 121-22, 233, 234, 237, 337-39
  See also Lungs
Revolving screens 251
Risk assessment 14, 15, 17, 19, 21, 29
Rodenticides 193-99, 223-25, 339, 487-509
Roll coating 274
Roll crushers 250
Roller mills 250
Rose v. Locke 400
Rotary dryers 247
Rotary presses 235
Rotating pans 212
Rubber 93, 216, 217, 220-23, 227, 228, 296, 298, 300, 302, 304, 306, 308, 310, 312, 314, 316, 318, 320, 359, 360, 372
Rubberized fabrics 275
Rupture disks 242
Rural sewage 530

S

Safe Drinking Water Act 2, 33, 48, 426, 519-22
Safety and relief valves 242
Salt of Saturn 371
Sand 360
Sand mills 218
Sanitation workers 360
Scale hoppers 232
Screening 213, 217, 220, 222, 225, 229, 250-52, 282

Screw conveyors 257
Scrubbing 53
Secretary of Labor v. Amerace Corporation 399
Secretary of Labor v. Goodyear Tire & Rubber Company 391-93
Secretary of Labor v. Proko Company of Texas, Inc. 383
Secretary of Labor v. Republic Steel Corporation 413-16
Secretary of Labor v. Research-Cottrell, Inc. 395-97
Secretary of Labor v. Spencer Leathers 417-20
Seed grain 91
Selenium 49-50, 119, 320-21
Self-reaction 44-5, 109
Sensitizers 39-40, 107-08, 121, 123
Sequesting agents 232
Sewage sludge. See also Sludge. 39, 53, 88
Shaking screens 251
Sheeting dryers 247-48
Shellac 217, 360, 372
Shipping requirements 122
Shoes 226, 360, 372
Signal words 104
Signs 342
Silica 218, 245-63
Silicon dioxide 403-05
Silk 357, 359, 372
Silver 49-50, 356, 360-61
Silver nitrate 119
Skin cancer 81, 427
  See also Dermal toxicity
Slate 361
Sludge 82, 98
Small Business Administration 18
Small manufacturers 152, 160-61, 577-78
Smelters and smelting 81, 88, 355, 357
Smog 100
Smokeless flare systems 272
Soap 230, 358, 360-61
Society of the Plastic Industry v. Department of Labor 326
Soda ash 225
Sodium 119
Sodium bicarbonate 229
Sodium carbonate 232
Sodium hydroxide 230, 273, 413-16
Sodium nitrilotriacetic acid 233
Sodium salt 218
Sodium silicate 230, 232, 341
Sodium sulfate 232
Sodium sulfide 220-21
Sodium sulphydrate 417-20
Sodium tripolyphosphate 230-32
Solar radiation 328
Solderers 361, 372
Solid waste 82, 91-92, 430, 433-34
Solvent adhesives 227

Solvents 217
Source assessment 21
Soybean glues 226-28
Special permits 48
Spills and spillage 52, 428
Spray-finishing 223, 286, 296, 298, 300, 302, 304, 306,
  308, 310, 312, 314, 316, 318, 320
Spray painting 213, 217, 220, 223, 225, 229, 267-68,
  273, 361
Sputum examinations 369
Squamous cell carcinoma 89
Stabilizers 213
"Standard Method of Test for 'Index of Dustiness' of
  Coal and Coke" 254
Standards Completion Program 204
Starches 226
State grants 17, 526
Steam distillation 233-34
Steam tube rotary dryers 248
Stearic acid 221
Steel 361
Sterility 224
  See also Reproductive system
Stereotyping 235, 361
Stoddard solvent 308-09
Stokinger and Woodward method 53, 55
Stomach 78
Stone workers 361
Styrene 221, 226, 300-01, 361
Subacute toxicity 426
Submerged arc welding 277
Substantial risk 22
Sugar 362
Sugar beet 93
Sugden v. Department of Public Welfare 405
Sulfates 217
Sulfur 119, 221
Sulfur dioxide 345-63
Sulfur oxides 410-11, 427
Sulfuric acid 119, 318-19, 362, 417-20
Suppositories 211
Sulfuric acid 318-19
Surfactant manufacturing 232
Synergistic effects 24
Synthetic adhesives 227-28
Synthetic Organic Chemical Manufacturers v. Brennan
  325
Synthetic polymers 217

T

TRW Systems Group 39, 40, 53
Table of Chemical Hazards 104
Taconite 83, 84
Tagliabue open cup tester 42

Talc 331, 363, 424
Tanning industry 363, 372
Teratogenesis 129, 148, 429, 431
Terphenyls 300-01
Terpineols 233
Test marketing 154
Testing procedures and administration 9, 19, 22, 24, 37,
  39, 45-56, 121-31, 441-54
  acute studies 109-113, 122-26
  chronic studies 128-29
  class B poison 106
  economic considerations 135-40, 142-43, 147
  Federal Hazardous Substances Act 122
  insufflation 128
  pesticides 493
  prolonged studies 126-28
  special 116-18, 229-31
  volatile materials 266
Tetrachloroethane 217
Tetrachloroethylene 314-15
Tetrasodium pyrophosphate 230, 232
Tetrahydrofuran 306-07
Textiles 86, 273, 286, 296, 298, 300, 302, 304, 306,
  308, 310, 312, 314, 316, 318, 320, 362, 369, 372
Thermal spraying 277
Threshold limit value (TLV) 39, 53
Thermoplastic adhesives 227, 228
Thionyl chlorine 362
Thyroid gland disorders and cancer 81
Tile 362
Tin 320-21
Tires 221, 222
Tobacco 3-4, 440
Toluene 221
Toluene diisocyanate 229, 271, 345, 352, 354, 356, 357,
  358, 360, 362, 363
Torch brazing 277
Toxaphene 49, 51
Toxic gases 417-20
Toxic substances
  agriculture 223-25
  carcinogens 325-45
  classification systems 39-42, 43-45, 46-47, 48, 107-09,
    494, 567-70
  controls 10-11, 13, 84, 85, 87, 88, 90, 92, 93, 95, 96,
    97, 99, 100, 102-03, 224, 225, 237, 427, 433
  court cases 383-420
  definition of "hazardous" 39-55, 332-333, 564-67
  health risks. *See* Health considerations
  inventory of 150-63, 561-95
  management 212-30
  measurement of 336-40
  monitoring 338-39
  occupational hazards 344
  regulation of 3-55, 330-32, 429, 434, 454-60

research 279-321
"special case" 567-68
testing of 4, 9, 19, 35, 38, 45, 121-31, 142-43, 147
trademark 563
transport of 42-45, 223, 260, 262, 427, 505, 555
Toxic Substances Control Act 3-38, 437-485
  benefits of 145
  control of toxic substances 13, 29, 32
  data collection and dissemination 10, 11-13, 14, 17, 18-26, 33-36
  ecological concerns 16
  economic considerations 37, 38, 135-49
  employee protection 478-79
  enforcement 7, 471-72
  environmental concerns 36-37
  exports and imports 7-8
  goals of 9, 10-11
  gross national project 139-140
  highlights of 10
  impact of 36-38
  industry's role 11-13
  information collection and dissemination 150-80, 461-64
  intermediate chemicals 26-27
  inventory of chemicals 26-27
  manufacturing and processing notices 447-54
  marketing delays 138
  preliminary list 59-76
  premarket notification, review, screening 4-5, 13, 20, 26, 28 ,29-30 , 32
  priorities 13, 16
  recordkeeping and reporting 461-64
  regulatory authorities 12-13, 26, 33, 454-60
  reporting and recording 5-6, 19, 22, 24, 150-180
  research 8, 17-18, 465-466, 469
  response to urgent problems 32-33
  risk assessment 17, 19, 21
  state programs, grants, laws 7-8, 472-73
  testing requirements 4, 16, 17-18, 19, 20, 31, 32-34, 441-454
  training 466
Toxic Substances Control Act Interagency Testing Committee 79
Toxicity criteria and rating 39-55, 106-107, 121, 124
Trace contaminants 48
Trade associations 157
Trade secrecy 7, 19, 501
Trademark 563
Train, Russell 407-08
Training 17, 338, 427, 466, 487
Transport of toxic substances 42-45, 223, 427, 505, 555
Tray dryers 247
Treatment works 527
Triphenyl phosphate 220, 221
Triphenyl phosphite 220

Tris (2,3-Dibromopropyl Phosphate (TBPP) 101-102, 230, 232
Tritium 425
Trivalent arsenic 79, 82
Tuluene diisocyanate 214
Tunnel drying 247
Turpentine 120, 308-09
Typesetting 363-372

U

Union Electric Co. v. Environmental Protection Agency 411
United Mine Workers 193
United States Coast Guard 43
United States Congress 9-10, 24, 32, 438
United States Customs 468
United States Steel Corp. v. O.S.H.R.C. 400
United States v. Ball 405
United States v. Beckerman 405
United States v. National Dairy Products Corp. 401
Uniroll mill 218
University of North Carolina 221
Unstable materials 109
Use exposure documents 205-06

V

Vacuum devices 245-46
Vacuum flash tank 220
Vacuum rotary dryers 248
Vacuum tray dryers 248
Vandinite 371
Vanadium 363
Vaporous materials 43-44, 105, 111, 215, 209-21, 263
Varnishes 217-19, 363
Veneer 226
Ventilation 205, 212, 214-17, 220-22, 224-25, 237-39, 240, 241-45, 248-52, 281, 285, 287-89
  See also Air quality and Dust
Vibrating screens 251
Vibratory conveyors 257
Vinyl acetate 120
Vinyl chloride 2, 47, 91, 102-03, 243-45, 325, 331, 434-35
Vinyl chloride monomers 215-17, 245
Vinyls 217
Volatile materials 263-66
  See also Flammability
Vulcanization 220-22, 363

W

Warfare agents 512
"Warning" 104

Waste disposal 82
Waste materials 39, 427-28
Waste steam constituents 53
Wastes 553-60
Water pollution 45, 48, 273
Water purification 297, 299, 301, 303, 305, 307, 309, 310, 311, 313, 315, 317, 319
Water quality 425, 428, 430, 434, 525-45
Water transport 42
Water treatment 82, 86
Welding operations 275-78, 287, 363, 372
"Wet layup" 214, 217, 265
White lead 371
Wine 363
Wood pulp 227

Wood and woodworking industry 238-41, 296, 298, 300, 302, 304, 306, 308, 310, 312, 314, 316, 318, 320
Work clothing 341
See also Protective clothing
Workplace standards and environment 234, 345-63, 383-405, 428-29, 434-35
World Health Organization 88

X

Xylidine 318-39

Z

Zincs 76, 81, 88, 120, 218, 363
Zirconium 320-21